2nd International Conference on Frontiers in Automobile and Mechanical Engineering 2016

IOP Conference Series: Materials Science and Engineering
Volume 197

Chennai, India
7 - 9 July 2016

Editor:

J. Lilly Mercy

ISBN: 978-1-5108-4174-1
ISSN: 1757-8981

Printed from e-media with permission by:

Curran Associates, Inc.
57 Morehouse Lane
Red Hook, NY 12571

proceedings.com

Some format issues inherent in the e-media version may also appear in this print version.

Printed by Curran Associates, Inc. (2017)

For permission requests, please contact the Institute of Physics
at the address below.

Institute of Physics
Dirac House, Temple Back
Bristol BS1 6BE UK

Phone: 44 1 17 929 7481
Fax: 44 1 17 920 0979

techtracking@iop.org

Additional copies of this publication are available from:

Curran Associates, Inc.
57 Morehouse Lane
Red Hook, NY 12571 USA
Phone: 845-758-0400
Fax: 845-758-2633
Email: curran@proceedings.com
Web: www.proceedings.com

Table of contents

Volume 197

Frontiers in Automobile and Mechanical Engineering

7–9 July 2016, Sathyabama

University, Chennai, India

Accepted papers received: 2 May 2017

Preface

011001
OPEN ACCESS
Frontiers in Automobile and Mechanical Engineering

011002
OPEN ACCESS
Peer review statement

Materials

012001
OPEN ACCESS
Development of friction material by using precast prefired (pcp f) blocks

R Dineshkumar, E V V Ramanamurthy and Ch Krishnapavanteja.....1

012002
OPEN ACCESS
Comparison of the Effect of Curing on the Properties of E-Glass/Cyanate modified Epoxy Cross Plied Laminates

W Andrew Nallayan, K.R Vijayakumar and Usama Tariq Rasheed.....8

012003
OPEN ACCESS
Optimization of Profile and Material of Abrasive Water Jet Nozzle

K P Anand Bala Selwin and S Ramachandran.....21

012004
OPEN ACCESS
Analysis of copper mixed kerosene servotherm in EDM of Monel 400™

P.A. Anandakumar, Baya Molla, Fisha Biruke and S. Aravind.....31

012005
OPEN ACCESS
Optimality of cycle time and inventory decisions in a two echelon inventory system with exponential price dependent demand under credit period

Seelam Krugon and Dega Nagaraju.....38

012006
OPEN ACCESS
Comparative Study of Fault Diagnostic Methods in Voltage Source Inverter Fed Three Phase Induction Motor Drive

R. B. Dhumale and S. D. Lokhande.....49

012007
OPEN ACCESS
Design, Static Analysis And Fabrication Of Composite Joints

G. Mathiselvan, R. Gobinath, S. Yuvaraja and T. Raja.....63

012008
OPEN ACCESS
Mechanical Characterization of Baslat Based Natural Hybrid Composites for Aerospace Applications

J. Alexander and S.J Elphej Churchill.....71

012009
OPEN ACCESS
Recent developments in turning hardened steels – A review

V Sivaraman and S Prakash.....79

012010
OPEN ACCESS
Review on Microstructure Analysis of Metals and Alloys Using Image Analysis Techniques

Suganthini Rekha and V.K. Bupesh Raja.....88

012011
OPEN ACCESS
Optimization of Tensile Process Parameters of Bi-Axial Glass Fibre Reinforced Epoxycomposite Using Taguchi Method

I Infanta Mary Priya and B K Vinayagam.....93

Internal Combustion Engine

012012
OPEN ACCESS
Analysis of CO_2, CO and HC emission reduction in automobiles

K N Balan, T N Valarmathi, Mannem Soma Harish Reddy, Gireddy Aravinda Reddy, Jammalamadaka K M K Sai Srinivas and Vasan.....100

012013
OPEN ACCESS
Green fuel utilization for diesel engine, combustion and emission analysis fuelled with CNSO diesel blends with Diethyl ether as additive

Ashok Kumar, K. Rajan, K.R. Senthil Kumar, K. Maiyappan and Usama Tariq Rasheed.....106

012014
OPEN ACCESS
Experimental Investigation of Performance and emission characteristics of Various Nano Particles with Bio-Diesel blend on Di Diesel Engine

N Karthik, X Goldwin Xavier, R Rajasekar, P Ganesh Bairavan and S Dhanseelan.....116

012015
OPEN ACCESS
A Comprehensive Review of Effect of Biodiesel Additives on Properties, Performance, and Emission

S Madiwale, A Karthikeyan and V Bhojwani.....122

012016
OPEN ACCESS
Optimization of Rice bran biodiesel blends on CI engine and investigating its effects

J. Jayaprabakar, Biraj Dey, Krishanu Dey, Batchu Hareesh and M. Anish.....129

012017
OPEN ACCESS
Comparative study of performance and emissions of a CI engine using biodiesel of microalgae, macroalgae and rice bran

J. Jayaprabakar, A. Karthikeyan, K. Saikiran, N. Beemkumar and Nivin Joy.....138

012018
OPEN ACCESS
Experimental investigation on Performance and Emission Characteristics of J20, P20, N20 Biodiesel blends and Sound Characteristics of P20 Biodiesel blend Used in Single Cylinder Diesel Engine

R. rajasekar, N. karthik and Goldwin Xavier.....145

012019
OPEN ACCESS
Review on Enhanced Heat Transfer Techniques using Modern Technologies for 4S Air Cooled Engines

S Ramasubramanian and V K Bupesh Raja.....150

012020
OPEN ACCESS
Experimental Investigations on Diesel engine using Methyl esters of Jatropha oil and fish oil

A Karthikeyan, J Jayaprabakar and Richard Dude Williams.....155

012021
OPEN ACCESS
Performance and Emission Analysis of Diesel Engine by Copper Coating Over Piston Crown and Cylinder Head

S. Yuvaraja, G. Mathiselvan and R. Gobinath.....162

012022
OPEN ACCESS
Effects of Nano Additives in engine emission Characteristics using Blends of Lemon Balm oil with Diesel

J. Senthil kumar, S. Ganesan, S. Sivasaravanan, S. Padmanabhan, L. Krishnan and V.C. Aniruthan.....169

012023
OPEN ACCESS
Performance Test on Compression Ignition Engine by Blending Ethanol and Waste Plastic Pyrolysis Oil with Cetane Additive

S Padmanabhan, S Ganesan, J Jeswin Arputhabalan, Varun Chithrala and P Ganesh Bairavan.....178

012024
OPEN ACCESS
Emission and Performance Analysis of ZrO_2 And CeO_2 Coated Piston Using Refined Vegetable Oils

J Hemanandh, K.V. Narayanan and Vemuri Manoj.....185

012025
OPEN ACCESS
Influence of MgO on Performance and Emissions of DI Engine using blends of Castor oil

S Ganesan, s Padmanabhan, J Senthil kumar, Navakanth Polina and Sanivada Krishna kumar.....193

012026
OPEN ACCESS
I.C. Engine emission reduction by copper oxide catalytic converter

S.P. Venkatesan, Desai Shubham Uday, Borana Karan Hemant, Kagita Rajarshi Kushwanth Goud, G. Lakshmana Kumar and K. Pavan Kumar.....199

012027
OPEN ACCESS
Prospects of pyrolysis oil from plastic waste as fuel for diesel engines: A review

V L Mangesh, S Padmanabhan, S Ganesan, D PrabhudevRahul and T Dinesh Kumar Reddy.....208

Thermal

012028
OPEN ACCESS
Investigation on Nano Refrigeration

G Senthil kumar, K Saravanan and K Ajay Kumar.....213

012029
OPEN ACCESS
Electric Power from Cryo (Nano) Ice

A. Kandasamy and M. Chandran.....218

012030
OPEN ACCESS
Design and Analysis of a Two-Stage Adsorption Air Chiller

P Benrajesh and A John Rajan.....224

012031
OPEN ACCESS
Experimental investigations on effect of different materials and varying depths of one turn exhaust channel swiss roll combustor on its thermal performance

Sagar B. Mane Deshmukh, A. Krishnamoorthy, V. K. Bhojwani and Ashwini Pawane.....234

012032
OPEN ACCESS
Experimental study on productivity of modified single-basin solar still with a flat plate absorber

V Ramanathan, B Kanimozhi and V K Bhojwani.....245

012033
OPEN ACCESS
Thermal Analysis of Fluidized Bed and Fixed Bed Latent Heat Thermal Storage System

N Beemkumar, A Karthikeyan, Kota Shiva Keshava Reddy, Kona Rajesh and A Anderson.....252

012034
OPEN ACCESS
Review of Phase Change Materials Based on Energy Storage System with Applications

R. Thamaraikannn, B. Kanimozhi, M. Anish, J. Jayaprabakar, P. Saravanan and A. Rohan Nicholas.....260

012035
OPEN ACCESS
Modelling and Optimising the Value of a Hybrid Solar-Wind System

Arjun Nair, Kartik Murali, S.P Anbuudayasankar and C.V. Arjunan.....268

012036
OPEN ACCESS
Performance Improvement of Energy Storage System with nano-additivesin HTF

N Beemkumar, A Karthikeyan, B Saravanakumar and J Jayaprabakar.....281

012037
OPEN ACCESS
Recent developments in drying of food products

T.N. Valarmathi, S. Sekar, M. Purushothaman, S.D. Sekar, Maddela Rama Sharath Reddy and Kancham Reddy Naveen Kumar Reddy.....286

012038
OPEN ACCESS
Investigation of Sensible and Latent Heat Storage System using various HTF

N Beemkumar, A Karthikeyan, A Manoj, J S Keerthan, Joseph Paul Stallan and P Amithkishore.....296

012039
OPEN ACCESS
Analysis of Thermal Energy Storage Tank by ANSYS and Comparison with Experimental Results to Improve its Thermal Efficiency

N Beemkumar, A Karthikeyan, Kota Shiva Keshava Reddy, Kona Rajesh and A Anderson.....302

012040
OPEN ACCESS
Charging and Discharging Processes of Thermal Energy Storage System Using Phase change materials

Dr. B. Kanimozhi, Kasilanka Harish, Bellamkonda Sai Tarun, Pogaku Saty Sainath Reddy and Padakandla Sai Sujeeth.....311

012041
OPEN ACCESS
Experimental investigations of hybrid PV/Spiral flow thermal collector system performance using Al_2O_3/water nanofluid

R Gangadevi, B K Vinayagam and S Senthilraja.....318

Design

012042
OPEN ACCESS
Design and Static Analysis of Airlesstyre to Reduce Deformation

Nibin Jacob Mathew, Dillip Kumar Sahoo and E. Mithun Chakravarthy.....327

012043
OPEN ACCESS
Agricultural robot designed for seeding mechanism

Dr. K A Sunitha, G S G S Suraj, CH P N Sowrya, G Atchyut Sriram, D Shreyas and T Srinivas.....334

012044
OPEN ACCESS
Design and Analysis of Wind Turbine Blade Hub using Aluminium Alloy AA 6061-T6

S. Ravikumar, V. Jaswanthvenkatram, Y.J.N.V. Sai kumar and S.Md. Sohaib.....340

012045
OPEN ACCESS
Design and Analysis of Boiler Pressure Vessels based on IBR codes

B Balakrishnan and B Kanimozhi.....348

012046
OPEN ACCESS
Flexural Behavior of Aluminum Honeycomb Core Sandwich Structure

Vidyasagar Matta, J Suresh Kumar, Duddu Venkataraviteja and Guggulla Bharath Kumar Reddy.....356

012047
OPEN ACCESS
Design Intent for CAD Modeling Features Using Boolean Operations

Chandrakant R Sonawane and Ghadge Sujit.....367

012048
OPEN ACCESS
Design and Analysis of Drive Shaft using Kevlar/Epoxy and Glass/Epoxy as a Composite Material

P Karthikeyan, R Gobinath, L Ajith Kumar and D Xavier Jenish.....375

012049
OPEN ACCESS
Finite element analysis of chip formation usingale method

V. Jayaprakash.....383

012050
OPEN ACCESS
Design and Analysis of High Pressure Hydraulic Filter for Marine Application

Toshin Momin, RP Chandrasekar, S Balasubramanian and Dr. AM Junaid Basha.....390

Composites

012051
OPEN ACCESS
Experimental Analysis of Exhaust Manifold with Ceramic Coating for Reduction of Heat Dissipation

J Saravanan, T N Valarmathi, Rajdeep Nathc and Prasanth Kumar.....399

012052
OPEN ACCESS
Fabrication of SiC particulate reinforced polyester matrix composite and investigation

R. Selvam, S. Ravi and R. Raja.....405

012053
OPEN ACCESS
Effect of Steel slag on Wear Characterization of Aluminium Composite Using Taguchi Technique

K S Sridhar Raja, V K Bupesh Raja, F Aaro Rinold and L.S Abhilash.....413

012054
OPEN ACCESS
Computational analysis of frp composite under different temperature gradient

P. Gunasekar and S. Manigandan.....419

Production

012055
OPEN ACCESS
Evaluation of Thrust force in Drilling Woven roving Glass fibre reinforced Aluminium Sandwich laminates with TiAlN coated drill using Taguchi analysis

G. Ramya Devi and K. Palanikumar.....425

012056
OPEN ACCESS
Measurement and analysis of thrust force in drilling sisal-glass fiber reinforced polymer composites

M. Ramesh and A. Gopinath.....433

012057
OPEN ACCESS
Quality of Work Life – A Study

Dr. T.K Mily Velayudhan and M.D Yameni.....440

012058
OPEN ACCESS
Evolution and Recent Trends in Friction Drilling Technique and the Application of Thermography

Mathew Alphonse, V.K. Bupesh Raja, K. Logesh and N. Murugu Nachippan.....454

012059
OPEN ACCESS
Evaluation and Selection of Best Priority Sequencing Rule in Job Shop Scheduling using Hybrid MCDM Technique

Kalla Kiran Kumar, Dega Nagaraju, S Gayathri and S Narayanan.....468

012060
OPEN ACCESS
Evaluation and selection of material handling equipment in iron and steel industry using analytic hierarchy process

Sajja Varun, Raj Harshita, Sesha Pramod and Dega Nagaraju.....478

012061
OPEN ACCESS
Evaluation and Selection of Predicaments in Pharmaceutical Supply Chain using AHP under Fuzzy Environment

E.V.S. Nikhil, V. Sai Ram, V. Charan Yadav, Kalla Kiran Kumar and Dega Nagaraju.....487

012062
OPEN ACCESS
Reduction of Defects in Al-6061 Friction Stir Welding and Verified by Radiography

D Kumaravel, Dr V K Bupesh Raja, Chakravarthy Potnuru and Navakanth Polina.....494

012063
OPEN ACCESS
Experimental Investigation and Taguchi Optimisation of Drilling Properties on Teak Wood Reinforced Epoxy Resin

J. Lilly Mercy, Mohamed Shaqir Tanvir and K. Swaroopkanth.....499

012064
OPEN ACCESS
Process improvement by cycle time reduction through Lean Methodology

R Siva, Mahamed naveed khan patan, Mane lakshmi pavan kumar, M Purusothaman, S Antony pitchai and Y Jegathish.....505

012065
OPEN ACCESS
Plasma Nitriding of CP Titanium Grade-2 and Ti-6Al-4V Grade-5

J.R. Deepak, V.K. Bupesh Raja, J. Senthil Kumar, Subin Thomas and Thomas Raju Vithaiyathil.....512

012066
OPEN ACCESS
Salt Bath Nitriding of CP Titanium Grade-2 and TI-6AL-4V Grade-5

J. R. Deepak, V.K. Bupesh Raja, K. Arun Kumar, H Varun Radhakrishnan and Sonu Skariah Thomas.....518

012067
OPEN ACCESS
Experimental Investigation of Mechanical Properties of Welded Corten Steel A588 Grade Plate Using ER70S - 6 Filler Material for Construction Application

J.R. Deepak, V.K. Bupesh Raja, Mittapalli Janardhan Guptha, Palaparthi Hari Durga Prasad and V. Sriram.....524

012068
OPEN ACCESS
Experimental Investigation Nano Particles Influence in NPMEDM to Machine Inconel 800 with Electrolyte Copper Electrode

K Karunakaran and M Chandrasekaran.....532

012069
OPEN ACCESS
Investigation on the Effect of Friction Welding Parameters on Impact Strength in Dissimilar Joints

C. Shanjeevi, J.Jeswin Arputhabalan, Rohan Dutta and Pradeep.....547

012070
OPEN ACCESS
Optimization of Drilling Parameters for Reducing the Burr Height in Machining the Silicon Carbide Particle (SiCp) Coated with Multi Wall Carbon Nano Tubes (MWCNT) Reinforced in Aluminum Alloy (A 356) Using Meta Modeling Approach

M. Sangeetha and S. Prakash.....553

012071
OPEN ACCESS
Building Brand Power

S Lakshmi and S Dr. Muthumani.....563

Automobile

012072
OPEN ACCESS
Experimental Investigation on The Electromagnetic Clutch Water pump and Pneumatic Compressor for Improving the Efficiency of an Engine

R Kumarasubramanian, Goldwin Xavier, W Mary Nishanthi and R Rajasekar.....568

012073
OPEN ACCESS
Experimental Investigation of Thermal Performance in a Vehicle Cabin Test Setup With Pcm in the Roof

M. Purusothaman, Saichand kota, C Sam Cornilius and R Siva.....576

012074
OPEN ACCESS
CFD and Thermo Mechanical Analysis on Effect of Curved vs Step Surface in IC Engine Cylinder Head

S. Balaji, N. Ganesh and A. Kumarasamy.....585

012075
OPEN ACCESS
Analysis of Effect of Inlet Swirl In Four Stroke Single Cylinder Diesel Engine With Different Inlet Valve Geometries Using CFD

R. Gobinath, G. Mathiselvan and R. Kumarasubramanian.....592

Aeronautical

012076
OPEN ACCESS
Aerodynamic Analysis Over Double Wedge Airfoil

U S Prasad, V S Ajay, R H Rajat and S Samanyu.....605

012077
OPEN ACCESS
Aerodynamics and flow characterisation of multistage rockets

G Srinivas and M V S Prakash.....611

012078
OPEN ACCESS
Recent developments of axial flow compressors under transonic flow conditions

G Srinivas, K Raghunandana and B Satish Shenoy.....625

012079
OPEN ACCESS
Fault diagnosis of helical gearbox using acoustic signal and wavelets

SK Pranesh, Siju Abraham, V. Sugumaran and M. Amarnath.....635

012080
OPEN ACCESS
Numerical simulation of dimples in airfoil using MATLAB

P. Booma Devi and Dilip A Shah.....649

012081
OPEN ACCESS
Experimental and numerical analysis of convergent nozzlex

G Srinivas and Bhupal Rakham.....658

012082
OPEN ACCESS
Numerical investigation of mixing characterstics of chevron nozzle by passive controls method

J. Devipriya and Dr. Kanimozhi.....671

012083
OPEN ACCESS
Aerodynamic analysis of seamless horizontal stabilizer

S. Nithya and S. Kanimozhi.....679

012084
OPEN ACCESS
Analysis and Optimization of Central Processing Unit Process Parameters

R Kaja Bantha Navas, Budi Venkata Chaitana Vignan, Margani Durganadh and Chunduri Rama Krishna.....686

012085
OPEN ACCESS
Experimental and Transient Thermal Analysis of Heat Sink Fin for CPU processor for better performance

S. Ravikumar, Parisaboina Subash Chandra, Remella Harish and Tallapaneni Sivaji.....695

012086
OPEN ACCESS
Design and analysis of Air flow duct for improving the thermal performance of disc brake rotor

T Raja, G Mathiselvan, M Sreenivasulureddy and X. Goldwin Xavier.....701

012087
OPEN ACCESS
MRR and TWR evaluation on electrical discharge machining of Ti-6Al-4V using tungsten : copper composite electrode

J Prasanna, S Rajamanickam, O Amith Kumar, G Karthick Raj and P V V Sathya Narayanan.....705

IOP Conf. Series: Materials Science and Engineering **197** (2017) 011001 doi:10.1088/1757-899X/197/1/011001

Preface

The 2016, 2nd International Conference on Frontiers in Automobile and Mechanical Engineering was held at Sathyabama University, South India between 7th to 9th July. The conference addressed various advances in the fields of Mechanical and Automobile Engineering. Oral and poster sessions addressing different research areas like Materials, Production, Automobile, Aeronautical, Thermal, Design, Internal Combustion Engines, Composites etc., were conducted during the conference. Over 200 delegates from various parts of the globe attended the Conference.

The Eminent speakers for the Conference include Dr.P.Sivakumar, Director, CVRDE, Chennai, Dr.S.Gomathinayagam, Director General, NIWE, Chennai, Dr.K.Elangovan, Professor, Ibri College of Technology, Oman, Dr. Bale V.Reddy, Professor, University of Ontario Institute of Technology, Canada, Dr.Manoj Gupta, Professor, National University of Singapore, Dr.Kannan Sivaprakasam, Graduate Director, St.Cloud State University, USA, Dr.K.Palanikumar, Professor, Sairam Institute of Technology, Chennai, Dr.Eriki.A.Kumar, Nilai University, Malaysia, Dr.G.Nagarajan, Professor, Anna University, Chennai, Dr.G.Kumaresan, Associate Professor, Anna University, Chennai, Dr.P.Sathiya, Associate Professor, NIT, Trichy, Dr.T.Ram Prabhu, Assistant Director, DRDO, Bangalore, Dr.I.A.Palani, Assistant Professor, IIT, Indore, Dr.Om Kumar, Associate Professor, Anna University, Chennai.

Special thanks to Dr.P.Sivakumar, Director, of CVRDE for joining hands and supporting this great event. The credibility of the conference goes to the organisers, internal organising committee and reviewers for their consistent efforts. The selfless dedication and commitment of Ms.J.Lilly Mercy, the Chief Corresponding Editor has resulted in bringing out this volume of proceedings in such a grand manner.

Organising Secretary
Dr.T.Sasipraba, Pro Vice Chancellor, Sathyabama University

Organising Chair
Dr.P.Sivakumar, Director, CVRDE, Chennai

Convenors
Dr.S.Prakash, Dean, School of Mechanical Engineering, Sathyabama University
Dr.S.Ramachandran, Research Head, School of Mechanical Engineering, Sathyabama University,

Co Convenors
Dr.B.Kanimozhi, Professor, School of Mechanical Engineering, Sathyabama University
Dr.G.Senthil Kumar, Professor, School of Mechanical Engineering, Sathyabama University

Chief Corresponding Editor
Ms.J.Lilly Mercy, Assistant Professor, School of Mechanical Engineering, Sathyabama University

IOP Conf. Series: Materials Science and Engineering **197** (2017) 011002 doi:10.1088/1757-899X/197/1/011002

Peer review statement

All papers published in this volume of *IOP Conference Series: Materials Science and Engineering* have been peer reviewed through processes administered by the proceedings Editors. Reviews were conducted by expert referees to the professional and scientific standards expected of a proceedings journal published by IOP Publishing.

IOP Conf. Series: Materials Science and Engineering **197** (2017) 012001 doi:10.1088/1757-899X/197/1/012001

Development of friction material by using precast prefired (pcp f) blocks

R Dineshkumar[1] , E V V Ramanamurthy[2] and Ch Krishnapavanteja[3]

[1 2 3]Department of Mechanical Engineering, Sathyabama University, Chennai-119, India

Email : evvrm.mech@gmail.com & rpdineshkumar1992@gmail.com

Abstract. The braking system used to control and stop automobile system. The braking system converts the kinematic energy into heat energy by friction. The performance of the brake pad depends on composition of friction materials. The asbestos brake pads are carcinogenic nature and it makes so many health problems. The present research work is going to replacement of asbestos by new materials. The new material is made by fused ceramic materials from industrial wastage. In this study the industrial waste are recycled and conducted the suitable test to compare the performance of the new material with existing brake pad material. The wear test was conducted by pin on disc experiment. The non asbestos, nonfused, fused samples are represented by x1, x2 and x3. The new brake pad material is formed by non fused and fused ceramic materials. The brake pads are manufactured by powder compacting process.

Keywords:
Friction, Wear, PCPF Block, Fused PCPF Block, Binder, Pin on Disc.

1. Introduction

The brake pad have been maintain consistent friction coefficient and should not be decompose at high temperatures. Commercially three types of brake pads are available in application such as asbestos, metallic and ceramic brake pads. Asbestos brake pad provided good frictional property and thermal resistance. Asbestos brake pads are carcinogenic nature because of the manufacturer started to find the alternative materials. The ceramic brake pads reduce the vibration, dust and disc or drum wear. The metallic brake pads provide long service life and increased rotor wear [1].the brake pad have been maintain the friction coefficient range from 0.2 to 0.4 under various velocity, friction heat and various operating conditions such as mud and water [2]. The existing brake pads consist of more than 10 elements [3]. Addition of the lubricant increases the stability of friction coefficient and improving the thermal conductivity of brake pad. Although graphite is a important ingredients for maintaining desired hardness level [4]. The binder in the brake pads is used to hold all the particles. Furthermore the addition of filler reduces the cost with increasing productivity of brake pad [5]. The friction materials need the following properties such as fade resistance, wear resistance and corrosion resistance [6]. Sliding velocity is a most important factor for affecting coefficient factor [7]. The wear and friction is plays an important role in development of new material. The wear and friction coefficient is complicated to find because of the following parameters such as particle size, speed, surface temperature [8].the microstructure of the abrasive particles can affect the friction performance of the brake pad [9]. Most of the brake pad materials were conducted wear test with the cast iron disc [10]. Typically 20-30% percentage of solid lubricant required to maintain the coefficient of friction between 0.3-0.4. Commercially various types of metal sulphide used as a solid lubricant in the brake pad [11]. The additions of glass fibre improve the thermal conductivity and diffusivity of brake pad [12]. The friction materials are generally developed by trial and error method or experience helps to develop the product. Now a day the following optimization methods are used to develops the composition of friction material [13].

2. Composition of materials

The various types of brake pads are used in automobile system such as asbestos, metallic, ceramic brake pads. The non asbestos, nonfused and fused ceramic samples are represented by x1, x2, and x3. The synthetic resin is used as a binding agent in a fused ceramic brake pad. The synthetic resins are non toxic nature which prevents from health problems. The synthetic resin provides good mechanical strength of fused ceramic sample (X3). The composition of the materials are have been found by EDAX test. The composition of friction materials are shown in the below tables.

Table 1. EDAX result of non asbestos sample

Elements	Weight (Wt %)
C	28.97
O	20.98
Na	0.48
Mg	0.34
Al	0.53
Si	1.29
S	5.88
Cl	0.57
K	0.26
Ca	2.61
Fe	4.25
Ba	33.85
Total	100

Table 2. EDAX result of nonfused sample

Elements	Weight (Wt %)
C	5.30
O	51.70
Na	0.93
Al	35.85
Si	5.40
Ca	0.35
Fe	0.47
Total	100.00

Table 3. EDAX result of fused sample

Elements	Weight (Wt %)
C	6.97
O	29.29
Na	1.10
Al	4.72
Si	6.24
Ca	2.09
Fe	49.59
Total	100.00

Table 4. The composition of rotor disc

Elements	Weight (Wt %)
C	3.310
Si	1.910
Mn	0.790
S	0.060
P	0.057

3. Properties of friction material and wear test experiment

The wear test was conducted by the pin on disc experimental set up. The non asbestos sample (X1) is prepared from existing commercial brake pad. The new material samples are made by the powder compacting process. The dimensions of the samples are diameter (ϕ) 8 mm and length is 35 mm. The cast iron disc is used as a rotor in the wear test experiment. The samples are manufactured by powder compacting process. The power metallurgy process is economical process which is used increase the productivity. The pin on disc experiment was conducted in dry condition. The properties of the samples are shown in the table 5. The wear test conducted and absorbed result of the samples.

Figure 1. Non asbestos pin.

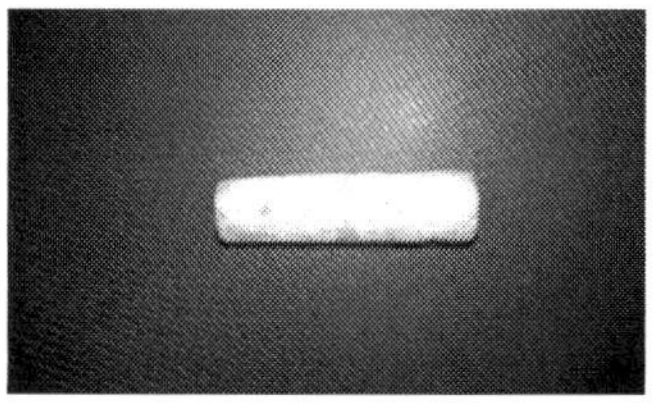

Figure 2. Nonfused ceramic pin

Figure 3. Fused ceramic pin.

Table 5. Physical properties of samples

Properties	X1	X2	X3
Density (g/mm^3))	0.0028	0.0032	0.0036
Hardness (Rockwell)	78	87.5	30.40
Compressive Strength(Mpa)	12.45	31.63	15.41

Figure 4. Pin on disc experiment set up.

IOP Conf. Series: Materials Science and Engineering **197** (2017) 012001 doi:10.1088/1757-899X/197/1/012001

4. Result and discussions

4.1 Friction and Wear Property Of The Material

The wear test was conducted on existing and new brake pad materials. The following graphs show the experimental result of the friction materials. The coefficient of friction and wear rate of a material was noted at every hundred meters. The wear tests of the samples are conducted by pin on disc experiment. The experimental set up was shown in the figure 4.

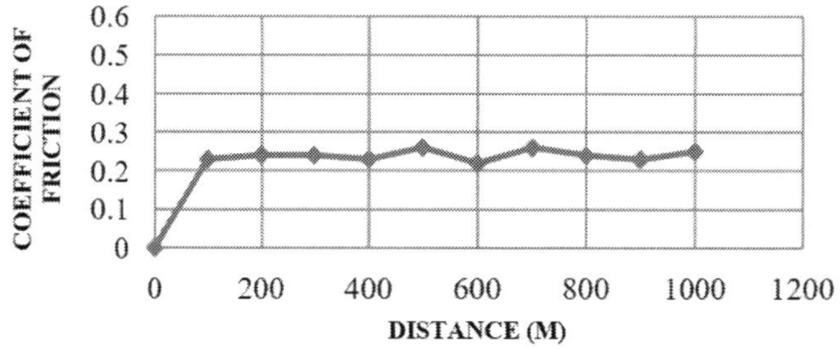

Figure 5a. Coefficient of friction v/s Distance for non asbestos material

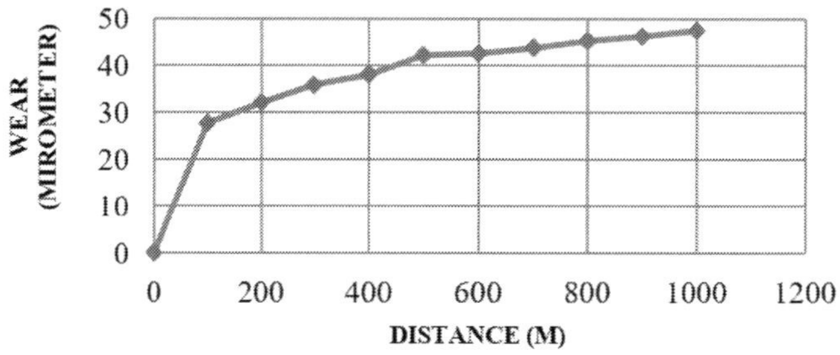

Figure 5b. wear v/s distance for non asbestos material

The non asbestos brake pad material maintains the friction coefficient between 0.2 to 0.3. The presents of graphite maintain consistent friction coefficient. The metallic brake pads are increasing the friction force with increasing applied force on brake pad. After reaching 500 m small fluctuation occurred in frictional force. The cause of fluctuation is adequate amount of graphite presented in the material. The wear rate is gradually increasing above 40 micrometer. The steel fibres increase the wear rate of rotor and increase the fluctuations of friction coefficient.

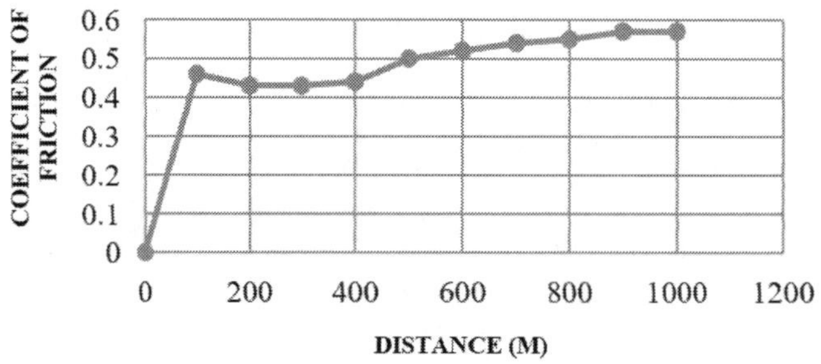

Figure 6a. Coefficient of friction v/s Distance nonfused for ceramic material

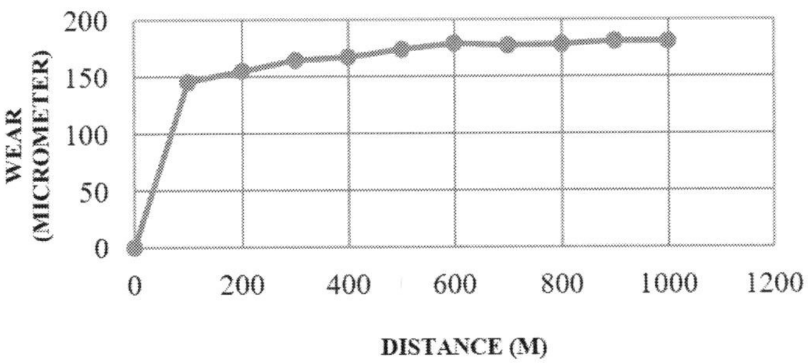

Figure 6b. Wear v/s Distance nonfused for ceramic material

The nonfused ceramic material maintains the friction coefficients between 0.4 to 0.6 shown in figure 6a. The presents of abrasive particle (SiO2) increase the friction force and increase the brittleness of the non fused pin. The amounts of SiO2 particles increase the wear of the rotor disc. The Al2O3 particles are reduced the compressibility of material during sintering process and also decrease the density of non fused ceramic pin. The metal oxides increase the hardness of nonfused friction material. The non fused ceramic sample maintain constant wear rate figure 6b. But the presents of aluminium particles increase the counterpart wear.

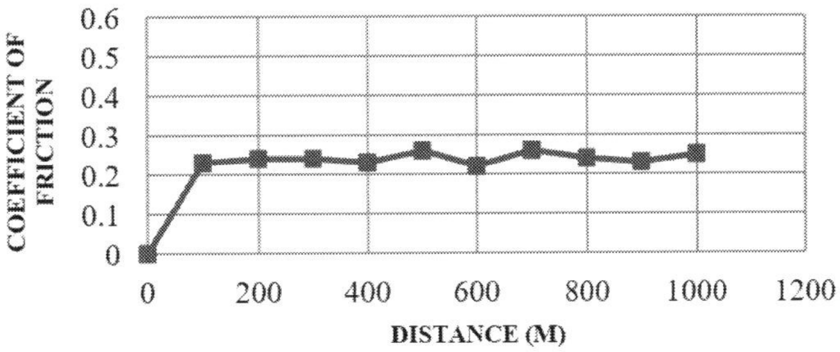

Figure 7a. Coefficient of friction v/s Distance for fused ceramic material

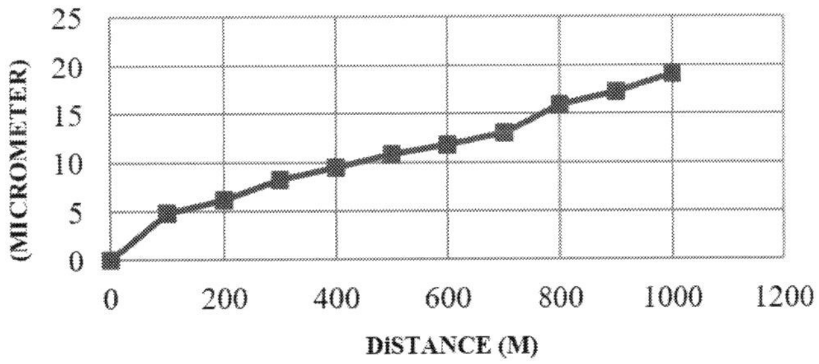

Figure 7b. Wear v/s Distance for fused ceramic material

The fused ceramic materials maintain stable friction coefficient between 0.35 to 0.45 shown in the figure 7a. The steel particles are increasing the mechanical strength of the brake pad. And it is reduce the wear rate of the brake pad. The wear rate of a fused ceramic is less than other materials such as non asbestos and nonfused ceramic materials shown in the figure 8b. The synthetic resin provides good structure rigidity under various operating conditions. The synthetic resins are non toxic nature.

IOP Conf. Series: Materials Science and Engineering **197** (2017) 012001 doi:10.1088/1757-899X/197/1/012001

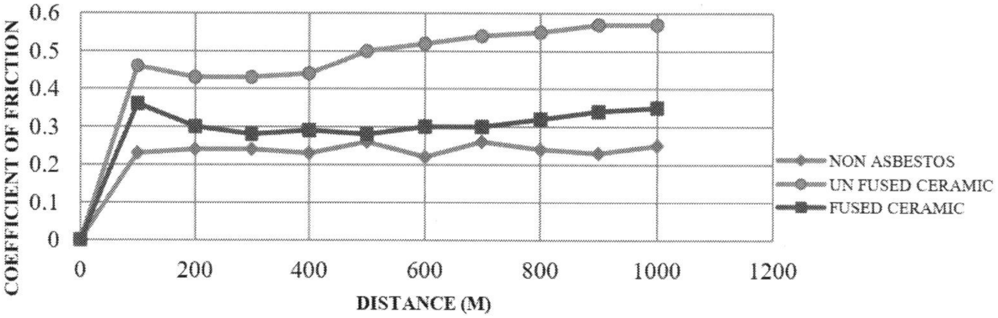

Figure 8a. Coefficient of friction v/s Distance

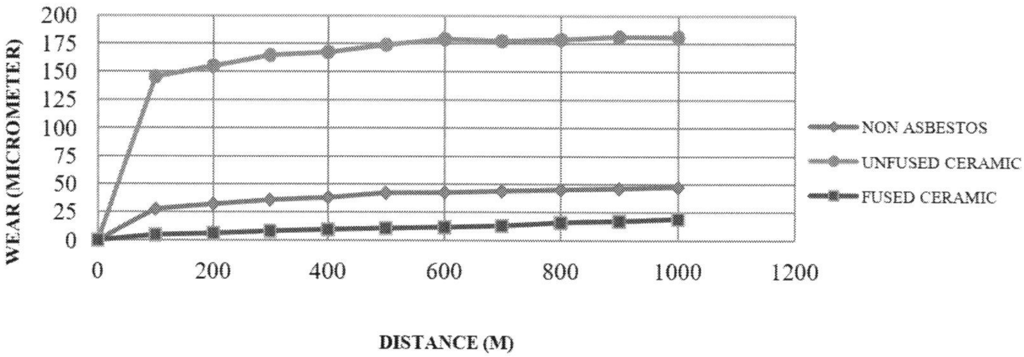

Figure 8b. Wear v/s Distance

The fused ceramic sample gives stable coefficient of friction compared to nonfused and non asbestos samples figure 8a. The SiO2 particles in the nonfused sample increase the friction coefficient and increase the counterpart wear. The commercial brake pad also gives consistent value of friction coefficient. But the wear rate of non asbestos material is more than fused ceramic material. The iron oxide particles reduce the wear rate of the fused ceramic material. The iron oxide particles are acting as an abrasive particle. The consistent friction coefficient shows sufficient amount of graphite and resin are presented in the fused ceramic pin. More amount of abrasive particles increases the inconsistency of friction coefficient and counterpart wear in a nonfused ceramic samples shown in the figure 8b.

The newly developed fused ceramic sample shows better wear resistance figure 8b. The wear of fused ceramic pin gradually increased to 25 micrometer. The non fused ceramic sample wear is increased above 175 micrometer. After reaching 600m the wear of nonfused ceramic pin gradually degreased. The presents of solid lubricant provide the friction stability and reduced the wear of the friction materials. The sufficient graphite content in the friction material maintain stable friction coefficient.

5. Conclusion

The brake pad should be provide consistent friction coefficient under various working condition such as various temperature, water, mud, etc., The fused ceramic material shows better compressive strength. The performance of the brake pad depends on the proper composition of ingredients. The friction coefficient of fused ceramic material is better than other sample. The fused ceramic sample is maintaining constant friction coefficient. The wear rate of fused ceramic pin is less than 25 micrometer. The presents of more SiO2 particles in a nonfused pin increases the coefficient of friction

IOP Conf. Series: Materials Science and Engineering **197** (2017) 012001 doi:10.1088/1757-899X/197/1/012001

is shown in the figure 6a. The metallic brake pad maintains consistent coefficient of friction. Metallic brake pad material maintains the coefficient between 0.2 to 0.3. Compared other friction materials fused ceramic material (x3) provide better braking performance.

Reference

[1] Chan D and Stachowiak G.W April 2004 Review of automotive brake friction materials, *Proc. Instn Mech. Engrs* Vol. **218** Part D: J. Automobile Engineering, pp 957.

[2] Yuji Enomoto and Takashi Yamamoto 1998 New materials in automotive tribology, *Journal of tribology, Tribology letters* **5**, pp 13-24.

[3] Shin M.W, Cho K.H and Lee W.K 2010 Tribological characteristics of binder resins for brake friction materials at elevated temperatures, *Tribol lett* **38**: pp 161-168.

[4] Bhane A.B, Amira Sellami, Mohamed Kchaou, Riadh Elleuch, Anne-Lise Cristol, Yannick Desplanquesl September 2014 Investigation of tribological properties for brake pad material : a review, *International Journal of Emerging Technology and Advanced Engineering*, Volume **4**, Issue 9.

[5] Amira sellami, Mohamed Kchaou, Riadh Elleuch, Anne-Lise Cristol, Yannick Desplanques 2014 Study of the interaction between microstructure, mechanical and tribo-performance of a commercial brake lining material, *Materials and design* **59**, pp 84-93.

[6] Popa N, Onescu C 2008 About the Tribological behaviour of ceramic materials, *Tribology in industry*, Volume **30**, No. 3&4.

[7] Zhencai Zhu, Guoan Chen, Yilei Li June 2011 Analysis on tribological properties of potentially new friction material with response surface method, *Journal of wuhan university of technology-mater.sci.ed*, Volume **26**, Issue 3, pp 499-503.

[8] Nagesh S.N, Siddarsju C, Prakash S V, Ramesh M R 2014 characterization of brake pads by variation in composition of friction materials, *International conference on advances in manufacturing and materials engineering*, AMME.

[9] Sung Soo Kim, Hee Jung Hwang, Min Wook Shin, Ho Jang 2011 Friction and vibration of automotive brake pads containing different abrasive particles, *Wear* **271**, pp 1194-1202.

[10] Liew K.W, Umar Nirmal 2013 Frictional performance evaluation of newly designed brake pad materials, *Materials and design* **48**, pp 25-33.

[11] Osterle.W, Deutsch.C, Gradt.T, Orts-Gil.G, Schneider.T, Dmitriev.A.I 2014 Tribological screening tests for the selection of raw materials for automotive brake pad formulations, *Tribolgy international* **73**, pp 148-155.

[12] Bakloutia.M, Cristolc.A.L, Desplanquesc.Y, Elleucha.R 2015 Impact of the glass fibers addition on tribological behavior and braking performances of organic matrix composites for brake lining, *Wear* **330-331**, pp 507-514.

[13] Han I, Huang L, Zhang J, Lu Y 2006 Optimization of ceramic friction materials, *Compos Sci Technol:* **66**, pp 2895-2906.

Comparison of the Effect of Curing on the Properties of E-Glass/Cyanate modified Epoxy Cross Plied Laminates

Andrew Nallayan W[1&2*], Vijayakumar K.R[2], Usama Tariq Rasheed[3]

[1]Research Scholar (Mechanical Department, St.Peter's University, 600054, Tamil Nadu, India)

[2]Professor (Mechanical Department, Dr. M. G. R. Educational and Research Institute, 600095, Tamil Nadu, India)

[3]Student (Mechanical Department, Dr. M. G. R. Educational and Research Institute, 600095, Tamil Nadu, India)

Email: wa_nallayan@yahoo.com

Abstract High performance polymer composite laminates that are used in Aerospace and Electronics industries requires laminates that are structurally rigid besides exhibiting high stiffness and good di electrical properties. They are required to be transparent to EM waves in order to transmit the signal with almost zero transmission loss. Response of the laminates under different loadings could hence establish a potent material combination with high structural strengths that could be used in sectors dealing with Signal transmissions. The results thus acquired can be used as a database for choosing relatively better materials for Radome and their advanced versions in the coming decades. To augment this, thin laminates with 4 plies with simple stacking configurations of 0/90/0/90 degrees as applicable to a cross plied laminates were fabricated with cyanate ester modified epoxy resin and 1200GSM E glass unidirectional fiber. Flexural and Impact strength were the properties identified for the accessing the structural responses of the Laminate as against room and oven curing conditions. FESEM images were applied to validate the experimental findings.

Index Terms—Cross ply, Cyanate loading, Structural properties, Radome.

1. SCOPE AND OBJECTIVE:

This work attempts to identify the relative assessment of few critical properties pertaining to two laminates done with Hand Lay and Compression molded technique. Laminates were cured under Room and Oven curing conditions respectively (1-2). The Properties that were identified for the assessment were Flexural and Impact strength and further after which they were scanned for FESEM morphologies. A simple 4 ply, cross ply configuration was used for laying the laminate under both the curing conditions. Signal processing divisions with Radar installations in a variety of disciplines needs a cover to protect them, wherein optimized mechanical and Di electrical properties besides thermal and environmental properties are mandatory criteria's. (3) Such covers called as Radome are made by these kinds of Laminates and a wide scope of developing them is a real possibility in the ever demanding high tech world of generation next Communication systems.

2. EXPERIMENTAL WORK

Laminate Fabrication. The material and optimization of the curing agent Epoxy resin, curing agent and unidirectional fibers of 1200 GSM were procured from Sackthi fibrer, Chennai, India and Bisphenol, the Cyanate ester was imported from Shangai Righton, Shanghai, China. Curing agent i.e., the hardener (Tetraethyleneamine), was optimized with regard to Cyanate ester resin (BACY) modified Epoxy resin (Di-glycidyl ether of Bisphenol A (DGEBA)). Optimization was done by varying the form, state and quantity of the curing agent, through number of trials till the defect free curing was observed in the resin mixture. Good potting period of around 15 to 30 minutes with zero defects was observed when 25.33% of the curing agent was mixed by uniform stirring for 10 minutes and later added to the blend of cyanate ester/Epoxy resin mix and allowed to cure. Before this optimization was achieved, defects like continuous foaming resulting sponginess in the cured blend, presence of glassy pin holes of varying sizes inside and open blow holes on the cured surface and presence of partially trapped undissolved cyanate were observed visually. On the addition of the un- optimized curing agent earlier and due to the obvious exothermic reaction, accelerated gelation of the blend was observed, thereby reducing the potable time of the resin mixture (6-7).

The laminate composites were initially fabricated from modified Epoxy resin loaded with 0%,5%,10% and 15% of Cyanate ester for Matrix and E–Glass unidirectional fiber of 410 GSM with four plies kept at the same symmetrical cross ply orientation of (0°/90°) reinforcement (8). Later after finding the optimized laminate from the above constraints, four more laminates of the same stacking configurations of (0°/90°)2 were made by Hand lay method using 1200 GSM unidirectional fiber. All precautions as per the ASTM guidelines were observed while fabricating the laminates.

Different types of loading were subjected on the specimen to find out the flexural and impact strength of the laminates. The Charpy impact test, also known as the Charpy V-notch test, was done in another apparatus consisting of a pendulum of known mass and length that is dropped from a known height to impact a notched specimen of material.

Fabrication of Laminates and Testing. A steel mould plate coated with silicone release agent and then a layer of resin was applied by a brush. The first layer of fiber (300X300 mm2) was placed on the resin and consolidated using roller. The process was repeated to construct four sets of 8 cross plied laminate with 0%, 5%, 10% and 15% of cyanate loadings. Four of the Laminates were cured at room temperature and were then demoulded after 10hours. And another four were compression moulded and latter oven cured. Using ASTM standards and by water jet cutting, specimens were prepared. UTM of associated scientific

IOP Conf. Series: Materials Science and Engineering **197** (2017) 012002 doi:10.1088/1757-899X/197/1/012002

Engg. Works, New Delhi, India, with a maximum testing load rate of 5 ton with a digital encoder with built up software of FIE make was used for performing the flexural tests. The speed of the testing was the relative to the rate of motion of the grips on test fixtures during the test. The grips were tightened evenly and firmly to prevent any slippage. The speed of testing was set at the proper rate and the machine was started. Specimens were subjected to different loading to find out the flexural Strength of the laminate.
The Charpy impact test, also known as the Charpy V-notch test, was done in another apparatus consisting of a pendulum of known mass and length that is dropped from a known height to impact a notched specimen of material.

Figure 1: Flexural testing apparatus

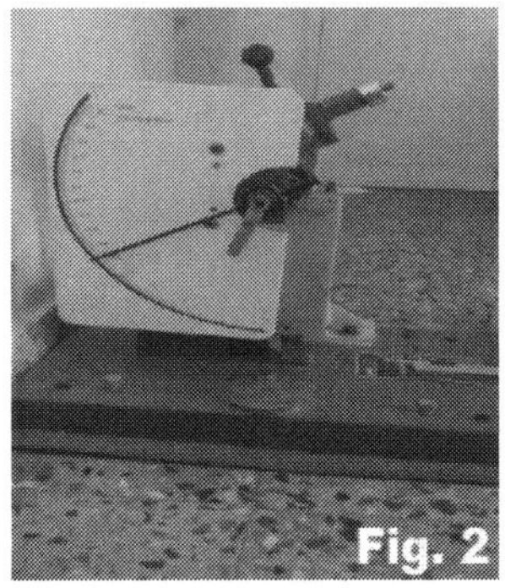

Figure 2: Impact Testing Apparatus

3. MATERIAL CONSTITUENTS OF THE LAMINATE

- Unidirectional Fiber 1200 GSM
- Epoxy blended with cyanate ester
- Hardener (Tetraethyleneamine)

4. METHODOLOGY

The very first and foremost step, before starting to prepare the laminate was to optimize the hardener. The 4, cross plied and balanced laminates were fabricated by using an optimized amount of cyanate ester (15%, found through number of trials). The laminates were put in compression molding (4) for 1 hour at 120^0C and further at 180^0C. Here the air pockets, structural voids, pores, etc. are removed due to pressure acting on the laminate. Compression allows the laminate to get pressed so that excessive amount of the resin blend flows out facilitating even distribution of liquid matrix material /resin. By eliminating the formation of voids and shifting of reinforcing fibers during the compaction process of the compression moulding, results in favorable changes of properties in the finally cured composites.

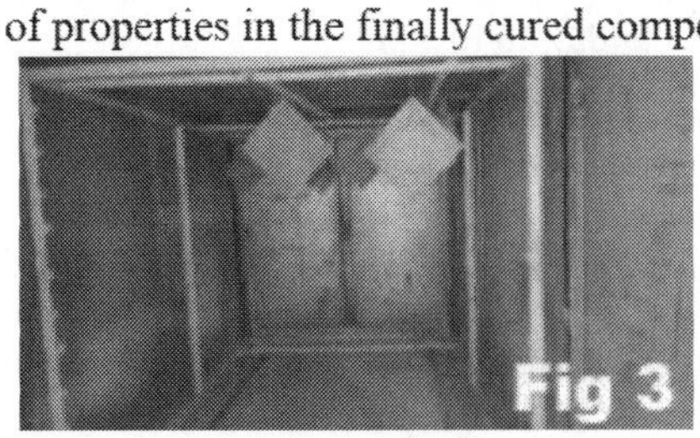

Figure 3: Oven Curing

5. ROOM AND OVEN CURING

Out of the eight samples, four were room cured and the other four were oven cured. The samples that were cured in convection oven were maintained at 2200^{0}C for two hours. After completion of two hours, samples were let to get cooled at room temperature. As a result of curing at elevated temperature the colour of composites turn yellowish or brownish. (5) Some of deformation like slight bending of the laminates was observed due to the relative thinness of the laminate.

6. TESTING AND CHARACTERISATION

Mechanical Test.

Flexural Properties. The flexural strength of a material is the strength value of resistance under bending deformation. For materials that deform significantly but do not break, the load at yield, typically measured at 5% deformation/strain of the outer surface, is reported as the flexural strength or flexural yield strength.

In a three-point flexural testing apparatus load is applied by one supported beam to the center of the specimen. In this study, flexural strength test is performed according to ASTM D 790-03 with three-point test apparatus.

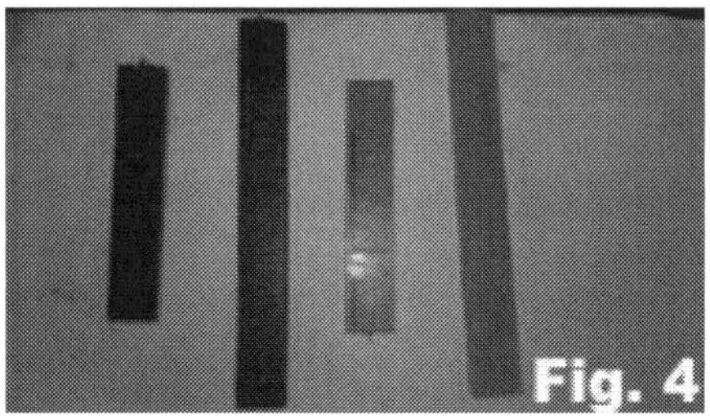

Figure 4: Few of the Flexural and Impact test specimens

Impact strength is the ability of material to withstand impact loading and it gives an idea about the brittleness of material. This property is determined by impact strength testing, measuring energy absorbed by material over a short period of time.

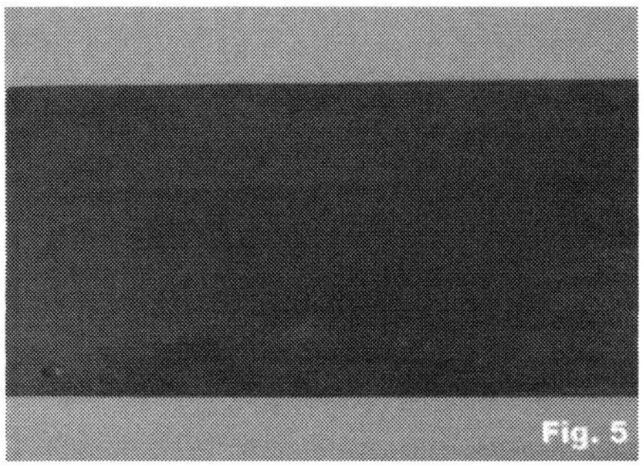

Figure 5: Samples of Compression Moulded Oven Cured

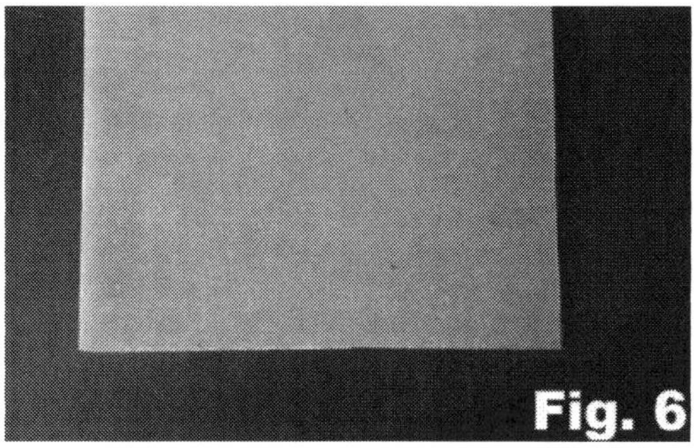

Figure 6: Samples of Compression Moulded, Room Cured

IOP Conf. Series: Materials Science and Engineering **197** (2017) 012002 doi:10.1088/1757-899X/197/1/012002

7. PROPERTY VARIATIONS OF THE OVEN AND ROOM CURED LAMINATES

Table 1 Oven Cured

Test Oven Cured	**Unidiretioal +Epoxy+0% +Cyanate Ester**	**Unidiretioal +Epoxy+5 %+Cyanate Ester**	**Unidiretioa l+Epoxy+10 %+Cyanate Ester**	**Unidiretioa l+Epoxy+15 %+Cyanate Ester**
Flexural Strength (KN)	0.570	0.625	0.90	0.950
Impact Strength (Joules)	7.7	8.4	10.3	10.2

Table 2 Room Cured

Test Room Cured	**Unidirection al+Epoxy+0 %+Cyanate Ester**	**Unidirection al+Epoxy+5 %+Cyanate Ester**	**Unidirection al+Epoxy+10 %+Cyanate Ester**	**Unidirectio nal+Epoxy+ 0%+Cyanat e Ester**
Flexural Strength (KN)	0.120	0.050	0.125	0.750
Impact Strength (Joules)	3	3.9	5	10.8

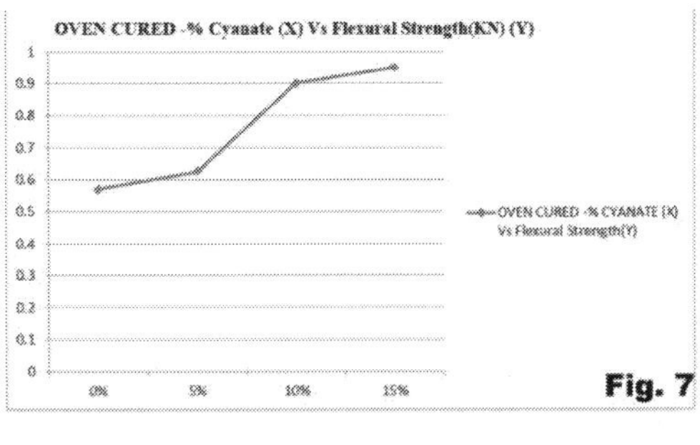

Figure 7: Cyanate Vs Flexural Strength Oven Cured Room Cured

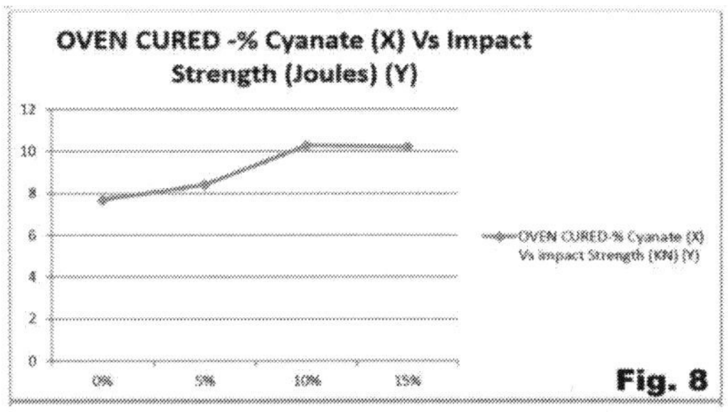

Figure 8: Cyanate Vs Impact Strength

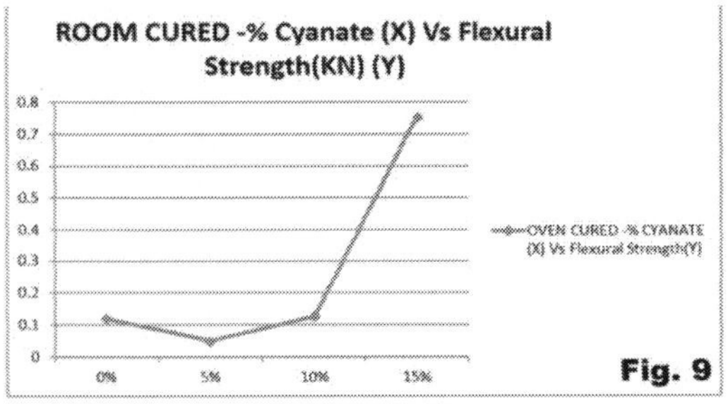

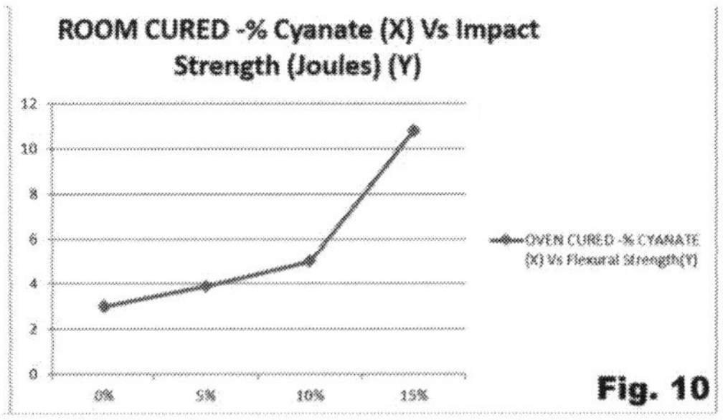

Figure 9: Graph of Room Cured: Cyanate Vs Impact

Figure 10: Graph of Room Cured: Cyanate Vs Flexural Strength

Strength

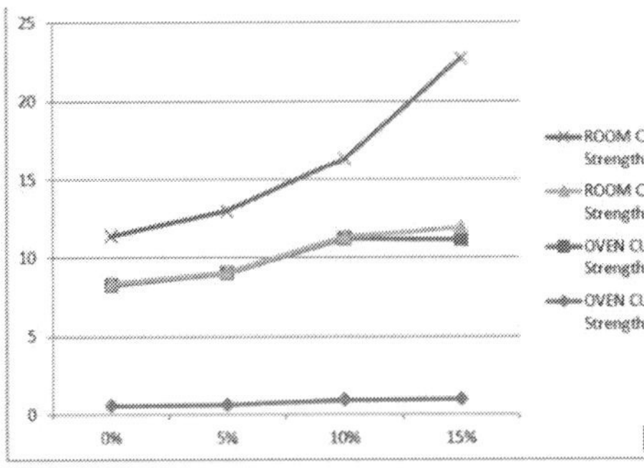

Fig. 11

Figure 11: Graphs of Oven Cured and Room Cured

CONCEPTS OF FESEM

The field emission scanning electron microscope (FE-SEM) is an analytical device used for investigating the surface of the samples through scanning. The scanning takes place by utilizing high-energy beam of electrons, these electrons are shot using a field emission gun. The sensitivity of this equipment is appreciably high when compared to the emission of tungsten filament it has 1000 times more power to generate the emissions but this can be achieved only under high vacuum conditions.

The functioning of this emission occurs obediently when the electrons beam exit the electron gun restraining the transaction of these electrons by metallic apertures and magnetic lenses. At the end, the potentials of positive and negative are isolated within the microscopes by detectors pertaining to both the type of charge emissions. The morphology of the particles in our samples was studied by Nova 200 Nano Lab Field Emission Scanning Electron Microscope (FE-SEM).

IMAGES AND DISCUSSIONS

Mechanical test results of laminates in general are dependent on both the fibre and matrix characteristics (7). The other main factors influencing are the bonding effectiveness of the fibre and matrix in transferring the loads and the laminate geometry of the composite panel. When it comes to the flexural test, the maximum tensile strength of the fibres present at the outer most layers of the laminate controls the strength and in the matrix front the polymer chemistry between the epoxy and the cyanate ester blend under the curing conditions play a decisive role in dictating the properties. When comparing the plain epoxy and cyanate ester modified version of the epoxy laminates, the latter is bound to show a higher value owing to increased toughness provided by the cyanate ester due to a higher cross linked density because of the formation of a network structure between cyanate ester and the epoxy matrix. Moreover, when the laminates are oven cured the cyanate increases this cross link density further.

Impact strength of the laminate is also influenced by the degree of toughness of the laminate in general and this also varies under room and oven curing conditions. As discussed above the main factor here will be the role played by the matrix chemistry besides to some extent the fibre geometry and the laminate configurations. The higher cross sectional area of the fibre besides the ability to behave in a ductile way by the laminate under the sudden loads could possibly decides the resultant impact strength.

The above discussions based on the theoretical concepts and earlier research works have been found to be proven under the present investigations.

IOP Conf. Series: Materials Science and Engineering **197** (2017) 012002 doi:10.1088/1757-899X/197/1/012002

The validations in terms of the FESEM images have been discussed as below:

In flexural under oven cured conditions, the higher magnification of 1000x to 1200x clearly shows few fibres getting broken in the outermost layers with elongated cross linked fibres in the inner layers in the cross plied layers of the laminate. Under lower magnification the straight line fracture with minimum bristles with almost a perfect resin coating are revealed as expected. Obviously the test also showed a value of 950 KN as against 750 KN under room curing conditions, shows a higher flexural strength for an oven cured specimen.

The presence of straight line mode fracture at the peripheral layers depict a relatively a fast brittle fracturing mode even under flexural loading. In the room curing conditions, under the higher magnifications, it is observed the presence of sheared matrix with not much fibre deformities. But the lower value of 750 KN is an indication of the underperforming cyanate under room temperatures. Though the matrix plays a second fiddle in load carrying task, the warp and weft nature of the observed cross linked fibres and a sheared resin could have played a role in bringing the value down.

Impact strength of the room cured specimen showed a higher value of 10.8 Joules than the oven cured one with 10.2 Joules. Though the cyanate ester enhances the strength better under a higher curing temperature by improving the toughness of the matrix, the presence of intact fibres ,the main load bearers along the loading directions under the higher magnification besides the presence of a partially sheared resin coat under low and latter a completely removed resin coat at a higher magnification reveals ultimately a ductile mode of fracture to justify a higher value of the impact strength for the room cured specimen.

8. OVEN CURED SPECIMEN /FLEXURALTEST/ ESEM IMAGES UNDER DIFFERENT MAGNIFICATIONS:

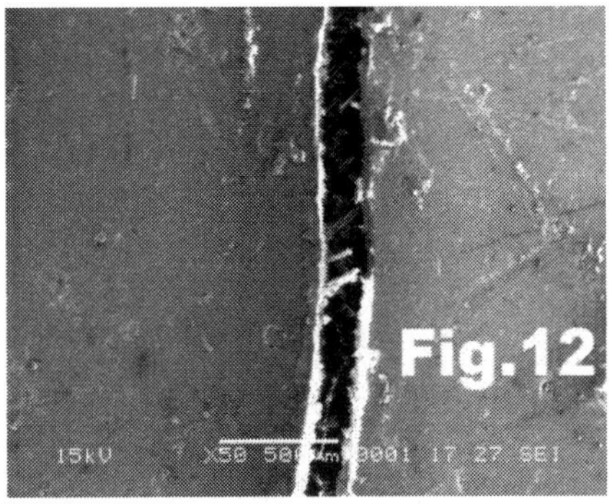

Figure 12

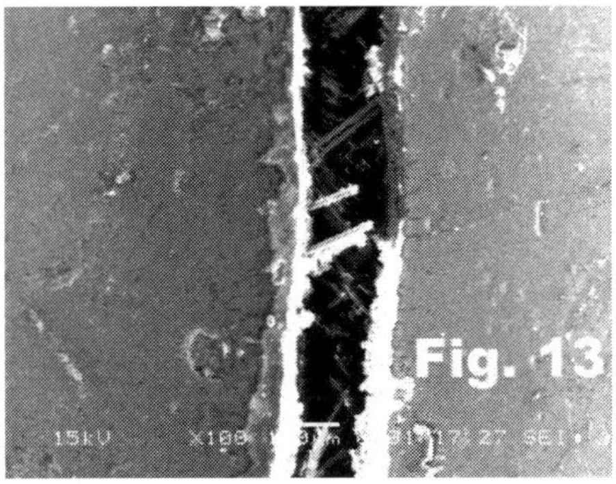

Figure 13

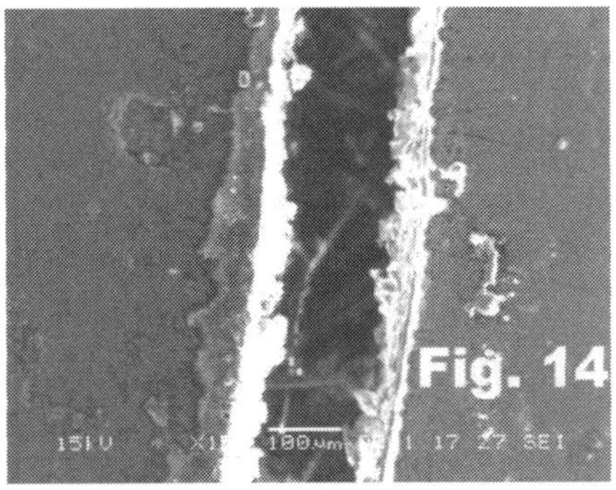

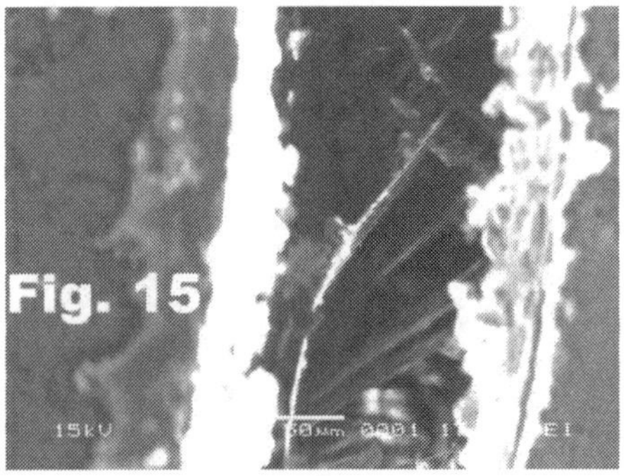

IOP Conf. Series: Materials Science and Engineering **197** (2017) 012002 doi:10.1088/1757-899X/197/1/012002

Figure 14

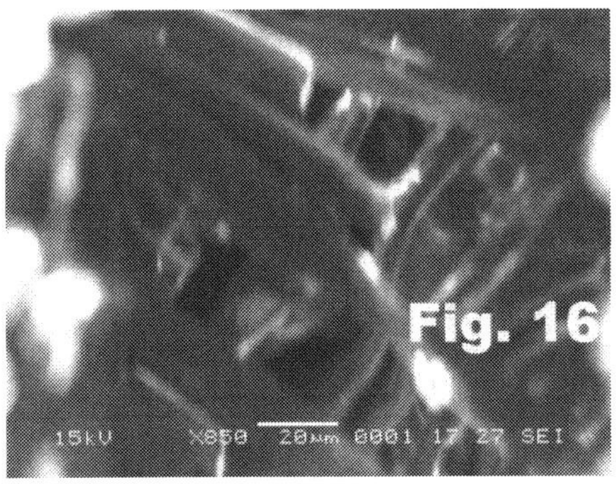

Figure 15

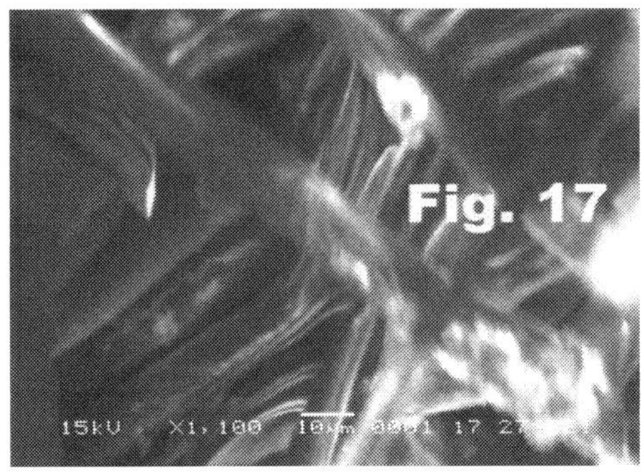

Figure 16

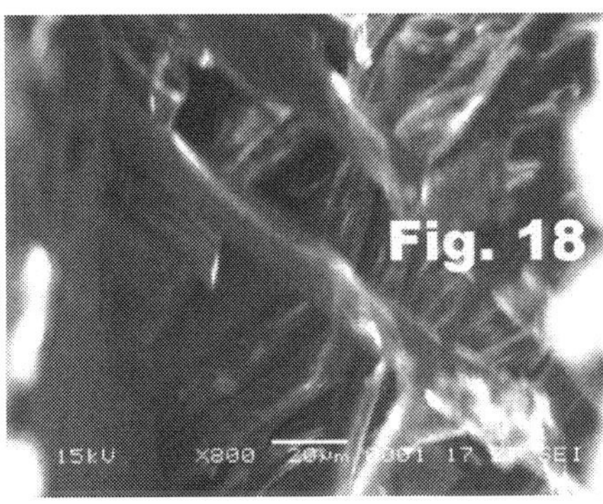

Figure 18

Figure 17

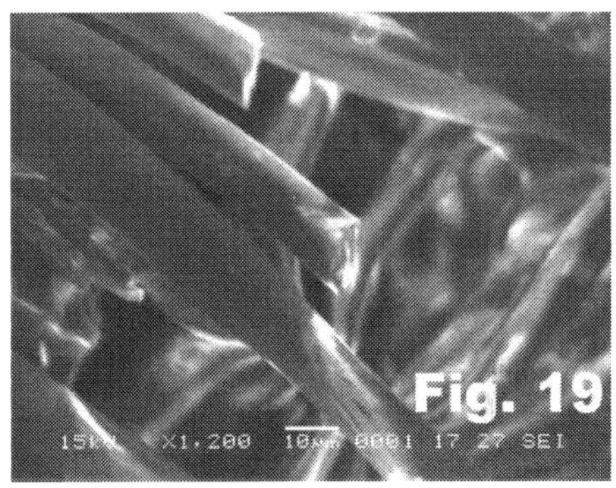

Figure 19

Figure 12: Shows fracture bonding is linear in nature typical of brittle Fracture. The fibre have cut with least bristles and no resin coating found flawed.

Figure 13: Higher magnification shows, the presence of spokes in the inner layer after fracture. However, the outer surface shows a (Straight line) fracture typical of Brittle fracture. The lower typical impact value is in correlation with the fracture nature.

Figure 14: Shows the top layer with the dominant resin layer that has fractured, and the inner layer shows the elongated fibre cross-linking.

Figure 15: Similar to the above but at the higher magnification of 200X showing the existence of fibre Cross linking

Figure16: Higher magnification of SEM at the inner fractured zone shows the cross linked fibre almost intact. The sandwiched fibre remains intact except elongated in the direction of force.

IOP Conf. Series: Materials Science and Engineering **197** (2017) 012002 doi:10.1088/1757-899X/197/1/012002

Figure 17, 18 and 19: The higher magnification of the fracture shows few fibres that have been cut at the peak load.

9. OVEN CURED SPECIMEN /IMPACT TEST/ FESEM IMAGES UNDER DIFFERENT MAGNIFICATIONS:

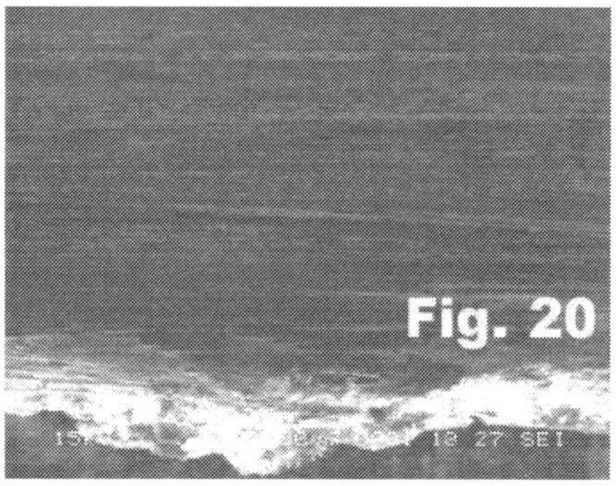

Figure 20

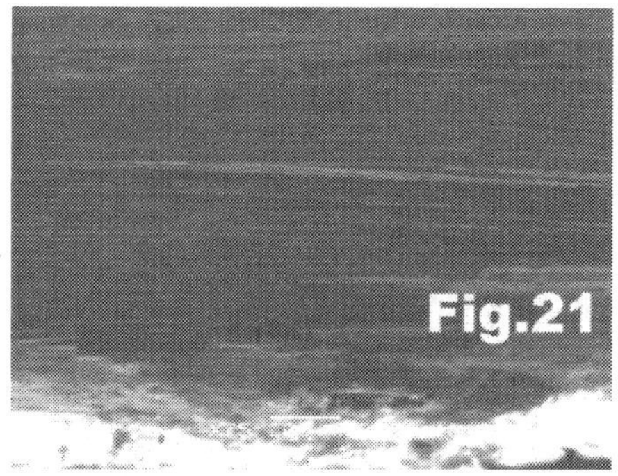

Figure 21

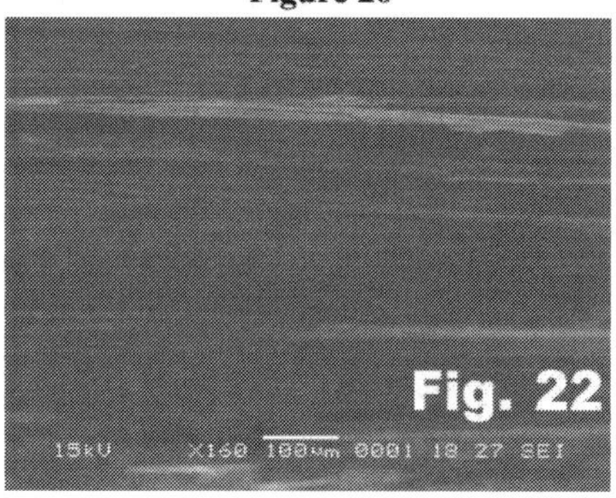

Figure 22

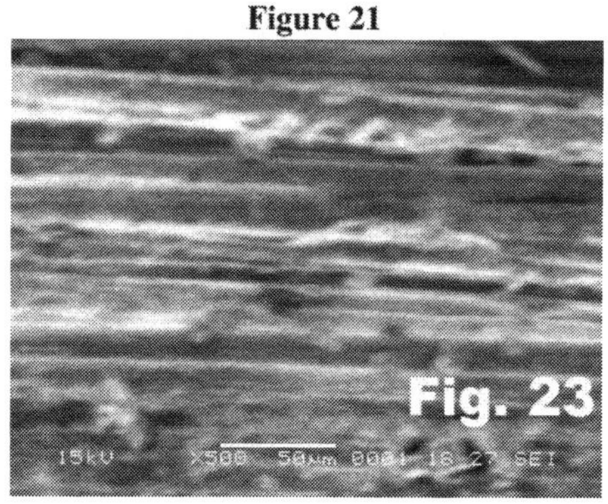

Figure 23

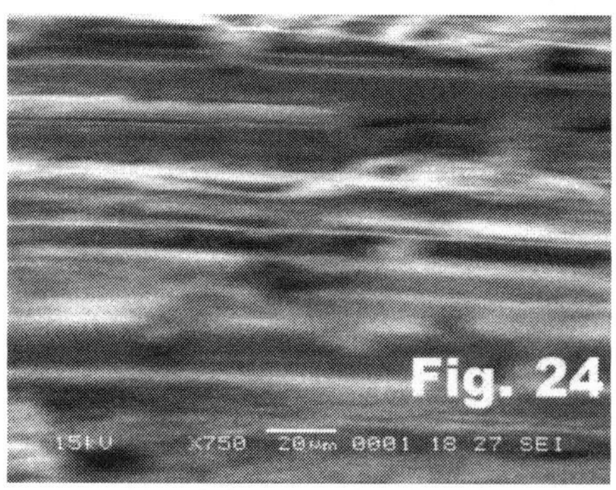

Figure 24

Figure 20: The SEM image shows the fracture edge with fine finished surface of the resin. The fracture is brittle type as under 40X.

IOP Conf. Series: Materials Science and Engineering **197** (2017) 012002 doi:10.1088/1757-899X/197/1/012002

Figure 21: Shows the fracture edge of the sample with dominant resin surface with no fibers observed. The magnification is 75X.
Figure 22: Shows the fractured zone with fiber free location and only resin mixture is seen at 160 X.
Figure 23: This fractured zone at 500 X shows the alternate layers of sandwiched fiber with resin.
Figure 24: Shows the same at higher magnification.

10. ROOM CURED SPECIMEN /FLEXURAL TEST/ FESEM IMAGES UNDER DIFFERENT MAGNIFICATIONS

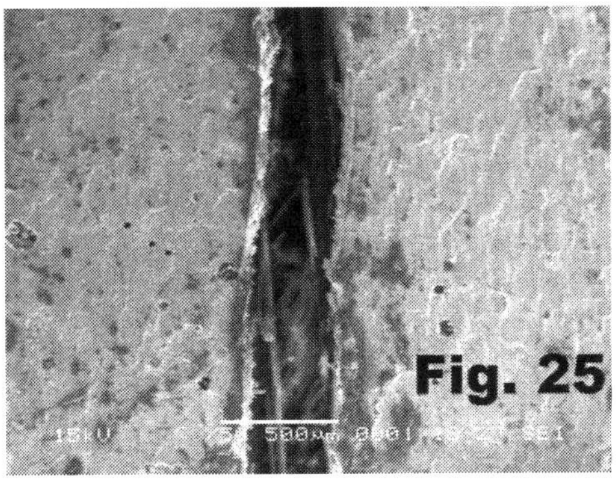

Figure 25

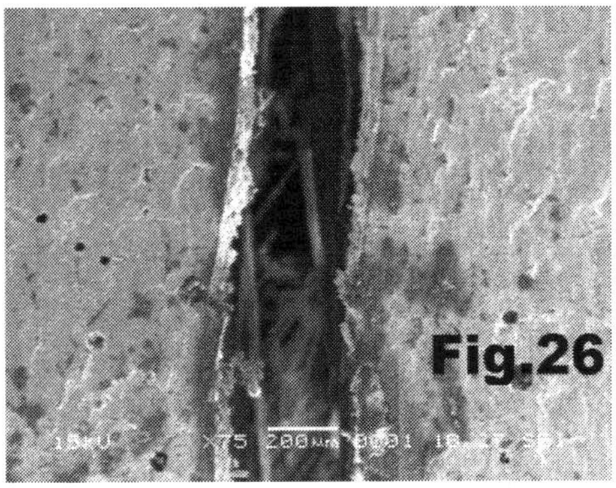

Figure 26

Figure 27

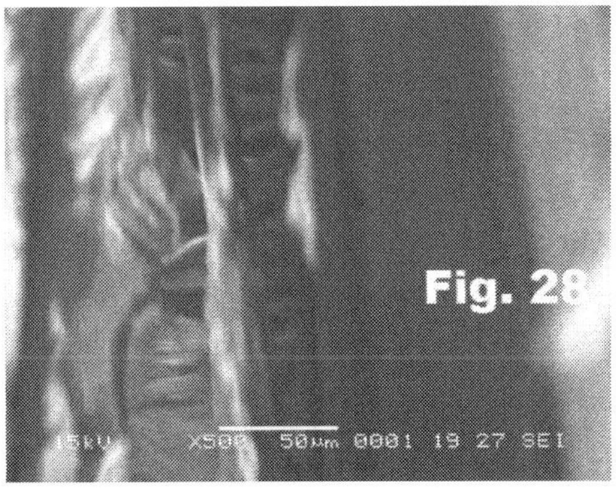

Figure 28

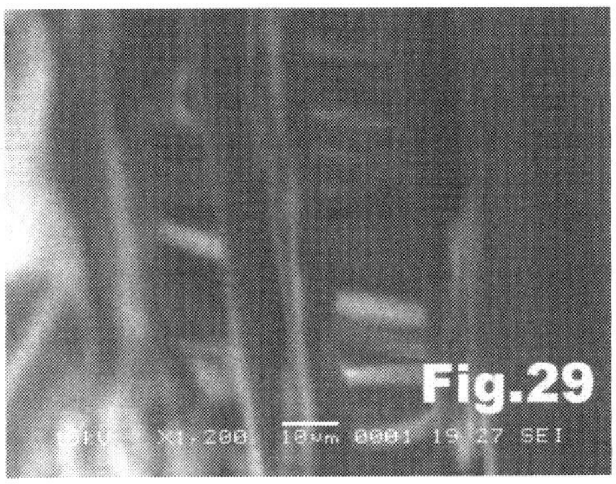

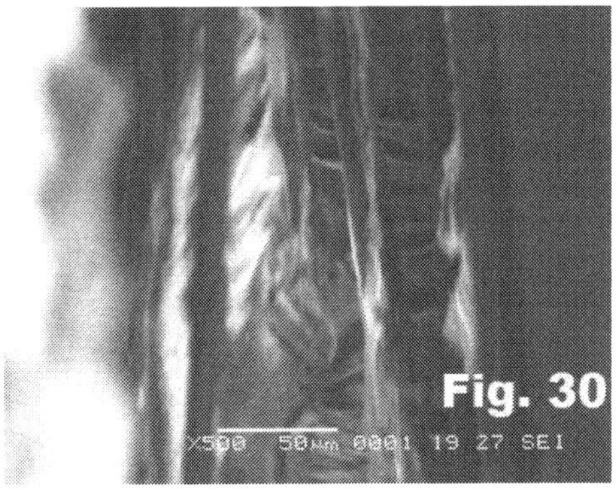

Figure 29 **Figure 30**

Figure 25: Shows fracture lines in the clear presence of fibers in the sub surface. Magnification here is 50X.
Figure 26: Similar to with top resin surface with brittle fracture. But the inner cross linked fiber not fractured.
Figure 27: Magnification at 250 X with alternate layers of cross linked fiber and resin alternate layers.
Figure 28: Same as but at higher magnification of 500X. With clear bonding of resin and fibers.
Figure 29: Same but at 1200 X showing the warp and weft of glass fiber cross linked.
Figure 30: Shows the resin layer and fiber. Resin sheared and fiber not sheared.

ROOM CURED SPECIMEN / IMPACT TEST/ FESEM IMAGES UNDER DIFFERENT MAGNIFICATIONS

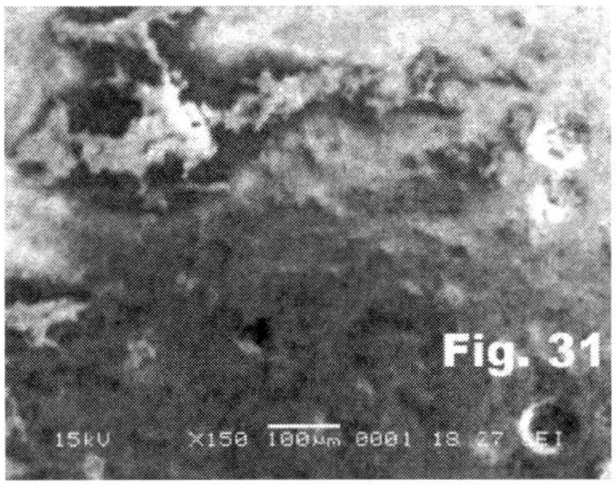

Figure 31

Figure 32

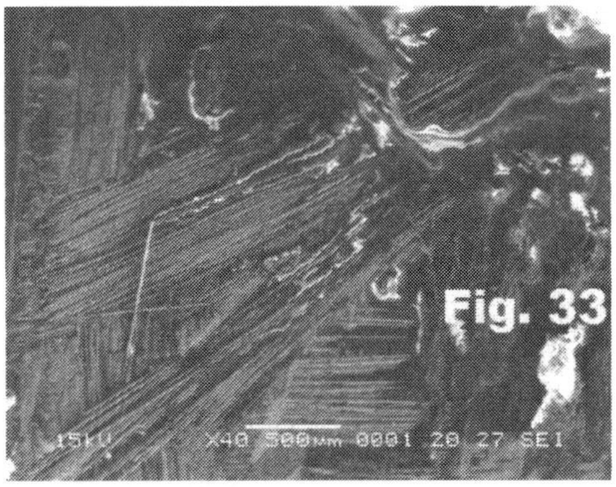

Figure 33

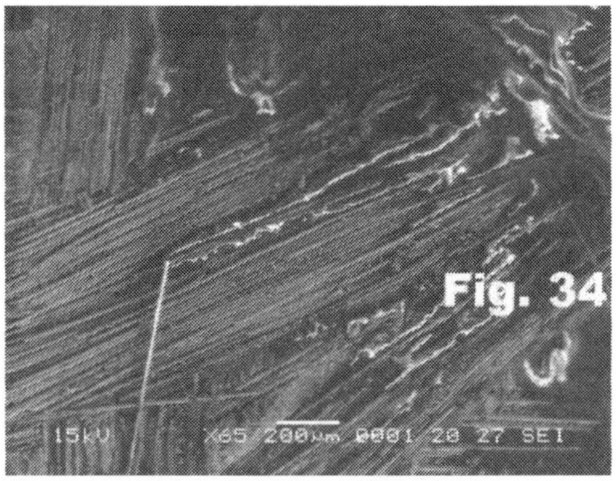

Figure 34

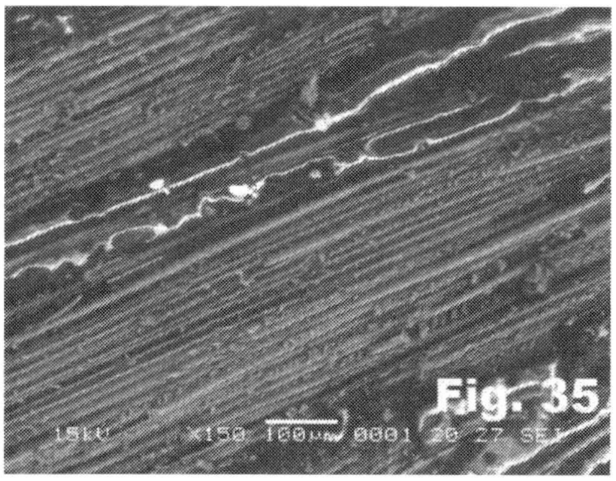

Figure 35

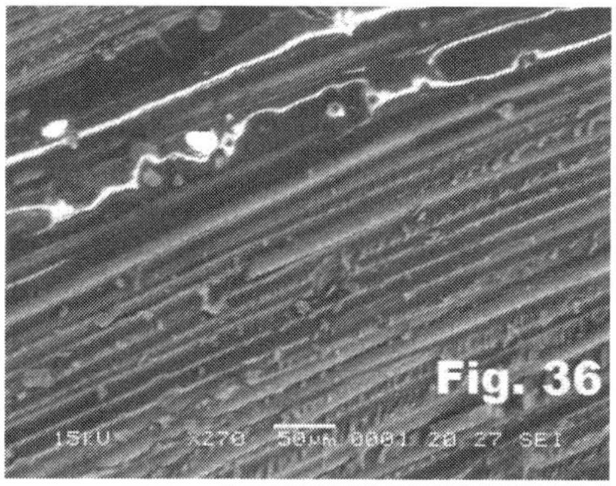

Figure 36

Figure 31: Shows the fractured surface with dominant resins and partially deformed resins. The fracture not seemed to be brittle. Magnification 150X.

Figure 32: Magnification 35 X Shows the fractured surface with resin showing ductile type fracture and sub surface glass fiber elongated.

Figure 33: This image shows complete fiber Cross link with resin completely fractured and removed.
Figure 34: As above at 65 X.

Figure 35: As above at 150 X the fiber part intact along the loading direction without resin bonding.
Figure 36: As above at 270 X

11. CONCLUSION

The role played by Composites in the field of Aerospace and allied areas of Electronics is quiet significant. A variety of material properties could be generated owing to the anisotropic nature of the composite laminates. In the past few decades the emergence of Cyanate ester, as a high performance thermosetting resin is also being investigated and getting proven in many disciplines of Engineering. The investigation in our work with regard to the Cyanate blended epoxy resin in E glass reinforcement in developing Laminates with better structural related properties., like Flexural and Impact strength has been validated with FESEM analysis.

The performance of the laminate was found to be much better under oven curing conditions than room curing and the 15% addition of cyanate ester was also found to exhibit better properties.

Prediction of the laminate for the other properties and hence the overall response apart from these critical properties alone could very well make this particular type of laminate an eligible candidate and a potent structural material for many relevant applications.

Signal processing divisions with Radar installations in a variety of disciplines needs a cover to protect them, wherein optimized Mechanical and Electrical properties besides Thermal and Environmental properties are mandatory criteria's. Such covers called as Radome are made by these kind of Laminates and a wide scope of developing them is a real possibility in the ever demanding high technology world of Generation Next Communication systems.

IOP Conf. Series: Materials Science and Engineering **197** (2017) 012002 doi:10.1088/1757-899X/197/1/012002

12. REFERENCES

[1] A. C, Meeks. (1974). Polymer, 15, 675

[2] R. J. Myslinski (2007).'Synthesis and characterization of cyanate epoxy composites High Performance Polymers'19:33-47

[3] Nikhil Khatavkara and Balasubramanian K.' Composite materials for supersonic aircraft radomes with ameliorated radio frequency transmission"-a review RSC Advances., issue 8, 2016

[4] B.Das,N.Raghupathi,TV Thimons."Process for providing glass reinforced thermoplastic compression molded materials and said molded material (1990) US Patent 4,950.532

[5] JM Berrong,RM Weed,IS Schwartz. "Color stability of selected Dual core composite resin cement"-Journal of Prosthodontics (1993)-Wiley online Library

[6] K.Shilpha "Damage & Degradation study of FRP Composites.(2010)-ethesis.nitrkl.ac.in

[7] LT Drzal 'Fiber matrix interphase structure and its effect on adhesion and composite mechanical properties (1990) Controlled interphase in CM -Springer

[8] Yi Xiao, Ri-Song Qin, Manchu Mahalingam "Effect of geometry and properties of fiber and Matrix on overall Composite Parameters, Journal of Applied science and Engineering Research, Vol. 3, Issue 4, June pp 792-803

IOP Conf. Series: Materials Science and Engineering **197** (2017) 012003 doi:10.1088/1757-899X/197/1/012003

OPTIMIZATION OF PROFILE AND MATERIAL OF ABRASIVE WATER JET NOZZLE

Anand Bala Selwin K P [a], Ramachandran S [b]

[a] Master degree student, Mechanical Engineering Department, Sathyabama University, Chennai, India

[b] Research Head, Faculty of Mechanical Engineering Department, Sathyabama University, Chennai, India

ABSTRACT

The objective of this work is to study the behaviour of the abrasive water jet nozzle with different profiles and materials. Taguchi-Grey relational analysis optimization technique is used to optimize the value with different material and different profiles. Initially the 3D models of the nozzle are modelled with different profiles by changing the tapered inlet angle of the nozzle. The different profile models are analysed with different materials and the results are optimized. The optimized results would give the better result taking wear and machining behaviour of the nozzle.

INTRODUCTION

Abrasive water jet machining is the process of cutting materials with good finish on the machined surface unconventionally. This process can be implemented in broad range of industries where both hard and brittle materials can be machined effectively using this process [2]. The problem associated with these high speed jet nozzles are wear along the nozzle walls [11]. The wear at the wall of the nozzle is mainly due to the complex function of particle impact and wall properties [3]. Also these complex phenomenons include the geometry of the nozzle and other operating parameters [4]. Jukti et al [5] has done major experiment on this abrasive water jet machining process and said that experiments in future had to be done with various ceramic materials.

The main objective of the work is to study the wear in the nozzle walls with different profile and materials. Junkar et al [6] has concluded from his work that an explicit Finite

Element Analysis can be used for understanding the influences of process parameters on it. The nozzle materials vary differently which depends on the operating pressure and nozzle design [7] and hence there are not large materials available for selection of nozzle material. Ye et al [8] in his paper had concluded that the internal shape of nozzle which is critical to the acceleration of abrasive particles and to the wear of the nozzle. The work also states that the optimized inlet angle can lead to less erosion along the walls. Deepak et al [9] has experimented on abrasive water jet machining technique and concluded that increasing the inlet pressure makes an increase in skin friction coefficient.

By taking these aspects into consideration to improve wear resistance and machining process, nozzle geometry and nozzle materials are taken as parameters. The optimized result gives the perfect result to improve the wear and machining behaviour of the nozzle.

EXPERIMENTS AND METHODS

The 3D model of the nozzle is modelled using the Catia V5 R20 software. The basic dimensions taken for modelling the nozzle are

Diameter - 1mm

Length - 20mm

Cylindrical Tube Length - 6mm

Four models of the nozzle is modelled using four different inlet angles respectively. The four different inlet angles used are 50^0, 60^0, 70^0, 80^0.

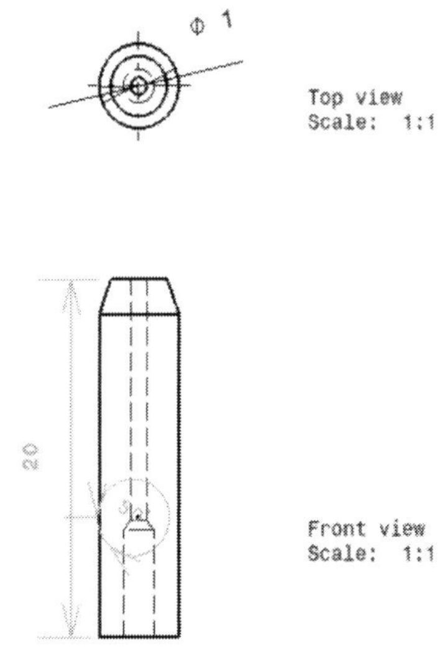

Fig 1 2D model of the abrasive water jet nozzle

In the selection process of the materials the main aspects considered are the wear properties of the material. The ability to withstand high pressure within the nozzle and the ability to withstand the abrasive particles striking the walls during the machining process are the main factor considered for selecting the material. The temperature does not affect the material selection as the whole of the process takes place approximately within the room temperature. The materials that are selected also do not react to the water as water is the carrier of abrasive particles. Based on these aspects, the major materials to be considered in this experiment are

- Hardened Steel (HS)
- Silicon Carbide (SiC)
- Titanium Carbide (TiC)
- Tungsten Carbide (WC)

For four different inlet angles and four different materials, the finite element analysis is done using the Ansys 14 software for a constant load. By performing the finite element analysis on the nozzle the total deformation on the nozzle and the maximum stress induced along the walls of the nozzle can be numerically obtained. Though the numerical analysis doesn't give perfect result when compared to practically made experimental results, it gives proper idea to the perfect results and helps to save cost and valuable time.

The constant load that is used in this project work is 3.5E5 Pa pressure which is assumed to be acting along the cylindrical tube of the abrasive water jet nozzle.

For the four different materials and four different inlet angles the orthogonal array will be 2^4, which is total number of 16 experiments.

The analysis is done and the total deformation and the stress values are noted in the experimental design tabulation for further optimization process.

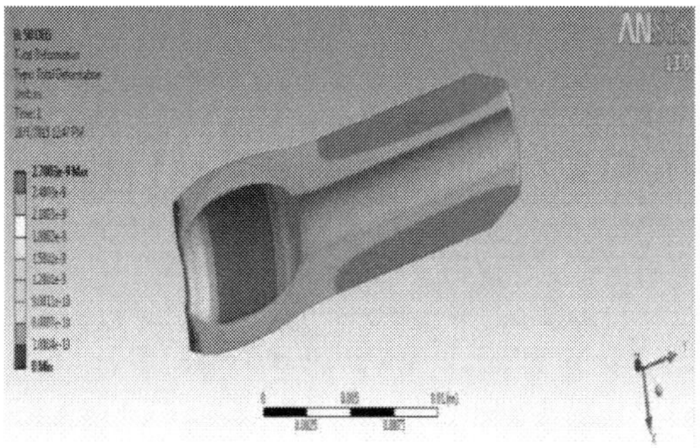

Fig 2: Total Deformation result for 60^0 inlet angle and Hardened steel material

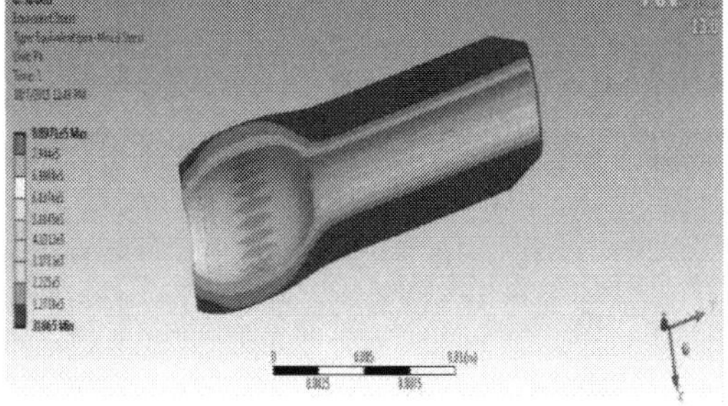

Fig 3: Stress result for 60^0 inlet angle and Hardened steel material

Table 1: Parameters level used in the experiment

Variables	Levels			
	1	2	3	4
Inlet Angle (deg)	50	60	70	80
Materals	HS	SiC	TiC	WC

Table 2: Experimental design with observed results

Trial	Inlet angle (deg)	Material	Deformation (m) $\times 10^{-9}$	Stress (pa) $\times 10^{5}$
1	50	HS	2.7003	8.8971
2	50	SiC	1.2621	9.4299
3	50	TiC	1.3223	9.2638
4	50	WC	0.9689	8.8581
5	60	HS	2.911	9.3033
6	60	SiC	1.3652	9.7185
7	60	TiC	1.4291	9.5809
8	60	WC	0.9992	9.2765
9	70	HS	2.6866	8.5684
10	70	SiC	1.2532	9.105
11	70	TiC	1.3143	8.9388
12	70	WC	0.9222	8.5282
13	80	HS	2.6338	8.8344
14	80	SiC	1.2245	9.3776
15	80	TiC	1.2862	9.2091
16	80	WC	0.9042	8.794

IOP Conf. Series: Materials Science and Engineering **197** (2017) 012003 doi:10.1088/1757-899X/197/1/012003

The optimization process to be used is the Taguchi-Grey relational method as there is more than one response in the experimental design.

The S/N ratio for the observed results are found out using the formula

$$Y_{ij} = -10\log(y^2)$$

which is the formula for smaller the better

Since the deformation and stress values have to be minimised, the approach that is used in this particular experiment is smaller the better for both the responses.

Table 3: S/N ratios for the responses

Trail	Deformation	Stress
1	173.372	-118.985
2	177.393	-119.490
3	177.57	-119.336
4	180.274	-118.947
5	170.718	-119.372
6	177.297	-119.752
7	176.899	-119.628
8	180.01	-119.348
9	171.414	-119.658
10	178.041	-119.186
11	177.62	-119.026
12	180.705	-118.617
13	171.588	-118.924
14	178.237	-118.442
15	177.815	-119.284
16	180.877	-118.884

The next step is to find out the normalised S/N ratio for the above S/N ratio. The formula for normalised S/N ratio for smaller the better is

$$X_{ij} = (\max Y_{ij} - Y_{ij}) / (\max Y_{ij} - \min Y_{ij})$$

Table 4: Normalised S/N ratio for the responses

Trail	Deformation	Stress
1	0.935	0.324
2	0.343	0.769
3	0.325	0.6334
4	0.059	0.291
5	1	0.665
6	0.352	1
7	0.392	0.890
8	0.085	0.644
9	0.931	0.036
10	0.279	0.501
11	0.320	0.360
12	0.0169	0
13	0.914	0.270
14	0.259	0.727
15	0.301	0.588
16	0	0.23

The next step is to find out the grey relational grade, which helps in indicating the relational degree between the sequences of the response. The grey relational coefficient can be calculated using the formula

$$GC_{ij} = (\Delta_{min} + \Psi\Delta_{max}) / (\Delta_{ij} + \Psi\Delta_{max})$$

After calculating the grey relational coefficients, the grey relational grade value is calculated by calculating the average value of the grey relational coefficient of both the responses. Grey relational grade gives the relational degree between the sequences.

Table 5: Grey relational coefficient and Grey relational grade

Trial	Grey coefficient (deformation)	Grey coefficient (stress)	Grey relational grade G_i
1	0.897	0.597	0.768
2	0.603	0.812	0.707
3	0.597	0.732	0.665
4	0.515	0.585	0.55
5	1	0.749	0.875
6	0.607	1	0.804
7	0.622	0.9	0.761
8	0.522	0.737	0.629
9	0.935	0.509	0.722
10	0.581	0.667	0.624
11	0.595	0.609	0.602
12	0.504	0.5	0.502
13	0.921	0.578	0.749
14	0.574	0.785	0.679
15	0.589	0.708	0.648
16	0.5	0.566	0.533

RESULTS AND DISCUSSION

Grey relational grade gives the relational degree between the sequences. After calculating the entire grey relational grade, a response table is created by providing the grey relational grade for each level of the variable.

Table 6: Response table for Grey relational grade

Variables	Level 1	Level 2	Level 3	Level 4
Angle	0.684	0.767	0.6425	0.652
Material	0.790	0.704	0.669	0.583

The above response table gives the value of the grey relational grade for each variables in every level.

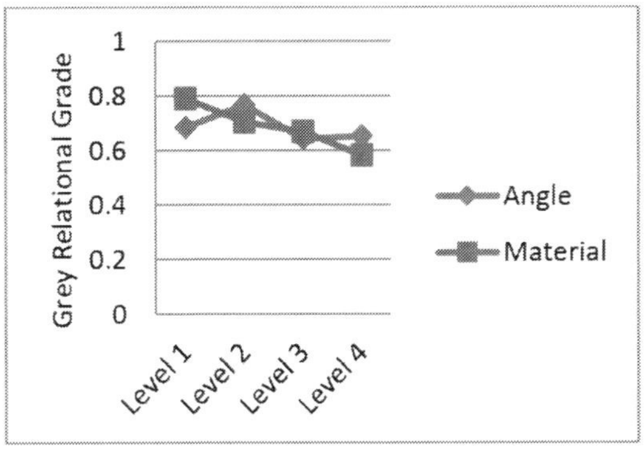

Fig: 4 Grey Relational Grade

CONCLUSION

The Taguchi grey relational analysis optimizes the result to be 70^0 tapered inlet angle and tungsten carbide material. From the experiment the change in wear properties by changing the profile of the nozzle and by changing the material is found out.

The optimised inlet angle and proper selection of material can reduce the wear along the walls of the nozzle. The nano abrasive particles with the nanotechnology can be used for abrasive particles to reduce erosion along the walls of the nozzle and future work can be done in the process of using nano particles for abrasives.

REFERENCE

1. Umang Anand, Joseph Katz, "Prevention of nozzle wear in abrasive water suspension jets (awsj) using porous lubricated nozzles", Journal Of Tribology, 2003
2. Bharanitharan P, "Design and analysis of nozzle in ajm process using computational fluid dynamics", Star journal, 2015
3. Mostafa G, Kwak Yong Kil, Ahn Jung Hwan,"Computational fluid analysis of abrasive water jet cutting head", Journal of Mechanical Science and Technology 24 (2010) 249~252
4. Saurabh Verma, Mishra S K, Moulick S K, "CFD analysis of nozzle in abrasive water suspension jet machining", International Journal of Advanced Engineering Research and Studies E-ISSN2249–8974, 2015
5. Jukti Prasad Padhy, Kanhu Charan Nayak, "Optimization and effect of controlling parameters on ajm using taguchi technique", Int. Journal of Engineering Research and Applications, ISSN : 2248-9622, Vol. 4, Issue 3(Version 1), March 2014
6. Junkar M, Jurisevic B, Fajdiga M, Grah M, "Finite element analysis of single-particle impact in abrasive water jet machining", International Journal of Impact Engineering 32 (2006) 1095–1112
7. Wright D, Wolgamott J, Zink G, "Water jet nozzle material types", 2003 WJTA American Water jet Conference August 17-19, 2003
8. Ye J, Kovacevic R, "Turbulent solid-liquid flow through the nozzle of premixed abrasive water jet cutting system", Proc Instn Mech Engrs Vol 213, 1999
9. Deepak D, Anjaiah D, Vasudeva Karanth K, Yagnesh Sharma N, "CFD simulation of flow in an abrasivewater suspension jet: the effect of inlet operating pressure and volume fraction on skin friction and exit kinetic energy", Advances in Mechanical Engineering Volume 2012, Article ID 186430, 2012
10. Rajeev Kumar, Gurdeep Singh Deol, Kalra C S, Vijay Sharma, "Analysis on performance of different parameters during abrasive jet machining by taguchi method", International Journal of Emerging Research in Management &Technology ISSN: 2278-9359 (Volume-3, Issue-8), August 2014
11. Conn, "A review of the 10th international symposium of jet cutting technology", International Journal of water jet technology, pg 135-149, 1991.

Analysis of copper mixed kerosene servotherm in EDM of Monel 400™

P.A. Anandakumar[1], Baya Molla[2]Fisha Biruke[3] S.Aravind[4]

1. Lecturer, Department of Mechanical Engineering, Dilla University, Dilla, Ethiopia
2. Lecturer & Head, Department of Mechanical Engineering, Dilla University,Dilla University,Dilla, Ethiopia
3. Lecturer & Department of Mechanical Engineering, Dilla University,Dilla University,Dilla, Ethiopia
4.Assistant professor, Department of mechanical Engineering, VelTech Dr.RR .Dr.SR Engineering college.
E-mail : anandme24@gmail.com

Abstract : Powder mixed electro discharge machine process (PMEDM) is a hybrid machine process where a conductive powders is mixed to the dielectric fluid to facilitate effective machining of advanced material. This present study focused on performance of copper mixed kerosene servotherm as dielectric medium in EDM of Monel 400™. The ratio of kerosene sevothermis 75 : 25.The copper powder was mixed with dielectric medium of kerosene servothem of 6g, 8g and 10 g respectively. This mixture was analyzed using different current rate of 8 amps, 10 amps and 12 amps to know the performance characteristics by using material removal rate, tool wear rate, diameter overcut, surface finish and dimensional accuracy. Based on the experimental investigation it is concluded that copper powder of 10g with 10 amps performed well than that of all other parameters.
Keywords : Electrical discharge machining, copper powder, material removal rate,tool wear rate, diameter overcut

Introduction

It is used for produced automobile components and in clinical equipment's. Edm tool is not directly contact with the work piece due to this one it can reject the mechanical stresses chattering and vibrant problems in the time of machining [1]. We can machine any harden and conductive type material only can be machined in Edm machine [2]. The Edm machine working in the principle of thermoelectric energy. It is formed between the work and tool submerged in the dielectric fluid with the supplying electric current. In this process the energy concentrated and repetitive spark discharge is maintain between work piece and tool electrode causing melting and evaporation work piece.Electrode travel towards the work piece till the opening is small enough so that the fascinated voltage is great enough to ionize the insulator medium [3]

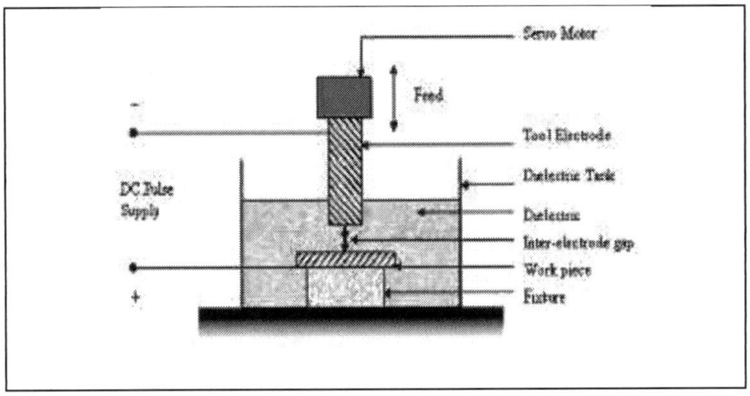

Fig 1

To considered the result of silicon powder adding into dielectric fluid ,the surface finish of H-13 die steel. Machining with adding up of silicon powder mixed with dielectric fluid produces fine and corrosion-resistant surfaces having roughness of 2 μm.[4]. In this EDM process kerosene and distilled water as a dielectric medium. In the operation MRR is more and TWR is less in the time of distilled water as dielectric medium compare to the kerosene in Ti-6Al-4V.[5].

In the present work, the Ti–6Al–4V alloy was machined by electrical discharge machining process with different machining conditions and electrode materials. Summarizing the mean features of the results, the following conclusions may be drawn. The surface integrity of ED Ti-6Al-4V includes roughening because of decomposition of recast layer on the surface, surface microcracks, debris and melted drops. Below the recast layer a slightly softening or tempered layeris occurring due to the low thermal conductivity of Ti6Al4V. The microstructure of white layer may be hexagonal martenstic a zero based on the cooling diagram.

The hardness of hite layer is significantly higher than the bulk material due to the Ti24C15 carbides. The material removal rate, surface roughness and electrode wear are increasing with process parameters for each electrode material except the prolonged pulse duration of 200 ms. Graphite electrode gives the highest material removal rate, followed by electrolytic copper and aluminium. Graphite exhibits the lowest wear rate due to higher melting point at all the applied condition. Aluminium electrode exhibits the best performance with regard to surface finish[6].

In machining parameters for EDM of 94WC-Co.In the case of the Ra parameter the most influential factors were intensity, followed by the pulse time factor, while the duty cycle factor was not significant at the considered confidence level. When either intensity or pulse time were increased, the roughness value also increased. Furthermore, a significant interaction effect was observed between the intensity and pulse time factors.Therefore, in order to obtain high values of material removal rate for the case of tungsten carbide, within the work interval considered in this study, one should use, above all, high values for intensity and duty cycle. Furthermore, although to a lesser extent, low values of the pulse time factor should also be used.[7]

The results of the experiments on EN-31 tool steel with different electrode materials, the following conclusion are arrived at For the EN-31 work material, copper and aluminium electrodes offer higher MRR• Diameteral overcut produced on EN-31 is comparatively low

when using copper and aluminium electrodes, whichmay be preferred for En-31 when low diameteral overcut (higher dimensional accuracy) is the requirement. Copper and copper–tungsten electrodes offer comparatively low electrode wear for the tested work material.
Aluminium electrode also shows good results while brass wears the most, of all the tested electrodes. Of the four tested electrode materials, Cu and Al electrodes produce comparatively high surface roughness for the tested work material at high values of currents. Copper–tungsten electrode offers comparatively low values of surface roughness at high discharge currents giving good surface finish for tested work material.Copper is comparatively a better electrode materials as it gives better surface finish, low diameteral overcut, high MRR and less electrode wear for En-31 work material,and aluminium is next to copper in performance, and may be preferred where surface finish is not therequirement [8].
From the literature review, it has been observed that more extensive work has been done with different tool electrode materials on the Monel 400^{TM}. There exists a great need for investigating the effects of various electrode materials and pulsed discharge currents on material removal rate, electrode wear and surface roughness in electric-discharge machining of Monel 400^{TM}.

Experimental Investigation
The experiments were carryout that EDM spark erosion machine mount on custom built dielectric cyclic system. The experiment carried out the following conditions

Table 1, Experimental conditions

Work piece	: Monel 400^{TM}
Dielectric type	: Copper powder mixed Kerosene + Servotherm oil,
Electrode materials	: Copper
Polarity	: Standards for positive on work piece, negative on electrode
Flushing	: Side flushing
Supply voltage	: 110 V
Average current	: 0.1 to 500 A
Machining Time	: 45 Minutes
Spark gap	: 0.013 to .13 mm

Figure : 2

The chemical composition of the work material is as follows: C – 0.30 max, Mn – 2.00 max, S – 0.024 max, Si – 0.50 max, Ni – 63.0 min, Cu – 28.0-34.0, Fe – 2.50 max. The dimension of the work material was 75mm in diameter and 6mm in thickness. The mean hardness value of the one end work material was 32 HRC. The EDM experiments were performed on NC EDM Glory Engineering. The concentration of the copper powder is varied 6 g / lit, 8 g / lit and 10 g / lit

The chemical composition of MONEL 400 is given in Table 2

Element	Ni	Fe	Si	Mn	C	Cu	S
%	66.50	2.50	0.50	2.00	0.300	31.00	0.024

The machining tests were performed with copper electrode, The tool and material were fixed in the machine and immersed about 5 cm in copper powder mixed kerosene + servotherm (75: 25) as dielectric mediums. Machining test was carried out for 3 mm depth of cut at three different current settings, 8A, 10A and 12A. The mass lost was measured in every current setting. The mass lost from the electrode and work material was weighed using a digital weighing pocket scale and recorded. Machining procedures were repeated for different current ratingand various g / lit of copper powder mixed dielectric mediums

Metal is removed from work in the form of crater due to machining and can be estimated for single pulse rate.

Material removal rate (MRR) = (A-B)/T g/min

Where,

A – Initial weight of work piece

B – Weight of work piece after machining

T – Machining time

The tool wear rate is required to account for tooling cost, machining accuracy and estimate the time of machining for desired depth of cut.
However, there are three types of wear observed in an electrode, end, corner and side.
Tool wear rate (TWR) = (D-E)/T g/ min.
Where,
D – Weight of tool electrode before machining
E – Weight of tool electrode after machining
T – Machining time

Result and Discussion

It was observed that the energy intensity and size of the spark increases with an increase in current, which ultimate by affects the surface finish and dimensional accuracy. The material removed rate and tool electrode wear have been measured by using Fig 3. It shows the influence of pulsed discharge current and liquid dielectric media on MRR for the Monel 400™ using copper tool electrode. From these figure, generally, the trend of the curves is resemblance i.e material removal rate increase with increasing the discharge current is due to fact that the high energy intensity is the spark. The copper electrodes acquire the best material removal rate with the 10 g / lit copper mixed kerosene + servotherm liquid dielectric compared to 6 g / lit and 8 g / lit copper mixed kerosene + servotherm liquid dielectric.

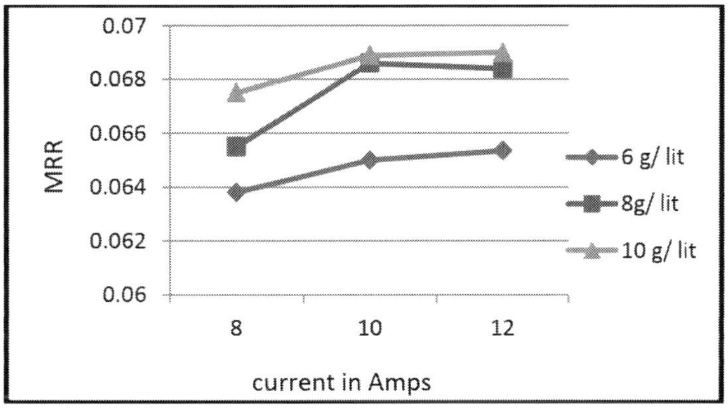

Figure : 3

Fig- 4 show the effect of discharge current and dielectric media on wear of copper tool electrodes while machining of Monel 400™. Generally tool wear rate is due to impingement of positive ions which are being generated during ionization of dielectric and also melting and evaporation of work material, when straight polarity is impressed.
From the four plots in Fig- 4 it laws observed that the tool wear generally increase with increasing discharge current due to enhancement in population density of positive ions and the type of dielectric also influences the tool wear. The tool electrode have minimal wear with 10 g/ lit copper mixed kerosene and servotherm liquid dielectric at all value of discharge current where acquire maximum wear compared to to 6 g / lit and 8 g / lit copper mixed kerosene + servotherm liquid dielectric.
From the above arguments, it is clear to have minimal tool wear, the pair 10 g / lit copper mixture servotherm and kerosene dielectriis recommended for copper tool electrodes for Monel 400™

From the four curves in Fig-4it was observed that the brass tool electrode have minimal wear with mixture kerosene of servotherm liquid dielectric, followed by kerosene.

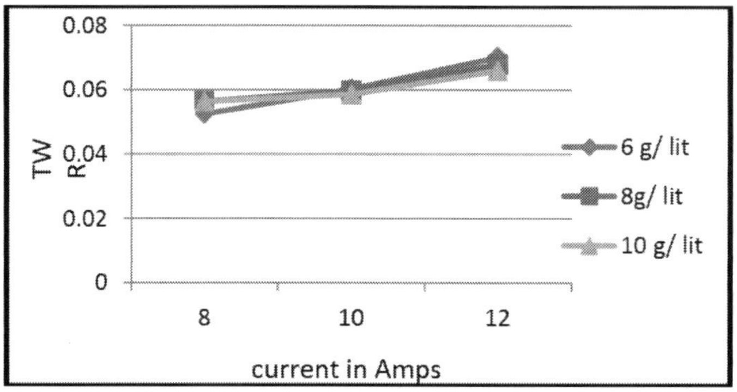

Figure - 4

Figure 5 indicate the effects of discharge current and dielectric media on surface roughness for Monel 400^{TM} using copper tool as electrode. By comparing the three plots in figure 5 it is observed the trends of the curved for 10 g / lit copper mixed kerosene + serovotherm oil low surface roughness. The surface roughness is low compared to to 6 g / lit and 8 g / lit copper mixed kerosene + servotherm liquid dielectric.

Fig 5 indicate the effects of discharge current and dielectric media on dimensional accuracy achieved for Monel 400tm when using copper electrode .

From the curves in figure 5 dimensional accuracy is decreases as the value of discharge current increases in the 6 g / lit and 8 g /lit copper powder mixed kerosene + servothermoil. Due to the fact that the increase causing enlarging the diameter of the crater formed on the surface of the work piece the dimensional accuracy achieved is good for the pair of copper tool and 10 g / lit copper powder mixed Kerosene + Serovotherm dielectric medium.

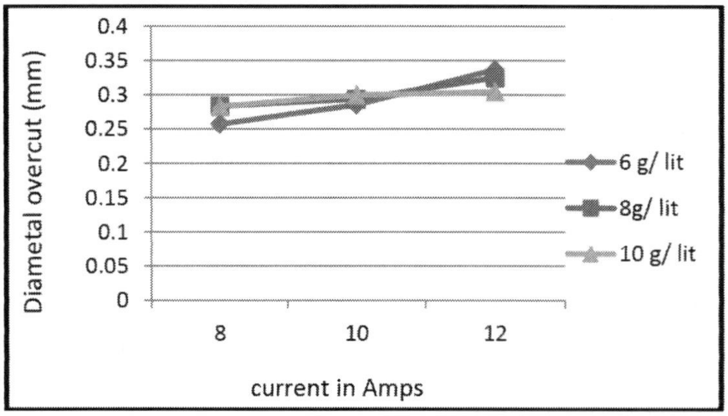

Figutr -5

Conclusion

The influence of copper powder mixed to the dielectric on material removal rate, tool wear rate, surface roughness and dimensional accuracy was studied, and the conclusions were drawn as follows,

- The copper electrodes acquire the best material removal rate with the 10 g / lit copper mixed kerosene + servotherm liquid dielectric compared to 6 g / lit and 8 g / lit copper mixed kerosene + servotherm liquid dielectric.
- The minimal tool wear, the pair 10 g / lit copper mixture servotherm and kerosene dielectriis recommended for copper tool electrodes for Monel 400^{TM}
- The surface roughness is low compared to to 6 g / lit and 8 g / lit copper mixed kerosene + servotherm liquid dielectric.
- The dimensional accuracy achieved is good for the pair of copper tool and 10 g / lit copper powder mixed Kerosene + Serovotherm dielectric medium.

References

1. K.H.Ho, S.T. Newman, State of the art Electrical discharge machining (EDM), International journal of Machine Tools & Manufacture 43 (2003) 1287-1300.
2. H.Ramaswamy, L.Blunt, Effect of EDM process parameters of 3D surface topography, journal of Materials Processing Technology 148 (2004) 155 -164
3. Singh S., S.Maheswari and P.C.Panday 2004, Some investigations in to the electrical discharge machining of harden tool steel using different electrode materials, J.Mater process technol: 49 , 272:277
4. KittimaSillapasa, SawaiDanchaaivijit, and KuljiraSujirote. (2005), 'Effects of Nickel Powder Size on the Processing of Reaction-BondedSilicon Nitride', - Journal of Metals, Materials and Minerals Vol.15 No.2pp.97-102.
5. S.L.Chen ,B.H.Yan . F.Y, Huang , 1999 influence of kerosene and distilled water as dielecric on the electrical discharge machining characteristics of Ti-6Al-4V,journal of materials processing technology 104 :171-179
6. Ahmet Hascalik,and Ulas Caydas,(2007), Electrical discharge machining of titanium alloy(Ti-6AI-4V),Applied surface science253:9007-9016
7. I. Puertas, Etal,Analysis of the influence of EDM parameters on surface quality, MRR and EW of WC–Co, Journal of Materials Processing Technology 153–154 (2004) 1026–1032
8. Shankar Singh, S.Maheswari, P.C.Pandey Some investigations into the electrical discharge machining of hardened tool steel using different electrode materials, (2004)

IOP Conf. Series: Materials Science and Engineering **197** (2017) 012005 doi:10.1088/1757-899X/197/1/012005

Optimality of cycle time and inventory decisions in a two echelon inventory system with exponential price dependent demand under credit period

Seelam Krugon[1], **Dega Nagaraju**[2]

[1]Mechanical Engineering, Bapatla Engineering College, Bapatla-522101, A.P., India.
[2]Manufacturing Division, School of Mechanical Engineering, VIT University, Vellore-632014, Tamilnadu, India
E-mail:seelamkrugon@gmail.com

Abstract. This work describes and proposes an two echelon inventory system under supply chain, where the manufacturer offers credit period to the retailer with exponential price dependent demand. The model is framed as demand is expressed as exponential function of retailer's unit selling price. Mathematical model is framed to demonstrate the optimality of cycle time, retailer replenishment quantity, number of shipments, and total relevant cost of the supply chain. The major objective of the paper is to provide trade credit concept from the manufacturer to the retailer with exponential price dependent demand. The retailer would like to delay the payments of the manufacturer. At the first stage retailer and manufacturer expressions are expressed with the functions of ordering cost, carrying cost, transportation cost. In second stage combining of the manufacturer and retailer expressions are expressed. A MATLAB program is written to derive the optimality of cycle time, retailer replenishment quantity, number of shipments, and total relevant cost of the supply chain. From the optimality criteria derived managerial insights can be made. From the research findings, it is evident that the total cost of the supply chain is decreased with the increase in credit period under exponential price dependent demand. To analyse the influence of the model parameters, parametric analysis is also done by taking with help of numerical example.

1.Introduction

The environmental factors have led to an enlarged emphasis on supply chain management over the past several decades; they have also been important drivers in the increasing importance and varying role of the purchasing function within organisations. Globalization of business has been increased and the resulting, competitive pressures have created from a seller's to a buyer. As a result, the difficulty of buying decisions has increased and the role of the purchasing organisation in managing the buyer-seller interface has become more critical. New product development activities have been improved, due to increased rate of technological change and introduction of new product have become more resulted in greater demands on suppliers in support of their customers. The company's operations are the control of current assets being procured or produced with certain inventory management policies. Hence coordination will come in picture between two players. In this paper the coordination mechanism are introduced by manufacturer to the retailer. The coordination mechanism can be considered as credit period under the demand function as expressed in terms of exponential price dependent. This research work is presented as follows. Review of literature of the problem is carried out in Section 2. Section 3 explains development of the mathematical model comprises of the features and assumptions, notations used. In Section 4, a numerical investigation is presented. Finally, Section 5 is narrating the conclusions of the present work and indicating future scope of the work.

2.Literature Review

Many of the authors' have demonstrated different approaches for supply chain coordination. Saha et al (2015) modeled supply chain coordination contracts with inventory level and retail price dependent demand. They classified contracts as joint rebate contract, wholesale price discount contract and cost sharing contract under stock and price induced demand for two echelon supply chain coordination perspective. Du et al (2013) explored coordination of two-echelon supply chains using wholesale price discount and credit option. Chakraborty et al (2015) proposed multi-item integrated supply chain model for deteriorating items with stock dependent demand under fuzzy random and bifuzzy environments. The authors considered the non linear function as production set up cost, which is expressed in terms of production rate, suppliers transportations cost, retailer procurement cost. Under the fuzzy random and bifuzzy environment models are framed, to evaluate optimized values of credit periods and total supply chain time with the constraints of budget and space. Jaikumar et al (1990) designed a distribution systems requires two essential decisions, one is number of levels between the customer and the producer and the other is trade discounts in channel management policies. Sana (2011) proposed a dynamical system for an EOQ model for salesmen's initiatives, stock and price sensitive demand of similar products. The author considered the demand rate of product is dependent on salesmen's initiatives and selling price where display space is limited. Also, the rate of replenishment depends on the level of stock of the items. Rahdar et al (2014) proposed coordination mechanism for a deteriorating item in a two-level supply chain system. The author considered the model as several buyers and a manufacturer and assumed the inventory level decreases. Jindala et al (2016) invented integrated supply chain inventory model with quality improvement involving controllable lead time and backorder price discount. The author assumed that lead time follows normal distribution. The total cost is minimized by continuous review by taking number of shipment, order quantity, reorder point, lead time, process quality and back order price quality. Jaggi et al (2016) proposed effects of inflation and time value of money on an inventory system with deteriorating items and partially backlogged shortages. Xiao et al (2010) proposed coordination of a supply chain with consumer return under demand uncertainty. The author is design to find the profitability of the player by using buyback/markdown money contract to coordinate the supply chain under partial refund policy. Sana et al (2014) proposed a three layer supply chain model with multiple suppliers, manufacturers and retailers for multiple items. Recently the impact of future price increase on ordering policies for deteriorating items under quadratic demand are proposed by Shah et al (2016). The author proposed two scenarios about the model as 1. Special order time coincides with retailers' replenishment time and 2. Special order time falls during the retailer's sales period.In this work, two-echelon supply chain is modelled under exponential price dependent demand environment. This research attempt aimed at evaluating the optimal values of decision variables like cycle time, retailers' replenishment quantity, shipment frequencies and the total relevant cost of the supply chain with trade credit.

3. Mathematical Model Development

In the proposed model development, a two-echelon inventory system, comprising of a single manufacturer supplying a single kind of product to a single retailer is considered.

Notation

The following notations are used in the development of the model.

λ Annual demand rate of the retailer (units/year) $(\because \lambda = ae^{-bP_R})$ where a>0, b>0 & a>>b

E_R Retailer ordering cost (in Rs./order)

ρ_R Fixed transportation cost of the retailer for receiving a shipment from the manufacturer (in Rs./shipment)

R_R Unit cost at the retailer (in Rs./unit)

P_R Unit selling price at the retailer (in Rs./unit)

W_e Interest rate earned (in Rs./ Re./year)

W_P Interest rate paid (in Rs./Re./year)

t or j Permissible credit period

E_m Fixed production setup cost (in Rs./batch)

M_m Unit cost at the manufacturer (in Rs./unit)

ρ_m Fixed transportation cost of the manufacturer for shipping a shipment quantity to retailer (in Rs./shipment)

q Shipment quantity from manufacturer to the retailer in each shipment (in units) δ Number of shipments from manufacturer to the retailer

T or D Length of the cycle time at retailer (in years)

φ_R Annual total relevant cost of the retailer (in Rs.)

φ_m Annual total relevant cost of the manufacturer (in Rs.)

φ_S Annual total relevant cost of the supply chain (in Rs.)

3.1. Features and assumption

In the mathematical model formulation, the following features and assumptions are considered.

a) Demand rate is exponential price dependent
b) Infinite Production Rate
c) Instantaneous replenishment rate
d) Manufacturer's inventory level is integer multiple of retailer's inventory level
e) Shortages are not permitted
f) Manufacturer provides the trade credit to the retailer

3.2. Model Formulation

For the model formulation, it is assumed that the manufacturer provides the trade credit to the retailer. Inventory associated cost factors like ordering cost, carrying cost and transportation costs incurred at the retailer and manufacturer are incorporated in the model development as shown below.

3.2.1. Retailer Point

Case I: $(D \geq j)$

Annual replenishment cost of the retailer $= \frac{E_R}{D}$

Annual holding cost of the retailer $= \frac{ae^{-bP_R} DR_R W_p}{2}$ $(\because \lambda = ae^{-bP_R})$

Annual transportation cost of the retailer $= \frac{\rho_R}{D}$

Interest paid by the retailer per cycle $= \frac{R_R ae^{-bP_R} (D-j)^2 W_p}{2}$

Interest paid by the retailer per year $= \frac{R_R ae^{-bP_R} (D-j)^2 W_p}{2D}$

Interest earned by the retailer per year $= \frac{P_R ae^{-bP_R} j^2 W_e}{2D}$

Annual total relevant cost of the retailer is expressed by subtracting the interest earned per year from the sum of annual ordering cost, holding cost, transportation cost and Interest paid per year.

$$\phi_R(D) = \frac{E_R}{D} + \frac{ae^{-bP_R} DR_R W_p}{2} + \frac{\rho_R}{D} + \frac{R_R ae^{-bP_R} (D-j)^2 W_p}{2D} - \frac{P_R ae^{-bP_R} j^2 W_e}{2D} \quad (1)$$

Case II: (j>D)

Interest earned by the retailer per year $= P_R ae^{-bP_R} W_e \left(j - \frac{D}{2} \right)$

Annual total relevant cost of the retailer is obtained by subtracting the interest earned per year from the sum of annual ordering cost, holding cost, transportation cost and Interest paid per year.

$$\phi_R(D) = \frac{E_R}{D} + \frac{\rho_R}{D} + \frac{ae^{-bP_R} DR_R W_p}{2} - P_R ae^{-bP_R} W_e \left(j - \frac{D}{2} \right) \quad (2)$$

3.2.2. Manufacturer Point

Annual replenishment cost of the manufacturer $= \frac{E_m}{\delta D}$ $(\because Q_m = \delta q_R)$

Annual holding cost of the manufacturer $= \frac{(\delta - 1) ae^{-bP_R} DM_m W_p}{2}$

Annual transportation cost of the manufacturer $= \frac{\rho_m}{D}$

Annual total relevant cost of the manufacturer is obtained as the sum of annual ordering cost, holding cost, transportation cost.

$$\phi_m(\delta, D) = \frac{E_m}{\delta D} + \frac{\rho_m}{D} + \frac{(\delta - 1) ae^{-bP_R} DM_m W_p}{2} \quad (3)$$

3.2.3. Total Supply Chain

If the manufacturer and retailer are agreed to follow the joint optimal inventory decision making policy, the expression for the total relevant cost of the entire supply chain is obtained by adding the expressions representing annual total relevant cost of the retailer and manufacturer.

Case I: $(D \geq j)$

The annual total relevant cost of the total supply chain is expressed as:

$$\phi_S(\delta, D) =$$

$$\frac{1}{D}\left(E_R + \rho_R + \frac{E_m}{\delta} + \rho_m\right) + \frac{ae^{-bP_R} DW_p}{2}\left(R_R + (\delta-1)M_m\right) + \frac{ae^{-bP_R}(D-j)^2 R_R W_p}{2D} - \frac{P_R ae^{-bP_R} j^2 W_e}{2D} \quad (4)$$

Optimality Criterion:

For the given value of shipment frequency (δ), the expression representing the annual total relevant cost of the supply chain is convex in terms of optimal cycle time (D). The optimal cycle time is obtained by taking the first order and second-order partial derivatives of equation (4) with respect to cycle time. By equating the first-order derivative to zero, the optimal cycle time is obtained. Hence, $\frac{\partial}{\partial D}(\phi_S(\delta, D)) = 0$

$$\left\{2\left(E_R + \rho_R + \frac{E_m}{\delta} + \rho_m\right) + ae^{-bP_R} j^2\left(R_R W_P - P_R W_e\right)\right\} = D^2 ae^{-bP_R} W_p\left(2R_R + (\delta-1)M_m\right) \quad (5)$$

$$\frac{\partial^2}{\partial D^2}(\phi_S(\delta, D)) = \frac{2}{D^3}\left(E_R + \rho_R + \frac{E_m}{\delta} + \rho_m\right) + ae^{-bP_R}\left(\frac{j}{D^3}\right)\left(R_R W_p - P_R W_e\right) \quad (6)$$

As the second order partial derivative, $\frac{\partial^2}{\partial D^2}(\phi_S(\delta, D)) > 0$ for all values cycle time and shipment frequency and other model parameters, the cycle time becomes optimal. From equation (5), the optimal cycle time is obtained as

$$D^* = \left(\frac{2\left(E_R + \rho_R + \frac{E_m}{\delta} + \rho_m\right) + ae^{-bP_R} j^2\left(R_R W_P - P_R W_e\right)}{ae^{-bP_R} W_p\left(2R_R + (\delta-1)M_m\right)}\right)^{0.5} \quad (7)$$

Similarly, for the given value of cycle time (D), the expression representing the annual total relevant cost of the supply chain is convex in terms of shipment frequency (δ). The optimal value of shipment frequency satisfies the following two-inequality conditions.

$$\phi_S(\delta^*) \le \phi_S(\delta^* - 1) \quad \text{and} \quad \phi_S(\delta^*) \le \phi_S(\delta^* + 1)$$

Upon substitution of relevant values in equation (4) for the condition $\phi_S(\delta^*) \le \phi_S(\delta^* - 1)$

$$\frac{1}{D}\left(E_R + \rho_R + \frac{E_m}{\delta} + \rho_m\right) + \frac{ae^{-bP_R} DW_p}{2}\left(R_R + (\delta-1)M_m\right) + \frac{ae^{-bP_R}(D-j)^2 R_R W_p}{2D} - \frac{P_R ae^{-bP_R} j^2 W_e}{2D} \le$$

$$\frac{1}{D}\left(E_R + \rho_R + \frac{E_m}{\delta-1} + \rho_m\right) + \frac{ae^{-bP_R} DW_p}{2}\left(R_R + (\delta-2)M_m\right) + \frac{ae^{-bP_R}(D-j)^2 R_R W_p}{2D} - \frac{P_R ae^{-bP_R} j^2 W_e}{2D}$$

After simplification and rearranging the terms, the following inequality is obtained as

$$\delta^*\left(\delta^* - 1\right) \le \frac{2E_m}{ae^{-bP_R} D^2 M_m W_P} \quad (8)$$

Similarly, upon substitution of relevant values in equation (4) for the condition $\phi_S(\delta^*) \le \phi_S(\delta^* + 1)$

$$\frac{1}{D}\left(E_R + \rho_R + \frac{E_m}{\delta} + \rho_m\right) + \frac{ae^{-bP_R} DW_p}{2}\left(R_R + (\delta-1)M_m\right) + \frac{ae^{-bP_R}(D-j)^2 R_R W_p}{2D} - \frac{P_R ae^{-bP_R} j^2 W_e}{2D} \le$$

$$\frac{1}{D}\left(E_R + \rho_R + \frac{E_m}{\delta+1} + \rho_m\right) + \frac{ae^{-bP_R} DW_p}{2}\left(R_R + (\delta)M_m\right) + \frac{ae^{-bP_R}(D-j)^2 R_R W_p}{2D} - \frac{P_R ae^{-bP_R} j^2 W_e}{2D}$$

After simplification and rearranging the terms, the following inequality is obtained as

$$\frac{2E_m}{ae^{-bP_R}D^2M_{m}W_P} \le \delta^*\left(\delta^*+1\right) \quad (9)$$

Combining equations (8) and (9), the inequality satisfying optimality for shipment frequency is obtained as $\delta^*\left(\delta^*-1\right) \le \frac{2E_m}{ae^{-bP_R}D^2M_{m}W_P} \le \delta^*\left(\delta^*+1\right)$ (10)

Case II: (D<j)

The annual total relevant cost of the total supply chain is expressed as:

$$\phi_S(\delta,D)=\frac{1}{D}\left(E_R+\rho_R+\frac{E_m}{\delta}+\rho_m\right)+\frac{ae^{-bP_R}DW_p}{2}\left(E_R+(\delta-1)M_m\right)-P_Rae^{-bP_R}W_e\left(j-\frac{D}{2}\right) \quad (11)$$

For the given value of shipment frequency (δ), the expression representing the annual total relevant cost of the supply chain is convex in terms of optimal cycle time (D). The optimal cycle time is obtained by taking the first order and second-order partial derivatives of equation (11) with respect to cycle time. By equating the first-order derivative to zero, the optimal cycle time is obtained. Hence, $\frac{\partial}{\partial D}\left(\phi_S(\delta,D)\right)=0$

$$2\left(E_R+\rho_R+\frac{E_m}{\delta}+\rho_m\right)=D^2ae^{-bP_R}\left\{W_P\left(R_R+(\delta-1)M_m\right)+P_RW_e\right\} \quad (12)$$

$$\frac{\partial^2}{\partial D^2}\left(\phi_S(\delta,D)\right)=\frac{2}{D^3}\left(E_R+\rho_R+\frac{E_m}{\delta}+\rho_m\right)$$

As the second-order partial derivative, $\frac{\partial^2}{\partial D^2}\left(\phi_S(\delta,D)\right)>0$ for all values cycle time, shipment frequency and other model parameters, the cycle time becomes optimal. From equation (12), the optimal cycle time is obtained as, $D^*=\left(\frac{2\left(E_R+\rho_R+\frac{E_m}{\delta}+\rho_m\right)}{ae^{-bP_R}\left\{W_P\left(R_R+(\delta-1)M_m\right)+P_RW_e\right\}}\right)^{0.5}$ (13)

For given value of cycle time (D), the expression representing the annual total relevant cost of the supply chain is convex in terms of shipment frequency (δ). The optimal value of shipment frequency satisfies the following two inequality conditions.

$\phi_S\left(\delta^*\right)\le\phi_S\left(\delta^*-1\right)$ *and* $\phi_S\left(\delta^*\right)\le\phi_S\left(\delta^*+1\right)$

Upon substitution of relevant values in equation (11) for the condition $\phi_S\left(\delta^*\right)\le\phi_S\left(\delta^*-1\right)$

$$\frac{1}{D}\left(E_R+\rho_R+\frac{E_m}{\delta}+\rho_m\right)+\frac{ae^{-bP_R}DW_p}{2}\left(R_R+(\delta-1)M_m\right)-P_Rae^{-bP_R}W_e\left(j-\frac{D}{2}\right)\le$$

$$\frac{1}{D}\left(E_R+\rho_R+\frac{E_m}{\delta-1}+\rho_m\right)+\frac{ae^{-bP_R}DW_p}{2}\left(R_R+(\delta-2)M_m\right)-P_Rae^{-bP_R}W_e\left(j-\frac{D}{2}\right)$$

After simplification and rearranging the terms, the following inequality is obtained as

$$\delta^*\left(\delta^*-1\right)\le\frac{2E_m}{ae^{-bP_R}D^2M_{m}W_P} \quad (14)$$

Similarly, upon substitution of relevant values in equation (11) for the condition $\phi_S\left(\delta^*\right)\le\phi_S\left(\delta^*+1\right)$

$$\frac{1}{D}\left(E_R+\rho_R+\frac{E_m}{\delta}+\rho_m\right)+\frac{ae^{-bP_R}DW_p}{2}(R_R+(\delta-1)M_m)-P_Rae^{-bP_R}W_e\left(j-\frac{D}{2}\right)\le$$
$$\frac{1}{D}\left(E_R+\rho_R+\frac{E_m}{\delta+1}+\rho_m\right)+\frac{ae^{-bP_R}DW_p}{2}(R_R+(\delta)M_m)-P_Rae^{-bP_R}W_e\left(j-\frac{D}{2}\right)$$

After simplification and rearranging the terms, the following inequality is obtained as

$$\frac{2E_m}{ae^{-bP_R}D^2M_mW_P}\le\delta^*\left(\delta^*+1\right)(15)$$

Combining equations (14) and (15), the inequality satisfying optimality for shipment frequency is expressed as

$$\delta^*\left(\delta^*-1\right)\le\frac{2E_m}{ae^{-bP_R}D^2M_mW_P}\le\delta^*\left(\delta^*+1\right)$$

Case III: (j = D)

When the optimal cycle time becomes equal to the credit period, the annual total relevant cost of the total supply chain is obtained as:

$$\phi_S(\delta,D)=\frac{1}{D}\left(E_R+\rho_R+\frac{E_m}{\delta}+\rho_m\right)+\frac{ae^{-bP_R}DW_p}{2}(R_R+(\delta-1)M_m)-\left(\frac{P_Rae^{-bP_R}DW_e}{2}\right)(16)$$

For given value of cycle time (D), the expression representing the annual total relevant cost of the supply chain is convex in terms of shipment frequency (δ). The optimal value of shipment frequency satisfies the following two inequality conditions.

$$\phi_S\left(\delta^*\right)\le\phi_S\left(\delta^*-1\right)\quad and\quad \phi_S\left(\delta^*\right)\le\phi_S\left(\delta^*+1\right)$$

Upon substitution of relevant values in equation (16) for the condition $\phi_S\left(\delta^*\right)\le\phi_S\left(\delta^*-1\right)$

$$\frac{1}{D}\left(E_R+\rho_R+\frac{E_m}{\delta}+\rho_m\right)+\frac{ae^{-bP_R}DW_p}{2}(R_R+(\delta-1)M_m)-\left(\frac{P_Rae^{-bP_R}DW_e}{2}\right)\le$$
$$\frac{1}{D}\left(E_R+\rho_R+\frac{E_m}{\delta-1}+\rho_m\right)+\frac{ae^{-bP_R}DW_p}{2}(R_R+(\delta-2)M_m)-\left(\frac{P_Rae^{-bP_R}DW_e}{2}\right)$$

After simplification and rearranging the terms, the following inequality is obtained as

$$\delta^*\left(\delta^*-1\right)\le\frac{2E_m}{ae^{-bP_R}D^2M_mW_P}(17)$$

Similarly, upon substitution of relevant values in equation (16) for the condition $\phi_S\left(\delta^*\right)\le\phi_S\left(\delta^*+1\right)$

$$\frac{1}{D}\left(E_R+\rho_R+\frac{E_m}{\delta}+\rho_m\right)+\frac{ae^{-bP_R}DW_p}{2}(R_R+(\delta-1)M_m)-\left(\frac{P_Rae^{-bP_R}DW_e}{2}\right)\le$$
$$\frac{1}{D}\left(E_R+\rho_R+\frac{E_m}{\delta+1}+\rho_m\right)+\frac{ae^{-bP_R}DW_p}{2}(R_R+(\delta)M_m)-\left(\frac{P_Rae^{-bP_R}DW_e}{2}\right)$$

After simplification and rearranging the terms, the following inequality is obtained as

$$\frac{2E_m}{ae^{-bP_R}D^2M_mW_P}\le\delta^*\left(\delta^*+1\right)(18)$$

Combining equations (17) and (18), the inequality satisfying optimality for shipment frequency, is expressed as

$$\delta^*\left(\delta^*-1\right)\leq\frac{2E_m}{ae^{-bP_R}D^2M_mW_P}\leq\delta^*\left(\delta^*+1\right)$$

4. Numerical Investigation

In the present section, the optimality of ordering policies and rate of shipment has been tested for a supply chain with the help of numerical data. Based on the following data, the numerical example is devised here to illustrate the model. To solve the model MATLAB program is written. Based on the step-by-step procedure developed above, the optimal values of decision variables and objective function are computed for two-echelon inventory model and the results are tabulated in Table 1.The inventory parameter values: a=6000, b=0.00009, e=0.009,

E_m = Rs. 2000/setup, E_R=Rs. 500/order, M_m=Rs. 100 per unit, R_R= Rs. 210 per unit, PR=Rs.240 per unit, λ = a*e^(-b*PR) units per year, W_P = 0.15 Rs./Re./year, W_e = 0.09 Rs./Re./year.

Table 1: Optimal Values of Decision Variables and Objective Function

Description	t = 10	t = 20	t = 30	t = 40	t = 50	t = 60
T^* (in Years)	0.0599	0.0619	0.1051	0.1051	0.1051	0.1051
q^* (in Units)	397.9	411.4	698.4	698.4	698.4	698.4
δ^* (Integer)	3.0	3.0	2.0	2.0	2.0	2.0
Q^* (in Units)	1193.9	1234.2	1396.7	1396.7	1396.7	1396.7
Φ_R^* (in Rs.)	19665.7	15082.9	13718.6	9733.0	5747.5	1761.9
Φ_m^* (in Rs.)	29615.5	29044.9	21883.2	21883.2	21883.2	21883.2
Φ_S^* (in Rs.)	49281.2	44127.8	35601.8	31616.2	27630.7	23645.1

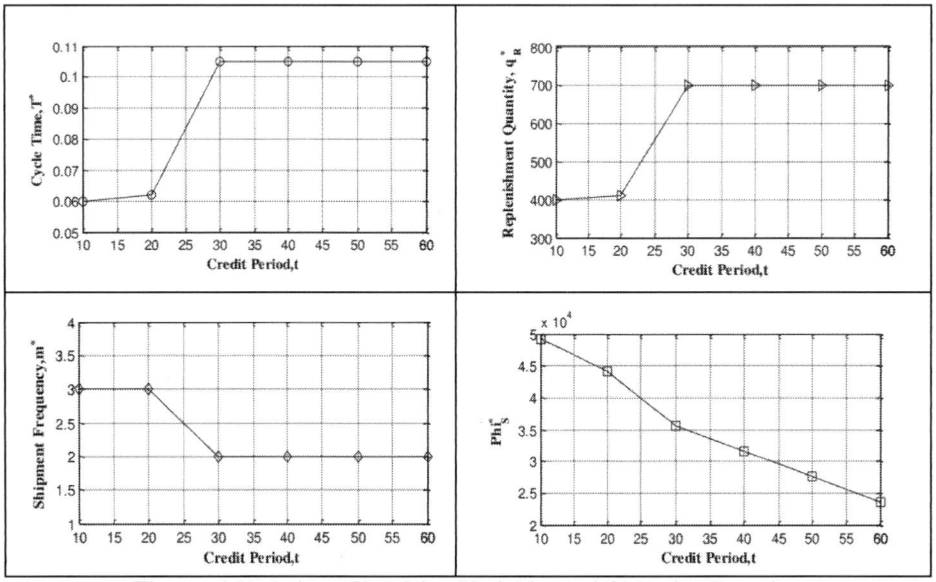

Figure. 1 Variation of Decision Variables and Objective Function w.r.t Credit Period

5. Conclusions

The values of optimal cycle time increases with increase in trade credit. By increasing the trade credit value beyond certain period, there is an sudden increase in optimal cycle time and remain constant. The optimal retailer's replenishment quantity is more, if manufacturer permits more number of days to delay the payments of the retailer. As trade credit increasing it is shown that shipment frequencies are decreases. The model analysis describes that, the total annual relevant cost of the supply chain is less for more trade credit values. From the Figure 2 sensitivity analysis is observed that total annual relevant cost of the supply chain increases with increase in ordering cost, manufacturer set up cost and unit cost at retailer and manufacturer. The retailer's replenishment quantity increases with increase in ordering cost and set up cost and decreases with increase in unit cost at retailer and manufacturer. Similarly, the total annual relevant cost of the supply chain decreases with increase in retailers selling price. A quantitative model for two echelon inventory system is proposed for optimal values of total annual cost of the supply chain, retailer's replenishment quantity, shipment frequencies and cycle time under trade credit with the influence of exponential price dependent demand. The model describes about the study of decision variables and objective function for a supply chain under trade credit with demand is expressed as function of exponential price dependent. A MATLAB program is written to derive the optimality of cycle time, retailer replenishment quantity, number of shipments, and total relevant cost of the supply chain. Also, the parametric analysis is done to observe the behavioral pattern of decision variables and objective function with respect to variation in model parameters.

Consumer goods industries are preferred for the novelty of the present model and research findings. Managerial decisions like cycle time, retailer's replenishment quantities, number of shipments and total annual cost of the supply chain can be decided with the help of this model. The present research work can be extended to a three- echelon supply chain with wide variations and assumptions in demand function.

IOP Conf. Series: Materials Science and Engineering **197** (2017) 012005 doi:10.1088/1757-899X/197/1/012005

Table 2: Sensitivity Analysis

		t=20					t=50					t=80				
		T^*	q^*	δ^*	Q^*	TVCs	T^*	q^*	δ^*	Q^*	TVCs	T^*	q^*	δ^*	Q^*	TVCs
E_m	-40%	0.06	372.1	4.0	1488.3	54009.6	0.11	752.2	2.0	1504.3	35279.8	0.11	752.2	2.0	1504.3	27308.7
	-20%	0.05	363.7	4.0	1454.8	52644.9	0.11	725.8	2.0	1451.5	33482.0	0.11	725.8	2.0	1451.5	25510.8
	+20%	0.06	385.8	3.0	1157.5	47572.8	0.10	669.8	2.0	1339.7	29674.2	0.10	669.8	2.0	1339.7	21703.1
	+40%	0.06	428.0	2.0	855.9	43866.8	0.10	640.1	2.0	1280.1	27645.8	0.10	640.1	2.0	1280.1	19674.7
E_R	-40%	0.06	415.5	3.0	1246.6	51748.7	0.11	725.8	2.0	1451.5	33482.0	0.11	725.8	2.0	1451.5	25510.8
	-20%	0.06	406.8	3.0	1220.5	50528.4	0.11	712.2	2.0	1424.4	32558.1	0.11	712.2	2.0	1424.4	24586.9
	+20%	0.06	388.9	3.0	1166.7	48005.0	0.10	684.3	2.0	1368.5	30655.4	0.10	684.3	2.0	1368.5	22684.2
	+40%	0.05	337.4	4.0	1349.6	48345.2	0.10	669.8	2.0	1339.7	29674.2	0.10	669.8	2.0	1339.7	21703.1
M_m	-40%	0.06	374.5	3.0	1123.6	52727.8	0.10	669.5	2.0	1339.0	33667.1	0.10	669.5	2.0	1339.0	25696.0
	-20%	0.06	385.7	3.0	1157.2	51030.6	0.10	683.5	2.0	1366.9	32652.5	0.10	683.5	2.0	1366.9	24681.3
	+20%	0.06	370.9	4.0	1483.8	48818.5	0.11	714.3	2.0	1428.5	30556.9	0.11	714.3	2.0	1428.5	22585.7
	+40%	0.06	389.0	4.0	1556.2	46276.1	0.10	636.3	3.0	1908.8	29297.0	0.10	636.3	3.0	1908.8	21325.8
R_R	-40%	0.05	324.8	4.0	1299.2	55577.1	0.10	641.5	2.0	1283.1	35829.3	0.10	641.5	2.0	1283.1	27858.2
	-20%	0.05	338.7	4.0	1354.9	53486.0	0.10	668.1	2.0	1336.3	33767.4	0.10	668.1	2.0	1336.3	25796.3
	+20%	0.06	424.9	3.0	1274.7	46448.0	0.11	733.1	2.0	1466.2	29363.0	0.11	733.1	2.0	1466.2	21391.9
	+40%	-	-	-	-	-	-	-	-	-	-	-	-	-	-	-
P_R	-40%	0.06	402.1	3.0	1206.3	49981.5	0.10	671.4	2.0	1342.8	28276.9	0.10	671.4	2.0	1342.8	16653.8
	-20%	0.06	400.1	3.0	1200.2	49632.4	0.10	684.1	2.0	1368.3	30021.3	0.10	684.1	2.0	1368.3	20259.3
	+20%	0.06	395.8	3.0	1187.3	48928.1	0.11	714.3	2.0	1428.6	33062.5	0.11	714.3	2.0	1428.6	26814.0
	+40%	0.06	393.5	3.0	1180.5	48573.2	0.11	732.3	2.0	1464.7	34360.2	0.11	732.3	2.0	1464.7	29768.3
W_P	-40%	0.05	341.2	3.0	1023.6	57378.5	0.09	618.9	2.0	1237.9	37719.4	0.09	618.9	2.0	1237.9	29748.2
	-20%	0.06	365.9	3.0	1097.8	53528.6	0.10	655.1	2.0	1310.1	34759.7	0.10	655.1	2.0	1310.1	26788.6
	+20%	0.07	441.7	3.0	1325.1	44504.4	0.11	751.6	2.0	1503.1	28249.7	0.11	751.6	2.0	1503.1	20278.5
	+40%	-	-	-	-	-	-	-	-	-	-	-	-	-	-	-
W_e	-40%	0.06	394.0	3.0	1181.9	49093.0	0.10	657.9	2.0	1315.8	28166.2	0.10	657.9	2.0	1315.8	17006.6
	-20%	0.06	396.0	3.0	1187.9	49186.9	0.10	677.2	2.0	1354.4	29913.0	0.10	677.2	2.0	1354.4	20347.7
	+20%	0.06	400.0	3.0	1199.9	49375.9	0.11	721.6	2.0	1443.3	33271.5	0.11	721.6	2.0	1443.3	26894.6
	+40%	0.06	401.9	3.0	1205.8	49471.0	0.11	747.4	2.0	1494.8	34873.9	0.11	747.4	2.0	1494.8	30091.2
a	-40%	0.05	473.0	3.0	1419.1	57153.0	0.09	826.3	2.0	1652.6	33952.8	0.09	826.3	2.0	1652.6	22793.2
	-20%	0.05	437.0	3.0	1310.9	53424.6	0.10	765.0	2.0	1530.0	32967.0	0.10	765.0	2.0	1530.0	23401.6
	+20%	0.07	355.1	3.0	1065.4	44588.5	0.12	624.6	2.0	1249.3	29783.8	0.12	624.6	2.0	1249.3	23406.9
	+40%	0.08	306.9	3.0	920.6	39120.9	0.14	540.9	2.0	1081.9	27273.3	0.14	540.9	2.0	1081.9	22490.6
b	-40%	0.06	406.3	3.0	1219.0	50181.5	0.10	712.7	2.0	1425.4	31931.7	0.10	712.7	2.0	1425.4	23629.5
	-20%	0.06	402.1	3.0	1206.4	49729.6	0.10	705.5	2.0	1411.0	31774.9	0.10	705.5	2.0	1411.0	23639.8
	+20%	0.06	393.9	3.0	1181.6	48836.2	0.11	691.3	2.0	1382.6	31455.9	0.11	691.3	2.0	1382.6	23645.3
	+40%	0.06	389.8	3.0	1169.3	48394.7	0.11	684.3	2.0	1368.6	31294.0	0.11	684.3	2.0	1368.6	23640.8

References

[1] Saha S, Goyal SK. Supply chain coordination contracts with inventory level and retail price dependent demand. International Journal of Production Economics. 2015,**161**,140-52.

[2] Du R, Banerjee A, Kim SL. Coordination of two-echelon supply chains using wholesale price discount and credit option. International Journal of Production Economics. 2013, **143(2)**, 327-34.

[3] Chakraborty D, Jana DK, Roy TK. Multi-item integrated supply chain model for deteriorating items with stock dependent demand under fuzzy random and bifuzzy environments. Computers & Industrial Engineering. 2015,**88**,166-80.

[4] Jaikumar R, Rangan VK. Price discounting in multi-echelon distribution systems. Engineering Costs and Production Economics. 1990, **19(1)**, 341-9.

[5] Sana SS. An EOQ model for salesmen's initiatives, stock and price sensitive demand of similar products–A dynamical system. Applied Mathematics and Computation. 2011,**218(7)**, 3277-88.

[6] Rahdar M, Nookabadi AS. Coordination mechanism for a deteriorating item in a two-level supply chain system. Applied Mathematical Modelling. 2014, **38(11)**, 2884-900.

[7] Shah N, Jani M, Chaudhari U. Impact of future price increase on ordering policies for deteriorating items under quadratic demand. International Journal of Industrial Engineering Computations. 2016, **7(3)**, 423-36.

[8] Jindala P, Solankib A. Integrated supply chain inventory model with quality improvement involving controllable lead time and backorder price discount. International Journal of Industrial Engineering Computations. 2016, **7(2)**, 463-80.

[9] Jaggi C, Khanna A, Nidhi N. Effects of inflation and time value of money on an inventory system with deteriorating items and partially backlogged shortages. International Journal of Industrial Engineering Computations. 2016, **7(2)**, 267-82.

[10] Xiao T, Shi K, Yang D. Coordination of a supply chain with consumer return under demand uncertainty. International Journal of Production Economics. 2010, **124(1),** 171-80.

[11] Sana SS, Chedid JA, Navarro KS. A three layer supply chain model with multiple suppliers, manufacturers and retailers for multiple items. Applied Mathematics and Computation. 2014,**229**,139-50.

Comparative Study of Fault Diagnostic Methods in Voltage Source Inverter Fed Three Phase Induction Motor Drive

R. B. Dhumale[1], S. D. Lokhande[2]

Department of Electronics and Telecommunication
Sinhgad College of Engineering
Pune, India. [1, 2]

rbd.scoe@gmail.com

Abstract. Three phase Pulse Width Modulation inverter plays vital role in industrial applications. The performance of inverter demeans as several types of faults take place in it. The widely used switching devices in power electronics are Insulated Gate Bipolar Transistors (IGBTs) and Metal Oxide Field Effect Transistors (MOSFET). The IGBTs faults are broadly classified as base or collector open circuit fault, misfiring fault and short circuit fault. To develop consistency and performance of inverter, knowledge of fault mode is extremely important. This paper presents the comparative study of IGBTs fault diagnosis. Experimental set up is implemented for data acquisition under various faulty and healthy conditions. Recent methods are executed using MATLAB-Simulink and compared using key parameters like average accuracy, fault detection time, implementation efforts, threshold dependency, and detection parameter, resistivity against noise and load dependency.

1. Introduction

In many industrial applications the Voltage Source Inverters (VSI) are used to convert DC power into AC power. In automation industries, VSIs are used to drive AC induction motor (ACIM). Different faults can occurs in ACIM, 38% faults in ACIM are caused due to failure of power devices, electroctrolytic capacitors, and other electronic components [1]. The breakdown of switching devices in power electronics like IGBTs or MOSFETs are broadly classified as [2]:

- Base or collector open fault,
- Misfiring fault and
- Short circuit fault

During IGBT collector or base open situation, IGBT residue in off condition. Thermic cycling is one of the reasons for incidence of base or collector open fault, which root cause is stirring of linking wires. Another reason of IGBT base or collector open fault is fault in driver circuit otherwise a short-circuit-fault. The electrical system is able to work in IGBT base or collector open faulty condition with less performance. The short circuit fault causes because of an incorrect voltage across gate. This can be reason of driver circuit break down or supplementary power supply breakdown. One more reason for short circuit fault is an intrinsic failure which may cause due by overvoltage/ avalanche stress or temperature overshoot. IGBTs can tolerate short circuit fault up to 10 μs [3]. The fault diagnostic methods are classified into Knowledge based systems, Model based systems, Signal based system, and Hybrid-Active fault diagnosis approach. In this paper Knowledge based systems and Hybrid fault diagnosis approaches are discussed. These methods are classified shown in Figure 1. The Discrete Wavelet Transform (DWT) and Fuzzy Logic (FL) are collectively applied for fault diagnosis of single IGBT for variable speed drive systems [9][10]. But this combination is not appropriate for fault diagnosis of multiple IGBTs for variable speed drive system. The features extracted from Detailed Coefficients (DC) of DWT are used to train Artificial Neural Network (ANN) in [11]. This method is suitable for fault diagnosis of multiple IGBTs and less than 5% errors are reported. The Fault Diagnostic Systems (FDSs) based on the combination of Park's Vector Transform (PVT) with the Fuzzy Logic (FL), ANN or Clusterive-Adaptive Fuzzy Neural Network (C-ANFIS) is proposed in [6], [7] and [8]. The methods proposed in [1] and [12] are more robust under variable load conditions but not able to diagnose short circuit faults. The combination of Entropy and ANN has been proposed for fault diagnosis of open circuit and short circuit fault diagnosis in [3]. The methods proposed in [3], [6], [8], [9] and [11] are called as NSADDFDS. The diagnostic variables (DV) and average current of 3Φ currents are used for fault classification of base or collector open fault in IGBTs for variable speed drive system [1] and [10]. The fault classifier is implemented using If-Then rules and known as Qualitative KFDM. The If-Then rules are implemented based on the expert knowledge. The Qualitative KFDM has low generality and low expandability [5].

In this paper recent methods which are commonly used in FDS are implemented using MATLAB-Simulink and compared using key parameters like average accuracy, fault detection time, implementation efforts, threshold dependency, detection parameter, resistivity against noise and load dependency.

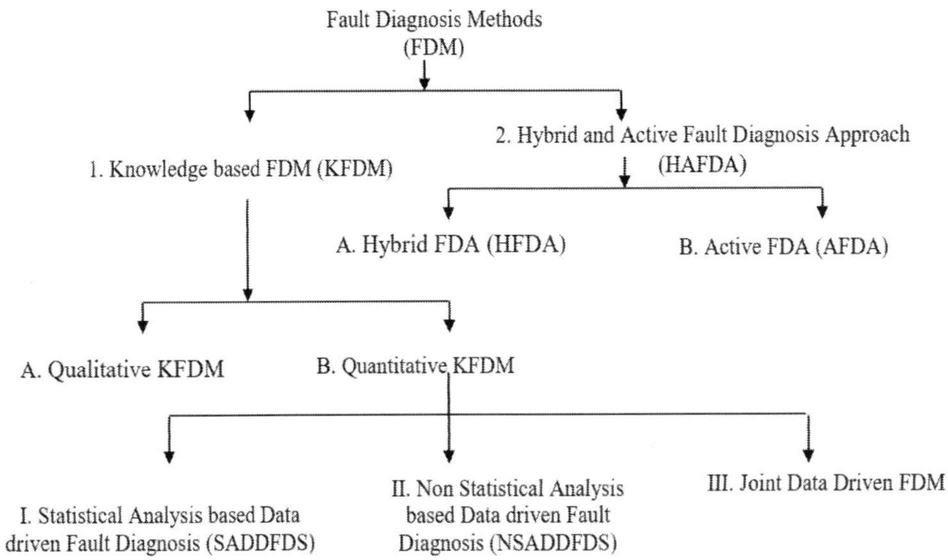

Figure 1. Classification of fault diagnostic methods

2. Voltage Source Inverter

The fundamental arrangement of 3Φ inverter is shown in Figure 2. In 3Φ inverter is implemented using six IGBTs which work complimentary. For standard situation, 3Φ inverter provides entirely balanced 3Φ sinusoidal currents and voltages as given by equation 1 and equation. 2.

$$i_n = \begin{cases} i_R = I_m \sin(\omega_s t + \phi) \\ i_Y = I_m \sin\left(\omega_s t - \frac{2\Pi}{3} + \phi\right) \\ i_B = I_m \sin\left(\omega_s t + \frac{2\Pi}{3} + \phi\right) \end{cases} \quad 1$$

Where n = R, Y or B, is the maximum amplitude of current, frequency and initial phase angle are given by I_m $w_s t$ and ϕ.

$$v_n = \begin{cases} v_R = V_m \sin(\omega_s t + \phi) \\ v_Y = V_m \sin\left(\omega_s t - \frac{2\Pi}{3} + \phi\right) \\ v_B = V_m \sin\left(\omega_s t + \frac{2\Pi}{3} + \phi\right) \end{cases} \quad 2$$

Where maximum amplitude of voltage is given by V_m.

Different FDSs are listed below and detail explanation will be given in Section III. Performance of the FDS is compared in Section IV.

A. DWT-FL [10],
B. DWT-ANN [13],
C. Diagnostic Variables and Average Value of 3Φ Current [12],

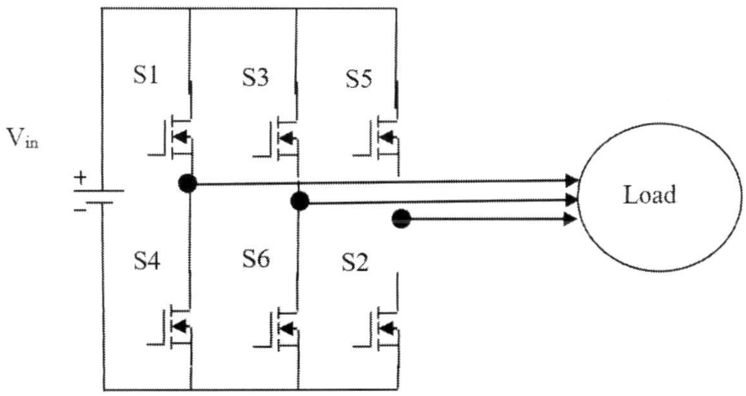

Figure 2. Three Phase VSI

3. Different Methods of Fault Diagnosis

3.1. Discrete Wavelet Transform - Fuzzy Logic (DWTFL)

The block diagram of DWTFL is shown in Figure 3. This method is implemented using three phase currents i_R, i_Y *and* i_B. Three phase currents are converted into data packets, each data packet consist of samples for 72° of fundamental current signals. The three phase currents are passed through DWT to compute *detail coefficients* (DCs) and *approximate coefficients* (ACs). h_n and g_n are responses of *Low Pass Filter* (LPF) and *High Pass Filter* (HPF) in DWT calculated using equation 3 and equation 4 respectively. Minimum *(min)* and maximum *(max)* values of DCs are extracted at various load conditions. K_d and K_f are selected as upper and lower threshold values respectively observing different values of DCs. The *min and max* values are compared with K_d and K_f to generate *fault diagnosis signatures* as given in equation 5. The fault signatures are created as shown in Table 1. Figure 4 shows the stator currents and *detail coefficients* of S1,S4, S1S6 or S1S4 open-circuit fault. DB4 mother wavelet is used for analysis. IGBT open circuit fault established the direct current offset in stator currents, which is clearly observed in Fig. 4. A higher direct current offset is noticed in phase in which IGBT is faulty, as compared to the healthy phases. Also, the polarity of direct current offset in faulty phase is opposite with the other two phases. A *Fuzzy Inference System (FIS)* is implemented observing fault signatures.

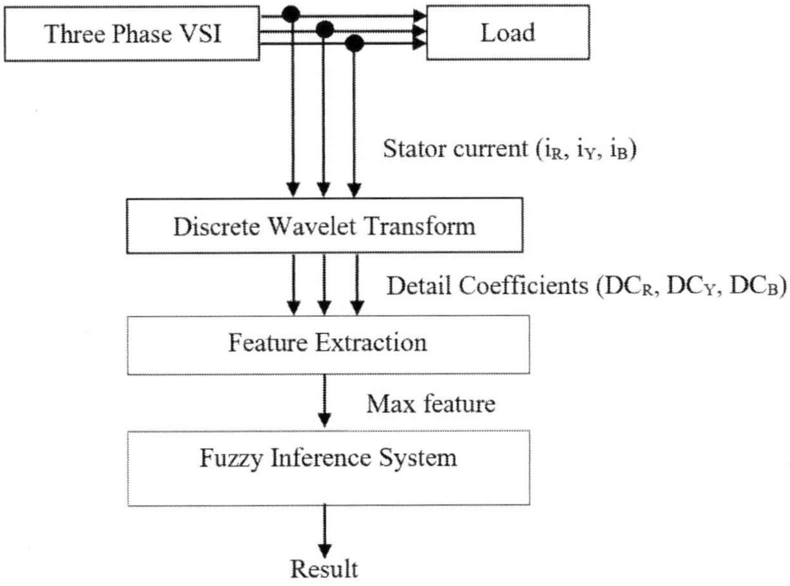

Figure 3. Discrete Wavelet Transform - Fuzzy Logic

$$DC_n = \sum_{k=0}^{l} i_n(k) \times h(n-k) \quad 3$$

$$AC_n = \sum_{k=0}^{l} i_n(k) \times g(n-k) \quad 4$$

Where l is number of current samples in one data packet.

Table 1 Diagnostic signatures of DWTFL

Faulty switches	Diagnostic variable		
	DC_R	DC_Y	DC_B
S1	Neg	Pos	Pos
S4	Pos	Neg	Neg
S3	Neg	Pos	Neg
S6	Pos	Neg	Pos
S5	Pos	Peg	Neg
S2	Neg	Neg	Pos

IOP Conf. Series: Materials Science and Engineering **197** (2017) 012006 doi:10.1088/1757-899X/197/1/012006

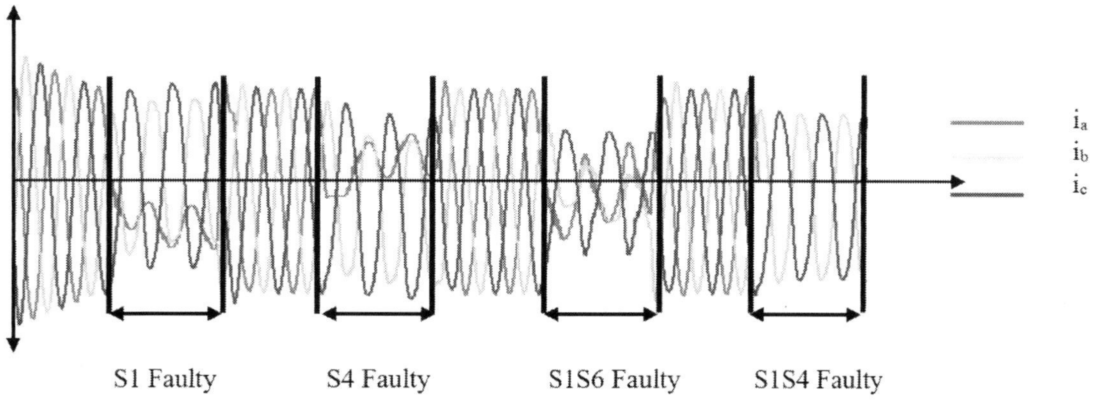

a) Three phase current waveform during healthy and faulty conditions

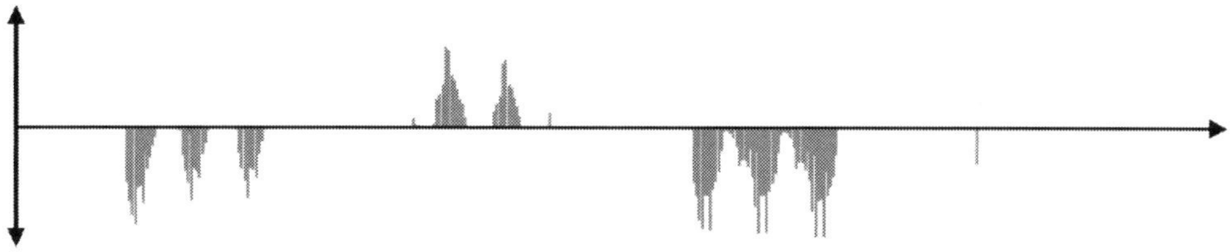

(b) Detail coefficient (DC) of ia

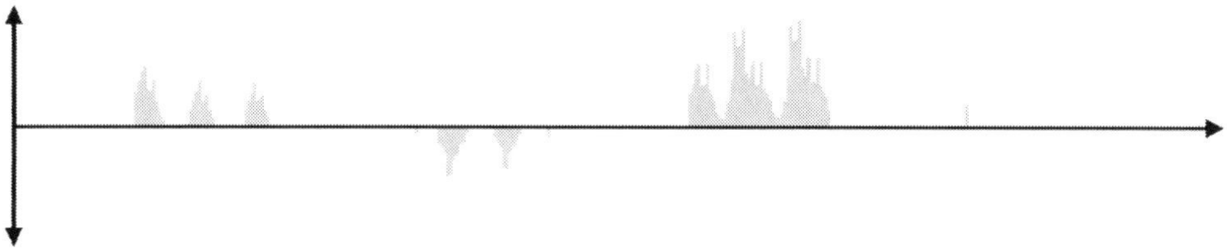

(c) Detail coefficient (DC) of i_b

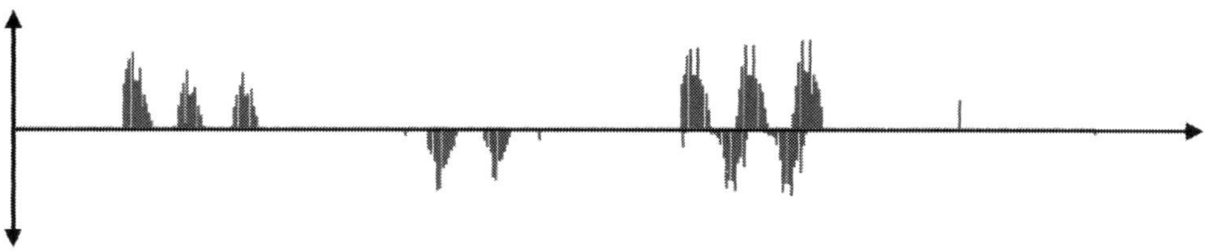

(d) Detail coefficient (DC) of i_c

Figure 4. Detail coefficients during healthy and faulty conditions

$$S_n = \begin{cases} \text{Neg if } DC_n < k_f \\ \text{Pos if } k_f \leq DC_n < k_d \\ \text{Dbl if } k_d \leq DC_n \end{cases} \quad 5$$

3.2. Discrete Wavelet Transform - Artificial Neural Network (DWTANN)

The block diagram of DWTANN is shown in Figure 5. This method is implemented using three phase currents i_R, i_Y *and* i_B. Three phase currents are converted into data packets, each data packet consist of samples for 72° of fundamental current signals. The three phase currents are passed through DWT to compute *detail coefficients* (DCs) and *approximate coefficients* (ACs) which are given by equation 3 and equation 4 respectively. Minimum *(min)*, Maximum *(max)*, Median *(med)*, Mean *(mean)* and Standard Deviation *(std)* features are extracted. The features are used to train *Feed Forward Back Propagation Neural Network* (FFBPNN). The structure of FFBPNN is decided changing parameters during training of ANN. Neurons in hidden layer, learning rate and training data size are changed to find structure of ANN. ANN with 5 input nodes, 1 hidden layer and 22 output nodes have structure like *15-24-22* is trained and tested for different faulty conditions which gives better performance with *less Mean Square Error (MSE)*.

3.3. Diagnostic Variables and Average Value of 3Φ Current (DVAC)

The d-q transformation or *Parks Vector Approach (PVA)* is mathematical transformation. This technique is applied to transform *3Φ Current* of VSI i_R, i_Y *and* i_B into two phase currents i_d *and* i_q. The Parks vector components are given by equation 6 and equation 7.

$$i_d = \sqrt{\frac{2}{3}} i_R - \frac{1}{\sqrt{6}} i_Y - \frac{1}{\sqrt{6}} i_B \quad 6$$

$$i_q = \frac{1}{\sqrt{2}} (i_Y - i_B) \quad 7$$

The normalization of *3Φ Current* currents is necessary *to keep away from the difficulty of machine mechanical operating condition dependency*. This method employs the dq transformation. By using Parks vector approach *Parks Vector Modulus (PVM)* can be calculated as equation 8,

$$|\overline{is}| = \sqrt{i_d^2 + i_q^2} \quad 8$$

The normalization is done by dividing *3Φ* currents by PVM. The normalized *3Φ* currents is given by equation 9,

$$I_{pvm} = \frac{i_l}{|\overline{is}|} \quad 9$$

Since this normalization method, the normalized phase currents I_{pvmN} will forever to obtain values inside the range of ± *0.8164966*, which is not dependent of the measured currents amplitude and given by equation 10.

$$I_{pvmN} = \begin{cases} i_{RN} = \sqrt{\frac{2}{3}}\sin(\omega_s t + \phi) \\ i_{YN} = \sqrt{\frac{2}{3}}\sin(\omega_s t - \frac{2\Pi}{3} + \phi) \\ i_{YN} = \sqrt{\frac{2}{3}}\sin(\omega_s t + \frac{2\Pi}{3} + \phi) \end{cases} \quad 10$$

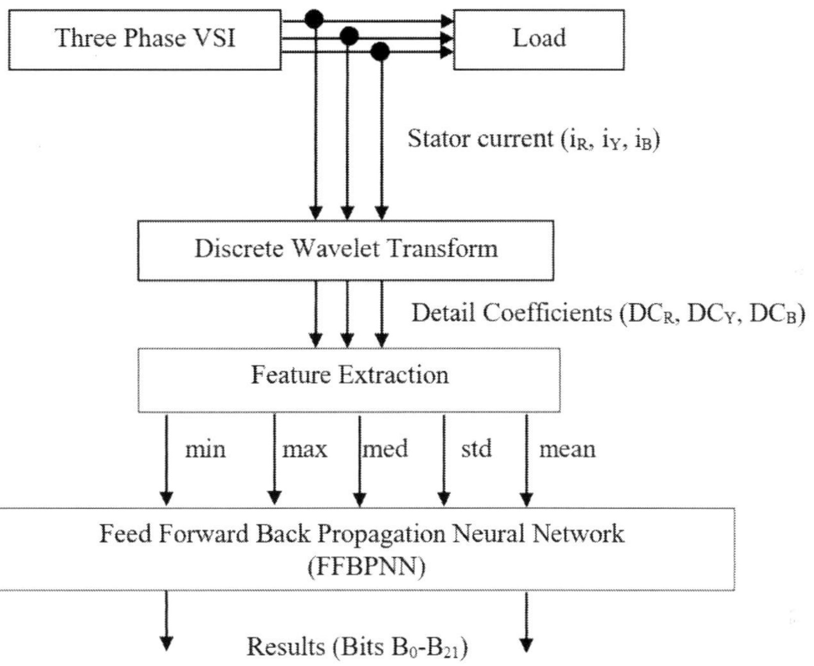

Figure 5. Wavelet Transform – Artificial Neural Network based fault diagnostic system

Average absolute values of normalized *3Φ* currents of VSI are applied to calculate diagnostic variables. For *3Φ*, three diagnostic variables are needed to be calculated. These three diagnostic variables e_{nN} are obtained from the errors of the normalized currents average absolute values and given by equation 11.

$$e_n = \xi - |I_{pvmN}| \quad 11$$

Where ξ is a stable rate. The rate of ξ is equal to the average absolute value of the normalized *3Φ* currents of inverter under healthy working conditions which is given by equations. 12, 13 and 14.

$$\xi_a = 0.5144 \quad 12$$
$$\xi_b = 0.5173 \quad 13$$
$$\xi_c = 0.5287 \quad 14$$

These three diagnostic variables have specific characteristics which are very useful for fault diagnosis. When inverter is in normal operating condition, all three diagnostic variables take approximately zero value. If *Base or collector open fault* is occurred in 3Φ inverter, as a minimum one of the diagnostic variables turn into positive. If fault is set up in IGBT S1, then diagnostic variable of phase which contains that IGBT takes positive value. In such way, diagnostic variables give information about the faulty phase. But it does not give any information about the faulty switches. For this, average current of three phases is used with diagnostic variables. Diagnostic variables detects faulty phase and average current identifies faulty switch from that faulty phase. From equations equation. 15 and equation 16, E_n and M_n are calculated which are used in fault diagnostic process. As the technique is normalized, it is not necessary to change these values for various load and speed conditions.

$$E_n = \begin{cases} F \text{ if } e_n < k_f \\ Y \text{ if } k_f \leq e_n < k_d \\ D \text{ if } k_d \leq e_n \end{cases} \quad 15$$

$$M_n = \begin{cases} N \text{ if } \langle I_{pvmN} \rangle < 0 \\ P \text{ if } \langle I_{pvmN} \rangle > 0 \end{cases} \quad 16$$

These two equations create fault diagnostic signatures. Value of K_d is decided by observing maximum value from e_a, e_b or e_c; when single switch is faulty. Similarly, K_f is minimum positive value from e_a, e_b or e_c; when two switches are faulty in different phases. The values of K_d and K_f are 0.20210 and 0.00553 respectively for 10 kW active power as shown in Table 2. These values are decided on experimentation basis observing Table 2. The diagnostic signatures are shown in Table 3.

4. Different Methods of Fault Diagnosis

Lists of evaluation and performance indices are given below:

Effectiveness: Diagnostic system is effective if it effectively pin point the faulty IGBT. Effectiveness of the fault diagnostic systems is calculated using Eq.17.

$$\text{Effectiveness (\%)} = \frac{\text{True Positive Output}}{\text{Total Number of Input Samples}} \text{ X } 100 \quad 17$$

Table 2 Diagnostic variables at 10 kW power

IGBT	e_a	e_b	e_c
S1	0.20048	-0.0354	-0.1579
S4	0.20173	-0.0385	-0.1566
S3	-0.1502	0.20192	-0.0456
S6	-0.1577	0.20210	-0.0397
S5	-0.0412	-0.1512	0.20139
S2	-0.0314	-0.1487	0.18745
S1S6	0.51974	-0.1487	0.18745
S3S4	-0.1873	0.51975	-0.1873
S5S2	-0.1873	-0.1873	0.51974
S1S2	0.12251	-0.082	-0.0261
S1S3	0.00553	0.19652	-0.2083

Table 3 Diagnostic signatures

Faulty switches	Diagnostic variable			Average current		
	E_a	E_b	E_c	M_a	M_b	M_c
S1	Y	F	F	N	-	-
S4	Y	F	F	P	-	-
S3	F	Y	F	-	N	-
S6	F	Y	F	-	P	-
S5	F	F	Y	-	-	N
S2	F	F	Y	-	-	P
S1S6	D	-	-	-	-	-
S3S4	-	D	-	-	-	-
S5S2	-	-	D	-	-	-
S1S2	Y	Y	F	N	N	P
S1S3	Y	F	Y	P	P	P
S1S5	F	Y	Y	P	P	P
S1S4	Y	Y	F	P	P	P

Table 4 Effectiveness of different diagnostic methods (in %) under variable load condition

Fault	DWTANN			DWTFL			DVAC		
	5000W	1000W	1500W	5000W	1000W	1500W	5000W	1000W	1500W
S1	70.73	82.63	66.6	100	100	86.6	100	100	100
S2	42.22	62.25	58.33	93.33	82.35	58.33	100	100	100
S3	37.5	57.87	49.89	100	88.23	100	100	100	97.26
S4	38.1443	59.73	43.75	88.23	100	93.75	100	100	100
S5	48.8095	57.89	51.98	85	85.71	100	100	100	100
S6	38.1443	54.45	85.71	85.71	86.36	85.71	100	100	100
S1-S6	65.28	85.37	73.98	100	99.99	99.99	100	100	100
S1-S4	76.19	87.23	81.98	99.99	99.99	99.99	100	100	100
S1-S3	66.13	85.23	72.93	89.46	100	100	96.35	100	100
S3-S2	66.89	87.81	73.29	100	84	83.33	100	100	100
S3-S4	71.67	85.65	76.29	96	100	92.3	100	100	100
S3-S6	67.87	84.69	72.98	99.99	99.9	99.99	100	100	100
S5-S6	53.98	73.75	65.76	96.15	99.99	100	100	100	100
S5-S2	65.98	82.97	71.09	100	99.99	97.94	100	100	100
S5-S4	56.98	72.84	65.98	91.29	99.99	99.99	100	100	100

Effectiveness of fault diagnostic system is tested over 316 samples. Out of these 38 samples are collected for healthy condition and 278 samples for faulty conditions. Effectiveness of different methods is shown in Table 4. In case of DWTFL, by observing polarities of detailed coefficients, significant accuracy is achieved for classification of single switch open circuit fault. Multiple switch open circuit fault diagnostics system requires additional number of fuzzy rules. Some features of detailed coefficients at different faulty conditions are matching with each other or are slight different in coefficients. Hence in DWTFL fault diagnostic system false alarms are increased. The DWTANN shows good results for 10 kW loads, as it is trained for such condition. It is very difficult to collect training data set at variable load conditions. In some faulty cases as mentioned in DWTFL, features are matching with each other at different conditions. The accuracy of DWTANN depends on the training data and network parameters. It is difficult to train ANN for high and light load conditions. In DVAC fault diagnostic system, accuracy at variable load conditions depends on threshold values. Threshold values are carefully selected Remarkable accuracy of fault diagnostic system is achieved. This method of fault diagnosis is effective under variable load conditions.

Diagnostic System Classifier Performance Metrics:

This type of metrics is used to evaluate performance of diagnostic algorithm, normally considering multiple fault classes. Classifier Performance Metrics is a square metrics consisting of equal rows and columns where its entry corresponds to strength of correlation between its ith and j^{th} elements. It shows the diagnostic outcome into several classes. The classifier performance metrics for different methods are shown in Table 5 to Table 7 for DWTFL, DWTANN and DVAC In Table 7, all conditions of misclassifications and few conditions of good classifications are shown. The tables show the real deficiency as heading across the top of the table and how they were classified as heading down the first column. The diagonal elements shown in metrics are correctly classified faults, whereas subsequent elements in the column represent misclassification of faults. The confusion matrix can be created using percentage or real cases viewed. The probability of isolation or fault isolation rate (FIR) is calculated using equation 18.

$$FIR = \frac{A_T}{A_T + C_T} X\,100 \qquad 18$$

$$A_T = \sum_i A_i \qquad 19$$

$$C_T = \sum_i C_i \qquad 20$$

Where A_i and C_i are the isolated and non isolated faulty conditions. The FIR for DWTFL, DWTANN and DVAC are 57.53%, 76.15% and 99.30% respectively. These values are calculated for 21 faulty and a healthy condition.

Fault Detection Time (FDT): FDT is defined as time interval between instance of fault occurrence and instance of fault isolation. The fault detection time is reliant on the difficulty of the detection algorithm and its resistivity to noise, as the threshold is determined according to that. FDT of different methods is shown in Table 8. The fault detection time required is less in case of fuzzy based diagnostic system as compared to ANN based fault diagnostic systems. The selection of neural network, number of hidden layer, hidden nodes and training of neural network decide the fault detection time of ANN based system. The computational difficulty in case of ANN is extra as compared to fuzzy logic. The fault detection time of DVAC method is low and it is very close to DWTFL; as the computational complexity in both systems is same.

Implementation Efforts (IE): The Implementation efforts of fault diagnostic system depend on the ease of sensing the detection parameter like voltage, current, vibration etc. In above mentioned diagnostic methods current signals are used as detection parameter. Similarly, packet size in case of all mentioned diagnostic method is same i.e. 4 ms except DVAC it is 20 ms. These methods are trained or implemented for fix size of packet as shown in Table 9. This increases data acquisition burden which results into high implementation efforts. The implementation efforts of fault diagnostic system also depend on the mathematical operation and decision-making process involved.

Tuning Effort (TE): It is beneficial to have thresholds independent fault diagnostic system and tolerances to be kept as highest as feasible. The fault diagnostic method based on artificial intelligent are threshold independent but requires high efforts to train neural network or to build rules in fuzzy inference system. In case DVAC, threshold values are selected to cover overall power region.

Table 5 Confusion matrix used for diagnostic metrics of DWTFL methodology

Fault → / Classified as ↓	S1	S2	S3	S4	S5	S6	S2S4	S2S6	S3S5	S4S5	S4S6	S5S6	S5S4
S1	56	17	19	18	08	20	07			20	14	12	10
S2	8	44			11	13	06	12					04
S3	11	10	59	12		21		05	05				
S4		12		44	10				03	08	08	10	
S5			05		54	19							
S6	27	17	17	26	17	27	09	05		10			07
S2S4							70				05		
S2S6								78	05				
S3S5									79	04		06	
S4S5										32			
S4S6							08				68		
S5S6									08			65	10
S5S4										18	05	02	72

Table 6 Confusion matrix used for diagnostic metrics of DWTANN methodology

Fault → / Classified as ↓	S1	S2	S3	S4	S5	S6	S2S4	S2S6	S3S5	S4S5	S4S6	S5S6	S5S4
S1	96								1.2		5.04		
S2		98		2				14			1.00		
S3			97										
S4		2		98									
S5					100								
S6	4		3			94	2.46						
S2S4							97.54						
S2S6								78					
S3S5									98.2				
S4S5										100			
S4S6											95.96		
S5S6												100	
S5S4													100

Table 7 Confusion matrix used for diagnostic metrics of DVAC

Fault → / Classified as ↓	S1	S2	S3	S4	S5	S6	S1S3	S2S6	S3S5	S4S5	S4S6	S5S6	S5S4
S1	100												
S2		100											
S3			99.35										
S4				100			0.62						
S5			0.65		100								
S6						100							
S2S4							99.28						
S2S6								100					
S3S5									100				
S4S5										100			
S4S6											100		
S5S6												100	
S5S4													100

Table 8 Fault detection Time in millisecond

Faults	DWTFL	DWTANN	DVAC
S1	62.793	96	59.953
S2	64.264	138	74.978
S3	57.919	111	60.842
S4	70.231	126	52.946
S5	57.598	95	47.784
S6	62.37	99	61.277
S1-S6	59.375	91	69.351
S1-S4	62.203	44	58.703
S1-S2	67.669	97	55.597
S3-S2	66.606	92	75.191
S3-S4	48.957	87	63.561
S3-S6	64.074	95	53.416
S5-S6	67.78	98	68.672
S5-S2	68.54	94	59.097

Table 9 Implementation efforts based on different parameters

Parameters	DWTFL	DWTANN	DVAC
Fault detection parameter	Current	Current	Current
	Low	Low	Low
Packet Size	4ms	4ms	20 ms
	High	High	High
Mathematical operation	High	High	High
Decision-making process	High	High	High

Resistivity against False Alarms (R): False alarms can take place during light-load and transient conditions, due to increased noise, i.e., low signal-to-noise ratio. Resistivity against transients can be improved by providing a dead time during transients. However, this approach has the drawback of slower detection. The necessary dead time is an indicator of resistivity against false alarms. Resistivity against noise can be improved by having higher threshold. However, it leads to false alarms during light loads. This can be handled by making a method independent of load. The DWTFL and DWTANN methods are dependent on load as shown in Table 4. DVAC diagnostic method is load independent but threshold dependent. The comparison of fault diagnostic methods regarding tuning efforts and resistivity is shown in Table 10.

Table 10 Comparison with respect to tuning efforts and resistivity against noise

Methods	Parameter		
	Tuning Effort	Resistivity	
		Against noise	Against load
WTFL	High	Low	Low
WTANN	High	Low	Low
DVFL	High	High	High

5. Conclusion

Different methods of Fault diagnosis for voltage source inverter are compared in this paper. These methods are generally based on grouping of 3Φ current transformations from time to other domain and classifier. Three phase currents are generally transformed using Discrete Wavelet Transform or Park Vector Transform. Discrete Wavelet Transform is more robust against high frequency noise but required threshold values to diagnose faults under variable load conditions. This transformation is useful for *Base or collector open fault* diagnosis of IGBTs. Such transformation is generally used with combination of If-Then rules, Fuzzy Logic or Artificial Neural Network. Parks Vector Transform is appropriate for open fault diagnosis of IGBTs in variable speed drive as current is normalized. This method is suitable and threshold values are used which covers overall power region. Fault diagnostic signatures calculated using PVT is combined with If-Then rules. The diagnostic method based on diagnostic variable and average value of three phases current is more suitable for diagnosis of switching devices under different operating conditions.

References

[1] M. R Ubale, R. B. Dhumale, S. D. Lokhande, Open switch fault diagnosis in three phase inverter using diagnostic variable method, International Journal of Research in Engineering and Technology, Vol. 2, iss. 12, pp 636-640, 2013.

[2] Lalit Patil, R. B. Dhumale, S. D. Lokhande, Fault Detection and Diagnosis Technique gor Power Inverter using Discrete Wavelet Transform, International Journal of Electronics Circuits and systems, Vol. 3, iss. 2, pp 174-178, June 2014

[3] M. R Ubale, R. B. Dhumale, S. D. Lokhande, Method of Open Switch Fault Detection in Three Phase Inverter using Artificial Neural Network, International Journal of Research in Science and Advance Technology, Vol. 3, iss. 3, pp 78-82, June 2014.
[4] Aubert B.,Regnier J., Caux S., Alejo D.,Kalman-Filter-Based Indicator for Online Interturn Short Circuits Detection in Permanent-Magnet Synchronous Generators, IEEE Trans. Ind. Electron., Vol. 62 , no. 3, pp.1921-1930, Aug. 2014.
[5] Bin Lu, Santosh K. Sharma, A Literature Review of IGBT Fault Diagnostic and Protection Methods for Power Inverters, IEEE Trans. Ind. Appl., vol.45, no.5, pp. 1770-1777, Oct. 2009.
[6] F. Charfi, F. Sellami, and K. Al-Haddad, Fault diagnosis in power system using wavelet transforms and neural networks, in Proc. IEEE Int. Symp. Ind. Electron. 2006, pp. 1143-1148.
[7] M. A. Awadallah and M. M. Morcos, Diagnosis of switch open-circuit fault in PM brushless dc motor drives, in Proc. Large Eng. Syst. Conf. Power Eng., May 2003, pp. 69-73.
[8] J. H. Park, D. H. Kim, S. S. Kim, D. J. Lee, and M. G. Chun, C-ANFIS based fault diagnosis for voltage-fed PWM motor drive systems, in Proc. Annu. Meeting North Amer. Fuzzy Inform. Soc., 2004,pp. 379-383.
[9] M. R. Mamat, M. Rizon, and M. S. Khanniche, Fault detection of 3-phase VSI using wavelet-fuzzy algorithm, Amer. J. Appl. Sci., vol. 3, no. 1, pp. 1642-1648, 2006.
[10] R. B. Dhumale, S. D. Lokhande, N. D. Thombare, M. P. Ghatule, Fault Detection and Diagnosis of High Speed Switching Devices in Power Converters, International Journal of Research in Engineering and Technology, Vol. 4, iss. 2, pp 253-257, Feb 2014.
[11] R. B. Dhumale, S. D. Lokhande, "Diagnosis of multiple open switch faults in three phase voltage source inverter", Journal of Intelligent & Fuzzy Systems, vol. 30, no. 4, pp. 2055-2065, 2016.
[12] J. O. Estima, A. J. M. Cardoso, A New Algorithm for Real- Time Multiple Open-Circuit Fault Diagnosis in Voltage-Fed PWM Motor Drives by the Reference Current Errors, IEEE Trans. Ind. Electron., Vol. 60, no. 8, pp. 3496-3505, Aug. 2013.

IOP Conf. Series: Materials Science and Engineering **197** (2017) 012007 doi:10.1088/1757-899X/197/1/012007

Design, Static Analysis And Fabrication Of Composite Joints

G.Mathiselvan[1], R.Gobinath[2],S.Yuvaraja[3],T.Raja[4]
[1,2,3,4] Department of Automobile Engineering, Sathyabama University, Chennai, India
Email id: mathiselvan@yahoo.com

Abstract The Bonded joints will be having one of the important issues in the composite technology is the repairing of aging in aircraft applications. In these applications and also for joining various composite material parts together, the composite materials fastened together either using adhesives or mechanical fasteners. In this paper, we have carried out design, static analysis of 3-D models and fabrication of the composite joints (bonded, riveted and hybrid). The 3-D model of the composite structure will be fabricated by using the materials such as epoxy resin, glass fibre material and aluminium rivet for preparing the joints. The static analysis was carried out with different joint by using ANSYS software. After fabrication, parametric study was also conducted to compare the performance of the hybrid joint with varying adherent width, adhesive thickness and overlap length. Different joint and its materials tensile test result have compared.

Keywords: Mechanical fasteners, Composite joints, Tensile test

1. Introduction

Over the past three decades, application of composite materials are continuously increasing from traditional application areas such as vehicles body frame, ship body, aircraft body to various engineering fields including automobiles, robotic arms and even architecture. Due to its superior properties, composites have been one of the materials used for repairing the existing structures. In such applications and also for joining various composite parts together, they are fastened together either using adhesives or mechanical fasteners. Nowadays, a novel method called hybrid joint is also being employed, where a combination of both adhesive and mechanical fasteners. Composite materials have been widely used as structural elements in vehicles structures due to their superior properties. The structure consists of an assembly of sub-structures properly arranged and connected to form a load transmission path. Such load transmission path is achieved using joints. Joints constitute the weakest zones in the structure. Failure may occur due to various reasons such as stress concentrations, excessive deflections etc. or a combination of these. Ever increasing aerospace performance requirements make the high strength-to-weight ratios and cost efficiency associated with bonded joints attractive. However, bonding cannot be fully utilized without validated analytical methods to increase confidence in bonded designs and to reduce the expensive testing often necessary to certify bonded joints in critical locations. Current standard analysis methods are not capable of predicting all of the complex failure mechanisms associated with composite bonded joints. Most existing bonded joint analyses do not include shear deformation of the adherents and cannot account for peel failures at the end of the overlap which are often a primary cause of joint failure. In addition, they often truncate the adhesive stress-strain.

2. Composite Joints

Short description of the three types of joints used in the present paper namely, bonded, riveted and hybrid joints was given below.

a. Bonded joint

Bonded joints can be made by gluing together pre-cured laminates with the suitable adhesives or by forming joints during the manufacturing process, in which case the joint and the laminate are cured at the same time (co-cured). Here, load transfer between the substrates take place through a distribution of shear stresses in the adhesive. In general, there are numerous advantages of adhesive bonded joints over the traditional mechanical fastened joints.

b. Riveted joint

Riveted joints can be used quite successfully on laminates up to about 3mm thick and also where a tight fit, called interference fit, is necessary. The choice lies between solid & hollow types and whichever is chosen, care must be taken to minimize damage to the laminate during hole-drilling and closing of rivet. In addition to material and configurationally parameters, the behaviour of riveted joints is also influenced by rivet parameters such as rivet size, clamping force, hole size and tolerance.

c. Hybrid joint

Hybrid joints have a combination of adhesive bonding and mechanical fasteners. In the present case, rivet has been used as the mechanical fastener. The advantages of using a combined bonded-riveted design apply mainly in a repair situation. It is generally accepted that a bonded joint is stronger than a mechanically fastened joint and a well-designed bonded joint is stronger than a hybrid joint.

3. Literature Review

Hart-Smith (1986) conducted theoretical investigation of combined bonded/bolted stepped lap joints between titanium and Carbon Fiber Reinforced Plastic (CFRP). While no significant strength benefits were found in comparison to perfectly bonded joints, the combined bolted-bonded joint was found to be beneficial for repairing damaged bonded joints and limiting damage propagation. Under room temperature and ambient humidity conditions, 98% of the applied load was predicted to be transferred by the adhesive. Chan and Vedhagiri (2001) investigated that the use of bolted, bonded and combined bonded-bolted joints used in repair. The structural response of various configurations of single lap joint, namely, bonded, bolted and bonded-bolted joints, was analyzed by three-dimensional finite element method. For the case of hybrid joints, it was found that the bolts do not take an active role in load transfer before the initiation of the failure. However, the bolts in the hybrid joint actually reduce the in-plane axial stress near the edge of overlap. Gordon Kelly (2005) investigated that the load distribution in hybrid joints numerically through the use of finite element analysis. A three dimensional finite element model was developed including the effects of bolt-hole and nonlinear material behaviour and large deformation. From the study, it was found that load transferred by the bolt increases with increasing adherent thickness and adhesive thickness and decreases with increasing overlap length, pitch distance and adhesive modulus. Fu and Mallick (2001) investigated that the static and fatigue strength of hybrid joints in a Structural Reaction Injection Molded (SRIM) composite materials. The authors performed an experimental investigation on a single lap joint considering the effect of different washer designs. It was concluded that the performance of the hybrid joints was dependent upon the washer design which affected the distribution of the bolt clamping force. The hybrid joints were shown in higher static strength and longer fatigue life than adhesive bonded joints for the studied material system. Jin-Hwe Kweon et al., (2006) conducted tests to evaluate the strength of carbon composite to aluminium double lap joints with two different adhesive materials, film and paste types.

4. Methodology

Previous studies on hybrid joints have considered fixed joint geometries and material systems. Limited consideration has been given to the prediction of load distribution in the joints. This study was focused on the analysis of stress distribution in three prominent joining methods namely, bonded, bolted and hybrid joints. FEA is used to study the stress distribution in the members involved under various

IOP Conf. Series: Materials Science and Engineering **197** (2017) 012007 doi:10.1088/1757-899X/197/1/012007

design conditions. In the present project, an attempt is made to analyze the stress distribution in 3-D models of three configurations of single lap joint, namely, bonded, riveted and hybrid joints.

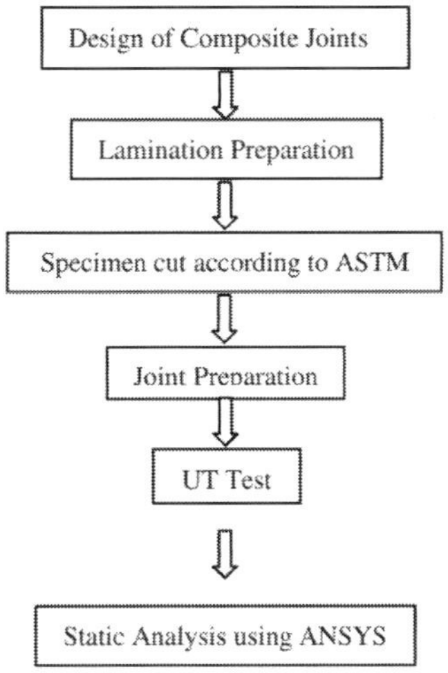

Figure 1. Methodology for Composite Joint Analysis

5. Design of Composite Joints

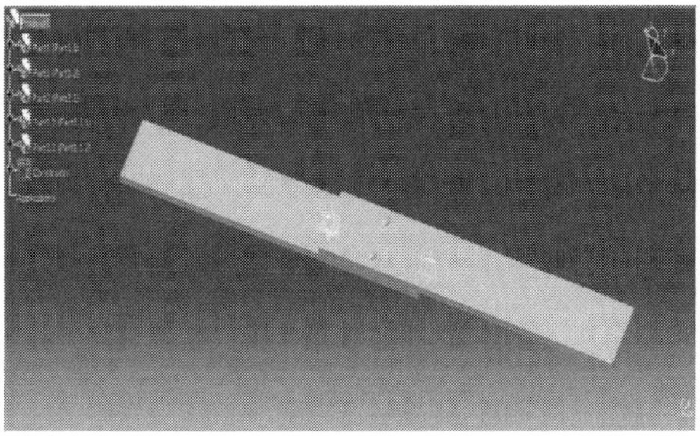

Figure 2. Design for Hybrid Composite Joint

In this laminate, REINFORCEMENT - Glass reinforced plastic fiber (bi-directional type) MATRIX-Epoxy. Correct ratio of resin and hardener is 10:1 Resin : LY556 Hardener : HY951. For Hybrid Composite Joint Design is as shown in figure 2. Similarly we done bonded and riveted design through ANSYS.

6. Fabrication of composite Joints

Even though the method has been replaced with automated techniques, the lay-up of preimpregnated material by hand is the oldest and most common fabrication method for advanced composite structures. Furthermore, the basic features of the method remain unchanged. Each step must follow in successive fashion in order to obtain a high-quality composite laminate after final processing. A description of these steps follows.

Step 1

Make the surface of the glass fiber material clean, if the dust or unwanted particles on the material will affect lamination preparation as shown in figure 3.

IOP Conf. Series: Materials Science and Engineering **197** (2017) 012007 doi:10.1088/1757-899X/197/1/012007

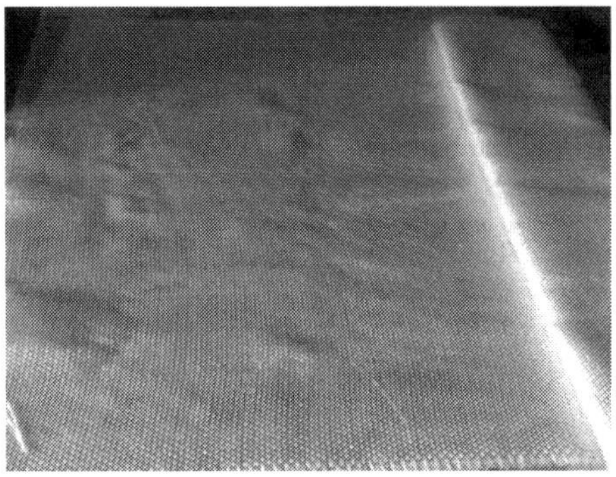

Figure 3. Lamination for Glass Fibre Material

Step 2

Cut the glass fiber material according to the required dimension for lamination preparation, into six layers for getting 3mm thickness.

Step 3

Prepare the epoxy resin by mixing the adhesive (LY556) and hardener (medium HY951) in the ratio 10:1 and mixing it continuously for lamination.

Step 4

Placing the plastic layer of cover on the cover glass fiber layer of material is pasted one by one by using the epoxy resin, repeating the pasting process for six layers, make sure that there should be no vacuum created in between the laminations. press each layer after applying the adhesive for better lamination.

Step 5

Cover the laminated surface by using plastic layer of cover and keep the glass plane on the lamination for distribution of even application of load and apply load about 4 to 5kg on the plane. Keep in observation about 48 hours.

Step 6

Take the lamination and cut the outer surface of the lamination because size and surface of the boundary layer will differ from the centre part. Make the marking on the lamination for cutting the exact dimension for getting the specimen(150mm*25mm).

Step 7

Filing the edges of the cutter specimen because while cutting the specimen by using hand saw some improper dimension will occur in the specimen. Filing the surface of specimen from one side about 50mm for joining the specimen in bonded and hybrid types of joint.

Step 8

Joining the specimen only by using epoxy resin on the 50mm filed surface of the specimen .Place any material under the specimen for the balancing support and apply load on the specimen for getting the composite structure. Keep it in observation about 48 hours and file the specimen to remove excess adhesive particles.

Step 9

For the riveted joint, drill the hole about 3mm diameter according to the respective dimension in the model of the specimen. Joining the specimen by using aluminium rivet by hammering the rivet in to the respective holes of the specimen.

Step 10

For making hybrid joints, bond the specimen by using epoxy resin and make the hole by drilling according to the respective dimension at 3mm diameter and make the second joint by using aluminium rivet by hammering in the respective holes as shown in figure 4.

Figure 4. Specimen for Hybrid Joint

Step 11

The cure of epoxies is the conversion of the liquid resin and hardener components to a solid high-performance plastic material. Cure is only initiated once the components are metered in the correct ratio to one another and are physically mixed together. The cure of all epoxies is an exothermic process where heat is liberated as a natural consequence of the chemical reaction. Success in using epoxies most efficiently is dependent upon handling the product in the correct way in order to avoid wastage and premature cure, and this can be achieved by some understanding of the basic chemistry and the various stages of the chemical transformation.

7. Mechanical Testing

After the fabrication of specimen, the specimen should undergo through a mechanical testing to take the experimental evaluation. The testing were made namely, tensile test. This type of tensile machine has two crossheads; one is for adjust the length of the specimen and other is for driven to apply tension .There are four main parameters: force capacity, Speed, precision and accuracy. Force capacity refers to the machine must able to generate enough force to fracture the specimen. The machine must be accurately and precisely measure the gauge length and force applied. A standard specimen is prepared in a round or a square section along the gauge length, depending on the standard used. Both ends of the specimens should have sufficient length and surface condition so they are firmly gripped during testing. The testing process involves placing the test specimen in the testing machine and slowly extending it until it factures. During this process, the elongation of the gauge section is recorded against the applied force.

8. Results and Discussion

Figure 5 shows the normal stress induced in specimen under the action of 1500N load. The maximum stress is induced at the fixed end of the laminate its maximum value is 302.05 MPa. Red zone indicates the area of maximum stress and blue zone indicates the area of minimum stress.

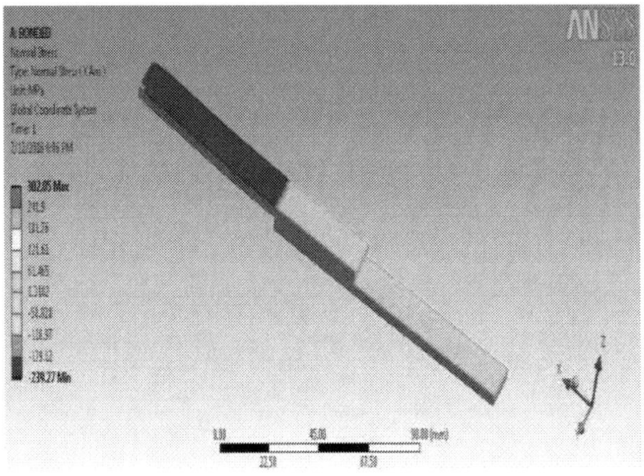

Figure 5. Normal stress for bonded joint

Stress intensity, total deformation analysis were done for bonded joint. Similarly Normal stress, stress intensity, total deformation analysis were done for riveted joint and hybrid joints. The test has taken at macro lab which is ISO associated (IS/ISO/IEC 17025).It is located at SP-101,2nd Main Road, Ambattur Industrial Estate,Chennai.

Result 1

For bonded joint, the figure 6 shows the maximum load of 6.1KN is obtained as maximum extent at this point bonded joint breaks. The bonded joint attains the maximum stress level of about 65Mpa at full tensile load.

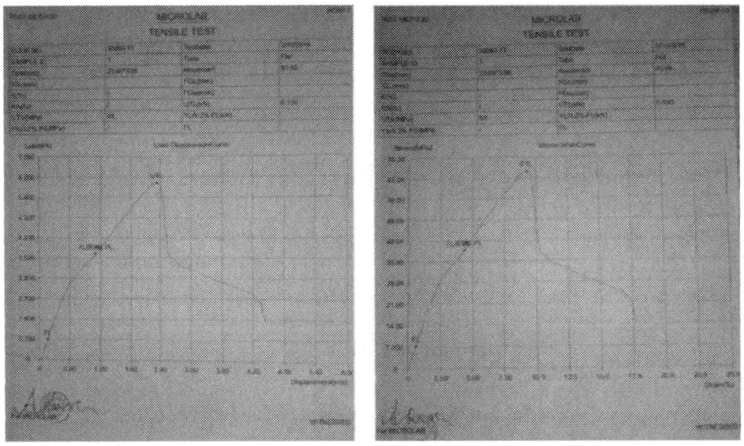

(a) Load-Displacement Curve (b) Stress-Strain Curve

Figure 6. Test report of bonded joint

Result 2

For riveted joint, the figure 7 shows the maximum load of 1.67KN is obtained as maximum extent at this point riveted joint breaks. The riveted joint attains the maximum stress level of about 19Mpa at full tensile load. The rived joint has very lower tensile strength when compare to both bonded and hybrid joint.

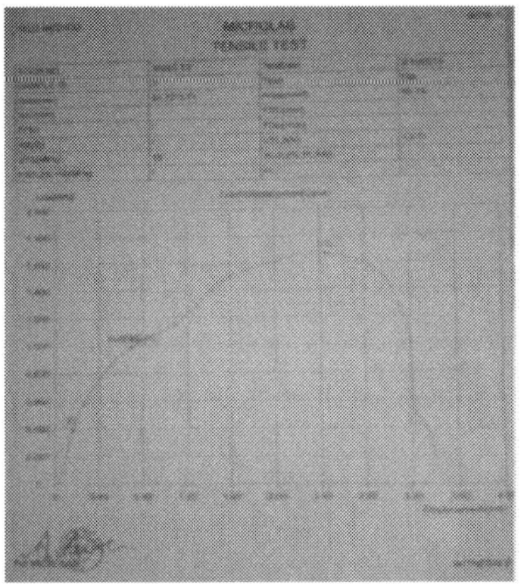

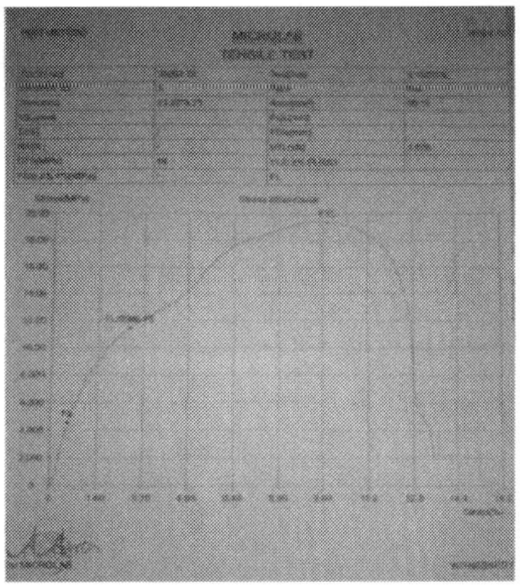

(a) Load-Displacement Curve (b) Strain- Stress Curve

Figure 7. Test report of riveted joint

Result 3

For hybrid joint, the figure 8 shows the maximum load of 7.125KN is obtained as maximum extent at this point hybrid joint breaks. The hybrid joint attains the maximum stress level of about 82Mpa at full tensile load. When compare to bonded and riveted joint the hybrid joint withstand more tensile stress and has higher tensile strength.

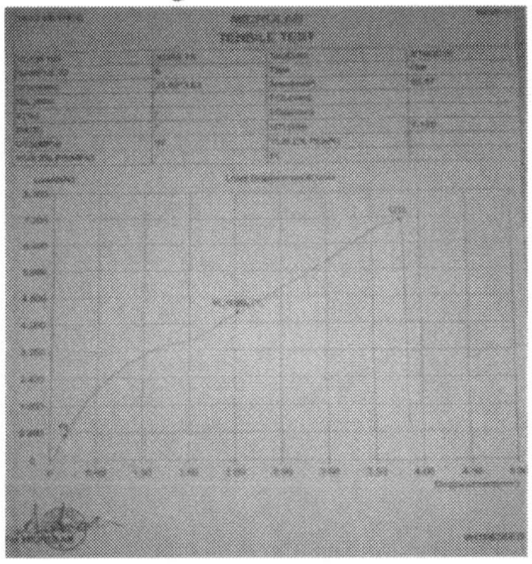

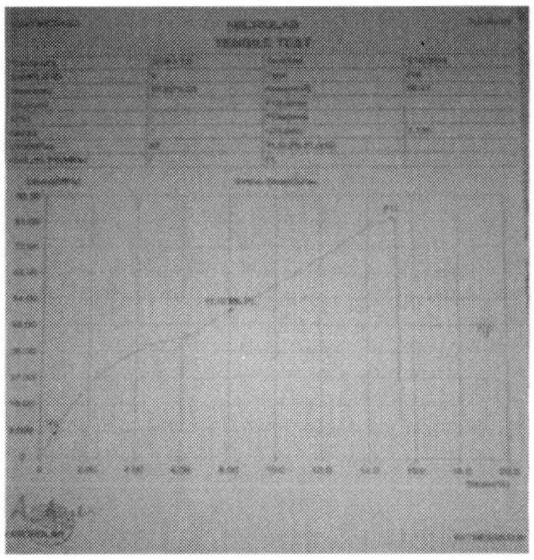

(a) Load-Displacement Curve (b) Strain-Stress Curve

Figure 8. Test report of hybrid joint

9. Conclusion

From the results, it is observed that the hybrid joint gives better result in tensile strength on joints under static load condition. In the present work, Numerical and experimental for the prediction of stress distribution in bonded, riveted and hybrid joints has been carried out. 3-D models were created and analyzed using ANSYS FEA software. Von Misses stress was used to compare the results. Parametric study has been performed to reveal the effect of various geometric parameters on the stress distribution of hybrid joint. Thus from the present study, it was found that a well-designed hybrid joint is very efficient when compared to bonded or riveted joints n the case of repair situation in aircraft structures.

IOP Conf. Series: Materials Science and Engineering **197** (2017) 012007 doi:10.1088/1757-899X/197/1/012007

References

[1] FU, M and P.K. MALLICK"fatique of hybrid joints in SRIM composites. Int.J. Adhesion Adhesives, 21: 145-159.

[2] Gordon Kelly, "Load transfer in hybrid (bonded/bolted) composite single-lap joints," Composite Structures 69 (2005) 35–43.

[3] Hart-Smith, L.J., "theoretical investigation of combined bonded/bolted stepped lap joints between titanium and Carbon Fiber Reinforced Plastic".

[4] Wen S. Chan S. Vedhagiri "Analysis of Composite Bonded/Bolted Joints Used in Repairing" Journal of Composite Materials June 2001 vol. 35 no. 12 1045-1061.

IOP Conf. Series: Materials Science and Engineering **197** (2017) 012008 doi:10.1088/1757-899X/197/1/012008

MECHANICAL CHARACTERIZATION OF BASLAT BASED NATURAL HYBRID COMPOSITES FOR AEROSPACE APPLICATIONS

J.Alexander[1], Elphej Churchill.S.J[2]
Sathyabama University, OMR, Chennai-119,India

[1]E- mail: vsjalexander@gmail.com

Abstract: Advanced composites have attracted aircraft designers due to its high strength to weight ratio, high stiffness to weight ratio, tailoring properties, hybridization of opposites etc. Moreover the cost reduction is also another important requirement of structural components. Basalt fibers are new entry in structural field which has excellent properties more or less equivalent to GFRP composites. Using these basalt fibres, new hybrid composites were developed by combining basalt fibres with natural fibres. The mechanical and thermal properties were determined and compared with BFRP and GFRP composites. Results proved that hybrid composites have some good qualities.

1. INTRODUCTION

High strength to weight ratio and high stiffness to strength are the major design factors for aircraft components. The aircraft undergoes through various loads, especially air loads. These loads are static and dynamic. A suitable high strength, lightweight and low cost materials are required. In this point of veiew, composite materials are selected. This composite materials can be fabricated using artificial and natural fibres. Advanced composites are derived using artificial fibres like Kevlar fibre, corbonfibre , Kevlar fibre and basalt fibres. To vary the strengths, based on applications, hybrid composites are preferable. [1]M.Sakthivel et al have done a research on hybrid composites with natural fibres like sisal, banana fibres, hemp and jute. They proved that polymer banana reinforced natural composites is the best natural composites among the various combinations. [2] Alexander.J et al have investigated about basalt fibre reinforced composites at various curing conditions and proved that basalt fibres re the best composites that can be used for aerospace applications.. [3] Alexander et al also have done a study on tress analysis of BFRP composites. [4]Kuruvila joseph et al reviewed the sisl fibre composite andconcluded their results with there will be a good scope and better future for sisal fibre. Therefore addition of natural fibre with artificial fibre improves the quality of the composites as well as reduce the cost.Thiswork or project we have selected three different fibers (Basalt, Sisal and Glass) and matrix Epoxy (LY556). The laminates were prepared by using hand layup process and then the specimens were cut as per ASTM standards. The tensile, bending, impact and DMA tests were conducted to get the mechanical properties of the composites.

2. EXPERIMENTAL

IOP Conf. Series: Materials Science and Engineering **197** (2017) 012008 doi:10.1088/1757-899X/197/1/012008

Three laminates were prepared for conducting experimental tests and all of them were done in the Aircraft structures lab of Aeronautical department, at Sathyabama University. Laminates were prepared by hand layup.7 layers of basalt and 2 layers of sisal laminate,5 layers of glass and 2 layers of sisal laminate and 4 layers of basalt,2 layers of glass and 2 layers of sisal laminate with resin named epoxy LY556 with their hardener. The basalt fiber is bought from the Germany, Sisal and glass fibersare from, Chennai. Laminates were prepared by using hand lay up process.

2.1 TENSILE TEST

This test was conducted using the UTM machine (1000 KN capacity) in the Aircraft structures lab of Aeronautical department, Sathyabama University. This testing helped in determining the mechanical characteristics of the material. The specimens were cut for tensile test according to the ASTM standards (D3039) shown in fig 2.1.

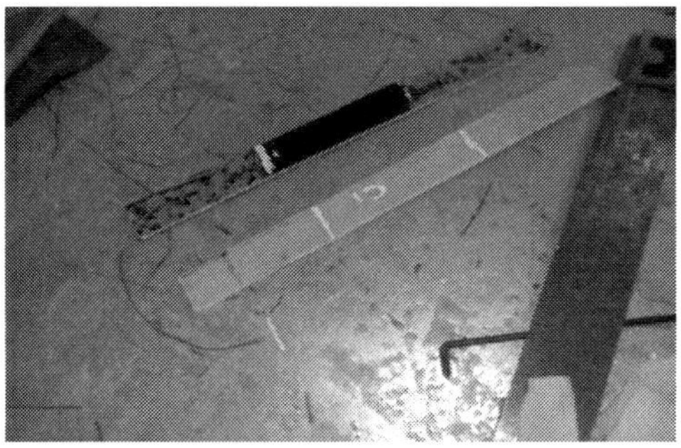

Fig: 2.1Tensile test Specimen as per ASTM standards

2.2 BENDING TEST

This test was conducted using the UTM machine in the Aircraft structures lab of Aeronautical department, Sathyabama University. The specimens were cut for bending test according to the ASTM standards(D790). The specimen was kept under 3 point setup at the edge and a compressive load was made to act at the middle from the top. The values were transferred to the system through the serial ports and they were taken down with the help of WIN UTM software. All the results were tabulated.

2.3 IMPACT TEST

The specimens were cut for impact test according to the ASTM standards(D790).The specimen was fixed in the machine and the handle is pulled down and the readings from the dials are noted.

Range : 0J to 14J.

Dead weight in the oscillator : 8Kg

Model& Make : K-1.4 & KrystalEquipments, India.

The un-notched specimens were kept in a cantilever position, and a pendulum has swung around to break the specimen. The impact energy (J) was calculated using a dial gauge that was fitted on the machine. Five samples were taken for each test, and the results were averaged.

2.4 DYNAMIC MECHANICAL ANALYSIS

Dynamic mechanical analysis also known as dynamic mechanical spectroscopy is a system used to study and illustrate materials property. It is mainlyhelpful for studying the viscoelastic performance of polymers. A sinusoidal stress is applied and the strain in the material is measured, allowing individual to verify the complex modulus. The temperature of the sample or the frequency of the stress are often varied, leading to variations in the complex modulus; this advance can be used to position the glass transition temperature of the material, as well as to recognize transitions corresponding to other molecular motions.

2.4.1 STORAGE MODULUS (E'):

The viscoelastic property of a polymer is calculated by dynamic mechanical analysis where a sinusoidal force (stress σ) is applied to a material and the consequential displacement (strain) is calculated. For a perfectly elastic solid, the resultant strain and the stress will be perfectly in phase. For a purely viscous fluid, there will be a 90 degree phase lag of strain with respect to stress.Viscoelastic polymers have the characteristics in between where some phase lag will occur during DMA tests.

$$\text{Stress: } \sigma = \sigma_0 \sin(t\omega + \delta)$$
$$\text{Strain: } \varepsilon = \varepsilon_0 \sin(t\omega)$$

Where,
ω →is frequency of strain oscillation,
t →is time,
δ →is phase lag between stress and strain.
The storage modulus measures the stored energy, in lieu of the elastic portion, and the loss modulus measures the energy dissolute as heat, in lieu of the viscous portion.

IOP Conf. Series: Materials Science and Engineering **197** (2017) 012008 doi:10.1088/1757-899X/197/1/012008

$$E' = \frac{\sigma_0}{\varepsilon_0} \cos \delta$$

2.4.2 GLASS TRANSITION TEMPERATURE (T_g):

In a composite material, the temperature at which the matrix material changes from a glassy to rubbery state or vice versa is termed the glass transition temperature . This change in condition may effect in a two to three orders of extent change in modulus. The transition is characteristic of the detailed molecular structure of the matrix polymer. As well as modulus the heat capacity and thermal expansion coefficient also change at the glass transition temperatures. As shown in the schematic diagram, the transition is not sharp but occurs progressively over a temperature range which may be up to some tens of degrees wide. The quoted is usually the mid-point of the range. Different methods of measuring may y give somewhat different values, so the method used should always be quoted.

2.4.3 LOSS MODULUS (tan δ):

Polymer matrices will absorb moisture and become plasticised. Plasticisation can reduce the leading to a 'wet' value. The reduction in depends on the initial crosslink density of the matrix material. Measurement of for a composite matrix material requires the specimen to be heated, which in turn, removes the absorbed moisture and hence affects the wet value by increasing it slightly. It is possible to seal the specimens to prevent moisture loss. Also in thick specimens, moisture may be lost on the outside surfaces resulting in a wider glass transition range perhaps with two inflection points representing the inner wet material and the drier outer material.Polymer matrices in composites are rarely used at temperatures above their because they cannot effectively transfer load in their rubbery state

$$E'' = \frac{\sigma_0}{\varepsilon_0} \sin \delta$$

3. RESULTS AND DISCUSSIONS

Mechanical static tests like Tensile test and Bending Test were conducted for all three types of hybrid composites and the results are tabulated in Table 3.1 and table3.2.

Table 3.1 Result of tensile test results of various hybrid composites

Hybrid composite name	Tensile Strength (N/mm^2)	Tensile Modulus (N/mm^2)	% Of Strain
sisal/glass/epoxy	583	116	6.7
sisal/basalt/epoxy	402	193	7.5
basalt/sisal/glass/epoxy	414	236	5.9

Tensile strength of Sisal/Glass/epoxy hybrid composites is found to be excellent .But the tensile modulus of Basalt/glass/sisal/epoxy composite is found to better when compare with other combinations. This is due to the cross linking between the fibers and matrices are improved due to different fiber combinations. The bending strength and Interlaminar shear strength are very high for sisal/Basalt/epoxy composites. This shows a very good bond is generated between the basalt fibre and epoxy and sisal. This is due to the surface behavior of the fibre mats. Bond between the glass and sisal is weaker than this.

Other important dynamic property composites are Impact strength. Specimens are prepared as per the ASTM standard. Charpy impact machine was used. The initial energy of the material at the beginning of the test was noted down. Then the Impact load were applied and the amount of Energy Absorbed during the breakage of a specimen in a standardized test was noted down, and using the area of the specimen the impact strength of that specimen was determined with the help of the following formula and were shown in fig 3.1. Impact strength of sisal/basalt/epoxy composites are much better than the other combinations. From the SEM images taken for all types, the bond between the fibres and the matrix for sisal/basalt/epoxy is good. This improved the bending strength.

$$\text{Impact strength} = \frac{E}{A} \qquad (3.1)$$

Where

E = Initial Energy-Final Energy
A =Width x Thickness

IOP Conf. Series: Materials Science and Engineering **197** (2017) 012008 doi:10.1088/1757-899X/197/1/012008

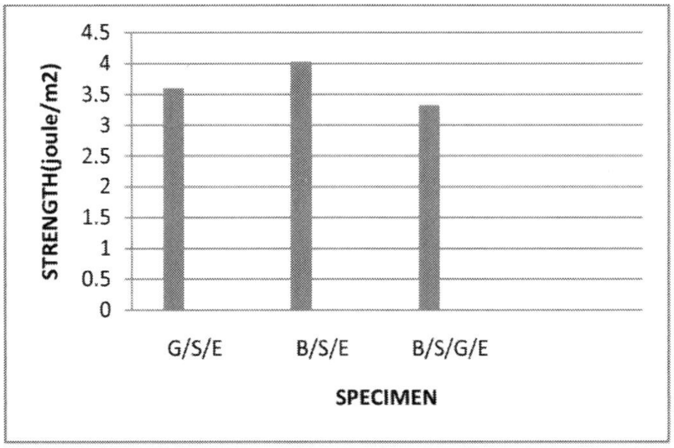

Fig 3.1 Impact test strength of various Hybrid composites

3.1 DYNAMIC MECHANICAL ANALYSIS RESULTS

From the dynamic mechanical Analysis the storage modulus, loss modulus and glass transition temperature of any materials can be determined. The results of all composites are shown in fig 3.2 -3.4. Table 3.2 gives the tangent modulus, loss modulus and glass transition temperature of all materials.

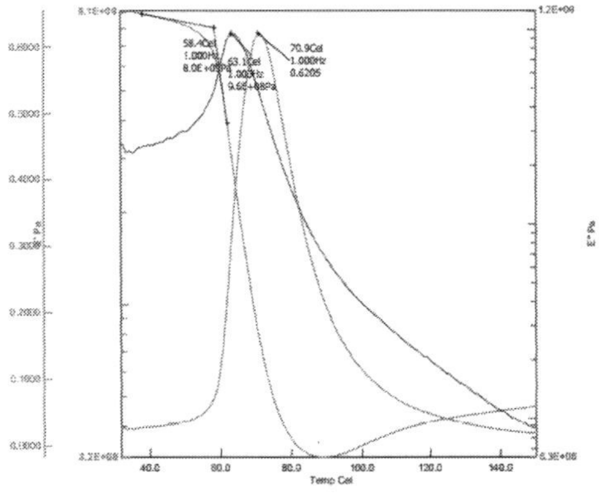

Fig 3.2 DMA result for glass/sisal/epoxy laminate

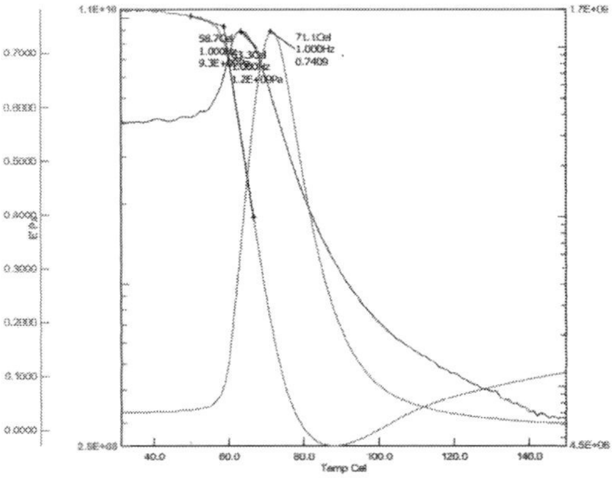

Fig 3.3 DMA result for basalt/sisal/epoxy laminate

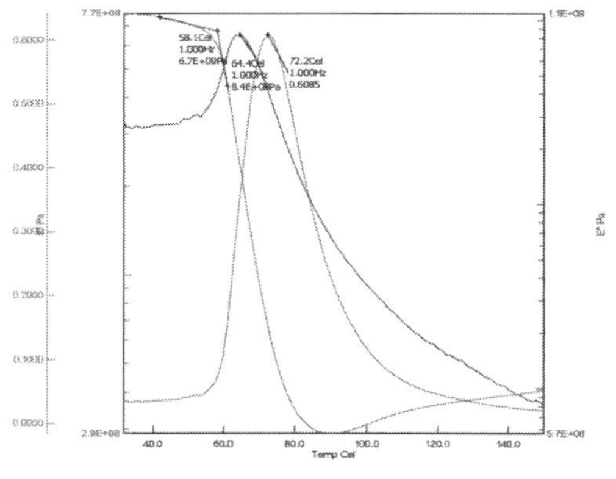

Fig 3.4 DMA result for basalt/sisal/glass/epoxy laminate

The results shows even the thermal properties also good for Sisal/basalt/epoxy composites. The storage modulus and loss modulus are higher than the other two composites. Even glass transition temperature is higher than sisal/ glass/ epoxy composites.

Table 3.2 DMA test results for different materials

Composite name	Storage modulus e (gpa)	Loss modulus tan δ (gpa)	Glass transition temperature T_g^0
sisal/glass/epoxy	8.009	0.030845	70.9
sisal/basalt/epoxy	9.309	0.032281	71.1
basalt/sisal/glass/epoxy	6.705	0.023961	72.2

When a material is operated beyond glass transition temperature, all the mechanical properties will decrease. Mechanical properties will not always increases with glass transition. Mechanical properties vary with respect to bond strength between the matrix and the fibre.

4.CONCLUSION

Mechanical and thermal properties of sisal/basalt/epoxy, sisal/glass/epoxy and sisal/glass/basalt/epoxy composites were determined and the corresponding strengths were determined. The following observations were made.

1.Addition of natural fibres with advanced fibres like glass and basalt improves the quality of the composites in all respect.

2. When the sisal fiber is added with basalt and glass separately, the bond strength between the sisal and basalt is better than that of sisal/glass due to the cross linking between the basalt epoxy sisal fiber is better than that of glass and epoxy sisal.

3. Thermal behavior of baslt, sisal combination is excellent than that of glass, sisal combination.

4. Combining basalt with sisal fibers gives a good performance. Therefore sisal/basalt/epoxy composites can be used in aircraft structural applications.

References

1.M.Sakthivel1, S.RAMESH, (2013),Mechanical Properties Of Natural Fibre (Banana, Coir, Sisal) Polymer CompositessciencePARK,
2.Alexander.J, Dr.BSM.Augustine (2016), "influence of microwave post curing on the mechanical and thermal properties of basalt / epoxy composites for aerospace applications" ,Journal of the Balkan Tribological Association, Vol. 22(1), pp.220–234.
3.Alexander.J, Dr.BSM.Augustine (2015), "Inter ply stress analysis of basalt fibre reinforced plastic laminated composites" International Journal of Applied Engineering Research, Vol. 10(11), pp.10448-10450.
4.Kuruvilla Joseph, Romildo Dias TolêdoFilho, Beena James, Sabu Thomas & Laura Hecker De Carvalho,(1999),A Review On Sisal Fiber Reinforced Polymer,Composites,RevistaBrasileira De EngenhariaAgrícola E Ambiental, V.3, N.3, P.367-379.

IOP Conf. Series: Materials Science and Engineering **197** (2017) 012009 doi:10.1088/1757-899X/197/1/012009

Recent developments in turning hardened steels – A review

Sivaraman V[1,a], Prakash S[2,b]

[1] Research scholar, Sathyabama University, Jeppiar Nagar, Chennai, India.
[2] Faculty Head, Faculty of Mechanical Engineering, Sathyabama University, Jeppiar Nagar, Chennai, India.

[a] vsivaramancbe@gmail.com, [b]prakash_s1969@yahoo.com

Abstract - Hard materials ranging from HRC 45 – 68 such as hardened AISI H13, AISI 4340, AISI 52100, D2 STL, D3 STEEL Steel etc., need super hard tool materials to machine. Turning of these hard materials is termed as hard turning. Hard turning makes possible direct machining of the hard materials and also eliminates the lubricant requirement and thus favoring dry machining. Hard turning is a finish turning process and hence conventional grinding is not required. Development of the new advanced super hard tool materials such as ceramic inserts, Cubic Boron Nitride, Polycrystalline Cubic Boron Nitride etc. enabled the turning of these materials. PVD and CVD methods of coating have made easier the production of single and multi layered coated tool inserts. Coatings of TiN, TiAlN, TiC, Al_2O_3, AlCrN over cemented carbide inserts has lead to the machining of difficult to machine materials. Advancement in the process of hard machining paved way for better surface finish, long tool life, reduced tool wear, cutting force and cutting temperatures. Micro and Nano coated carbide inserts, nanocomposite coated PCBN inserts, micro and nano CBN coated carbide inserts and similar developments have made machining of hardened steels much easier and economical. In this paper, broad literature review on turning of hardened steels including optimizing process parameters, cooling requirements, different tool materials etc., are done.

1. Introduction

In the modern world with fast developing technologies, manufacturing industries are striving hard to survive with economical and shorter time production methodologies. Based on the functional requirements, some parts demands
high hardness. Conventional procedure for machining these hard materials is
(a) Annealing the work piece.
(b) Turning the work piece to near net shape.
(c) Performing the required hardening processes.
(d) Finally grinding the material to the required dimension.
Finish grinding process involves in removing the material in micron level and consumes more time.

Until 1970, turning these types of hard materials are considered as difficult process. Recent developments in the designing of rigid machineries, tool holders, newer super hard tool materials, work pieces are directly turned to correct dimensions after hardening, thus eliminating annealing process, the need to turn the work piece to near net shape and finish grinding process. The direct turning of hard material is made feasible by the arrival of newer super hard tool materials such as cemented carbide, Cubic Boron Nitride(CBN), Polycrystalline Cubic Boron Nitride(PCBN), coated carbide inserts, CBN coated carbides, micro and nanocomposite coated tools. Also, Physical Vapor Deposition (PVD) and Chemical Vapor Deposition (CVD) coating technologies have made hard turning an easier process.

Implementation of hard turning results in reduction in set up time, process steps and increased productivity yielding more cost saving. Other major advantage of hard turning is that it can be machined without or minimum lubrication. Handling of the lubricants also adds cost to the process and excluding the lubricant makes the process more economical. Absence of lubricants makes hard turning an environment friendly system. The recent developments in turning hardened steels are reviewed in this paper, the areas of research done, scope for further research are discussed.

2. Cutting tool materials

A cutting tool material must possess the following qualities.

(a) Hardness greater than work piece material.
(b) Hot hardness – Retain the hardness at elevated temperatures.
(c) Toughness – Should not fracture or chip during cutting operation.
(d) Wear resistance – Able to resist wear leading to longer tool life.
(e) Chemical affinity – Should not react with work piece material and coolant.
(f) Impact resistance – Withstand impact loads.

None of the cutting tool materials incorporates all these characteristics. Based on the specific requirements trade-offs occur in tool materials.

The history of tool material starts from carbon tool steels followed by High speed steel (HSS), cemented carbide, ceramics, cermets, coated carbides, coated ceramics, Poly Crystalline Diamond, Cubic Boron Nitride (CBN), Poly Cubic Boron Nitride, CBN coated carbide tools.

In 1923 Osram Lamp Works developed cemented carbide tool a composite material having tungsten carbide (WC) as the hard aggregate and Co as the metallic binder. Introduction of cemented carbide tools increased the cutting speed to approximately 4 times the speed of high speed steel. The addition of TiC to cemented carbide increased the tool life. Later TaC, VC, Cr_3C_2 and other carbides were added to cemented carbide based on the requirement and the performance of cemented carbide has increased.

For hard turning, alumina based ceramics plays an important role. Though alumina has most of the required properties such as high hot hardness, increased wear resistance and chemical inertness, it is highly brittle leading to shorter tool life [1]. To overcome this, alumina based ceramic tools are generally reinforced with TiC, TiN, (W,Ti)C, Ti(C, N), ZrO_2, TiB2, SiCw, SiCp [2].

PCBN tools are very expensive as it needs high temperature and pressure in the manufacturing compacting process. In consequence, an alternate tool material, with near equivalent quality but with less price have gained importance. Fulfilling both requirement ceramic tools replaces the costly PCBN [3, 4, 5].

Abhijeet et al. [6] compared CBN-TiN composite coated WC – Co insert with PCBN tool in turning hardened AISI 4340 steel. Though PCBN tool performance was better than CBN-TiN composite coated WC-Co tool in terms of tool life, cutting force, CBN-TiN composite coated tool takes an edge as it is cheaper than the PCBN tool.

In 1986, Brandt [7] observed better flank wear resistance in mixed alumina ceramic tool when compared to oxide alumina ceramic tools. Later Senthil kumar et al. [8] experimented EN 24 steelwith hardness HRC 40 to HRC 45 with two types of cutting tools, Ti[C,N] mixed alumina ceramic cutting tool and Zirconia toughened alumina ceramic cutting tool. Comparing these two tools, the author found that with respect to adhesive wear Zirconia toughened alumina ceramic cutting tool was better and best surface finish was recorded by Ti[C,N] mixed alumina ceramic cutting tool.

Luis Henrique Andrade Maia et al.[9] have made study on AISI 4340 hardened steel with uncoated, AlCrN coated and nano AlCrN coated WC inserts. The author utilized acoustic emission methodology to collect wear rate and power spectral density (PSD) technique and auto-covariance coupled PSD

technique to analyze. In his study he found that compared to uncoated WC insert and ordinary AlCrN coated WC insert, nano structured AlCrN coated WC insert performed superiorly.
Usage of Polycrystalline cubic boron nitride (PCBN) cutting tool in hard turning is attributable to its ultra high hardness, lower chemical affinity towards work piece and extreme wear resistance[10 -19]. Researchers have indicated that highly precision, rigid machine and rigid tool clampingare required to use CBN tools as it is brittle in nature and chips easily [20, 21]. Many researchers have made research about the composition, wear and tool temperatures of CBN cutting tools [22 -24]. Prior studies in turning AISI 52100 hardened steel revealed that low content CBN tools generates good surface finish compared to high content CBN tools [22, 24].
Pradeep Kumar and Chauhan [25] experimented hard turning of hardened AISI H13 die tool steel using CBN tools and investigated cutting forces and surface roughness of three work pieces with different hardness values (45 HRC, 50 HRC and 55 HRC). The authors considered five parameters cutting speed, depth of cut, feed rate, nose radius and hardness of work piece for their study. They concluded that turning work piece with higher hardness lead to better surface finish. Larger nose radius, high cutting speed resulted in reduced surface roughness and increase in cutting speed reduced the cutting force.
Shalaby et al. [26] investigated about the tool life and cutting forces in Hard turning of D2 tool steel with three different tools mixed alumina ceramic tool, PCBN tool and coated PCBN tool. He concluded that mixed alumina ceramic tools resultedin less cutting forces and increased tool life compared to PCBN tools, this can be accredited to its less chemical affinity to work piece, thermal stability. But, the brittle nature limits its application.
Ranjit Babu B. G and Sonachalam M [27] experimented turning of AISI 52100 hardened steed using nano coated tool insert and optimized the process parameters using Taguchi technique. Vereschaka A. A et al.[28] successfully developed nano-scale multilayered composite coatings on cutting tool using chemical vapor deposition (CVD) and physical vapor deposition (PVD) and stated that nano coated tool have more wear resistance. Escobar C et al.[29] investigated tribological and wear behavior of Hafnium Nitride (HfN) and Vanadium Nitride (VN) nano multilayer coated tool. The authors concluded that turning work piece with nano multilayer coated tool shows high tribological performance and have high wear resistance. Also the nano coated tools have 24% higher tool life than uncoated WC tool.
Several researches are going on to develop newer cutting tool material, nano coated and nano composite coated tool insert aiming at reducing the tool wear, cutting force, surface roughness and increasing the tool life, metal removal rate and productivity. Emerging researchers have an ample of scope in developing newer tool material.

3. Tool geometry

One of the major factors in tool life is cutting edge geometry, a tool with poor edge geometry fails quickly [30-31]. To accomplish high productivity with economy, cutting tool geometry has to be optimized. Heat generated in the cutting process is related to the cutting edge geometry. Cutting edge significantly affects the surface integrity of the work piece [22, 32-35]. As a whole, cutting tool edge preparation influences on the performance and reliability of the cutting tool.
Tugrul Ozel et al.[36] studied the effects of cutting edge geometry on surface roughness and cutting forces in turning of hardened AISI H13 steel and concluded that cutting edgegeometry on surface roughness is significant and also cutting edge geometry influences the cutting forces. The author also states that honed edge geometry produces better surface finish. Authors Muammer Nalbant et al. [37] and Huges et al. [38] in their study stated that, in metal cutting process, cutting tool edge geometry significantly influences tool wear, tool life, chip formation, surface roughness, cutting force and cutting temperature.
In hard turning, tool geometry influences the following tool wear, surface finish, chip form, residual stress, cutting forces, heat generation, white layer and variations in micro hardness [39]as shown in the fig. 1.

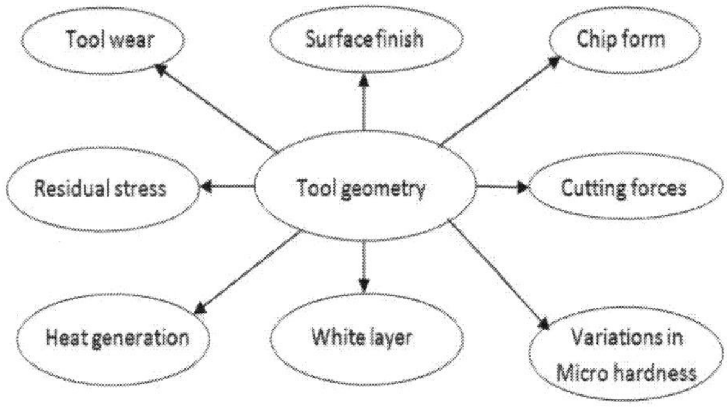

Figure. 1. Effect of tool geometry in hard turning.

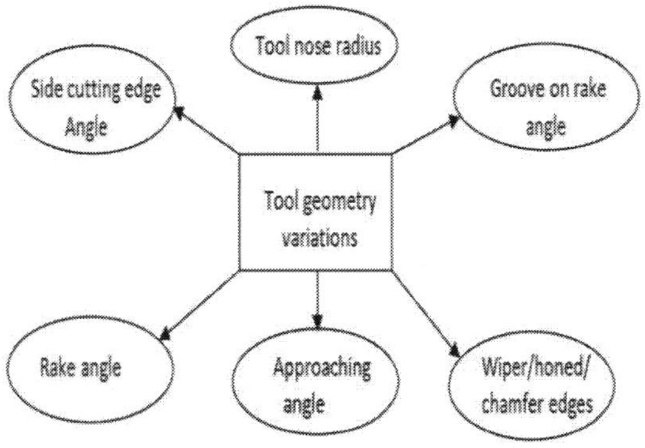

Figure. 2. Cutting tool geometry variations.

Variation in tool geometry is one of the prime factor in enhancing surface integrity and tool life. The probable variations that can be made in the cutting tool are shown in the fig. 2. [39]. From the above literature study, it is obvious that by adopting proper tool cutting geometry variation, productivity is much improved in turning of hardened steel.

4. Minimum quantity lubrication

Generally hard turning is done in dry condition as it is believed that, due to high temperature it reduces the cutting force and in turn the power consumption of the machine tool. At the same time, surface integrity and tool life are affected by dry cutting. Recent survey on European automotive industries exposed that the cost of cooling lubricant nears 20% of thetotal manufacturing cost [40], which is more than the cost of cutting tools (7.5%). This factor forces the need to reduce the consumption of cutting fluids. Also, the governmental initiatives on environmental aspect, lead to a tighter regulations to force the industries to take pollution preventing measures to achieve eco-friendly manufacturing.

Finding a solution to satisfy both the extreme conditions of dry cutting and flooded wet cutting, emerges the concept of Minimum Quantity Lubrication (MQL). The usage of lubricant in MQL is very less, in the order of 50 to 500 ml/hour. This near dry lubrication is environmental friendly and reduces the hazards caused by the flooded cutting fluids in the shop floor. MQL reduces the handling cost of lubricant leading to reduction in cost and time.

As a recent advancement, using nano-lubricants in MQL proved to be an effective method to minimize the friction between the contact surfaces in machining. Nanoparticles dispersed in the coolant oil easily penetrate into the rubbing surface and gives better elasto-hydrodynamic lubrication effect [41].Mohd Sayuti et al.[42] succeeded in using SiO_2 nano-lubricant in turning of AISI 4140 hardened steel. The author reports minimum tool wear and better surface finish achieved in the study.

5. Cryogenic cooling

Turning hard material involves high speed machining generating more heat, ending in shorter tool life and poor surface finish. Increase in tool work interface being the main cause for reduced tool life. The heat generated is removed by flooded coolant or minimum quantity lubrication (MQL) and thus reducing tool wear and increasing tool life. The use of cutting fluids is nowadays restricted due to the adverse health and environmental hazards, and also high cost of coolant, handling cost and disposal cost directed the researchers to look for other methods of cooling. Alternate to the traditional and MQL methods, usage of cryogenic fluids like liquid nitrogen (LN_2) emerged as new technology. Liquid nitrogen is inert, odorless, non toxic, non-combustible gas satisfying the environmental requirements and also capable of achieving very low temperature (-196 °C).

The method of spraying liquid nitrogen directly on the cutting area, while machining, yielded better tool life [43, 44]. Many researchers reported reduction in cutting temperature [45-47] as the main advantage resulting in higher surface integrity and tool life. Increase in tool life [48-49] is observed by several authors as there is reduction in tool wear due to cryogenic cooling. Ghost et al. [49] investigated the effects of cryogenic cooling in machining of A2 hardened tool steel by alumina ceramic tool.

Comparing to flooded and dry cooling systems cryogenic cooling excelled in performance. Sunil Magadum et al. [50] evaluated the tool life and cutting forces in machining EN -24 steel having hardness 45 HRC with coated carbide insert in cryogenic cooling environment. The authors found that tool life increased up to 38.60% and the cutting forces are reduced up to 15.42%.

Garcia Navas et al.[51] studied about the surface integrity of AISI 4150 steel while turning in dry, wet and cryogenic cooling conditions and concluded that cryogenic coolingresulted in increased tool life and good surface integrity. Several authors are still doing research in cryogenic cooling and there is a lot of scope for further research.

6. Laser assisted hard turning

Laser assisted hard turning involves in heating the workpiece using laser and removing the material simultaneously. Heating the workpiece with laser beam takes place first and is followed by the cutting tool by a short gap. The material is softened by the laser irradiation and the tool that follows removes the material easily.

Ding H and Shin YC studied about laser assisted machining of hardened steel [52].Laser assisted turning has its own limitations, that for removing the material a high power laser is required to induce the thermal softening and integrating a laser beam into the machine have become the challenging tasks.

7. Laser tempering based turning

The above process of laser assisted turning is decoupled into two steps in the laser tempering based turning, first the surface of material is softened by laser treatment through tempering process and then after cooling to room temperature, the material is removed by turning. Since the outer layer of the material is soft due to tempering, higher material removal rate can be achieved with the hard CBN tool. Otherwise a low cost ceramic tool can be used for turning.

Satyanarayanan Raghavan et al [53]. experimented laser assisted turning of AISI 52100 hardened steel (63 HRC) with two cutting tools (a) Mixed alumina ceramic tool and (b) CBN tool and made the following conclusions. Ceramic tools can be used in place of high cost CBN tools resulting in almost same tool life and metal removal rate. In turning the hardened steel based on laser tempering high cutting speeds is possible leading to high metal removal rate. The authors also reported lower cutting forces in laser tempering based turning process.

8. Acoustic emission controlled hard turning

Several studies about tool wear were reported, and when the tool wears, it has to be replaced by a new tool. The time of changing the tool plays a major role in the production industries. To change a tool at right time needs monitoring of the tool wear. Monitoring the tool wear can be done in two ways, online or continuous method and offline or intermittent method.

Tool wear is accurately monitored by offline method, otherwise called direct method, but practically it has its limitations such as cutting fluid usage, illumination and possibility of accessing. Online or indirect methods of measuring are not accurate but practically possible to measure it. Some of the direct or offline methods of measuring tool wear are

(i) Optical
(ii) Radioactive and
(iii) Electrical resistance.

Indirect or online method involves in measuring the parameters such as cutting force, surface roughness, sound, vibration, temperature and acoustic emission (AE).

8.1*Acoustic emission (AE)*

The spontaneous release of energy from the work material, when it undergoes a change from its original condition produces sound. Metal removing, corrosion, wear, shear are some of the acoustic emission generating processes. Acoustic Emission is one of the foremost techniques in monitoring tool wear [9].

Li X [54] explained acoustic emission as a wave of tension travelling through the material when there is a sudden release of tensile energy. Hase A et al.[55] correlated AE signals with wear mechanisms.

It is evident from the literature survey that AE monitored hard turning will be helpful in the advancement of the hard turning technology and budding researchers have plenty of scope in developing this area.

9. Conclusion

It is evident from the review that, introduction of latest tools, cooling methods, monitoring systems, newer modeling techniques, tool geometry hard machining process attained new high in technological development. Still there is a possibility of developing newer tools with micro and nano coating technologyresulting in more tool life, less economic and increased productivity. Dry and cryogenic cooling system have led to the pollution free environment, increased performance and reduced cost. Ultrasonic assisted turning, and AE monitored machining have given more control over the production process leading the metal cutting industries to a new higher level of performance.

In spite of these developments, turning of hardened materials such as D2, D3, M7, M35, M42, O1, O2 tool steels, which have hardness value above 60 HRC pose a challenge to manufacturing industries and there is a lot of scope in doing further research in the field of turning hardened materials.

10.References

[1]R. C. Dewes and D. K. Aspinwall. (1997). A review of ultra high speed milling of hardened steels. J Mater Process Technol. 69, pp. 1–17.

[2] Ersan Aslan, Necip Camuscu and Burak Birgoren. (2007) Design optimization of cutting parameters when turning hardened AISI 4140 steel (63 HRC) with Al_2O_3 + TiCN mixed ceramic tool. Materials and Design. 28, pp. 1618–1622.

[3] K. Aslantas, I. Ucun and A. Cicek. (2012). Tool life and wear mechanism of coated and uncoated Al_2O_3/TiCN mixed ceramic tools in turning hardened alloy steel. Wear. 274, pp 442–451.

[4] V.N. Gaitonde, S.R. Karnik, Luis Figueira and J.Paulo Davim (2009). Machinability investigations in hard turning of AISI D2 cold work tool steel with conventional and wiper ceramic inserts. Int. J. Refract. Met. Hard Mater. 27 (4), pp. 754–763.

[5] M. Elbah, M.A. Yallese, H. Aouici, T. Mabrouki and J.F. Rigal. (2013). Comparative assessment of wiper and conventional ceramic tools on surface roughness in hard turning AISI 4140 steel. Measurement. 46 (9), pp. 3041–3056.
[6] Abhijeet S. More, Wenping Jiang, W.D. Brown and Ajay P. Malshe. (2006). Tool wear and machining performance of cBN–TiN coated carbide inserts and PCBN compact inserts in turning AISI 4340 hardened steel. J of Mater Process Technol. 180, pp. 253–262.
[7] G. Brandt. (1986).Flank and crater wear mechanisms of alumina based cutting tools when machining steel. Wear. 112(1), pp. 39–56.
[8] A. Senthil Kumar, A. Raja Durai and T. Sornakumar. (2003). Machinability of hardened steel using alumina based ceramic cutting tools. Int. J. of Refract. Met. & Hard Materials. 21, pp. 109 – 117.
[9] Luis Henrique Andrade Maia, Alexandre Mendes Abrao, Wander Luiz Vasconcelos, Wisley Falco Sales, and Alisson Rocha Machado. (2015). A new approach for detection of wear mechanisms and determination of tool life in turning using acoustic emission. Tribology International. 92, pp. 519–532.
[10] X.L. Liu, D.H. Wen, Z.J. Li, L. Xiao and F.G. Yan. (2002). Experimental study on hard turning hardened GCr15 steel with PCBN tool. J.Mater. Process. Technol. 129 (1-3), pp. 217–221.
[11] T. Hodgson, P. H. H. Trendler and G. F. Michelletti. (1981). Turning hardened tool steels with cubic boron nitride inserts. CIRP Ann. - Manuf. Technol. 30(1), pp. 63–66.
[12] Roland Meyer, Jens Köhler and Berend Denkena. (2012). Influence of the tool corner radius on the tool wear and process forces during hard turning. Int J Adv Manuf Technol, 58 (9), pp. 933–940.
[13] Li Qian and Mohammad Robiul Hossan. (2007). Effect on cutting force in turning hardened tool steels with cubic boron nitride inserts. J. Mater. Process. Technol. 191, pp. 274–278.
[14] H. K.Toenshoff, C. Arendt and R. Ben Amor. (2000). Cutting of hardened steel. Ann CIRP 49(2), 1–19.
[15] R. M'Saoubi, T. Czotscher, O. Andersson and D. Meyer. (2014). Machinability of powder metallurgy steels using PcBN inserts. Procedia CIRP. 14, pp. 83–88.
[16] M. A. El Hakim, M. D. Abad, M. M. Abdelhameed, M. A.Shalaby and S. C. Veldhuis. (2011). Wear behavior of some cutting tool materials in hard turning of HSS. Tribol Int. 44, pp. 1174–81.
[17] F. Klocke and G. Eisenblatter. (1997). Dry cutting. Ann CIRP. 46(2), pp. 519–526.
[18] J.C. Camargo, D.S. Dominguez, E.O. Ezugwu and A.R.Machado.(2014). Wear model in turning of hardened steel with PCBN tool. Int. J. Refract. Met. Hard Mater. 47, pp. 61–70.
[19] W. Koenig, A. Berktold and F. Koch. (1993). Turning versus grinding – a comparison of surface integrity aspects and attainable accuracies. Ann CIRP 42(1), pp. 39–43.
[20] Y. Matsumoto, F. Hashimoto and G. Lahoti. (1999). Surface integrity generated by precision hard turning. Ann CIRP 48(1), pp. 59–62.
[21] M. A. Davies, T. J. Burns and C. J. Evans. On the dynamics of chip formation in machining hard metals. Ann CIRP 46(1), pp. 25–30.
[22] Y. K. Chou, C. J. Evans and M. M. Barash. (2003). Experimental investigation on CBN turning of AISI 52100 steel. J Mater Process Technol. 134, pp. 1–9.
[23] T. Ueda, M. Al Huda, K. Yamada and K. Nakayama. (1999). Temperature measurement of CBN tool in turning high hardness steel. Ann CIRP 48(1), pp. 63–66.
[24] Y. K. Chou and C. J. Evans. (1997). Tool wear mechanism in continuous cutting of hardened tool steels. Wear. 212, pp. 59–65.
[25] Pradeep Kumar and S. R. Chauhan. (2016) An Investigation on Cutting Forces and Surface Roughness during Hard Turning of AISI H13 Die Tool Steel with CBN Inserts using RSM. Int. J. Adv. Engineering Research and Applications, 1(9). pp. 345-356.

[26] M.A. Shalaby, M.A. El Hakim, M.M. Abdelhameed, J.E. Krzanowski, S.C. Veldhuis and G.K. Dosbaeva. (2014). Wear mechanisms of several cutting tool materials in hard turning of high carbon–chromium tool steel. Tribol. Int. 70, pp. 148–154.
[27] B. G. Ranjit Babu and M. Sonachalam. (2014). Machining Optimization of Nano Coated Tool Insert in Hard Turning using Taguchi Method. International Journal of Engineering Research & Technology. 3 (4), pp. 1686-1691..
[28] A. A. Vereschaka, S. N. Grigoriev, A. S. Vereschaka, A. Yu. Popov and A. D. Batako. (2014). Nano-scale multilayered composite coatings for cutting tools operating under heavy cutting conditions. Procedia CIRP. 14, pp. 239 – 244.
[29] C. Escobar, M. Villarreal, J. C. Caicedo, W. Aperador and P. Prieto . (2014). Tribological and wear behavior of HfN/VN nano-multilayer coated cutting tools. Ingeniería e Investigación. 34(1), pp. 22-28.
[30] N. Fang and G. Fang. (2007). Theoretical and experimental investigations of finish machining with a rounded edge tool. Journal of Materials Processing Technology. 191(1-3), pp. 331-334.
[31] Jeffrey D. Thiele, Shreyes N. Melkote, Roberta A. Peascoe and Thomas R. Watkins. (1999). Effect of cutting edge geometry and workpiece hardness on surface residual stress in finish hard turning of AISI 52100 steel. J. Manuf. Sci. Eng 122(4), pp. 642-649.
[32] Jeffrey D. Thiele and Shreyes N. Melkote. (1999). Effect of cutting-edge geometry and workpiece hardness on surface generation in the finish hard turning of AISI 52100 steel. J. Matl. Proc. Tech. 94(2-3), pp. 216-226.
[33] T. Özel.(2003). Modeling of hard part machining: effect of insert edge geometry in CBN tools. J Mater Process Technol. 141, pp. 284–293.
[34] G. Poulachon, A. Moison and I. S. Jawahir. (2001). On Modeling the influence of thermo-mechanical behavior in Chip Formation During Hard Turning of 100Cr6 Bearing Steel. Annals of the CIRP. 50(1), pp. 31-36.
[35] Y. B. Guo and C. R. Liu. (2002). Mechanical properties of hardened AISI 52100 steel in hard machining processes. J Manuf Sci Eng 124, pp. 1–9.
[36] Tugrul O zel, Tsu-Kong Hsu and Erol Zeren. (2005). Effects of cutting edge geometry, workpiece hardness, feed rate and cutting speed on surface roughness and forces in finish turning of hardened AISI H13 steel. Int J Adv Manuf Technol. 25, pp. 262–269.
[37] M. Nalbant, A. Altin and H. Gokkaya. (2007). The effect of cutting speed and cutting tool geometry on machinability properties of nickel-base Inconel 718 super alloys. Materials and Design. 28, pp. 1334-1338.
[38] J. I. Hughes, A. R. Sharman and C. Ridgwayk. (2006). The effect of cutting tool material and edge geometry on tool life and workpiece surface integrity. Proceedings of the Institution of Mechanical Engineers. Part B. Journal of Engineering Manufacture. **220**(2), pp. 93-107.
[39] M. Dogra, V. S. Sharma and J. Dureja. (2011). Effect of tool geometry variation on finish turning – A Review. Journal of Engineering Science and Technology Review. **4**(1), pp. 1-13.
[40] T. Brockhoff and A. Walter. (1998). Fluid Minimization in Cutting and Grinding. Abrasives. 38-42.
[41] D. X. Peng, Y. Kang, R. M. Hwang, S. S. Shyr and Y. P. Chang. (2009). Tribological properties of diamond and SiO2 nanoparticles added in paraffin. Tribol. Int. **42**(6), pp. 911-917.
[42] Mohd Sayuti, Ahmed A.D. Sarhan and Faheem Salem. (2013). Novel uses of SiO2 nano-lubrication system in hard turning process of hardened steel AISI4140 for less tool wear, surface roughness and oil consumption. Journal of Cleaner Production. pp. 1-12.
[43] Franci Pusavec, Peter Krajnik and Janez Kopac. Janez Kopac. (2010). Transitioning to Sustainable Production – Part I: Application on Machining Technologies. Journal of Cleaner Production. **18**(2), pp. 174-184.
[44] Z. Wang and K. P. Rajurkar. (2000). Cryogenic Machining of Hard-to-Cut Materials. Wear. **239**(2), pp. 168-175.

IOP Conf. Series: Materials Science and Engineering **197** (2017) 012009 doi:10.1088/1757-899X/197/1/012009

[45] N. R. Dhar and M. Kamruzzaman. (2007). Cutting temperature, tool wear, surface roughness and dimensional deviation in turning AISI-4037 steel under cryogenic condition. International Journal of Machine Tools and Manufacture. **47**(5), pp. 754-759.

[46] B. Dilip Jerold and M. Pradeep Kumar. (2012). Experimental comparison of carbon-dioxide and liquid nitrogen cryogenic coolants in turning of AISI 1045 steel. Cryogenics. **52**(10), pp. 569-574.

[47] Y. Kaynak, H. E. Karaca, R. D. Noebe and I. S. Jawahir. (2013). Analysis of Tool-wear and Cutting Force Components in Dry, Preheated, and Cryogenic Machining of NiTi Shape Memory Alloys. Procedia CIRP. **8**, pp. 498-503.

[48] M. J. Bermingham , J. Kirsch, S. Sun, S. Palanisamy and M. S. Dargusch. (2011). New Observations on Tool Life, Cutting Forces and Chip Morphology in Cryogenic Machining Ti-6Al-4V. International Journal of Machine Tools & Manufacture. **51**, pp. 500-511.

[49] R. Ghosh, Z. Zurecki and J. H. Frey. (2003). Cryogenic machining with brittle tools and effects on tool life. Proceedings of ASME IMECE2003-42232.

[50] Sunil Magadum, S. Arun Kumar, V. G. Yoganath, C. K. Srinivasa and T.GuruMurthy. (2014). Evaluation of Tool life and Cutting forces in Cryogenic Machining of Hardened Steel. Procedia Materials Science. **5**, pp. 2542-2549.

[51] V. Garcia Navas, D. Fernandez, A. Sanda, C. Sanz, S. Suzon and T. Fernandez de Mendiola. (2014). Surface integrity of AISI 4150 (50CrMo4) steel turned with different types of cooling-lubrication. Procedia CIRP. **13**, pp. 97 – 102.

[52] H. Ding and Y. C. Shin. (2010). Laser-assisted machining of hardened steel parts with surface integrity analysis. International Journal of Machine Tools and Manufacture. **50**(1), pp. 106–114.

[53] Satyanarayanan Raghavan, Shreyes Melkote and Fukuo Hashimoto. (2013). Laser tempering based turning process for efficient machining of hardened AISI 52100 steel. Journal of Manufacturing Processes. **15**, pp. 318–328.

[54]Xiaoli Li. (2002). A brief review: acoustic emission method for tool wear monitoring during turning. Int J Mach Tools Manuf. **42**(2), pp. 157–165.

[55] Alan Hase, Hiroshi Mishina and Masaki Wada. (2012). Correlation between features of acoustic emission signals and mechanical wear mechanisms. Wear. **292–293**, pp. 144–150.

REVIEW ON MICROSTRUCTURE ANALYSIS OF METALS AND ALLOYS USING IMAGE ANALYSIS TECHNIQUES

[1]Suganthini Rekha, [2]V.K.Bupesh Raja*

[1]Professor&Head, Department of Automobile Engineering, Sathyabama University, Chennai, Tamil Nadu, India

[2]Assistant Professor, Department of Mechanical Engineering, Dr.M.G.R Educational & Research Institute University, Chennai, Tamil Nadu, India

[1]suganthinirekha@gmail.com, [2]bupeshvk@gmail.com

*Corresponding author: bupeshvk@gmail.com

Abstract: The metals and alloys find vast application in engineering and domestic sectors. The mechanical properties of the metals and alloys are influenced by their microstructure. Hence the microstructural investigation is very critical. Traditionally the microstructure is studied using optical microscope with suitable metallurgical preparation. The past few decades the computers are applied in the capture and analysis of the optical micrographs. The advent of computer softwares like digital image processing and computer vision technologies are a boon to the analysis of the microstructure. In this paper the literature study of the various developments in the microstructural analysis, is done. The conventional optical microscope is complemented by the use of Scanning Electron Microscope (SEM) and other high end equipments.

Keywords: Microstructure, image processing, mechanical properties, SEM.

1. INTRODUCTION

Many of the manufacturing industries have a constant pressure to manufacture a high quality product with a less cost. Thus they required a new technologies in increasing the production with a high quality. Hence they prefers to change their machining tools and manufacturing materials so that they can increase their efficiency in their manufacturing. The replacement among the tools will progressively reducing the tool wear. Reduction of tool wear is placing a major significance in tool designing industries. Wherever the aluminium alloys placing a better shaping and cutting process over the metals. These effect can be found by analysing the difference of metals and alloys [1].

Hard materials were commonly utilized for its wear application and the machining process. Due to the various reasons the material were tested to highly fatigue metals. Among the past years they have developed to improve hardness, bending strength and the fracture toughness. In order to design fatigue proof tools or structural components made of hard metals, mechanical and design engineers require material data like yield strength or fatigue strength [4].

Cast Al alloys are important structural materials for the light weighting of cars and, consequently, reduction of greenhouse gases emission and pollution. The microstructure and properties of cast Al alloys could be further improved by means of grain refinement, practise which cannot efficiently be performed with common Ale TieB grain refiners used for wrought Al alloys [2].

The intelligent atomization processing is the on-line in situ sensing of powder size and velocity. The principles of operation and current analogous applications of several sensors make them potential candidates for consideration. In addition to meeting the specific measuring requirements, viable sensors must be non-intrusive, and must work reliably in high ambient temperature, high ambient light intensity, and high dust particle density environments. Usually non-intrusive sensors are based on electric, electromagnetic or optical principles, the latter being most developed in fields obviously related to atomization. Optical methods for particle sizing can be divided into imaging and non-imaging techniques. The imaging techniques include microphotography, holography and cinematography; non-imaging techniques can be further divided into ensemble or multi-particle techniques and single particle counting techniques [3].

L. Bolzoni has studied where a commercial Ale12Si alloy was inoculated with different level of NbþB addition to assess the grain refining potency of NbþB inoculation it can be concluded that insitute formed Nb-based intermetallics compounds are potent heterogeneous nucleation substrates with high potency for the refinement of AleSi cast alloys [2].

2. STUDIES ON MICROSTRUCTURE ANALYSIS

Microstructure analysis is used among the various industries to find the structure of the material at various stages of testing. Thus the analysis explains about the structure of the materials. Generally microstructure analysis is done by using the optical microscope but now a days there is development done with a digital analysis using Image progression methods. By using this analysis we can find the yield strength and ultimate tensile or compressive strength are not as well established in the literature mainly due to the high hardness and the low ductility that makes the determination of reliable information of material as a challenging task. So the general values of ultimate tensile strength and yield strength are higher in bending than in uniaxial testing conditions due to the differences in tested effective volume [5] and due to the non-linear stress distribution in the bent specimens which are partially deformed plastically before fracture [6]. The cyclic behaviour of hard metals was investigated by several authors [7]. The results from literature on microstructure and properties can be summarized. Roebuck finds significant higher fatigue life in hard metals at 300°C testing temperature compared to room temperature tests [8]. Varying testing temperatures up to 700°C encounters lifetimes significantly below the values determined at room temperature. The behaviour of hard metals under cyclic loads cannot be deduced from their behaviour under static loading conditions [9]. Has studied the microstructure of friction welded cu/al joint and its weld interface characterization were identified and discussed and also the correlation between the microstructure and mechanical properties has been investigated[10]. The full advantage that these materials can provide is strongly dependent on composition and microstructure [11].

3. STUDY ON IMAGE ANALYSIS

The solidification process itself is analysed using the UMSA device by appliance of the Derivative Thermo Analysis. The influence of the cooling rate on the alloy microstructure was investigated using computer aided image analysis, in this work also the content of particular phases was analysed, as well the percentage of pinholes compared to the chosen cooling rate [12]. The aluminum images have inconsistent distribution of illumination due to light concentration from the metallograph image acquisition system. This is particularly apparent in the corners of the image. Since the main objective was to isolate the dark and gray blobs based on their intensity levels, a background correction algorithm was performed to compensate for the inconsistent pixel intensity of the background. This minimized the possibility of dark and gray blobs being misidentified [4]. The microstructure of an engineering material depends on its properties. For the reason of the dependence settlement between the microstructure and properties there should be performed an image analysis. The process of image analysis can be defined as a image investigation for object identification and its contribution valuation on the properties. This process consist of data accusation from the image as well the analysis through: identification, classification, measurement and valuation of the physical state of the sample, analysis of the pattern observed in the image as well the spatial relation between the features of a given sample. In case of data acquisition about metal microstructure, the manual, stereological or conventional image analysis methods are often time-consuming and not always user friendly. As a solution for this problem there are applied computer aided qualitative and quantitative metal microstructure analysis [13]. The alloy precipitates present in the image are represented by the blobs which are fairly distinctly distinguished into two shades of gray – the white blobs which are the lighter ones and the darker or grey blobs. The objective is to identify the two types of blobs and statistically characterize them for comparison purposes [4].

4. STUDY ON METALS

All Metallic materials are broadly of two kinds – ferrous and non-ferrous materials. This classification is primarily based on tonnage of materials used all around the world. Ferrous materials are those in which iron (Fe) is the principle constituent. All other materials are categorized as non-ferrous materials. Another classification is made based on their formability. If materials are hard to form, components with these materials are fabricated by casting, thus they are called cast alloys. B.K Prasad has investigated on the observations pertaining to the effects of specimen and slurry compositions as well as traversal distance on the slurry wear response of a zinc-based alloy [14]. among the different kinds of alloys the zinc alloy has a better coating and it scarifying to protect steel structures against the corrosion taken place in the environment. Zinc-based alloys have been shown to be more effective as protective coatings and sacrificial anodes than pure zinc in more corrosive environments [15]. It has been observed that the corrosion resistance of zinc-based alloys increases with Al content [16] while addition of Si to the alloy system has been proved to be still more beneficial in this context [17]. Studies suggest that Zn-Al alloys behave in a manner similar to that of pure Zn from corrosion resistance standpoint since Al remains practically unaffected [18]. Steels are alloys of iron and carbon plus other alloying elements. In steels, carbon present in atomic form, and occupies interstitial sites of Fe microstructure. Alloying additions are necessary for many reasons including: improving properties, improving corrosion resistance, etc. Arguably steels are well known and most used materials than any other materials. Low carbon steels: These are arguably produced in the greatest quantities than other alloys [19]. Carbon present in these alloys is limited, and is not enough to strengthen these materials by heat treatment; hence these alloys are strengthened by cold work. Their microstructure consists of ferrite and pearlite, and these alloys are thus relatively soft, ductile combined with high toughness. Hence these materials are easily machinable and weldable. Typical applications of these alloys include: structural shapes, tin cans, automobile body components, buildings, etc [20]

Magnesium alloys: The most sticking property of Mg is its low density among all structural metals. Mg has HCP structure, thus Mg alloys are difficult to form at room temperatures. Hence Mg alloys are usually fabricated by casting or hot working. As in case of Al, alloys are cast or wrought type, and some of them are heat treatable. Major alloying additions are: Al, Zn, Mn and rare earths. Common applications of Mg alloys include: hand-held devices like saws, tools, automotive parts like steering wheels, seat frames, electronics like casing for laptops, camcoders, cell phones etc.

Titanium alloys: Ti and its alloys are of relatively low density, high strength and have very high melting point. At the same time they are easy to machine and forge. However the major limitation is Ti's chemical reactivity at high temperatures, which necessitated special techniques to extract. Thus these alloys are expensive. They also possess excellent corrosion resistance in diverse atmospheres, and wear properties. Common applications include: space vehicles, airplane structures, surgical implants, and petroleum & chemical industries [20].

5. CONCLUSION

The implementation over the microstructural analysis is explained for the testing of different metals and alloys. We also gathered the information of analyzing the microstructures of metals and alloys. Thus the structure explains about the various strength of the materials. The different kinds of alloys were discussed in this paper.

References

[1] Mohammed E. Hoque, Ralph M. Ford, and John T. Roth Penn State Erie, Automated Image Analysis of Microstructure Changes in Metal Alloys, School of Engineering and Engineering Technology Erie, PA 16563-1701.

[2] L. Bolzoni, M. Nowak, N. Hari Babu, On the effect of Nb-based compounds on the microstructure of Ale12Si alloy (vol-162 (2015) 340e345), Brunel University London, Institute of Materials and Manufacturing, Uxbridge, Middlesex, UB8 3PH, UK.

[3] G. Jiang, H. Henein, and M. W. Siege, Intelligent Sensors for Atomization Processing of Molten Metals and Alloys, The Robotics Institute Carnegie-Mellon University Pittsburgh, PA 15213 June 1988.

[4] Thomas Klunsnera,, Stefan Marsoner, Reinhold Ebnera,Reinhard Pippan, Johannes Glatzle, Arndt Puschel, Effect of microstructure on fatigue properties of WC-Co hard metals. Procedia Engineering 2 (2010) 2001–2010.

[5] Quinn GD. Weibull effective volumes and surfaces for cylindrical rods loaded in flexure. *J Am Ceram Soc* 2003;**86**[3]:475–9.

[6] Lackner JM. Entwicklung eines Zugversuches fur hochstfeste Werkzeugstahle. Leoben, Diploma Thesis, 2001.

[7] Kursawe S, Pott P, Sockel HG, Heinrich W, Wolf M. On the influence of binder content and binder composition on the mechanicalroperties of hardmetals. *Int J Refract Met Hard Mater* 2001;**19**:335-40.

[8] Roebuck B, Maderud CJ, Morrel R. Elevated temperature fatigue testing of HM using notched testpieces. Int J Refract Met Hard Mater 2008;26:19-27.

[9] Kindermann P, Schlund P, Sockel HG, Herr M, Heinrich W, Gorting K, et al. High-temperature fatigue of cemented carbides under cyclic loads. Int J Refract Met Hard Mater 1999;17:55-68.

[10]CH.Muralimohan, S.Haribabu, Y.Hariprasada Reddy, V.Muthupandi, K.Sivaprasad, Evaluation of microstructures and mechanical properties of two dissimilar materials by friction welding,

[11] Ulrike Taffner, Veronika Carle, and Ute Scha¨fer, Max-Planck-Institut fur Metallforschung, Stuttgart, Germany Michael J. Hoffmann, Institut fur Keramik im Maschinenbau, Universitat Karlsruhe, Germany Preparation and Microstructural Analysis of High-Performance Ceramics

[12] M. Krupiński, K. Labisz, L.A. Dobrzański, Z. Rdzawski, Image analysis used for aluminium alloy microstructure investigation, VOLUME 42 ISSUES 1-2 September-October 2010.

[13] J.P. Anson, J.E. Gruzleski, The quantitative discrimination between shrinkage and gas microporosity in cast aluminum alloys using spatial data analysis, Materials Characterization 43 (1999) 319-335.

[14] B.K. PRASAD, O.P. MODI, Slurry wear characteristics of zinc-based alloys: Effects of sand content of slurry, silicon addition to alloy system and traversal distance, Advanced Materials and Processes Research Institute (CSIR), Bhopal-462026, India Received 16 April 2008.

[15] KUBEL E J. Expanding horizons for ZA alloys [J]. Advanced Materials and Processes, 1987, 132: 51−57.

[16 DAFYDD H, WORSLEY D A, MCMURRAY H N. The kinetics and mechanism of cathodic oxygen reduction on zinc and zinc-aluminium alloy galvanized coatings [J]. Corrosion Science, 2005, 47: 3006−3018

IOP Conf. Series: Materials Science and Engineering **197** (2017) 012010 doi:10.1088/1757-899X/197/1/012010

[17] CHEN W, LIU Q, ZHU L, WANG L. A combinatorial study of the corrosion and mechanical properties of Zn-Al material library fabricated by ion beam sputtering [J]. Journal of Alloys and Compounds, 2008, 459: 261−266.
[18] G. E. Dieter, Mechanical Metallurgy, Third Edition, McGraw-Hill, New York, 1986.
[19]William D. Callister, Jr, Materials Science and Engineering – An introduction, sixth edition, John Wiley & Sons, Inc. 2004.
[20] ASM Handbook, Heat treating, Vol. 4, ASM International, Materials Park, OH, 1991.

IOP Conf. Series: Materials Science and Engineering **197** (2017) 012011 doi:10.1088/1757-899X/197/1/012011

OPTIMIZATION OF TENSILE PROCESS PARAMETERS OF BI-AXIAL GLASS FIBRE REINFORCED EPOXYCOMPOSITE USING TAGUCHI METHOD

I Infanta Mary Priya[1] and B K Vinayagam[2]

[1]Asst Professor, Mechanical Department, SRM University

[2]Professor, Mechatronics Department, SRM University

E-mail: infanta.i@ktr.srmuniv.ac.in

Abstract: The aim of this work is to optimize the parameters namely load, elongation and thickness in tensile test of glass fibre reinforced polymer(GFRP) composites. In the present work, experiments were carried out as per the Taguchi experimental design and an L9 orthogonal array was used to study the influence of various combinations of parameters on stress and strain factors of the composite using MINITAB 17. Analysis of variance (ANOVA) test was conducted to determine the significance of each parameter on stress and strain value of the composite. The results indicate thatload is the most significant factor influencing the stress, and also it is the most significant factor inducing the strain of the composite. This work is useful in selecting the optimum values of various parameters that would, not only maximize the stress in the composite but also reduces the strain to a minimum level and improve the strength of the composite by acquiring a higher load bearing capacity.

Key Words: Glass Fibre Reinforced Polymer (GFRP), L9 orthogonal array, analysis of variance (ANOVA), stress, strain.

1. Introduction

Composite materials are widely used in many fields of automobiles and aerospace industries due to their high strength to weight ratio, lightweight, high stiffness, corrosive property and low thermal expansion, etc. Most of the literature [1-3] reported on Taguchi experimental design and an L9 orthogonal array was used to study the influence of various parameters. According to Taguchi method, engineering optimization of a process must be carried out in three different stages: system, parameter and tolerance design [4]. Also Taguchi method uses orthogonal arrays to study a large number of variables with a minimum number of experiments by using design of experiments theory. Therefore orthogonal arrays reduce the experimental configurations in number that is to be studied. The conclusions drawn by these small number of experiments are within the entire experimental region by the usage of control factors and the settings[5]. Taguchi modified the usage and provided standard orthogonal arrays in tabulated sets with the corresponding linear graphs to fit specific conditions of the project[6].

In this work, process parameters are optimized in the tensile test of GFRP composites using Taguchi design of experimental technique and the results are analysed using ANOVA technique to know the percentage contribution of each parameter on stress and strain of the composite. In this present work, statistical analysis software MINITAB 17 is used for the design and analysis of experiments to perform the Taguchi and ANOVA analysis and also to establish regression models.

2. Methodology

2.1. Taguchi Design of Experiment

Taguchi design of experiment consists of orthogonal arrays to study all the parameters of the project with a minimum number of experiments only. Then the results of the experiments are transformed into a signal-to-noise (S/N) ratio. It considers the S/N ratio to be a measure of quality characteristics that are deviating from or nearing to the required values. There are three different stages of quality characteristics in the analysis of the S/N ratio, i.e. the lower the better, the higher the better, and the nominal the better. The formula used for calculating S/N ratio is given below.

Smaller the better: It is used where the smaller value is required.

$$\frac{S}{N} ratio(\eta) = -10 \log_{10}\frac{1}{n}\sum_{i=1}^{n} {y_i}^2$$

Where y_i = observed response value and n = number of replications.

Nominal the best: It is used where the nominal or target value and variation about that value is minimum.

$$\frac{S}{N} ratio(\eta) = -10 \log_{10}\frac{\mu^2}{\sigma^2}$$

Where μ = mean and σ = variance.

Higher the better: It is used where the larger value is required.

$$\frac{S}{N} ratio(\eta) = -10 \log_{10}\frac{1}{n}\sum_{i=1}^{n} \frac{1}{{y_i}^2}$$

Where y_i = observed response value and n= number of replications.

2.2 Selection of process parameters

In this study, tensile test parameters like load, elongation and thickness of the specimen were considered. According to the Taguchi's design of experiments for three parameters and three levels L9 Taguchi orthogonal array was selected. The number of factors and their corresponding levels are shown in table 1.

Table 1 – Selected variable levels for the tensile test of GFRP

Code	Variable	Level1	Level2	Level3

1	Load(KN)	2.5	5.5	8.5
2	Elongation(mm)	0.839	2.573	7.226
3	Thickness(mm)	2	3	4

2.3. ANOVA Analysis

The ANOVA technique identifies the important parameters and calculates the influence of percentage of each parameter on different quality characteristics. The S/N ratios are calculated for the stress and strain responses and it is represented in table 2 and table 3 respectively.

Table 2 – Response table for S/N ratios for stress

Level	Load(KN)	Elongation(mm)	Thickness(mm)
1	21.98	27.18	28.63
2	32.66	31.37	31.00
3	37.45	33.54	32.46
Delta	15.47	6.36	3.84
Rank	1	2	3

Analysis of Variance

Source	DF	Adj SS	Adj MS	F-Value	P-Value
Load(KN)	2	5379.23	2689.62	884.73	0.001
Elongation(mm)	2	505.01	252.50	83.06	0.012
Thickness(mm)	2	5.35	2.68	0.88	0.532
Error	2	6.08	3.04		
Total	8	5895.67			

Model Summary

S	R-sq	R-sq(adj)	R-sq(pred)
1.74357	99.90%	99.59%	97.91%

Table 3 – Response table for S/N ratios for strain

Level	Load(KN)	Elongation(mm)	Thickness(mm)
1	9.1382	-0.9389	-4.9295
2	-13.7873	-6.9035	-7.6916
3	-20.3110	-17.1177	-12.3390
Delta	29.4492	16.1788	7.4095
Rank	1	2	3

Analysis of Variance

Source	DF	Adj SS	Adj MS	F-Value	P-Value
Load(KN)	2	142.838	71.4191	330.90	0.003
Elongation(mm)	2	46.304	23.1519	107.27	0.009
Thickness(mm)	2	7.179	3.5895	16.63	0.057
Error	2	0.432	0.2158		
Total	8	196.753			

IOP Conf. Series: Materials Science and Engineering **197** (2017) 012011 doi:10.1088/1757-899X/197/1/012011

Model Summary

S	R-sq	R-sq(adj)	R-sq(pred)
0.464575	99.78%	99.12%	95.56%

2.4. Experimentation

In this work, bi-axial glass fibre reinforced plastic composite laminate specimens with 50% fibre volume ratio were prepared with E-glass fibre using epoxy resin by vacuum infusion process. The specimen size of 250 x 25 x 2 mm was cut from a laminate using abrasive water jet cutting. The tensile test was carried our according to the ASTM D3518 standards. The experimental setup along with the software is shown in figure 1.

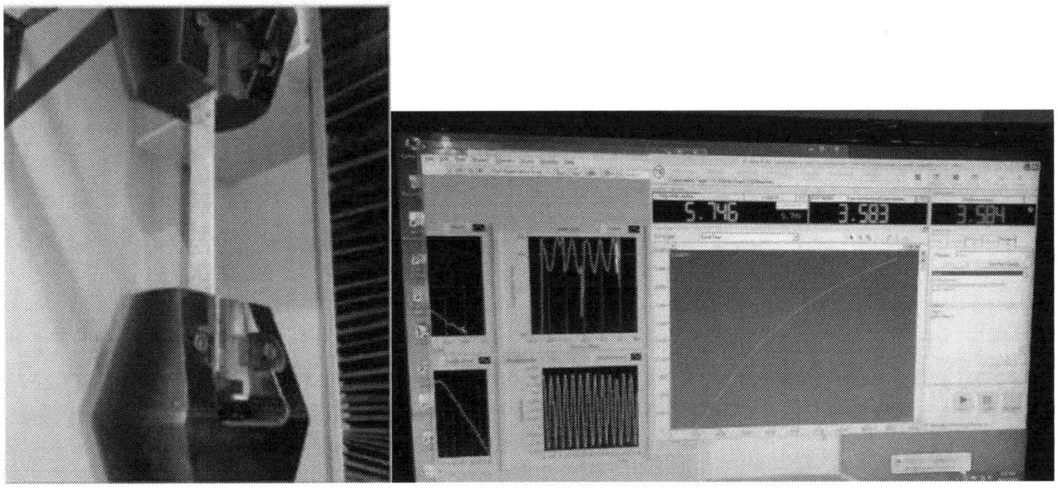

Figure 1 - Experimental set up

Tensile test was conducted according to Taguchi's L9 orthogonal array as shown in table 4 on the bi-axial GFRP specimens.

Table 4 – Taguchi L9 orthogonal array

Expt.No	Control factors and their levels		
	A	B	C
1	1	1	1
2	1	2	2
3	1	3	3
4	2	1	2
5	2	2	3
6	2	3	1
7	3	1	3
8	3	2	1
9	3	3	2

3. Results and discussion

The experimental results for all the responses using the S/N ratio values are presented in table 2 and table 3. Similarly the main effect for the mean and S/N ratio is plotted in figure 2 and figure 3.

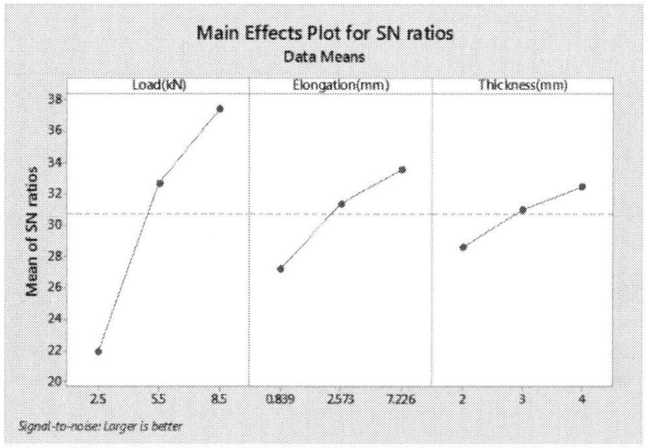

Figure 2 – Main effects plot for S/N ratio (Stress)

From the figure 2 and the table 2, the most influencing parameters on the stress are obtained as A3, B3, C3. It is clearly understood from the response table for stress, that the maximum values are achieved at level 3 for all the three factors. Also from the figure 3 and table 3, the influencing parameter on the strain of the component are obtained as A1, B1, C1. It is very clear that for strain of the specimen, the minimum values are achieved at level 3 for load, elongation and thickness.

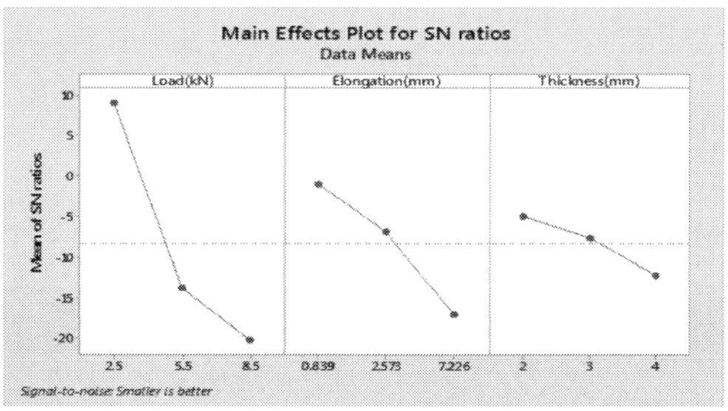

Figure 3 – Main effects plot for S/N ratio (Strain)

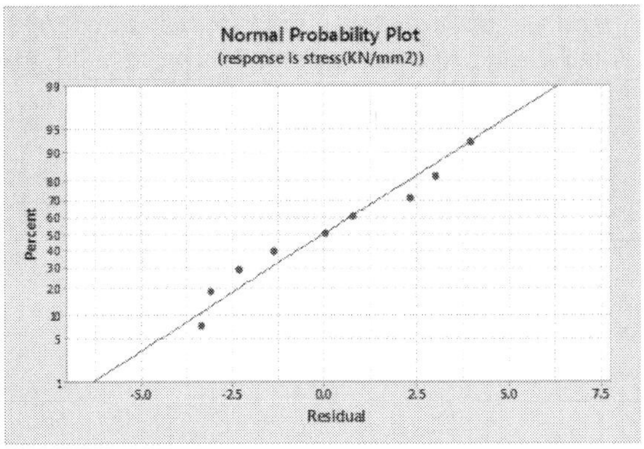

Figure 4 – Normal probability plot of residuals for stress

After conducting ANOVA analysis, to verify the normality assumption of the residual, residual plots were plotted. Figures 3 and 4 represents the normal probability plots of the residuals. From both the figures it is revealed that almost all the residuals follow a straight line pattern.

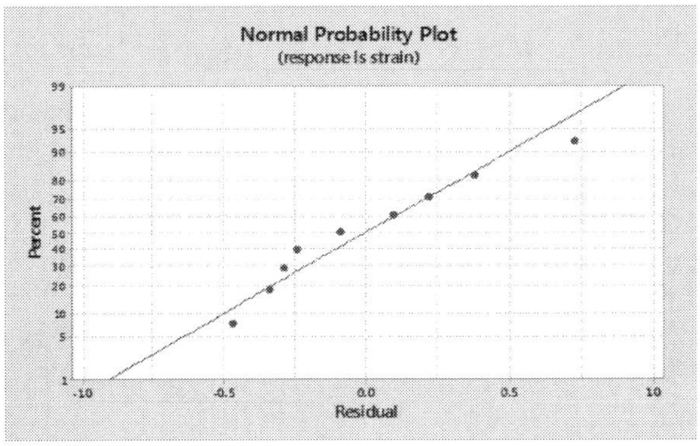

Figure 5 – Normal probability plot of residuals for strain

The regression equation for the stress and strain are given below

Stress (KN/mm^2) = -22.26 + 9.976 Load (KN) + 2.648 Elongation (mm) + 0.84 Thickness (mm)

Strain = -2.27 + 1.5872 Load (KN) + 0.8031 Elongation (mm) - 1.157 Thickness (mm)

Table 5 - Error prediction of Stress

Load(KN)	Elongation(mm)	Thickness(mm)	Stress(KN/mm^2)	Predicted stress(KN/mm^2)	%Error
2.5	0.839	2	6.24	6.581672	-5.475512821
2.5	2.573	3	13	12.013304	7.589969231
2.5	7.226	4	25.22	25.174448	0.180618557
5.5	0.839	3	35.04	37.349672	-6.59152968
5.5	2.573	4	45.12	42.781304	5.183280142
5.5	7.226	2	50.06	53.422448	-6.716835797
8.5	0.839	4	65.02	68.117672	-4.764183328
8.5	2.573	2	75	71.029304	5.294261333

8.5	7.226	3	85	84.190448	0.952414118

Table 6 - Error Prediction for Strain

Load(KN)	Elongation(mm)	Thickness(mm)	Strain	Predicted strain	% Error
2.5	0.839	2	0.055	0.0578009	-5.09255
2.5	2.573	3	0.29	0.2933763	-1.16424
2.5	7.226	4	2.67	2.8732006	-7.61051
5.5	0.839	3	3.4	3.6624009	-7.71767
5.5	2.573	4	3.592	3.8979763	-8.51827
5.5	7.226	2	9.578	9.9488006	-3.87138
8.5	0.839	4	7.396	7.2670009	1.744174
8.5	2.573	2	10.418	10.9735763	-5.33285
8.5	7.226	3	14.45	13.5534006	6.20484

4. Conclusion

This work presents optimizing the parameters namely the load, elongation and thickness of the specimens in the tensile test of glass fibre reinforced polymer composites using the Taguchi design of experiments and followed by ANOVA analysis. The following conclusions are drawn:

- Stress is maximum at load 8.5KN, elongation of 7.226mm and 3mm of thickness specimen
- Strain is minimum under 2.5KN load, elongation of 0.84mm and 2mm thickness specimen
- The percentage of error between the experimental stress value and the predicted stress is below 10%. It is also the same, while comparing the experimental strain and the predicted strain value from the table 5 and 6.
- From the main effect plot it is clear that the most influencing parameter of the tensile test is the load.
- ANOVA analysis reveals more than 95% confidence level for the responses.

References

[1] Manna A and Bhattacharyya B 2006 *Int J Mach MachMater* **14** 488-99
[2] Abrao AM, Faria PE, Campos Rubio J C and Paulo Davim J 2007J Mater Process *Technology* **186** 1–7
[3] Gopalsamy B, Mondal B and Ghosh S 2009 *J SciInd Res* **68** 686–95
[4] Taguchi G 1986 *Introduction to quality engineering* Asian Productivity Organization Dearborn MI: Distributed by American Supplier Institute Inc
[5] Phadke S M 1989 *Quality engineering using robust design* Englewood Cliffs N.J: Prentice Hall
[6] ASI 1989 *Taguchi methods: implementation manual* Dearborn MI USA: American Supplier Institute Inc

IOP Conf. Series: Materials Science and Engineering **197** (2017) 012012 doi:10.1088/1757-899X/197/1/012012

Analysis of CO_2, CO and HC emission reduction in automobiles

K N Balan[1]*, T N Valarmathi[2], Mannem Soma Harish Reddy[3], Gireddy Aravinda Reddy[4], Jammalamadaka K M K Sai Srinivas[5], Vasan[6]

[1] Assistant Professor, Faculty of Mechanical Engineering, Sathyabama University, Chennai, India

[2] Assistant Professor, Faculty of Mechanical & Production Engineering, Sathyabama University, Chennai, India

[3, 4, 5] UG Students of Mechanical Engineering, Sathyabama University, Chennai, India

[6] UG Students of Mechanical & Production Engineering, Sathyabama University, Chennai, India

Email id: knb5463@gmail.com

Abstract. In the present scenario, the emission from automobiles is becoming a serious problem to the environment. Automobiles, thermal power stations and Industries majorly constitute to the emission of CO_2, CO and HC. Though the CO_2 available in the atmosphere will be captured by oceans, grasslands; they are not enough to control CO_2 present in the atmosphere completely. Also advances in engine and vehicle technology continuously to reduce the emission from engine exhaust are not sufficient to reduce the HC and CO emission. This work concentrates on design, fabrication and analysis to reduce CO_2, CO and HC emission from exhaust of automobiles by using molecular sieve 5A of 1.5mm. In this paper, the details of the fabrication, results and discussion about the process are discussed.

1. Introduction

Pollution is becoming a challenging problem in our environment. It is not an exaggeration to say that the advancement in science and technology had caused an uncontrollable change in the environment along with an alarming increase in the pollution. Among all the pollutions, air pollution is considered as a major constituent which is causing diseases, damaging food crops and causing ecological imbalance. Air pollution can be caused by both anthropogenic and natural sources. Air pollution is causing stratospheric ozone depletion which is recognized as a threat to human health as well as to the Earth's ecosystem. Air pollution is defined as emission of harmful gases like CO_2, CO, HC, SO_2 and NOx from the industrial chimneys, automobiles and aircrafts. Worldwide air pollution is caused due to the emission of hydrocarbons, nitrogen oxides, carbon monoxide and carbon dioxide. Emission of these gases takes place mainly due to the incomplete combustion of fuel and air mixture.

IOP Conf. Series: Materials Science and Engineering **197** (2017) 012012 doi:10.1088/1757-899X/197/1/012012

Fuel + Air ⟹ Unburned Hydrocarbons + Nitrogen oxides + Carbon dioxide + Carbon monoxide + water

1.1 CO_2 sequestration process

It is a process involving the carbon dioxide capture and long term storage of CO_2. It has been proposed as a way to slow the atmospheric and marine accumulation of greenhouse gases, which are released by burning fossil fuels. There are two major types of CO_2 sequestration: Terrestrial and geological.

- Terrestrial sequestration
- Geological sequestration

2. Literature review

Shuichi Yamamoto [1] conducted experiments on CO_2 sequestration and found that sequestration in deep aquifers is one of the options for reducing the emission of green house gases an axis symmetric horizontal aquifer- cap rock system is modeled, to study various factors such as injection rate, permeability, stiffness and stability of the cap rock and aquifer. Koji Kadono [2] in collaboration with Kansai electric power and Mitsubishi heavy industries limited developed an economical process to reduce the CO_2 emissions by carbon capture choosing MEA as solvent. It will capture approximately 150000 tons of CO_2 annually at a CO_2 capture rate of over 90%. Nikolett Sipocz [3] conducted experiments on low temperature CO_2 capture using MEA 30% Wt solvent, in the desorbed the absorbed CO_2 is heated up and stripped from the solution. Eventually the purified CO_2 is sent for compression while the regenerated solvent is sent back to absorber.

Jillian Dickinson [4] determined that sequestration process is a two stage process which consists of absorption and desorption columns. The process relies on the reversible reaction between the CO_2 and MEA, at lower temperature CO_2 is absorbed and at higher temperatures CO_2 is released. Xu [5] studied sequestration process on by injecting CO_2 into high permeability sand stone and found that there is sharp decrease in Ph value with an increase in HCO_3. The CO_2 solubility will increase enhancing the solubility trapping with a decrease in density of aqueous phase. Anggit Raksajati [6] studied sequestration using phase change solvent by considering few samples of L-CLASS solvent and S- CLASS solvent which will form carbonate ions and bicarbonate ions on reacting with CO_2. Takashi Nakamoto [7] conducted experiments on aqueous solutions of amine in high pressure conditions to study CO_2 absorption among three samples RH-X, MDEA aqueous solution and DEPG solutions, he found that RH-X has the highest CO_2 absorption performance.

3. Experimental work

Literature suggests that CO_2 sequestration process is one of the methods to reduce CO_2 emission rate from thermal power station and industries. However, limited information is available in contribution of CO_2 sequestration process. The present study has chosen to investigate the interactions of various gas reductions from automobile exhaust. The various gases include CO_2, CO and HC.

3.1. Experimental setup

The assembled exhaust unit consists of stainless steel cylinder, fan, SS mesh. Inside the cylinder a stainless steel mesh is provided at the middle height from the base, in which the mesh is to support the chemical and it doesn't allow falling down. Base of the cylinder is fitted with a fan in which it circulates air equally in all direction inside the cylinder. At the base of the cylinder is tapered to reduce the pressure. Assembled exhaust unit

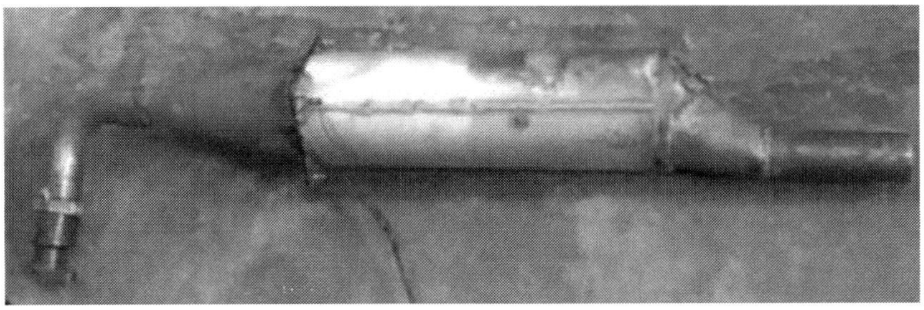

Figure 1. Assembled exhaust unit

3.2. Plan of experiment

NETEL auto exhaust/multi gas analyzer offers the greatest benefit of having the need to conduct the experiment and to study the measurement of CO, CO_2, HC gases. Hence it is used to measure the relative volumes of certain gaseous constituents in the exhaust of automobile vehicles. The gases are Carbon monoxide (CO), Carbon dioxide (CO_2), Hydrocarbons (HC). In this study, experiment has conducted on three control parameters (CO, CO_2, and HC) at seven levels.

Figure 2. NETEL Auto exhaust / multi gas analyzer

Figure 3. Exhaust unit attached to the petrol engine

3.2.1 Molecular sieve 5A of 1.5mm

The aim of the present study is to propose a chemical to reduce the emission rate from auto mobile exhaust and compare its effectiveness against the normal exhaust silencer is taken into under consideration. The chemical used in this study is molecular sieve 5A of 1.5mm.

Upon including the chemical of molecular sieve 5A of 1.5mm, a series of experiment was conducted by inserting the chemical in the assembled exhaust unit, in which the experiment was carried out for the consideration of molecular sieve 5A of 1.5 mm and compared with normal engine exhaust.

Figure 4. Molecular sieve 5A crystals of 1.5mm

4. Results and discussion

By using the NETEL auto gas analyzer, two experiments were carried out for the two wheeler, four stroke petrol engine without using molecular sieve exhaust unit and by using molecular sieve 5A of 1.5mm and their results are recorded at relative volume at a speed range of 1000 rpm to 7000 rpm shown in table :

Table 1. Emissions at different speeds with and without using Molecular sieve 5A of 1.5mm.

S.no	Speed in rpm	Without using Molecular sieve 5A of 1.5mm			Using Molecular sieve 5A of 1.5mm		
		HC in ppm vol	CO in % vol	CO_2 in %vol	HC in ppm vol	CO in % vol	CO_2 in %vol
1	1000	240	1.3	1.1	110	1.20	1.6
2	2000	260	1.5	1.4	130	1.38	2.1
3	3000	280	1.7	1.8	160	1.69	2.3
4	4000	341	3.7	3.9	179	2.0	2.5
5	5000	478	7.9	8.2	190	3.0	2.9
6	6000	639	8.9	8.9	250	4.1	3.2
7	7000	752	9.8	9.5	271	4.4	3.3

4.1 Comparison of CO emission rate

The comparison of the CO emission rate for both experimental values obtained using molecular sieve and without using molecular sieve for a speed range of 1000 to 7000 rpm shown in Table 1.

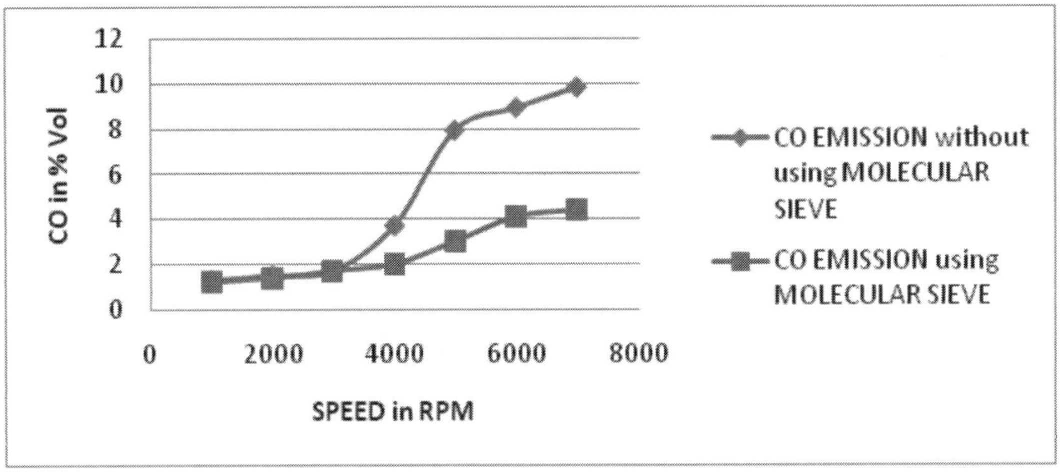

Figure 5. Comparison of CO emission rate

4.2 Comparison of CO_2 emission rate

The comparison of the CO_2 emission rate for both experimental values obtained using molecular sieve and without using molecular sieve for a speed range of 1000 to 7000 rpm shown in above Table 1.

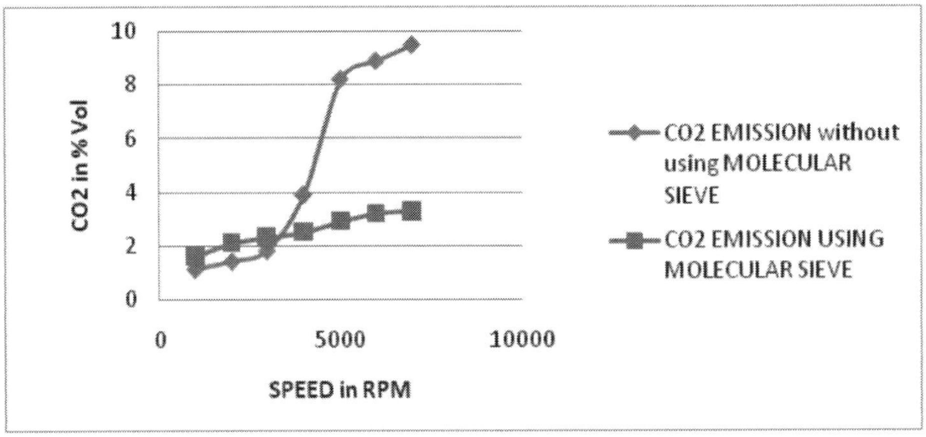

Figure 6. Comparison of CO_2 emission rate

4.3 Comparison of HC emission rate

The comparison of the HC emission rate for both experimental values obtained using molecular sieve and without using molecular sieve for a speed range of 1000 to 7000 rpm shown in above Table 1.

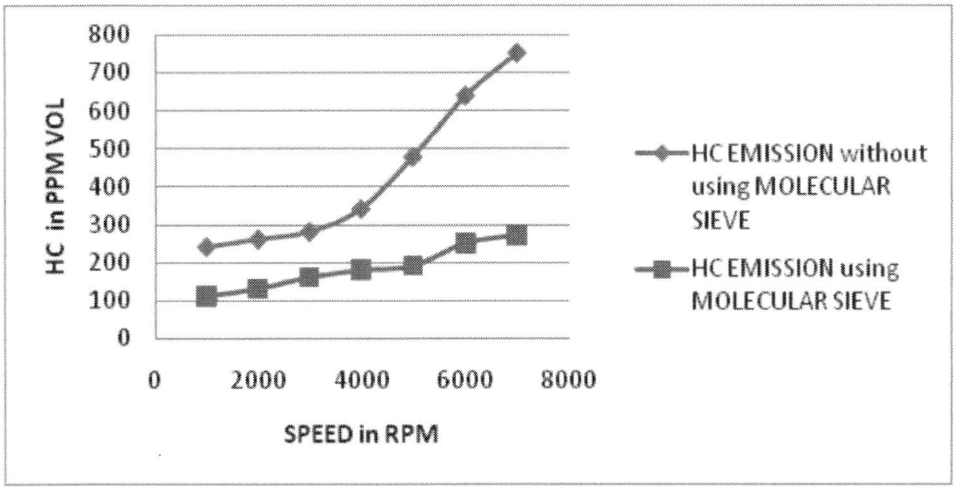

Figure 7. Comparison of HC emission rate

From the above graphs it is concluded that the emission of CO_2, CO and HC was optimized successfully.

5. Conclusions

In this work it has been presented a dynamic effort to reduce HC, CO and CO_2 emissions through Carbon capture and storage mechanisms. Adsorption technique is followed to control the Carbon emissions from the exhaust gas. The solid adsorbent used in this work is molecular sieve 5A of 1.5mm, in which it locks and holds the carbon molecules from the exhaust. The carbon capture storage is successively applied for automotive emission control. It is the action taken for the automobile sector for controlling HC, CO, CO_2emission rate from the automobile exhaust. Finally it is concluded that the emission of HC, CO, CO_2 gases for four stroke petrol engine was reduced successfully by using molecular sieve 5A.

References

[1] Shuichi Yamamoto, Satoru Miyoshi, Shin Sato, Kenichiro Suzuki 2013 *Energy Procedia* p 3989.
[2] Koji Kadono, Asao Suzuki, Masaki Iijima, Tsuyoshi Ohishi, Hiroshi Tanaka, Takuya Hirata, Masami Kondo 2013 *Energy Procedia* p 1785.
[3] Nikolett Sipocz, Alvaro Hernandez-Nogales, Miguel A.Gonzalez-Salazar, Roger Shisler, Vitali Lissianski 2013 *Energy Procedia* p 1228.
[4] Jillian Dickinson, Graeme Puxty Andrew Percy, T.vincent Verheyen 2015 *IFAC-Papers Online* p 216.
[5] Xu,T and Li.J 2013 Procedia Earth and Planetary Science p 912.
[6] Anggit Raksajati, Minh T.Ho, Dianne E.Wiley 2014 *Energy Procedia*, Pages 2280–2288.
[7] Takashi Nakamoto, Toshinori Muraoka, Shin Yamamoto and Takayuki Higashii 2014 *Energy Procedia* p 1940.

IOP Conf. Series: Materials Science and Engineering **197** (2017) 012013 doi:10.1088/1757-899X/197/1/012013

Green fuel utilization for diesel engine, combustion and emission analysis fuelled with CNSO diesel blends with Diethyl ether as additive

Ashok Kumar[1]*, K. Rajan[2], K.R. Senthil Kumar[3], K. Maiyappan[4], Usama Tariq Rasheed[5]
[1,2,4 & 5] Department of Mechanical Engineering, Dr. MGR Educational and Research Institute University, Chennai 600095, India
[3]Department of Mechanical Engineering, R.M.K.Engineering College, Chennai, India
*ashok.simme@gmail.com

Abstract
The experimental investigation is conducted to evaluate the effects by using Diethyl ether (DEE) as an additive. The Cashew Nut Shell Oil diesel blends (CDB) are tested in a 4-stroke single cylinder DI unmodified diesel engine, rated power is 4.4 kW at a speed of 1500 rpm. The effect of combustion analysis of test fuels on net heat release rate, cylinder pressure, engine power, BSFC, BTE, EGT were observed by the performance tests. The combustion and emission characteristics of a diesel engine with an additive of high cetane number is utilized with CDB and thus investigated. The influence of blends on CO, CO_2, HC, NOx and smoke opacity is investigated by emission tests. Initially, the experiment was conducted with different blends of CDB diesel blends like 10%, 20%, & 30% by volume basis in a diesel engine. Among this blends B20 shows reasonable result and heat dissipation rate at full load conditions. The BTE of B20 is 27.52% whereas base diesel fuel is 29.73%. Addition of the DEE by 5%, 10% and 15% by volume basis with B20 which is a base fuel has resulted with improved estimates. The result shows that at full load conditions BTE of B20D10 is 28.96% which is close to the base fuel i.e. B20. The emissions like CO_2 shows reducing trends while HC emission rises with increase in CNSO blends. The HC in diesel corresponds to 30ppm and in B20 it is 34ppm, but addition of DEE shows a decreasing trend as in B20D5 has 29ppm and B20D15 has 23ppm respectively. NOx also shows increasing trends with CNSO blend, after addition of DEE it shows declining trend. The NOx for diesel, B20, B30, B20D5, B20D10 and B20D15 emits 1195, 1450, 1511, 1327, 1373 and 1200ppm respectively. The smoke emission is 3.96, 3.38, 3.15 FSN of B20, B20D15 and diesel respectively.
Index Terms— CNSO diesel blends; Diesel Engine; Diethyl ether

NOMENCLATURE

BTE	Brake thermal efficiency	DI	Direct injection
SFC	Specific fuel consumption	DEE	Diethyl ether
B20	80%Diesel and 20% CNSO	EGT	Exhaust gas temperature
bTDC	Before top dead center	FSN	Filter smoke number
B20D5	95% B20and 5% Diethyl ether (v/v).	kPa	Kilo Pascal
CNSO	Cashew nut shell oil	kW	Kilo watt
CI	Compression ignition	NOx	Oxide of nitrogen
CDB	CNSO Diesel blends	ppm	Parts per million (by vol.)
cSt	Centi stoke	rpm	Revolutions per minute
CO	Carbon monoxide	UHC	Unburnt hydrocarbon
D100	Neat diesel	HRR	Heat release rate

1. INTRODUCTION

The vulnerability of fossil fuels globally has led to create a demand for finding an alternative fuel. The fossil fuels have a crucial role in the human survival. In order to meet the energy demands of the world, CNSO is abundantly available and it one of the low-cost resources [1]. The blend was chosen from a variety of options available from the best alternate sources for minimizing greenhouse effect. So to increase the resource of alternative fuel, the importance of vegetable oil is essential. Vegetable oils are available everywhere in the world, they are easily renewable. The plant used for producing the oil seeds are planted with ease, also vegetable oils which are friendly to the environment and contain less sulphur as quoted by Raheman and Phadatare in 2004. Due to the above advantages the vegetable oils are suitable substitute to the diesel fuel [2].There are various methods to reduce the viscosity like pyrolysis, dilution, trans-esterification and blending. As we know that the vegetable oils have high viscosity and less volatility, this leads in poor combustion and atomization of fuel in diesel engine [3]. The research on a number of oils like Jatropha, sunflower, soybean oil, palm oil, karanja oil, etc. was done for making this paper, the results of the researchers have been compared and summarized CNSO is relatively clean burning alternative fuel, produced from domestic and renewable resources. [4] CNSO contains no petroleum components, but it can be blended in any given amount with diesel to create a biodiesel blend. It can be used in CI engines with little or no modifications [5]. CNSO is biodegradable, nontoxic, and essentially free from sulphur and aromatics. CNSO is most promising alternative solution to protect the world from global warming, gases, emissions, tail pipe particulate matter, HC, CO and other air toxics. Biodiesel improves lubrication and reduces premature wearing of fuel pumps [6]. Fuels are recommended for the better enhancement of saving the fossil fuels and improve the emissions. The oxygenated alternative fuels have gained massive attention recently [7]. The savings on any alternative fuel regarding the economical consumption by one person at the rate of savings of 20% in single refuelling the amount saved is upto 6 litres of diesel per month i.e. if consuming 30 litres per month. Therefore, if this exercise is performed by 1000 people it saves 6000 litres of diesel per month. A comparative study of CBD in diesel engine was made at different loading conditions. The engine combustion parameters are seen as peak pressure, time of occurrence of peak pressure and heat release rate which were found specifically to CNSO [8-11]. Thus, in India alone a great quantity of diesel can be saved as it is the second largest population on earth. The figure 1 explains BS-IV (2017) regulations are subjected to improve upon the same and implement BS-VI directly in 2020. Therefore, CNSO blends can be used in CI engine in rural area for meeting energy requirement in various agricultural operations such as irrigation, harvesting, etc. The objective of this project work is to analyse the effect of DEE as an additional oxygenated fuel in the ratio of 5%, 10% and 15% by volume basis along with B20.

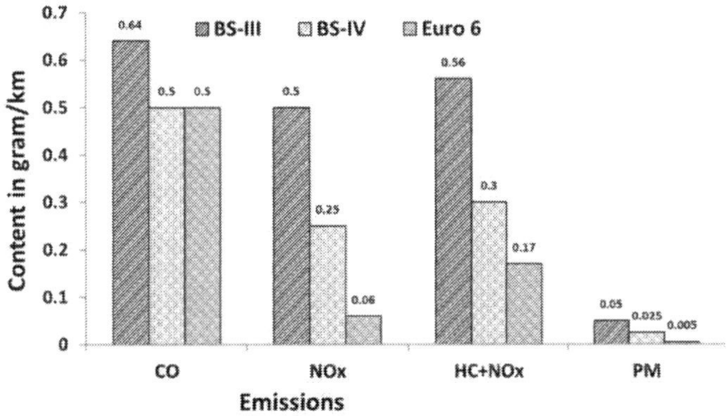

Figure 1: BS-IV Norms for 2020

2. TESTED PROPERTIES OF CNSO

The important chemical and physical properties of CNSO were determined by standard methods as compared with diesel. The analytical data are shown in Table 1, 2 and 3.

Table 1: Analytical data of Diesel, CNSO and Diethyl Ether

Properties	Diesel	CNSO	DEE
Density (gm/cc) at 40 °C	0.843	0 .930	0.713
Kinematic viscosity (cSt) at 40 °C	3.12	30.24	0.21
Cetane No.	51	38.00	125
Flash point at °C	53	222	16
Calorific value MJ/kg	43	39.94	33.9

Table 2: Properties of CNSO blends

IOP Conf. Series: Materials Science and Engineering **197** (2017) 012013 doi:10.1088/1757-899X/197/1/012013

CNSO-Diesel Blends	Density (kg/m³)	Kinematic Viscosity (Cst)	Calorific value (kJ/kg)	Flash Point (°C)
B10	836.6	4.18	42029	83.3
B20	845.86	5.59	42149	94.6
B30	855.12	7.01	42269	105.9

Table 3: Properties of CNSO blends with DEE

DEE with CDB	Density (kg/m³)	Kinematic Viscosity (Cst)	Calorific value (kJ/kg)	Flash Point (°C)
B20D5	830.27	4.04	41583.5	76.8
B20D10	839.53	5.45	41703.5	88.1
B20D15	848.79	6.86	41823.5	99.4

3. DIETHYL ETHER (DEE)

DEE is also known as ethyl ether or simply ether or ethoxyethane. It is an organic compound with the chemical formula $(C2H5)_2O$. It can be mixed in any proportion in fossil fuel or vegetable oils. DEE is good for cold- start aid for engines. Southwest Research Institute (SWRI) reported that the cetane number of DEE is higher which about 125 is and it is used as a preliminary fluid. When combined with petroleum it distillates for gasoline and diesel engines because of high volatility and low flash point.

4. PHOTOGRAPHIC VIEW OF APPARATUS

A photographic view and schematic diagram of the experimental setup of the engine are shown in figure 2 and 3 respectively. Before starting the experimental reading, first run the engine to let it reach the operating temperature. Thus, the peak pressure and heat release rate can be calculated using pressure-crank angle diagram and heat release diagram.

Figure 2: Photographic View

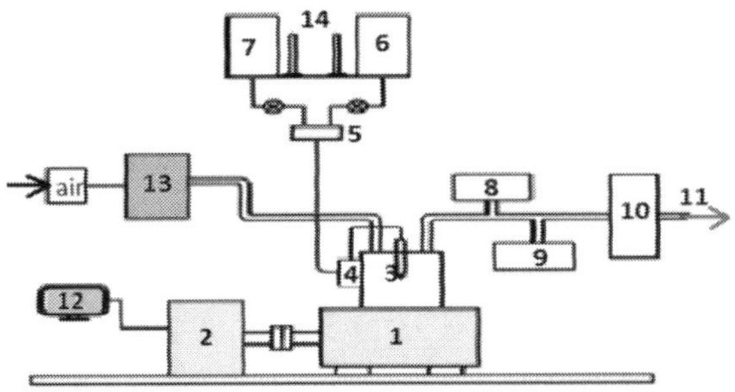

Figure 3: Schematic Diagram of the Experimental Setup

1. Engine; 2.Eddy current dynamometer; 3. Fuel injector; 4. Fuel pump; 5. Fuel Filter; 6. Diesel tank 7. B20 tank; 8. AVL gas analyser; 9. AVL smoke meter; 10. Calorimeter; 11.Exhaust silencer; 12. Monitor; 13. Air stabilizing tank; 14. Burette.

Table 4 Engine specification

Parameters	Description
Engine type and make & number of cylinder	4- stroke water cooled single cylinder CI engine, Kirloskar TAF-1
Bore Diameter	87.5 mm
Stroke length	110 mm
Maximum power	4.4 kW
Compression ratio	17.5:1
Speed	1500 rpm
Displacement volume	553 cm3

5. ENGINE TEST RESULTS AND DISCUSSION

The experiments were conducted with CDB with 5%, 10% and 15% DEE blends at different loads. The measured values were analysed and compared with those of diesel fuel.

A.HEAT RELEASE RATE

The duality of HRR (heat release rate) in accordance to crank angle at full load conditions for CBD is shown in Fig. 4 (a) and 4 (b). The cooling effect of the fuel vaporing as it is injected into the cylinder, the HRR is slightly negative during the injection period after the combustion is started, and this becomes positive. The initial phase of combustion is called the premixed combustion which is very rapid because of the combustion of a fuel that has mixed with it during ignition delay. This phase is known as mixed- control combustion. It is observed that peak HRR of neat diesel is higher than the any blends of diesel. From the result it is observed that HRR during the ignition delay period in B20, B20D10 and diesel is 66.335, 70.615 and 70.732 KJ/m^{3}deg at full load condition. Thus increases in HRR 6.22% with respect to diesel and B20. Also it is observed that HRR B20D10 is much closed to the diesel. This may be due to the DEE which acts as an ignition improver for CDB and hence it increases the premixed combustion rates, thus resulting in increased HRR.

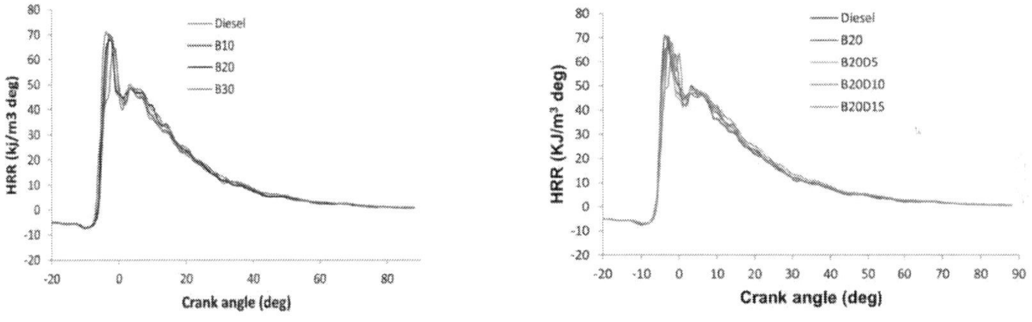

Figure 4(a) **Figure 4(b)**

Figure 4(a) and 4(b): Variation of HRR with crank angle at different blends.

B.PRESSURE CRANK ANGLE

The variations in peak cylinder pressure with crank angle for all test fuels are shown in Figure 5(a) and 5(b). The peak pressure mainly depends upon the combustion rate in the initial stages, which is influenced by the fuel taking part in uncontrolled heat release phase. A similar kind of result was obtained by the addition of DEE with CDB. The peak pressure increases with the addition of DEE with neat diesel. It has been seen that the peak cylinder pressure for B20D10 is considerably increased as compared to B20 under full load conditions. Indicating an improved combustion process thus, resulting due to DEE combustion. Especially when the engine is operating at low load condition fuel to air equivalence ratio is low. The addition of DEE can promote the formation of combustible mixture for a better combustion. However, with further increase in the blend ratio, the volatility effect of DEE is traded off by the significant reduction in calorific value. As a result, the peak cylinder pressure reduces reliably. This may be due to the higher cetane number and high flammability of DEE which increases the premixed combustion phase resulting in higher peak pressure. The peak pressure for CNSO with adding 10% and 15% DEE is 68.922 and 68.360 bars, respectively at full load. While for diesel and B20 the values are 69.479 and 67.879 bars at full load condition. The peak pressure reduces at high power output with the introduction of DEE along with CNSO due to reduction in ignition delay as compared with diesel fuel.

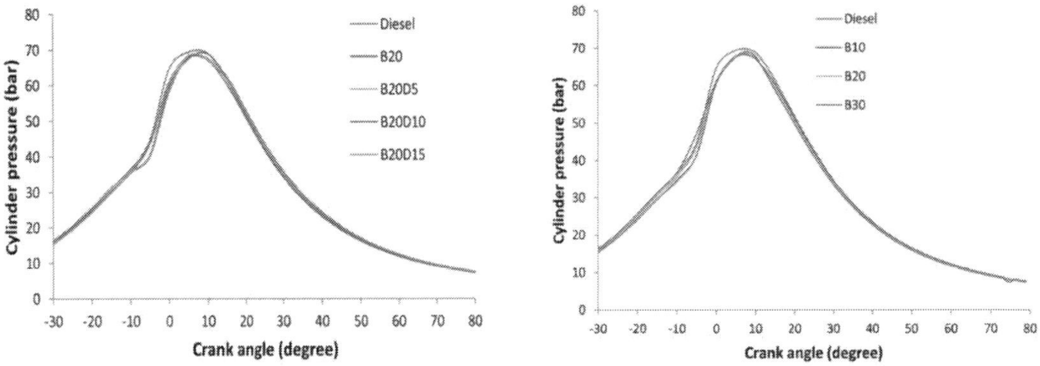

Figure 5(a) **Figure 5(b)**

Figure 5(a) and 5(b): Cylinder Pressure variation with respect to crank angle with various blends

C. BRAKE THERMAL EFFICIENCY (BTE)

The BTE with respect to load for diesel and various blends of CNSO and DEE are shown in Figure 6(a) and 6(b). In all cases, BTE has increased with increase in load. This is due to reduction in heat losses and increase in brake power with increase in load. The BTE increase slightly for lower blends of DEE but with the increasing concentration of DEE blends its reducing trends. This is due to little more amount of oxygen present in the blends, which gives enhanced combustion as compared to pure diesel. The variation of BTE at full load operation is 28.37, 27.52, and 29.73% for B20D15, B20 and diesel respectively. It is observed that BTE B20D10 is closed to the diesel fuel. BTE of B20D10 is increased by 5.3% with respect to B20 at full load conditions. Table 5(a) and 5(b) shows the relative differences of the engine BTE between diesel fuel and that of CNSO blends.

Table 5(a) and 5(b): Relative differences of BTE at different loading conditions for different fuel blends (%)

Table 5(a)

Engine load (%)	Diesel	B10	B20	B30
0	-	-	-	-
25	-	-4.1	-8.1	-17.4
50	-	-0.12	-4.1	-9.4
75	-	-2.6	-4.8	-8.8
100	-	-3.7	-7.4	-12.7

Table 5(b)

Engine Load (%)	Diesel	B20	B20D5	B20D10	B20D15
0	-	-	-	-	-
25	-	-8.1	-7.6	-6.3	-6.3
50	-	-4.1	-7.1	-5.7	-5.9
75	-	-4.8	-2.9	-2.1	-4.2
100	-	-7.4	-3.2	-2.6	-4.6

D. BRAKE SPECIFIC FUEL CONSUMPTION (BSFC)

It shows variation in BSFC with respect to load for CNSO in figure 7 (a) and 7(b). If load increases BSFC decreases, mainly due to lower calorific value and high viscosity of CNSO. BSFC of CNSO is higher than that of diesel at all loading conditions. BSFC for diesel is 0.202 kg/Kw-hr and for various blends such as B10; B20 and B30 are 0.216, 0.237, and 0.248 kg/Kw-hr respectively. In next stage, B20 is considered as a base fuel and blended with DEE as 5%, 10%, and 15% by volume. It shows that for various blending of DEE, BSFC shows reducing trend for B20D10 and B20D15 blends which is, 0.226 and 0.231 kg/Kw-hr respectively. The results show BSFC for B20D10 is closed to the B20. Table 6(a) and 6(b) shows the relative BSFC differences.

Table 6(a) and 6(b): Relative differences of BSFC at different loading conditions for different fuel blends (%)

Table 6(a)

Engine load (%)	Diesel	B10	B20	B30
0	-	-	-	-
25	-	8.7	12.1	20.2
50	-	-1.2	1.3	9.3
75	-	- 4.7	0.47	5.7
100	-	6.9	17.3	22.7

Table 6(b)

Engine load (%)	Diesel	B20	B20 D5	B20 D10	B20 D15
0	-	-	-	-	-
25	-	12.1	7.3	-4.9	7.1
50	-	1.3	11.8	7.9	10.6
75	-	0.47	7.1	2.4	3.3
100	-	17.3	13.7	10.6	12.5

E. EXHAUST GAS TEMPERATURE (EGT)

The EGT is high with neat CNSO due to low HRR and poor combustion, so a part of the combustion extends into the exhaust stroke, thereby increasing the combustion duration. Higher ignition delays result in a delayed combustion and higher exhaust temperature. The maximum EGT is 409°C for B30 blends and 378°C for diesel at full load conditions. Result shows that the EGT is relatively low at initial load and then high at full load of B30 as compared to diesel and CBD, which shown in Fig.8 (a) and 8(b). Using DEE with B20 blends gives 398°C. Table 7(a) and 7(b) shows the relative differences of the engine EGT.

Table 7(a) and 7(b): Relative differences of EGT at different loading condition for different fuel blends (%)

Table 7(a)

Engine load (%)	Diesel	B10	B20	B30
0	-	3.5	6.7	8.7
25	-	4.4	7.5	9.3
50	-	4.5	8.3	10.6
75	-	5.6	8.9	10.5
100	-	3.4	5.3	8.2

Table 7(b)

Engine load (%)	Diesel	B20	B20D5	B20D 10	B20D 15
0	-	6.7	4.6	2.6	1.5
25	-	7.5	5.3	0.4	-0.9
50	-	8.3	6.1	4.2	1.9
75	-	8.9	6.8	6.2	4.6
100	-	5.3	3.7	2.6	5.1

6. EMISSION CHARACTERISTICS

The emissions prevailing from the exhaust gas analyser are processed using data processing and calculations which were conducted by the analyser system to determine the various emissions.

A. CARBON MONOOXIDE (CO)

The tested CO emissions with varying loading conditions are shown in Figure 9(a) and 9(b). CO emission of CI engine is low due to lean mixture operations. It depends on the air fuel ratio. The CO emission is higher for neat CNSO than its blends and diesel, due to poor atomization of CNSO. From the result it is observed that CO percentage decreases in B20 with respect to B20D15 which is 11.11%. Also in diesel and B20 it increases by 23.8% by volume. Presence of DEE in the blends enhances the combustion process, which is reflected in the reduction of CO emissions. Table 8(a) and 8(b) shows the relative differences of the engine CO emissions between diesel fuel and that of CNSO blends.

Table 8(a) and 8(b): Relative differences of CO at different loading conditions for different fuel blends (%)

Table 8(a)

Engine load (%)	Diesel	B10	B20	B30
0	-	6.7	-16.7	83.3
25	-	28.6	-14.3	71.3
50	-	9.1	-27.3	36.4
75	-	6.7	-6.7	20
100	-	4.8	-23.8	19.1

Table 8(b)

Engine load (%)	Diesel	B20	B20 D5	B20 D10	B20D 15
0	-	-16.7	16.7	33.3	50
25	-	-14.3	14.3	28.6	57.2
50	-	-27.3	9.1	27.3	18.2
75	-	-6.7	20	40	0
100	-	-23.8	4.8	19.1	-14.3

B. CARBON DIOXIDE (CO_2)

The result of CO_2 emissions with different loading conditions is shown in figure 10(a) and 10(b). It is inferred the percentage of CO_2 by volume of diesel, B10, B20 and B30 blends are 7.1, 6.5, 5.9 and 5.6% respectively, which is a reducing trend than neat diesel fuel at various load conditions. But adding DEE the result for B20D5, B20D10 and B20D15 blends are. 6.2, 6.4 and 6.8% respectively. We observed that percentage decrease of CO_2 with respect to diesel and B20 is 16.9% and 9.81% viz. with respect to diesel and B20D10 blends by volume basis. Table 9(a) and 9(b) shows the relative differences of the volume basis engine CO_2 emission between diesel fuel and that of CNSO blends.

Table 9(a) and 9(b): Relative differences of CO2 at different loading conditions for different blends (%)

Table 9(a)

Table 9(b)

Engine load (%)	Diesel	B10	B20	B30
0	-	-15	-30	-40
25	-	-11.5	-20	-31.4
50	-	-8.9	-17.8	-26.7
75	-	-5.2	-12.1	-18.9
100	-	-8.5	-16.9	-21.2

Engine load (%)	Diesel	B20	B20 D5	B20 D10	B20 D15
0	-	-30	-20	-10	-5
25	-	-20	-11.4	-5.7	-2.9
50	-	-17.8	-11.1	-6.7	-2.2
75	-	-12.1	-10.4	-6.9	-3.5
100	-	-16.9	-12.7	-9.9	-4.3

C. HYDRO CARBON (HC)

The tested results of HC emissions at different loading conditions are shown in Fig.11 (a) and 11(b). Result shows that the HC emission is increased for neat CDB and it is decreased with adding DEE in the CDB at all load conditions. The percentage increases of HC with respect to diesel and B20 is 13.33% and percentage decreases of HC with respect to B20 and B20D15 is 32.35% by volume. This decrease in HC emission for DEE blends due to improper combustion.

D. OXIDE OF NITROGEN (NO_X)

The tested results of NOx emissions at different loading conditions are shown in Figure 12(a) and 12(b). NOx emissions of CDB are higher than that of diesel fuel as there is reduction in HRR and combustion temperature. This will result in high NOx concentration in the exhaust. It is observed that NOx for diesel fuel is 1195 ppm and B10, B20, and B30, blends are 1316, 1450 and 1511ppm respectively, which is more than neat diesel fuel. It is observed that the next stage after addition of DEE, NOx for B20 is 1450 ppm which is higher than that of B20D5 (1327ppm). It means that the addition of DEE in B20 blends reduces the NOx emission. Table10 (a) and 10(b) shows the relative differences of the NOX.

Table 10(a) and 10(b): Relative differences of NO_X at different loading conditions for different fuel blends (%)

Table 10(a)

Engine load (%)	Diesel	B10	B20	B30
0	-	2.4	6.8	7.97
25	-	-3	7.9	18.3
50	-	11.7	19.7	25
75	-	10.4	19.3	24.63
100	-	10.13	21.4	26.44

Table 10(b)

Engine load (%)	Diesel	B20	B20D 5	B20 D10	B20 D15
0	-	6.8	9.96	9.2	2
25	-	7.9	5.94	6.2	3.82
50	-	19.7	14.97	3.2	1.6
75	-	19.3	8.7	7.3	1.4
100	-	21.4	11.1	6.53	0.42

E. SMOKE OPACITY

The tested results of smoke level at different loading conditions are shown in Figure 13(a) and 13(b). CNSO has prolonged aromatic chain-like structure and it has poor combustion. Thus it results in more amounts of smoke level increases it may be due to poor atomization of the CNSO. It is observed that the smoke opacity of CNSO blend is higher than diesel. Result shows at full loading condition smoke level of B20 is 3.96 FSN whereas for diesel its level is 3.15 FSN. After addition of DEE with B20 it is reduced. It has been found that Smoke level of B2OD15 is 3.38 FSN. This is because improved viscosity resulting improve combustion rate of the fuel blends. Table 11(a) and 11(b) shows the relative differences of the Smoke emission between diesel and CDB.

Table 11(a) and 11(b): Relative differences of Smoke level at different loading condition for different fuel blends %.

Table 11(a)

Engine load (%)	Diesel	B10	B20	B30
0	-	166.7	133.3	158.3
25	-	90.9	72.73	86.4
50	-	97.4	55.3	79
75	-	28.5	15.2	27.3
100	-	11.11	25.7	33.33

Table 11 (b)

Engine load (%)	Diesel	B20	B20D 5	B20D 10	B20D 15
0	-	133.3	125	100	41.7
25	-	72.73	63.64	50	27.3
50	-	55.3	34.2	21	5.3
75	-	15.2	3	-1.82	-12.1
100	-	25.7	18.4	13.65	7.3

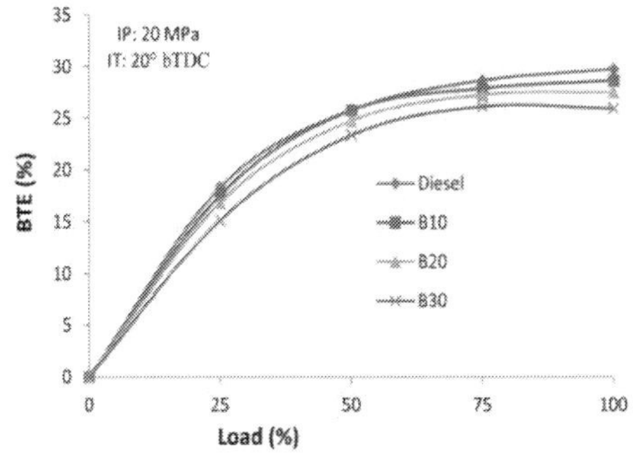

Figure 6(a)

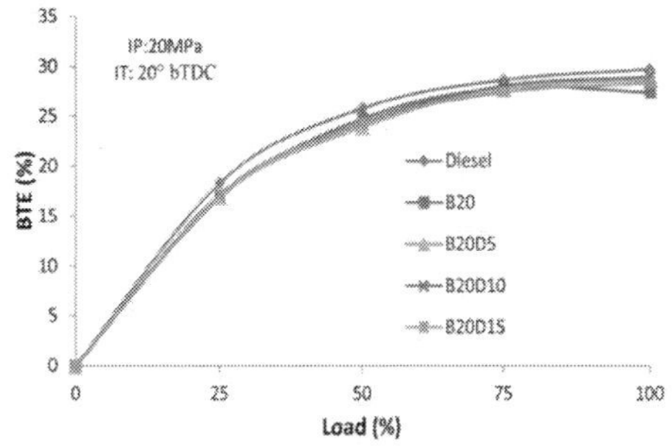

Figure 6(b)

Figure 6(a) and 6(b): Evaluation of BTE of engine at various loads and blends

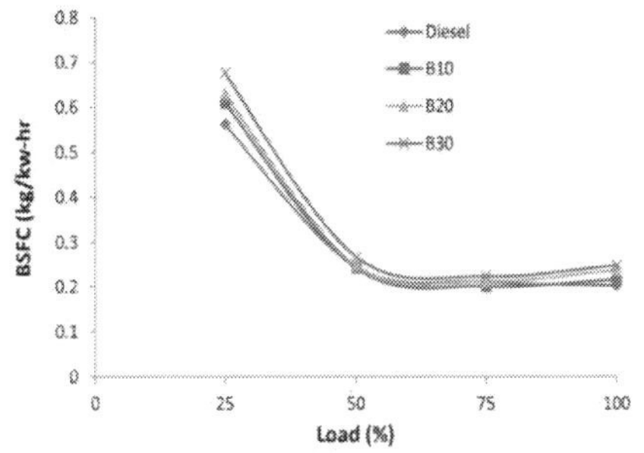

Figure 7(a)

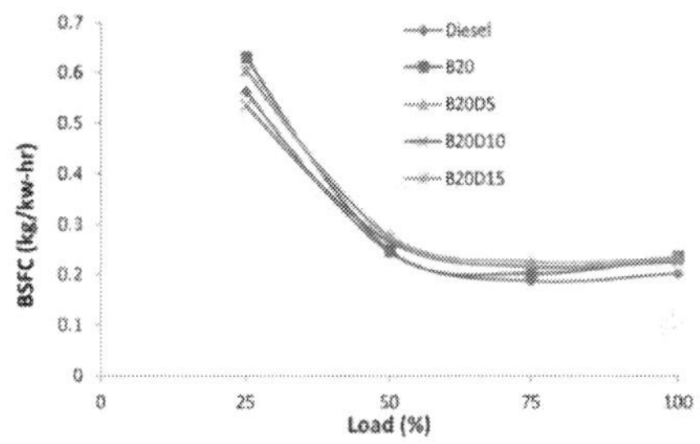

Figure 7(b)

Figure 7(a) and 7(b): Comparison of BSFC of engine with various loads for various blends

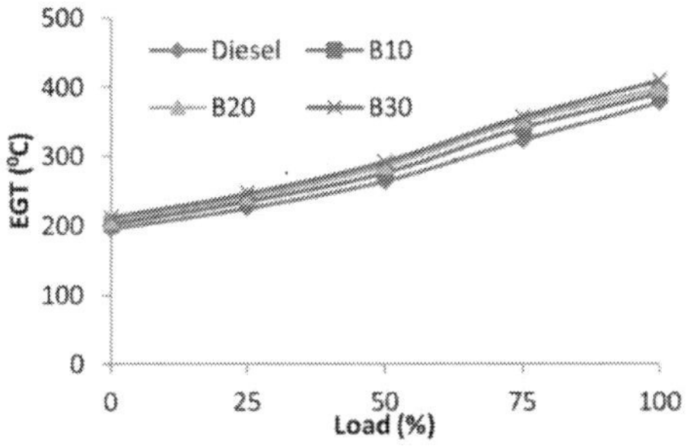

Figure 8(a)

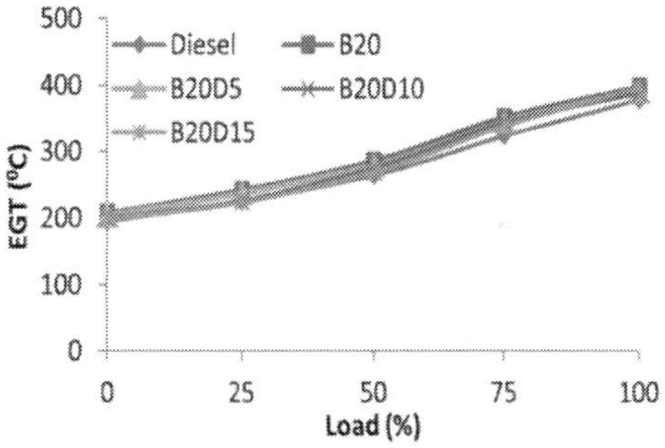

Figure8(b)

Figure 8(a) and 8(b): Comparison of EGT of engine with various loads for various blends

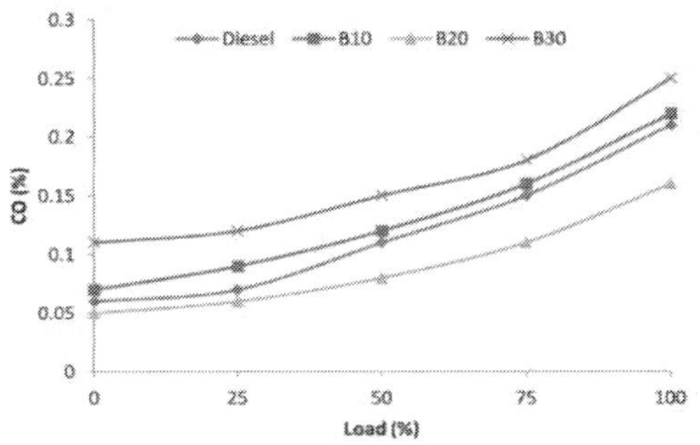

Figure 9(a)

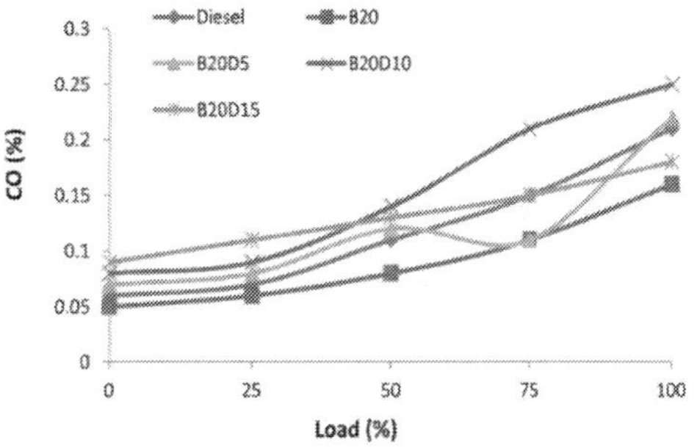

Figure 9(b)

Figure 9(a) and 9(b): Comparison of carbon monoxide of engine with various loads for various blends

IOP Conf. Series: Materials Science and Engineering **197** (2017) 012013 doi:10.1088/1757-899X/197/1/012013

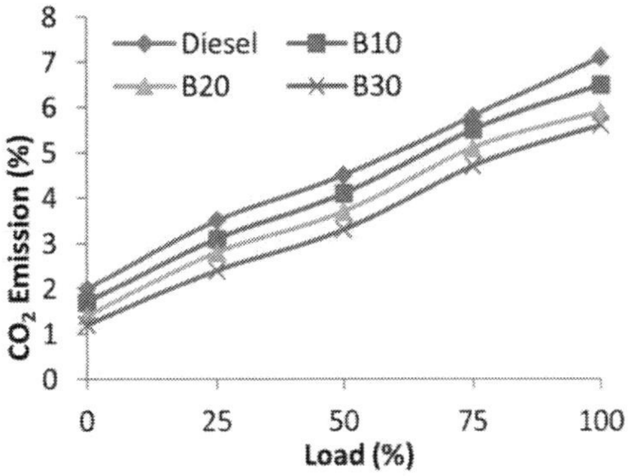

Figure 10(a)

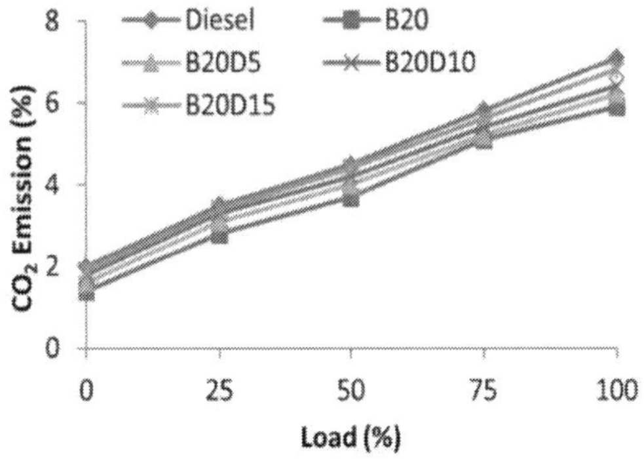

Figure 10(b)

Figure 10(a) and 10(b): Comparison of CO_2 emissions of engine with various loads for various blends

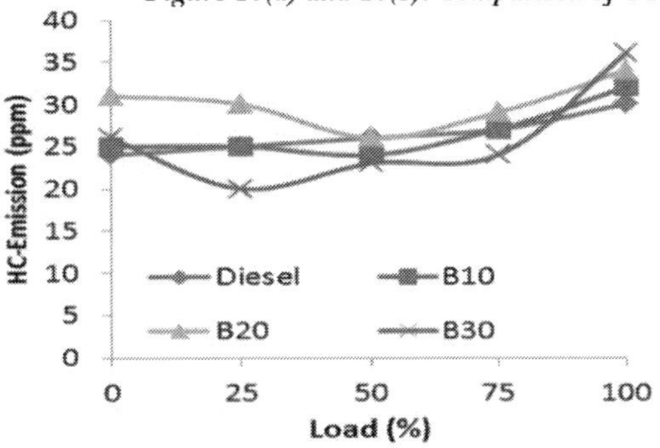

Figure 11(a)

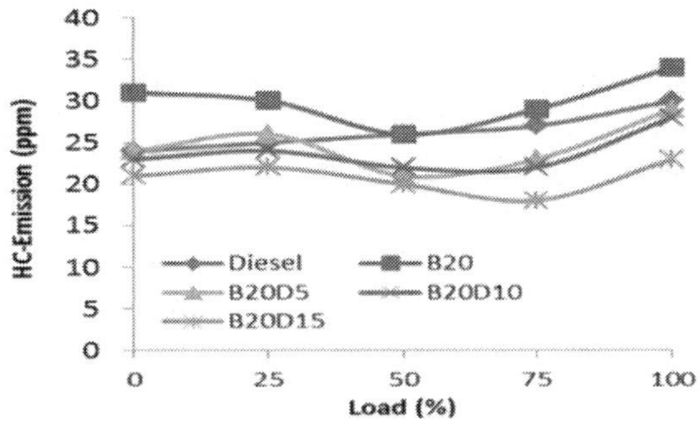

Figure 11(b)

Figure 11(a) and 11 (b): Evaluation of HC emission with various loads and blends

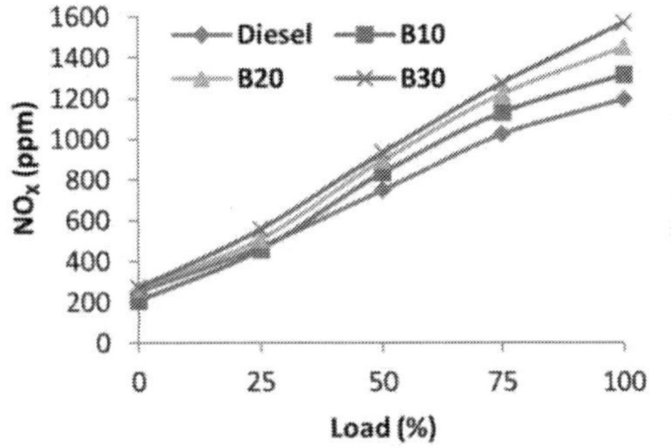

Figure 12(a)

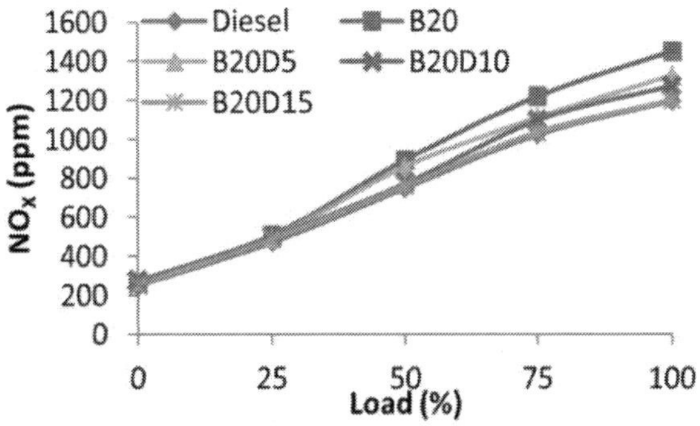

Figure 12(b)

Figure 12(a) and 12(b): Comparison of NOx emission of engine with various loads for various blends

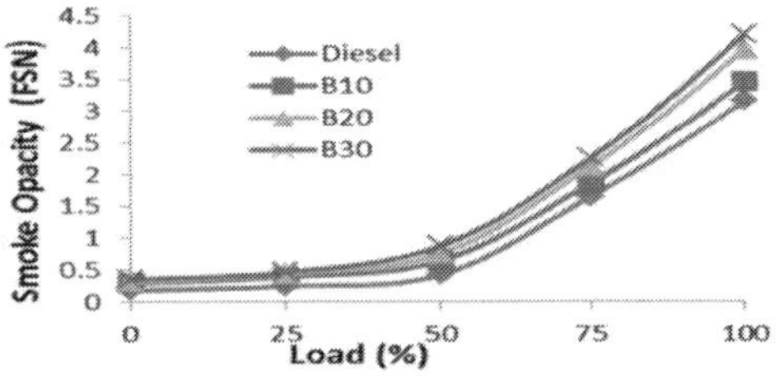

Figure 13(a)

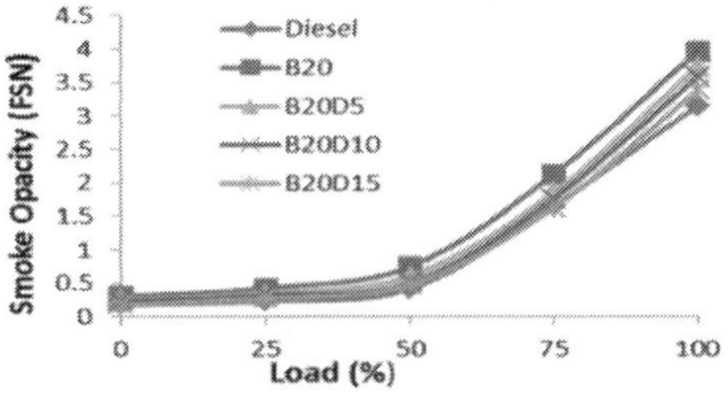

Figure 13(b)

Figure 13(a) and 13(b): Evaluation of smoke opacity of engine with various loads and blends

CONCLUSION

An extensive experimental study is conducted to evaluate and compare the use of DEE as an additive to the CBD in a diesel engine. The CNSO is cheaper than other kinds of vegetable oils, which is an important advantage for biodiesel production. In addition, DEE has high oxygen content and cetane number (125) it decreases the density and the viscosity of the blends. The BTE increases of blends B20D5, B20D10 and B20D15 which are slightly higher than B20. The B20D5, B20D10 and B20D15 have higher oxygen content than B20. The EGT increases for all the blends with increase in the applied load. It is inferred that the CO emissions for pure CNSO are infeasible as being higher than the blend of CNSO and Diesel.The NO_X emission increases with increased proportion of blends and also with higher EGT. This result is achieved as the presence of oxygen in CNSO, leads to improved oxidation at higher temperature and it is responsible for more NO_X emission. The NOx of B20 is increased by 21.34% with respect to diesel at full load conditions and with respect to diesel and B20D5 it is 9.94%. Prediction for CO_2 emission decreases when we use DEE as an additive in B20, but HC emissions will be higher in comparison with diesel. Smoke opacity increases in B20 especially at high engine load using DEE with B20 it shows a decline. DEE has good solubility in the diesel fuel and this blend can be effectively used in diesel engines without any further modifications in the engine.

REFERENCES

[1] Gerpen JV. Biodiesel Processing and Production. Fuel Process Technology 2005; 86:1097-107.

[2] Midilli A Dincer I, Rosen MA. The role and future benefits of green energy. Int J. Green Energy 2007; 4(1):65-87

[3] S Saka, K Dadan, Transesterification of rapeseed oils in supercritical methanol to biodiesel fuels, in: R.P. Overend, E. chomet (Eds.) Processing of the 4th Biomass Conference of the Americans, Oakland, CA, 1999 (Aug.29-Sept. 2)

[4] Ashok kumar, K.Rajan, Performance and emission characteristics of a direct injection diesel engine fuelled with Cashew Nut Shell Oil (CNSO)-diesel blends with Diethyl ether as additive, Applied Mechanics and Materials.,787 (2015) 746-750.

[5] Piyali Das., Anuradda Ganesh., 2003. Bio Oil from Pyrolysis of CNSO, Biomass and Bioenergy, vol.36.

[6] K.Rajan, Senthil Kumar K.R., Performance and Emission Characteristics of Diesel Engine with Internal Jet Piston using Biodiesel, International Journal of Environmental Studies., 2010, 67 (4), 556- 567.

[7] John B Heywood., Internal combustion engine fundamentals Mc Graw Hill Publishers. New York, USA.1998.

[8] Rakopoulos CD, Antonopoulos KA, Rakopoulos DC. Experimental heat release analysis and emissions of a HSDI diesel engine fueled with ethanol-diesel fuel blends. Energy 2007; 32(10):1791–808.

[9] Kegl B. Influence of biodiesel on engine combustion and emission characteristics. Appl Energy 2011; 88(5):1803–12.

[10] D.N. Mallikappa, Rana Pratap Reddy and Ch.SN. Murthy, "Performance and emission characteristics of stationary CI engine with cardanol bio fuel blends", Int. J.Scientific and Eng Research, vol. 2, no.2, pp. 1-6,2011.

[11] Velmurugan, A., Loganathan, M., 2011. Performance and emission characteristics of a DI diesel engine fuelled with CNSL-diesel blends, World Academy of Science, Engineering and Technology 58, p. 884.

Experimental Investigation of Performance and emission characteristics of Various Nano Particles with Bio-Diesel blend on Di Diesel Engine

N Karthik[1], X Goldwin Xavier [2], R Rajasekar[3], P Ganesh Bairavan[4], S Dhanseelan[5]

[1, 3] Assistant Professor, Faculty of Mechanical Engineering, Sathyabama university, Chennai
[2] Assistant Professor, Department of Mechanical Engineering, Loyolo Institute of Technology, Chennai
[4,5]Students, Department of Automobile Engineering Sathyabama University, Chennai
Email: nkarthikme@gmail.com

Abstract. Present study provides the effect of Zinc Oxide (ZnO) and Cerium Oxide (CeO_2) nanoparticles additives on the Performance and emission uniqueness of Jatropha. Jatropha blended fuel is prepared by the emulsification technique with assist of mechanical agitator. Nano particles (Zinc Oxide (ZnO)) and Cerium Oxide (CeO_2)) mixed with Jatropha blended fuel in mass fraction (100 ppm) with assist of an ultrasonicator. Experiments were conducted in single cylinder constant speed direct injection diesel engine for various test fuels. Performance results revealed that Brake Thermal Efficiency (BTE) of Jatropha blended Cerium Oxide (B20CE) is 3% and 11% higher than Jatropha blended zinc oxide (B20ZO) and Jatropha blended fuel (B20) and 4% lower than diesel fuel (D100) at full load conditions. Emission result shows that HC and CO emissions of Jatropha blended Cerium Oxide (B20CE) are (6%, 22%, 11% and 6%, 15%, 12%) less compared with Jatropha blended Zinc Oxide (B20ZO), diesel (D100) and Jatropha blended fuel (B20) at full load conditions. NOx emissions of Jatropha blended Cerium Oxide is 1 % higher than diesel fuel (D100) and 2% and 5% lower than Jatropha blended Zinc Oxide, and jatropha blended fuel.

1. Introduction

Due to higher kinematic viscosity, reduced smoke emission and rich oxygen content, Biodiesel is considered as promising alternative fuel for CI engines. The 10% of oxygen content in biodiesel provides better combustion of fuel [3]. On the contrary, The NOx emissions are higher due to the formation of high local temperature. Nearly 12% of increase in NOx emission for biodiesel fuel noted when compared with petrol and diesel fuel [4].To control such NOx emission numerous methodologies have been trailed by analysts around the nations, for example, blending of biodiesel, changes in engine specifications, fumes gas decrease strategies and adjustment in fuel definitions [2]. Among them, fuel plan systems are considered as the most useful method for controlling the level of toxins at the engine fumes. In continuation of such methods, the nanoparticle addition with biodiesel results in pollution reduction and increase in engine performance are noticeably due to the higher surface area to volume ratio [6]. In order to determine the performance and emission characteristics of Jatropha bio diesel, the experiments were conducted in a constant speed single cylinder 4 stroke CI engine. The Jatropha biodiesel causes reduction in carbon monoxide (CO), unburned hydrocarbon (HC) emissions, but they cause increases in nitrogen oxides (NOx) emissions [1]. Brake thermal efficiency of Jatropha biodiesel was lower when compared with neat diesel for the same test condition [5]. An analyst was conducted few experiments on diesel and Jatropha biodiesel fuels with cerium oxide nano particle (at 20, 40 and 60 ppm (parts per million)) as an additive. The results showed that there was 30% reduction in NOx emission and 40% reduction in hydrocarbon emission and 1.5%

increase in brake thermal efficiency [7]. In this study, experiments were conducted in a constant speed(1500rpm) 4stroke single cylinder direct injection diesel engine for diesel(D100), B20(Jatropha 20%+Diesel 80%), B20ZO (Jatropha 20%+Diesel 80%+Zinc oxide), B20CE(Jatropha 20%+Diesel 80%+Cerium Oxide) to investigate the performance and emission characteristics.The Nano particles (100 ppm) were dispersed in the fuel by an ultrasonicator. The tests were conducted with the condition of 220 bar injection pressure and 21° BTDC ignition timing. The specifications of the engine are given in Table1.The properties of test fuels are given in Table 2. The experimental setup has been shown in Figure 1.An AVL model gas analyser is used to measure and carbon monoxide (CO) in percentage volume (%vol) and Unburnt hydro carbon (HC) and Nitric Oxide (NOx) in parts per million(ppm).

Table 1.The test engine specifications

Parameters	Values
Bore	87.5mm
Stroke	110.0mm
Speed	1500(constant speed)
Compression ratio	17.5:1
Rated power	4.4 kW
Number of cylinders	One
Type of cooling	Air cooled - eddy Current dynamometer
Injector opening	21°BTDC
Pressure	220 bar
No. of stroke	4 stroke

Table 2 Properties of fuels used in the experiment

Property	Diesel	Jatropha Oil
Sp. Gravity	0.74	0.96
Viscosity at 40^0C(mm2/s)	4.15	4.4
Calorific Value (KJ/kg)	42000	39500

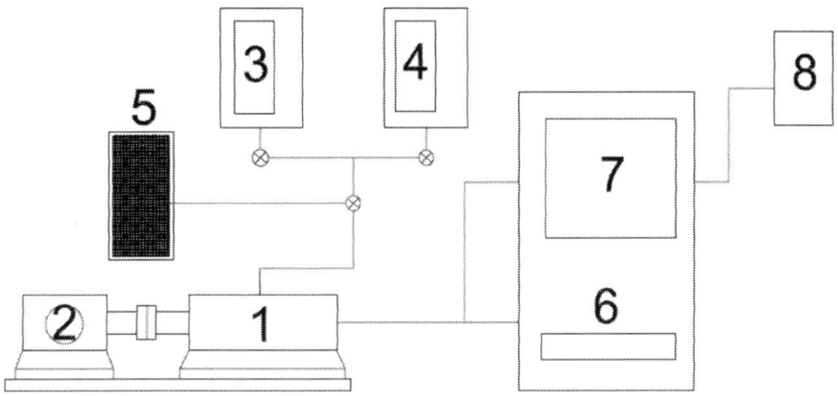

Figure 1. Experimental setup

1. Engine 2.Alternator 3.Diesel Tank 4. Air inlet 5. Biodiesel tank 6. AVL meter 7.Display 8.Smoke Meter

2. Result and discussion:

2.1 Variation of Brake Thermal Efficiency

The variation of Brake Thermal Efficiency with load is shown in figure 2.The results show that the brake thermal efficiency of the diesel engine is improved by the addition of Zinc oxide and cerium oxide nanoparticle in the Jatropha biodiesel. The nano particle presents in the fuel act as an oxygen buffer and lead the combustion to complete. Lower brake thermal efficiency noted for B20 biodiesel compared to D100 due to its lower calorific value, higher density and poor atomization property. The observed values of brake thermal efficiency for D100 and B20 are 32%, 27.8% at full load condition. However, improvement in brake thermal efficiency is observed for B20ZO, B20CE test fuels as 29.9% & 30.75% respectively at full load condition. Due to its high surface area to volume ratio of the nanoparticle increases the atomization of biodiesel and brake thermal efficiency is improved near to that of diesel fuel.

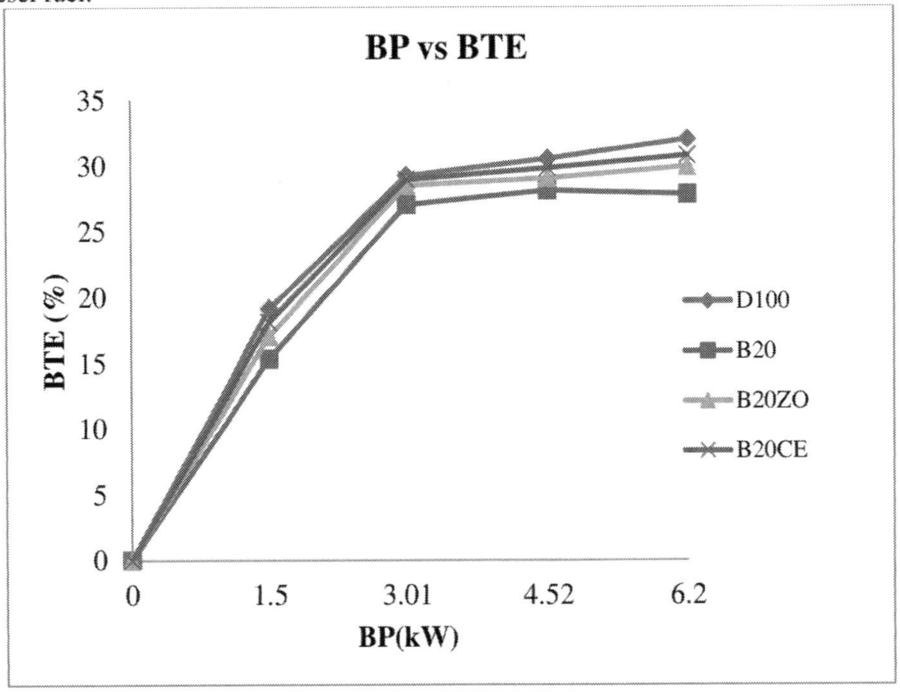

Figure 2. Brake Power vs. Brake Thermal Efficiency

2.2Variation of Hydrocarbon (HC) Emissions

The variation of hydrocarbon emission with engine load is shown in figure 3.Formation of lean charge during the delay period is the primary reason for hydrocarbon (HC) emission. The observed values are 88ppm, 80ppm, 76ppm, 72ppm for D100, B20, B20ZO, and B20CE respectively at full load condition and the percentage reduction Unburned HC for B20ZO, and B20CE are 14% and 18% respectively compared with D100 and 9% and 5% respectively compared with B20.The zinc oxide and cerium oxide nanoparticles in the biodiesel lowers the carbon combustion activation temperature and enriches the hydrocarbon oxidation. Thus unburned HC emissions lowered.

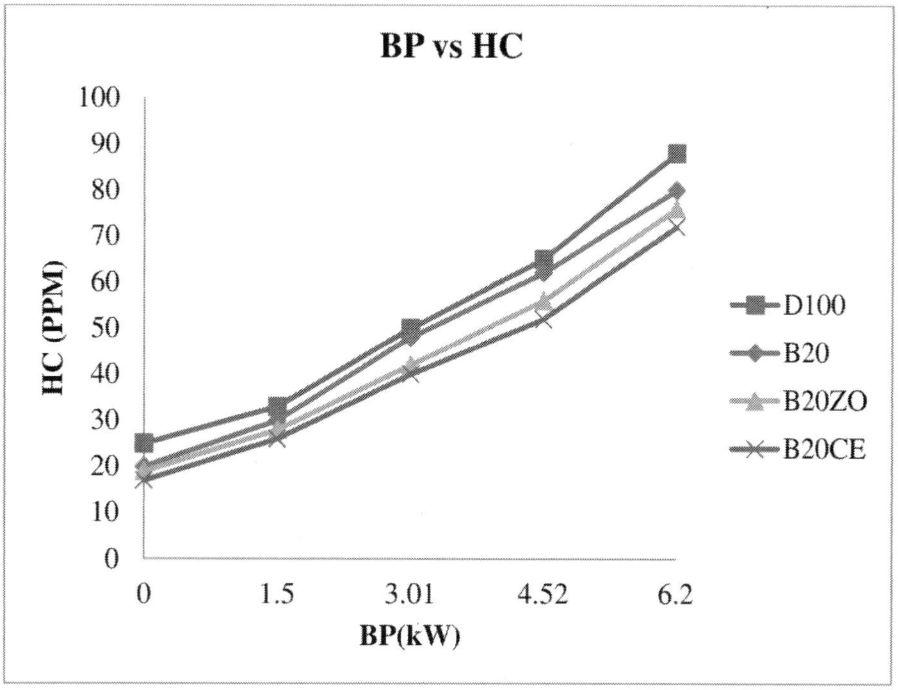

Figure 3. Brake Power vs. Hydrocarbon

2.3 Variation of Carbon Monoxide (CO) Emissions

Figure 4 illustrates the effect of test fuels on CO, under different load conditions. CO emission is formed mainly due to partial combustion of fuel and also by inadequate supply of air during combustion. Lower CO is witnessed for Jatropha biodiesel (B20) when compared with neat diesel (D100), owing to its rich oxygen content that helps to burn the charge completely. The observed CO emission for D100, B20, B20ZO, and B20CE are 0.38% Vol, 0.37% Vol,0.35% Vol,0.33% Vol respectively at full load conditions and the percentage reduction of CO emission for B20ZO, and B20CE are 8 % and 13 % respectively, when compared with diesel(D100), and 5% and 11% respectively, when compared with 20% Jatropha biodiesel(B20).

2.4 Variation of Nitrogen Oxide (NOx) Emissions

Figure 5 illustrates the effect of test fuels on NOx under different load conditions. Biodiesel being an oxygenated fuel enrichesthe rate of oxidation during combustion and results in high exhaust temperatures. At higher temperature, the nitrogen presentin air combines with oxygen and forms a series of gas phasereactions resulting in NOx emission. The observed NOxemission for D100(diesel), B20 (Jatropha biodiesel), B20ZO, and B20CEare 1226 ppm,1300 ppm,1260 ppm,1240 ppm, respectively at full load conditionand the percentage reduction of NOx emission for B20ZO, and B20CE are 3 % and 5 % respectively, when compared with Jatropha biodiesel and 3% and 1% higher than diesel(D100).

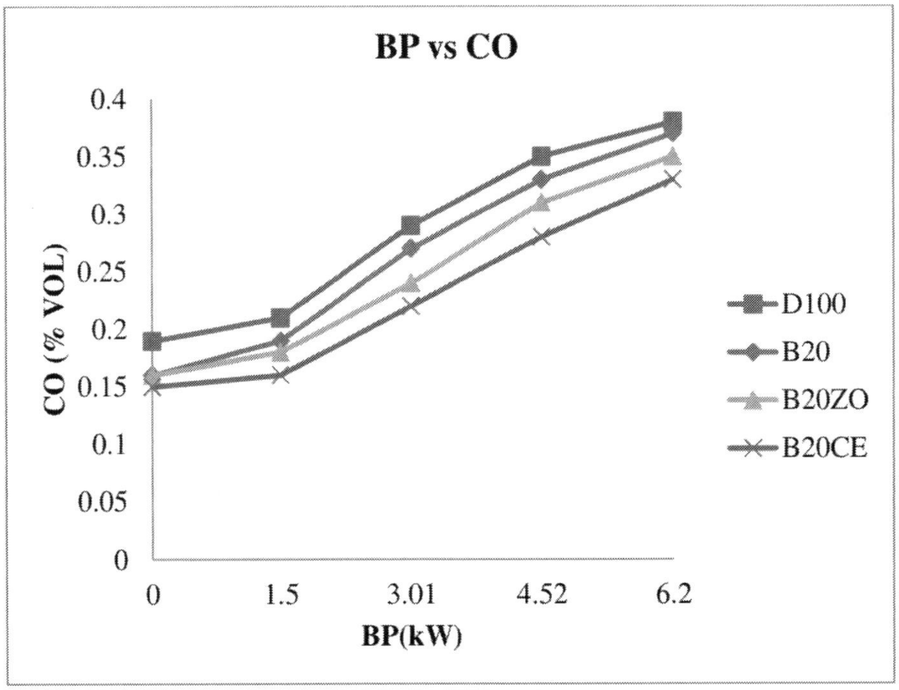

Figure 4. Brake Power vs. Carbon Monoxide

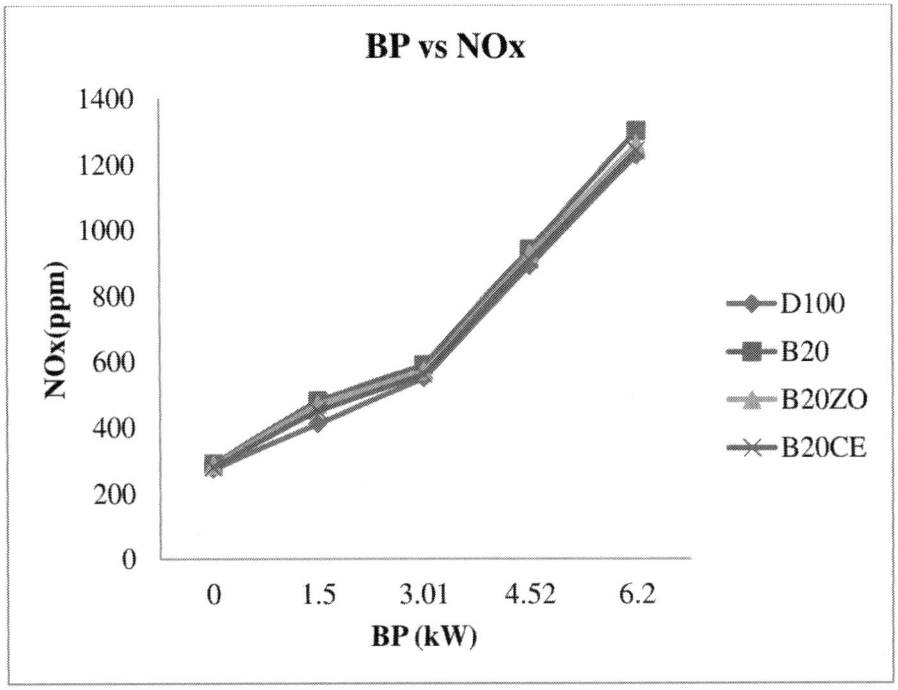

Figure 5. Brake Power vs. Nitrogen oxide

IOP Conf. Series: Materials Science and Engineering **197** (2017) 012014 doi:10.1088/1757-899X/197/1/012014

3. Conclusion

The following conclusions are drawn from the experimental investigation of Zinc oxide and cerium nanoparticle as addition in Jatropha biodiesel.

1. For B20ZO test fuel - Percentage reduction of NOx by 3 %, Unburned HC by 9 %, and CO by 5 % are observed when compared to Jatropha biodiesel (B20) and Percentage increase of NOx by 3 %, reduction in Unburned HC by 14 %, and CO by 8 % are observed when compared to neat diesel(D100).
2. For B20CE test fuel - Percentage reduction of NOx by 5 %, Unburned HC by 9 %, and CO by 11 % are observed when compared to Jatropha biodiesel (B20) and Percentage increase of NOx by 1 %, reduction in Unburned HC by 18 %, and CO by 13 % are observed when compared to neat diesel(D100).
3. 8% and 11% increase in BTE is observed for B20ZO, B20CE test fuels compared to Jatropha biodiesel (B20) and, due to the improvement in atomization.

References

[1] Ajay V Kolhe, Shelke R E and Khandare S S 2014 *Jordan Journal of Mechanical and Industrial Engineering* **8**pp 7-12

[2] Karthikeyan S, A Elango , A Prathima 2014 *Journal of scientific and industrial research***73** pp 187-190.

[3] Kyunghyun Ryu, 2010 *Bioresource Technology*, **101**pp 78–82.

[4] Melissa, A Hess, Michael, J Haas, Thomas, A Foglia 2007 *Fuel Processing Technology*, **88**pp 693–699.

[5] R.Rajasekar, Vijith Chiranjiv R 2015 *International Journal of Applied Engineering Research***10**, No 3 pp 7245-7251.

[6] Sadhik Basha J, R B Anand 2013 *Journal of Brazilian society of Mechanical Sciences and Engineering***35** pp 257-264.

[7] Sajith V, C B Sobhan, G P Peterson 2010 *Advanced Mechanical Engineering*pp 1- 6.

IOP Conf. Series: Materials Science and Engineering **197** (2017) 012015 doi:10.1088/1757-899X/197/1/012015

A Comprehensive Review of Effect of Biodiesel Additives on Properties, Performance, and Emission.

S Madiwale[1], A Karthikeyan[2] and V Bhojwani[3]

[1]Research Scholar of Sathyabama University,Chennai 600119,Tamilnadu,India.
[2]Professor & Head, Department of Automobile Engineering, Sathyabama University, Chennai 600 119,Tamilnadu,India
[3]Professor, Department of Mechanical Engineering, JSPM's ,Jayawantrao Sawant College of Engg.,Hadapsar,Pune 411028,Maharashtra,India.

Email id: shrikantmadiwale@gmail.com.

Abstract. Objectives: -To presents the literature review on effect of biodiesel additives on properties, performance and on emission. Method:-In the current paper reviews are taken from previous years paper which necessitates the need of addition of additives in the blends of biodiesel and studied the its effect on properties , performance and emissions. Emissions from the diesel powered vehicles mostly damaged the earth's environment and also increased the overall earth's temperature. This attracts the need of alternative fuels in the field of transportation sector. Past inventions and research showed that Biodiesel can be used as an alternative fuel for the diesel engine. Biodiesel have good combustion characteristics because of their long chain hydrocarbon structure. However biodiesel possesses few disadvantages such as lower heating value, higher flow ability, much high density and not able to flow at low temperature. Higher rate of fuel consumption is identified and higher level of NOx emissions when biodiesel used in an engine as an alternative fuels. Findings:-Different additives such as antioxidants, improvers for cetane number, cold flow properties improver ,etc were investigated by the many researcher and scientists and added in the different feedstock of biodiesel or blends of biodiesel with diesel in different proportions. Directly or indirectly fuel additives can improve the reduction in the emissions, improve the fuel economy, and reduce the dependency of the one's nation on other. Performances of biodiesel vehicles were drastically improved because of additioninthe blends of biodiesel with diesel fuel in specific percentages to meet the international emission standards. Addition of additives in the biodiesel or in the blends of biodiesel basically changes the high temperature and low temperature flow properties of blends of biodiesel. Current paper finds and compares properties of different additives and its effect on blends of biodiesel properties, performance and on emissions from diesel engines. Improvement:-This paper presents the literature review on effect of biodiesel additives on properties, performance and on emission.

1. Introduction

Diesel engines are largely used in the transportation sector. However the reserves of conventional fuels reduced day by day. Therefore many researcher and scientists are now investigating other sources of fuel as an alternative fuel to the conventional petroleum fuel [1]. Biofuels can be produced from animal fats, plant oils or waste from agriculture, hotel or from industry. One of the known alternative fuels is biodiesel. The ester of long chain FFA(free fatty acid) used with the help of transesterification process produced from non conventional raw feedstock such as vegetable oil, animal fat etc. During the transesterification process the methanol is added with KOH as a catalyst which forms methyl esters and residual as a glycerin [2]. Different feed stocks of plant oil and animal fats such as Jatropha oil, Soybean oil, Palm oil, Cottonseed oil, Waste vegetable oil, Marine fish oil etc are being investigated as an alternative resource for conventional petroleum fuels [3]. Biodiesel is biodegradable and non toxic. The harmful emissions from biodiesel such as carbon and smoke were drastically reduced as compared to conventional petroleum [4]. But on the different side lowheating value, higher viscosity , higher density and poor cold weather flow properties such as cloud point , pour point and cold filter plugging point limits the use of biodiesel which tends to have poor atomization of fuel, injector clogging , narrow spray pattern of fuel in the combustion chamber and incomplete combustion [5-8]. Many researchers investigated that without doing modification in the standard engine, the fuel properties such as the high temperature and low temperature properties of biodiesel can be enhanced by using different fuel additives such as oxygenated additives, metal based additives, antioxidant additives, cold flow properties improver, lubricity improver additives and cetane number improver additives [9-10].The literature review emphasis mainly on use of different biodiesel additives on the properties of fuel , engine performance and engine emissions.

2. Biodiesel Additives

Additives play a very important role in meeting the international fuel standards and real time problems which are associated with biodiesel. With the help of additives the fuel properties can be enhanced and biodiesel in blend can be used with diesel in the vehicle which able to increase the performance and reduction in exhausts emissions from the engine. The selection of biodiesel additives is mainly based on different properties of additives such as flash point, fire point, viscosity, density, calorific value, solubility etc [11]. The use of additives in biodiesel also solves many technical problems which limits the acceptability of biodiesel as an alternative fuel in all conditions. Table 1 shows the different properties of biodiesel additives.

Table 1 Properties of different bio-additives.

Additive	Kinematic Viscosity at 40^0C (cst)	Density (kg/m^3)	Calorific value (kJ/g)	Cetane number	Flash point (^0C)	Ref.
Ethanol	1.14*	791*	27.33	5-8	-	12
n-Butanol	3.00	812	34.33	25	35	12
Diethyl Ether	0.22	712	33.89	25	-	12
Methanol	0.59	790	19.62		11	13

(*- measured at 20^0C)

Ethanol contains s 34% higher oxygen by weight. Biomass can be converted to the ethanol by using the process of fermentation from the different feedstock the ethanol can be produces such as sugarcane, sugar beet, sorghum, potatoes, sunflower, molasses, corn , wheat, cotton etc[14]. Oxygenated fuels are generated because of addition of ethanol in diesel fuels. Ethanol with diesel is also named as Diesohol. Many investigators documented that addition of ethanol in small percentage in the blends of biodiesel increases the brake thermal efficiency, increases the rate of heat release, reduces the viscosity and reduces the smoke in the exhaust of the engine. Also the physical and thermal properties of ethanol-biodiesel –diesel blends improved drastically which helps the combustion [15].

Diethyl ether is also an oxygenated additive which is biomass based produced from the ethanol. It is colourless fuel. It possesses high volatility. It also have high tendency to respond to low temperature because of high flammability. It has high cetane index. But on the other side the heating value of diethyl ether is lower than conventional diesel. But it used to mix with biodiesel or diesel easily. It has good solubility characteristics. Many investigators reported that addition of diethyl ether with diesel–biodiesel blend improves the properties of the blend and enhanced the performance and reduced the engine pollutants[16].

In past few years n-butanol is also emerged as an additive for the biodiesel – blend because of its higher percentage of O2 content. It is also refer as 1-butanol and produced from fermentation of different feedstock of biomasses. It has straight chain with OH group and has lower hydrophilic nature. High cetane number, good solubility and higher heat content attract many researchers to consider n-butanol as an additive to biodiesel-diesel blends [16].

Due to the poor cold flow properties of biodiesel-diesel blend it is not possible to use the blend in cold weather condition without the additives. Methanol is also a very promising alternative additive higher level of oxygen which is basically responsible to enhance the rate of combustion in the cylinder. It is an analysis grade anhydrous ethanol of purity 99.7%. The latent heat of methanol is 1109kJ/kg, which is considerably higher than the conventional fuel [17]. Because of all such type of properties, methanol is the strong competitor as an additive in the biodiesel–diesel blend.

3. Additives and Fuel Properties

Investigators have documented their results which show improvement in the fuel propertiesofdifferent feedstock of biodiesel with diesel by using different additives in the blends.In[18] studied the properties of biodiesel with 5% addition of methanol as an additive and reported that addition of additive decreases the density of blend from 845 kg/m^3 to 829 kg/m^3, viscosity from 4.5 cSt to 3.3 cSt and flash point from 100^{0}C to 92^{0}C. In[19] investigated and reported the properties of cottonseed biodiesel with diesel with an addition of ethyl levulinate as additive, and found the significant improvement in low temperature flow properties of blend. The cloud point was lowered from 4-5^{0}C, the pour point 3-4^{0}C, and cold filter plugging point was reported as 3^{0}C. In[20]had done the analysis and study of effect of ethanol and butanol on the cyclic variation in the diesel engine. In[21]investigated the effect of kerosene and ethanol as an additive in the Mahua oil biodiesel. They reported that addition of kerosene and ethanol in the blend of B20 biodiesel drastically enhanced the cloud point and pour point. In[12] studied the engine performance and exhaust gas emission of blends of biodiesel with n-butanol and diethyl ether as an additive, in[16]studied the effect of n-butanol and diethyl ether as additives on engine characteristics of a multiplecylinder diesel engine fuelled with diesel–Jatropha biodiesel blend and reported that kinematic viscosity of blend changes from 3.36 cSt to 3.16 cSt, lower calorific value 43.69 to 43.10 MJ/kg and flash point was improved from 96.5 to 71.5^{0}C. Table 2 gives the effect of addition of additives on the fuel properties of feedstock of biodiesel.

TABLE 2 Fuel additives and fuel properties of different feedstock of biodiesel

Feed stoke of biodiesel	% of Additives	Density at 20 ^{0}C (kg/m^3)	Viscosity at 40 ^{0}C (cSt)	Flash point ^{0}C	Pour point ^{0}C	Cetane number	Calorific value (MJ/kg)	Ref
Palm oil (B20)	5% Ethanol	833	3.23	84.5	-	46	43.08	12
Palm oil (B20)	5% n-butanol	833	3.39	85.5	-	47	43.43	12
Palm oil (B20)	5% DEE	832	3.37	81.5	-	52	43.41	12
Jatropha Oil (B30)	10% ethanol	832.87	2.80	14	-3	50	39.93	22
Palm oil (B20)	2% Methanol	837	2.31	13	–	–	40.21	23
Palm oil (B20)	2% Ethanol	838	2.37	18	–	–	40.53	23
Palm oil (B20)	2% DEE	840	2.35	15	–	–	39.65	23
Neem oil (B100)		-	4.5	165	-	51	41	24
Neem oil (B100)	5% Ethanol	–	4.2	153	–	49	40.4	24
Rice bran oil (B100)		874.2	4.63	165	3	56.2	37.9	25
Rice bran oil (B100)	5% Ethanol	782	1.35	22	$\leq$ 35	10	–	25
Cottonseed oil (B100)		871	5.28	-	-	51	-	26
Cottonseed oil (B100)	5% Ethanol	867	-	-	-	-	-	26

4. Biodiesel Additives Performance and Emissions

Researchers analyzed the effect of additives on the performance and the emissions from the engines. In[25] studied the performance of single cylinder 4S diesel engine with rice bran oil biodiesel with 2.5% ethanol as an additive, and reported that the 6.98% increase in BSFC for ethanol blend and increased in BTE up to 27.82%.In[27] studied the effect of additives on performance and on emissions when fueled with biodiesel, the investigation reported that there was a 9% increase in brake power, 6% lower BSFC and NOx,CO2 and Co emissions were reduced. In[28]reported performance of engine, exhaust emission and combustion and heat release of a common-rail diesel engine fueled with bio-ethanol as a fuel additive in coconut oil biodiesel blends.Paper reported that there wasa increase in BTE and BSFC when operated with additives. Emissions from the engine also lowered the NOx, smoke, carbon monoxide as compared to diesel. In[29]investigated that when the engine is fueled with soybean biodiesel with ethanol as a additive then emissions from the engine such as NOx, CO, CO2, PM reduces. In[30] tested the feedstock of biodiesel from tall oil with magnesium and nickel based additives at various loads. They reported that BSFC was increased by 6%. Exhaust emissions from biodiesel with additives also decreases, CO was decreased by 64.28% and smoke by 30.91%. Also there was a significant reduction in the NOx level. In[31] studied the ethanol biodiesel and diesel blend at different proportion and at different temperature. Investigation shows that there was a significant reduction in the CO and HC but NOx started to increase at higher engine load. In[32]studied emission parameters of a diesel engine with methanol and ethanol as a biodiesel additive. NOx and PM were reduced from the engine when fueled with additives. Alcohol increased the carbon monoxide and hydrocarbon in the exhaust. In[33]investigated and analyzed the exhaust emissions from a CI engine operated with biodiesel and methanol as an additive. Investigation reported that there was a reduction in CO2 and NOx and PM[34]. All experiments were performed on the constant speed engine at different load.

5. Conclusion

The comprehensive study framed the effect of additives on the biodiesel properties, biodiesel performance and emissions when used as an alternative fuel in the conventional diesel engine. The following conclusions are drawn from this review. Different additives with more oxygen contents are available. If such type of additives added in to the blends of biodiesel with diesel then they can enhance the quality of combustion and leads to complete combustion. Diethyl ether, ethanol, methanol,and n-butanol can be used as biodiesel additive because of their higher oxygen content. The use of additive enhanced the performance of engine and reduced the BSFC. Emissions such as nitrogen dioxide carbon monoxide, particulate matter etc will be reduced when fueled with biodiesel additives.

6. References

[1].Rashedul H.K , Masjuki H.H, Kalam M.A, Ashraful A.M, Ashrafur Rahman S.M, Shahir S.A. The effect of additives on properties, performance and emission of biodiesel fuelled compression ignition engine . Energy Conversion and Management. 2014; 88,: 348-364.

[2].Mohammadi P., Nikbakht A. M., Tabatabaei M., Farhadi Kh.A Novel Diesel Fuel Additive to Improve Fuel Properties and to Reduce Emissions. International Journal of Automotive Engineering.2012 ; 2 (3) :156-162.

[3].Swarup Kumar Nayaka, Bhabani Prasanna Pattanaika. Experimental Investigation on Performance and Emission Characteristics of a Diesel Engine Fuelled with Mahua Biodiesel Using Additive . Energy Procedia.2014 ; 54: 569-579,.

[4].Abbaszaadeh A, Ghobadian B, Omidkhah MR, Najafi G. Current biodiesel production technologies: a comparative review. Energy Convers Manage.2012; 63:138-148.

[5].Zhang X, Reece D, Haws R, Moller G. Biodegradability of biodiesel in the aquatic environment. ASAE.1998; 1423:41-50.

[6].SMA Rahman, HH, Masjuki MA Kalam, MJ Abedin, A, Sanjid ,H Sajjad. Impact of idling on fuel consumption and exhaust emissions and available idle reduction technologies for diesel vehicles – a review. Energy Convers Manage.2013; 74:171-82.

[7].Murayama T. Low carbon, lower buildup, low smoke, and efficient diesel operation with vegetable oils by conversion to mono-esters and blending with diesel oil or alcohols. SAE . 1984; 841161:292-302.

[8].Demirbas A. Importance of rural bioenergy for developing countries. EnergyConvers Manage.2007;48: 2386-2398.

[9].Varatharajan K, Cheralathan M, Velraj R. Mitigation of NOx emissions for a jatropha biodiesel fuelled DI diesel engine using antioxidant additives. Fuel.2011; 90:2721-2725.

[10].Lapuerta M, Herreros JM, Lyons LL, Garcia-Contreras R, Briceno Y.Effect of the alcohol type used in the production of waste cooking oil biodiesel on diesel performance and emissions. Fuel.2008; 87:3161-3169.

[11].Silitonga AS, Masjuki HH, Mahlia, Ong HC, Chong WT, Boosroh MH.Overview properties of biodiesel diesel blends from edible and non-edible feedstock. Renew Sustain Energy Rev.2013; 22: 346-360.

[12].Imtenana S., Masjukia H.H., Varmana M., Arbaba M.I., Sajjada H., Rizwanul I.M. Fattaha,M.J. Abedina, Abu Saeed Md. Hasibb. Emission and performance improvement analysis of biodiesel-diesel blends with additives . Procedia Engineering. 2014;90: 472- 477.

[13].D.H. Qi , H. Chen, L.M. Geng, Y.ZH. Bian, X.CH. Ren .Performance and combustion characteristics of biodiesel–diesel–methanol blend fuelled engine . Applied Energy . 2010; 87:1679-1686.

[14].Mofijur a,n M., Rasul a M.G., Hydea J., Azada A.K., Mamatb R.,Bhuiya mmk. Role of biofuel and their binary (diesel–biodiesel) and ternary (ethanol–biodiesel–diesel) blends on internal combustion engines emission reduction. Renewable and Sustainable Energy Reviews. 2016; 53 :265-278.

[15].Shahir n S.A., Masjuki nn H.H., M.A. Kalam, Imran A., Ashraful Centre A.M.Performance and emission assessment of diesel–biodiesel–ethanol/ bioethanol blend as a fuel in diesel engines: A review. Renewable and Sustainable Energy Reviews , 2015;48:62-78.

[16].Imtenan S., Masjuki H.H., Varman M., Rizwanul Fattah I.M., Sajjad H.,. Arbab M.I.Effect of n-butanol and diethyl ether as oxygenated additiveson combustion–emission-performance characteristics of a multiple cylinder diesel engine fuelled with diesel–jatropha biodiesel blend.Energy Conversion and Management . 2015;94:84–94.

[17].D.H. Qi, H. Chen, L.M. Geng, Y.ZH. Bian, X.CH. Ren. Performance and combustion characteristics of biodiesel–diesel–methanol blend fuelled engine. Applied Energy .2010; 87:1679-1686 .

[18].Mat Yasin M, Yusaf T, Mamat R, Fitri Yusop A.Characterization of a diesel engine operating with a small proportion of methanol as a fuel additive in biodiesel blend. Appl Energy.2014;114:865-873.

[19].Joshi H, Moser BR, Toler J, Smith WF, Walker T.Ethyl levulinate: a potential bio-based diluent for biodiesel which improves cold flow properties. Biomass Bioenergy.2011; 35: 3262–3266.

[20].Obed M. Alia, Nik Rosli Abdullahb, Rizalman Mamata, Abdul Adam Abdullaha. Comparison Of The Effect Of Different Alcohol Additives With Blended Fuel On Cyclic Variation In Diesel Engine . Energy Procedia. 2015;75: 2357- 2362.

[21].Bhale PV, Deshpande NV, Thombre SB.Improving the low temperature properties of biodiesel fuel. Renew Energy.2009; 34:794–800.

[22].Mofijur M. ,. Rasul M.G, Hyde J. Recent Developments on Internal Combustion Engine Performance and Emissions Fuelled With Biodiesel-Diesel-Ethanol Blends. Procedia Engineering, 2015 ;105: 658-664.

[23].Vedaraman N, Puhan S, Nagarajan G, Velappan K. Preparation of palm oil biodiesel and effect of various additives on NOx emission reduction in B20: an experimental study. Int J Green Energy.2011; 8: 383–397.

[24].Sivalakshmi S, Balusamy T.Influence of ethanol addition on a diesel engine fuelled with neem oil methyl ester. Int J Green Energy, 2012;9:218–228.

[25].Venkata Subbaiah G, Raja Gopal K.An experimental investigation on the performance and emission characteristics of a diesel engine fuelled with rice bran biodiesel and ethanol blends. Int J Green Energy. 2011;8: 197-208.

[26].Anbarasu A, Saravanan M, Loganathan M.The effect of ethanol addition in a biodiesel operated DI diesel engine on combustion, performance, and emission characteristics. Int J Green Energy, 2013;10: 90–102.

IOP Conf. Series: Materials Science and Engineering **197** (2017) 012015 doi:10.1088/1757-899X/197/1/012015

[27].Shahabuddina M.,Masjuki H.H., Kalam M.A.,. Mofijur M,.Hazrat M.A,.Liaquat A.M.Effect of Additive on Performance of C.I. Engine Fuelled with Bio Diesel. Energy Procedia vol. 2012;14 :1624 - 1629.
[28].Howa H.G., Masjukia H.H., Kalama M.A., Teoha Y.H.,b a Centre. Engine performance, emission and combustion characteristics of a common-rail diesel engine fuelled with bioethanol as a fuel additive in coconut oil biodiesel blends. Energy Procedia.2014; 60:1655-1659.
[29].Mario L. Randazzo , Jose R. Sodre .Exhaust emissions from a diesel powered vehicle fuelled by soybean biodiesel blends (B3–B20) with ethanol asadditive (B20E2–B20E5). Fuel .2011;90: 98-103.
[30].Keskin A, Gürü M, Altıparmak D. Biodiesel production from tall oil with synthesized Mn and Ni based additives: effects of the additives on fuel consumption and emission. Fuel.2007;86:1139-1143.
[31].Kwanchareon P, Luengnaruemitcha A, Jai-In S. Solubility of a diesel–biodiesel–ethanol blend, its fuel properties, and its emission characteristics from diesel engine. Fuel, 2007; 86:1053–1061.
[32].L Zhu, CS Cheung, WG Zhang, Z Huang. Emissions characteristics of a diesel engine operating on biodiesel and biodiesel blended with ethanol and methanol. Sci Total Environ.2010;408,4:914–921.
[33].Cheng CH, Cheung CS, Chan TL, Leec SC, Yao CD, Tsang KS.Comparison of emissions of a direct injection diesel engine operating on biodiesel with emulsified and fumigated methanol.Fuel .2008 ;87:1870–1879.
[34].P. Bridjesh, N. Prabhu Kishore, M. V. Mallikarjuna, N. Alekhya,Performance Analysis of Variable Compression Ratio Diesel Engine using Calophyllum inophyllum Biodiesel ,Indian Journal of Science and Technology.2016; 9(35)

IOP Conf. Series: Materials Science and Engineering **197** (2017) 012016 doi:10.1088/1757-899X/197/1/012016

Optimization of Rice bran biodiesel blends on CI engine and investigating its effects

J.Jayaprabakar*, **Biraj Dey, Krishanu Dey, Batchu Hareesh & M. Anish**

[1]Department of Mechanical Engineering, Sathyabama University, Chennai, India.
E-mail: jp21tn@gmail.com

Abstract. Bio-diesel can be produced from various plant oils like soybean, sunflower or rice bran. Here the focus is on converting the rice bran oil into bio-diesel which is produced by transesterifying the rice bran oil with a low molecular weight alcohol (methanol) and a non-conventional catalyst (lipase). Using a lipase based catalyst brings down the cost of bio diesel production significantly by reducing the number of washing cycles and its ability to be reused further. Four different blends of B10, B20, B30, B40 and straight diesel are tested in a single cylinder, fourstroke, vertical air cooled Kirloskar Diesel Engine having ignition timing of 23° before Top Dead Centre (TDC). As compared to straight diesel the Brake Thermal Efficiency (BTE) value for all the blends are higher. The Specific Fuel Consumption (BSFC) values for most of the blends are less as compared to diesel. Emissions of CO, CO2 and HC for all the blends decreased quite significantly. As a summary, the blend B20 records better performance as well as emission characteristics as compared to diesel.

1. Introduction

In the current context, compression ignition engines are aimed to deliver high thermal efficiency and lowered specific fuel consumption. Due to recently imposed stringent emission norms, several tools and techniques were developed all over the world to limit the CI engine emissions within the specified limits [1].Biodiesel, a form of diesel fuel produced from vegetable oils and animal fats. It is proven to safe, biodegradable and emits less air pollutants than conventional diesel. Biodiesel is defined as a liquid fuel made as fatty acid methyl esters, fatty acid alkyl esters or the long chain mono alkyl esters. As purest form (B100) or blended with pure diesel the biodiesel can be used. Fuel properties of the blend will vary according to the amount of biodiesel in the blend. The smaller volume percentage of biodiesel with diesel in the blend performs better. Biodiesel for diesel engines is produced by chemically reacting a vegetable oil or animal fat with an alcohol such as methanol or ethanol. This reaction takes place for the oil in the presence of a catalyst, normally a conventional catalyst such as sodium or potassium hydroxide, and produces new compounds called methyl esters. Non conventional catalyst like enzymes may also be used. Because the carbon in the oil from carbon dioxide present in the atmospheric air, biodiesel is considered to contribute much less to global warming than petroleum fuels. A CI engine operated with biodiesel emits lower emissions of carbon monoxide, unburnt hydrocarbons, particulate matter, and air toxics than whenoperated on Diesel fuel. Biodiesel have advantages like, provides a market for huge production of animal fats and vegetable oils, and decreases a country's dependence on imported Diesel.

Rice bran oil, a non conventional, inexpensive and low grade is also source of high value added by-product are derived from the crude rice bran oil and the resultant oil is used as a feedstock for biodiesel. Rice bran oil is the oil extracted from the germ and inner husk of rice. It is popular as

cooking oil in several Asian countries including Japan and China. Several reports were already supports the use of Rice bran oil in CI engine. Calorific value of the Rice bran biodiesel is very much nearer to diesel value and it is possible to convert high FFA rice bran oil into biodiesel [2].Biodiesel fuel blend with a small portion of Rice bran biodiesel was technically feasible for a CI engine. The lesser amount of biodiesel will improve the Specific Fuel Consumption characteristic [3].Biodiesel derived from Ricebran oil may be effectively used as a blended fuel in CI engine as its properties are within the suggested standard fuel limit. The biodiesel blends B20, B40 produced from rice bran oil are considered to be a better alternative to diesel without any modification on the CI engine [4].In a test conducted on a single cylinder CI engine the peak pressures for the Ricebran oil biodiesel blends occurred within 1°-9° crank angle after TDC and it is very nearer to with that of diesel [5].Viscosity of Rice bran biodiesel is normally higher. When its concentration increases in the blend, the viscosity of blend increases [6].While the engine running with Rice bran Biodiesel, during the initial stage of combustion better intermixing of air and fuel were attributed [7].It was tested and proven with Rice bran Biodiesel that the maximum yield of ester can be obtained working with an initial concentration of catalyst and operation temperature of 40°C [8].
The aim of the project is to produce biodiesel from rice bran and mixing it with straight diesel in the four proportions of B10 to B40 blends. Rice bran oil derived biodiesel's fuel properties are observed and tested by following standard procedures. It is found that rice bran oil properties are almost similar to that of other commonly used bio-diesel. Then an experimental Test rig is used to investigate the combustion , performance and emission of rice bran oil biodiesel blends (B10, B20, B30 and B40). The standard injection timing of 23° before top dead center is adopted for all the blend testing. The various parameters are measured and calculated. The emission graphs are plotted against the varying load. Finally comparisons are made between the blends and with the standard value of the diesel

2. Experimental Setup and Methodology

.

Table 1. Properties of Rice Bran oil

Properties	Results
Free Fatty Acid	0.592 %
Kinematic Viscosity at 40°C	37.548 cSt
Density	0.910 gram/cc
Moisture Content	0.002 %
Flash point (PMCC method)	278°C
Fire point (PMCC method)	307°C
Copper Strip Corrosion at 60° C for 3 hours	1a (slight tarnish)
Could point ° C	16°C
Pour point ° C	1°C
Calorific value	3808.356 cal/gram

First 3000 ml of rice bran oil is obtained from a commercial store under the brand name POORNA RICE BRAN OIL. The various fuel properties of the raw rice bran oil were tested at Centre for Waste Energy and Management, Sathyabama University, Chennai. The Properties of the raw Ricebran oil are presented in Table 1. Here the FFA content of the raw rice bran oil is less than 4%. Therefore, no further esterification is needed; it can be now directly transesterified to form biodiesel. The catalyst used in this work is Lipase (Steapsin), it is a digestive enzyme of pancreatic juices that catalysis the hydrolysis of fats to fatty acids and glycerol and it is a fat splitting enzyme. The advantages of lipase are it can be used again and again without loss in its effectiveness, the biodiesel which is produced after the transesterification process is not needed to be washed thereby reducing the cost of the biodiesel. The amount of catalyst used for this study is 3% of the total weight of oil samplThe oil sample is checked for its FFA.Oil is heated to 50°C and maintained thereafter till completion of the reaction. Lipase is added and stirred continuously with a magnetic stirrer. Methanol is added stage wise (every half an hour) to avoid deactivation of the catalyst. After the reaction completed, the contents are transferred into a separating funnel; two layers were seen in the container, the upper layer is biodiesel and lower layer is Glycerol. Lipase is settled down at the bottom. The variables in the transesterification process are presented in Table 2. The biodiesel obtained after the transesterification process is tested for various fuel properties and are presented in Table 3.

Table 2. Transesterification variables report

Parameters	**Value**
Amount of Lipase added	3% by weight of oil sample
Total Amount of Methanol added	210 ml for 1000ml oil sample
Amount of Methanol Added in every	
half an hour	52.5 ml
Reaction Temperature	50°C
Reaction time	2 hours
Magnetic Stirrer Speed	1000 rpm
Washing	No
Yield	96%

Table 3. Properties of Rice bran oil Biodiesel

Properties	**Values**
Kinematic Viscosity at 40°C	3.906 cSt
Density	0.873 gram/cc
Moisture Content	0.001%
Flash point (PMCC method)	166°C
Fire point (PMCC method)	198°C
Copper Strip Corrosion at 60°C for 3hours	1a(slight tarnish)
Cloud point °C	10° C
Pour point °C	-3° C
Calorific value	8214.692 cal/gram

The Engine Test rig used in this work is a Kirloskar, Four stroke, Single Cylinder, Vertical Air Cooled Diesel Engine having ignition timing of 23° before TDC.the specifications of the engine is shown in the Table 4.After property testing, the samples are then taken to next process for performance, combustion and emission test. Four different blends B10, B20, B30, B40 and pure diesel are tested on a four stroke, single cylinder, vertical air cooled CI engine. While the engine is in running condition, load is increased gradually (0%, 25%, 50%, 75% and 100%).The exhaust gases temperature coming out of the engine is noted down for different load conditions after 10g of fuel is consumed. The time taken for consuming 10g of fuel for a particular load is noted down at the same time the difference in manometer reading is also noted. The engine is set to run till the full load condition and all the readings are noted down.After this the engine is stopped. The performance results are calculated from the various readings which are noted down during engine testing. On the other hand the emission results are calculated with the help of supporting system.

Table 4. Specifications of the engine

Rated power	4.4 kW
Rated speed	1500 rpm
Bore Diameter (D)	87.5 mm
Stroke (L)	110 mm
Compression ratio	17.5 : 1
Orifice Diameter	13.6 mm
Coefficient of discharge (C_d)	0.6
Loading Device	Swing field Electrical Dynamometer

3. Results and Discussion

The combustion, performance and emission characteristics of straight diesel and tested rice bran blends B10, B20, B30 and B40 are discussed in this section.

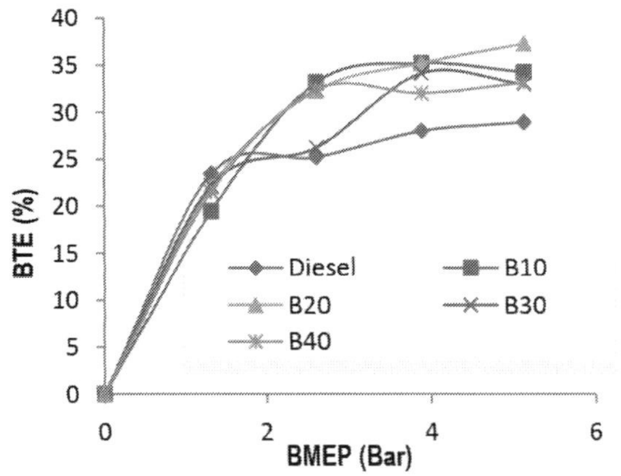

Figure 1. BMEP vs. BTE

3.1. Brake thermal efficiency

The Figure 1. illustrates the comparison of Brake thermal efficiency values of biodiesel blends and straight diesel against BMEP. Most of the blends have higher BTE than diesel. This is due to the inhumed mixing of biodiesel-air along with better combustion results leading to maximum thermal efficiency. The BTE of the diesel increases initially but as the BMEP keeps on increasing; the BTE decreases and gradually has the least BTE value. The BTE values for all the blends vary drastically except for the blend B20 which increases continuously.

3.2. Brake specific fuel consumption

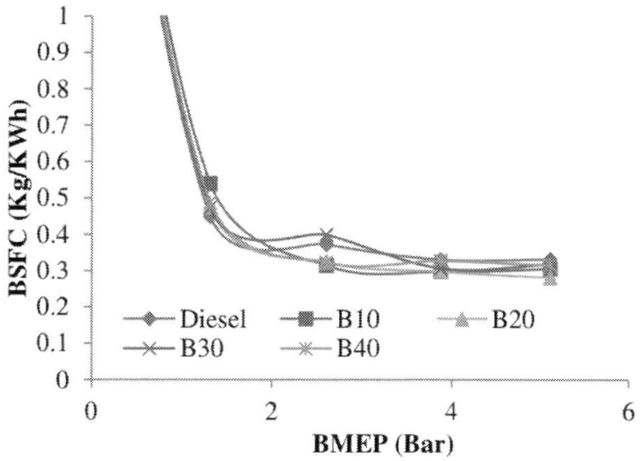

Figure 2. BMEP vs. BSFC

Figure 2. Depicts the variation of BSFC values against BMEP for the tested biodiesel blends. When brake mean effective pressure of the cylinder increases, specific fuel consumption will decrease. The specific fuel consumption values of blends normally decreases for all the blends and straight diesel with increase in the BMEP values. The blend B30 and diesel show sudden increase in fuel consumption at pressure between 2 to 3 bar. The reason is at higher pressure fine atomized fuel mixes homogenously with air and additives and hence less fuel is needed.

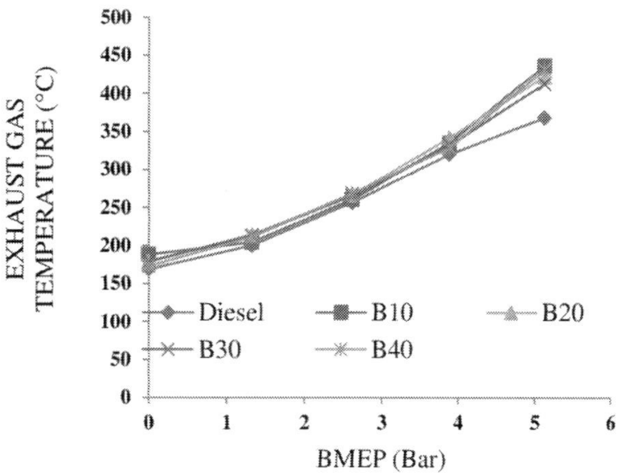

Figure 3. BMEP vs. Exhaust Gas Temperature

3.3. Brake specific fuel consumption

The Figure 3. delivers the comparison of exhaust gas temperature values of tested fuel blends against BMEP. The exhaust gas temperature is the temperature of the gas which comes from the engine. The EGT of the engine is considered as a most important parameter which measures utilization of heat energy of the fuel. The graph shows that the exhaust gas temperature values for all biodiesel blends and diesel increases with increase in BMEP. A gradual increase in the exhaust temperature for all the blends was identified but the exhaust gas temperature for the straight diesel is least at BMEP of 5 bar.

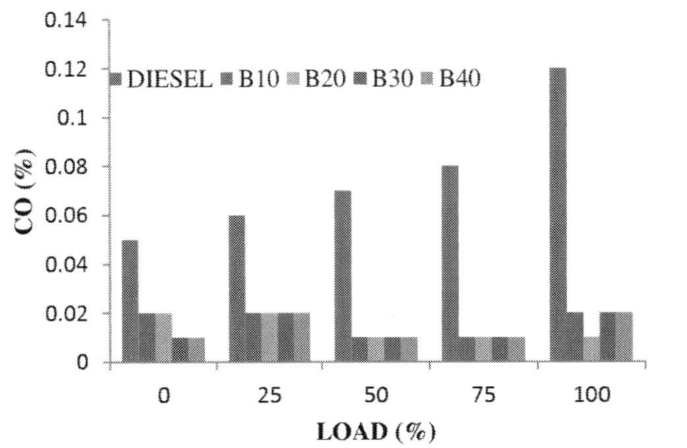

Figure 4. CO Emission chart for all blends and diesel

3.4. Carbon monoxides

The CO emission of various biodiesel blends and straight diesel at different load conditions are shown in the figure 4. The carbon monoxide emission is found to be maximum for diesel among all tested fuels at all loads. The reason for this is due to the incomplete combustion of diesel fuel in the engine cylinder. When the engine load increases the CO emission for diesel increases but this trend could not be seen in the biodiesel blends. At No load the CO emission is least for the blends B30 and B40 and at load full load condition the CO emission is least for the blend B20.

3.5. Hydro carbons

The emission of unburnt hydro carbons in the exhaust gasses has many adverse effects on the atmosphere which leads to global warming. The Figure 5. Illustrate the HC emission at different loads for diesel and biodiesel in conventional engines. Diesel records the highest HC emission at all loads. The other blends have fluctuation values at different loads. At load 0 the HC emission is least for B30 and B40. At load 50 the emission is least for B40 and for load 100 the emission is least for B40.

3.6. Carbon di oxides

The Figure 6. Presents the variation of CO2 emissions in the exhaust gases of tested blends and straight diesel. The straight diesel has almost same emission as that of the biodiesel at various loads with little variations. This indicates that the combustion pattern of the biodiesel blends is similar to that of the straight diesel. At No load condition the CO2 emission is less for the biodiesel blend B40 and at the maximum load condition the CO2 emission is least for the blends B10 and B20.

3.7. Nitrogen oxides

Figure 7. Furnishes the NOX emission values for the biodiesel blends and diesel against rated loads. NOX emission is least for diesel among all the tested fuels at all loads. The highest value of NOX emission was recorded by the blend B40 at all applied loads.

IOP Conf. Series: Materials Science and Engineering **197** (2017) 012016 doi:10.1088/1757-899X/197/1/012016

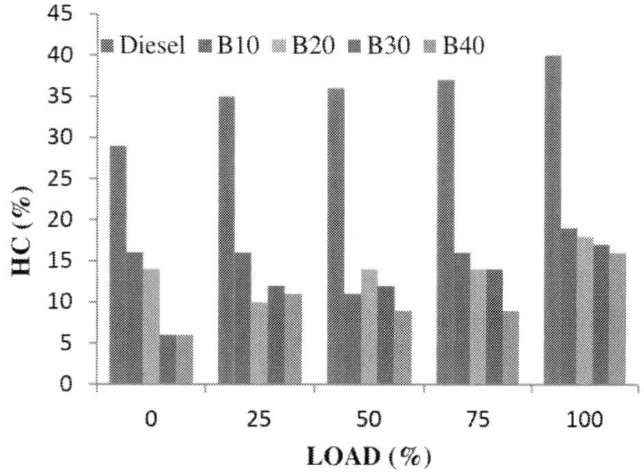

Figure 5.UHC Emission for the blends and diesel

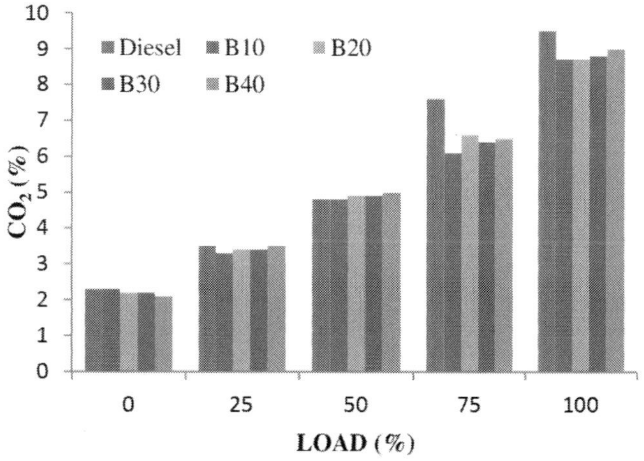

Figure6. CO_2 Emissions for all blends and straight diesel

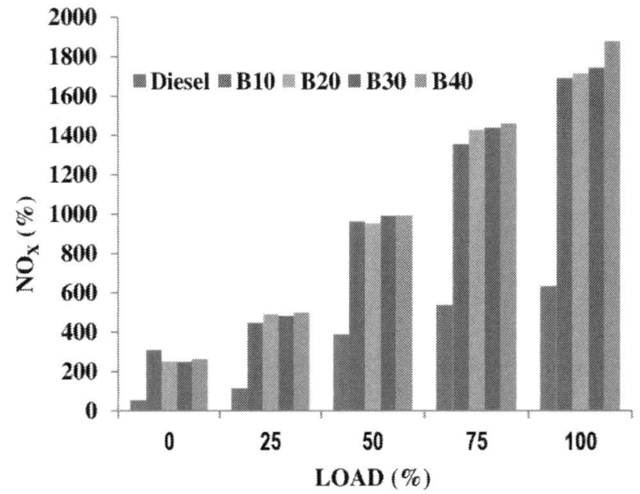

Figure 7. NO_xemission chart for all blends and diesel

4. Conclusion

The summarized conclusions drawn from the combustion, performance, emission and characteristics investigation conducted for the four blends of Rice bran with straight diesel are discussed here.

➢ When the Brake Mean Effective Pressure increases the Brake Thermal Efficiency value increases for all blends B10, B20, B30 and B40.

➢ The fuel consumption is less for all the tested blends than straight diesel except for the blend B30 at pressure between 2 to 3 bar. The Blend B20 shows a remarkable decrease in the fuel consumption values.

➢ The Exhaust Gas Temperature values are is higher for all the biodiesel blends than the straight diesel at all pressure. The Exhaust temperature increases gradually with the increase in pressure for the all the fuel types.

➢ The CO, HC, CO_2 emission values for straight diesel is quite higher than the biodiesel blends at all loads.

➢ The NO_X emission values for the biodiesel blends are higher than the straight diesel at all loads.

References

[1] D.C. Rakopoulos, C.D. Rakopoulos, E.C. Kakaras, E.G. Giakoumis 2008 Effects of ethanol–diesel fuel blends on the performance and exhaust emissions of heavy duty DI diesel engine, Energy Conversion and Management **49** pp 3155–3162.

[2] Subhan Kumar Mohanty 2013 A production of biodiesel from rice bran oil and experimenting on small scale capacity diesel engine, IJMER, **3**, pp 920-923.

[3] Wahome A, Ngunjiri G.M.N, Shitanda D and Ogola W.O 2013 Alternative Diesel Engine Fuel from Kenyan Pishori Rice Bran, International Journal of Engineering and Science **2** pp 75-79.

[4] Jayaprabakar J, Karthikeyan A, Aashwin Josiah, Anu Shajan 2015 Experimental Investigation on the performance and emission characteristics of a CI engine with rice bran and micro algae biodiesel blends Journal of Chemical and Pharmaceutical Sciences, **special issue 7** pp 19-22.

[5] Jayaprabakar. J, Karthikeyan. A and Rameshkumar.V 2017 Effect of injection timing on the combustion characteristics of rice bran and algae biodiesel blends in a compression-ignition engine International Journal of Ambient Energy, **38(2)** pp 116–121.
[6] Anandkumar S Malipatil and Manjunath Bandi 2014 Experimental study on performance and emission characteristics of diesel engine using rice bran oil as fuel, International Journal of Research in Aeronautical and Mechanical Engineering **2(9)** pp 23-33.
[7] Mohit Vasudeva, Sumeet Sharma and S.K Mohapatra 2013 Optimizing the Compression ratio of compression ignition engine fuelled with esters of crude rice bran oil **2(3)** pp 131-134.
[8] El Boulifi N, Bouaid A, Martinez M, Aracil J 2013 Optimization and oxidative stability of biodiesel production from rice bran oil, Renewable Energy**53** pp 141–147.
[9] Jincheng Huang , Yaodong Wang , Jian-bin Qin, Anthony P. Roskilly 2010 Comparative study of performance and emissions of a diesel engine using Chinese pistache and jatropha biodiesel, Fuel Processing Technology **91** pp 1761–1767.
[10] A. Demirbas 2009 Production of Biodiesel from Algae Oils, Energy Sources, **Part A, 31** pp 163–168.
[11] Ekrem Buyukkaya 2010 Effects of biodiesel on a DI diesel engine performance, emission and combustion characteristics, Fuel **89** pp 3099–3105.
[12] Enweremadu, C. C., and Rutto, H. L 2010, Combustion, Emission and Engine Performance Characteristics of Used Cooking Oil Biodiesel —A Review, Renewable Sustainable Energy Rev. **14(9)** pp 2863–2873.
[13] Agarwal, A. K., 2007 Biofuel (Alcohol and Biodiesel) Applications as 369 Fuels for Internal Combustion Engines," Prog. Energy Combust. Sci., **33(3)** pp 233–271.

IOP Conf. Series: Materials Science and Engineering **197** (2017) 012017 doi:10.1088/1757-899X/197/1/012017

Comparative study of performance and emissions of a CI engine using biodiesel of microalgae, macroalgae and rice bran

J. Jayaprabakar[1*], A. Karthikeyan[2], K. Saikiran[1], N. Beemkumar[1], Nivin Joy[1]

[1]Dept. of Mechanical Engg. Sathyabama University, Chennai, India.
[2]Dept. of Automobile Engg. Sathyabama University, Chennai, India

E-mail: jp21tn@gmail.com

Abstract. Biodiesel is an alternative and safe fuel to replace conventional petroleum diesel. With high-lubricity and clean-burning ability the biodiesel can be a better fuel component for use in existing diesel engines without any modifications. The aim of this Research was to study the potential use of Macro algae oil, Micro algae oil, Rice Bran oil methyl ester as a substitute for diesel fuel in diesel engine. B10 and B20 blends of these three types of fuels are prepared by transesterification process. The blends on volume basis were used to test them in a four stroke single cylinder diesel engine to study the performance and emission characteristics of these fuels and compared with neat diesel fuel. Also, the property testing of these biofuels were carried out. The biodiesel blends in this study substantially reduces the emission of unburnt hydro carbons and smoke opacity and increases the emission of NOx emission in exhaust gases. These biodiesel blends were consumed more by the engine during testing than Diesel and the brake thermal efficiency and volumetric efficiency for the blends was identical with the Diesel.

1. Introduction

Diesel engines are considered as widely used prime movers since hundred years before; these have been used in many areas like generators, automobiles ranging from cars to trucks, ships and locomotives. A diesel engine has high thermal efficiency, less fuel consumptions, reliability and higher durability [1]. The recent uncertainties in petroleum availability and increases in cost of petroleum products made an interest towards vegetable oil derived biofuels for diesel engines [2]. Biodiesel is defined as the fuel which is derived from plant oils such as soybean, rapeseed, Jatropha or animal fats like beef tallow. The most popular method used to produce biodiesel from vegetable oils is the transesterification method [3]. Transesterification is defined as a reaction between the oil and an alcohol with the presence of a catalyst at specified temperature, the reaction produces biodiesel and glycerol. Availability of feedstock , the cost, resulting by products, the government incentives on environment and economical issues are considered to be important factors which decides the competitiveness of any biodiesel to be substitute for the diesel [4]) Biodiesel produced from biomass have already available through commercial markets as blends with diesel up to 20% in many countries [5]. It is always not advisable to use an edible oil source for biodiesel as it affects the food chain. In India more than 50% of the imported oil is used for cooking purpose. In this context the current need is to have an economic and eco friendly biodiesel source. The Non edible oil sources like Jatropha, Pongamia and other non edible oils are treated as potential feed stocks in India. Compared to these non edible oil sources Algae oil and Rice bran oil are considered to be important emerging non edible biodiesel sources due to their low cost and less land area requirement. Algae are photosynthetic organisms that can grow in any aquatic region with size ranges from micrometers to meters. Macro algae are large (size in inches) multi cellular organisms growing in water bodies. Micro algae are very small (size in micrometers) unicellular organisms normally grow in suspension within a water body [6]. The development of biodiesel from Algae biomass is at emerging stage in stage in India [7]. Paddy, the feedstock for rice bran oil is the leading crop in India and also it has greater potential for producing Rice bran oil. When considering the improved income through rural employment and cleaner environment, Rice bran can be considered as a potential biodiesel resource [8]. A study on Rice bran oil biodiesel revealed that, Rice bran oil biodiesel blends recorded improved brake thermal efficiency and reduced pollutants when compared to Diesel [9].This study involved in testing the pure Diesel and its two blends by volume from each fuel type;MAME10 (90% Diesel with 10% Micro algae oil biodiesel), MAME20 (80% Diesel with 20% Micro algae oil biodiesel), AME10 (90% Diesel with 10% Macro algae oil biodiesel), AME20 (80% Diesel with 20% Macro algae oil biodiesel), RME10 (90% Diesel with 10% rice bran oil biodiesel), RME20 (80% Diesel with 20% rice bran oil biodiesel). The performance parameters brake thermal efficiency, volumetric efficiency, brake specific fuel consumption and the unburnt hydro carbon, smoke opacity and nitrogen oxide emissions are measured and discussed in comparison with the pure Diesel.

2. Experimental Setup and Methodology

The required amount of Macro algae biomass was collected from sea shore near Chennai, the micro algae biomass was collected at a fresh water open pond situated at Sathyabama University, Chennai. The collected biomass were cleaned and dried in open sun for 2-3 days. Then dried biomass was pulverized and the powder was crushed again and again with hexane and isoproponal solvent mixture for extracting the lipid in them. Extracted lipid was separated from the biomass and heated for removing the solvent in it. The remaining oil is subjected to transesterification process. Neat refined rice bran oil was procured for the experimental purpose and it was subjected to transesterification process.

Transesterification process is the method used to prepare biodiesel from the oil and it will reduce the viscosity of oil. This work used sodium hydroxide as catalyst for the reaction and methanol as the alcohol. The reaction was held between these two with the three types of oil separately, at 65°C for 3 hours duration and the solution is made to mix well during the reaction. The reaction resulted biodiesel as top layer and glycerol as bottom layer. The top layer was separated after filtering the glycerol by using separator. The prepared biodiesel from the oils were tested for its properties. The properties of the three biodiesel sample were tested by adapting ASTM testing protocols for biodiesel and they are presented in the Table 1.

Table 1. Properties of Biodiesel

Property	Diesel	MAME	AME	RME
Viscosity(cst) at 40 °C	3.9	3.51	4.84	5.045
Cetane Number	49	57	45	46
Calorific Value (kJ/kg)	43200	40362	33300	37081
Flash Point (°C)	58	92	130	124
Fire Point(°C)	64	124	139	147
Specific Gravity	0.804	0.792	0.795	0.8899
Sulfur content	-	-	18	Less than 20 ppm

The Test rig used in this study is a diesel engine of constant speed, four stroke, vertical, air cooled type. The fuel consumption is measured by using a graduated burette for specific time. The normal injection timing and pressure were maintained during the test were 23° BTDC and 200 bar. For measuring the air consumption rate an orifice meter with U-tube manometer is provided along with an air tank on the suction line. A Smoke meter (AVL415) is provided for measuring FSN of exhaust gases. The test rig is installed with AVL software for obtaining various curves and results during operation. The detailed specification of the test rig was presented in the Table 2. The test was done with conventional Diesel and its two blends (B10 and B20) of each biodiesel.5 sets of loading was used for testing from 0% to 100%.

Table 2. Engine specifications

Type	**Kirlosker, Four stroke, single cylinder,**
Rated power	4.4 kW
Rated speed	1500 rpm
Bore dia	87.5 mm
Stroke	110 mm
Compression ratio	17.5:1
Orifice dia	13.4 mm
Coefficient of discharge	0.6
Loading device	Eddy current dynamometer
Make	Benz systems
Dynamometer constant	2000
Supply voltage	240 ±10 % AC, 50 Hz, 1 ϕ
Maximum excitation current	6 to 8 Amp

3. Results and Discussion

The performance and emission characteristics of straight diesel and tested biodiesel blends B10 and B20 are discussed in this section.

3.1. Brake thermal efficiency

Brake thermal efficiency gives the amount heat energy that converted to useful work. This parameter purely depends on calorific value of the fuel and the viscosity. The variations in the brake thermal efficiency values against engine loading for the fuels Diesel, MAME10, MAME20, AME10, AME20, RME10 and RME20 is presented in the figure 1(a).The Average BTE Values for Diesel, MAME10, MAME20, AME10, AME20, RME10 and RME20 at all loads are 26.38, 26.19, 26.14, 20.83, 21.6, 23.52 and 24.23% respectively. For all the fuel types the BTE values reaches its maximum value at 75% load and then they started decreasing. The Biodiesel blends followed similar trend of pure diesel. In some stages the micro algae biodiesel blends MAME 20 and MAME 10 recorded slightly high BTE values than diesel, the kinematic viscosity of MAME blends is less than Diesel as indicated by the property test, it improved the fuel atomization and thus leads to better combustion. The BTE value for this fuel type is almost identical with the pure diesel. The other biodiesel blends have low volatility, slightly superior viscosity, and lower heating value it had shown lower brake thermal efficiency than conventional diesel. The macro algae biodiesel blends AME 10, AME 20 recorded less BTE values than other fuel types. The rice bran biodiesel blends RME 10, RME 20 recorded slightly higher BTE values than AME blends and also the trend of this fuel type is found identical with MAME blends and diesel. At 100% load all the fuel types recorded almost same BTE values.

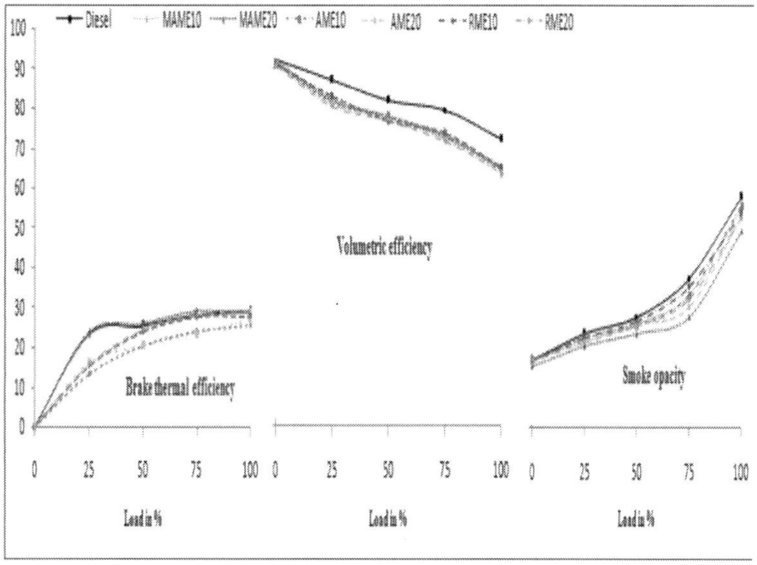

Figure 1. a) Brake thermal efficiency; **b)** volumetric efficiency; and **c)** smoke opacity

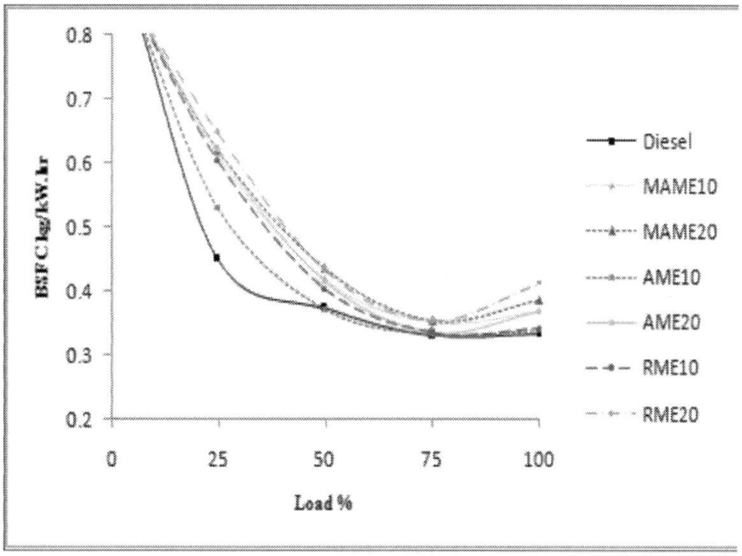

Figure 2. Brake specific fuel consumption in kg/kW-hr against load in %

3.2. Volumetric efficiency

Volumetric efficiency indicates the breathing ability of any air borne engine. It relates the actual and theoretical amount of mixture sucked to engine. The variations in the volumetric efficiency values against engine loading for the fuels Diesel, MAME10, MAME20, AME10, AME20, RME10 and RME20 is presented in the figure 1(b). The Average Volumetric Efficiency Values for the fuels Diesel, MAME10, MAME20, AME10, AME20, RME10 and RME20 at all loads are 82.31, 78.14, 77.62, 77.57, 76.83, 77.93 and 76.69% respectively. The Volumetric Efficiency for all the fuel types decreases with increase in load. The minimum volumetric efficiency value for diesel is 71.83% at 100% load the same for MAME10, MAME20, AME10, AME20, RME10 and RME20 at 100% load are 65.58, 65.07, 64.28, 63.5, 64.97 and 64 % respectively. The maximum Volumetric Efficiency for diesel is 92% occurred at 0% load the same for MAME10, MAME20, AME10, AME20, RME10 and RME20 at 0% load are 91.85, 91.34, 90.87, 90.9, 91.25 and 90.55% respectively. Due to its higher heating value, Diesel has got higher Volumetric Efficiency values than the other fuel types. All the fuel types other than diesel have almost same Volumetric Efficiency values. The difference between their Volumetric Efficiency values is less than 1%.

3.3. Smoke Opacity

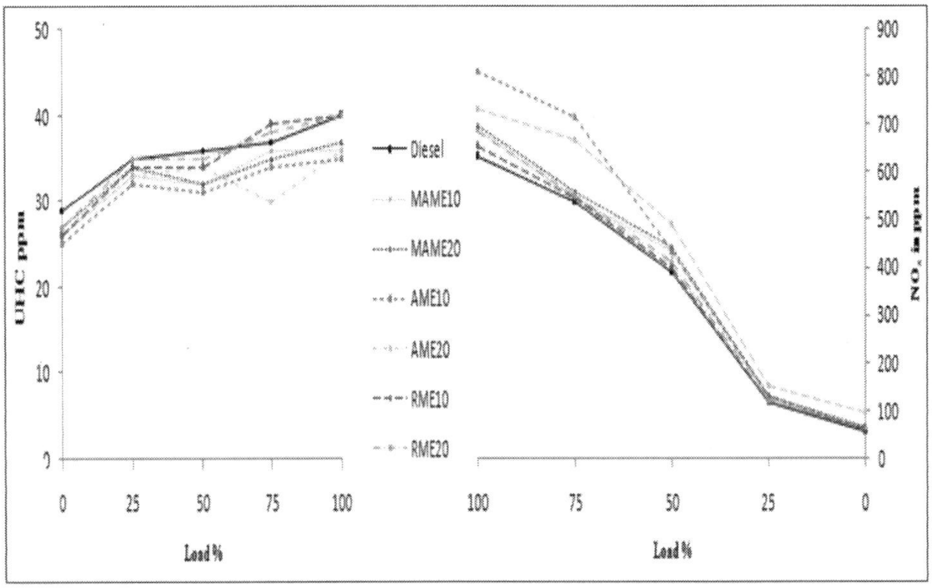

Figure 3. Emission in ppm against load in %, **a)** unburnt hydro carbons and **b)** nitrogen oxides

Smoke in the engine emission indicates how completely the combustion was held. These variations in percentage against engine loading for the fuels Diesel, MAME10, MAME20, AME10, AME20, RME10 and RME20 are presented in the figure 1(C).The Average Smoke Opacity Values for the fuels Diesel, MAME10, MAME20, AME10, AME20, RME10 and RME20 at all loads are 30.6, 28.62, 25.5, 28.8, 2.5,29 and 27.8 % respectively. The smoke opacity for all the fuel types increases with increase in engine load. At 0% load the smoke opacity values for all the fuel types are almost same and those were less than 17%. When the load increases the smoke opacity starts increase and reached to a maximum at 100% load. The average maximum smoke opacity values for all the fuel types is 50.25% .The biodiesel blend MAME 20 has recorded lesser smoke opacity values than the other fuel types at all loads. The RME biodiesel blends show higher smoke opacity values among the six blends but it is still lesser than diesel. Due to the high oxygen content in the biodiesel which helped for complete combustion the blends could record less smoke opacity values than Diesel.

3.4. Brake specific fuel consumption

Brake specific fuel consumption is the amount of fuel consumed to produce one unit power at unit time. It mainly depends upon the fuel property heating value. The variations of BSFC in kg/kW-hr against engine loading for the fuels Diesel, MAME10, MAME20, AME10, AME20, RME10 and RME20 are presented in the figure 2. The Average BSFC Values for the fuels Diesel, MAME10, MAME20, AME10, AME20, RME10 and RME20 at all loads are 0.37, 0.44, 0.45, 0.39, 0.43, 0.42 and 0.46 kg/kW –hr respectively. Compare to the biodiesel blends diesel shows less BSFC values at all the loads due to its high heating value and low density. The BSFC values for all the fuel types decreases with increase in the load. At higher loads the difference in BSFC values between diesel and its blends with three biodiesel types are lesser. The differences between the BSFC values of all the fuel types are greater at lower loads. When comparing between biodiesel blends MAME Blends were consumed more by the engine than the other blends. The AME 10 and RME 10 were very close to diesel. The BSFC values of MAME 10 and MAME 20 blends were almost identical. The BSFC values range from 0.33 kg/kW–hr, 0.64 kg/kW–hr for all the fuel types tested at all loads.

3.5. Un-burnt hydro carbon(UHC) Emissions

The unburnt hydro carbon emission measures the combustion ability of any fuel. The variations of UHC in ppm against engine loading for the fuels Diesel, MAME10, MAME20, AME10, AME20, RME10 and RME20 are presented in the figure 3(a). The average UHC emission values for Diesel, MAME10, MAME20, AME10, AME20, RME10 and RME20 at all loads are 35.4, 32.6, 33, 31.4, 31.8, 34.6 and 35 ppm respectively. These emissions increases as increase in engine load for all the fuel types. As no carbon content in the AME and MAME blends they emitted less UHC emissions than the RME blends. The UHC emissions for RME blends are approximately identical with that of pure diesel. The increase in UHC emission when increasing the engine load is of very small amount. The UHC emissions range from 25ppm to 40 ppm for all the types of fuels at all loads. The lower UHC emissions of the biodiesel blends indicates they had better and complete combustion than the Pure Diesel.

3.6. Nitrogen oxide(NO_x) emissions

The nitrogen oxide emission of any fuel is greatly affected by the oxygen content in the fuel, Cylinder temperature and the fuel retention time during combustion. The variations of NO_x emission in ppm against engine loading for the fuels Diesel, MAME10, MAME20, AME10, AME20, RME10 and RME20 are presented in the figure 3(b).The average NO_X emission values for the fuels Diesel, MAME10, MAME20, AME10, AME20, RME10 and RME20 at all loads are 346,371,379,430,428,355 and 367ppm respectively. Oxygen content in the plant oil derived biodiesel blends are more. When biodiesel is used in the engine, more fuel is sucked and it needs more time to burn it completely, also the complete combustion increases the cylinder temperature. Because of these reasons the NO_x emissions for the biodiesel blends are high when compared to pure diesel. The least NO_X value was found to be 55 ppm emitted by diesel at 0% load. The highest NO_X value is 812 ppm emitted by the AME 10 blend at 100% load. The average values of RME blends are very close to the diesel values. At lower loads all the fuel types emitted almost same NO_X emissions especially at 0% load and 25% load, because at lower engine load cylinder temperature will not be that much higher. The Diesel, RME and MAME blends followed identical trend at all loads.

4. Conclusion

The performance and emission characteristics of the micro algae, macro algae and rice bran oil biodiesel blends in the tested engine prove that the carbon emissions of the biodiesel blends are less compared to Diesel except NOX emission. The engine consumed more fuel when the biodiesel blends are used. The volumetric efficiency values of the blends were slightly lesser than Diesel. The performance test revealed that the biodiesel blends exhibited complete and improved combustion. This study reveals that these types of biodiesel blends can be a better alternate to the pure diesel fuel to be used in a CI Engine without any modification.

References

[1] Jincheng Huang, Yaodong Wang, Jian-bin Qin, Anthony P. Roskilly 2010 Comparative study of performance and emissions of a diesel engine using Chinese pistache and jatropha biodiesel. Fuel Processing Technology 91 pp 1761–1767.

[2] Demirbas A 2009 Production of Biodiesel from Algae Oils. Energy Sources. Part A, 31 pp 163–168.

[3] Ekrem Buyukkaya 2010 Effects of biodiesel on a DI diesel engine performance, emission and combustion characteristics. Fuel 89 pp 3099–3105.

[4] Hanna M A, Loren Isom and John Campbell 2005 Biodiesel:Current perspectives and future. Journal of Scientific and Industrial Research 64 pp 854-857.

[5] Padma Vasudevan, Satyawati Sharma and Ashwani Kumar. Liquid fuel from biomass:An Overview. Journal of Scientific and Industrial Research. 2005; 64, 822-831.

[6] Chang H 2009 Marine Biodiversity and Systematics. Research Programmes. http://www.niwa.cri.nz/rc/prog/marinebiodiversity/obj7.

[7] Mukesh Kumar, Sharma M P, Gaurav Dwivedi 2013 Algae Oil as Future Energy Source in Indian Perspective. International Journal of Renewable Energy Research. 3(4) pp 913-921.

[8] Janahiraman Krishna Kumar, Karupannan V S, Venkatachalapathy and Sellappan Elancheliyan 2008 Technical aspects of Biodiesel production from vegetable oils. Thermal Science 12(2) pp159-169.

[9] Kandu Kalpatti Chinnaraj Velappan 2001 Less NOx biodiesel: CI engine studies fuelled with rice bran oil biodiesel and its five blends. Journal of scientific and Industrial Research 66 pp 60-71.

IOP Conf. Series: Materials Science and Engineering **197** (2017) 012018 doi:10.1088/1757-899X/197/1/012018

Experimental investigation on Performance and Emission Characteristics of J20, P20, N20 Biodiesel blends and Sound Characteristics of P20 Biodiesel blend Used in Single Cylinder Diesel Engine

R.rajasekar[1*], N.karthik[2], Goldwin Xavier[3]

[1*, 2] Assistant Professor, Faculty of Mechanical Engineering, Sathyabama university, Chennai
[3] Assistant Professor, Department of Mechanical Engineering, Loyola Institute of Technology, Chennai
*sekarauto@gmail.com

Abstract-Present work provides the effect of biodiesel blends and Sound Characteristics of P20 Biodiesel blend compared with Performance and emission Characteristics of diesel. Methods and analysis biodiesel blends was prepared by the Transesterification Process. Experiments were conducted in single cylinder constant speed direct injection diesel engine for various test fuels. Research is mainly focused on pongamia oil .It was observed that a 20% Pongamia oil blends and its properties were similar to diesel. The results showed that 20% Pongamia oil blends gave better performance, less in noise and emission compared with ester of Jatropha and neem oil blends. Hence Pongamia blends can be used in existing diesel engine without compromising the engine performance.

1. Introduction:

The demand of petrodiesel is resend days due to depletion of fossil fuels. The substitute source of vegetable oil biodiesel blends can be used in CI engines. The various properties and performance of different biodiesel blends of cotton seed methyl ester and neem oil methyl ester compared to diesel fuel. C20 biodiesel blend results are closer performance to diesel [1]. Biodiesel has been found as an alternative fuel for compression ignition engine because of increased performance and reduced emission. The waste cooking oil methyl ester (WCO) biodiesel has similar results to that of diesel. Brake thermal efficiency, carbon monoxide emission, hydrocarbon emission, and smoke emission are lower compared to pure diesel and specific energy consumption and oxides of nitrogen is higher compared to diesel [2].The depleting petroleum fuel resources increase the price of fuel continuously. The various biodiesel fuels identified and tested successfully. The performance and emissions of different rubber seed oil (RSO) biodiesel blend compared with diesel. B30 biodiesel blend are closer performance and lower emission compared to diesel [3].Direct use of neem oil is problem for diesel engine because of very high viscosity. Esterified neem oil different biodiesel blends are tested in compression ignition engine. The experiment concluded that B20 biodiesel blend gave better performance and lower emission because of extra amount of oxygen present in the blend [4] [5] various properties of B20 biodiesel blends shown in table1.

IOP Conf. Series: Materials Science and Engineering **197** (2017) 012018 doi:10.1088/1757-899X/197/1/012018

Table1. Properties of Diesel and Bio-diesel Oils

PROPERTIES	DIESEL	J20	P20	N20
Calorific Value(kJ/kg)	43000	41600	41300	35125
Flash Point (0c)	50	57	56	68
Viscosity(cst)	3.9	4.3	4.1	4.7
Density (kg/m3)	840	860	850	920
Cetane number	50	54	52	57

1.1 Procedure For Transesterification Process:

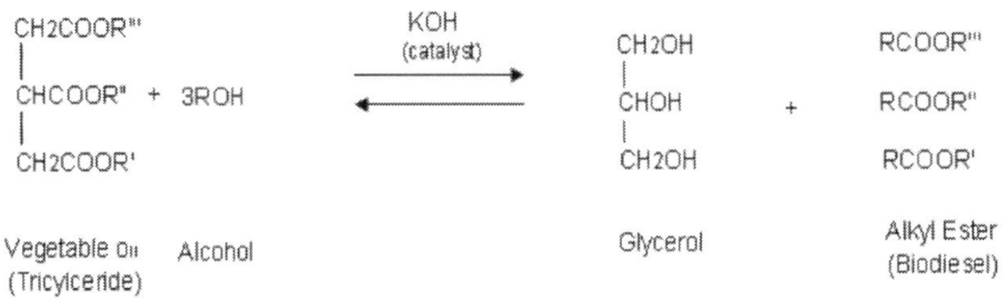

The required amount of catalyst pellets (NaOH / KOH) is quickly weighed, shielding it as much as possible from atmospheric moisture and carbon di-oxide. The pellets are quickly transferred to the dry mixture grinder to convert into powder form. The powdered form of catalyst and solvents (methanol) solution is then vigorously shaken in a conical flask for homogeneous mixing. At this point dissolved catalyst (NaOH) is presumed to have been converted into potassium/sodium/ ethoxide/ methoxide. A fixed quantity of solvent and catalyst solution is poured into the oil to mark the beginning of the reaction. Once the reaction reaches the preset reaction time, heating and stirring are stopped. Once the reaction period is over, the product is permitted to settle overnight. Two different liquid phases are identified. Crude esters at the top and glycerol phase at bottom. That is bio diesel process turns the oil into esters, extrication out the glycerol. The glycerol sinks at the bottom and the bio diesel float on top and can be siphoned off. The crude ester phase is separated and glycerol phase is then washed by warm de-ionized (double distilled) water for four times and water becomes clear. So no need for further wash. The excess methanol and water in the ester phase is removed (or recycled) by evaporation under atmospheric condition after being measured for product yield calculation.

1.2 Description of Engine setup:

Figure 1. Experimental setup

The Kirloskar engine is one of the widely used engines in agricultural, transport-vehicles, pump sets, small and medium scale commercial purposes. Further, the necessary modification on the cylinder head and piston crown can be easily carried out in this type of engine. Hence this engine is selected for present research work.

A single cylinder, four stroke compression ignition engines with a displacement volume of 661 cc, compression ratio of 17.5:1, and developing 5.9kW at 1800 rpm was used for the present study. The specifications of the engine are given in table 2. Variable load tests are conducted for no load, 25%, 50%, 75% and 100% power output at a constant rated speed of 1800 rpm, with fuel injection pressure of 200 bars and cooling water exit temperature at 600 C. The injection timing recommended by the manufacturer was 270 bTDC. The governor was used to maintain constant speed under varying load conditions. Experimental setup is shown in fig1.2 and engine specifications shown in table 2.

Table2. Technical specifications of engine

Manufacturer	Kirloskar oil engines ltd, pune, india
Type of Engine	Vertical,4Stroke cycle, Single acting, Single cylinder, high speed compression ignition diesel engine
Speed	1800 rpm
Rating at 1500 rpm	5.2 kw
Bore (B) & Stroke (S)	87.5 & 110
Compression Ratio	17.5:1
Fuel Injection Timing	27° BTDC
Method of Cooling	Water cooling
Injection pressure	200-205 bar

2. Result and Discussion:

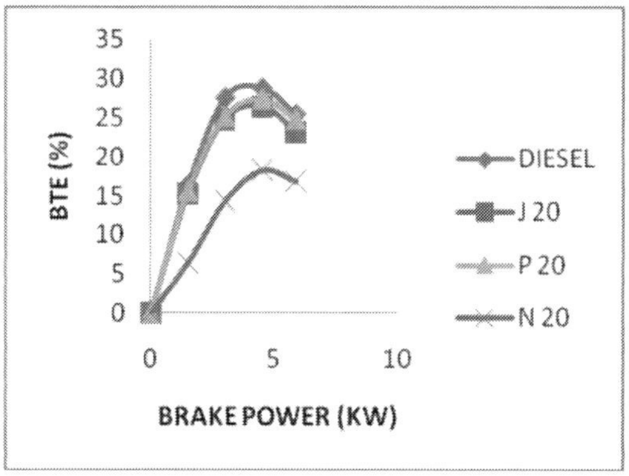

Figure 2. Variation of BTE with BP

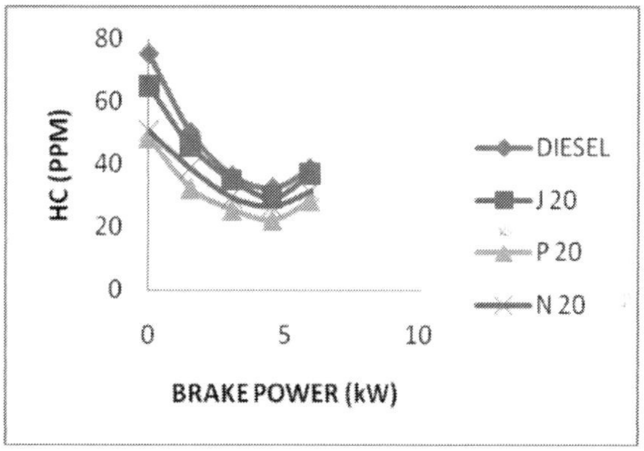

Figure 3. Variation of HC with BP

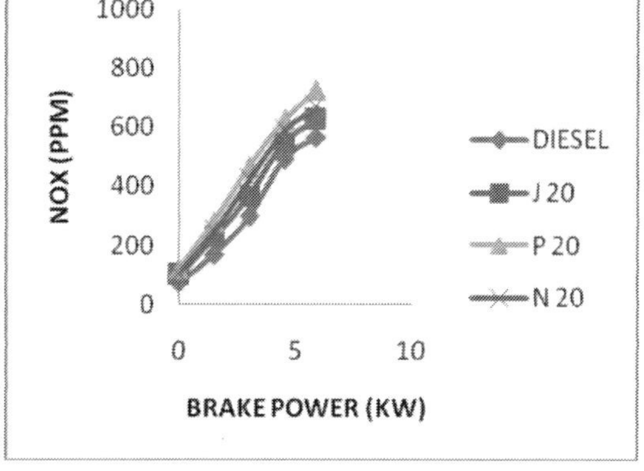

Figure 4. Variation of NOX with BP

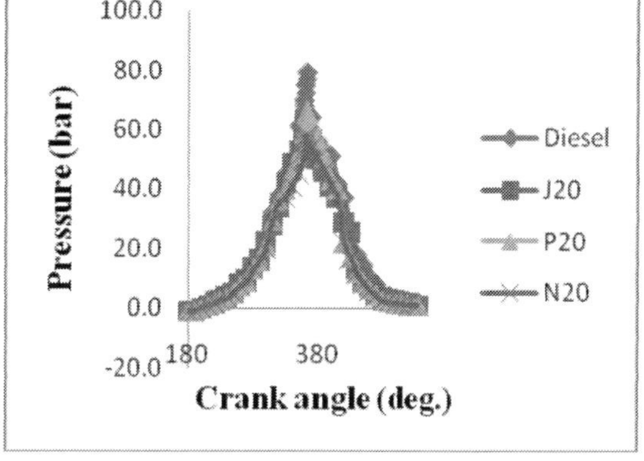

Figure 5. Comparison of Pressure with crank angle

In Figure 2, a slight drop in efficiency was found with methyl esters (biodiesel) when compared with pure diesel. This drop in thermal efficiency must be attributed to the poor combustion characteristics of methyl esters due to high viscosity. It was experimental that the brake thermal efficiency of B20 methyl ester of Pongamia methyl ester (MEOP) had better brake thermal efficiency than compared with the methyl esters of Jatropha and Neem. So B20 can be suggested as best blend for biodiesel preparation with Pongamia oil.
Figure 3 shows the comparison of HC emissions for diesel, J20, P20 and N20 at standard engine specification. The experimental results show that for all the blends, the HC emissions were lower at part load, but increases with increase in load. The HC emission for pure diesel at maximum load was 38.7 ppm, where it was decreased by 4.7%, 26.1% and18.61% for J20, P20and N20 blend. This may be due to the fact that all the biodiesel contain oxygen in their chemical composition. This favors comparatively better combustion for biodiesel when compared with diesel
Figure 4 shows the comparison of NOX emissions for pure diesel, J20, P20 and N20 at standard engine specification. The experimental results show that for all the blends, the NOX emissions were lower at part load, but increases with increase in load. As stated, the presence of oxygen in the biodiesel has led to complete combustion of biodiesel better than diesel. As a result, the adiabatic flame temperature inside the cylinder is more in case of biodiesel than pure diesel. Hence this catalyzes the reactions for oxidation of nitrogen and hence NOx emissions are more for diesel. The NOx emission for diesel at maximum load was 567ppm, where as it is more by 10.58%, 28.04% and 17.98 % for J20, P20 and N20 blend. This is due to the fact that all the biodiesels contain little nitrogen when compared with pure diesel.
Figure 5 shows the comparison of cylinder pressure for pure diesel, J20, P20 and N20 at standard engine specification. The Peak pressure for neat diesel was 79.24 bars, whereas at decreases by 20.24%, 16.45% and 26.57% for J20, P20 and J20 blend. This is due to the increased calorific value and decreased viscosity of diesel compared to biodiesel which leads to complete combustion. Among the bio diesel blends, the peak pressure was obtained for P20 blend. This may be due to the reduced viscosity and increased calorific value which leads to better combustion compared to other biodiesel blends.

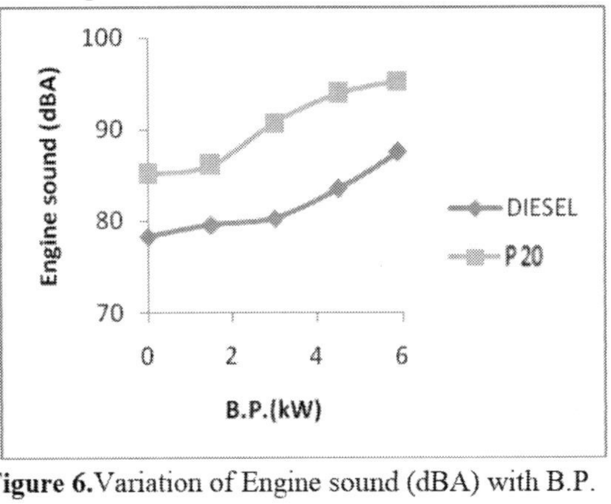

Figure 6. Variation of Engine sound (dBA) with B.P.

Figures 2.5 show the variations of engine sound in degree dBA with brake power for MEOP and its blends with diesel at an injection timing of 270bTDC .In general, the engine sound level is increased with increase in load. The engine sound level at 270bTDC is 95.3 dBA. This may be due to large drop let size of the fuel compared to diesel.

3. Conclusions:

From the conclusions of present investigation are drawn, Brake thermal efficiency of P20 bio-diesel is very close to pure diesel because fuel properties like viscosity, flash and fire point, calorific value are closer to diesel. HC emission for P20 blend is very less when compared to pure diesel. This may be due to higher oxygen content of bio-diesel. NOx emission for P20 blend is very high. This may be due to high adiabatic flame temperature at the end of combustion and higher oxygen content compared to pure diesel. Peak pressure for P20 blend is very close to diesel because of slightly higher viscosity and lower calorific value compared to pure diesel. Noise of P20 blend is lower compared to diesel, because of slower heat release rate and lower max pressure rise compared to pure diesel.

IOP Conf. Series: Materials Science and Engineering **197** (2017) 012018 doi:10.1088/1757-899X/197/1/012018

References:

[1] K.Dilip Kumar, P.Ravindra Kumar 2012. Experimental Investigation of Cotton Seed Oil and Neem Methyl Esters as Biodiesel on CI Engine, *International Journal of Modern Engineering Research* **2**, pp-1741-1746

[2] K.Nantha Gopal, Arindam Pal et al, 2014.Investigation of emissions and combustion characteristics of a CI engine fueled with waste cooking oil methyl ester and diesel blends, *Alexandria Engineering Journal* **53**, 281–287

[3] B.Dinesh Babu1, R.Vidya Sagar Raju et al, 2014. Experimental Investigation of Rubber Seed Oil as Biodiesel on C.I. Engine, *International Journal of Engineering Research* **3**, pp: 58-64

[4] Nishant Tyagi1, Ambuj Sharma, 2012. Experimental Investigation of Neem Methyl Esters as Biodiesel on C.I Engine, *International Journal of Engineering Research and Applications* **2**, pp.1673-1679.

[5] N. Karthik and S. Mahalingam Emission,2014. Characteristics on Single Cylinder Constant Speed Diesel Engine using Diesel Blends with Jatropha and Rubber Seed Oil, *International Journal of Applied Engineering Research* **21** pp. 11373-11380.

Review on Enhanced Heat Transfer Techniques using Modern Technologies for 4S Air Cooled Engines

S Ramasubramanian [1] and V K Bupesh Raja[2]

[1] Department of Mechanical Engineering, Vels University, Chennai, India
[2]Professor and Head, Department of Automobile Engineering, Sathyabama University Chennai, India.

E-mail: raams87@gmail.com

Abstract: Engine performance is a biggest challenge and a vital area of concern when it comes to automobiles. Researchers across the globe have been working decades together meticulously improvising the performance of engine in terms of efficiency. The durability of the engine components mainly depends on the thermal stress it undergoes over the period of operation. Air cooling of engine is the simplest and most desirous technique that has been adopted for ages. In this regard fins or extended surfaces are employed for effective cooling of the cylinder while in operation. The conductive and convective heat transfer rate from the cylinder to the fins and in turn from the fins to surrounding ambience determines the effective performance of the engine. In this paper an attempt is made to review and summarize the various researches that were conducted on the Fins in terms of profile geometry, number of fins, size, thickness factor, material used etc., and to bring about a long term solution with the modern technologies like nano coatings and nano materials.

1. Introduction

Automobiles are a fascinating and most attractive industry around the world. The competitiveness in developing new technologies with modern innovations for meeting the standards and expectation of the user has reached varied horizons. Performance of the automobile mainly depends on the efficiency of the engine. The thermal stresses developed on the engine have great impact on all the components that is associated with the combustion chamber. To increase the durability of the engine components and there by an efficient performance of the engine it become very essential to quickly dissipate the heat generated during combustion. Typical air cooled engines use fins or the extended surfaces along its periphery for this purpose.[1] Heat is removed from the inside of the cylinder through conductive heat transfer to the fins and from fins to the surrounding through convective heat transfer. Hence the heat transfer rate plays a very important role in maintaining the required engine temperature. Extensive works carried out in the past shows significant results in achieving better heat transfer rate. The availability of software like Pro-E ANSYS, MATLAB, has greatly helped in understanding the thermal distribution for various fin geometry with different material compositions enabling effective optimization of fins for a particular engine. In addition, the recent advancements in nano materials and its suitability to extend its application for engine has opened a new dimension for the researchers working on the engine cooling systems.

2. Literature Review

G. Babu and M. Lavakumar [2], have worked on a heat transfer analysis for optimization of engine cylinder fins by varying the geometry and material. Experimentation has been conducted by designing a cylinder fin for the 100cc Hero Honda Motorcycle and modeling was done in parametric 3D modeling software Pro/Engineer. The fin material Aluminum alloy 204 was replaced with Aluminum alloy 6061 and magnesium alloy. The profile of the fin was modified from rectangular to circular and curve shaped. The default thickness of fin was reduced from 3mm to 2.5mm. It is found that efficiency is increased by reducing the thickness and the fin profile is changed to curve shape, thereby reducing the weight of the fin body. The weight of the fin body is said to have reduced further with the use of Magnesium alloy. Thermal analysis on the fin is made by varying materials, geometry and thickness. From the analysis results it was observed that circular fin made of Aluminium alloy 6061 and thickness of 2.5mm is better since heat transfer rate is more. But by using circular fins the weight of the fin body was found increasing. In terms of weight reduction curved fins are found effective than other geometries. So it was concluded that aluminium alloy 6061 with reduced thickness of 2.5 mm and curved shape fins are very suitable for the increased heat transfer rate of the selected engine. The Figure1&2 shows the variation of the temperature distribution over the rectangular and circular fin profile for the same engine configuration.

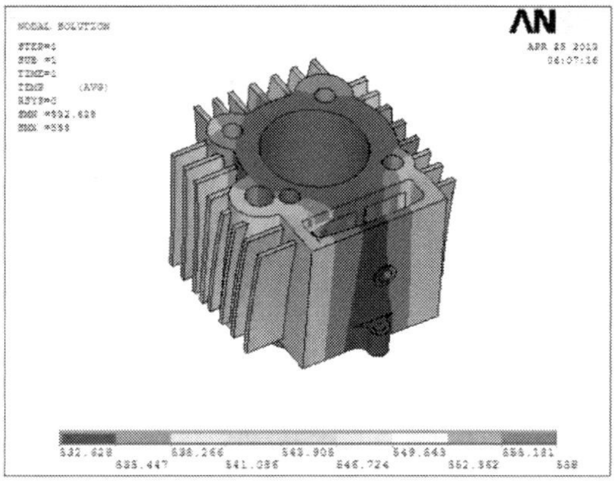

Figure 1. Temperature distribution over a rectangular fin

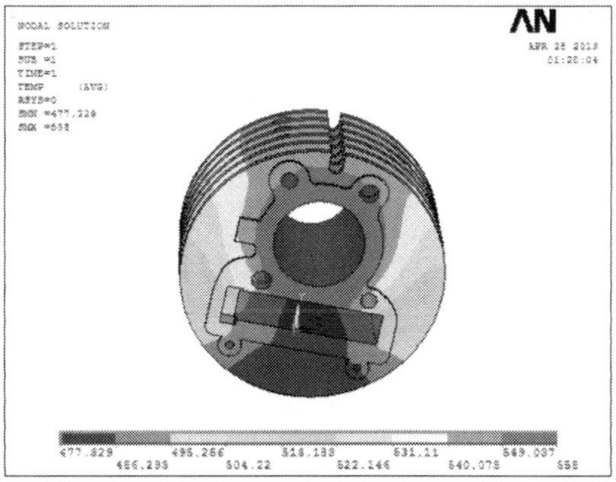

Figure 2. Temperature distribution over a circular fin

Mohsin A .Ali, Prof. S.M Kherde [3] have concluded from their review on design modification of two wheeler fins that, design of fin plays an important role in heat transfer between the cylinder to the surroundings. The fin geometry and the area of cross section affect the heat transfer coefficient to a larger extent.

In high speed vehicles, thicker fins produced swirls which increased the heat transfer and thereby increasing the efficiency of the fin. Also large number of fins with lesser thickness in high speed engines than thick fins with lesser number helped in higher heat transfer by inducing greater turbulence. Contact time between air flow and fin is also found to be an important factor for such heat transfer.

Denpong Soodphakdee et.al [4] has compared the heat transfer performance by adopting various fin geometries. The fin geometry taken for study is plate fins or pin fins. The plate fins can be continuous parallel plates or staggered. A circular array of 1mm diameter pin fins with a 2mm pitch is taken as a basis for comparison. It is observed that the ratio of solid to fluid thermal conductivity for aluminium and air is greater, this led to modeling of fins as isothermal surfaces than the conjugate solid. The two-dimensional CFD simulations were conducted on planes having symmetry parallel to the flow. The air approach velocity is in the range of 0.5 to 5m/s. It has been observed that the staggered plate fin geometry has higher heat transfer for the given combination of pressure gradient and flow rate.

N. Phani Raja Rao et.al. [5] typically analyzed the thermal properties by varying thickness, geometry and material of cylinder fins. From the results the thickness is reduced and the fin shape is changed to circular which increased the efficiency and heat transfer rate of fin. By using Aluminum Alloy 6061 as circular fin material the heat transfer rate is better with higher fin efficiency and effectiveness.

S.S. Chandrakant et.al.[6] has made extensive experiments on rectangular and triangular fin profiles with a varied air velocities ranging from 0 to 11 m/s. Experiments carried out and CFD simulation result proves that annular fins with rectangular fin profiles have optimal heat transfer when compared to triangular fin profiles. It is observed that at different air velocity the triangular fin surface temperature is greater than rectangular fin. The rectangular fin has higher heat transfer coefficient than triangular fin. The comparison resulted in rectangular profile being more efficient than the triangular profile.

Pulkit Agarwal et.al [7] in their experimental investigation has simulated the heat transfer in motor-cycle engine fins using CFD analysis. During investigation it was found that, the surrounding ambient temperature played a important role in engine efficiency. The heat transfer rate was very higher at low ambient temperatures, which in turn increased the heat transfer to a larger extent which caused over cooling of the. This resulted in excess fuel consumption and hence need to reduce the air velocity striking the engine surface. In this aspect a diffuser is placed in front of the engine to reduce the relative velocity of the air stream thus decreasing the heat loss.

Bassam A and K Abu Hijleh [8] has made investigation of a cross flow forced heat transfer for a horizontal cylinder that has permeable fins which are equally spaced multiple layout on the outer surface. These fins exhibit higher heat transfer rate than the traditional fins of similar structure. The aerodynamic and thermal wakes caused in permeable fins is reduced by downstream fins, when θ is below 90°. A better convection heat transfer from the cylinder can be obtained by using a long permeable fin.

L. Dialameh et. al. [9] have conducted a numerical study to predict natural convection from an array of aluminum horizontal rectangular thick fins of 3 mm < t < 7 mm with short lengths (L 6 50 mm) attached on a horizontal base plate. Finite volume scheme is used to solve the three-dimensional elliptic governing equations of laminar flow and heat transfer. Two types of flow patterns in the channel of the fin arrays were observed based on the model verified, fluid flow and thermal structure. Effect of various fin geometries and temperature differences on the convection heat transfer from the array was determined for Rayleigh numbers based on fin spacing of 192–6784. Then correlations were developed to predict the Nusselt numbers with reference to non-dimensional parameters. It is concluded that natural convection heat transfer coefficient increases with increasing temperature differences and increases with fin spacing and decreases with fin length.

Mr. A. Raj Kumar et. al. [10] has made an investigation in which the transient analysis is performed for a engine running at 6000rpm for a minute. In the initial stage temperature distribution in the fin area is analyzed and thermal analysis is done. In the next stage the obtained thermal load is used for doing the structural analysis. It is found that the maximum stress of 386.094 MPa for the material A413.0 and maximum stress of 363.354 MPa for the material C443.0. The material B390.0 has more Factory of Safety better than the other two materials, so it is the best material.

Wladyslaw Mitianiec et. al. [11] have dealt with calculation of the thermal loads and temperature distribution of the cylinder and cylinder head of the two-stroke engine 115 cm^3 capacity cooled by air at mean engine load. The meshing is done in CATIA and simulation is carried out in ANSYS program. The simulation results were verified with the experimental results. The results showed that the cooling heat in the air-cooled two-stroke engine increases with the engine speed and cooling energy amounts above 30% of the total energy delivered with fuel. Also, spark plug and the outer ribs which experience the highest temperature in SI two-stroke can be lessened in order to decrease the weight of the parts.

Mohd Faizal Mohideen Batcha et. al. [12] in their investigation on Mist cooling, have found dissipating heat fluxes with low coolant mass fluxes. The temperature distribution is analyzed for plate fins. The fins are analyzed by giving actual conditions on a cylinder head by subjecting to constant heat flux and there by cooling by applying airflow at velocities in the range of 21.5 m/s to 40.4 m/s. The amount of heat transferred is measured from the temperature distribution in the fin at different locations. On repeated experimentation with the injection of mist a comparison is made for the augmentation of heat transfer. The heat transfer coefficients increased between 200 and 400%, the surface temperatures were depressed. Therefore it is concluded that during critical conditions mist-cooling is adopted to enhance the heat transfer and lower the temperatures in air-cooled systems.

Benjamin pineel [13] in his experimental investigation have analyzed the heat-transfer processes of an air-cooled engine for the purpose of determining the manner in which the various engine and cooling conditions combine to determine the cylinder temperature. Investigation was conducted on the cylinders of Pratt & Whitney 1535 and a Pratt & Whitney 1340H. It is concluded that the values of the effective gas temperatures were practically independent of the engine speed and brake mean effective pressure. The normal range of operation were equal to 1,150° F. for the head and 600° F for the barrel for both the Pratt & Whitney 1535 and the 1340-U cylinders. A suggestion to incorporate turbulence devices in front of the cylinder to increase in the heat-transfer coefficient of the order of 30 percent for the same pressure drop is made.

S. Pashah, Syed et al [14] have studied the effect of coating layer on variable profile composite annular fins using a non-dimensional finite element formulation. The non-linear relationship between the temperature and relative humidity is done by piecewise linear approximation. The coating effects are completely governed by a single dimensionless parameter Bic. A limiting value of Bic ¼ 1 is identified for which the coating effects become negligible. Bic 1 results show a minor increase in fin thermal performance. Further, finite element formulation can be used to study any arbitrary fin profile and fin type under fully or partially wet condition.

S. Pashah , A.F.M. Arif ,et al [15] have studied the effects of coating layer and interface resistance on thermal performance of variable thickness composite annular fins. The obtained results are in dimensionless form, by using the governing parameters for dimensionless system the coating layer effect is governed by single dimensionless parameter. This dimensionless parameter groups important material and geometry parameters of coating and fin substrate. The coating degrading effects is insignificant for contact Biot number. The interface resistance effects are governed by a single dimensionless parameter Bi which groups the important material and geometry parameters of interface and fin substrate. For identical value of Biot number the interface resistance effect is more important than the coating layer resistance effect.

3. CONCLUSION

From the literature review done it is very predominant that the conductive and convective heat transfer rate of the fins varies with the material, profile geometry, number of fins, pitches, coating provided etc., in other hand changing the existing fin profiles with complicated geometry involves tedious techniques. No significant research has been conducted so far with the fins coated with nano materials. The complexity in design and inadequate technology in the past to enable nano coating on the fins could be a reason for the researchers to keep away from this aspect. With recent advancements in the field of nano technology, the plasma spray coating techniques have come as a solution for such complex applications. The nano coated fins with suitable high thermal conducting material could thus become a long standing solution and a most effective means of enhanced heat transfer method for air cooled engines.

4. REFERENCES

[1] Arvind S, Sorathiya A , Sathishkumar N Parar and Prof.Pravin P Rathod 2014 Review paper on Effect of

IOP Conf. Series: Materials Science and Engineering **197** (2017) 012019 doi:10.1088/1757-899X/197/1/012019

Cylinder Block Fin Geometry on Heat Transfer Rate of air – cooled 4S SI Engine 2(1) 2347-6435

[2] Babu G and Lavakumar M 2013 Heat transfer analysis and optimization of engine cylinder fins of varying geometry and material *IOSR Journal of Mechanical and Civil Engineering (IOSR-JMCE)* **7(4)** *24-29*

[3] Mohsin A .Ali, Prof. Kherde S M 2014 Design Modification and Analysis of Two Wheeler Cooling Fins- A Review **1**.

[4] Denpong Soodphakdee, et al. 2001 A Comparison of Fin Geometries for Heat sinks in Laminar Forced Convection Part 1 - Round, Elliptical, and Plate Fins in Staggered and In-Line Configurations *The International Journal of Microcircuits and Electronic Packaging* **24(1)**.

[5] Rao N P R and Vardhan T V 2013 Thermal Analysis of Engine Cylinder Fins By Varying Its Geometry and Material *International Journal of Engineering* **2(8).**

[6] Chandrakant S S et al. 2013 Numerical and Experimental Analysis of Heat Transfer through Various Types of Fin Profiles by Forced Convection *International Journal of Engineering Research & Technology (IJERT)* **2(7)**.

[7] Agarwal P et al. 2011 Heat Transfer imulation by CFD from Fins of an Air Cooled Motorcycle Engine under Varying Climatic Conditions *Proceedings of the World Congress on Engineering*.

[8] Bassam A and Hijleh K A 2003 Enhanced Forced Convection Heat Transfer from a Cylinder Using Permeable Fins *ASME Journal of Heat Transfer* **125**.

[9] Dialameh L, Yaghoubi M and Abouali O 2008 Natural convection from an array of horizontal rectangular thick fins with short length *Applied Thermal Engineering* **28.**

[10] Raj Kumar A, Janardhana Raju G and Amarendar Rao G 2013 Heat Transfer Analysis In The Cylinder Head Of A Four-Stroke Si Engine *International Journal of Engineering Research & Technology (IJERT)* **2(5)**

[11] Wladyslaw Mitianiec and Konrad Buczek 2007 Analysis of thermal loads on air cooled engine, Journal of KONES Powertrain and Transport **14(3)**

[12] Mohd Faizal Mohideen Batcha, Amer Isyraqi Hussin and Vijay Raghavan R 2008 Engine cylinder head cooling enhancement by mist cooling- A simulation study *Proceedings of International Conference on Mechanical & Manufacturing Engineering (ICME2008)*

[13] Benjamin pineel (1937) Heat transfer process in air cooled engine *Langley Memorial aeronautical laboratory, National advisory committee for aeronotics Langley Field.*

[14] Pashah S, Syed M Zubair and Arif A F M Study of coating effects on variable profile annular fins when subjected to dehumidifying operating conditions *International Journal of Refrigeration*

[15] Pashaha S, Arif A F M, Syed M. Zubair and Mustafa M T The effect of coating and interface resistance on thermal performance of variable thickness annular composite fins

IOP Conf. Series: Materials Science and Engineering **197** (2017) 012020 doi:10.1088/1757-899X/197/1/012020

Experimental Investigations on Diesel engine using Methyl esters of Jatropha oil and fish oil

A Karthikeyan[1*], J Jayaprabakar[2]& Richard Dude Williams[2]

[1]Dept. of Automobile Engg., Sathyabama University, Chennai, India.
[2]Dept. of Mechanical Engg., Sathyabama University, Chennai, India

E-mail: naveenkarthik2004@gmail.com

Abstract. The aim of the study is to use fish oil methyl ester (FME) and Jatropha oil methyl ester (JME) as a substitute for diesel in compression ignition engine.Experiments were conducted when the engine was fuelled with Diesel, Fish oil methyl ester and Jatropha oil methyl ester. The experiment covered a range of loads. An AVL smoke meter was used to measure the smoke density in HSU (Hatridge Smoke Unit). The exhaust emissions were measured using exhaust gas analyzer. High volume sampler was employed to measure the particulate matter in exhaust.The performance of the engine was evaluated in terms of brake specific fuel consumption, brake thermal efficiency. The combustion characteristics of the engine were studied in terms of cylinder pressure with respect to crank angle. The emissions of the engine were studied in terms of concentration of CO, NOx, particulate matter and smoke density.The results obtained for Fish oil methyl ester, Jatropha oil methyl ester, were compared with the results of diesel. Bio-diesel, which can be used as an alternate diesel fuel, is made from vegetable oil and animal fats. It is renewable, non-toxic and possesses low emission profiles.

1. Introduction

Fluctuations in the fuel price and limited number of oil reserves are the causes of for the arise of alternative fuel technology. In the current context, India is producing only 30% of petroleum fuels requirement. The rest 70% is being imported1. Biodiesel is a form of diesel fuel manufactured from vegetable oils and animal fats. It is safe, biodegradable and produces less air pollutants than petroleum-based diesel. Biodiesel is a liquid fuel made up of fatty acid alkyl esters, fatty acid methyl esters (or) long-chain mono alkyl esters. Like conventional diesel, biodiesel is used in compression-ignition engines. Any fatty acid source may be used to prepare biodiesel. Thus, any animal or plant lipid should be a ready substrate for the production of biodiesel. The use of edible vegetable oils and animal fats for biodiesel production has recently been of great concern because they compete with food materials.

There are several countries has already started to substitute biodiesel for the conventional diesel. India is also one among them. So many plant sources are available to get oil from it for the production of Biodiesel. Some of them are Jatropha, Maize, Soybean, UCO (used cooking oil), Cotton seed and Rape seed. Biodiesel- the substitute has the fuel properties very close to the conventional Diesel. There are literatures which reported that the output power for biodiesel is almost equal to diesel. But the Specific fuel consumption for biodiesel is higher than diesel. This is due to its lower heating value [2]. Transesterification is one the popular techniques used to convert the oil in to biodiesel, it also enhance the properties of plant oils to meet the requirements of a diesel engine [3,11].

IOP Conf. Series: Materials Science and Engineering **197** (2017) 012020 doi:10.1088/1757-899X/197/1/012020

Jatropha oil is having promising career as a biodiesel source. Tests were already conducted on the CI engines running with Jatropha oil blends with Diesel. While using as dual fuel mode with Diesel it has recorded brake thermal efficiency very closer to pure Diesel as well less smoke [4]. Most research on biodiesel has focused on using plant based oils as feed stocks. There has been much less works on converting animal-based oils into biodiesel [5]. One potential source of oil is fish oil. A large amount of fish by-products are generated from the fishing industry. Some of these products are converted into fish meal and oil, but approximately 60% are not utilized [6,9]. The oil contained in the discarded fish parts can be extracted and better converted in to fish oil methyl ester (Biodiesel) by transesterification process [7].

2. Experimental Setup and Methodology

The work used crude fish oil extracted from the lipid content present in the soap stock of discarded parts of different marine fish. The soap stock was treated by series of processes, i) cooking it thoroughly in boiling water, ii) squeezing it and iii) centrifugally separating the product. The crude fish oil of the marine fish was brown and contains many impurities, such as water, fish residue and saline compounds. To remove these impurities the lipids was refined by a set of pre-treatment processes. These processes included the absorption of the fish residue by active clay, winterizing at 4°C for 2 hrs, centrifuging for 10 min at 3000 rpm to remove the solid impurities, washing with water by 5 vol. % distilled water for 15 min, and heating to 105°C for 30 min [6,7]. The refined fish oil then used to produce biodiesel.

Table 1. Fuel properties

Properties	Diesel	Fish oil methyl ester	Jatropha oil methyl ester
Kinematic viscosity at 40°C (Cst)	3.52	4.96	5.4
Density at 15°C (kg/m^3)	830	850	870
Flash point (°C)	49	162	169
Cetane number	50	51	53
Calorific value (kJ/kg)	43000	37800	38450
Total sulphur (% by mass)	0.01	0.05	Nil
Distillation (% by volume)	-	90 at 333°C	90 at 330°C
Ash (% by mass)	0.01	Nil	0.03
Oil ester (biodiesel) %	-	89.96	90.69
Carbon residue	0.1	0.47	0.35

The refined fish oil was transesterified with methyl alcohol to produce biodiesel. Sodium methoxide was prepared by mixing 25% (by volume of oil) of pure methanol and 6.25g (Per liter of oil) of sodium hydroxide (NaOH). The sodium methoxide, which played a role of enhancing the transesterification process was poured in to a reaction vessel and mixed with the refined fish oil. To prevent the methanol from being vaporized by the reactant mixture, the reacting temperature was fixed at 60ºC, which is just below the boiling point of methanol at 63ºC [8]. The mixing was done for 50 to 60 minutes. Then the mixture was poured in to a separating funnel and it was allowed to rest for 12 hours. The glycerin was formed at the bottom and the methyl ester at the top. Methyl ester was separated from the glycerin which is called biodiesel [10]. Finally, it was washed and dried to remove the excess alcohol. The properties of the tested fuel are presented in Table 1.

A vertical water cooled single cylinder four stroke, direct injection diesel engine was used for this study. The engine was coupled with eddy current dynamometer for load measurement. The smoke density was measured using AVL smoke meter. Exhaust emissions were measured using exhaust gas analyzer. Particulate matter was measured using high volume sampler. Experiments were conducted with pure diesel, fish oil methyl ester (FME) and Jatropha oil methyl ester (JME). The experiment covered a range of loads.

3. Results and Discussion

The Combustion, performance and emission parameters of the engine with the considered biodiesel blends are analyzed here. The results of Fish oil and Jatropha oil biodiesel blends are compared with the results of diesel.

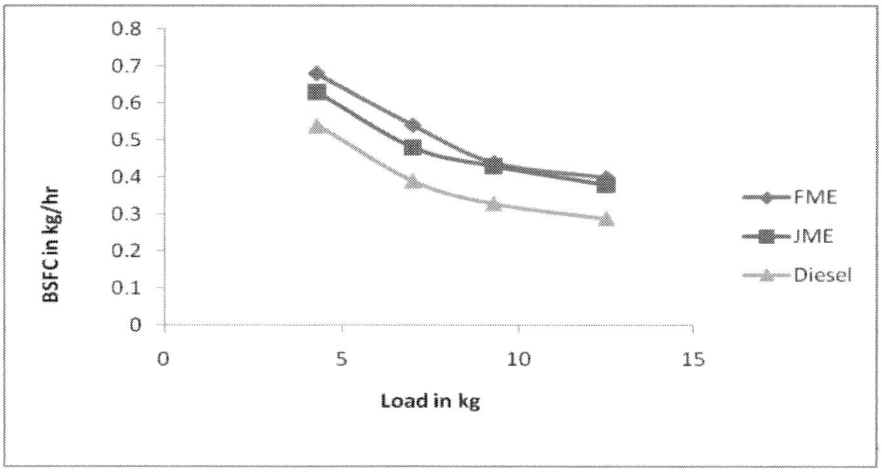

Figure 1. Load Vs BSFC

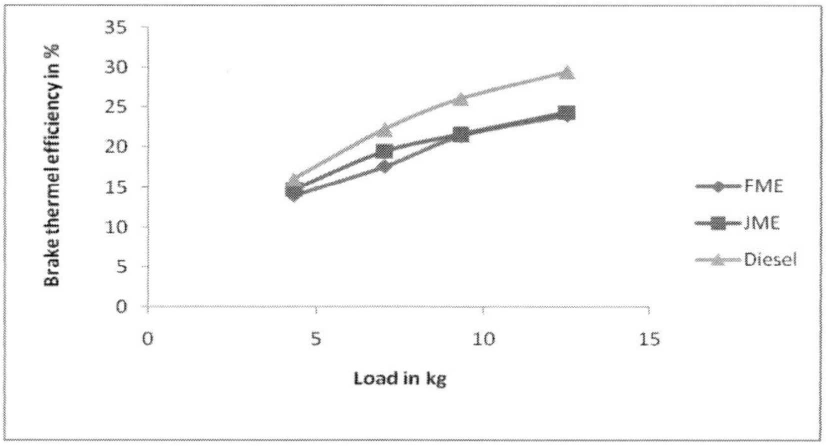

Figure 2. Load Vs Brake Thermal Efficiency

3.1. Brake specific fuel consumption

Brake specific fuel consumption of the engine is defined as the ratio between mass of fuel consumed to produce one unit of brake power for one hour. The BSFC results for the tested fuels are presented in the figure 1. For all the tested fuels the BSFC values decrease with increase in the engine load. The differences between the values of BSFC for FME and JME are small. The minimum BSFC values calculated are 0.29 kg/kW hr, 0.38 kg/kW hr and 0.40 kg/kW hr for Diesel, JME and FME respectively. The higher BSFC values of the FME and JME indicates their lower calorific value and the higher fuel flow rate due to high density. More quantity of fuel is consumed to maintain the engine speed constant.

3.2. Brake thermal efficiency

BrakeThermal efficiency of the engine is defined as the ratio between the power output and heat energy supplied through fuel injection. The BTE results for the tested fuels are presented in the figure 2. For all the tested fuels the BTE values increase with increase in the engine load. The brake thermal efficiency values are higher for pure diesel compared to JME and FME at all loads. At higher loads the brake thermal efficiency values of JME and FME are almost equal. The brake thermal efficiency values at full load are 29.55%, 24.37 % and 24.04 % for pure diesel, JME and FME respectively. The low brake thermal efficiency values of methyl esters are due to their low calorific value and increase in fuel consumption.

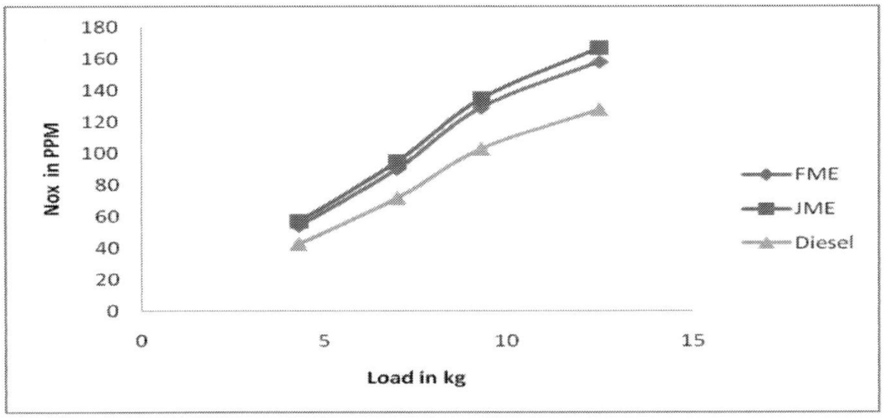

Figure 3. Load Vs NO_x

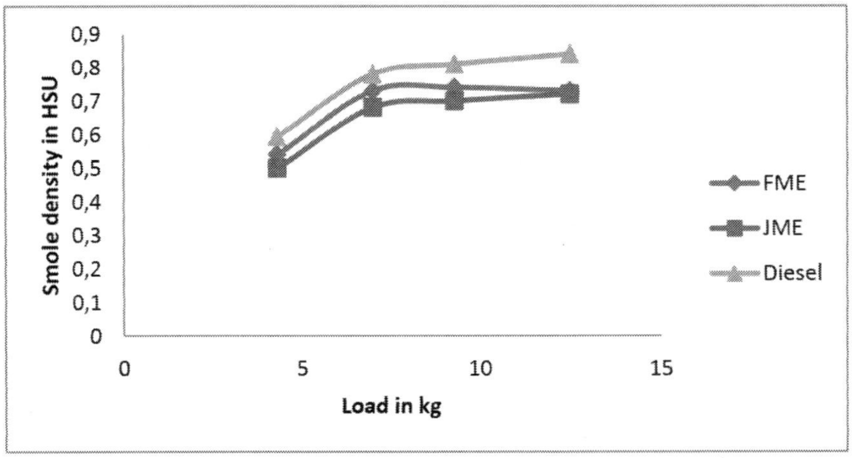

Figure 4. Load Vs Smoke Density

3.3. NO_x emission

The formation of NO_x is mainly dependent upon the availability of oxygen during combustion and flame temperature. Figure 3. Indicates the variation of NO_X emission with load for the tested fuels. It is clear from the figure that the NO_x emission value decreases in the order of Diesel, FME and JME. It is observed that FME shows 25% higher value of NO_x emissions when compared with diesel. It is observed that JME shows around 30-32% higher value of NO_x emissions when compared with diesel. The nitrogen oxide emission increases with increase in loads due to the increase in cylinder gas temperature.

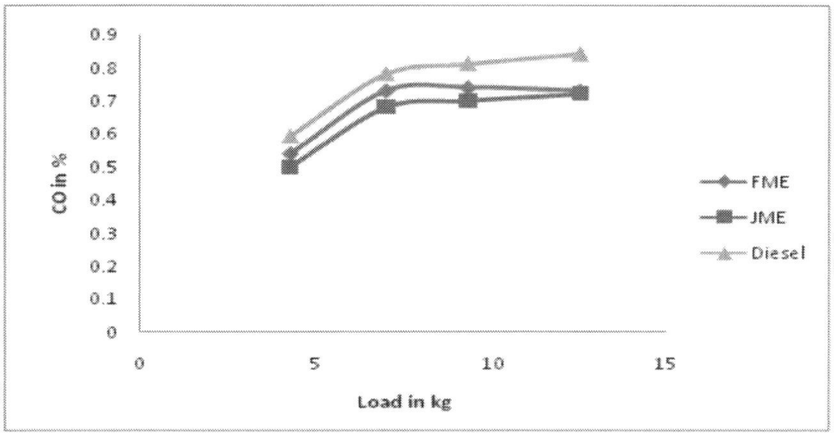

Figure 5. Load Vs Carbon monoxide

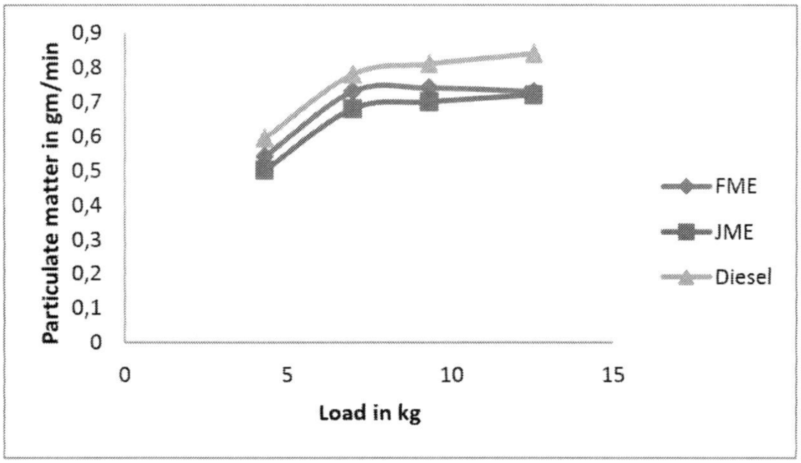

Figure 6; Load Vs Particulate matter

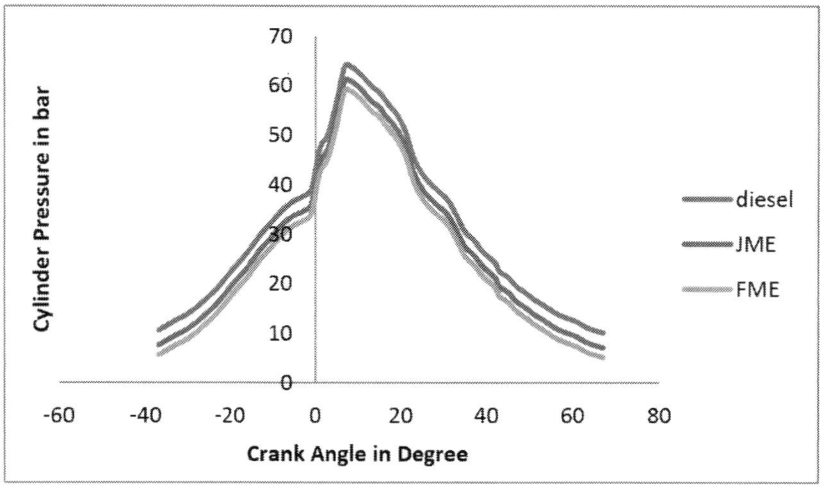

Figure 7. Crank angle Vs Cylinder Pressure

3.4. Smoke density

The formation of smoke is primarily resulted from the incomplete burning of the hydrocarbon fuel and partially reacted carbon content in the liquid fuel. Figure 4 shows the variation of smoke density with load for various test fuels. It is observed that FME shows 18% lower value of smoke density when compared with diesel. It is observed that JME shows 25% lower value of smoke density when compared with diesel. This may be due to improved combustion characteristics of methyl esters. The smoke density is found to be higher for FME compared with JME. The reason may be due to its larger quantity of carbon residue compared with JME.

3.5. CO emission

Figure 5 shows the variation of CO emission with load for various test fuels.CO is predominantly formed due to the lack of oxygen. Since methyl esters are oxygenated fuels, it lead to better combustion of fuel resulting in decrease in CO emission. It is found that the burning of FME and JME produced fewer CO emissions than the diesel. The CO emission is found to be lower for JME and its blends compared to other two fuels. It is found that FME shows 60% reduction and JME shows 65% reduction when compared with diesel at all loads.

3.6. Particulate matter

Brake Particulate emissions are as a result of incomplete combustion. It was clear from the figure 6 that the Particulate matter emissions decrease in the order of Diesel, FME and JME. It is observed that FME shows around 8-10% lower value of particulate matter emission when compared with diesel. It is observed that JME shows around 14% lower value of particulate matter emission when compared with diesel. Since methyl esters are oxygenated fuels, they promote better combustion and result in reduction of particulate matter emissions.

3.7. Pressure variation with crank angle

The pressure variation in the cycle is important in the analysis of the performance characteristics of any fuel. Figure 7 shows the variation of cylinder pressure with crank angle for various fuels at rated load. Similar trends are observed for other loads also. JME and FME exhibit slightly lower pressure at all crank angles compared to diesel. The peak pressure values are 64 bars for diesel, 61 for JME, 59 bars for FME.

4. Conclusion

From the above analysis it can be concluded that fish oil methyl ester produce lower smoke, CO, Particulate matter emissions compared to diesel but slightly higher compared to Jatropha oil methyl ester. The NOx emissions of fish oil methyl ester are higher than diesel and lower than Jatropha oil methyl ester. The fish oil methyl ester is having satisfactory performance and combustion characteristics which are comparable with diesel and Jatropha oil methyl ester.

References

[1] Murugesan AUmarani C Subramanian R Nedunchezhian N 2009 Bio-diesel as an alternative fuel for diesel engines—A review. Renewable and Sustainable Energy Reviews. 13 pp653–662.
[2] Qi D H, Geng L M, Chen H, Bian Y ZH, Liu J, Ren X CH 2009 Combustion and performance evaluation of a diesel engine fueled with biodiesel produced from soybean crude oil. Renewable Energy 34 pp 2706–2713.
[3] Emmanuel I. Bello, Taye S. Mogaji and Makanju Agge 2011 The effects of transesterification on selected fuel properties of three vegetable oils. Journal of Mechanical Engineering Research3(7) pp 218-225.
[4] Senthilkumar M, Ramesh A and Nagalingam B 2001 Complete vegetable oil fueled Dual fuel compression ignition engine. SAE 01-153.
[5] Fangrui M A, Hanna M A 1999 Biodiesel production: A review. Bio source Technology70 pp 1-15.
[6] Chiou B S, El-mashad H M, Avena-Bustillos R J, Dunn R O, Bechtel P J 2008 Biodiesel from waste salmon oil. American society of Agricultural and Biological Engineers51(3)pp 797-802.
[7] Cherng Yuan Lin, Rong Ji Li 2009 Fuel properties of biodiesel produced from the crude fish oil from the soap stock of marine fish. Fuel processing Technology 90pp130-136.
[8] El-mashad H M, Zhang R, and Avena-Bustillos R J 2008 A two-step process for biodiesel production from salmon oil. Biosystems Eng. 99(2)pp 220-227.
[9] Karthikeyan A, Durgaprasad B 2009 Experimental Investigation on Diesel Engine Using Fish oil Biodiesel and its Diesel Blends. International Journal of Applied Engineering Research4(7) pp 1139-1149.
[10] Agarwal A K, Das L M 2000 Biodiesel development and characterization for use as a fuel in C.I Engine. Journal of Engineering, Gas turbine and power (ASME)123 pp440-447.
[11] Ramadhas, Jayaraj S, Muraleedharan C 2004 Use of vegetable oils as I.C engine fuels: A Review. Renewable Energy29 pp727- 742.

IOP Conf. Series: Materials Science and Engineering **197** (2017) 012021 doi:10.1088/1757-899X/197/1/012021

PERFORMANCE AND EMISSION ANALYSIS OF DIESEL ENGINE BY COPPER COATING OVER PISTON CROWN AND CYLINDER HEAD

S.YUVARAJA[1],G.MATHISELVAN[2], R.GOBINATH[3]
[1,2,3,4] Department of Automobile Engineering, Sathyabama University, Chennai, India
Mail: yuvar_1985mech@yahoo.co.in mathiselvan@yahoo.com gobinath32@ymail.com

Abstract: The major toxins emitted from SI engine are carbon monoxide (CO) and unburnt hydrocarbons (UHC). These are harmful and create health problems to human beings, and hence control of these pollutants calls for instant attention. It has been recognized from the literature review that copper coating inside the cylinder head and over the piston crown will reduce the emission and increase the overall performance. Hence, in this Project work piston crown and engine head are coated using copper and experiment are conducted. A Kirloskar AV1 engine is used for conducting experiment. The copper coated piston crown and engine head is used to reduce the emission (HC, CO, O_2, and CO_2). The performance and characteristics of the copper coated engine has been studied.

1.Introduction

Exhaust gas or flue gas is produced as a result of the combustion of fuels such as natural gas, gasoline, petrol, biodiesel mixtures, diesel fuel, fuel oil, or coal. According to the type of engine, it is discharged into the atmosphere through tail pipe, flue gas stack, or propulsive nozzle. It usually disperses downwind in a pattern called an exhaust plume. It is a main component of motor vehicle emissions (and from stationary internal combustion engines), which can also include:

- Crankcase blow-by
- Evaporation of unused gasoline

Motor vehicle emissions contribute to pollution and area unit a significant ingredient within the creation of smogginess in some giant cities. A 2013 study by MIT indicates that fifty 3000 early deaths occur per annum within the us alone due to vehicle emissions. in keeping with another study from identical university, traffic fumes alone cause the death of 5000 folks per annum simply within the UK.Over the past 3 decades, application of composite materials is incessantly increasing from ancient application. the most important a part of most combustion gas is gas (N2), water vapor (H2O) (except with pure-carbon fuels), and greenhouse gas (CO2) (except for fuels while not carbon); these aren't venomous or pestilent (although greenhouse gas may be a greenhouse emission that contributes to world warming). a comparatively tiny a part of combustion gas is undesirable pestilent or venomous substances, appreciate CO (CO) from incomplete combustion, hydrocarbons (properly indicated as CxHy, however usually shown merely as "HC" on emissions-test slips) from unburnt fuel, gas oxides (NOx) from excessive combustion temperatures, and particulate (mostly soot). When a chilly engine is started, the fuel doesn't vaporize fully, making higher emissions of hydrocarbons, nitrogen oxides and monoxide, that diminishes solely because the enginereaches in operation temperature. The length of this start-up section has been reduced by advances in materials and technology, along with computer-controlled mechanical system, shorter intake lengths, and pre-heating of fuel and/or inducted air. Catalytic converters square measure terribly inefficient till up to their in operation temperature. This point has been a lot of reduced by moving the device nearer to the manifold and even additional therefore inserting a little nonetheless quick-to-heat-up device directly at the manifold. The little device handles the start-up emissions that permit enough time for the larger main device to heat up. Additional enhancements are often complete in some ways, as well as electrical heating, thermal battery, chemical process preheating, flame heating and superinsulation.

2.Coating

Copper could be a substance with image copper metallic element (from Latin: cuprum) and number 29. It's a soft, malleable and ductile metal with terribly high thermal and electrical conduction. A freshly exposed surface of pure copper features a reddish-orange color. It's used as a conductor of warmth and electricity, as an artifact, and as a constituent of assorted metal alloys. Copper is found as a pure metal in nature, and this was the supply of the primary metal to be employed by humans, ca. 8,000 BC; it absolutely was the primary metal to be smelted from its ore, ca. 5,000 BC; it absolutely was the primary metal to be solid into a form in an exceedingly mould, ca. 4,000 BC; and it absolutely was the primary metal to be purposefully alloyed with another metal, tin, to make bronze, ca. 3,500. Copper was mined as Cyprus in the Roman era.The name originated from aes cyprium (metal of Cyprus), later transferred to cuprum, from that to copper (English), Koper (Dutch), cuivre (French), and Kupfer (German) got derived. Its compounds area unit unremarkably encountered as copper (II) salts, which regularly impart blue or inexperienced colors to minerals like azurite, mineral and turquoise and are wide used traditionally as pigments. Bailiwick structures engineered with copper corrode to convey inexperienced verdigris (or patina). Ornamental art conspicuously options copper, each by itself and within the style of pigments.

3.Methodology

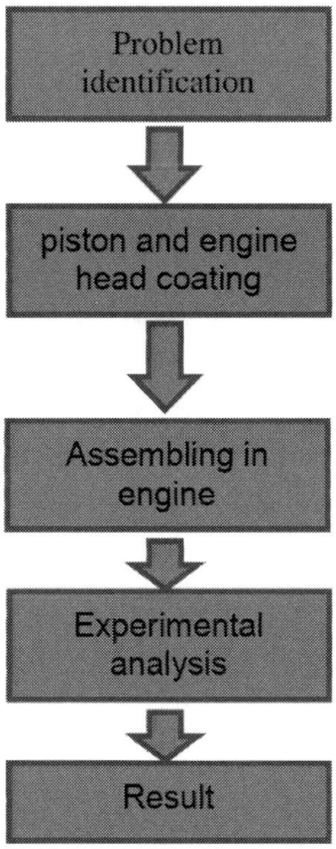

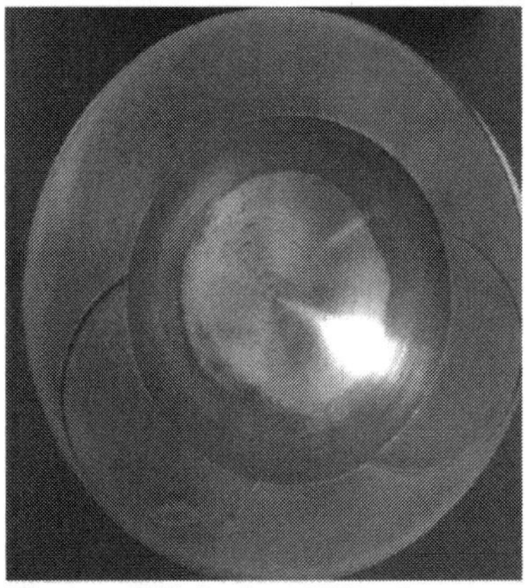

Figure 1. Piston of Kirloskar AV1 after copper coating

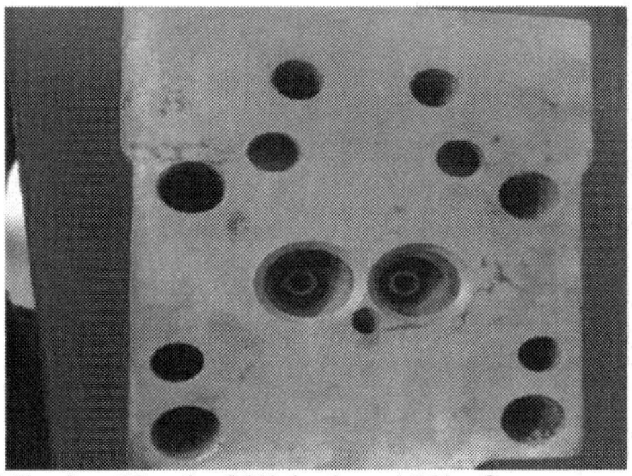

Figure 2. Engine Head of Kirloskar AV1 after copper coating

4. Experimental Setup

The experimental setups were dispensed on single cylinder four stroke diesel (Kirloskar 500cc).

- The Exhaust emission like (HC, CO, dioxide & O2) was measured by exploitation Automotive Emission Gas analyzer.
- The main pollutants that are objectionable and are to be reduced are HC, CO & dioxide by exploitation chemical action coating together of the techniques wide accustomed cut back emission and improve performance

Table 1. Engine Specifications

Make	Kirloskar
Model	AV1
No. of Cylinders	1
Cubic capacity (ltr)	0.553
Rated Speed (rpm)	1500
Bore Diameter	80 mm
Stroke length	100 mm
BHP	5
Orifice Diameter	14 mm
Fuel	Diesel

Calorimeter	Exhaust gar calorimeter
Cooling	Water cooled engine
Starting	Hand cranking
Compression ratio	16.09:1, variable from 13.51 to 19.69

The AVL Smoke Meter could be a filter-type smoke meters for measurement the soot content within the exhaust of diesel and GDI engines. The variable sampling volume and thermal exhaust learning assures large applications vary, e.g. measurements throughout engine development or DPF standardization.AVL Di gas analyzer moveable Emission Measuring System for measure the HC, NO/NO2, CO, CO2 and O2 concentrations among the exhaust gas of diesel and gasoline vehicles.

5.Perfomance and emisson characteristics

The experimental were conducted on Single cylinder four stroke diesel standard engine and copper coated engine with varying load and the results are tabulated in the table. Emission of HC, CO2, O2 and CO was also noted. Standard engine and copper coated engine emissions readings are tabulated

Table 2. Emission parameters

Gases	Before coating	After coating
HC (in PPM)	13.2	11.41
CO (% of volume)	0.05	0.046
CO2 (% of volume)	1.42	1.33
O2 (% of volume)	18.66	16.45

Table 3. Performance Parameters

S.No.	FP (kw)	BP (kw)	τ (Nm)	IP (kw)	TFC	SFC	$\eta mech$
1.	3	0	0	3	0.504	-	0
2.	3	0.741	4.74	3.741	0.572	0.771	19.80
3.	3	1.482	9.56	4.482	0.677	0.456	33.06
4.	3	2.223	14.42	5.223	0.851	0.382	42.56
5.	3	2.964	19.37	5.964	0.960	0.323	49.69

IOP Conf. Series: Materials Science and Engineering **197** (2017) 012021 doi:10.1088/1757-899X/197/1/012021

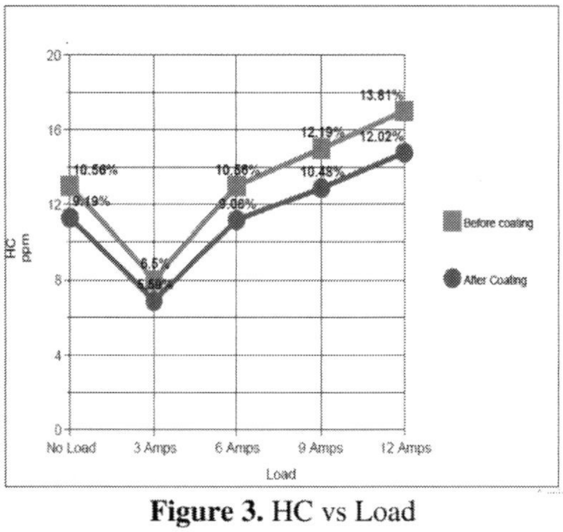

Figure 3. HC vs Load

From the above graph, the emission level of the Hydro Carbon gas with the standard and the copper coated engine was observed with various loads. The emission level of hydrocarbon is reduced 1.7% after the copper coating.

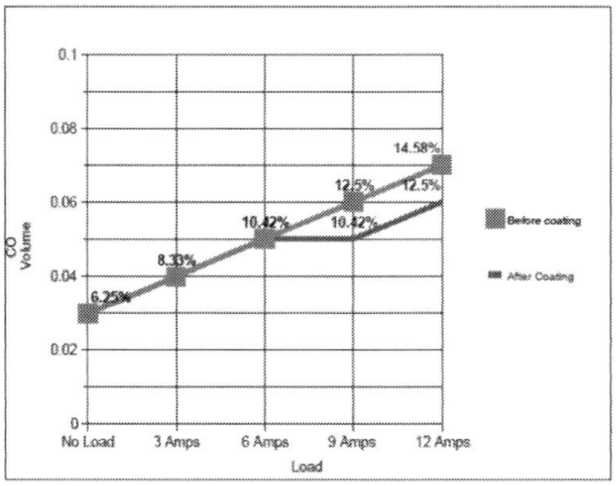

Figure 4. CO vs Load

From the above graph, the emission level of the Carbon Monoxide gas with the standard and the copper coated engine was observed with various loads. The emission level of carbon monoxide is reduced 2.2% after the copper coating.

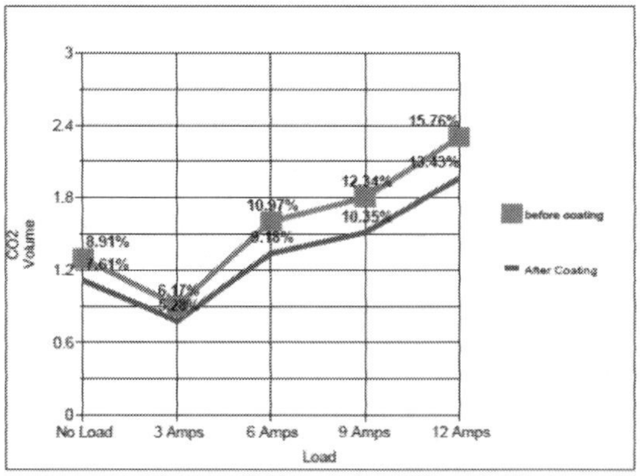

Figure5.CO_2 vs Load

From the above graph, the emission level of the Carbon monoxide gas with the standard and the copper coated engine was observed with various loads. The emission level of carbon monoxide is reduced 2.3% after the copper coating

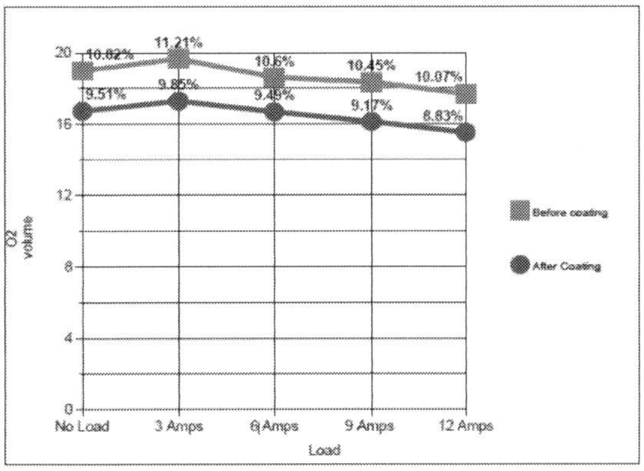

Figure 6.O_2 vs Load

From the below graph, the emission level of the Oxygen (O2) gas with the standard and the copper coated engine was observed with various loads. The emission level of oxygen is reduced 2.4% after the copper coating.

6.Results and discussion

From the above calculation and the emission analysis test of the engine with the piston crown and the engine head coated with copper shows that the variation in emission level of the HC, CO, CO2, O2 from the copper coated engine.

From the experimental investigation of copper coated engine the following conclusion could be arrived:

- From the experimental results at constant speed with various loads, the average HC emission standard engine is 13.2 ppm, whereas HC emission for copper coated engine is 11.41 ppm. (i.e.) 13.56 % of emission is reduced for copper coated engine compared to standard engine. It is represented in graph 3.
- From Graph 4. CO emission standard engine is 0.05 % of volume whereas copper coated engine gives 0.046 % of volume. From the results the average amount is 8 % of emission is reduced in copper coated engine at idle speed with various load condition.
- From the Graph 5. CO2 emission in standard engine is 1.42 % of volume whereas copper coated engine gives 1.33 % of volume. From the results is 6.33 % of emission is reduced in copper coated engine with various load at idle speed condition
- From the Graph 6. O2 emission in standard engine is 18.66 % of volume whereas copper coated engine gives 16.45 % of volume. From the results is 11.84 % of emission is reduced in copper coated engine with various load at idle speed condition.

Thus, the emission of the exhaust gases namely HC, CO, CO2 and O2 have been reduced to 11.41 ppm, 0.046 %, 1.33 % and 16.45 % respectively using Automotive Emission Gas Analyzer. This can be further reduced by implementing further additional technologies.

References

[1] P. N. Shrirao, A. N. Pawar. Evaluation of Performance and Emission characteristics of Turbocharged Diesel Engine with Mullite as Thermal Barrier Coating, International Journal of Engineering and Technology Vol.3 (3), 2011, 256-262.
[2] K. Kokini, Y.R. Takeuchi, and B.D. Choules. Surface thermal cracking of thermal barrier coatings owing to stress relaxation: zirconia vs mullite, Surf.Coat. Technol., vol. 82, 77–82,1996.
[3]Berndt CC (1985), Plasma spray coating, Department of Materials Engineering, Clayton, Victoria: Monash University
[4] Jian Gao, Deming Jiang, Zuohua Huang. Spray properties of alternative fuels: A comparative analysis of ethanol–gasoline blends and gasoline, Journal of Fuel 2007, 86, 1645–1650.
[5] Owen Keith, Coley Trevor, Automotive fuels reference book, 2nd Edition. New York: SAE. 1995
[6] Serdar, H. Yu¨cesu, Adnan Sozen, Tolga Topgu, Erol Arcakliog, Comparative study of mathematical and experimental analysis of spark ignition engine performance used ethanol–gasoline blend fuel,Journal of Applied Thermal Engineering 2007, 27, 358–368.
[7]Wei-Dong Hsieh, Rong-Hong Chen, Tsung-Lin Wu, Ta-Hui Lin. Engine performance and pollutant emission of an si engine Using ethanol–gasoline blended fuels, Atmospheric Environment 36 (2002) 403–410.
[8]C.Ananda Srinivasan, C.G.Saravanan. Emission reduction in SI engine using ethanol – gasoline blends on thermal barrier coated pistons, International journal of energy and environment *Volume 1, Issue 4, 2010,715-726*
[9] Rice, R.W., Sanyal, A.K., Elrod, A.C., Bata, R.M., 1991.Exhaust gas emissions of butanol, ethanol and methanol– gasoline blends. Journal of Engineering for Gas Turbine and Power 113, 337–381.
[10] P Lawrence, P. Koshy Mathews, B. Deepanraj. Experimental Investigation on Performance and emission characteristics of Low Heat Rejection Diesel Engine with Ethanol as Fuel, American Journal of Applied Sciences 8 (4): 348-354, 20

IOP Conf. Series: Materials Science and Engineering **197** (2017) 012022 doi:10.1088/1757-899X/197/1/012022

Effects of Nano Additives in engine emission Characteristics using Blends of Lemon Balm oil with Diesel

J.Senthilkumar[1], S. Ganesan[2], S.Sivasaravanan[3] , S.Padmanabhan[4] , L.Krishnan[5]andV.C.Aniruthan[6]
[1]Research scholar, Faculty of Mechanical Engineering, Sathyabama University, Chennai, India,
[2]Mechanical Engineering, Chennai Institute of Technology, Chennai, India.
[3-6]Faculty of Mechanical Engineering, Sathyabama University, Chennai, India,
j.senthilkumar1981@gmail.com[1], gansuma@gmail.com[2], sivakriya16@gmail.com[3], krishnalaxman25@gmail.com[1], and caniruth999@gmail.com[1]

Abstract

Economic growth in developing countries has led to enormous increase in energy demand. In India the energy demand is increasing at a rate of 6.5% every year. The crude oil demand of country is meet by bring in of about 70%. Thus the energy safety measures have become key issue for our country. Bio diesel an eco-friendly and renewable fuel alternate for diesel has been getting the consideration of researcher's entire world. The main aim of this paper is to evaluate the engine parameters using blend of pure lemon balm oil with diesel. Also nano Additives is used as a catalyst with blends of bio fuel to enhance the Emission Characteristics of various effective gases like CO2 ,NOx , CO and UHC with various levels of engine process parameters.
Keywords – Bio diesel, nano Additives, Emissions, lemon balm oil.

Introduction

In the transportation and agriculture sector, various renewable liquid bio-fuels derived from biological resources have proved to be good alternatives for fuels derived from crude oils. For the problems of environmental effects, energy safety, restricting imports, village people employment and agricultural wealth, these bio- fuels are gaining global acceptance as a solution. The smartest technologies deliver benefits to multiple interests, including improved economy, and a positive impact on the environment and governmental policies. One of the greatest advantages to biofuel is that it can be used in existing engines, vehicles and infrastructure with practically no changes. Bio diesel can be pumped ,stored and burned just like petroleum diesel fuel, and can be used pure,or in blends is practically identified to petroleum diesel fuel , and year round operation can be achieved by blending with diesel fuelMuthukumar et al [1] Calopylluminophyllum oil as alternate fuel for diesel engine creaked by fly ash.Then the oil is mixed with pure diesel of 75% in volume this reduce the emission of nitrogen oxide but the efficiency will be reduced by increasing the amount of Calopylluminophyllum oil in the biodiesel mixture.

Ganesan .S et al [2] used alternative fuel as lemongrass oil. An attempt was made to optimize the engine parameters such as Fuel Consumption, Nano Catalyst Proposition and Injection Pressure. Experiments were carried out based on Taguchi's Design of Experiments approach method to limit the

Number of experiment by the formation of orthogonal array, yet without sacrificing momentous information. On comparison, the predicted values from Taguchi and the experimental values are nearer, but the experimental values are higher than the predicted values. The results are well supported by the confirmatory test.

J.Senthilkumar et al [3] used jatropha and rubber seed oil with pure diesel as a alternative fuel .He compared B20,B40,B60 and found B20 is having the lower fuel consumption and also found that CO and HC emission were reduced with increase in blends.But NOx increases with increase in blends.

Viswanath K. Kaimal et al [4] studied the different blends of plastics waste oil and diesel such as P25,P50,P75 and found that P25 can replace pure diesel which increases the peak cylinder pressure, Ignition delay, combustion timing, heat released and also found that thermal efficiency of the engine reduced by increasing the blends.

S. Vedharaj et al [5] used KME (kapok methyl ester) – diesel blends of B25,B50,B75,B100 with two different combustion chamber geometries such as TRCC (trapezoidal combustion chamber) and TCC (toroidal combustion chamber) and found that B50 with TCC gives the maximum output which is increase in break thermal efficiency of 5.2% and reduces the CO and smoke by 15 and 8 percentage respectively.

M.S. Shehata et al [6] used corn and sugarcane oil as alternative fuel with different pressures 180,190,200 Bar At 200 Bar pressure the break thermal efficiency increased and specific fuel consumption decreased by 15% compared to pressure of 180 Bar.

Jo-Han Ng et al [7] The author proposed that In this study, a 0D/1D gas dynamics numerical model of a single-cylinder, 4-stroke diesel engine is used to simulate entire engine cycle from air intake to exhaust product. The engine model is successfully validated against experimental data of both palm biodiesel and fossil diesel for pertinent combustion parameters such as pressure trace, rate of heat release and ignition delay period.

K. Nanthagopal et al [8] They suggested the effect of Calophylluminophyllum methyl ester on diesel engine performance, emission and combustion characteristics at different injection pressures. Experimental investigations with varying injection pressures of 200 bar, 220 bar and 240 bar have been carried out to analyse the parameters like brake thermal efficiency, specific fuel consumption, heat release rate and engine emissions of direct injection diesel engine fuelled with 100% biodiesel and compared with neat diesel. The experimental results revealed that brake specific fuel consumption of C. inophyllum methyl ester fuelled engine has been reduced to a great extent with higher injection pressure. Significant reduction in emissions of unburnt hydrocarbons, carbon monoxide and smoke opacity have been observed during fuel injection of biodiesel at 220 bar compared to other fuel injection pressures.

Fuel Preparation

Sonication process was the second stage of the project where different fuel preparation takes. Various blends of biodiesel along with different proportion of CeO2 were prepared using an ultra-sonication process. The different mixture of fuel was prepared in 1000ml beaker and was kept in ultra-sonication machine for 45 minutes to prepare the homogenous mixture of various combinations. This allows to for uniform mixing of nanoparticles in fuel. Various blends of biodiesel along with different proportion of CeO2 were prepared using an ultra-sonication process. The different mixture of fuel was prepared in 1000ml beaker and was kept in ultra-sonication machine for 45 minutes to prepare the homogenous mixture of various combinations. This allows to for uniform mixing nanoparticles in fuel. The different mixture of fuel was prepared in 1000ml beaker and was kept in ultra-sonication machine for 45 minutes to prepare the homogenous mixture of various combinations. This allows to for uniform mixing of nano particles in fuel during the process the amount of nano particle to be mixed is being calculated by ultra-weighing machine with the accuracy of 0.001mg

Experimental Setup and Procedure

The performance test was conducted on a single cylinder CI engine to study the performance, emission, combustion characteristics at various load condition.The diesel engine is a high speed four stroke, vertical, air cooled type. The loading is by means of an electrical dynamometer. The fuel tank is connected to a graduated burette, to measure the quantity of the fuel consumed in unit time. An orifice meter with u-tube manometer is provided along with an air tank on the suction line for measuring air consumption. An AVL145 smoke meter is provided for measuring FSN of exhaust gases. The rig is installed with AVL software forobtaining various curves and result during operation. A five gas analyzer is used toobtain the exhaust gas composition.

The filters of the engine were replaced and the injectors were cleaned and calibrated according to the desired pressure. Then the AVL gas analyzer and smoke meter were installed. The input to the gas analyzer was taken from the exhaust port of the engine. The fuel tank was then filled with diesel and the engine was run. After the fuel is filled to the engine it was run at various loads of the dynamometer like 5,10,15,20,25kgs and respective readings were taken for fuel consumption/ sec. The readings of gas analyzer and smoke meter were noted in each case. After all the readings were taken, the leftover diesel was drained out of the tank and emulsion was poured. Same steps were taken and the readings were noted down for the emulsion. Before using the next emulsion the engine was again run with diesel so that the results are not biased. After taking all the observations finally graphs were plotted to compare the performance characteristics and emission characteristics of the engine in case of diesel and emulsion.Before using the next emulsion the engine was again run with diesel so that the results are not biased. The table 1 shows engine specifications.

Table 1. The Test Engine Specifications

Engine type	Four stroke, single cylinder, vertical air cooled diesel engine.
Brake power	4.4 KW
Speed	1500 (constant speed)
Bore Diameter (D)	80 mm
Stroke (L)	110 mm
Number of Cylinders	1
Type of loading	Electrical rheostat
Orifice diameter	13.4 mm
Cd of orifice	0.62
Number of Stroke	4 Stroke
Manufacturer	Kirloskar oil engine
Model	Model

Design of Experiment

Different optimization techniques have been discussed in the coming sections. Methods like Taguchi Technique, grey relational analysis have been discussed in detail. Once the experimental design has been determined and the trials have been carried out, the measured performance characteristic from each trial can be used to analyses the relative effect of the different parameters. Design of Experiments (DOE) techniques enables designers to determine simultaneously the individual and interactive effects of many factors that could affect the output results in any design. Table shows the factors and levels for present experiment.

The Taguchi method utilizes orthogonal arrays from theory of design of experiments to study a large number of variables with a small number of experiments. Using orthogonal arrays significantly reduces the number of experimental configurations to be studied. Figure 1 shows the orthogonal array for the present experiment.

Table 2. Design Factors and their Level

Factors	**Level 1**	**Level 2**	**Level 3**
Lemon balm oil - % (A)	10	20	30
Injection Pressure - Bar (B)	200	220	240
BP - KW (C)	1.5	3	4.4
MgO - ppm (D)	15	30	45

To select an appropriate orthogonal array for the experiments, the total degrees of freedom need to be computed. The degrees of freedom are defined as the number of comparisons between process parameters that need to be made to determine which level is better and specifically how much better it is.

Orthogonal Array	No of experiments	No of factors	Max no of factors at theses levels			
			Level 2	Level 3	Level 4	Level 5
L-4	4	3	3	-	-	-
L-8	8	7	7	-	-	-
L-9	9	4	-	4	-	-
L-12	12	11	11	-	-	-
L-16	16	15	15	-	-	-
L-'16	16	5		-	5	-
L-18	18	8	1	7	-	-
L-25	25	6	-	-	-	6
L-27	27	13	-	13	-	-
L-32	32	31	31	-	-	-
L-'32	32	10	1	-	9	-
L-36	36	23	11	12	-	-
L-'36	36	16	3	13	-	
L-50	50	12	1		-	11
L-54	54	26	1	25	-	-
L-64	64	63	63	-	-	-
L-'64	64	21	-	-	21	-
L-81	81	40	-	40	-	-

Fig: 1. Orthogonal array

For example, a two-level process parameter counts for one degree of freedom. Once the degrees of freedom are known, the next step is selecting an appropriate orthogonal array to fit the specific task. . The degrees of freedom are defined as the number of comparisons between process parameters that need to be made to determine which level is better and specifically how much better it is. Figure shows L9 orthogonal array of the present study.

Table 3.L9 Array with Design Factor

Ex. No	L.B oil(%)	IP(Bar)	BP(KW)	Mgo(ppm)
1	10	200	1.5	15
2	10	220	3	30
3	10	240	4.4	45
4	20	200	3	45
5	20	220	4.4	15
6	20	240	1.5	30
7	30	200	4.4	30
8	30	220	1.5	45
9	30	240	3	15

Experimental data using Taguchi Methodology

In the present experiment , using Taguchi methodology of L 16 orthogonal array, after the total number of experiment was done, the array with design factors, levels and responses for engine performance are given below in table 3.

Table 4. Experimental Results Taguchi Methodology

Ex. No	L.B oil (%)	IP (Bar)	BP (Kw)	MgO(ppm)	BTE (%)	SFC(kg/kw hr)	HC (ppm)	CO (%)	NOx(ppm)
1	10	200	1.5	15	28.9	0.575	30	12.7	750
2	10	220	3	30	22.95	0.38	32	14.98	821
3	10	240	4.4	45	28.52	0.27	34	13.43	780
4	20	200	3	45	28.27	0.313	40	12.67	700
5	20	220	4.4	15	22.05	0.41	17	15.04	745
6	20	240	1.5	30	29.4	0.626	21	16.88	428
7	30	200	4.4	30	23.67	0.381	33	15.46	722
8	30	220	1.5	45	26.8	0.6	32	16.82	430
9	30	240	3	15	28.9	0.575	30	12.7	750

Analysing Experimental Data

To determine the effect each variable on the output, the signal-to-noise ratio, or the SN number, needs to be calculated for each experiment conducted. The calculation of the SN for the first experiment in the array above is shown below for the case of a specific target value of the performance characteristic.

There are 3 Signal-to-Noise ratios

- Smaller the better
- Larger the better

Smaller-The-Better

This is usually the chosen S/N ratio for all undesirable characteristics like defects etc. for which the ideal value is zero. Also, when an ideal value is finite and its maximum or minimum value is defined then the difference between measured data and ideal value is expected to be as small as possible.The S/N ratio for Smaller -The-Better is calculated by using the following formulae.

$$S / N \ ratio \ (\eta) = -10\log 10\left(\frac{1}{n}\sum_{i=1}^{n} y_{ij}^{2}\right)$$

In this current above formulae is used to calculate the smaller S/N Ratio for SFC, CO,HC and NOX.

Larger-The-Better

This case has been converted to smaller-the-better by taking the reciprocals of measured data and then taking the S/N ratio as in the smaller-the-better case. These are used for analyzing the output of the industries and study of different property enhancing experiments. The S/N ratio for Larger – The – Better is calculated by using the following formulae.

$$S / N \ ratio \ (\eta) = -10\log 10\left(\frac{1}{n}\right)\sum_{i=1}^{n} \frac{1}{y_{ij}^{2}}$$

In this current above formulae is used to calculate the Larger S/N Ratio for BTE.S/N Ratio for the all the responses mention below the table.

Table.5.S/N ratio

BTE (%)	**SFC(Kg/Kw hr)**	**HC (ppm)**	**CO (%)**	**Nox (ppm)**
29.218	4.8066	-29.54	-22.08	-57.5
27.216	8.4043	-30.1	-23.51	-58.29
29.103	11.373	-30.63	-22.56	-57.84
29.027	10.089	-32.04	-22.06	-56.9
26.868	7.7443	-24.61	-23.54	-57.44
29.367	4.0685	-26.44	-24.55	-52.63
27.484	8.3815	-30.37	-23.78	-57.17
28.563	4.437	-30.1	-24.52	-52.67
29.218	4.8066	-29.54	-22.08	-57.5

RESULTS AND DISCUSSION

Response Curve Analysis for BTE

The effects of the following combinations were plotted, as shown in the figure. It was observed that the Brake Thermal Efficiency escalates at the points A1, B3, C1 and D3. Graph A highlights the optimal increase of BTE when B10 oil blend is compared with B20 and B30, the average value remaining relatively similar. Graph B shows the responses observed when the Injection Pressure is varied, and shows maximum efficiency at point B3 which is at 240 bars. Brake Power of 1.5KW, point C1, increases the efficiency to around 29%, and 45ppm of MgOnano additive also shows an increase in the BTE. Hence, according to the observations made, the best combination of parameters turns out to be A1-B3-C1-D3 shown in fig.2.

Response Curve Analysis for SFC

The observations made in the prediction stage, to get optimal engine output values, the lowest value in each of the graph has to be considered .It was observed that B30 oil blend, 240 bars of injection pressure, 1.5KW Brake Power and 15 ppm of MgO gives the best combination of parameters in order to achieve the minimum SFC value. The variations in each graph are quite high except the graph plotted for the Injection Pressure in which the values obtained are almost similar as show in fig.3.

Response Curve Analysis for HC

The graph plotted below shows the main effects of LB oil, injection pressure, brake power and parts per million of nano additive added in order to obtain a minimum value of HC. It was observed in the prediction stage, that B30 level of LB oil decreases the HC to a value around -30 which is comparatively the least among the other values. The second graph indicates that the optimum pressure to be 200 bars. The third and fourth graph shows that 3.0KW and 45 ppm of MgO gives the lowest required value of HC in the experiment and hence the best combination of parameters being A3-B1-C2- D3 as shown in fig.4.

Response Curve Analysis for CO

The graph given below compares the different combinations of bio diesel blend which points that the A3-B2- 4C1-D2 combination reduces the maximum amount of CO gas. the successful combination is 10% of mint oil and 90% of pure diesel at220 bar pressure and the break power is 1.5KW then finally 30 ppm of MgO as shown fig.5.

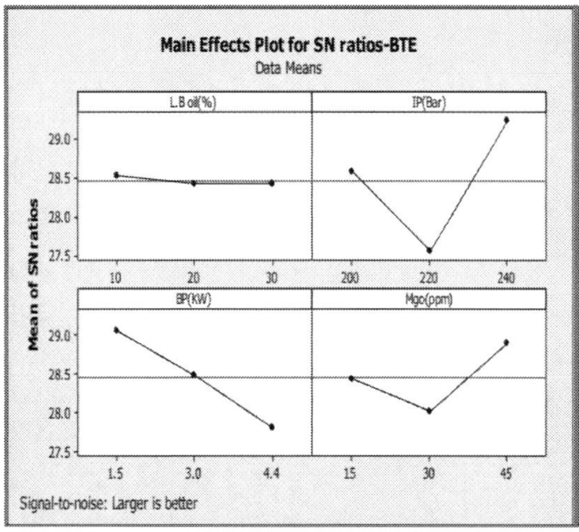

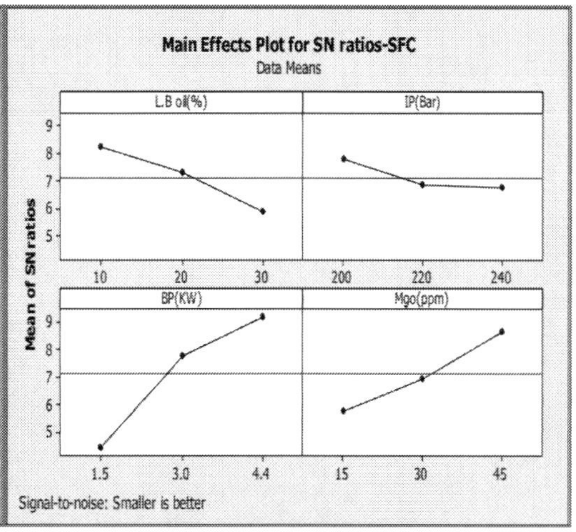

Fig.2: Response curve analysis for BTE **Fig.3:** Response curve analysis for SFC

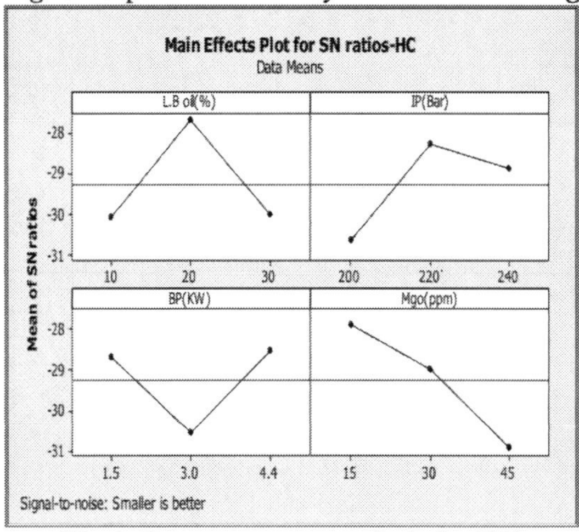

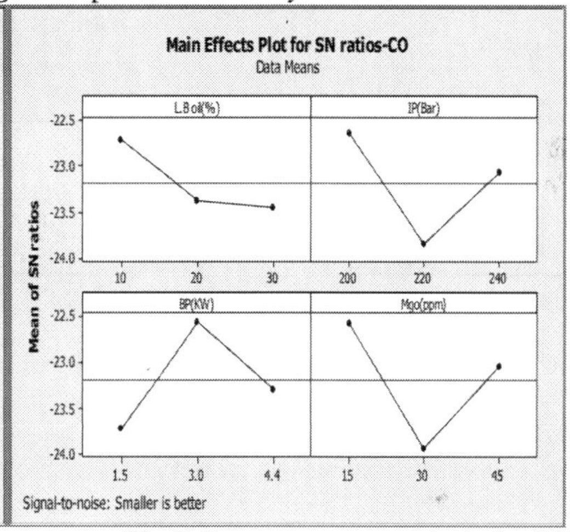

Fig.4: Response curve analysis for HC **Fig.5:** Response curve analysis for CO

Response Curve Analysis for NOx

The graph given below compares the different combinations which points that the A1-B1-C2-D1 combination reduces the maximum amount of NOx gas. the successful combination is 10% of mint oil and 90% of pure diesel at200 bar pressure and the break power is 3KW then finally 15 ppm of MgO as shown in fig.6.

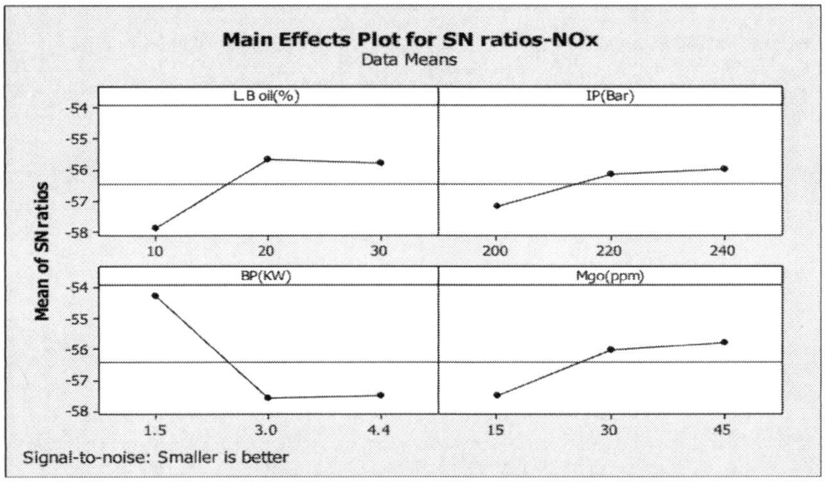

Fig.6: Response curve analysis for CO

Conclusion

From this experiment it is seen that biodiesel is going to be the natural choice for our future transport fuel. The foregoing combination of experience and reasoning it is possible to distil some conclusions about biodiesel as an automotive fuel particularly in a developing country like India. The following facts have been established through the present work. From the above experiment the multi-response parameters (engine performance and emission study) are optimized using grey relational analysis and converted into a single response. Then we have used Taguchi's methodology to analyze the experimental data.

For Optimum BTE, Larger the better S/N ratio consider. The engine operating condition A1B3C1D3.For Better Specific fuel consumption, Smaller the better S/N ratio consider. The predicted optimized Engine parameter of A3B3C1 and D1. Better HC, Smaller the better S/N ratio consider. The predicted optimized Engine parameter A1B1C2D3.In prediction stage an experiment should be conducted with MgO in 45 ppm, BP in 3KW, 200 bar of injection pressure and L.B in 10% levels to obtain optimal values .For Better CO, Smaller the better S/N ratio consider. The engine operating condition A3B2C1D2. For Better Specific fuel consumption, Smaller the better S/N ratio consider. The predicted optimized Engine parameter of A3B3C1 and D1.Better HC, Smaller the better S/N ratio consider. The predicted optimized Engine parameter A1B1C2D3.In prediction stage an experiment should be conducted with MgO in 45 ppm, BP in 3KW, 200 bar of injection pressure and L.B in 10% levels to obtain optimal value. For Better CO, Smaller the better S/N ratio consider. The engine operating condition A3B2C1D2. For Better NOx, Smaller the better S/N ratio consider.The predicted optimized Engine parameter A1B1C3D1.In prediction stage an experiment should be conducted with MgO in 15 ppm, BP in 4.4 KW, 200 bar of injection pressure and L.B in 10% levels to obtain optimal values.The predicted optimized Engine parameters were obtained from Response curve Analysis .Nox, CO, HC and Smoke emissions are lower for MgO nanoparticles blended fuels.

References

[1] MuthukumaranN, SaravananCGa, Prasanna Raj S Yadav, Vallinayagam R, VedharajS and RobertsWL 2015 Synthesis of cracked Calophylluminophyllum oil using fly ash catalyst 4 for diesel engine application, *Fuel*,Article in press.

[2] Ganesan S, Elango A, EluriVamsi Krishna and BalajiA 2014 Influence of Nano catalyst on Emission Characteristics of Di Diesel Engine with Blends of lemon grass oil Using Taguchi Approach in *International Journal of Applied Engineering Research*, Research India publication, **9, 21**, 9941-9949, ISSN: 0973-4562.

[3] Senthilkumar J and Mahalingam S 2015Effect of Injection Timing on Performance and Emission Analysis of Single Cylinder Diesel Engine Fuelled with Dual Bio-Fuels*International Journal of Applied Engineering Research*, **2**5041-5047.

[4] Viswanath KKaimal and Vijayabalan P 2015A detailed study of combustion characteristics of a DI diesel engine using waste plastic oil and its blends, *Energy Conversion and Management*,**105**, 951-956.

[5] VedharajS, Vallinayagam R, Yang WM, SaravananCG and Lee PS 2015Optimization of combustion bowl geometry for the operation of kapok biodiesel – Diesel blends in a stationary diesel engine,*Fuel*, **139**, 561-567.

[6] ShehataMS, Ali MA,Attia SM and Abdel RazekR Corn and soybean biodiesel blends as alternative fuels for diesel engine at different injection pressures*Fuel***161**49-58.

[7] Jo-Han Ng, Jing Huey Khor, Kang Yao Wong, SritharRajoo and Cheng Tung Chong 2015, Statistical Analysis of Engine System-Level Factors for Palm Biodiesel Fuelled Diesel Engine ResponsesThe 7th International Conference on Applied Energy – ICAE2015*Energy Procedia***75**99-104.

[8] NanthagopalK, AshokB and ThundilKaruppa Raj R2016Influence of fuel injection pressures on Calophylluminophyllum methyl ester fuelled direct injection diesel engine *Energy Conversion and Management***116**165-173.

Performance Test on Compression Ignition Engine by Blending Ethanol and Waste Plastic Pyrolysis Oil with Cetane Additive

PadmanabhanS[1*], GanesanS[1c],
JeswinArputhabalan J[1d], ChithralaVarun[2a], Ganesh BairavanP[2b],

[1]Associate Professor, School of Mechanical Engineering, Sathyabama University, Chennai.
[2]Department of Automobile Engineering, Sathyabama University, Chennai.
*corresponding author email: padmanabhan.ks@gmail.com
Co-author email: [2a]chithralavarun07@gmail.com; [1c]gansuma@gmail.com;
[1d]jeswin.j.john@gmail.com; [2b]gangun005@gmail.com;

Abstract: The demand for diesel fuel is higher than that of petrol throughout the world hence seeking alternative to mineral diesel is a natural choice.Alternative fuels should be easily available at lower cost, environment friendly and fulfill energy needs without modifying engine's operational parameters. Waste to energy is the trend in the selection of alternate fuels. In this work, Waste Plastic Pyrolysis oil (WPPO), Ethanol, Diesel blend with Cetane additive has been attempted as an alternative fuel. A Twin cylinder, Direct Injection engine was used to assess the engine performance and emission characteristics of waste plastic pyrolysis oil with cetane additive.Experimental results of blended plastic fuel and diesel fuel were compared.
Keywords:*cetane additive, diesel engine performance, emission characteristics, ethanol,waste plastic pyrolysis oil*

1. Introduction

An efficient energy conservation and management has become an important action in industrial field as energy is considered as a major element in the production cost. This positive approach has changed the pattern of energy use all over the world. Hence the developed countries have shown raise in their productivity, maintaining the unchanged energy consumption levels.Alternative fuels should be available abundantly and at low cost, be environment friendly and fulfil energy security needs without sacrificing engine's operational performance. Waste to energy is the development in the choice of alternate fuels. Fuels like alcohol, biodiesel, fuel from waste plastics are several of the tried out alternative fuels for the internal combustion engines.

The demand for diesel fuel is higher than that of gasoline throughout the world hence seeking alternative to mineral diesel is a natural choice. Plastics have turn intoanvitalpart in day today life, due to their lightweight, durability, energy efficiency, coupled with a faster rate of production and design flexibility, this process such as thermalpyrolysis andcatalytic pyrolysis can be used to securelychangeinto waste plastics into hydrocarbon fuels that can be used in diesel engines. Disposal of the waste plastics poses a great hazard to the environment and an effective method has not yet been implemented. Due to its non-biodegradable nature, the plastic waste contributes significantly to the problem of waste management also. Panda et al [1]stated that the production of liquid fuel from plastic waste is a better alternative as the calorific value of the plastic pyrolysis oil is almost to the values of

diesel fuels.Alan Hasen et al[2], in this review the properties and specifications of ethanol blended with diesel fuel are discussed.CenkSayin et al [3] in his performance study, ethanol was blended with diesel fuel from 0% to 15% and with an increment of 5%, and the injection timings were varied. It was observed that ethanol in the blend increases the levels of NO_x and CO_2 emissions and decreases the CO and UHC emissions considerably.
Viswanathetal[4] carried out experimental investigation using waste plastic pyrolysis oil with silica as catalyst. Waste plastic pyrolysis oil was mixed with diesel in proper ratios and used as fuel in diesel engine. Results showed increased exhaust temperature and higherNOx. Mani et al[5]had tried waste plastic pyrolysis oil produced from thermal pyrolysis as fuel in diesel engines and observed that NOx higher at peak loads and higher unburnt hydrocarbon and increased smoke levels. Osami Nishida etal[6] had tried waste plastic pyrolysis oil obtained from thermal pyrolysis as fuel in marine diesel engines. Since waste plastic oil has almost nil sulphur, SOx production been almost nil compared to Heavy Fuel Oil used in marine engines.Ozcanlietal[7] carried thermal pyrolysis of waste plastics and subsequently distillate those to observe better performance as fuel. In emission, increased in NOx has been observed even in light phase oil. Ahmad RozaimeeMustaffaetal[8]carried out cultivation of microalgae using sungaisurawater source for biodiesel production.Bridjeshetal[9] conceded experimental investigation of variable compression ratio diesel engine using Calophylluminophyllum Biodiesel. Ramaraoetal[10], done the experimental investigation on performance and emission characteristics of CI engine using nanoadditives withbiodiesel.
In this research paper, the diesel engine performance was evaluated by blending ethanol with waste plastic oil and diesel at definite mixing ratios of (Ethanol: WPPO: Diesel) 5:5:90, 10:10:80 and 15:15:70 respectively. The cetane additive AC2010Aadded to blended fuel is 1ml per 1000ml ofblended fuel. The objectives were followed as, to evaluate CI engine performance and emissions characteristics with blended diesel, Waste Plastic Pyrolysis oil (WPPO) and Ethanol and to observea substitutefuel by comparing the blended fuel results with base diesel.

2. Experimental Setup
A Twin cylinder, constant speed, Direct Injection engine was used to assess the engine performance and emission characteristics of plastic pyrolysis oil. The diesel runs under different load conditions at a constant speed of 1800 rpm with the different plastic oil proportions. The diesel engine was directly attached with an eddy current dynamometer for varying the loads from zeroload (0%) to full load (100%). Based on the engine power produced, the engine load range is varied from zero load condition of 0%, 25%, 50%, 75% and full load condition of 100%. The engine loads are varied manually with help of an eddy current dynamometer. AVL smoke meter was attached for measuring the smoke opacity and exhaust gas temperatures. The test rig was installed with AVL indimicro software to obtain various readings and results during operation. A five gas analyzer was used to measure the emission characteristics such as UHC, CO, NO_x, CO_2 and O_2 values from the exhaust gas.

2.1. Benefits of Cetane additive
Additives can increase the viscosity to make sure adequate lubrication of the injection pump, also it stabilizes the mixture in the presence of a high-water content, to ensure fuel is homogeneous under all conditions. It has good detergent properties to ensure fuel injection equipments remain clean and provide good atomization. Prime benefit of using the additive includes:

- Improved engine power and torque.
- Improved cetane number.
- Reduced emissions of carbon dioxide.
- Reduced corrosion.
- Reduced engine noise.

Table 1. Properties of Diesel, Plastic oil and Ethanol

Sl.No	Properties	Diesel Fuel	Plastic Oil	Ethanol
1	Density (kg/m^3)	852	794	790
2	Calorific Value (kJ/kg)	46450	44344	27710
3	Kinematic Viscosity (cSt)	3.10	2.13	1.05
4	Flash point °C	53	42	16.5
5	Fire point °C	57	46	98
6	Cetane Number	55	52	9
7	Sulphur (%)	< 0.037 %	< 0.023 %	-

Blended fuel in the blended ratio of Ethanol 10%; WPPO 10%; Diesel 80% was tested and its properties are tabulated in Table 2.

Table 2. Properties of Blended Fuel

Sl.No	Properties	Blend ofE/WPPO/D	E/WPPO/D withcetane additive
1	Kinematic Viscosity (cSt)	2.34	2.34
2	Cetane Number	51	58
3	Carbon Residue (%)	0.01 %	0.01 %
4	Sulphur (%)	0.04 %	0.04 %

3. Results and Discussion

The engine performance and emission analysis test of Waste Plastic Pyrolysis oil, Ethanol, Diesel with a cetaneadditive with different blends were tested and evaluated for varied load condition.

3.1. Variability of SFC Vs Load

The variability of Specific Fuel Consumption (SFC) for varied loads (kW) applied on the engine for various ratios of waste plastic pyrolysis oil, ethanol and diesel blends with AC 2010A Additive were shown below in the Figure 1.

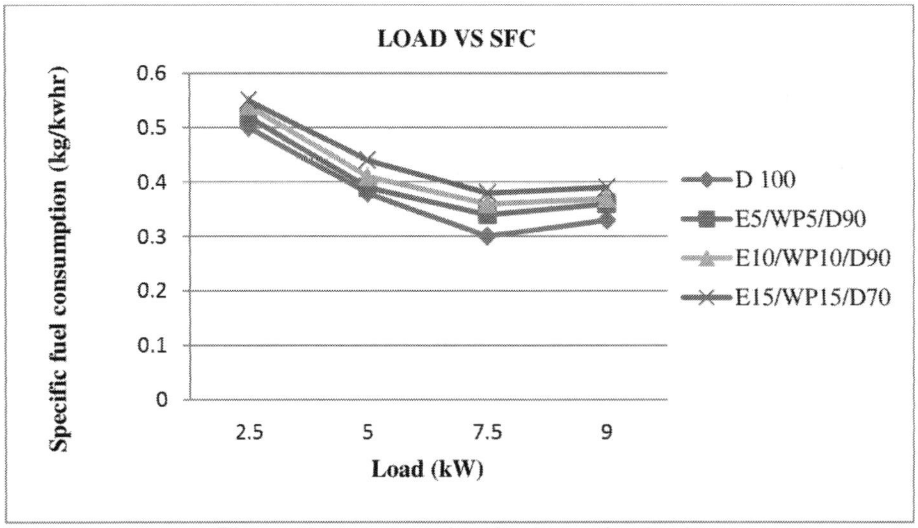

Figure. 1. Variation of SFC with engine loads

IOP Conf. Series: Materials Science and Engineering **197** (2017) 012023 doi:10.1088/1757-899X/197/1/012023

The Specific fuel oil consumption of blended oil is slightly higher than diesel oil. The E5/WPP5/D90 blend has specific fuel oil values much nearer to diesel. From the graph, it is derived that the diesel has a lower SFC because of high calorific value, with blended fuels such as E5 and E10 the equivalent SFC was very closer but higher than that for the pure diesel fuel. This was observed due to the slightly lower calorific value and higher viscosity of the plastic pyrolysis oil.

3.2. Variation of Brake Thermal Efficiency Vs Load

The variation of the brake thermal efficiency (BTE) for different loads with blended fuel at different ratios of diesel, waste plastic pyrolysis oil, ethanol and cetane additive blends were shown in figure 2. The brake thermal efficiency of blended oil is slightly lower than diesel oil. The E5/WPP5/D90 blend has brake thermal efficiency values almost like diesel.

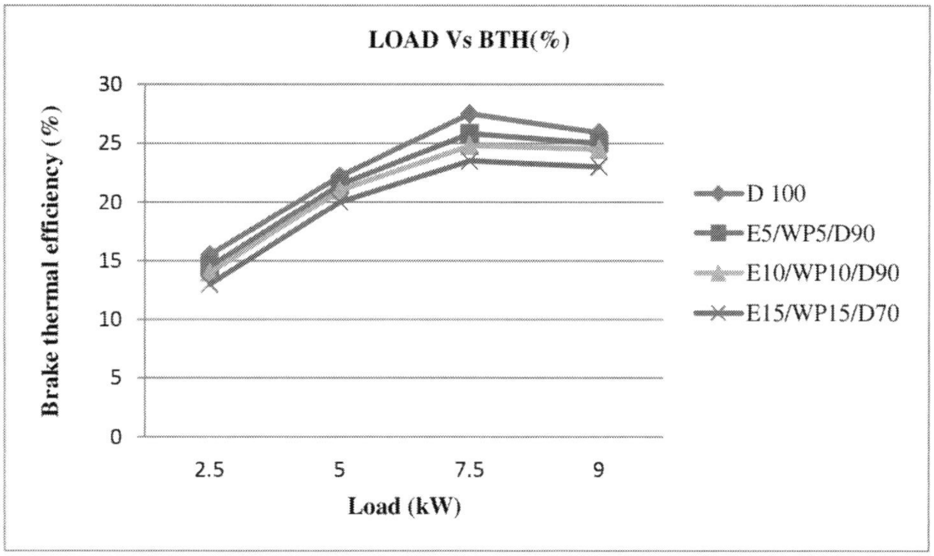

Figure. 2: Variation of BTE with engine loads

The higher viscosity of the blended fuel reduced the brake thermal efficiency. The BTE of the E5 blended fuel was similar to that of the diesel performance. Itis observed that BTE of 26% at maximum load condition and it was only 2% variation from that of the pure diesel performance.

3.3. Variation of Load with NOx

The variation of nitrogen oxide emission from diesel, waste plastic pyrolysis oil, ethanol and cetane blends are slightly lower than the pure diesel fuel performanceat maximum loads and it was shown in figure 3.

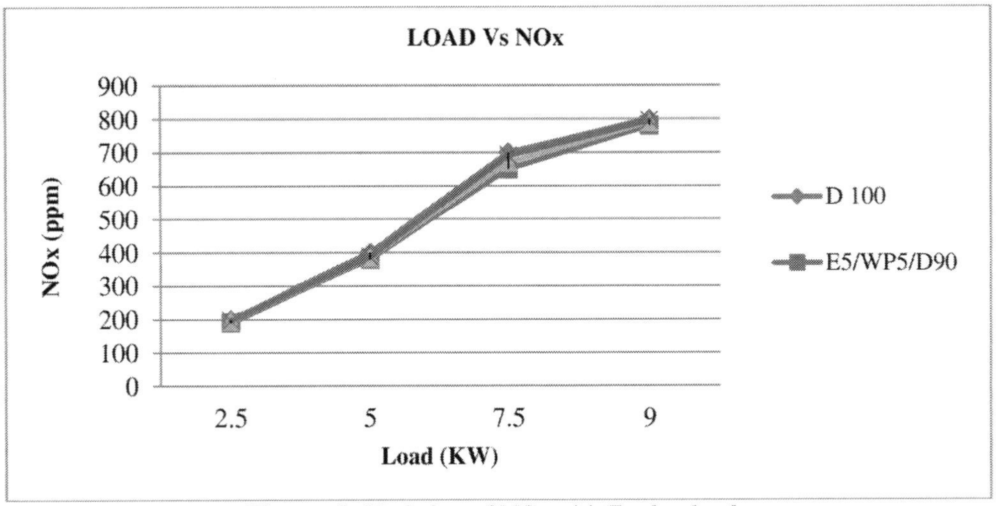

Figure. 3: Variation of NO_x with Engine loads

Because ofmore air action, much elevated than stoichiometric values, diesel engines produce more NO_x at part load conditions and at higher loads. Diesel has higher calorific value than that of waste plastic oil, so a lesser amount of diesel was injected into the combustion chamber. Due to this more heat was liberated during the combustion and it was increases the combustion chamber temperature.

3.4. Variation of Load with CO

The various Carbon Monoxideemissions for varying blends were shown in figure 4. The Carbon Monoxide emissions from blended fuel area bitmore than the normal diesel fuel at all loads. Diesel engines produce little amount of CO when compared to NO_x and particulate emission as the engine was not loaded. It also concluded that blended fuel emission of CO is like that of pure diesel.

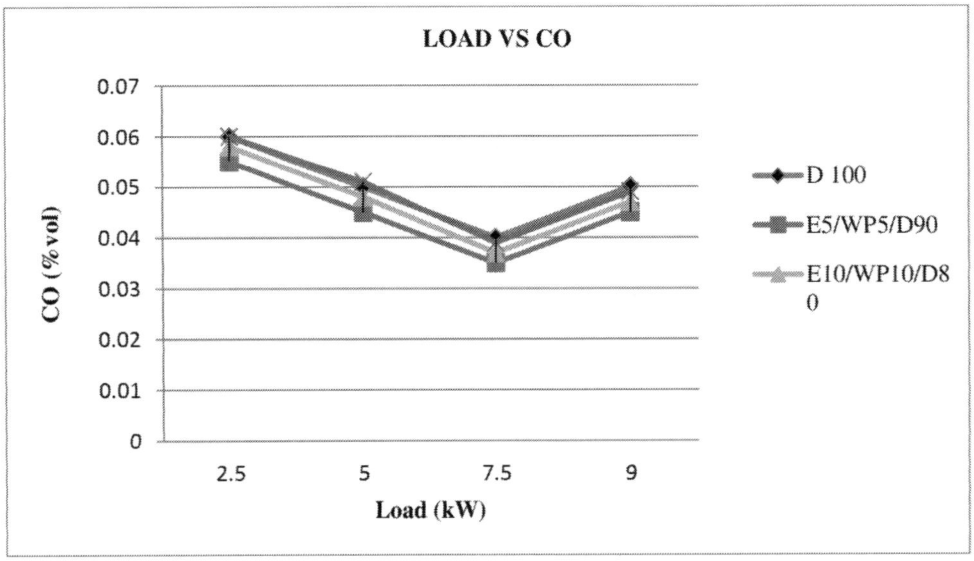

Figure. 4: Variation of CO with engine loads

3.5. Variation of Load with CO_2

The various Carbon Dioxide levels for varying blends were shown in figure 5. From the readings, it is inferred that emission of Caron dioxide is like that of pure diesel.The CO emission can be reduced by providingsurplus oxygen in to the combustion chamber. There by converting CO into CO_2this is possible by complete combustion.

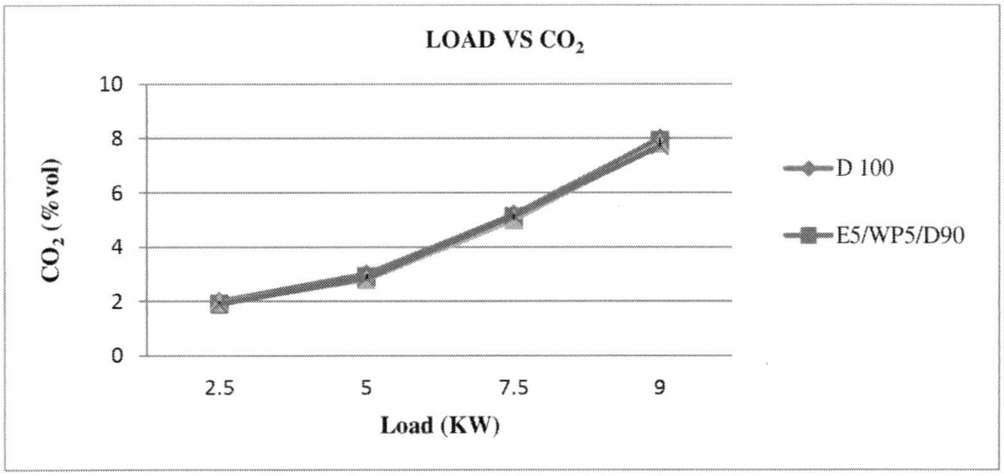

Figure. 5: Variation of CO_2 with engine loads

Carbon dioxide will appear in the exhaust of the diesel engine at full load condition. From the results of emission characteristics, it wasobserved that the Carbon dioxide emission for allblended ratio of blended fuel was more likely compared to diesel performance at all load conditions.

3.6. Variation of Load with UHC

The various hydrocarbonemissions for different blends were shown in figure 6. Excluding at higher loads the hydrocarbon emissions are little lower than diesel.

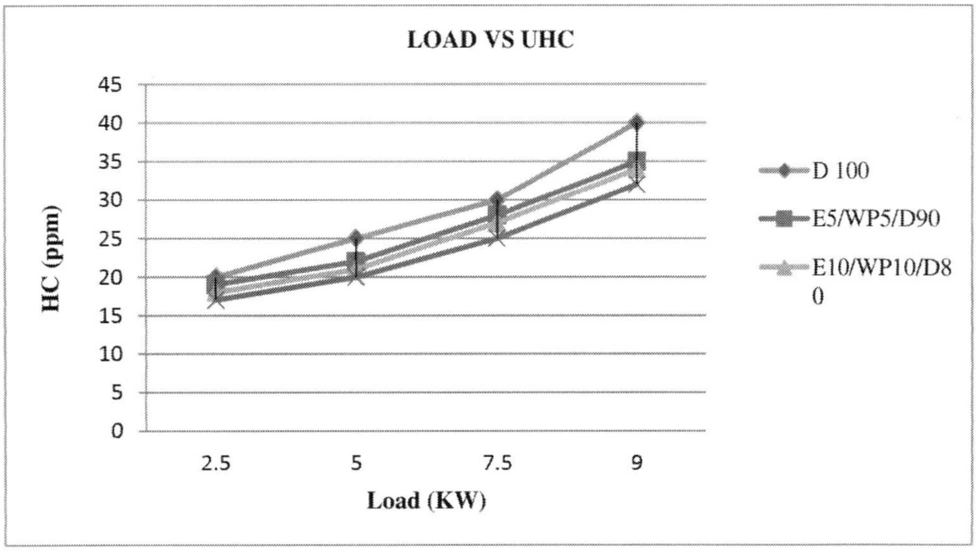

Figure. 6: Variation of UHC with engine loads

At different load conditions, the unburned hydrocarbon emission was lower for the diesel fuel with respect to the blends E5 to E15. This is because of the higher calorific value of diesel and due to which less amount of fuel was injected when compared to plastic oil blends. This builds the engine to function at leaner state. Because of oxygen enrich environment, combustion is complete. Hence lower unburned hydrocarbon emission wasobserved with diesel fuel.

4. Conclusion

Alternative fuels should be available at ease and at low cost, should be environment friendly and provide energy security needs without reducing engine's operational performance. Waste to energy is the trend in the selection of alternate fuels. Plastics waste has become a significantelement in resourcefulwaste. In this work, Waste Plastic oil Ethanol, Diesel blend with Cetane additive has been attempted as an alternative fuel. The experiments were conducted without any modification on the engine. CI engine performance tests were conducted with various blend ratios of diesel-WPPO-ethanol with cetane additive and the best combination of blends and additives were studied.Based on the engine performance and emission characteristic test of the Waste Plastic Pyrolysis oil, Ethanol, Diesel and Cetane Improver blend symbolizeaadmirablesubstitute fuel which gives better performance and similar emission characteristics results compared with base pure diesel. From this study, blend of diesel 90%;waste plastic pyrolysis oil 5%; ethanol 5% with cetane additivewas performance better when compared to other blends.

REFERENCES

[1] Panda A K, Singh RK and Mishra DK 2010, Thermolysis of waste plastics to liquid fuel A suitable method for plastic waste management and manufacture of value added Products-A world prospective, *Renewable and Sustainable Energy Reviews*, **14**233-248.

[2] Alan C Hasen, Qin Zhang and Peter WL Lyne2004, Ethanol Diesel fuel blend, Aerosol Science and Technology,**362–369**1521-7388.

[3] CenkSayin and MetinGumus 2011 Impact of compression ratio and injection parameters on the performance and emissions of a DI diesel engine fuelled with biodiesel-blended diesel fuel, *Applied Thermal Engineering*, **31**3182-3188.

[4] Viswanath KKaimal and Vijayabalan P2015 A detailed study of combustion characteristics of a Direct Ignition diesel Engine using waste plastic oil and its blends, Energy conversion and Management,**105** 951-956.

[5] Mani M, Nagarajan G and Sampath S 2011 Characterisation and effect of using waste plastic oil and diesel Fuel blends in compression ignition engine, *Energy***36** 212-219.

[6] Osami Nishida, Hirotsugu Fujita, WataruHarano, PhangHai VU,Tsuyoshi Fukui, Hitomi Saito and Norihiko Toyoshima 2001 Application of Waste Plastic Disposals to Marine Diesel Engines, *Bulletin of the JIME*, **29**128 -34.

[7] Ozcanli M 2013 Light and heavy phases derived from waste polyethylene by thermal cracking and their usage as fuel in DI diesel engine, *Journal of Scientific and Industrial Research*, **72** 198-202.

[8] Ahmad RozaimeeMustaffa, Ku Halim Ku Hamid, Mohibah Musa, JailaniSalihon andRoslinaRamli 2016, Cultivation of Microalgae using Sungai Sura Water Source as a Medium for Biodiesel Production, *Indian Journal of Science and Technology*, **9** 9, Doino:10.17485/ijst/2016/v9i9/88732.

[9] Bridjesh P, Prabhu KishoreN, MallikarjunaMV andAlekhyaN 2016 Performance Analysis of Variable Compression Ratio Diesel Engine using Calophylluminophyllum Biodiesel *Indian Journal of Science and Technology***9** 35, Doino:10.17485/ijst/2016/v9i35/95577.

[10] Ramarao K, RaoCJ andSreeramuluD 2015 The Experimental Investigation on Performance and Emission Characteristics of a Single Cylinder Diesel Engine using Nano Additives in Diesel and Biodiesel,*Indian Journal of Science and Technology***8**29, Doino:10.17485/ijst/2015/v8i29/75718.

IOP Conf. Series: Materials Science and Engineering **197** (2017) 012024 doi:10.1088/1757-899X/197/1/012024

Emission and Performance Analysis of ZrO_2 And CeO_2 Coated Piston Using Refined Vegetable Oils

J Hemanandh [1,*], K.V.Narayanan[2], Vemuri Manoj[3]

[1]Asst Professor, [2,3]Student, School of Mechanical Engineering,
Sathyabama University, Chennai-600119, Tamilnadu, India.
Email.Id: hemanandh123@gmail.com, Contact No:+91 9444167753.
[*] Corresponding author

Abstract. Increase in global warming and pollution leads to look for an alternative fuel. The aim of this paper to improve the performance and to reduce the emissions in DI diesel engine. The 80% of ZrO_2 and 20% of CeO_2 were mixed and coated on the piston head using plasma spray method. The B10 fuel of various refined vegetable oil methyl esters were used as fuel. The test was conducted in the 4-stroke DI diesel engine at a constant speed of 1500 rpm. The results show that the brake thermal efficiency, NO_x and BSFC was increased. The CO and HC were decreased.

1. Introduction

Sivakumar and Senthil Kumar [1] carried out an experimental investigation by using YSZ as the thermal barrier material on the piston and fuelled with diesel by varying the loads the results revealed that TE,CO and CO2 are increased, BSFC and HC are decreased. Selman Aydınet al., [2] has conducted an experimental study on a diesel engine coated with ZrO_2 and fuelled with biodiesel produced from waste cooking oil and conducted performance tests on both coated and uncoated engines and found that the modification of engine resulted in lower BSFC. The emissions were lowered excluding NOX emission and the results of the coated engine are better in gas pressure ,HRR and HR. Bahattin ISCAN et al., [3] has done an experimental investigation on the diesel engine coated with the ZrO_2 and fuelled with the different blends of cotton seed oil with PD and compared the obtained results with the results obtained by employing the number 2 fuel in uncoated engine and found that all the results are improved for the coated engine except NO_x emission. M. Ekström et al., [4] conducted a experimental study on the coatings of the exhaust manifold with seven different coatings (mullite, forsterite, $La_2Zr_2O_7$, 8YSZ, nanostructured 8YSZ) and two sol-gel composite (one sprayed and one dipped)) and were examined for their thermal insulation properties and oxidation and spoliation resistance and found that the reduction of temperature up to 50°C by using TBC UPTO 0.2 mm with thermal conductivity close to 0.1 W/m K, or a 3–6 mm thick TBC with thermal conductivity 1.5–3W/m K. Huseyin Aydin [5] has done an experimental investigation on the diesel in which some of its parts are coated with ZrO_2 and fuelled with the vegetable oils (cottonseed oil and sunflower oil) blended with PD and the results stated that the decrease in the BSFC,CO,HC and smoke opacity. Sunflower oil gives better performance than cotton seed oil fuel. Selman Aydin et al., [6] has conducted an experimental study on a 510 model Lombardini CI engine coated with ZrO_2 , MgO, $Al_2 O_3$, NiCrAl and fuelled with biodiesel produced from residual frying oil of cotton seed and the fuels D2, B5, B100. The performance, Combustion and the emission tests were carried out and the results show that there is partial increase in the B.P, exhaust manifold temp., engine noise and partial decreases were seen in BSFC and the emissions of CO,HC and smoke opacity while NO_x increases. Naveen.P et al [7] has done an experimental investigation on a diesel engine which was coated with the Al_2O_3 and TiO_2 fuelled with Honge BioDiesel and compared the results with the test conducted on the standard engine with same operating conditions and found that there was reduction in BSFC and increase in Thermal Efficiency in the coated piston. Vedharaj et al., [8] conducted an experimental study on a diesel engine coated with the PSZ and subsequently fuelled with the blends of kme B25

IOP Conf. Series: Materials Science and Engineering **197** (2017) 012024 doi:10.1088/1757-899X/197/1/012024

AND B50.Then the performance and emission tests were conducted and from the results they found that B50 shows increase in BTE, decrease in HC,CO and smoke opacity however there is increase in cylinder temperature and NO_x emissions when compared to uncoated engine. In order to reduce NO_x emission Urea based SNCR system was incorporated in the exhaust pipe through which the NO_x emissions are decreased. Ahmet Necati Ozsezen et al., [9] have conducted a experimental study on the direct injection diesel engine fuelled tentatively with the canola oil methyl ester and waste palm oil methyl ester. The test is conducted at constant speed at full load condition and found that there is reduction in CO, HC, CO_2 and B.P and increase in BSFC and NO_x when compared to PBDF. Muralidharan et al., [10] have done an experimental investigation on a diesel engine working at the constant speed, at constant load by varying the compression ratios. This engine if fuelled with the waste sunflower oil blended with the PD. The various blends were used as fuel and the blend B40 is identified as the best one in terms of thermal efficiency. The usage of these blends has resulted in the reduction of CO, HC emissions and increase in the NO_x emissions. Ertan Alptekin et al., [11] have conducted an experimental study on the DI diesel engine running at the constant speed and the varying load. The biodiesel produced from the pilot plant were blended with certain amounts of diesel fuel and bio ethanol. The tests were conducted on the engine with the biodiesel, diesel and the blends. The results state that BSFC of the biodiesel and the blends containing Bioethanol were higher when compared to the Diesel. The start of the combustion occurred at the earlier crank angles for biodiesel and occurred at later crank angles for bio ethanol when compared to diesel. biodiesels emitted lower carbon monoxide (CO) emissions, higher carbon dioxide(CO_2) and oxides of nitrogen (NO_x) emissions as compared to diesel fuel. For the blends the HC emissions increased and CO_2 emissions reduced. M. Mofijur et al [12] have carried out an comparative evaluation biodiesels. They found that many large number of literatures from highly rated journals in scientific indexes are reviewed including the most journals state that the usage of biodiesel reduces the emission of HC, CO, CO_2, PM but results in higher NO_x emissions compared to PD. Biodiesel lowered the B.P, BTE but increased the BSFC compared to diesel. S.M. Palash et al [13] conducted a study on the NO_x reduction techniques in the combustion of biodiesels. Increasing energy demand and reduction in the fossil fuels and the effect of fossil fuels on the environment has prompted for alternative fuels. As an alternative fuel biodiesel reduces the emissions but the NO_x emissions are high compared to PD. In biodiesels the NOx emissions is high due to many factors such as physicochemical properties and molecular structure of biodiesel, flame temperature, ignition delay , injection timing and engine load conditions etc. The NO_x emissions can be reduced by the both pre and post combustion techniques such as exhaust gas recirculation technique (EGR) and retarded injection timing . Among these EGR is the most effective one which decreases the NO_x by 5-25% and slightly decreasing the HC, CO. I.M. Rizwanul Fattah et al., [14] conducted a study on the biodiesels and reviewed the facts and utilization of biodiesels. Biofuels mainly soybean, rapeseed, palm, jatropha and cottonseed oils reduce engine exhaust gas, noise and petro dependency. Through this study the authors have found that the biofuels have many benefits like reduction of greenhouse gas emissions and many harmful pollutants along with noise emission.. Sakthivel et al., [15] have conducted an experimental study on a diesel engine running at constant speed, at variable loads fuelled with the biodiesel produced from the fish oil blended with diesel and the results have shown the decrease in the NO_x, HC, CO with increase in CO_2,smoke and BTE when compared to diesel .

Nomenclature —

CO	-	Carbon monoxide
CO_2	-	Carbon Dioxide
HC	-	Hydro Carbons
PD	-	Pure Diesel
RPM	-	Refined Palm oil
RCF	-	Refined Corn Oil
BSFC	-	Brake Specific Fuel Consumption

RSF	-	Refined Sunflower oil
BTE	-	Brake Thermal Efficiency

2. Materials and methods

2.1 Plasma Spray Coating

Plasma spray process is a thermal resistance technique .The thermal conductivity of this process ranges from 0.5w/mk to 1.5w/mk .This process is mainly used to improve the performance characteristics of various components which are subjected to high temperature in this process Argon and Hydrogen are used as arc gases .A high power source is required for this process. The mixture of ZrO2 and CeO2 80% and 20% are fed to the arc gun in the form of powder which are supplied to the piston surface .Arc is generated between the electrodes. Due to the high temperature attained by the arc the powder supplied gets melted and the splat formation takes place which is followed by condensation, solidification and adhesion. The formation of the coating on the substrate is due to the KE.

2.2 Transesterification process

The reaction of a triglyceride which is a combination of fat or oil with an alcohol to form glycerol and esters is called transesterification. During this reaction process a catalyst like Sodium Hydroxide (NaOH) is used. The alcohol reacts with fatty acids to form methyl esters and glycerol. In most processes methanol is used. A successful transesterification reaction is identified by the separation of the methyl ester and glycerol layers after the reaction time. The heavier, co-product, glycerol settles down. First vegetable oils which are brought from the local market are checked for the FFA If the free fatty acids are less than 5% then the one litre of refined Vegetable oil is treated with 250ml of methanol and 2.5gms of NaOH as catalyst. At first the oil is preheated up to 20 °C to 40 °C and is allowed to cool down naturally. At cold temperature i.e atmospheric temperature or lower than that Methanol is added to the catalyst in the preheated oil While performing Transesterification process of oil the temperature is raised to 70 °C to 80°C the process performed to reduce high viscosity and give pure methyl esters without any soap content.

After the completion of the process the glycerol and the methyl esters are separated even after separation obtained methyl esters contains small amount of water. so for removal of water from these esters they are heated up to 1hr at 20°C and then esters and water are separated This process is followed for PM ,CF , SF. Then these oils are blended with the pure diesel and the blends obtained are B10 , B25 ,B40.These blends are studied for the emission characteristics in the kirloskar diesel engine.

Table 1. Specification of Test Engine

Type	:	Kirloskar Vertical, 4S, Single acting, Diesel Engine
Combustion	:	Direct Injection
Rated Power	:	4.3 kW
Rated Speed	:	1500 rpm
Compression Ratio	:	17.5: 1
Injector type	:	Single 3 hole jet injector
Fuel injection pressure	:	210 bar

Table 2. Details of Measuring Systems

Pressure Transducer	-	GH 12 D
Data Analyzer from Engine	-	AVL PIEZO CHARGE AMPLIFIER
To measure pressure	-	AVL 364 Angle Encoder
Smoke meter	-	AVL 437 C Smoke

IOP Conf. Series: Materials Science and Engineering **197** (2017) 012024 doi:10.1088/1757-899X/197/1/012024

5 Gas Analyzer (NOx, HC, CO, CO2, O_2) - AVL DIGAS 444 Analyzer

2.3 Experimental Setup

The table shows the experimental setup. A kirloskar 4-Stroke, Direct Injection Diesel Engine is used to evaluate the Emission and performance characteristics of the various Refined Vegetable oils at various injection pressure and loading. by using this setup we measure the emissions of various vegetable oils from the engine at different loads. The emission parameters that we evaluate are (CO, HC, NOx, CO2) and performance of the Brake specific fuel consumption, and Brake thermal Efficiency. These parameters are measured by the Gas analyzer (AVL DIGAS 444 Analyser).The eddy current dynamometer is coupled to the engine to obtain the various loads (0%, 25%, 50%, 75%, 100%).

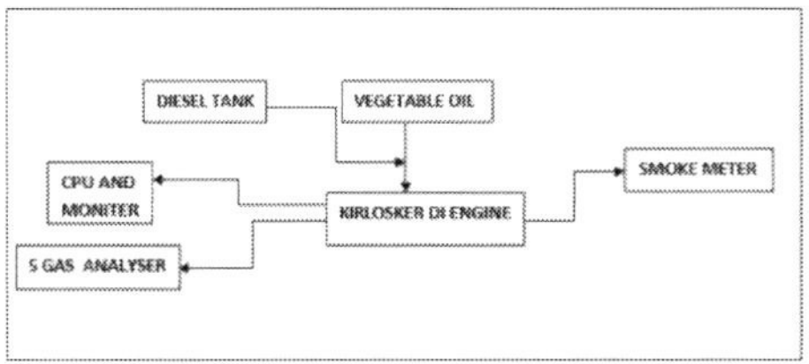

Figure 1. Schematic diagram of the Experimental Set- up

2.4 Test Procedure

This experiment is conducted by running the engine at a constant speed of 1500rpm. Initially the engine is run at no load condition for ten minutes before applying load for every vegetable oil. The various blended oils are injected at various injection pressures (180 bar, 210 bar, 240 bar) and then the load is increased gradually (up to 100%) and the various emission and performance characteristics are measured using the 5 Gas analyzer at the exhaust stream of the engine. The properties of vegetable oils are compared with Diesel, ASTM and BIS standards as given in Table -3.

Table 3. Comparison of properties of Diesel, Biodiesel standards & Vegetable oils

S.No.	Properties	Diesel	BIS standard Bio Diesel	ASTM D – 6751(IS15607: 2005	PM	CF	SF
1.	Cetane Index(Min)	46	51	-	36	35	38
2.	Density at 15°C kg/m^3	820 -845	860-900	860 – 900	874	923	917
3.	Kinematic Viscosity at 40°C cst	2 – 4.5	2.5 – 6	1.9 – 6	4.9	5.02	4.98
4.	Flash point °C min	35°C	262°C	130°C	220	162	163
5.	Calorific Value kJ/kg	42,000	-	-	37,037	36,824	39,8284

3. Results and discussion

3.1 Carbon Mono Oxide (CO)

The fig. 2 shows the CO emissions of the RPM, RSF , RCF, PD at various load conditions. When the emissions of the vegetable oils are compared with the petroleum diesel the emissions of the CO. At the full load condition is reduced by 35.71% by using RCF but RSF reduce by 21.42% and RPM reduce by 28.57% . This may be due to the adequate supply of oxygen.

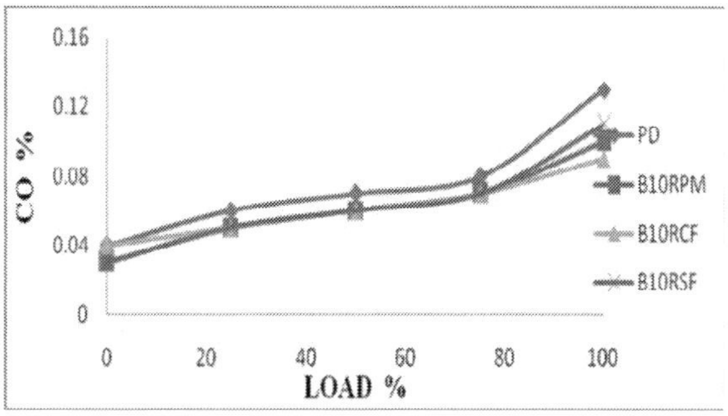

Figure 2. Variation of CO with respect to various vegetable oils

3.2 Hydrocarbon (HC)

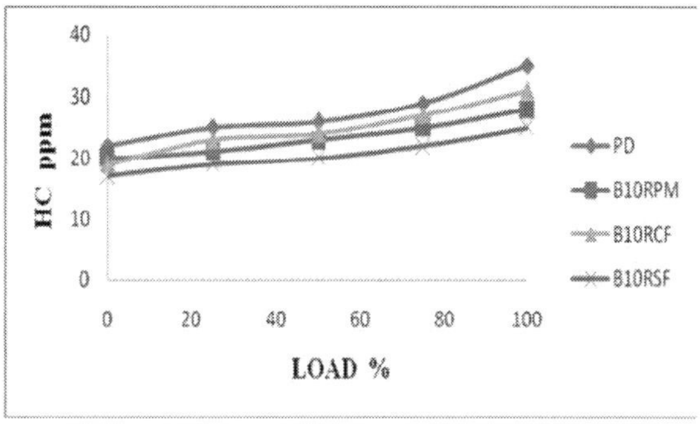

Figure 3. Variation of HC with respect to various vegetable oils

The fig.3 show the emissions of HC in the various vegetable oils and petroleum diesel. At the full load condition the emission of HC is reduced when compared to PD and the rate at which they reduce are in RPM by 20%, in RCF by 14.28% ,in RSF by 37.14%.There is reduction in the amount of HC in vegetable oils due to the difference in viscosity values of PD and vegetable oils.

3.3 Carbon Di Oxide (CO_2)

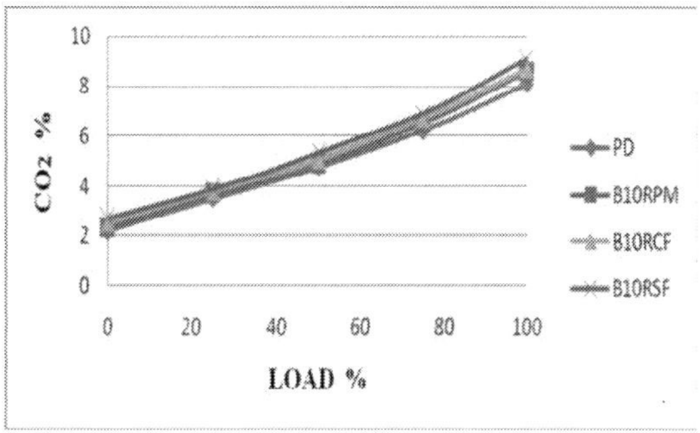

Figure 4. Variation of CO2 with respect to various vegetable oils

The fig. 4 show the emission of CO2 for PD and the 3 vegetable oils. There is marginal increase in the emission of CO2.This may be due to the incomplete combustion.

3.4 Nitrogen Oxide (NO_x)

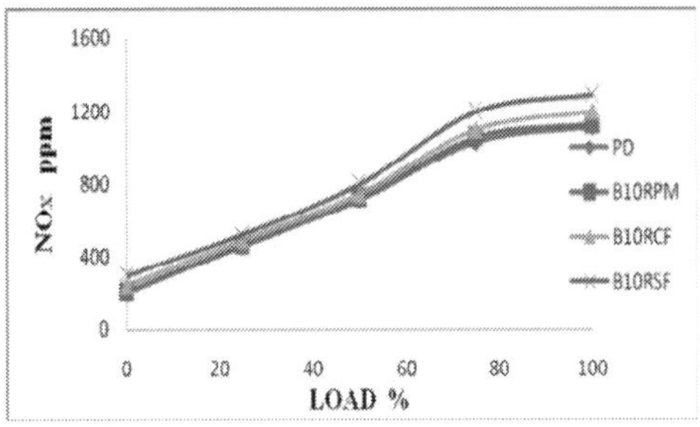

Fig. 5 Variation of NOx with respect to various vegetable oils

The fig.5 shows the emission of NOx at various loads of different oils. At full load condition the emissions of NOx is increased by 4.5%,8.33%,12% for RPM , RCF and RSF respectively. The emission of NOx is more in vegetable oils the reason for this may be due to the higher peak temperature.

3.5 Break Specific Fuel Consumption (BSFC)

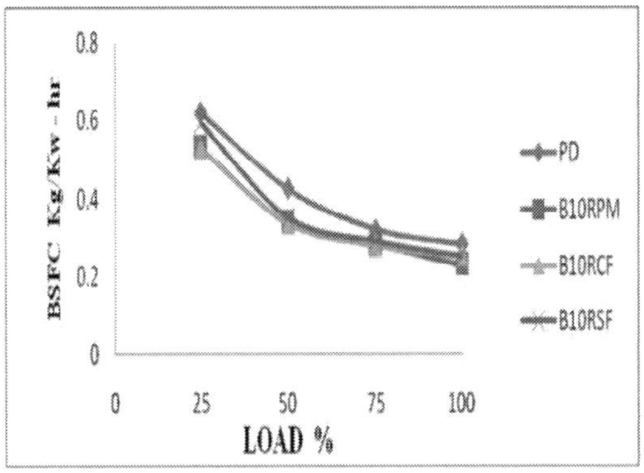

. **Figure 6.** Variation of BSFC with respect to various vegetable oils

The fig.6 shows the BSFC of various vegetable oils at different load conditions. The BSFC decreases marginally for vegetable oils when compared to PD. This may be due to the calorific value of the blends.

3.4 Break Thermal Efficiency (ηbt)

The fig.7 shows the BTE of various vegetable oils at the different load conditions. The BTE keeps increases as the load on the engine increases. The BTE of the vegetable oils are more than the PD . Among the RPM produces the maximum BTE i.e 23.07%.The amount of the BTE is more for biodiesels than the Diesel due to the better combustion of biodiesels

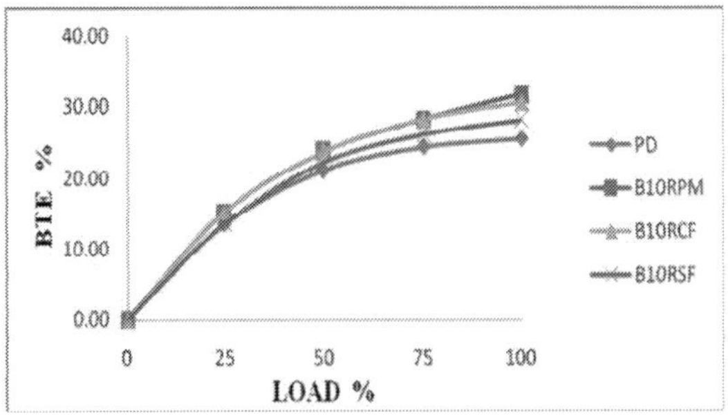

Figure 7. Variation of BTE with respect to various vegetable oils

IOP Conf. Series: Materials Science and Engineering **197** (2017) 012024 doi:10.1088/1757-899X/197/1/012024

4. Conclusions

The Engine was tested with various blends by varying load at constant speed. The properties of the biodiesel were analyzed and the following conclusion were arrived from the graph.

1. It was found that for a blend of diesel of 10% RPM, at full load , the CO, HC decreased by 28.57%, 20.03% and there is marginal increase in NOX and CO2 than PD.

2. The BTE is increased by 23.07% and there is marginal decrease in BSFC when compared to PD. The performance and emission characteristics for RPM has given better results than other two vegetable oils.

References

[1] G. Sivakumar, S. Senthil Kumar, "Investigation on effect of Yttria Stabilized Zirconia coated piston crown on performance and emission characteristics of a diesel engine"; Alexandria Engineering Journal ;2014;53, pp. 787–794

[2] Selman Aydın, Cenk Sayın, "Impact of thermal barrier coating application on the combustion, performance and emissions of a diesel engine fuelled with waste cooking oil biodiesel–diesel blends"; Fuel ;2014; 136 pp. 334–340)

[3] Bahattin _IS¸ CAN, "Application of ceramic coating for improving the usage of cottonseed oil in a diesel engine"; Journal of the Energy Institute;2015 pp. 1-8

[4] M. Ekström, A.Thibblin, A.Tjernberg ,C. Blomqvist ,S. Jonsson, "Evaluation of internal thermal barrier coatings for exhaust manifolds"; Surface & Coatings Technology;2015; pp. 198–212.

[5] Huseyin Aydin, "Combined effects of thermal barrier coating and blending with diesel fuel on usability of vegetable oils in diesel engines"; Applied Thermal Engineering;2013; 51 pp. 623-629.

[6] Selman Aydin et al, "Investigation of the usability of biodiesel obtained from residual frying oil in a diesel engine with thermal barrier coating"; Applied Thermal Engineering; 2015;80, pp. 212 - 219.

[7] Naveen. P et al,, "Effect of Titanium Oxide Coating on Performance Characteristics of Bio-Diesel (Honge) Fuelled C.I. Engine"; International Journal of Modern Engineering Research 2012; Vol.2, pp. 2825-2828

[8] S. Vedharaj et al, Reduction of harmful emissions from a diesel engine fueled by kapok methyl ester using combined coating and SNCR technology, Energy Conversion and management;2014;79 pp. 581-589.

[9] Ahmet Necati Ozsezen, Mustafa Canakci, "Determination of performance and combustion characteristics of a diesel engine fuelled with canola and waste palm oil methyl esters", Energy Conversion and Management; 2011;52 pp. 108-116.

[10] K. Muralidharan , D. Vasudevan, "Performance, emission and combustion characteristics of a variable compression ratio engine using methyl esters of waste cooking oil and diesel blends"; Applied Energy;2011;88 pp. 3959-3968.

[11] Ertan Alptekin et al, "Using waste animal fat based biodiesels–bioethanol–diesel fuel blends in a DI diesel engine"; Fuel;2015;157 pp. 245-254.

[12] M. Mofijur et al, "A study on the effects of promising edible and non-edible biodiesel feed stocks on engine performance and emissions production: A comparative evaluation" Renewable and Sustainable Energy Reviews; 2013; 23 pp. 391-404.

[13] S.M. Palash et al, "Impacts of biodiesel combustion on NOx emissions and their reduction approaches, Renewable and Sustainable" Energy Reviews; 2013 pp. 473-490.

[14] I.M. Rizwanul Fattah et al, "Impact of various biodiesel fuels obtained from edible and non-edible oils on engine exhaust gas and noise emissions, Renewable and Sustainable Energy reviews", 2013;18 pp. 552-567.

[15] G. Sakthivel et al, "Comparative analysis of performance, emission and combustion parameters of diesel engine fuelled with ethyl ester of fish oil and its diesel blends", Fuel; 2014;

IOP Conf. Series: Materials Science and Engineering **197** (2017) 012025 doi:10.1088/1757-899X/197/1/012025

Influence of MgO on Performance and Emissions of DI Engine using blends of Castor oil

Ganesan.S[1] , Padmanabhan.s[2],Senthil kumar.J[3],Polina Navakanth[4], Sanivada Krishna kumar[5],

[1,2,3]Faculty of Mechanical Engineering, Sathyabama University, Chennai, India
[45,]Department of Mechanical Engineering, Sathyabama University, Chennai, IndiaEMAIL ID:
Email id: [a]gansuma@gmail.coml, [b]Navakanthpolina@gmail.com,[c] sanivada1995@gmail.com

Abstract: With the technological development, the research over alternate fuels is increasing day by day in order to help the upcoming generation with a bright and greener future. In order to preserve the existing petroleum resources for future generation, it is necessary to soon switch to any alternate source which is easily available, renewable as well as environment friendly. In this paper I would like to highlight upon the usage of Diesel, Castor Oil and Nano Particles for a compression ignition engine and study the emission characteristics of this fuel at different mixing ratios and analyze the different levels of residue particles.

Introduction

The various research works going on the use of bio fuels, one of the researcher has performed a comparative experimental study for determining the effect of fuel properties on the constructive characteristics of some pieces of a current common rail injection system used in light duty diesel vehicles.Results showed that the use of the ethanol–biodiesel–diesel blend produced a similar effect on the durability on the injection pump components studied and on the injector nozzle as that produced by diesel fuel. [1]
Biodiesels are used in diesel engines without making any alterations to the engine. the ethanol mix biodiesel is widely researched and experimented with various compositions and their performance is noted down. Isopropanol is also being added to mixture. Experimental results showed that the ethanol addition reduced CO, soot and SO2 emissions, but it increased the NOx emission and power reductions. It was also found that increasing the injection pressure decreased CO and the smoke emissions. [2]
Ethanol being widely used as a blend in biofuels, it is being reseaarched extensively. During analysis, special emphasis is placed on the factors such as stability, viscosity and lubricity, safety and materials compatibility critical to the potential commercial use of these blends. The effect of the fuel on engine performance, durability and emissions is also considered. But maintaining vehicle safety with these blends may entail fuel tank modifications and work is required in specifying acceptable fuel characteristics, confirming the long-term effects on engine durability, and ensuring safety in handling and storing ethanol–diesel blends. [3]
In the ongoing research for searching a suitable biofuel in order to replace the non renewable resources, in 1996 an oil-soluble organo-metallic iron combustion catalyst was developed for turbine engines. It included oil-soluble magnesium to reduce vanadium deposits and corrosion which resulted in significant reduction of smoke in the exhaust of engines operating in steady state and non-equilibrium start-up conditions. The same biofuel when used for steam boilers and compression-

ignited reciprocating engines, it was noted that it promotes complete and more efficient combustion in the engine, resulting in increased power, improved fuel economy and radically reduced smoke emission. The iron-magnesium combination acts synergistically to give greater activity than expected. [4]

In the research works it is also been seen the influence of a ferrous picrate based homogeneous combustion catalyst on fuel consumption and smoke emission of a laboratory diesel engine. on analysis, it was noticed that the brake specific fuel consumption and smoke emission decreased as the dosing ratio of the catalyst doped in the diesel fuel increased. At a certain point, the brake specific fuel consumption was reduced by from 2.1% to 2.7% and the smoke emission was reduced by from 6.7% to 26.2% at the full engine load. The results indicated that the potential of the fuel saving at light load. [5]

In a recent study where ethanol with Palm Stearin Methyl-Ester oil as additive has been carried out in a CI diesel engine to study engine power, torque, break specific fuel consumption, break thermal efficiency and exhaust gas temperature with the diesel – ethanol blend with addition of small amount of biodiesel(PSME). The study was tested using diesel blended with ethanol at certain mixing ratios. From the experiment it was shown that the brake thermal efficiency of the engine increased for B40 blend for medium load capacity and showed that the exhaust gas temperature for B10 ratio is near the diesel fuel. The exhaust temperature for diesel fuel was higher compared to any mixing of the blended fuel. Brake specific fuel consumption of all ethanol Methyl-ester, diesel blends were lower compared with diesel at full load. [6]

Apart from the various blends which are being tested day by day with different additives, different methods of producing them are also been found. Some of the biodiesel production techniques are supercritical methanolysis, ultrasonication method and microwave technique by which maximum biodiesel can be produced. The new approach of using nano particle in biodiesel shows very good results in reducing the level of pollutant gases in the engine exhaust and increased performance without any engine modification. [7]

During the usage of Biodiesel, the particulate and high emissions of $NO_x$are the main problems. So, a mixture of acetal and regular diesel fuel has been tested in a heavy-duty diesel engine. The effect was a marked decrease in particle number and the estimated particle mass. However, the net effect was nevertheless a reduction in the emission of CO_2. The emissions of HC, CO, NOx, some aldehydes, and hydrocarbons were only slightly affected by the new fuel composition. An exception was the emission of acetaldehyde, which was almost quadrupled, probably reflecting the decomposition and oxidation of acetal to acetaldehyde. [8]

Apart from the various blends which are being tested day by day with different additives, different methods of producing them are also been found. Some of the biodiesel production techniques are super critical methanolysis, ultrasonication method and microwave technique by which maximum biodiesel can be produced. The new approach of using nano particle in biodiesel shows very good results in reducing the level of pollutant gases in the engine exhaust and increased performance without any engine modification. [9]

In order to increase the performance characteristics of bio fuels, many Nano particles are being added. In one such experiment, the nanoparticle formation during exhaust analysis. Growth rates of nanoparticles at different exhaust dilution ratios and temperatures have been determined by monitoring the evolution of particle size distributions in the first stage of the dilution system. Thus, as carbon emissions from engines are reduced, nanoparticle formation and growth becomes more likely unless emissions of sulfuric acid and hydrocarbons are correspondingly reduced. [10]

EXPERIMENT:

The experiments were conducted on a single cylinder Kirloskar build direct injection four stroke cycle diesel engine. Water cooled eddy current dynamometer was used for the experiments. The engine is equipped with crank angle sensor, cylinder pressure sensor, thermocouples to measure the temperature of water, air and gas, Rotameter to measure the water flow rate and manometer to measure air flow and fuel flow. The various parameters that govern the performance characteristics are Brake Power, total fuel consumption, exhaust gas temperature and Brake thermal efficiency

The various samples of biodiesel are as follows:

Sample 1- 80% diesel + 200% Castor oil + 50ppm Magnesium Oxide nanoparticles (200 Bar)
Sample 2-80%diesel + 20% castor oil + 50ppmMagnesium Oxide nanoparticles (220 Bar)
Sample 3-80%diesel + 20% castor oil + 50ppmMagnesium Oxide nanoparticles (240 Bar)

Results and discussion

After the detailed experimental analysis for the emission characteristics of the biodiesel at different composition ratios the following graphs shall be plotted as the emission curves.

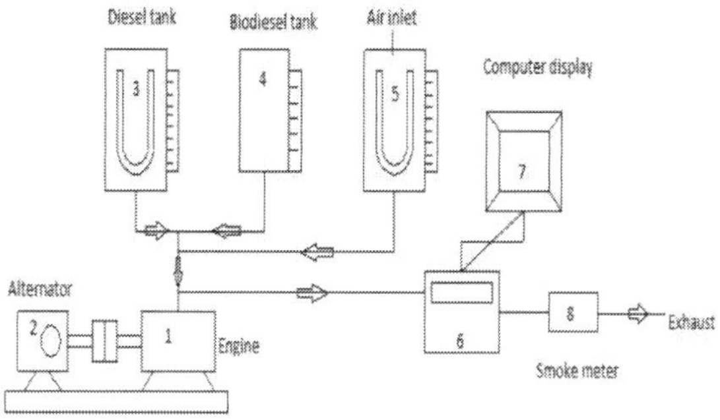

A.Brake power vs carbon monoxide

When all the three samples were compared on the basis of their amount of carbon monoxide released based on the experimental analysis it was seen that in sample one the brake power is higher when compared to sample two and three. Because the CO emission can be reduced by supplying excess oxygen in to the combustion chamber. By increasing the pressure from 200bar to 240bar,, the CO emission was very high than that of the diesel fuel. This can be reduced by supplying excess oxygen during combustion or increasing the oxygen content in bio-fuels.

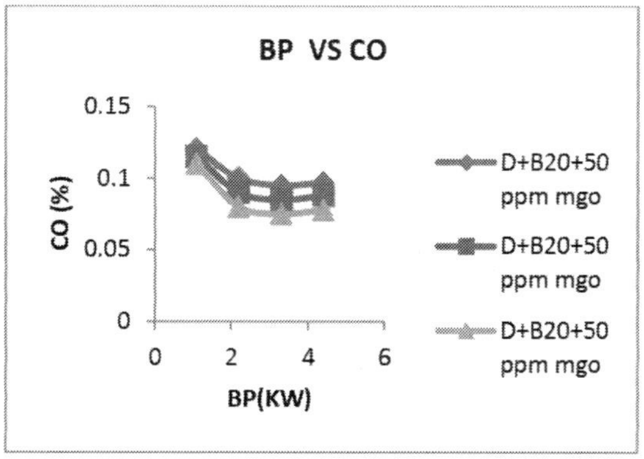

Figure 1. Brake Power Vs. Carbon Monoxide

B.Brake power vs carbon dioxide

It can be observed that the amount of carbon dioxide released based on the experimental analysis it was seen that in sample three the amount of CO2 released is maximum when compared to sample one and two. This leaves a remarkable significance over the emission characteristics of the biofuel.

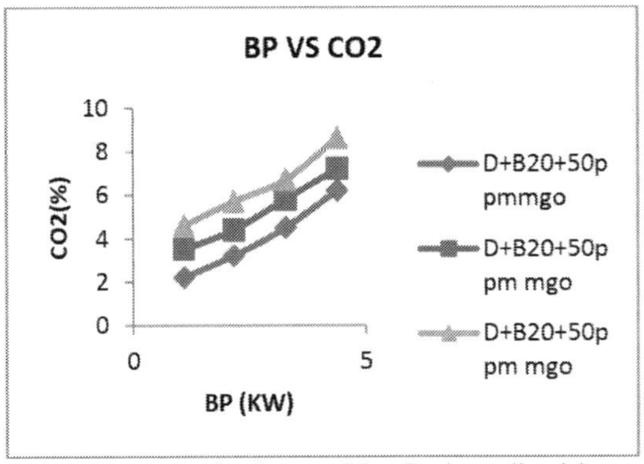

Figure 2. Brake Power Vs. Carbon dioxide

C.Brake power vs. Hydrocarbon

The Figure 6.20 shows the Unburned Hydro Carbon emission for various pressure levels from no load to full load conditions. When all the three samples were compared on the basis of their amount of Hydrocarbon released based on the experimental analysis it was seen that in sample three at 240bar, it is higher when compared to sample one and three at 200 and 220bar. For pure diesel, the HC varies from 10 ppm to a maximum of 28 ppm at full load level. In this investigation the minimum of 26ppm HC obtained at 200bar pressure compare to 220 and 240bar.

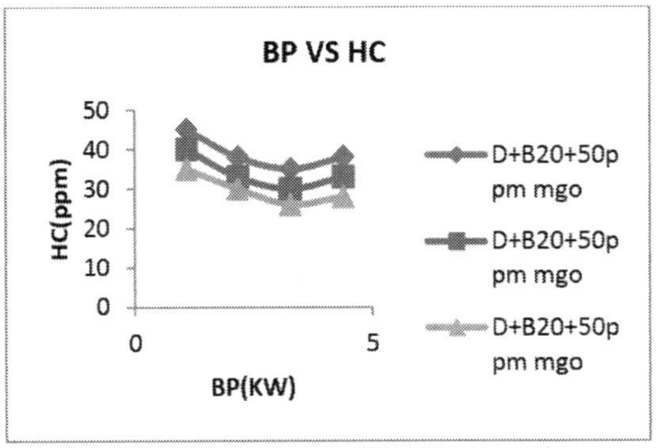

Figure. 3 Brake Power Vs. Hydrocarbon

D.Brake power vsNO_x

When all the three samples were compared for the variation of NO_x with brake power. It can be observed that NO_x decreases more for sample 1 at 200bar compare to sample two and sample three in the pressure levels of 220barand240bar.it was seen that in sample one the amount of O_2 released is more when compared to sample one and two. This leaves a remarkable significance over the emission characteristics of the biofuel.The pure diesel fuel emits higher NOx from 100 ppm to 750 ppm from no load condition to full load condition compared with other blended fuels. The maximum NO_x250ppm obtained at 240bar.

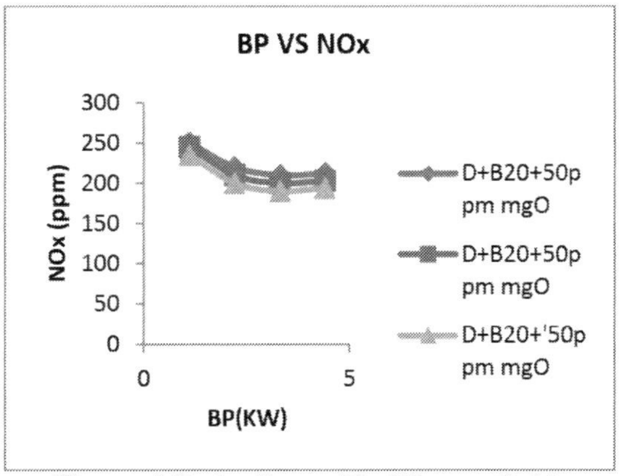

Figure 4.Brake Power Vs. Nitrous Oxide

E.Brake power vs Smoke opacity

Fig 6 represent the smoke opacity at 200,220 and 240 bar pressure levels.When all the three samples were compared on the basis of their smoke opacity values, based on the experimental analysis it was noted to be higher in sample three at 240bar when compared to sample one and two. This leaves a remarkable significance over the emission characteristics of the biofuel.

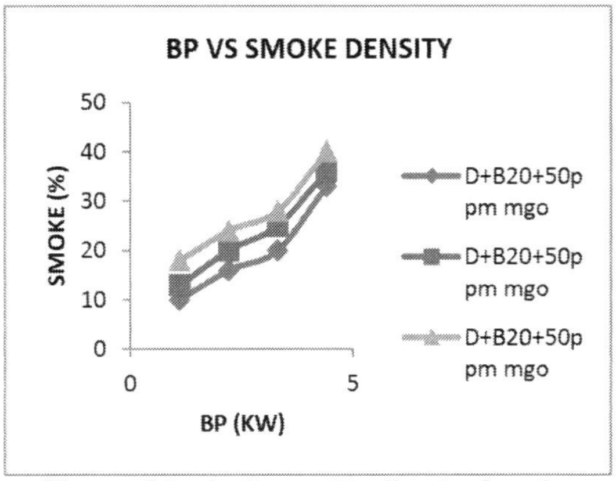

Figure 5.Brake Power Vs. Smoke Opacity

Conclusion

After the successful experimental analysis of the biofuel of various pressures, the best sample which gave the most positive outcomes is sample 2.

Comparision of emission characteristics of pure diesel with that of sample one (80% diesel + 20% Castor oil + 50ppm Magnesium Oxide) at 200bar gives lower value of NO_xUn burnt hydrocarbon than pure diesel. The lower value of Hydrocarbon content in the emission shows efficient burning of fuel. The value of CO emission has also been reduced, which also corresponds to the effective burning of the fuel.

IOP Conf. Series: Materials Science and Engineering **197** (2017) 012025 doi:10.1088/1757-899X/197/1/012025

REFERENCES

[1] Armas O, Martínez-Martínez S and Mata C, Effect of an ethanol–biodiesel–diesel blend on a common rail injection system.
[2] Ozer Can, IsmetCelikten and NazımUsta, Effects of ethanol addition on performance and emissions of a turbocharged indirect injection Diesel engine running at different injection pressures.
[3] Alan C. Hansen, Qin Zhang and Peter WLLyne, Ethanol–diesel fuel blends—a review.
[4] Walter R May, Houston and Edward A Hirs, Catalyst for Improving the Combustion Efficiency of Petroleum Fuels in Diesel Engines.
[5] Mingming Zhu, Yu Ma and Dongke Zhang, An experimental study of the effect of a homogeneous combustion catalyst on fuel consumption and smoke emission in a diesel engine.
[6] Mayur DBawankure, Prashant APotekar, Bhushan A Nandane and VivekRGandhewar, Performance Evaluation of Compression Ignition Engine by Diesel Ethanol-Methylester Blend.
[7] Prabu A and Anand RB, Production and Application of Biodiesel – A Case Study.
[8] Kente Nord and Danhaupt, Reducing the Emission of Particles from a Diesel Engine by Adding an Oxygenate to the Fuel.
[9] Ganesan S, Elango A, Eluri Vamsi Krishna and Balaji A 2014Influence of Nano Catalyst on Emission Characteristics of Di Diesel Engine with Blends of Lemon Grass Oil Using Taguchi Approach *International Journal of Applied Engineering Research* **9**9941-9945
[10] GanesanS and Elango A 2013 Influence of nanocatalyst in the performance of Direct injection CI Engine using blends of castor oil ICANMEET 2013, IEEE Xplore,645-647.
[10] Fred Brear, Nanoparticle Growth During Dilution and Cooling of Diesel Exhaust: Experimental Investigation and Theoretical Assessment

IOP Conf. Series: Materials Science and Engineering **197** (2017) 012026 doi:10.1088/1757-899X/197/1/012026

I.C. Engine emission reduction by copper oxide catalytic converter

S.P. Venkatesan[1*], Desai ShubhamUday[2], Borana Karan Hemant[3], KagitaRajarshiKushwanth Goud[4], G. Lakshmana Kumar[5], K. Pavan Kumar[6]

[1]Associate Professor, [2,3,4,5,6]U.G. Scholar
[1,2,3,4,5,6]Department of Mechanical Engineering,
Sathyabama University, Chennai-600119, Tamilnadu, India.
*E-mail address: spvsathyabama@gmail.com

Abstract— The toxic gases emitted from diesel engines are more than petrol engines. Predicting the use of diesel engines, even more in future, this system is developed and can be used to minimize the harmful gases. Toxic gases include NO_X, CO, HC and Smoke which are harmful to the atmosphere as well as to the human beings. The main aim of this work is to fabricate system, where the level of intensity of toxic gases is controlled through chemical reaction to more agreeable level. This system acts itself as an exhaust system; hence there is no needs to fit separate the silencer. The whole assembly is fitted in the exhaust pipe from engine. In this work, catalytic converter with copper oxide as a catalyst, by replacing noble catalysts such as platinum, palladium and rhodium is fabricated and fitted in the engine exhaust. With and without catalytic converter, the experimentations are carried out at different loads such as 0%, 25%, 50%, 75%, and 100% of maximum rated load. From the experimental results it is found that the maximum reduction is 32%, 61% and 21% for HC, NOx and CO respectively at 100% of maximum rated load when compared to that of without catalytic converter. This catalytic converter system is cash effective and more economical than the existing catalytic converter.

1. Introduction

An exhaust system is a gadget utilized as a part of vehicle to control emanations by changing over dangerous results from ignition (happening in the I.C engine) to less harmful substances by moving synthetic responses through catalyzed. The responses change contingent on the sort of substrate or catalyst altered. Albeit exhaust systems are essentially utilized as a part of exhaust system in cars, trains, planes, generator sets, mining hardware, and other motor fitted gadgets. They are likewise utilized on some wood stoves to control discharges. There are three sorts of exhaust system 1)reduction catalytic converter 2) oxidization catalytic converter 3) three-way catalytic converter (TWC). An oxidation catalyst is a gadget that is set on the end of fumes pipe of an auto. The oxidation catalyst is the second phase of the catalytic converter. It diminishes the hydrocarbons and carbon monoxide by oxidizing them over a platinum and palladium impetus[1]. This catalyst conveys the response of the CO and HC with the rest of the oxygen in the fumes gas. The reduction catalyst is utilized to control NO_X outflow and can be utilized as a different framework as a part of expansion to the oxidation exhaust system. The reduction catalyst is fitted at the underlying position of the oxidation framework. The reduction catalyst is the principal phase of the exhaust system. At the point when such particles interact with the catalyst, the catalyst evacuates the nitrogen iota out of the atom

IOP Conf. Series: Materials Science and Engineering **197** (2017) 012026 doi:10.1088/1757-899X/197/1/012026

and clutches it, liberating the oxygen as O_2. The nitrogen molecules bond with each other that are likewise adhered to the catalyst forming N_2.

$$2NO = N_2 + O_2$$

3-way catalytic converter (TWC) have the extra preferred standpoint of controlling the discharge of nitric oxide and nitrogen dioxide (both together shortened with NO_X and not to be mistaken for nitrous oxide). The diminishment and oxidation catalyst are in one finish single shell of the reactant convertor; in any case, in a few cases, they might be housed independently. A three-way exhaust system has three synchronous errands.Reduction of nitrogen oxides to nitrogen and oxygen[2].

$$2NO_X \rightarrow XO_2 + N_2$$

Oxidation process of carbon monoxide:

$$2CO + O_2 \rightarrow 2CO_2$$

Oxidation process of unburnt hydrocarbons (HC):

$$C_xH_2x+2 + [(3x+1)/2]O_2 \rightarrow xCO_2 + (x+1)H_2O.$$

The above responses happen in reactant convertor; for the most part when the temperature of fumes discharge is somewhat over the stoichiometric point. 3-way converters are viable when the engine is worked inside the scope of air-fuel proportions close to the stoichiometric point, with the end goal that the fumes gas arrangement sways between rich (abundance fuel) and incline (overabundance oxygen)[3]. Shut circle engine control frameworks are fundamental for successful operation of 3-way exhaust systems as nonstop adjusting is required for viable NOX lessening and HC oxidation.

2. Materials Used in Catalytic Converters

The catalytic converter is comprised of different materials. The catalyst substrate or monolith fluctuates as indicated by the vehicle. For example, in car catalytic converters the monolith is normally an ceramic monolith with a honeycomb like structure. At the point when created in vast amounts, artistic centers can be extremely modest. Metallic foil monoliths are made of iron-chromium-aluminum blend, and utilized as a part of different applications. Metallic centers are less costly when made for little creation runs, for example, in sports cars in which low back weight and solid consistent high load is vital[4]. Every one of these materials are intended to give a high surface range to bolster the catalyst shell. Cordierite ceramic substrates are additionally utilized as a part of most catalytic converters. The catalyst washcoat is a transporter for the catalytic materials, which is utilized to coat the materials over a high surface territory. Titanium dioxide, aluminum oxide, silicon dioxide or a blend of both silica and alumina can likewise be utilized[5]. The catalytic materials are suspended in the washcoat before settling it to the center. Washcoat materials have unpleasant, sporadic surface to expand surface zone, which augments the catalyst dynamic surface accessible to respond with the motor discharges. The catalyst utilized as a part of the converter is for the most part a valuable metal, for example, platinum, palladium and rhodium[6]. Platinum is utilized as a lessening catalyst and in addition an oxidation catalyst. Despite the fact that platinum is an exceptionally dynamic catalyst and broadly utilized, it is extremely costly and not reasonable for all applications. Rhodium is utilized as a decrease catalyst, while palladium is utilized as an oxidation catalyst. Sometimes, cerium, iron, manganese and nickel are likewise utilized[7]. In any case, a few organizations prohibit the utilization of some of these. In this present work copper oxide is utilized as a catalyst which is exceptionally savvy catalyst when contrasted and other catalyst like platinum, rhodium and palladium etc.

2.1. Preparation of honeycomb/ monolith

A monolith/honeycomb is a ceramic structure comprising of an extensive number of little straight and parallel directs as appeared in the Figure 1. The monoliths are made by expulsion. An extraordinary blend of additives and claybinder is pushed through a modern color to make the monolith structure[8]. The material is dried sliced to the required length and let go at high temperatures. The monolithic structure has a width of around 15 cm and can have distinctive shapes. The distance across of the channels ranges from 0.5 to 10 mm and the length of the honeycomb/monolith can be up to 1m long. On the surface of these channels a catalyst can be connected in which concoction responses can occur. As a result of the expansive number of channels the contact zone between the catalytic layer and the fumes gas that goes inside the channels is extensive[9]. Encourage the channels are straight and parallel so that the stream is not deterred.

Figure 1. monolith/honeycomb structure

2.2. Applying copper oxide coating over honeycomb

A major consideration for Manual Dispersion coating process is that the coating is done at a necessary thickness, and a number of different techniques are used to get this control, ranging from a simple brush for painting a wall, to some various machinery applying coatings in the coating industry. A further measure for 'non-all-over' coatings is that control is needed according to thickness of coating to be applied. This process is followed by manual dipping of the honeycomb structure into the solution of copper oxide and the acid solution and then air jet spraying over the honeycomb/monolith[10]. Hence, this is the cost effective coating process compared to all above process, this is the process which is used for coating of honeycomb.

Flow chart of manual dispersion technique is shown in the Figure 2.

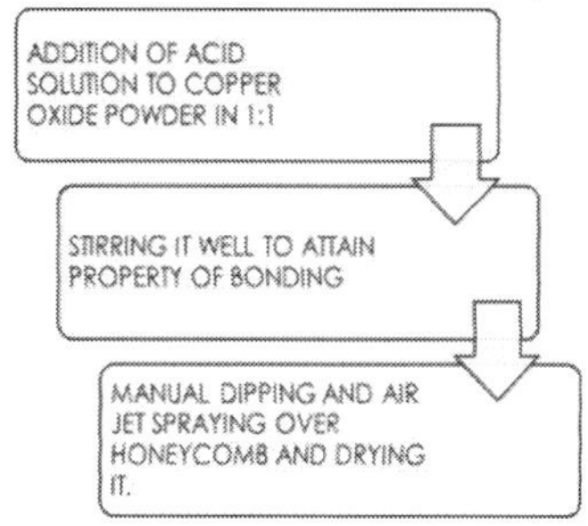

Figure 2. Flow chart of manual dispersion technique

2.3. Preparation of coating material
Copper oxide powder of 480 grams was mixed with the acid solution of dilute nitric acid in the ratio 1:1 until and mixed well to obtain bonding.

2.4 Process of coating
The process used for the coating process is called as manual dispersion technique, where the honeycomb is manually dipped and is subjected to air jet spraying so the pores of the honeycomb structure are clear to allow the passage of smoke. Later, it is dried well for 48 hours.

2.5 Assembling of catalytic converter
Catalytic converters are typically assembled by inserting a coated substrate assembly with copper oxide into a catalytic converter outer shell. The substrate assembly is nothing but insulating mat surrounding catalytic substrate. The mat is then held in place by adhesive. Force is applied to the substrate assembly to compress the mat around the catalytic substrate and stick to adhesive. An outer diameter of the substrate assembly is measured during application of the force. The outer diameter of the converter outer shell is then determined based on this outer diameter measurement of the substrate assembly. The substrate assembly is then fitted into the converter outer shell and the converter outer shell is subjected to subsequent welding operations to reduce the converter outer shell to the determined outer diameter. The subsequent welding operations utilizes electrode welding to pack the converter substrate into the shell. Further, each final assembled catalytic converter which is shown the Figure 3 should have a desired density characteristic, back pressure and other gas flow characteristics are calculated.

Figure 3. Assembled Catalytic Converter

3. Experimental Setup

Figure 4 demonstrates the Catalytic Converter of width 150 mm and a spine measurement of 40 mm fitted Kirloskar engine. Before settling of catalytic converter the outflows originating from engine were watched .Then the spine of the catalytic converter was altered as needs be to the rib of the engine with the assistance of tightens and dashes as demonstrated the figure. The determination of motor is appeared in the Table 1. The motor is made to run utilizing diesel of Calorific Estimation of 45000 KJ/kg. Legitimate measures are taken to abstain from dismantling of the rib of catalytic converter and engine. At first the motor is made to keep running at no load condition. After the enactment of catalytic converter i.e. 2 hours, load was expanded bit by bit and emanation of NOx, CO, HC and Smoke are noted and definite investigation is given underneath. The gulf of the AVL smoke meter appeared in the Figure 5 was associated with the fumes of the catalytic converter to quantify the emanation. The particular of engine and AVL 437 smoke meter are appeared in the Table 1 and Table 2.

Figure 4. Catalytic converter fitted test engine

Table 1. Engine Specification

PART	SPECIFICATION
Product	Single cylinder , 4 stroke, multi-fuel VCR (computerised)
Product code	240PE, Make: apex innovation pvt. Ltd.
Engine	Single cylinder, 4 stroke, water cooled, bore 87.5mm, capacity 661cc. Diesel mode: power 3.5 kw, speed 1500 rpm, Injection variation: 0 – 25 degree bTDC power 3.5 kW @1500 rpm, speed range 1200-1800 rpm,
Dynamometer	Eddy current i.e water cooled, with loading unit Model iag10 of Sai test plant pvt. Ltd.
Fuel tank	Capacity 15 lit. Type: dual compartment, with fuel metering pipe of glass
Piezosensor	Combustion: range 5000 psi, Diesel line: range 5000 psi
Crank angle sensor	Resolution 1degree, speed 5500 rpm with TDC pulse
Data acquisition device	Ni usb-6210, 16 bit, 250 ks/s
Engine control hardware	Fuel injector, fuel pump, ignition coil, idle air
Temperature sensor	Type rtd, pt100 and thermocouple, type k
Load indicator	Digital, range 0-50 kg, supply 230 v ac
Load sensor	Load cell, type strain gauge, range 0-50 kg
Software	“Enginesoft” engine performance analysis software

Table 2. Specification Of AVL 437 Smoke Meter

ITEM	DETAIL
Measurement principle	Measurement of filter paper blackening
Measured value output	FSN (filter smoke number) or mg/m3 (soot concentration)
Measurement range	0 – 10 fsn
Detection limit	0.002 fsn or 0.02 mg/m3

Resolution	0.001 fsn or 0.01 mg/m3
Exhaust pressure ranges	(-300*) – 100 to 400mbars (-500*) – 200 to 750 mbars 0 – 3000 mbars with the high pressure option (*) with activate altitude simulation
Maximum exhaust temperature	600° C with standard 340 mm sample probe
Interfaces	2 serial rs 232 interfaces with a k protocol Digital via instrument 6approx.6per 4210 1 ether net interface with in port option install
Power supply Power consumption	100 – 115V ac or 230V ac, 50/60 hz 700 va
Compressed air	150i/min during purge Grades 1.1.1 to 1.4.1 according to ISO 8573.1:2001I Recommended connection pressure on the AVL smoke meter: 5 to 8 bars at the measurement device input
Weight	<40 kg
Dimensions(w x h x d)	560 x 620 x 300 mm
Sample flow	10 i/min
Ambient condition	5 to 55° C /max. 95 rh; without condensation Sea level -500 to +5000m
Repeatability	Standard deviation is=+- (0.005 fsn+3% of the measured value@ 10 sec intake time)

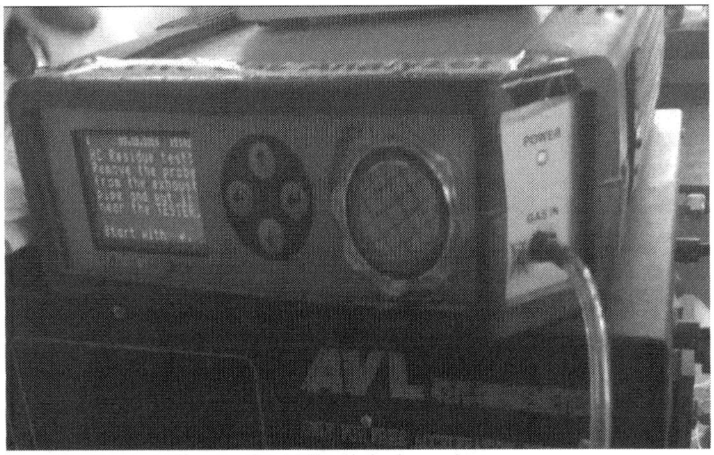

Figure 5. AVL digital smoke meter

4. Result and Discussions

4.1 Study of NO_x emission

The reduction of NO_x emission with and without using catalytic convertor is shown in the Figure 6. The reduction of NO_x considered the main reaction.Idea depends on fuse of a water gas shift and steam transforming responses .The reduction of NOx is approx.61% with using copper oxide based catalytic converter. Nitrogen Oxides (NO_x) are reduced into nitrogen (N_2) and carbon dioxide (CO_2).

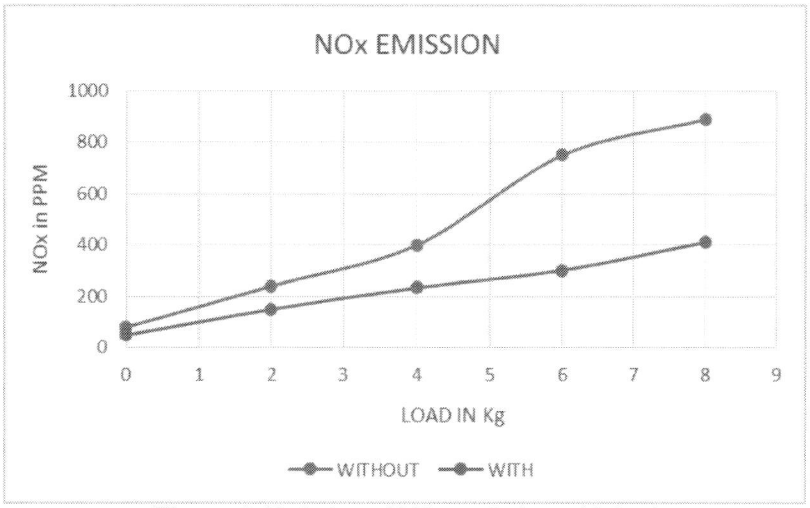

Figure 6. Variation of NOx emission with load

4.2 Smoke analysis

The Figure 7 shows the reduction of Smoke emission by using Catalytic Convertor. Soot or smoke made up of particles in the micrometer size range: Particulate matter causes negative health effects, including but not limited to respiratory disease and cancer. The Density of Smoke is Approx. 38 kg/m^3. The reduction of Smoke is approx. 32% by using Copper based Catalytic convertor.

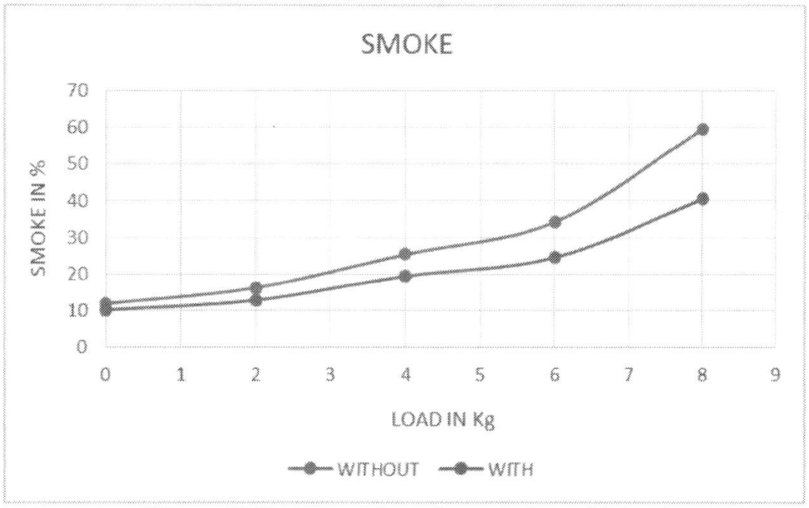

Figure 7. Variation of smoke emission with load

4.3 Analysis of CO emission

The reduction of CO emission with and without using catalytic convertor is shown in the Figure 8. The reduction of CO is approx. 21% with using copper oxide based catalytic converter. Carbon mono oxide is oxidized to carbon dioxide. Copper oxide when reacts with carbon mono oxide reduces it to carbon dioxide and deposition of copper, due to reduction property CO.

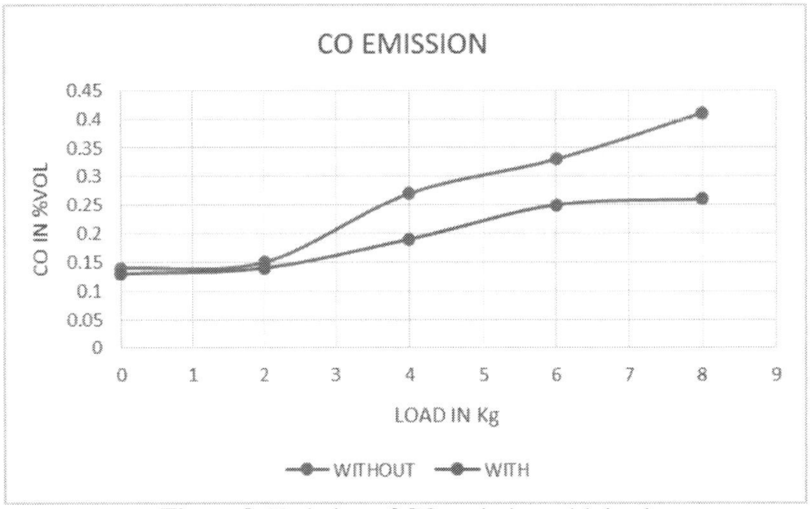

Figure 8. Variation of CO emission with load

4.4 Study of HC emission

The reduction of HC emission by using Catalytic Convertor is shown in the figure 9.Reducing of hydrocarbons by catalytic converter is due to Small amount of fuel in large mass of air causes smaller values of emission both CO and HC, but engine worked steady during experimental tests.The reduction of HC is approx. 32% by using Copper based Catalytic Convertor. Hydrocarbons (HC) are oxidized to create water (H_2O) and carbon dioxide (CO_2).

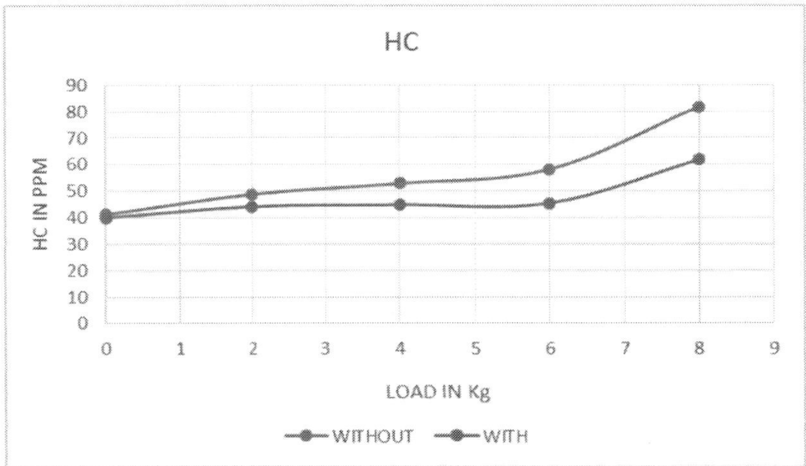

Figure 9. Variation of HC emission with load

5. Conclusion

The experimental investigation findings on emissions of catalytic converter fitted diesel engine are given below

- The catalyst (copper oxide) used in this catalytic converter is economically cost effective when compared to presently using noble catalysts
- It is found that NO_X is significantly reduced when the engine is allowed to run with a catalytic converter at all loads. The result shows that the maximum reduction of 61% in NO_X emission is recorded at full load when compared with that of without catalytic converter.

- From the emissions results, it is noticed that HC emission is reduced for copper oxide based catalytic converter when compared with that of without catalytic converter at all loads. The maximum reduction in HC is observed as 32% at full load.
- Copper oxide catalytic converter shows a remarkable reduction in smoke and CO emission when compared to those of without catalytic converter. The maximum reduction in smoke and CO is observed as 32% and 21% at full load.

As the catalytic activity of copper oxide on the engine exhaust gas which is passing over it is found to be similar to the noble catalyst, it is a better alternate for noble catalysts.

References

[1]. Balakrishna B, Srinivasarao Mamidal. Design optimization of catalytic converter to reduce particulate matter and achieve limited back pressure in diesel engine by cfd. *International Journal of Current Engineering and Technology*. 2014 February, Special Issue 2, pp. 651-658.
[2]. Prashant Katara. Review paper on catalytic converter for automotive exhaust emission. *International Journal of Science and Research.* 2016 September, 5 (9), pp. 30-33.
[3]. Dorit Adams, Elyse Dumas. Improving the effectiveness of catalytic converters via reduction of cold start emissions. University of Pittsburgh, Swanson School of Engineering, 2013.
[4]. Julie M Pardiwala, Femina Patel, Sanjay Patel. Review paper on catalytic converter for automotive exhaust emission. *International Conference on Current Trends in Technology, NUiCONE, Ahmedabad, India*, 2011, pp.1-6.
[5]. Karuppusamy P, Senthil R. Design, analysis of flow characteristics of catalytic converter and effects of back pressure on engine performance. *International Journal of Research in Engineering & Advanced Technology*. 2013 March, 1 (1), pp. 1-6
[6]. Narendrasinh R. Makwana, Chirag M. Amin, Shyam K. Dabhi. Development and performance analysis of nickel based catalytic converter. *International Journal of Advanced Engineering Technology*. 2013 June, 4 (2), pp.10-13.
[7]. Promit Choudhury, SrishaDeo. An innovative approach for emission control using copper plate catalytic converter. *International Journal of Advanced Science, Engineering and Technology*. 2014 May, 3 (2), pp. 19-23.
[8]. Srinivasa K. Chalapathi, Dr. Ch.Bhavanarayana Murthy, Dr. B. SudheerPrem Kumar. Development of automobile catalytic converter during last four decades. *International Journal for Research in Applied Science & Engineering Technology*. 2014 November, 2 (11), pp.321-333.
[9]. Syed C. Aalam, Dr. C. G. Saravanan, C. Mohamed Samath. Reduction of diesel engine emissions using catalytic converter with nanoaluminium oxide catalyst. *International Journal for Research in Emerging Science and Technology*. 2015 July, 2 (7), pp. 17-22.
[10]. Wojciech Marek, Wladyslaw Mitianiec. Modeling and research analysis of catalytic converter in a small SI two-stroke engine. *Journal of KONES Internal Combustion Engines*. 2002, 3 (4), pp.176-184.

IOP Conf. Series: Materials Science and Engineering **197** (2017) 012027 doi:10.1088/1757-899X/197/1/012027

Prospects of pyrolysis oil from plastic waste as fuel for diesel engines: A review

V L Mangesh[1], S Padmanabhan[2], S Ganesan, D PrabhudevRahul[3] and T Dinesh Kumar Reddy[3]

[1]Research Scholar, Faculty of Mechanical Engg, Sathyabama University, Chennai-600119, India.

[2]Associate Professor, Faculty of Mechanical Engg, Sathyabama University, Chennai-600119, India.

[3]UG Student, Faculty of Mechanical Engg, Sathyabama University, Chennai-600119, India.

vlmangesh@gmail.com, padmanabhan.ks@gmail.com

Abstract. The purpose ofthis study is to review the existing literature about chemical recycling of plastic waste and its potential as fuel for diesel engines.This is a review covering on the field of converting waste plastics into liquid hydrocarbon fuels for diesel engines. Disposal and recycling of waste plastics have become an incremental problem and environmental threat with increasing demand for plastics. One of the effective measures is by converting waste plastic into combustible hydrocarbon liquid as an alternative fuel for running diesel engines. Continued research efforts have been taken by researchers to convert waste plastic in to combustible pyrolysis oil as alternate fuel for diesel engines.An existing literature focuses on the study of chemical structure of the waste plastic pyrolysis compared with diesel oil. Converting waste plastics into fuel oil by different catalysts in catalytic pyrolysis process also reviewed in this paper. The methodology with subsequent hydro treating and hydrocracking of waste plastic pyrolysis oil can reduce unsaturated hydrocarbon bonds which would improve the combustion performance in diesel engines as an alternate fuel.

Keywords: Waste Plastics, Pyrolysis oil, Catalysts, Hydro treating and Hydro cracking.

1. Introduction

The usage of plastic has exploded worldwide leading to increase in researches to search for an alternate application of used plastics. Plastics are non-biodegradable polymers mostly containing carbon, hydrogen, and few other elements. Due to its non-biodegradable nature, the plastic waste contributes significant problem for waste management. Waste plastics like polyethylene used for domestic and industry purpose are not recycled for about 43% of its volume. Most of the plastics are recycled and sometimes they are not done so due to lack of sufficient market value and energy consumption. Ever increasing diesel consumption, large outflow of foreign exchange and concern for environment have prompted developing countries like India to search for a suitable environmental friendly alternative to diesel fuel. The countries have to simultaneous address the issues of energy insecurity, increasing oil prices and large-scale of unemployment. This paper deals with an objective to convert plastic waste in to liquid hydrocarbon fuel. The different process of conversion has been studied and compared. The performance of catalytic pyrolysis waste plastic oil as fuel in diesel engines have been reviewed and compared with thermal pyrolysis fuel. This existing literature is on hydro treating and hydro cracking of waste plastic material for converting in to fuel for diesel engine. The various studies and testes are made to understand the chemical composition of processed pyrolysis fuel from the existing literature work in which, the Bromine test was more advantageous to indicate the unsaturated bonds in the oil. The usefulness of Bromine test as an indicator of unsaturated bonds in the oil has been highlighted.

2. Conversion of Waste Plastic into Fuel Oil.

Chemical recycling is a series of processes in which polymers are broken down by heat and/or chemical agents which produces a large variety of products depending on the process employed.Thermal and catalytic pyrolysis processes are the two processes which are economically used to chemically convert waste plastics into fuel oil. They generate at moderate temperatures a wide variety of products (solid, liquid and gaseous fractions) with a broad range of potential applications.

2.1 Thermal Pyrolysis

It is the process of heating of plastic material in an inert atmosphere, a process that decomposes the organic part of the plastic polymer material and generates liquids and gases that are used as fuels and/or raw materials. This method is suitable for complex polymer waste that contains a number of highly mixed materials and prior separation step is not needed. Thermal pyrolysis is done at temperatures ranging between 400 to 800deg C. J.Walendiewskiet al., 2001 shows that fuel oil yield is highest at reaction temperatures between 400 to 450degC.

2.2 Catalytic Pyrolysis

It is the process of heating of the material in an inert atmosphere in the presence of catalysts, a process that decomposes organic part of the plastic polymer material and generates liquids and gases that are used as fuels and /or raw materials. This method is done at temperatures ranging between 350 to 550 °C. AguadoJ et al., 2008 states that the catalytic pyrolysis has following advantages over thermal pyrolysis, (a) Reduction of time and temperature of the reaction, with the consequent decrease in energy consumption, (b) Better selectivity to desired compounds depending on the characteristics of the catalyst, (c) Avoiding the formation of undesired compounds, for example, inhibiting the formation of halogenated compounds from samples containing PVC. J. Walendziewskiet al., 2001 had stated the disadvantages of using catalysts as follows (a) cost of the catalysts (b) recoverability of the catalysts (c) reusability of the catalysts.

2.3 Catalytic Hydrogenation Pyrolysis

J.Walendziewskiet al., 2001 has conducted experiments using hydrogen atmosphere along with catalytic conversion to increase saturated compounds in resulting fuel oil.

3.Performance of Thermal Pyrolysis Oil as Fuel in Diesel Engines.

M.Maniet al., 2009 had tried on waste plastic pyrolysis oil obtained from thermal pyrolysis as fuel in diesel engines. The following observation had been given, (a) Ignition delay higher (b) higher heat release rate (c) NOx higher by 25% at peak loads (d) Higher unburnt hydrocarbon and increased smoke levels. OsamiNishsida et al., 2001 had tried waste plastic pyrolysis oil obtained from thermal pyrolysis

IOP Conf. Series: Materials Science and Engineering **197** (2017) 012027 doi:10.1088/1757-899X/197/1/012027

as fuel in marine diesel engines. Delayed ignition, Increased exhaust temperature and there by a 19% increase in NOx were observed. Since waste plastic oil has almost nil sulphur, SO_xproduction been almost zero compared to Heavy Fuel Oil used in marine engines.M.Ozcanliet al., 2013 carried thermal pyrolysis of waste plastics and subsequently distillate those to observe better performance as fuel. It is found that increase in NOx, even in light phase oil. Viswanath.K.Kaimalet al., 2015 made an experiment on waste plastic pyrolysis oil and used silica as catalyst in the production of waste plastic oil. Waste plastic pyrolysis oil was blended with diesel and used as fuel in diesel engine. The Cylinder peak pressure was found higher along with increased heat release rate. The ignition delay was found increasing with higher blending ratios.

4. Performance of Catalytic Pyrolysis Oil as Fuel in Diesel Engines.

Sachin Kumaret al.,2011 had attempted using waste HDPE plastic pyrolysis oil obtained from catalytic pyrolysis as fuel, blended with diesel in diesel engines. Treated Kaolin has been used as catalyst in the experiment. The following observation has been given, (a) NOx is higher by 30% when compared to diesel fuel at peak loads. (b) Unburnt hydrocarbons and smoke were observed at higher levels. Preetham Reddy Churkuntiet al., 2016 used commercially derived fuel (CynDiesel™) made from a blend of waste plastics through catalytic pyrolysis was tested as a blend with Ultra Low Sulphur Diesel. Increased PM emissions and decreased NOx emissions were observed. Better results were attributed to increased saturated bonds in CynDiesel™.

5. Evolution and Developments in Catalytic Pyrolysis Process of Waste Plastics and Conversion to Liquid Fuels.

The thermal degradation process of waste plastics is further enhanced by using suitable catalysts for obtaining commercially valuable products. Common catalysts used in this process are Zeolite, alumina, silica, FCC Catalyst, reforming catalyst etc. Ahmad RahmanSongip and Kenji Hashimotoet al., 1993 conducted catalytic cracking using several solid acid catalysts such as HY, REY, HZSM-5 zeolites and silica-alumina (SA). REY Zeolite was found to be the most suitable catalyst for the reaction having the highest research octane number (RON) of 67 and a highest gasoline yield of 48wt %. A.Marcilla et al., 2008conducted catalytic pyrolysis of LDPE and HDPE plastic waste with HZSM-5 catalyst in a batch reactor and observed that the yield and composition of the products depends on the type of polyethylene used. HDPE gave higher N-paraffin's and iso-paraffin's whereas LDPE gave higher olefins at lower temperatures. Guohua et al., 2000 conducted catalytic cracking of HDPE and PP with Silica/Alumina catalysts in a powder particle fluidized bed. Over 86 wt. % conversion of liquid fuel was obtained. Kyong-Hwan et al.,2003 conducted catalytic cracking of HDPE, LDPE and PP using spent Fluidic catalytic cracking (FCC). It was observed that a yield of 89% yield was obtained and LDPE and PP degraded faster than HDPE. AchyutK.Pandaet al., 2013 conducted catalytic pyrolysis using kaoline and silica alumina as catalysts. Yield up to 91% was obtained and silica alumina was found better as compared to kaolin in liquid yield. Nnamso.S.Akpanudoh et al., 2005 conducted catalytic conversion of plastic waste using Y zeolite. The effect of plastic polymer to catalyst ratio-mentioned as the acidity content of the polymer/catalyst system – was studied on the creation of liquid hydrocarbons. J.Shahet al., 2005 used Lead sulphide as a catalyst for pyrolysis. The author observed more than 70% yield of gas and liquid fraction with boiling point up to 350 °C.

6. Hydro Treating and Hydro Cracking of Waste Plastic into Liquid Fuels.

H.S.Jooet al., 1998carried out on liquid plastic pyrolysis oil the sequential process of hydro treating, hydrocracking and distillation. The gasoline range product produced from upgrading the residual plastics pyrolysis liquid was characterized by higher paraffin's, lower aromatics, olefins and nitrogen. J. Walendziewskyet al., 2001 carried out thermal pyrolysis using catalyst and hydrogen inside closed autoclaves. The liquids obtained in the presence of hydrocracking catalyst are characterised by lower unsaturated hydrocarbons in comparison to liquid fractions obtained using cracking catalyst of thermal cracking.

7. Chemical Composition Tests for Pyrolysis Oil

7.1 FTIR

Sachin Kumaret al., 2011 states that Fourier Transform Infrared Spectroscopy (FTIR) is an important analysis technique that detects various characteristic functional groups present in oil. Upon Interaction of infrared light with oil, chemical bonds can absorb infrared radiation in specific wavelength ranges regardless of the structure of the rest of the molecules.

7.2 GC-MS

Gas chromatography–mass spectrometry is a testing method which includes the features of gas-chromatography and mass spectrometry to identify different substances within a test sample.

7.3 Bromine Test

Olefins are unsaturated hydrocarbon bonds and they are of very low presence in diesel. But Olefins are formed during the catalytic and thermal cracking. The presence of excess olefins causes the delayed ignition delay and higher heat release in diesel engines. Hydrocracking can substantially reduce the unsaturated bonds present in waste plastic pyrolysis oil. J.Walendziewski et al., 2004 stated that the presence of higher olefins is the reason for polymerisation and oxidation resulting in production of gums and deposits. Light gas oil obtained by pyrolysis had a bromine number of 88.48 g Br/100g and subsequently reduced to 0.54 g Br/100g by hydrogenation. Bhaskar.Met al., 2001 had observed bromine number for diesel is 4.5 shown in table.1

Table 1: Bromine Number of Hydrocarbons and Different Chemistry Olefins

Hydrocarbons	Purity %	Bromine Number
Paraffins	>99	0-0.1
Straight chain Olefins	>99	63-235
Branched Chain Olefins	>99	58-235
Cyclic Olefins	>99	134-237
Diolefins	>99	185-352
Monocyclic Aromatics	>99	0-0.7
Polycyclic Aromatics	>99	0-12

Conclusion

Though, significant progress has been made in chemical recycling of waste plastics its use as fuel in diesel engines is still facing many problems. Catalytic pyrolysis with subsequent Hydro treating and hydro cracking can reduce unsaturated hydrocarbon bonds and thereby improve combustion performance in diesel engines. A literature review of the works in the field of waste plastic pyrolysis oil as fuel for diesel engines is conducted and the results of the review has been presented in this paper. Catalytic pyrolysis has been found by researchers as a better process than thermal pyrolysis with only the cost and recovery of catalyst as areas of concern. Literature review of performance of fuel derived from thermal / catalytic pyrolysis has been done in this paper. Problems regarding delayed ignition, higher heat release rate and increased NO_x remains to be addressed to make this fuel suitable for diesel engines. Future research work should focus on reducing the unsaturated hydrocarbons in pyrolysis fuel and thereby improve the combustion performance on par with diesel. Hydro treating and hydrocracking are promising much better results in reduction of unsaturated hydrocarbons.

References

[1] WalendziewskiJ and SteiningerM 2001Thermal and catalytic conversion of waste polyolefins *Catal* Today **65**323-330.

[2] AguadoJ, Serrano D P and EscolaJ M2008 Fuels from waste plastics by thermal and catalytic processes: A Review *Ind.Eng.Chem.Res***47** 217982-7992.

[3] WalendziewskiJ 2004 Continuous flow cracking of waste plastics *Fuel Processing*.

[4] MMani C Subash G Nagarajan 2009 Performance, emission and combustion characteristics of a diesel engine using waste plastic Oil *Applied Thermal Engineering*.

[5] Osami Nishida HirotsuguFujitWataruHarano PhangHaiVU, Tsuyoshi Fukui Hitomi Saito and NorihikoToyoshima 2001 Application of waste plastic disposals to marine diesel engines, B*ulletin of the JIME***29**(1).

[6] OzcanliM 2013 Light and heavy phases Derived from waste polyethylene by thermal cracking and their usage as fuel in direct injection diesel engine *Journal of Scientific and Industrial Research***72** 198-202.

[7] Viswanath k kaimal and Vijayabalan2015 *Energy conversion and management***105** 951-956

[8] Sachin Kumar and R K Singh 2011 Brazilian journal of chemical engineering **28** 4 659-667

[9] Preetham Reddy Churkunti J Mattson Christopher Depcik Ger Devlin 2016 *Fuel Processing Technology***142**212-218.

[10] Ahmad RahmanSongip Takao MasudaHiroshi Kuwahara KenjiHashimoto 1993*Applied CatalysisB:Environmental***2** 2-3 153-164.

[11] MarcillaA,BeltranM I and NavarroR 2008Evolution of Products generated during the dynamic Pyrolysis of LDPE and HDPE over HZSM5 *Energy and Fuels***22** 2917-2924.

[12] GuohuaLuo TomohikoSuto SatomiYasu and Kunio Kato 2000 Catalytic degradation of high density polyethylene and polypropylene into liquid fuel in a powder- particle fluidized bed **70**97-102.

[13] Kyong-Hwan Lee and Dae-Hyun Shin2003 Catalytic degradation of waste polyolefinic polymer using spent FCC catalyst with various experimental variables *Korean J.ChemEng***20**1, 89-92.

[14] Achyut K Panda and SinghRK 2013Experimental optimization of process for the thermal Catalytic Degradation of waste polypropylene to liquid fuel *Advances in Energy Engineering*13

[15] NnamsoS Akpanudoh Karishma Gobin and George Manos 2005Journalof Molecular Catalysis A: Chemical**235** 1-2 67-73.

[16] ShahJ and Jan Z HussainM R 2005Catalytic pyrolysis of Low density polyethylene with lead sulphide into fuel oil *Polymer degradation and stability***87** 329-333.

[17] JooH S and Guinn J A 1998 *Fuel Process Technologies* 25-40.

[18] Bhaskar M 2001*Petroleum Science and technology- from Taylor and Francis Group* 18, 851.

IOP Conf. Series: Materials Science and Engineering **197** (2017) 012028 doi:10.1088/1757-899X/197/1/012028

Investigation on Nano Refrigeration

G Senthilkumar[1,*], K Saravanan[2], K AjayKumar[3]

[1]Associate Professor, [2]U.G. Scholar, [3]U.G. Scholar
[1,2]Department of Mechanical Engineering,Sathyabama University, Chennai-600119, Tamilnadu, India.
[3]GKM College of Engineering, Chennai, Tamilnadu, India.

* *E-mail address*: tosenthilgs79@gmail.com

Abstract— Boiling heat transfer is imperative in the refrigeration and air conditioning systems. R22 is the mostly widely used alternative refrigerant in refrigeration equipment such as domestic refrigerators and air conditioners. Though the global warming up potential of R22is relatively high, it is affirmed that it is a long alternative refrigerant in lots of countries. By addition of nano particles to the refrigerant results in improvements in the thermo-physical properties and heat transfer characteristics of the refrigerant, thereby improving the performance of the refrigeration system. The results indicate that cuo nano refrigerant works normally and safely in the refrigeration system.. The results indicate that heat transfer coefficient increases with the usage of nanocuo. Thus using cuo nano refrigerant in refrigeration system is found to be feasible. Objective of this project is to study the pre and post effects of addition of NRs in refrigeration and to predict COP and power consumption reduction.

1. Introduction

The problems encountered in freezers using R134a, R22 and R12 as the refrigerant are the critical issues as they impact directly on human life[7,8]. R152a and hydrocarbon mixtures may be considered as substitute for the above mentioned refrigerants. The use of Nano refrigerants can enhance the COP to a considerable extend[1]. The thermal impact of replacement of R12 with R134a on the performance of a vapour compression refrigeration system was studied[2,9]. R32 yielded undesirable characteristics, such as high pressure with minimum COP[3,4].

2. Literature Review

Saterile, M. S et al. (2012) studied the thermal analysis of the impact of replacement of R12 with the geotropic mixture R413a on the performance of a domestic vapour compression refrigeration system. They found that the evaporator outlet temperature ranged from 15o C to -10o C[9]. B.O. Bolaji et al. (2011) analyzed the performances of the three ozone Hydro fluorocarbon (HFC) refrigerants (R32, R134a and R512a) in a vapour compression refrigeration system experimentally. Based on the results, it was concluded that R32 yielded undesirable characteristics, such as high pressure and low COP[3].

3. Experimental Procedure

The figure 1 shows the P-H diagram of simple Vapour Compression Refrigeration(VCR) system and figure 2 describes the flow diagram of the VCR. The properties of copper oxide nano particles are Chemical Formula- CuO, Cu (% by wt) = 79.87% O (% by wt) = 20.10%, Melting point= 1201 ^{0}C, Boiling point= 2000 ^{0}C, Density= 6.31 g/cm^3, Molar mass=79.55 g/mol. The flow diagram is shown in figure 2.

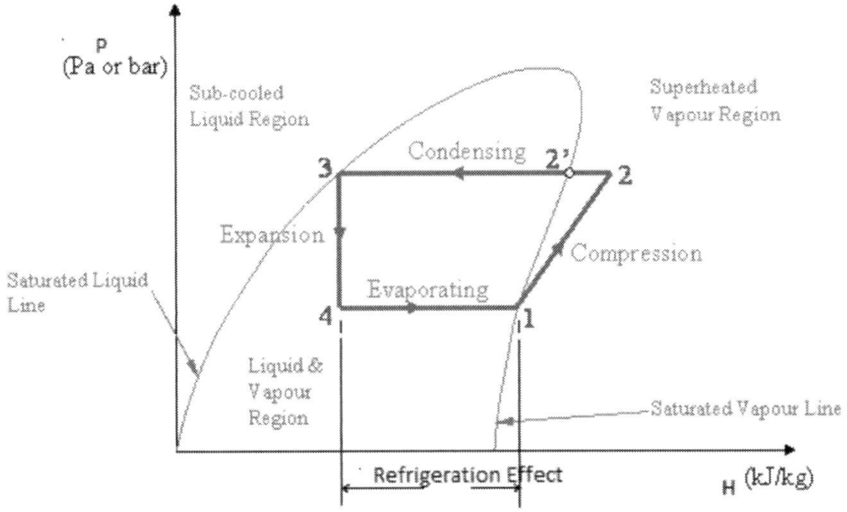

Figure 1. P-h diagram of VCR cycle

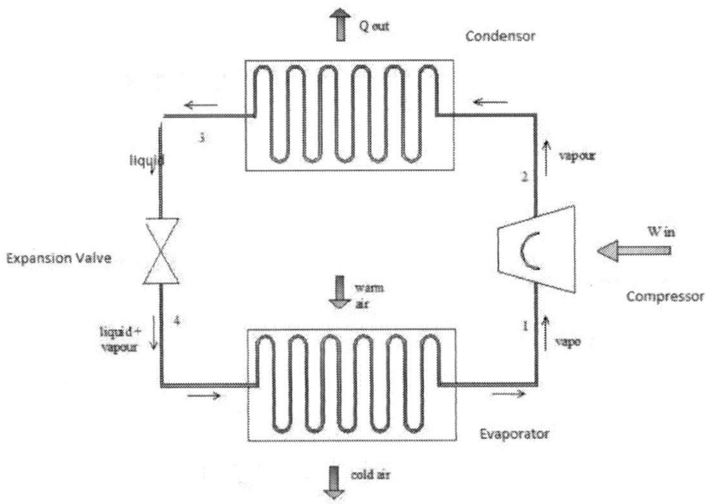

Figure 2. Flow Diagram of VCR System

The experimental set up consists of a compressor, condenser, expansion(Capillary) device and an evaporator as shown in figure 3. The evaporator is of serpentine coil type which is loaded with water. Service ports are provided at the inlet of expansion device and compressor to charge the refrigerant [5,6]. The mass flow rate is measured with the help of flow meter fitted in between expansion device and drier unit.

Figure 3. Experimental setup

Table 1 Experimental Readings

Descriptions	Compressor suction pressure(bar)	Compressor delivery pressure(bar)	*Temperature at junctions (^0C)*				DBT(^{0}C)		WBT(^{0}C)		Velocity (m/s)
			T_1	T_2	T_3	T_4	T_{in}	T_{out}	T_{in}	T_{out}	
Before adding Cu O nano-fluid	1.4	20	27.5	60.5	28.7	-2.4	28	26	22	20	3
After adding CuO nano-fluid	1.4	20	31	55	25	-3.1	26	24.2	19	17.8	3

The above table 1 illustrates the observations made in the experimental set up with and without the CuO nano particles on the performance of the system.

4. Results and Discussion

From the results, it is concluded that the COP of refrigeration system as Freon along with CuO (nano particle) additive is greater than that of the one which used conventional refrigerant as given in Table 2.

Table 2 Comparison of COPs

Refrigerant	$COP_{Theoretical}$	COP_{Actual}	$COP_{Relative}$
R22	16.05	9.32	0.58
R22+CuO	18.72	11.51	0.62

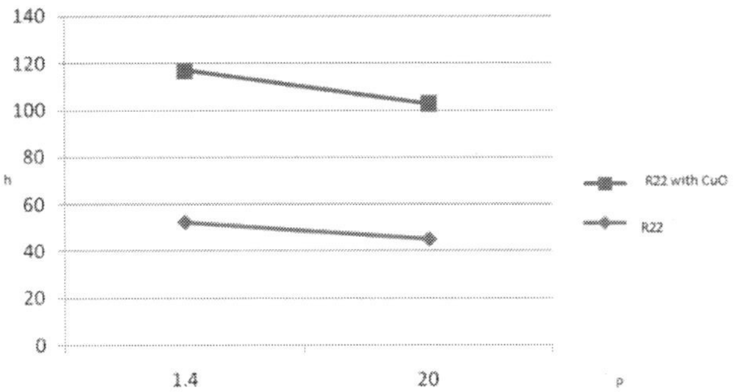

Figure 4. P-h variation with nano Refrigerant addition

From the above table we could clearly see that coefficient of performance of refrigeration system which used nano refrigerant as working fluid is significantly higher than that of conventional refrigerant system. The addition of nano particles, 0.05% by volume, has improved heat transfer properties of the refrigerant and reduced power consumption as shown in figure 4. From figure 5, it is evident that relative COP has increased from 0.58 to 0.62 when R22 with 0.05% concentration of nano-particles.

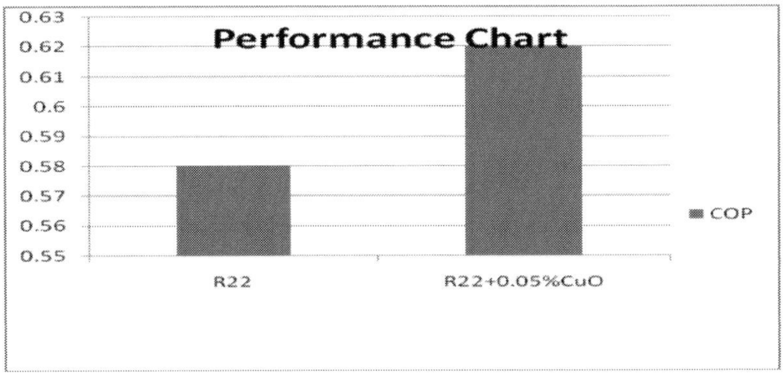

Figure 5. Performance chart

5. Conclusion

An experimental study conducted with the blend of R22 and the mixture containing 0.05%vol of nano particles (copper oxide) in refrigerant shows enhanced heat transfer rates. The nano refrigerant has further reduced compressor work done, which led to reduction in power consumption. Hence development for new refrigeration system with low refrigerant inventory is essential. Very limited pure alternatives are available. Therefore, the new refrigeration system designs should accommodate the non-linear property variation of environment friendly mixed refrigerants. Compatibility of the alternative refrigerants mixtures with lubricants and the construction materials is required to be studied further.

IOP Conf. Series: Materials Science and Engineering **197** (2017) 012028 doi:10.1088/1757-899X/197/1/012028

References

[1] Bane Agawam and Vipin shrinivastava 2010 Retrofitting of vapour compression refrigeration trainer by an eco-friendly refrigerant, *Indian Journal of Science and Technology* **4**

[2] Bolaji, B.O.2010 Experimental Study of R152a and R32 to replace R134a in a domestic refrigerator, *International Journal of Energy*, 35, pp 3793- 3798.

[3] Bolaji, B. O. . Akintunde M.A, and Falade T. O. 2011, Comparative Analysis of Performance of Three Ozone-Friends HFC Refrigerants in a Vapour Compression Refrigeration, *Journal of Sustainable and Environment*, **2**, pp 61-64.

[4] Ching-Song Jwo, Chen-Ching Ting and Wei-Ru Wang 2012, Efficiency analysis of home refrigerators by replacing hydrocarbon refrigerators, *International Journal of Measurement*, . **42**,pp 697-701.

[5] Ching-Song Jwo. 2008 Experimental Study on Thermal Conductivity of Lubricant containing Nano particles, *Rev Adv Mater Science*, **18**,pp 660-666.

[6] Wang J.J.,. Zheng R.T, Gao J.W., and Chen G. 2012 Heat conduction mechanisms in nano fluids and suspensions, *Nano Today* **7**,pp 124 -136

[7]. Sekhar S. J, and Lal D. M.. 2005 HFC 134a/HC600a/HC290 mixture a retrofit for CFC12 systems, International Journal of Refrigeration, **28**, pp 735-743.

[8] Joseph Sekhar S., Mohan Lal D, and Renganarayanan S.. 2004, Improved energy efficiency for CFC domestic refrigerators retrofitted with ozone-friendly HFC134a/ HC refrigerant Mixture, International Journal of Thermal Sciences **43**, pp 307-314

[9] Saterile, M. S., Sahin, H., Kavlicoglu, B., Liu, Y, and Graeve, 2012 O.A. *Nano Scale Res. Lett.* **6**, p 217.

ELECTRIC POWER FROM CRYO (NANO) ICE

A.Kandasamy[1] and M.Chandran[2],
Research Scholar Department ofMechanicalEngineering,Dr.M.G.R.E&RIUniversity.
Professor Department of Mechanical Engineering, Dr.M.G.R.E&RI University,
Email id: kandan.samy@gmail.com

Abstract. In this paper, the authors have studied experimentally the performance of cryocooler which is a mechanical device for producing very low temperature with significant cooling capacity.Nano particles were administrated to enhance the faster rate of cooling.. Electric power (energy) was produced from cryogenic (nano) ice with help of thermoelectric effect.The governing equations for energy conversions, cooling capacity, amount of electric power was also discussed.

Keywords: Ice energy, cryocooler, nano particles and electric power.

1.Introduction

Cryocooler research in the past 22 years has led to significant improvements in all the Potential problems with cryocoolers are reliability, efficiency, and weight, cool down time ,vibration, electromagnetic interference, heat rejection and cost.

1.1. Cryocooler

Cryocooler is also called a cooler which is a refrigeration machine used to cool and maintain cryogenic temperatures by cycling gases. Using cryocoolers a very low temperature of about 20K and below can also be achieved, with a small cooling capacity for useful application. Cryocooler is used for generation of low temperature due to compression and expansion of gas. It is a closed system that uses an oscillating pressure at one end to generate an oscillating gas flow in the rest of the system. Cryocooler has no moving parts in the low temperature region.

1.2. Types of cryocooler

There are five types of cryocoolers most commonly used to provide cryogenic temperatures for various applications are as follows
a).Joule–Thomson cryocooler,
b).Brayton cryocooler,
c).Sterling cryocooler,
d).Gifford–McMahon cryocooler,
e) Pulse tube cryocooler.

1.3 .Cryocoolers applications

IOP Conf. Series: Materials Science and Engineering **197** (2017) 012029 doi:10.1088/1757-899X/197/1/012029

Cryocoolers find many applications such as space, military,low temperature research, agriculture, transportation and security.

1.4. Thermo-electricity

When one junction has a different temperature than the other, an (E.M.F) electromotive force is produced in the circuit and current flows. The magnitude of the force or potential depends on the temperature difference between the two junctions.

2. Experimental set up and procedure

The Pictorial view of Experimental setup of the cryocooler with its components is shown in Fig.1. Water is filled in the collecting tank of the cryocooler, then the cryocooler is allowed to run for many hours to form ice and nano particle were added and sprayed in the tank to enhance and achieve faster cryogenic temperatures,the corresponding time and cryogenic temperatures were recorded. Once a large amount of ice was produced in the ice collecting tank with thick size and cryogenic temperatures ,after that thermoelectric device were connected and switched on to record the electric power produced. Three trials were conducted and readings were recorded in the Table-1 and Table-2

Figure 1. Pictorial view of Experimental set up

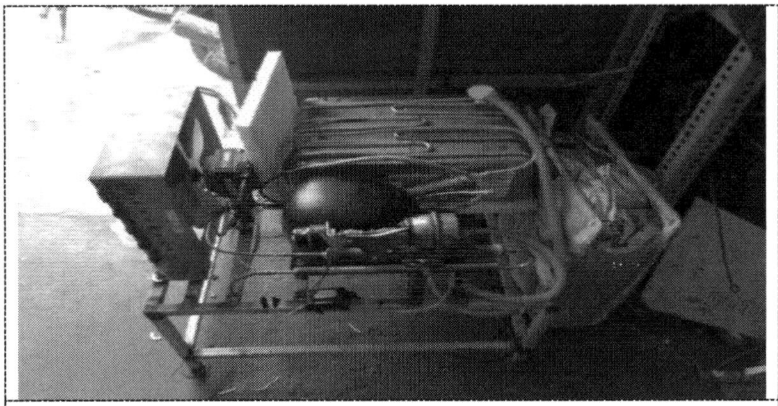

Figure 2. Pictorial view of Experimental set up

Ice energy received from cryocooler is converted in to electrical energy after passing through the thermoelectric device.

Sl. no	Name of component	Pictorial view
1	Condenser	
2	**Expansion valve**	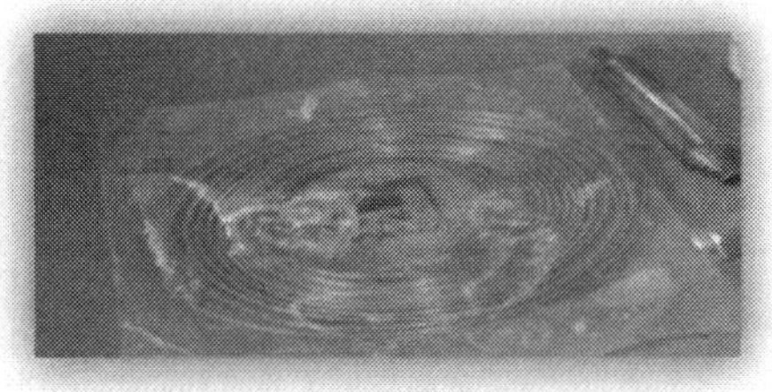
3	**Compressor**	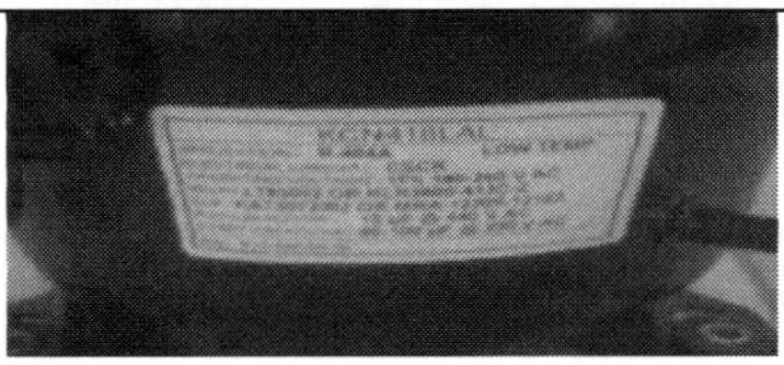

IOP Conf. Series: Materials Science and Engineering **197** (2017) 012029 doi:10.1088/1757-899X/197/1/012029

4	Evaporator	
6	Temperature indicator with reading	
7	Condenser fan and pressure gauges	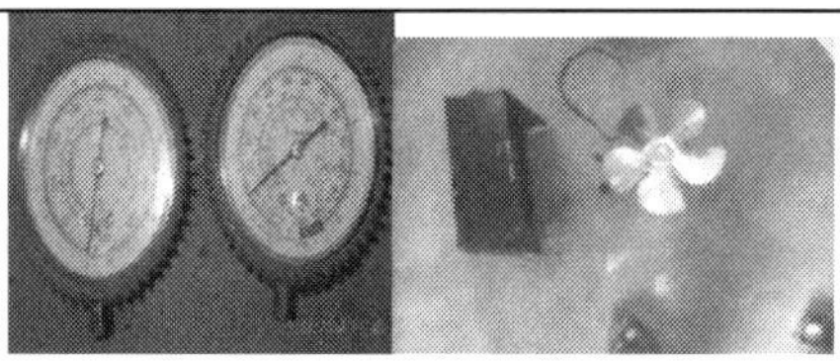

Figure 3. Description of the components of cryocooler

Table 1. Experimental investigation

Discharge Pressure (psi)	Suction Pressure (psi)	Temp(k)
201	100	303
201	102	298
201	104	293
202	107	288
203	110	283
204	110	278
204	110	273
203	110	272
203	110	263
203	110	258
203	110	253
203	110	248
203	110	246

IOP Conf. Series: Materials Science and Engineering **197** (2017) 012029 doi:10.1088/1757-899X/197/1/012029

Table 2. Trials

Trial-1		Trial- 2		Trial- 3	
Ice Formation time (sec)	**Cryogenic Temp(k)**	**Ice Formation time(sec)**	**Cryogenic Temp(k)**	**Ice Formation time (sec)**	**Cryogenic Temp(k)**
376	266	360	267	366	268
37	262	48	261	48	263
22	256	32	250	22	258
38	252	29	251	42	253
51	246	56	247	40	248
98	246	27	247	108	246

3. Results and discussion

The results obtained during the investigation are shown in the graphs.

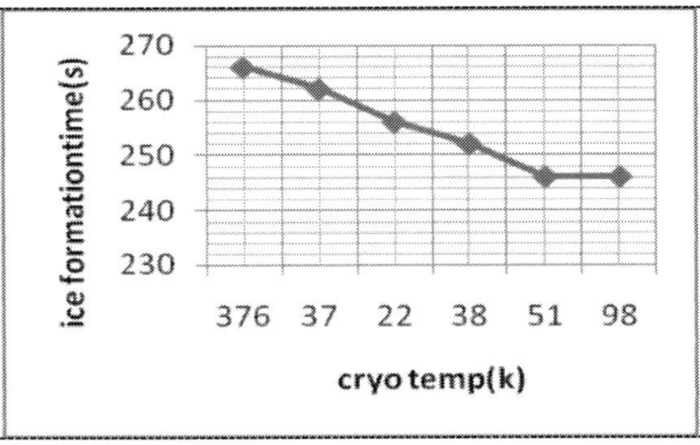

Figure 4. Time vs temperature. Trial - 1

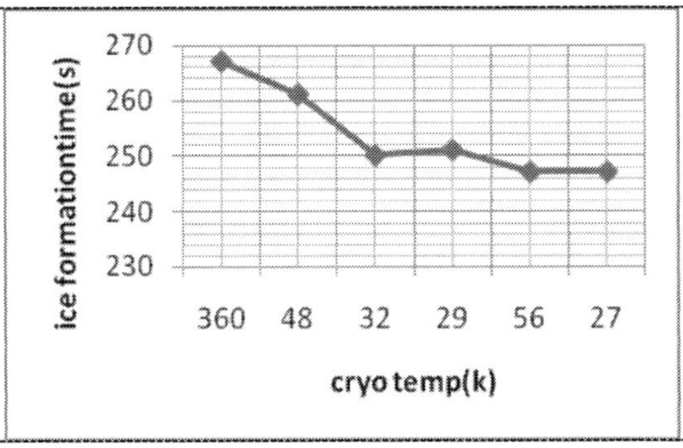

Figure 5. Time vs temperature. Trial - 2

IOP Conf. Series: Materials Science and Engineering **197** (2017) 012029 doi:10.1088/1757-899X/197/1/012029

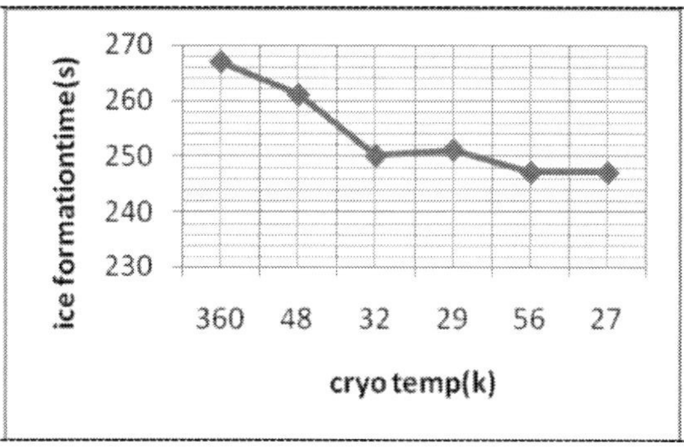

Figure 6. Time vs temperature. Trial - 3

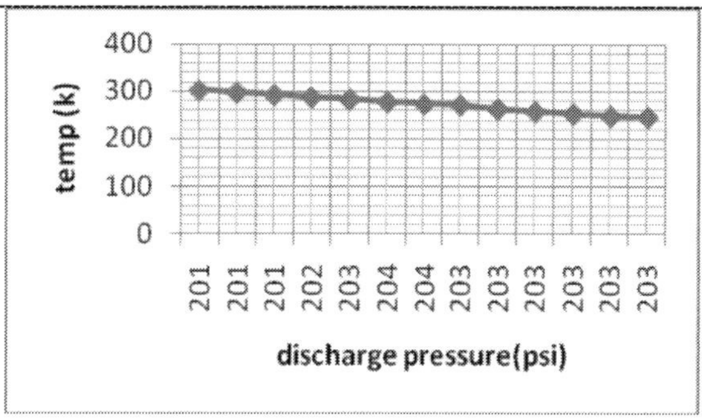

Figure 7. Temperature vs Discharge pressure

4. References

[1] Chandran. M, SenthilKumar .P, 2013, *International Journal of Mechanical & Mechatronics Engineering* **13** 06

[2] Gifford.W.E, Longsworth. R.C, 1964 , *.JIof Engg for industry*, **86** 264-268.

[3] Mikulin E.I, et al, 1984, *Adv. in Cryogenics Engg*. **29** 629-637.

[4] Garraway et al, 1996 *Int cryocooler conference*, 105-113.

5. Acknowledgement

A.kandasamy and Dr.M.Chandran. express sincere gratitude to the Management of Dr.M.G.R.Educational and research institute University for utilizing the research lab.

IOP Conf. Series: Materials Science and Engineering **197** (2017) 012030 doi:10.1088/1757-899X/197/1/012030

DESIGN AND ANALYSIS OF A TWO-STAGE ADSORPTION AIR CHILLER

P Benrajesh[1] and A John Rajan[2]

[1]Research Scholar, Sathyabama University, Chennai, Tamil Nadu, India.

[2] Professor, Department of Manufacturing Engineering, School of Mechanical engineering (SMEC), VIT University, Vellore, Tamil Nadu, India.

ben.rajesh@christuniversity.in

ABSTRACT

The objective of this article is to design and build a bio-friendly air-conditioner, by using adsorption method in the presence of 15% of calcium carbide in water. Aluminum sheet metals are used to form three identical tunnels, to pass the air for processing. Exhaust heat generated from the dairy sterilizing unit process is reutilized, for cooling the environment through this equipment. This equipment is designed, and the analysis is carried out to quantify the COP, SCP, and cooling power. Heat exchangers are designed; its Performance Parameters are quantified and correlated with the conventional designs. It is observed that the new adsorption chiller can produce the coefficient of performance of chiller as 1.068; the Specific cooling power of 10.66 (W/Kg); and the Cooling power of 4.2 KW. This equipment needs 0 to 15 minutes to reach the desired cool breeze (24°c) from the existing room temperature (29°c).

Keywords:

Adsorption Chiller, Cooling Equipment

1. INTRODUCTION:

Air Conditioning by non-conventional method evolved due to environmental regulations in Montreal protocol in 1987, Kyoto Protocol in 1997, and European Commission Regulation 2037/2000. These regulations decided to reduce the contribution of CFC, HCFC, and HFC-134a and to mitigate its effects in the depletion of ozone layer, global warming [1]. Amidst the total energy generation, eighty percent of energy is generated from fossil fuels. It helps the majority of legislative initiative to address towards the energy saving and use of renewable energy [2]. Eighty-five percent of the European population is working in cities, and it increases the scope of advanced solar cooling in European cities.

It can help to save the energy in the range of thirty to fifty Mtoe and it reduces the energy saving up to eight percent of total consumption [3].

2. LITERATURE REVIEW:

2.1 AUTOMOBILE AIR COOLING

The potential utility of adsorption air conditioners in the automobile is identified and validated by a virtual model test [4]. Adsorption air cooling helps to reduce the use of ozone-depleting substances such as CFC, HCFC. They are quiet, durable and inexpensive to maintain it. Working pair utilized was palm-derived activated carbon with methane and the cycle time of twenty minutes with an average chilled air of 22.6 degree Celsius. While the cooling coil temperature operates in the range of 9.5 to 14.7 degree Celsius, the Coefficient of performance and specific cooling power are 0.19 and 396.6 W/Kg [1].

An adsorption air conditioner was developed to cool the passenger cabin of a bus, by extracting the waste heat from the diesel engine. The activated carbon and ammonia are treated as the original working pair with the Coefficient of performance (COP) and cooling power of 0.16 and 2.58 KW. In the second stage, activated carbon is compressed to a small density of 900 Kilogram per cubic metre and then it is mixed with an additional adsorbent [5]. Another attempt to develop an adsorption air conditioner for locomotive driver cabin was made by using a working pair of Zeolite and water. It extracts the exhaust heat, and it generates the cooling power of 3.0 to 4.2 kW and Coefficient of performance (COP) of 0.18 to 0.21 [6]. An empirical test was conducted for the application of air conditioner in the electric vehicles, and it is observed that air conditioning by compression technology is having high supremacy than the adsorption air conditioner with Coefficient of performance (COP) of 0.28 which uses Zeolite and water as working pair[7]. An experiment of the intermittent cooling system was developed with Zeolite-13X, water as working pair and finned double tube heat exchanger as the absorber. It extracts the exhaust heat from the diesel engine with Coefficient of performance (COP) of 0.38 and Specific Cooling power (SCP) as 25.7 W/Kg [8].

2.2 SOLAR REFRIGERATION

Adsorption refrigeration was carried out by silica gel-water as working pair driven by waste heat and it operates closer to ambient temperature [9]. In a cold storage system, an adsorption principle was adopted with Zeolite and water as working pair with the Coefficient of performance (COP) of 0.18 to 0.21 and cooling power of 3.0 to 4.2 KW [10]. Dual power source refrigerator is developed by combined solar thermoelectric adsorption air cooler with activated carbon-methanol as working pair [11]. An effort was made to develop a refrigerator operates on the multi-stage adsorption method, which was several times larger than the conventional type and it is highly impossible to fit in the passenger cabin of automobile [12]. Because of sizing and cooling capacity limitations, no working prototype has been feasible to be fit in the automobiles [13]. Analysis was made on the energy efficiency and economic feasibility of solar refrigeration used in ice making and cold production [14].

A solar ice maker was developed with the cooling power of 4 to 7 Kilogram per square meter and with the Coefficient of performance (COP) of 0.2 to 0.6 [15]. Usually solar cooling plants are used for refrigeration; nevertheless, it has the feasibility to couple with traditional solar plants for producing warm water for industrial and sanitary use for heating [16]. Many solar collectors are used for cooling in European countries and since 2007 more than 100's are used for cooling [17]. To reduce the impact of fluctuating intensity of solar radiation, the working temperature of the working fluid gets reduced and it is obtained by flat plate collectors or thermal effluents [18]. Another design of adsorption refrigerator was developed with three stage adsorption cycles and it works with an optimum temperature of 90 degree Celsius [19]. Wang et al. developed a refrigerator, which operates within a range of -40 to +65 degree Celsius, which extracts heat from the low-level thermal source [20].

A seasonal cold storage indoors is developed with solar collectors, which operates only on broad daylight [21]. In a unique design of solar power absorption refrigeration system (SPAR), solar energy can be absorbed and stored as a chemical potential of the working fluid [22]. To improve the performance of solar air conditioner, a heat pump source, and flat plate collectors are added in which latter one as a replacement to vacuum tube collectors [23]. A solar refrigerator was developed by a team to process beef and pork in the food processing industries which can reduce the energy utility bill of 12000 pounds per year and it operates with the Coefficient of Performance (COP) of 1.15 [24]. Based on the environmental regulations, researchers from agriculture field started developing bio-friendly threshers [25].

3. DESIGN OF AN ABSORPTION CHILLER

3.1 EQUIPMENT USED

- Aluminum metal of grade 3003-H14 with the dimensions of 0.5 X 610 mm,
- 0.5 HP Self-Priming Motor,
- Bronze joints,
- Rivets, 10" Tin Cutter
- Two Exhaust fans,
- Miscellaneous (Burettes, Conical flasks, Calcium Carbides, Burette holders, Thermometers, Funnels, Cotton adhesive tapes 2", Rubber sheet black (1mm) -1.5 Kg, Rubber corks, Copper tubes, wire mesh, wire heating coil.)[26].

3.2 DESIGN DATA

Aluminium foils are rolled into four tunnels [Refer Fig(1)] with three similar sizes and one with dissimilar. Copper tubes (15mm) are used to construct exchange tubes.

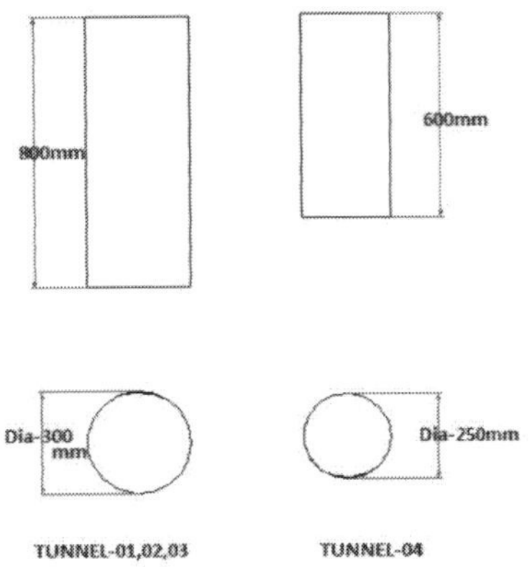

Figure 1: Tunnel Sizes [26]

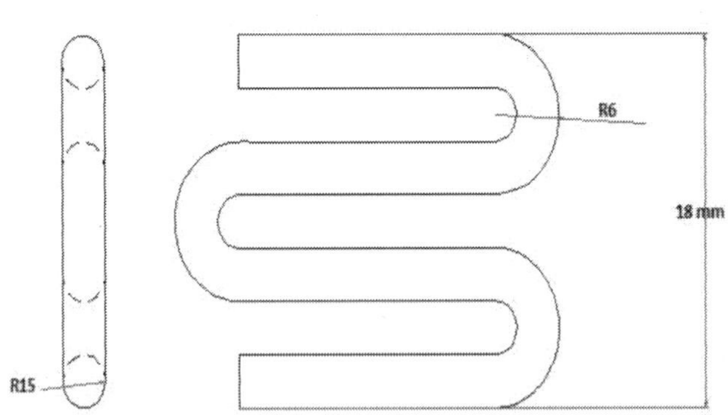

Figure 2: Heat Exchanger Design [26]

IOP Conf. Series: Materials Science and Engineering **197** (2017) 012030 doi:10.1088/1757-899X/197/1/012030

3.3 ASSEMBLY STEPS [26]:

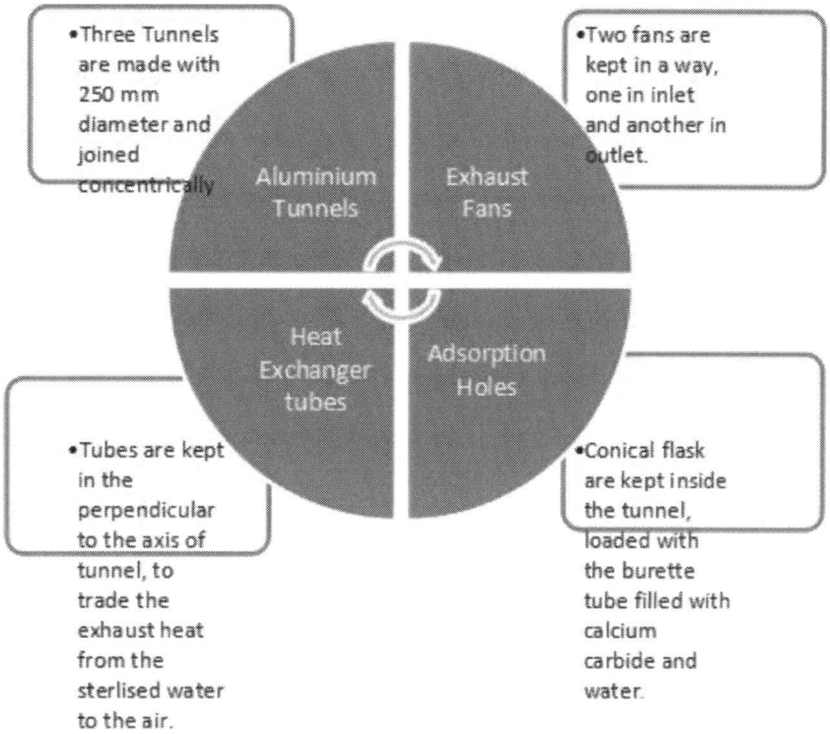

3.4 TESTIMONIALS

Figure 3 Heat exchanger tubes design [26]

Figure 4 Project Display [26]

IOP Conf. Series: Materials Science and Engineering **197** (2017) 012030 doi:10.1088/1757-899X/197/1/012030

4. ANALYSIS:

TABLE 1 Comparative Study of various adsorption designs.

SNO	INVESTI - GATORS	HEAT SOURCE	WORKING PAIRS	COP	SCP (W/Kg)	Cooling Power (Kw)
1.	Present Study	Exhaust heat	Calcium carbide + Water	1.068	10.66	4.2
2.	Leo sing lim[1]	Exhaust heat	Activated Carbon + Methanol	0.19	396.6	0.635
3.	Aceves [7]	Electric heater	Zeolite/Water	0.28	--	--
4.	Zhang [13]	Exhaust heat	Zeolite 13X + water	0.38	25.7	--
5.	Wang et al [5]	Exhaust Heat	Activated Carbon + ammonia	0.16	--	2.58
6.	Lu et al [6]	Exhaust heat	Zeolite/Water	0.18-0.21	--	3 to 4.2

4.1 FORMULAE

- COP (Coefficient of Performance) = $\{(T_c)/(T_H-T_c)\}=\{(24+273.15)/[(29+273.15)-(24+273.15)]\} = 1.068$

Specific Cooling Power (SCP) = $(Q_{ev})/(T_{cy}.M_{ac})$

Where Q_{ev} = Desired Room temperature after it pass through the evaporator.

T_{cy}=Cycle Time

M_{ac}= Mass of the adsorbed component

- SCP = $\{(24)/(15*0.15)\}$ =10.66 W/Kg

4.2 CALCULATIONS

Heat generation Amount by the customer machine=Q (Unknown)

Circulating fluid = water

Circulating fluid mass flow rate (Q_m) = $(\varrho *Q_v)/60$ Kg/sec

Circulating Fluid (density) ϱ = 1 Kg/dm^3

Circulating Fluid (volume) flow rate (Q_v) = 10 dm^3/min

Circulating fluid specific heat capacity (C) =4.2*10^3 J/Kg/k

Circulating Fluid Outlet temp(t_1)= 29°c= 302.15K

Circulating fluid Inlet temperature (t_2) = 24 °c= 297.15K

Circulating fluid return temperature (Δt) =5 °c=5K

Conversion factor : Minutes to seconds =60 S/minutes

- $Q = q_m * C * (T_2 - T_1)$
- $Q = \{(\varrho * Q_v * \Delta t)/60\}$ =3.5 Kw
- Cooling capacity = Consider the safety factor of 20%
  = 3500 X 1.2
  = 4200 W = 4.2 Kw

TABLE 2 Experiment Data [26]

SNO	TIME (MINUTES)	INLET AIR(°c)	OUTLET AIR(°c)	STEAM TEMP (°c)	COLD WATER TEMP (°c)
1.	05	29	27	95	19
2.	10	29	25	96	18
3.	15	29	24	98	17

5.GRAPHS:

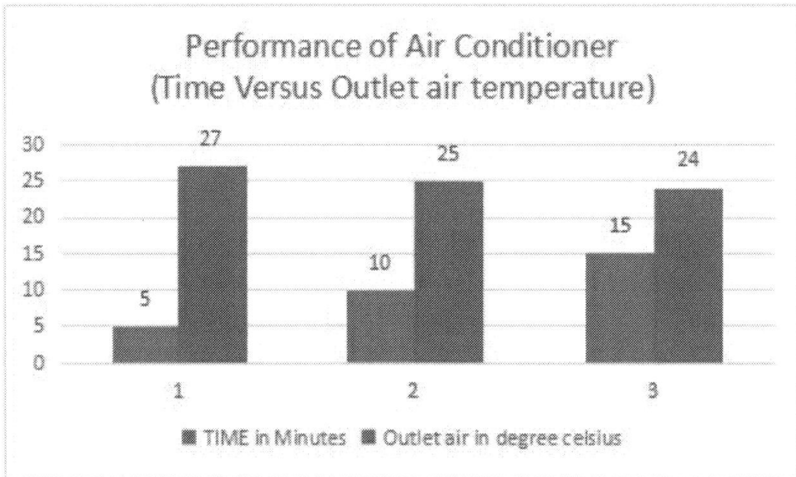

Figure 5: Adsorbing air conditioner Performance

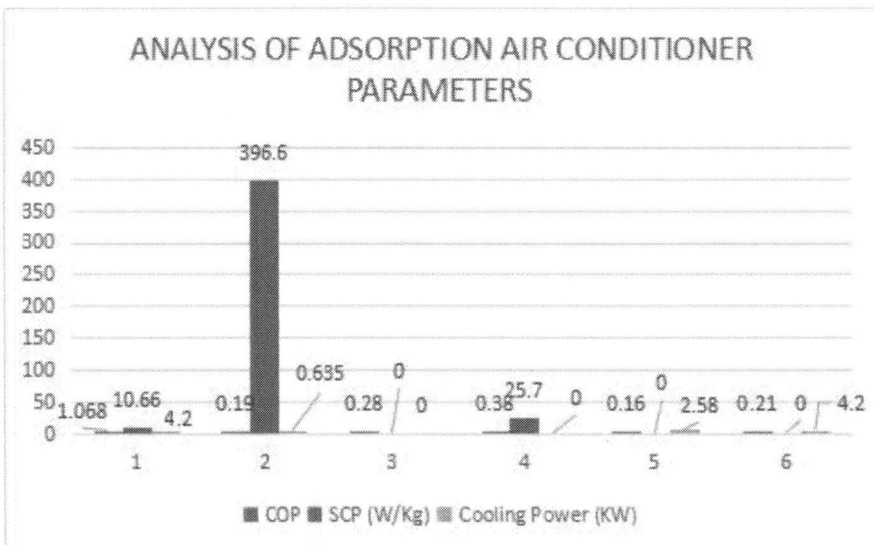

Figure 6: Plot for the Table-1 Values

6. RESULTS AND CONCLUSION:

The Graph-1 results, shown that decline in the outlet air temperature are directly proportional to operation time. It starts with 27 degree Celsius at the end of the 5th minute, and it progresses to 24 degree Celsius at the end of the 15th minute. The Graph-2 infers that the present study is executed with 10.66 W/Kg for the specific cooling power which is inferior to two designs of five, but it has the better performance in the cooling power. Thus, the adsorption air conditioner, have the ability to subside the heat waves without harming the ozone layer or by emitting poisonous gases. It complies with the environmental regulations made in the Montreal protocol in 1987, Kyoto Protocol in 1997, and European Commission Regulation 2037/2000. There are a lot of scopes to improve the COP, SCP, the Cooling power of adsorption air conditioner however without adding the size of the equipment.

APPENDIX:

- COP: Coefficient of Performance
- SCP: Specific Cooling Power
- HP: Horse Power
- VAC: Voltage of Alternate Current
- °C: Degree Celcius
- CFC – Chlorofluorocarbon
- HCFC –Hydro Chlorofluorocarbon.

7. REFERENCES:

[1] Leo Sing Lim, Mohammad Omar Abdullah, March 2010, *American Society of Heating, Refrigerating and Air-Conditioning Engineers, Inc.*, Experimental study of an automobile exhaust heat-driven adsorption air-conditioning laboratory prototype by using palm activated carbon-methanol, **16** 2

[2] ENEA, 2006, The Environment energy and environment report, *Italian National Agency for New Technologies Energy,* available from http://www.enea.it

[3] Santa Mouri. S M et al., 1999 Advanced Passive and Active solar cooling for European cities project, *European Commissioners ACTENER Programme* available from www.rehva.eu

[4] Suzuki. M, 1993, *Heat recovery systems &* CHP, Application of adsorption cooling systems to automobiles, **13**(4): 335-40

[5] Wang R Z et al. 2001, *Final report SJTU-UTRC Joint Research Program,* Shanghai Jiao Tong University (SJJU) and United Technologies Research Centre (UTRC) China Research and development on waste heat drove absorption bus conditioning system

[6] Lu Y Z, Wang R Z et al. 2004, *Appl. thermal engg.*, Practical Experiments on an adsorption air conditioner powered by exhausted heat from diesel locomotive **24**:1051-9

[7] Aceves S M 1996, *Energy resources technology, Transactions of ASME* Analytical Comparison of absorption and vapor compression air conditioner for electric vehicle applications **118** (1):16-21.

[8] Zhang L Z et al 2000, *Appl. thermal engg.,* Design and testing of an automobile waste heat adsorption cooling system **20**(1):103-14.

[9] Saha B B et al 2001, *The Prototype renewable energy,* Solar/Waste Heat driven two-stage adsorption chiller **23**(1):93-101.

[10] Lu Y.Z. et al, 2003 *Energy covers management* Adsorption cold storage system with Zeolite-water working paper used for locomotive air conditioning **44**(10):1733-43.

[11] Abdulla MO, 2003 *Energy Fuels* Cooling Performance of a combined solar thermoelectric adsorption cooling system: an experimental study **23**: 5677-83.

[12] Sato, Honda H S et al, *U.S. Patent,* Adsorptive type refrigeration apparatus, Patent No-5619866

[13] Abdullah M O et al., 2011, *Renewable, sustainable energy* Automobile adsorption air-conditioning system using oil palm biomass-based activated carbon: A review, **15**:2061-72.

[14] Kim D S et al, 2008, *Intl. J. of Refrigeration,* Solar refrigeration options-a state of the art **31**(1): 3-15.

[15] Wang R Z et al, 2006, *Prog. Energy combustible science* Adsorption refrigeration- an efficient way to make good use of waste heat and solar energy **32**: 424-58.

[16] Battisti R 2007, The Thermo hydrosanitary, *Courier* No **226**:2007 available from www.serviziolettori.it

[17] Henning H M et al, 2007, *A Handbook of planners, New York, Springer, Wien 2007* Solar assisted air conditioning in buildings.

[18] Ahachad M, 1993 *Intl. J. of Energy Research* Solar absorption heat transformer applications to absorption refrigeration machines **17** (1993):719-726.

[19] Venegas M, 2002 *Intl. J. of Energy Research* Thermodynamic study of multistage absorption cycles using low-temperature heat **26** (2002): 775-791.

[20] Wang J F et al 2000, *Intl. J. of Energy Research* An Improved absorption refrigeration cycle is driven by unsteady thermal sources below 100°c, **24**(2000):633-640.

[21] Dennis M et al, 2010, *Intl. J. of Refrigeration,* Use of variable geometry ejector with cold store to achieve high solar fraction for solar cooling, **34**(2010):1626-1632

[22] Henrik B, 1981, *Energy Research,* Absorption process for heating, cooling and energy storage- a historical survey **5**(1981) 43-59.

[23] Zhang Xingjuan et al., 2013 *Energy and Buildings,* Performance analysis of a new type of solar air conditioning system **60** (2013):280-285

[24] Umberto Desideri et al., 2009, *Appl. Energy,* Solar-Powered Cooling Systems: Technical and economic analysis of Industrial refrigeration and air conditioning applications **86** (2009) 1376-1386.

[25] Benrajesh P and Johnrajan A, Feb 2016, *Indian Journal of Science and Technology* Application of Green Technology in Agriculture Sector – A Solution for Sustainable Food Production **9**(6), doi: 10.17485/ijst/2016/v9i6/87675

[26] Benrajesh P, 2016 *A working Paper report to Christ University,* Design of an Adsorption chiller

Experimental investigations on effect of different materials and varying depths of one turn exhaust channel swiss roll combustor on its thermal performance

Sagar B. Mane Deshmukh[1], A.Krishnamoorthy[2], V. K.Bhojwani[3], Ashwini Pawane[4]

[1] Research Scholar of Sathyabama University, Chennai 600119, Tamilnadu, India
[2]Professor & Head, Department of Mechanical Engineering, Sathyabama University, Chennai 600 119, Tamilnadu, India
[3]Professor, Department of Mechanical Engineering, JSPM's, Jayawantrao Sawant College of Engg., Hadapsar, Pune 411028, Maharashtra, India
[4]Department of Mechanical Engineering, JSPM's, Jayawantrao Sawant College of Engg. Hadapsar, Pune 411028, Maharashtra, India
manedeshmukh_sagar@yahoo.co.in

Abstract. More energy density of hydrocarbon fuels compared to advanced batteries available in the market demands for development of systems which will use hydrocarbon fuels at small scale to generate power in small quantity (i.e. in few watts) and device efficiency should be reasonably good, but the basic requirement is to generate heat from the fuels like methane, propane, hydrogen, LPG and converting into power. Swiss roll combustor has proved to be best combustor at small scale. Present work is carried out on one turn exhaust channel and half turn of inlet mixture channel Swiss roll combustor. Purpose of keeping exhaust channel length more than the inlet mixture channel to ensure sufficient time for heat exchange between burned and unburned gases, which is not reported in earlier studies. Experimental study mentions effects of different design parameters like materials of combustor, various depths, equivalence ratio, mass flow rates of liquefied petroleum gas (LPG), volume of combustion space and environmental conditions (with insulation and without insulation to combustors) on fuel lean limit and fuel rich limit, temperature profile obtained on all external surfaces, in the main combustion chamber, in the channel carrying unburned gas mixture and burned gas mixture, heat loss to atmosphere from all the walls of combustor, flame location. Different combustor materials tested were stainless steel, Aluminum, copper, brass, bronze, Granite. Depths considered were 22mm, 15mm, 10mm and 5mm. It was observed that flame stability inside the combustion chamber is affected by materials, depths and flow rates. Unburned mixture carrying channel was kept below quenching distance of flame to avoid flash back. Burned gas carrying channel dimension was more than the quenching distance. Considerable temperature rise was observed with insulation to combustors. But combustors with more thermal conductivity showed more heat loss to atmosphere which led to instability of flame.
Keywords: Swiss roll combustor, Stable flame, heat loss, temperature profile, various materials and depths

1. Introduction

Recent innovations in the electrical energy generating devices have made it possible to generate electrical energy locally. There are regions around the globe where 24Hrs electricity is not available. New ideas for producing electrical energy would ensure generation and consumption of electricity locally. The macro energy generation is well proven in terms of distribution, performance and efficiency. Idea of producing energy from combustion of fuels is assumed to be the best way because of more energy density of the fuels containing hydrocarbons compared to more energy density batteries available in the society [1].Hence the development of energy generating device depends on the precise manufacturing of heat producing chamber (i.e. Combustion space/chamber).Applications of the small energy generating devices are remote sensors, micro turbines, future electronics, divert and attitude control systems, portable battle power, aerial vehicles, actuators, robots, rovers, small air conditioning devices, small space heating[8].Major Problems associated with small scale Combustion chambers are heat loss due to [1],[3],[6],[7]decreasing ratio of volume to surface area of combustion space, sealing and fabrication of assembly [3], limit on the dimension of the flame sustaining chamber (i.e. quenching dimension)[5],[6],instability of the flame/combustion for leaner mixtures [2],[10], instability of the flame for low grades fuels, more frictional losses for internal combustion engines[1],[3],[7],[15] and because of which engine design at small scale is not possible. Fabrication difficulty, material failure, destruction of the active radicals, stability of flame only at low flow rates of reactants, problems in measurement of parameters like temperature, pressure etc. are other major issues for small scale Combustion chambers. Success of the heat generating device mainly depends on establishing or achieving a stable flame for longer duration inside the chamber. All the issues discussed above also affect formation of stable flame. The above difficulties may be overcome by utilizing: high pressures and high temperatures of incoming Air/Fuel mixture [8], reduced heat loss[9], Catalytic activation of combustion [1],[10] by increasing residence time [7], avoiding moving parts (thermoelectric, photovoltaic, pyro electric [1], piezoelectric[4]), external heating [16], inactive combustor walls[16], high grade fuels like hydrogen, combustible fuel-air mixtures, insulating the combustion space to avoid heat loss, flame speed balanced with mixture velocity, excess enthalpy burners[1],[2],[3],[5],[7],[8],[9],[10],[14],[15]. The method of excess enthalpy burners (i.e. heat recuperation from products/burned gases and using it to heat reactants/unburned gases) is effective to improve flame stability by reducing heat loss with increased heat exchange from products to reactants, reduced frictional losses using thermoelectric systems, less emissions with reduced temperatures of products, use of low grade fuels, extended flame stability limits, improved thermal efficiency with increased total enthalpy of the reactants (i.e. chemical and thermal)etc. Previously attempts were made by use of excess enthalpy to improve extinction limits by the use of catalytic combustion[1],to increase leaner stability of the flame[2],to improve low temperature stability of the flame (i.e. to get flameless mode of the combustion) [3],to study flame characteristics of counter current excess enthalpy burner for different materials and scales of the burner [5],[15]to get a sublimit flames model [6],to develop model of first principle to judge heat loss through different locations like inside reactant and product dividing walls, heat loss to atmosphere and heat transfer [7],to develop combustor as a heater and to find out flash back and blow out limits [8], to check effect of the third dimensional heat loss on the reaction[9],to study temperature profiles and extinction limits[10].Limitations of the Swiss roll combustors as a heat producing device are not previously discussed preciously. Here in the models prepared one turn exhaust channel length was more than the inlet reactants channel length. In order to study effect of Swiss roll combustor as a heat generating device, models with different materials like Stainless steel with 22mm, 10mm, 15mm and 5mm depths and Aluminum, Brass, Bronze, Copper, Granite were prepared with 20mm depth and their effect on temperature profiles inside the combustion space, Channels and on surfaces is checked. Effect of equivalence ratio on temperature profile and heat loss for a specific material and depth is discussed.

2. Swiss roll models and experimental set up

Figure 1 shows the experimental set up used during testing on models under study. Compressor of 8bar capacity used to compress the air and then it is supplied to air storage tank. LPG tank with 14 bar pressure capacity used to store LPG. But all tests conducted at ambient conditions (i.e. at 1.013bar and 300K).Pressure regulating valves were placed after tanks to step down the pressure to ambient value. Flow meters used for LPG and air were corrected for 2% of the Full scale. Y mixer placed in connection to create a uniform mixture of air and LPG. Function of the flash trapping arrangement is was to avoid back flow of flame in the pipe to avoid accident. Just before inlet of the Swiss roll combustor intake manifold was placed to ensure entry of mixture in the combustor along the depth of combustor. High speed Camera used to capture images of flame for different velocities of the mixture and to find out location of the flame in Swiss roll model of the combustor. K-type thermocouples were used to measure temperatures on external surfaces and inside the channels with 2% accuracy. All the temperatures were measured at steady state. Figure 2 shows Swiss roll model with its different dimensions and surface temperature locations and table 1 gives idea about models under test and environmental conditions used. Reactants carrying channel, products carrying channel and combustion space dimensions were 2mm, 3mm and 10 mm x 7 mm (Refer figure 2). All the models were precisely manufactured by using Wire electro discharge machining. The dimension of the reactants carrying channel was purposefully kept below quenching dimension (i.e. 2mm) of the LPG at room pressure and temperature it helps in avoiding flash back. Exhaust channel dimension was 3mm to get easy push of products out of the combustion space. Combustion space dimension was selected in Meso range (i.e. from 1mm to 10mm) because of chances of flame instability increases with reduced value of the volume to surface area ratio at this scale [8]. Top plate was made of quartz material for all the trials for optical access. Bottom plate was made of the materials under study during all the trials. Insulation was provided for all the vertical surfaces only. Top plate being quartz (For flame visibility) and bottom plate allowing passage to thermocouples could not be insulated. Two electrodes of 0.5mm diameter were inserted through beads of ceramic materials which were placed in bottom plate to create a spark. Source of 50Hz, 6KV was provided to create spark Combustor as a heat source application is satisfied by stable flame inside combustion space. Initially stable flame obtained only after preheating of the combustor. After achieving stable flame for gaseous fuel quantity, air quantity is varied to decide the leaner and richer limits of combustion for different materials and depths. Readings of temperatures in the channels and on surfaces were collected for varying equivalence ratios. Equivalence ratio is a ratio of actual air-fuel ratio to the stoichiometric air-fuel ratio. To change the equivalence ratio quantity of air was varied during the tests by keeping LPG quantity constant, which in turn changes the mean velocity of the reactants. Mean velocity of the reactants is calculated by taking ratio of mass flow rate and product of density and cross sectional area at inlet reactant carrying channel. Cross sectional area is product of depth of the combustor and reactants channel width. Stable flame inside the combustor is a result of balance between flame velocity and reactants velocity.

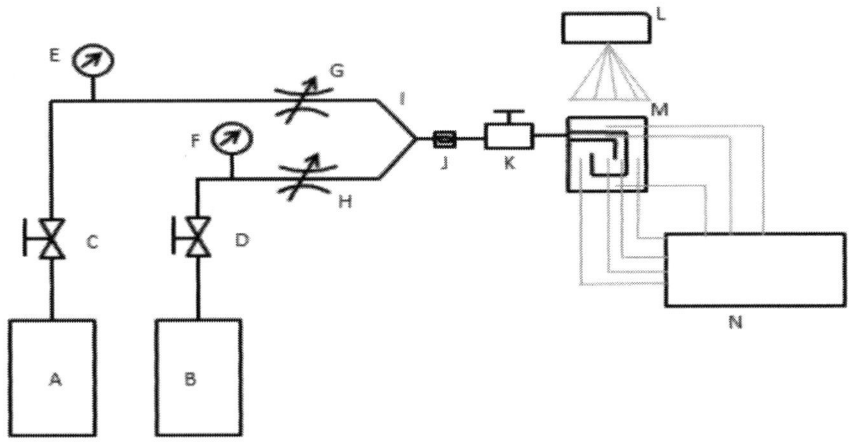

Figure 1 Experimental set up. A and B – LPG and air storage tanks, C and D – LPG and air control valves, E and F – Pressure regulating and indicating gauges, G and H – LPG and Air flow meters, I – Y Mixer, J- Flash trapping arrangement, K – Manifold, L –Camera, M – Combustor model, N – Temperature data acquisition arrangement.

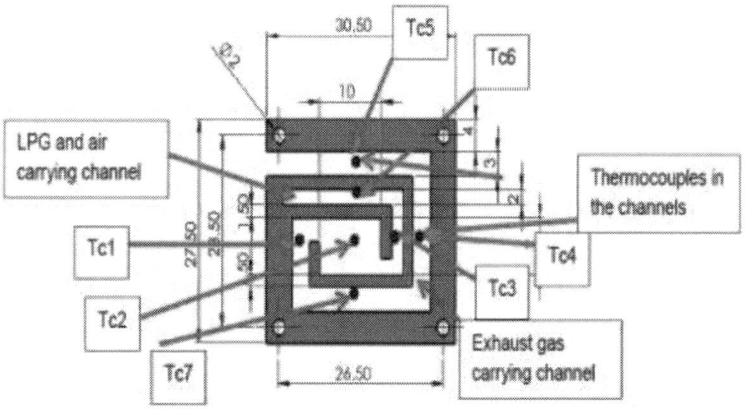

(a) Top view of one turn Exhaust Channel combustor with dimensions and thermocouples locations (i.e.Tc1, Tc2, Tc3, Tc4, Tc5, Tc6 and Tc7) inside the mixture carrying channel and exhaust gas carrying channel [17]

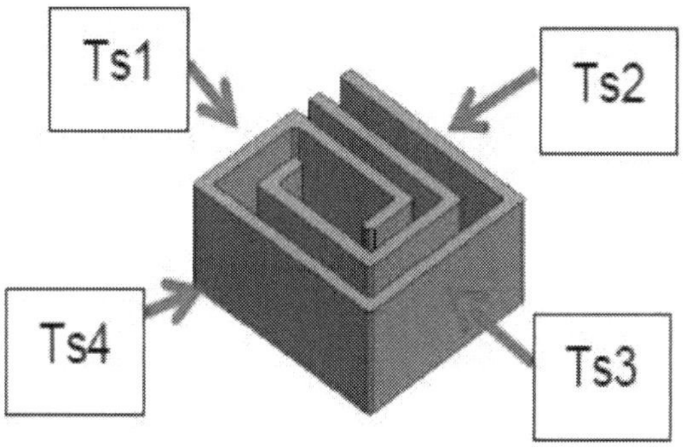

(b) 3D model of one turn exhaust channel combustor showing thermocouples Locations (Ts1, Ts2, Ts3 and Ts4) on the external surfaces S1, S2, S3 and S4 respectively

Figure 2 One turn exhaust channel swiss roll combustor.

Table 1. Swiss Roll Models Descriptions: For Different Materials and Depths with Different Environmental Conditions.

Sr. No.	Combustor Model No. (M)	model details	Environmental conditions		Combustor Volume,mm^3 (Vc)
			Insulation (I)	Without Insulation (WI)	
01	M1	SSD22Di2De3Dcs70	Y	Y	1540
02	M2	SSD15Di2De3Dcs70	Y	Y	1050
03	M3	SSD10Di2De3Dcs70	Y	Y	700
04	M4	SSD5Di2De3Dcs70	NY	Y	350
05	M5	BrD20Di2De3Dcs70	NY	Y	1400
06	M6	Brz D22Di2De3Dcs70	Y	Y	1540
07	M7	Cu D20Di2De3Dcs70	Y	Y	1400
08	M8	Al D22Di2De3Dcs70	Y	Y	1540
09	M9	Gr D22Di2De3Dcs70	NY	Y	1540

Table1 Gives idea about different models tested. 9 different combinations of models were made and tested by varying depths, materials and environmental conditions. Models are described as SSD22Di2De3Dcs70 where, SS - stainless steel(material),D22 – depth of combustor is 22mm, Di2 – reactants/mixture carrying channel width is 2mm,De3 –products/Exhaust gas carrying channel width is 3mm, Dcs – Cross Sectional area of combustion space (i.e. from top view) is70mm2(i.e.10mm X 7mm), similarly for different materials (i.e. Br - Brass, Brz - Bronze, Cu - Copper, Al - Aluminum, Gr – Granite)Y – Test Conducted successfully, NY – Test not conducted.

3. Findings from the experiments and discussions

At small scale of the combustor issue of the flame stability is major which is result of decreased volume to surface area inside the combustion space [8],[10],[5] and flame speed is to be balanced with the velocity of the mixture of LPG and air.

3.1. Flame Stability Limits for Different Values of the Equivalence Ratio and Mixture Velocities

Figure 3 shows variation of Mixture velocity with equivalence ratio for without insulating conditions of all the models and fuel less (i.e. fuel lean) and fuel more (fuel rich) limits for different models were obtained. Tests were conducted at approximately same mass flow rate. All the models tested were at higher mass flow rates (i.e. mass of LPG was 14.73 mg/s, which was more than the previously reported values of methane 0.4 mg/s, 1.04 mg/s by [10]). Amongst the nine models tested M9 model with granite material broke during the tests. Reasons for breaking of the granite material were high brittleness resulting into failure (i.e. beyond 1100K approximately). Fuel fewer limit gives idea about how much less fuel [2] can be used to get required output at a specific air quantity and fuel more limit gives idea about how much more heat can be released without extinction of the flame inside the combustion space which is important case for isothermal applications [1]. Mixture velocity of LPG and air decides back flow (flash back limit) of flame and blow out of the flame from the combustion space. Less velocity with sufficient preheating of the mixture of LPG and air gives back flow in mixture channel and more velocity results into blow out [5] in the exhaust carrying channel (i.e. due to insufficient residence time[3]). Out of the models tested model M8 gave wider flame stability limits because of lower velocities during stable flame. Minimum velocity flame stability was obtained at 2.33 m/s velocity at 0.44 equivalence ratio. But the maximum velocity limit was 6.83 m/s at equivalence ratio of 1.35. Maximum velocity flame stability was obtained for model M4 with 11.64 m/s, which was more than previously reported in available literature maximum 2 – 3 m/s by [5] and 6 – 7 m/s by [8]. Model M5 gave narrow flame stability limit because of its more thermal conductivity results into more heat loss to surrounding.

3.2. Flame Locations in the One Turn Exhaust Channel Combustor

Continuous stable flame exactly at the center of combustion space of the combustor is necessary to get a uniform heat generation and distribution [1] on the top surface, which is required by thermoelectric devices. In the combustion applications location of the flame plays significant role. Flame location depends on the equivalence ratio [12], mixture velocity and dimensions of the channel. Five locations were basically found during the experiments conducted: 1. In the mixture carrying channel (i.e. Flash back condition shown in Figure 4a this occurs because of decreased flow mixture velocity [13] compared to flame speed [6], because of more dimension of channels than the quenching dimension of flame at normal conditions [5] and more heat recirculation to mixture from exhaust gases which reduces flame thickness 2. At outlet of Mixture carrying channel (i.e. about to flash back shown in figure 4b this is result of decreased flow rate [11] and will result into deterioration of edge at outlet of mixture carrying channel because of continuous chemical reaction. 3.Continuously at center of combustion space (i.e. Stable flame shown in figure 4c result of balanced flame burning speed and mixture velocity [8], sufficient residence time, balance of ratio of heat loss to surrounding to heat generation in the combustion space [6], required surface to volume ratio [3], [5], [8], [9], [10] 4.Always cling to exhaust carrying channel (i.e. about to blow out condition shown in figure 4d which

IOP Conf. Series: Materials Science and Engineering **197** (2017) 012031 doi:10.1088/1757-899X/197/1/012031

is result of high mixture velocity 5.In the exhaust carrying channel (i.e. blowout condition shown in figure 4e result of large exhaust channel dimension than the quenching dimension.

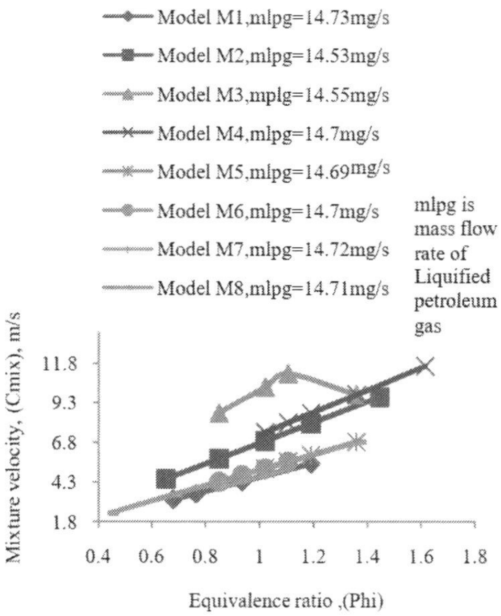

Figure 3. Flame stability limits for different equivalence ratios and mixture velocities and for different models at approximately same LPG quantity.

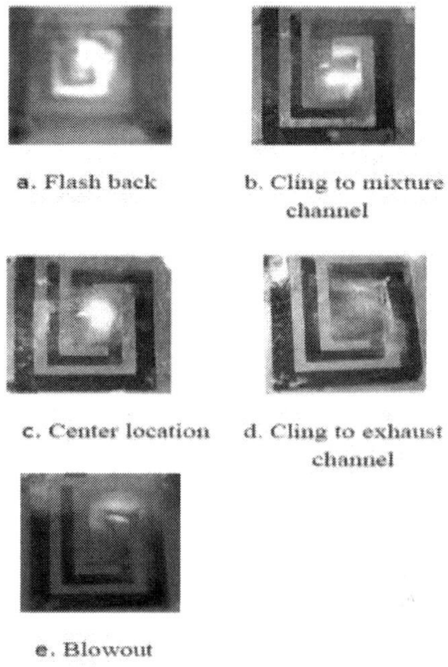

Figure 4. Actual Flame photos captured during tests indicating five locations.

3.3. Effect of an Equivalence Ratio on Temperature Profile

Figure 5 shows variation of external surface temperatures with equivalence ratio for model M1 when LPG was 14.73 mg/s. Locations of temperatures measured by K- type thermocouples mounted on the surfaces are shown in the figure 2. It was observed during the tests; temperatures on surface 2 measured by thermocouple Ts2 were higher at all the values of Phi and were maximum at Phi equal to unity. Surface 2 was close to combustion chamber and it was a second surface of wall separating exhaust carrying channel from the surrounding. Temperatures in descending manner were observed on Ts2, Ttop, Tbot, Ts1, Ts4 and Ts3. Highest temperature was observed at phi close to unity by Ts2 and it was 802.5 k. Model M1 has higher combustion volume which gives increased residence time [15] results into completing the chemical reaction. Results are shown for Model M1 because of higher temperature observance always on surfaces. Figure 6 shows variation of channel temperatures with equivalence ratio for model M1 and for without insulation conditions. Thermocouples placed at different locations in the channels gave maximum temperatures close to Phi equal to unity. Maximum temperatures were measured by thermocouple Tc2 as it was always exposed to flame inside the combustor. In case of blowout conditions only Tc1 measured maximum temperatures as it was exposed to flame directly. Maximum temperature was observed to be 1499 K at Phi equal to 1.19. On richer side of the mixture of LPG and air, proper heat exchange takes place between exhaust/ products and mixture/ reactants because of low velocity of mixture/reactants at quantity of LPG and air which give sufficient time to share heat, But for same LPG quantity as air quantity is increased mixture velocity also increases results into less time to share heat between products and reactants. As a result of which graphs for thermocouples Tc1 and Tc6 are converging on richer mixture side and diverging on leaner mixture side.

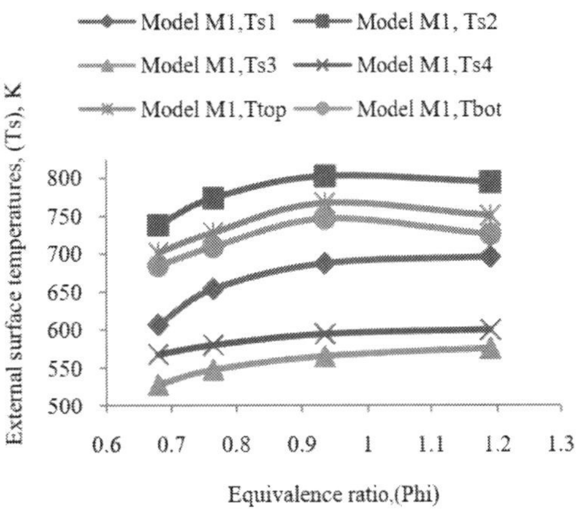

Figure 5 Variation of External surface temperatures with Equivalence ratio (Phi) for model M1 without insulation

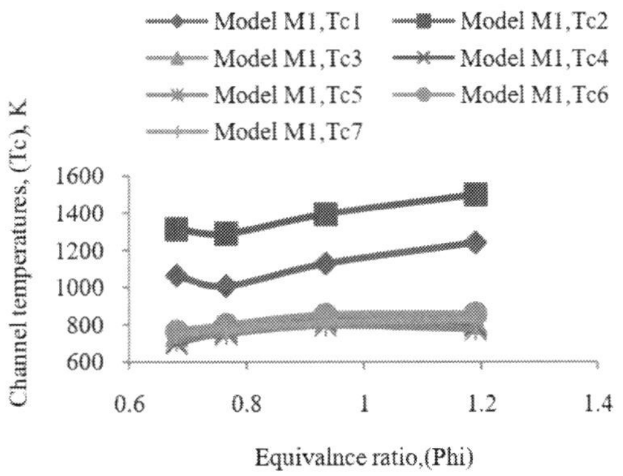

Figure 6 Channel temperatures (Tc) Vs. Equivalence ratio (Phi) for model M1 and without insulation conditions

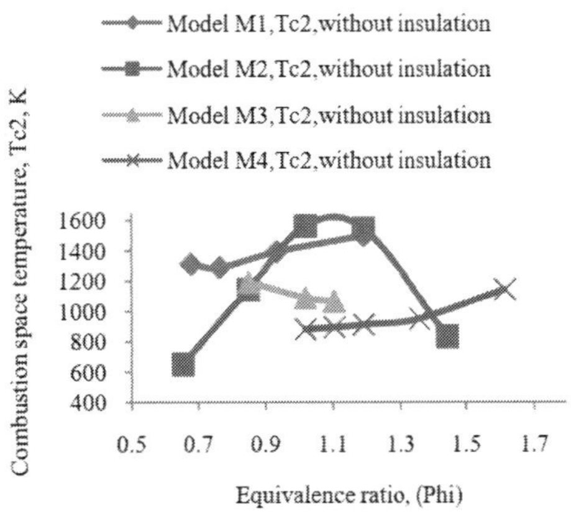

Figure 7 Combustion space temperature (Tc2) Vs equivalence ratio (Phi) for various depths for Stainless steel material

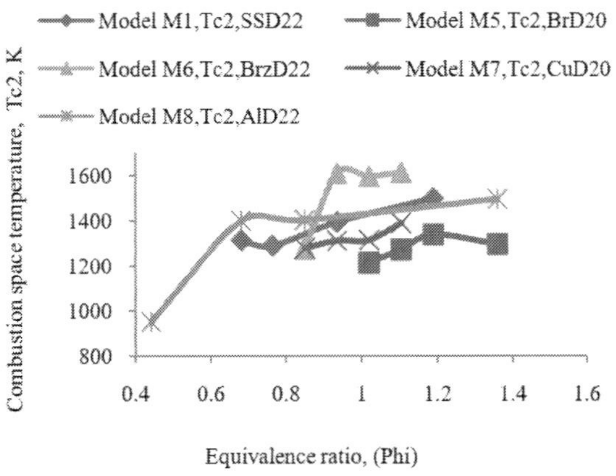

Figure 8 Combustion space temperature (Tc2) Vs equivalence ratio (Phi) for different materials.

3.4. Effect of an Equivalence Ratio on Temperature Profile

Low velocity limits are decided by the ratio of heat loss to surrounding to heat generated in the combustion space [3]. Variation of combustion temperature with Phi is shown in figure 7. Maximum temperature in the combustion space was 1563 K at Phi equal to unity. Least value of temperature was observed for model M2 at Phi equal to 0.65 and was 560 K because of incomplete combustion. Minimum values were measured for model M4 approximately for all equivalence ratios which is result of decreased depth and more heat loss from the top and bottom plate. figure 8 shows variation of the combustion space temperature with equivalence ratio for different materials. It was observed that

Model M6 combustor with bronze material gave higher temperatures and model M5 with brass material gave lower temperatures. But melting problems occurred with models M5, M6 and M8 on higher temperatures in the combustors. Model M8 with Aluminum material gave wider flame stability limits. Model M7 with copper material gave temperatures less than Models M1, M6, and M8 for same

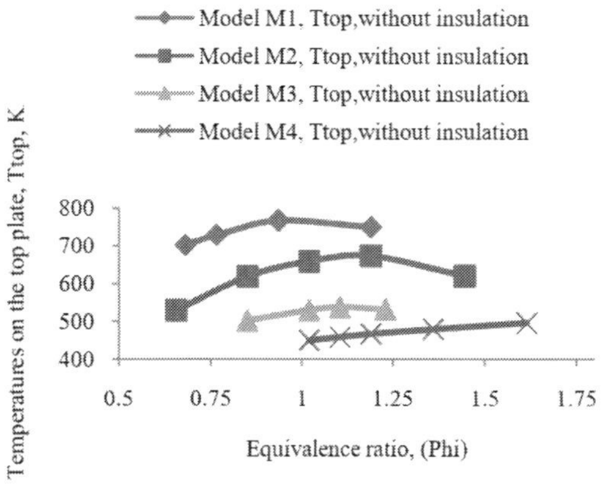

Figure 9 Temperatures on the top plate (Ttop) Vs Equivalence ratio (Phi) for different depths of stainless steel material

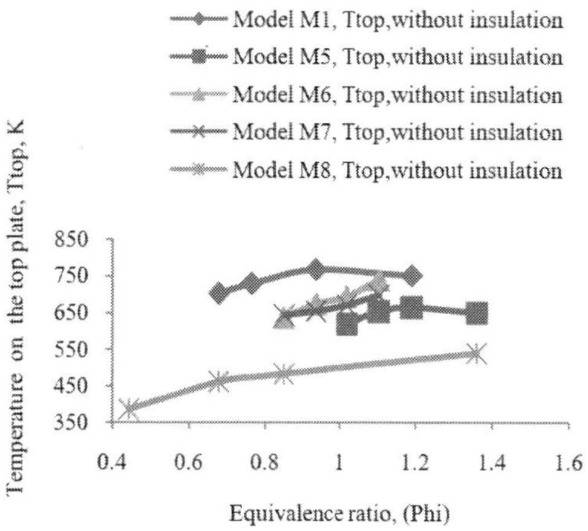

Figure 10 Temperatures on the top plate (Ttop) Vs Equivalence ratio (Phi) for different materials and same depth.

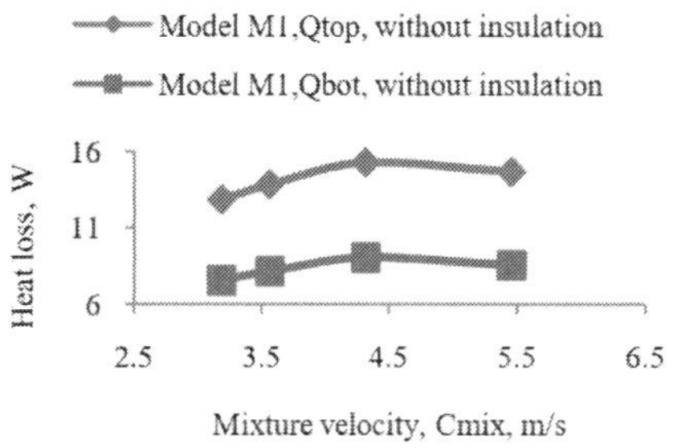

Figure 11 Heat loss through top quartz plate varying with mixture velocity

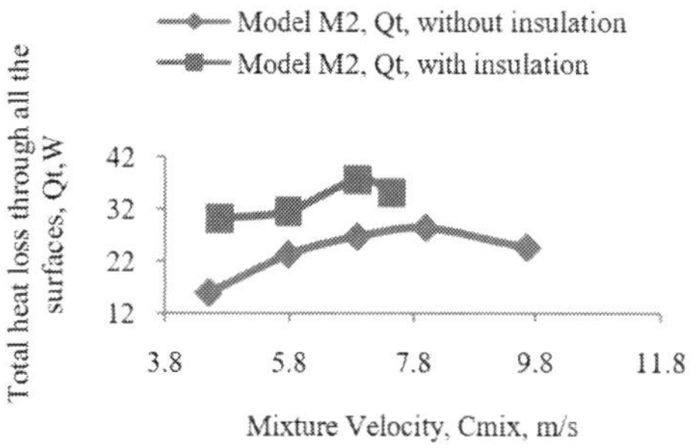

Figure 12 The Variation of total heat loss through all the surfaces Vs Mixture velocity for model M2

equivalence ratio but were more than the brass combustor. Amongst above all model M1 was stable for higher temperatures and not melted during the tests. Top surface temperature is very important from the view point of connecting combustor to thermoelectric device. Maximum temperature on the top plate will give maximum input to the thermoelectric device [4] to get increased output in the form of electrical energy, but more temperature on the top plate is indication of more heat loss to surrounding which results into flame extinction. Figure 9 shows variation of the temperatures on the top plate with equivalence ratio for different depths of stainless steel material. Model M1 with 22 mm depth showed higher temperatures for all the equivalence ratios compared to other depths. Temperatures on top plate were found decreasing with decreased depth. Highest temperatures were observed at equivalence ratio equal unity. Figure 10 shows variation of the Temperatures on the top plate with Phi for different materials and approximately same depths. Model M1 showed more temperatures on the top plate compared to other models. Lower temperatures were observed for M8 model (i.e. for aluminium material) because of more heat loss from the combustion space to all the locations inside due to higher thermal conductivity. Temperatures were found decreasing in sequence from Model M1, M6, M7, M5 and M8.

3.5 Effect of Mixture Velocity on Heat Loss through Top Plate

Maximum heat loss was observed for Model M1 and was through top plate for all the mixture velocities. Heat loss through top quartz plate and bottom plate is important to find out how much input can be given to thermoelectric devices to generate electric energy (i.e. maximum heat loss will give maximum input to the thermoelectric devices) and on contrary to find out limit of maximum heat loss at which flame is expected to be stable inside the combustion space. Third dimensional heat loss is important on lower mixture velocities [9]. Figure11 shows variation of the Heat loss through top plate [5], [8] and bottom plate with different mixture velocities. Maximum heat loss was found near the velocity of 4.5m/s and was high through top plate compared to bottom plate. Reference [8] used for Heat loss calculations. Figure 12 shows variation of the total heat loss variation with mixture velocity. Heat loss through all the surfaces was found to be increased because of insulation provided on all vertical surfaces. Least heat loss observed on vertical surfaces but improved heat loss through top and bottom plates resulted into increase in total heat loss. Maximum heat loss of 37.58 W was observed at 6.88 m/s for with insulation model. Consideration of adiabatic combustor is possible only with all the surfaces insulated.

3.6 Effect of Combustion Volume on total heat loss

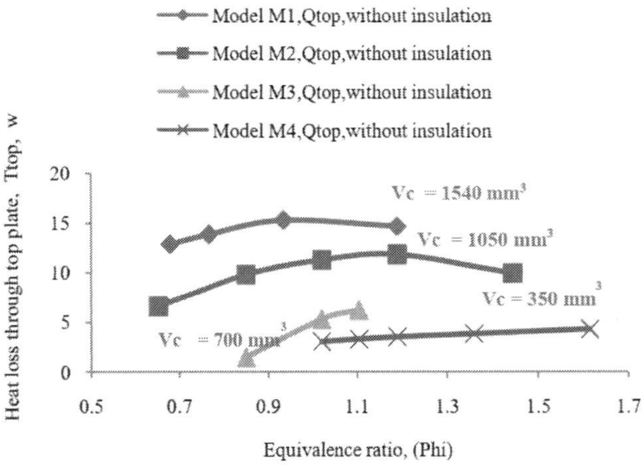

Figure 13 Heat losses through top plate for different volumes of stainless steel material without insulation.

Figure 13 shows the variation Model M1 with more combustion volume (i.e. Vc = 1540 mm3) results into increased heat loss to surrounding. Maximum heat loss was observed close to stoichiometry for all the models. On richer side of the mixture because of incomplete combustion temperatures measured were lower and hence less heat loss. Similarly on leaner side less heat loss was found because of less heat release during the leaner combustion. Decreased heat loss was observed ate leaner limit compared to richer limit.

4. Conclusion

Effects of the various depths and materials were compared to check the thermal performance of one turn exhaust channel combustor. Nine different models of the combustors were tested. Model M9 broke during the test because of brittleness and low temperature sustainability of granite material. Models M5, M6 and M8 melted during the tests on higher temperature sides. Minimum velocity at which flame stability was obtained was 2.33 m/s at 0.44 equivalence ratio. Flame was found stable at a maximum velocity of 11.64 m/s for model M4 (i.e. with 5 mm depth) and model M5 gave narrow flame stability limits (i.e. Brass can't be used as combustor material). Maximum temperatures were always obtained in the combustion space and maximum value of the temperature was found to be 1613 K. Amongst all the surfaces, quartz material top plate surface always showed more temperatures than other surfaces. Second highest temperature was always shown by surface 2, because it was adjacent to exhaust carrying channel. Heat exchange between exhaust carrying channel and mixture carrying channel was effective and more length of the exhaust channel improves heat sharing. Higher dimension of the exhaust channel results into blowout (i.e. it should be always below quenching dimension of the flame). High heat loss was observed through top and bottom plates for all the combinations. Total heat loss was maximum for model M1 with increased depth and combustion volume. Insulation on the materials showed improved performance in terms of increased temperatures.

References

[1] Ahn A J, Eastwood C, Sitzki L and Rooney P D 2005 Gas-phase and catalytic combustion in heat-recirculating burners *Proeedings of the combustion institute* 30 2463-2472
[2] Lloyd S A and Weinberg F J 1974 A burner for mixtures of very lowheat content *Nature* 251 47-49
[3] Sitzki L, Borer K, Schuster E and Ronney P D 2001 Combustion in micro scale heat recirculating burners *The Third Asia Pacific Conference on Combustion* June 24-27, Seoul, Korea.
[4] Jejurkar S Y and Mishra D P 2009 A review of recent patents on micro-combustion and applications *Recent Patents on Engineering* 3 194-209
[5] Lee M J, Cho S M, Choi B I and Kim N I 2010 Scale and materialseffects on flame characteristics in small heat recirculationcombustors of counter-current channel type *Applied Thermal Engineering* 30 222–2235
[6] Ju Y and Choi C W 2003 An analysis of sub-limit flame dynamics usingopposite propagating flames in mesoscale channels *Combustion and Flame* 133 483-493
[7] Ronney P D 2003 Analysis of non-adiabatic heat-recirculatingcombustors *Combustion and Flame* 135 421-439
[8] Kim N I, Kato S, Kataoka T, Yokomori T , Fujimori S T and Maruta K 2005 Flame stabilization and emission of small swiss rollcombustor as heaters *Combustion and Flame* 141 229–240
[9] Chen C H, Gowdagiri S, Kumar S and Ronney P D 2009 Numericaland experimental study in swiss roll heat-recirculating burner *PowerMEMS* 605-608
[10] Bei-Jing Z and Jian-Hua W 2010 Experimental study on premixedCH_4/air mixture combustion in micro swiss roll combustor *Combustion and Flame* 157 2222-2229
[11] Khandelwal B, Sahota G P S and Kumar S 2010 Investigations intothe flame stability limits in a backward step micro scalecombustor with premixed methane-air mixtures *Journal ofMicromechanics and Micro Engineering* 20 P. 095030.
[12] Akram M and Kumar S 2011Experimental studies on dynamics ofMethane - air premixed flame in Meso-scale diverging channels *Combustion and Flame* 158 915-924
[13] Jejurkar S Y and Mishra D P 2011 Flame stability studies in a Hydrogen - air premixed flame annular micro combustor *International Journal of Hydrogen Energy* 36 7326-7338
[14] Sanchez-Sanz M 2012 Premixed flame extinction in narrow channels with and without heat recirculation *Combustionand Flame* 159 3158-3167

[15] Chen C H and Ronney P D 2013 Scale and geometry effects onheat-recirculating combustors *Combustion Theory andModeling*
[16] Wang Y, Zhou Z, Yang W, Zhou J, Liu J, Wang Z and Cen K 2011 Instabilibilty of flame in micro-combustor under different external thermal environment *Experimental Thermal and Fluid Science* 35 1451-1457
[17] Mane-Deshmukh S B, Bhojwani V K, Krishnamoorthy A, Avati V, Patil S and Mandale E 2016 Experimental investigations of the characteristics of the heat recovering/recuperating one turn exhaust channel stainless steel and aluminum combustors *RAM – 2016*

IOP Conf. Series: Materials Science and Engineering **197** (2017) 012032 doi:10.1088/1757-899X/197/1/012032

Experimental study on productivity of modified single-basin solar still with a flat plate absorber

V Ramanathan[1], B Kanimozhi[1], V K Bhojwani[2]

[1]Department of Mechanical Engineering, Sathyabama University, Chennai, India

[2]Department of Mechanical Engineering, J.S.P.M. College of Engineering, Pune, India

vramnathan.sae@sinhgad.edu

Abstract. Solar still is an apparatus which uses solar energyto producedistilled water from saline water.This can be used in remote areas effectively wherein electricity is not available.The output from a conventional single basin solar still is found to be very low. Hence research is required to increase the productivity of the conventional solar still. This work is an attempt to increase the productivity of solar still. A flat mica plate is embedded in the conventional solar still to augment evaporation of the water from the input saline water. The flat plate absorber is placed in such a way that it is parallel to the glass cover of the solar still so as to maximize the absorption of solar radiations. By this modification, the maximum temperature of the absorber plate achieved was 95°C in comparison to 67°C of the conventional solar still. Experimental results of modified solar still were compared with conventional solar still. It was found that distillate output increased by 25% with a flat plate absorber when compared to conventional still.

Keywords—Solar Still; Flat Plate Collector; Distillation.

1. Introduction

A major amount of water on earth is either in the form of seawater or icebergs in the Polar Regions. About97% of this water is in sea. Only 1% of the fresh water is within human reach. Even this small fraction of water is said to be sufficient to support human life. The hydrological cycle of nature is a very large-scale process of solar distillation which naturally produces fresh water in the form of rain. Thishydrological cyclecan be explained in short as a process of production of water vapours above the surface of the seawater and cooling of air-vapour mixture through condensation and precipitation. The basin type solar still mimics this natural hydrological process on a small scale.

Solar distillation usessolar energy as a source of heat energy for obtaining fresh water from salty, brackish or contaminated water. It is a slow distillation process which allows only pure water to evaporate from the basin and condensate beneath the glass cover, leaving all particulate contaminants behind. Solar stills are suitable only for regions where freshwater demand is less than 200 m^3/ day [1]. Amongst the various types of solar stills available, the single-basin still has been much studied by various authors worldwide. One of the main setbacks for this type of desalination device is low thermal efficiency and productivity [2].

IOP Conf. Series: Materials Science and Engineering **197** (2017) 012032 doi:10.1088/1757-899X/197/1/012032

2. Historic Review

The solar distillation has been used by Arab alchemists as early as 1551. The French chemist Lavaosier has used wide glass lenses to concentrate solar rays to distillate water inside the flasks in 1862 [3]. The first known documented application of solar distillation was in the year 1872 at Las Salinas. It provided drinking water for animals used in nitrate mining [4]. Most of the solar stills have been built on a similar concept with some variations in geometry, materials, methods of construction and operations employed.

The thermal efficiency and productivity of the simple solar still has been improved by a number of both passive and active modifications. Various wick and porous materials were experimentally studied by Murugavel et al [5] for using in basin to improve the productivity. The results show that the still with black light cotton cloth as wick material is found to be more productive. The wick is a better option than porous materials for they maintain a thin water film which in turn improves the evaporation rate.

The performances of solar still with different sizes of sponges placed in the basin were studied by Bassam et al [6]. The productivity of the still was increased in the range of 18% to 273%. The optimal parameters were arrived at through series of experiments. However with increase in salinity, sponges were found to be less effective. Velmurugan et al [7] integrated fins at the basin of the still which resulted in productivity increase of the stillby 45.5%.

The surface heating of water mass was attempted by Valsaraj et al [8]. A perforated aluminum sheet is floated over the water surface. This sheet absorbs and dissipates the heat energy at surface of water layer. This prevents the whole water mass being heated by convection as in the normal still. The overall productivity was increased by 49.3 %.

The effect of suspended absorbers was investigated by Sebaii et al [9]. When aluminium plates were used it was found that the productivity increased by 20 % and 23% for mica plate as suspended absorber. The suspended absorber enables to increase the water mass contained in the basin thus increasing the nocturnal production of the still. Sahoo et al [10] studied the effect of blackened surface and insulation on the performance of a solar still. The efficiency of solar still got increased by 17% upon suitable modification of the solar basin with appropriate base liner and insulation.

The effect of varied depths of water in the basin was studied. The depth of water influences the internal heat and mass transfer and thereby the productivity of the still[11]. The maximum distillate output was achieved with water depth in still basin 2 cm. The maximum efficiency of the experimental still varies from 10% to 34%. The results indicate that an increase in depth of basin water, still productivity decreases [12].Patel et al [13] utilized the different semi-conducting oxides like CuO, PbO_2 and MnO_2 as photo-catalysts. The overall efficiency of conventional basin type solar still got enhanced. Metal oxides not only improve the efficiency of the process but the rate of production of desalinated water was also increased.

Bassam et al [14] analysed the Effect of water emissivity on solar still efficiency. The use of dye resulted in minimal improvement in still efficiency. The reduction in the water emissivity will reduce the radiation heat transfer from water to glass. This resulted in a substantial improvement in still efficiency and offers great potential for solar still usage.

The literature survey points out that output can be increased by decreasing the losses, increasing the temperature difference between the evaporating and condensing surfaces, making use of latent heat of condensation, and maximizing the driving force for evaporation by increasing the heat sink capacity. These considerations usually increase the cost of the system, since they require use of more materials.

The still productivity depends on parameters like solar radiation intensity, atmospheric temperature, basin water depth, glass cover material, thickness and its inclination, wind velocity and the heat capacity of the still. Amongst the above parameters, the inclination of the absorber surface and depth of the basin water are the main parameters that affect the performance of the still.

3. System Developement

In this work, a flat plate absorber plate (a mica plate of thickness 8 mm) was integrated to the basin of the still. The flat plate absorber was placed in such a way that it is parallel to the glass cover of the solar still so as to maximize the intensity solar radiations by reducing the angle of incidence. Depth of water on absorber is reduced for maximizing the absorption. This is achieved by supplying water on the absorber surface of solar still through a separate water tank by gravity. The primary objective of the work is to increase the productivity of single basin solar still with the help of low cost modifications.

The experimental investigations were performed using one of the simplest types of solar stills popularly known as asymmetric greenhouse type still (ASGHT). It has an effective area of 1 m^2. The frame is made of fiber reinforced plastic. It has a top cover of transparent glass, and the interior surface of its basin is blackened to enable absorption of solar energy to the maximum possible extent. The body is made of Fibre reinforced plastic, 3 mm thick; and shaped as an inclined box placed on a vertical mild steel stand. The whole assembly is air tight with help of rubber gasket and screws.

The unmodified still will be referred as conventional still henceforth. This conventional still serves as a standard for comparison with the modifications done in the solar still. The outputs of the two systems were compared to see the effect of the modification.

Figure. 1Conventional solar still

The conventional still is the still as in figure 1 is without any modification. Thermocouples were fixed at various points on the still to measure temperature. This still has served as the standard for comparison for the modified still.

Figure. 2 Modified solar still

The modified still as shown in figure 2 is geometrically same as that of the conventional solar still. Modifications like inclusion of a flat plate absorber and continuous flow of inlet water were made in the conventional still.

A flat absorber surface parallel to the glass will increase the efficiency of the system. A mica sheet 940 mm x 980 mm size was placed within the still parallel to the glass surface. The mica sheet is light in weight and can withstand up to 80-85°C. The modified still is of active nature as against a passive conventional still. Here a reservoir of 20 liter capacity is used to store and supply saline water to the modified still. The reservoir is mounted on a level higher than solar still.

The flat plate introduced in to the system is the new absorber surface as against a basin in a conventional still. The K-Type thermocouples were placed in the as shown in figure 3 for conventional still and figure 4 for modified still.

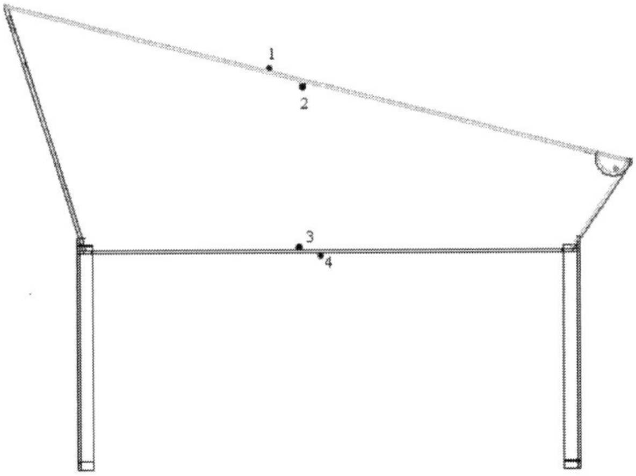

Figure. 3 Positioning of thermocouple in conventional still

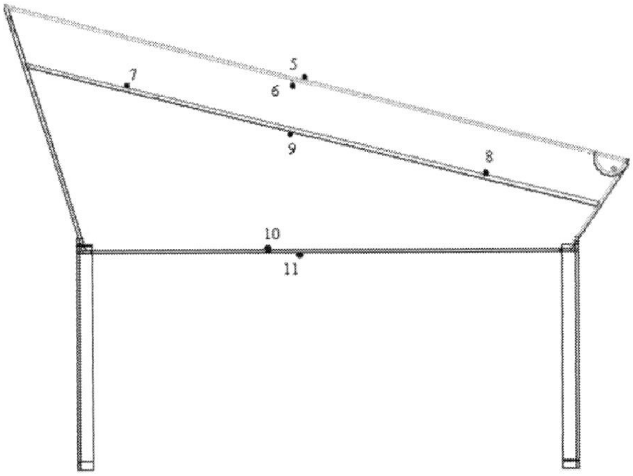

Figure. 4 Positioning of thermocouple in modified solar still

4. Experimental outcome

The main objective of this present work is to improve the distillate output. The modified system is tested for its performance. The conventional system serves as a standard to quantify the performance of the modified still. Both the stills were tested simultaneously under the same conditions. The experiments were conducted in actual outdoor conditions. The observations were recorded for the cover temperatures, basin water temperatures, plate surface temperatures, wind velocity, and ambient temperature.

Both the stills were supplied with 20 liters of water at same initial temperature and quality. In the conventional still, the water is filled at once in the basin trough before the experiment is started. In the modified still, 20 liters of water was filled in the water reservoir. The flow rate of the water entering the modified still can be fine adjusted using the control valve fitted to the base of reservoir. The

readings were recorded at an interval of one-hour. The distillates from both the stills were measured on hourly basis. The anemometer reading and pyranometer reading were recorded.
Initially the Standardisation of both the solar stills is done before any modification to ensure uniformity. Both the stills were subjected to the same conditions during the standardization test. The outputs of the two systems were measured and compared. Both the stills have shown nearly the same performance and variation in output is well within 2% as seen in graph.1.

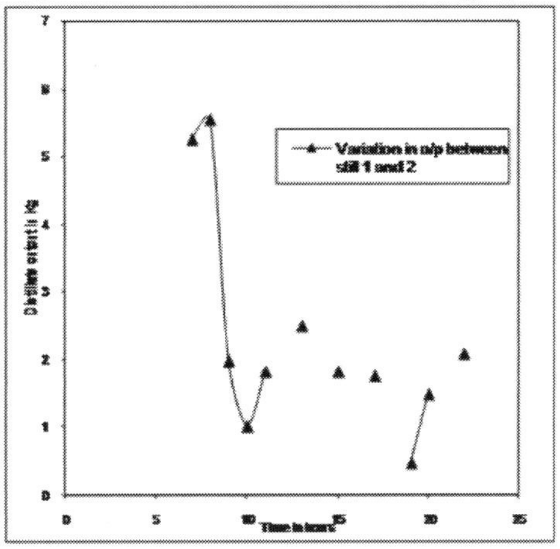

Graph.1 Standardisation of solar stills

The mica sheet is embedded in a solar still to make a modified still. This modified still is then tested against the conventional still.
From graph.2 it is noted that the maximum temperatures in a modified still is 95.5 0C on the flat absorber plate as against 67 0C in the basin trough of conventional still.

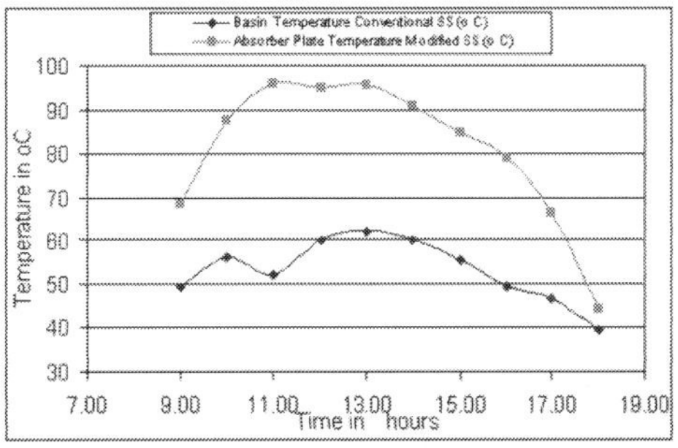

Graph.2 Temperature Plot of Conventional Still and Modified still

The distillate output of the modified solar still is increased by 25% conventional still as can be seen in graph.3

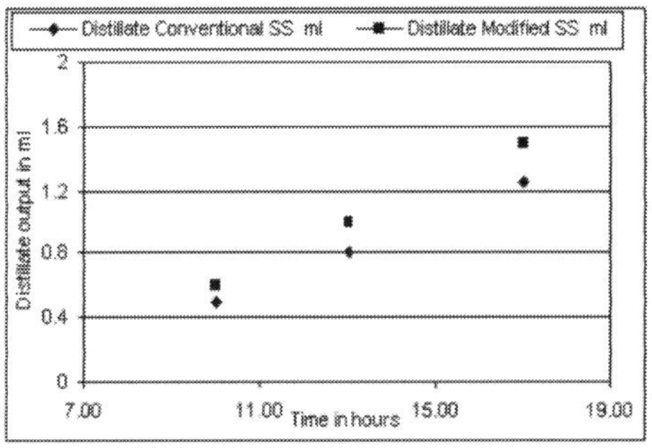

Graph 3. Distillate Output plot of Conventional Still and Modified Still

5. Conclusion

The modified solar still has an increased output by 25% compared to the conventional still. The increase in Thermal efficiency can be attributed to the following factors

1. The absorber plate is parallel to the glass surface and also approximately equal to the latitude of this place resulting in absorption of maximum solar radiations
2. Reduction in cavity volume between the absorber surface and glass surface: This led to enhancement of mass transfer i.e. more evaporation and condensation.
3. The water drained from the absorber plate gets accumulated in the basin. This further increases the heat mass capacity of modified still and enables condensation in nighttime also.
4. Due to accumulated water, the underneath of the absorber plate is not directly exposed to atmospheric temperature but to temperature of the vapor. This reduces the heat loss at the underside of absorber plate compared to the conventional still.

Thus it can be concluded that with the above simple modifications made in the solar still, it is possible to improve the output of the system significantly.

In this experimental work it is observed that the distribution of water is not uniform in the bare absorber plate. Although higher operating temperatures were achieved, a proportional increase in productivity was not achieved. The root cause of the problem was uneven distribution of water over the absorber surface. To overcome this problem the textile materials like jute, cotton cloth can be spread water over the absorber plate.

References

[1] Compain, P. (2012). Solar energy for water desalination, 1st international symposium on innovation and technology in the phosphate industry. Procedia Engineering, 46, 220–227.

[2] El-Samadony, Y. A. F., & Kabeel, A. K. (2014). Theoretical estimation of the optimum glass cover water film cooling parameters combinations of a stepped solar still. Energy, 68, 744–750.

[3] Janarthanan, B., Chandrasekaran, J., & Kumar, S. (2006). Performance of floating cum tilted-wick type solar still with the effect of water flowing over the glass cover. Desalination, 190, 51–62.

[4] Kaushal, A., & Varun. (2010). Solar stills: A review. Renewable and Sustainable Energy Reviews, 14, 446–453.

[5] Murugavel, K. K., Chockalingam, K. K. S. K., & Srithar, K. (2008). Progress in improving the effectiveness of the single basin passive solar still. Desalination, 220, 677–686.

[6] Bassam et al (2003) Experimental study of a solar still with sponge cubes in basin, Energy Conversion and Management, 44(9):1411-1418.

[7] Velmurugan, V., Deendayalan, C. K., Vinod, H., & Srithar, K. (2008). Desalination of affluent using fin type solar still. Energy, 33, 1719–1727.

[8] Valasaraj et al (2002) An experimental study on solar distillation in a single slope basin still by surface heating the water mass, Renewable Energy 25(4):607-612.

[9] El-Sebaii, A. A., Al-Ghamdi, A. A., Al-Hazmi, F. S., & Faidh, A. S. (2009). Thermal performance of a single basin solar still with PCM as a storage medium. Applied Energy, 86, 1187–1195.

[10] Sahoo, B. B., Sahoo, N., Mahanta, P., Borbora, L., Kalita, P., & Saha, U. K. (2008). Performance assessment of a solar still using blackened surface and thermocol insulation. Renewable Energy, 33, 1703–1708.

[11] M.K. Phadatare , S.K. Verma (2007) Influence of water depth on internal heat and mass transfer in a plastic solar still, Desalination, Volume 217, Issues 1–3, Pages 267–275

IOP Conf. Series: Materials Science and Engineering **197** (2017) 012032 doi:10.1088/1757-899X/197/1/012032

[12] Bilal A. Akash, Mousa S. Mohsen (2000) Experimental study of the basin type solar still under local climate conditions, Energy Conversion & Management 41 883-890
[13] Suresh G Patil, Shilpi Bhatnagar, Jitendra Vardia, Suresh C Aveta(2006) Use of photo catalysts in solar desalination, Desalination, vol. 189, pp. 287-291.
[14] Bassam et al (2003) Effect of water emissivity on solar still efficiency, International Journal of Sustainable Energy , 23(1):13-19.

IOP Conf. Series: Materials Science and Engineering **197** (2017) 012033 doi:10.1088/1757-899X/197/1/012033

Thermal Analysis of Fluidized Bed and Fixed Bed Latent Heat Thermal Storage System

N Beemkumar*, A Karthikeyan, Kota Shiva Keshava Reddy, Kona Rajesh, A Anderson

School of Mechanical Engineering, Sathyabama University, Chennai, India

E-mail: beem4u@gmail.com

Abstract. Thermal energy storage technology is essential because its stores available energy at low cost. Objective of the work is to store the thermal energy in a most efficient method. This work is deal with thermal analysis of fluidized bed and fixed bed latent heat thermal storage (LHTS) system with different encapsulation materials (aluminium, brass and copper). D-Mannitol has been used as phase change material (PCM). Encapsulation material which is in orbicular shape with 4 inch diameter and 2 mm thickness orbicular shaped product is used. Therminol-66 is used as a heat transfer fluid (HTF). Arrangement of encapsulation material is done in two ways namely fluidized bed and fixed bed thermal storage system. Comparison was made between the performance of fixed bed and fluidized bed with different encapsulation material. It is observed that from the economical point of view aluminium in fluidized bed LHTS System has highest efficiency than copper and brass.The thermal energy storage system can be analyzed with fixed bed by varying mass flow rate of oil paves a way to find effective heat energy transfer.

1. Introduction

Amount of conservation of energy is more than available resources, hence there is a need substitution for non-renewable resource source and so there should be a change over to the renewable source as they can be obtained continuously. Renewable resources like wind, solar radiation are continuously emitted, but only a small amount of this energy is being used. The present work has been undertaken to study the feasibility of store up thermal energy using PCM and utilizing this energy to heat fluid for domestic applications during night time [1]. This ensures that hot fluid is available throughout the day. The PCMs are the materials which has very less sensible heat storage rather the maximum heat storage capacity is due to their latent heat capacity which is due to change in its physical phase. The storage system consists of two at once functioning heat-absorbing units. One of them is a HTF and the other a heat storage unit consists of PCM [2]. PCM which is stored inside the encapsulation material .encapsulation which is capable of transferring heat to the PCM. Encapsulation material which is in orbicular shape with 4 inch diameter and 2 mm thickness orbicular shaped product of copper, brass and aluminium are used as encapsulation materials because of its high thermal conductivity. Arrangement of encapsulation material is done in two ways namely fluidized bed and fixed bed thermal storage system [3]. The PCM and HTF are used as latent heat and sensible heat storage system respectively. The mixed sensible and latent heat storage system can overcome the negative aspect of using either one of them separately. The PCM Charging and discharging process has been carry out for both fluidized and fixed bed thermal storage system. Charging is taken for every 10 minutes from the noted initial temperature. The change in temperature of heat transfer fluid is noted for every 10 minutes. This process continues during day time. Discharging takes place during night time.

Discharging time is also noted for every 10 minutes. Discharging is taken till the HTF reaches its room temperature. Comparative study is done between charging and discharging in fixed bed and fluidized bed latent heat thermal storage system. The experiments are done to find out the way for most proficient utilize of solar energy [4]. This experiment is to find out the heat exchange rate of HTF with D-Mannitol as PCM by using different encapsulation material in fixed bed and fluidized bed latent heat thermal storage system.

2. Experimental Setup and Methodology

First and foremost step in this experiment is to setup the solar collector. Solar collector is designed to track the intermittent energy given by sun. There are many types of collector to track the energy from sun. But the literatures are proved that the use parabolic trough collector (PTC) would give the better results which also pave a way to track the solar energy in the most efficient manner [5] [6]. Initially the HTF therminol-66 is circulated through the closed loop of thermal storage system using a slow speed high temperature pump. Therminol-66 is carrying heat to the encapsulated material during charging. Circulation of oil with the optimum mass flow rate for effective heat transfer was considered for investigations [7]. Thermal storage system is intentionally made for the storage of hidden (latent) heat for extended duration [8]. Reducing losses leads to efficient storage of thermal energy which is obtained by providing glass wool for insulation [9]. Material with high heat thermal conductivity is chosen as encapsulation material. Encapsulation materials are aluminium, brass and copper. Adding PCM inside the encapsulation material is followed by melting the PCM and then pouring inside the encapsulation material. PCM which chosen for this experiment is D-Mannitol. Limited amount of PCM is poured inside the encapsulation material by ensuring sufficient space of thermal expansion of PCM [10] [11]. Before starting the experiment connection of thermocouple, flow meter and heat pump has to attach to the setup. Minimum of 12 thermocouple is needed for the system for correct measurement of temperature in all tanks. Arrangement of encapsulation material inside the tank is to be carried in two methods such as fixed bed and fluidized bed. In fluidized bed encapsulated materials are arranged randomly in a container by ensuring without any air gap between the encapsulation materials. In fixed bed position of encapsulation materials are arranged in a certain order with help of each tray. The experiment was carried out in two modes -one for charging mode and another one for discharging mode. The process of heat absorbing mode (charging mode) and heat liberating mode (discharging mode) were carried out with PCM (D-Mannitol) and also with different encapsulation material (copper, aluminium, brass).charging and discharging is to be carried out with two position fixed bed and fluidized bed latent heat thermal storage system. Above two methods is to study the heat transfer rate between HTF and PCM in fixed bed and fluidized bed.

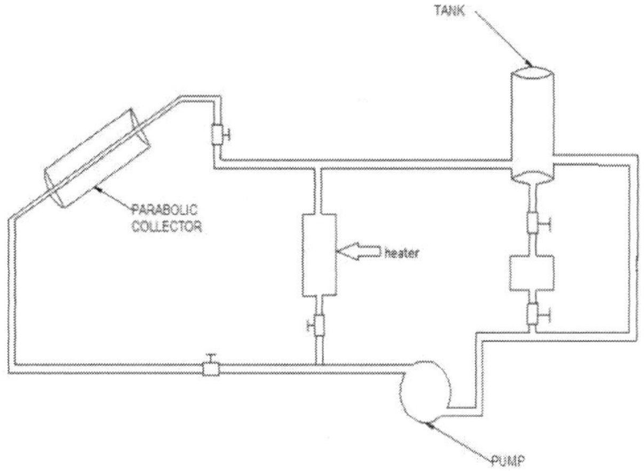

Figure 1. Experimental layout

Figure 2. PCM tank & fixed bed arrangement

Experiment setup is shown in the figure 1. The supply of oil to the remaining two tanks is closed with the help of check valve. The experimental set up consists of three thermal storage tanks, a heater and a circulating pump. The storage tank consists of both, the PCM and the HTF. Each cylindrical tank has a diameter of 256 mm and 300 mm length made of 6 mm thick MS plate. The storage tank and pipe lines are insulated with glass wool of thickness 0.10 m. A total quantity of 35 L of therminol-66 is filled in the storage tank and the circuit of pipe, and it is ensured that there is no leak in the oil flow path. An additional tank is also fabricated to house the heaters which will be used to provide the required heat energy. Valves are present at regular intervals to control the amount of HTF flow. This is used to reduce build-up of high pressure in the circuit, which could be dangerous [12]. The selection of PCM is depends on the required temperature range. The PCM chosen here is D-Mannitol, due to its stability and corresponding melting points of single tanks to use therminol-66 as HTF. Table 1 shows the properties of PCM and HTF. A gear pump maintains the HTF flow in the circuit, through the heater and storage tanks during charging and discharging process of PCM in which the heated oil i.e, HTF transfers heat to the PCM and vice-versa. The PCM is encapsulated in spherical balls of copper, aluminium and brass. The experimentation was done by altering different encapsulating material. The heater (supplementary source to the PTC) is switched on along with the pump. The temperature of HTF in all the tanks is noted after regular intervals of 10 minutes, this is called charging process of PCM. The heater is turned off, after attaining a set temperature. Keeping the flow rate on, the above process is repeated. This is the discharging process. Hence heat is subsequently stored and recovered in these processes. The above processes are conducted for all the three type of encapsulations material with D-Mannitol by using therminol-66 as HTF and the observations are noted down. PCM tank & fixed bed arrangement is shown in the figure 2.

Table 1. Properties of PCM &HTF.

PCM – D-Mannitol		**HTF – Therminol 66**	
Melting point (°C)	165	Density (kgm^{-3})	1005
Latent heat ($kJkg^{-1}$)	316.4	Specific heat ($kJkg^{-1}{}^{\circ}C^{-1}$)	1.495
Density ($kg\ m^{-3}$)	1490	Kinematic Viscosity (m^2s^{-1})	$29.64x10^{-6}$
Specific heat ($kJkg^{-1}{}^{\circ}C^{-1}$)	2.5	Boiling Temperature (°C)	359

3. Results and Discussion

The first tank is fitted with heater and the second tank is filled with brass encapsulation. During charging heater is kept on and the heat transfer fluid temperature increase gradually as well as the temperature of PCM increases simultaneously. This process goes on during charging. In discharging mode the heater and circulating pump is turned off. Temperature of working fluid and PCM decreases slowly and the temperature of HTF is noted for every 10 minutes during charging and discharging process.

3.1. Brass encapsulation analysis in fixed bed and fluidized LHTS system

Figure 3 shows the charging mode and discharging mode of HTF in storage tank when brass encapsulated balls are present in the tank. A comparison is made between charging and discharging in fixed and fluidized bed thermal energy storage system. The rate HTF temperature rise is quicker in both cases during charging process [13]; the heat is supplied by the external aid to the HTF. The temperature drop in both cases at slower rate in discharging process of PCM. In discharge process, the PCM liberates the heat to the HTF at slower rate. In fixed bed LHTS, the time to attain the temperature of 165°C from its room temperature is 190 minutes during charging process and in discharging process the time to attain its room temperature from 170°C is 280 minutes. In fluidized bed LHTS, the time to attain the temperature of 167°C from its room temperature is 170 minutes during charging process and in discharging process the time to attain its room temperature from 170°C is 340 minutes.

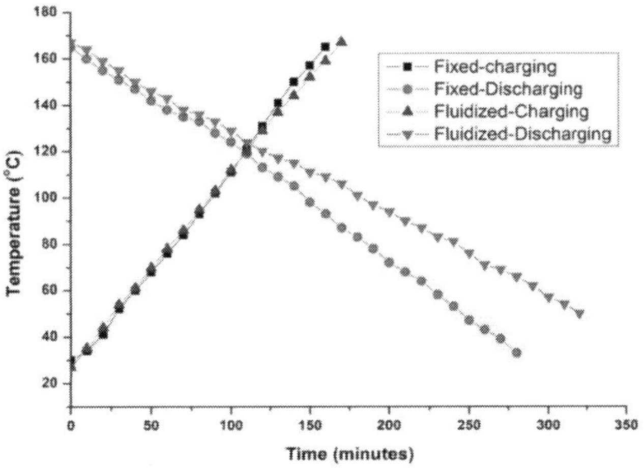

Figure 3. Brass encapsulation fixed vs. fluidized bed.

3.2. Copper encapsulation analysis in fixed bed and fluidized LHTS system

A comparison is made between charging and discharging in fixed and fluidized bed thermal energy storage system as shown in the figure 4. The D-mannitol PCM is encapsulated in copper balls are filled in the storage tank. Since copper has higher thermal conductivity property the heat absorbing and liberating of PCM was attained quickly in both cases (fixed and fluidized bed LHTS system).The time to attain the HTF temperature of 165°C from its room temperature is 160 minutes and during discharging the time to attain its room temperature from 165°C is 280 minutes in the fixed bed LHTS system. The time to attain HTF temperature of 170°C from its room temperature is 110 minutes and during discharging it's time to attain its room temperature from 170°C is 310 minutes in the fluidized bed LHTS system which shows the fluidized bed system has elevated energy density than fixed bed system [14].

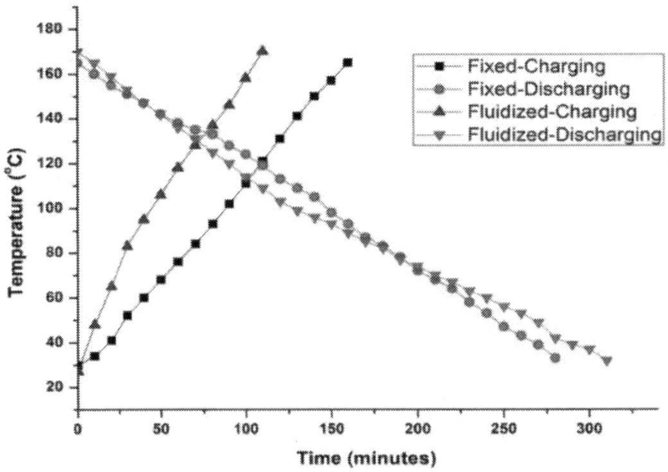

Figure 4. Copper encapsulation fixed vs. fluidized bed.

3.3. Aluminium encapsulation analysis in fixed bed and fluidized LHTS system

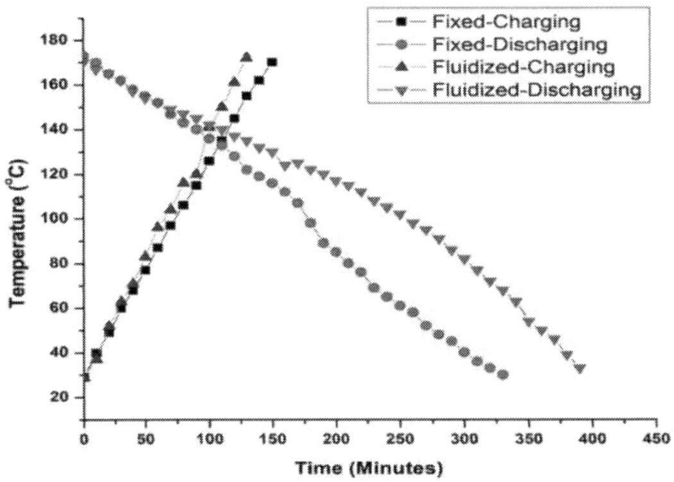

Figure 5. Aluminium encapsulation fixed vs. fluidized bed.

Figure 5 shows the charging mode and discharging mode of PCM inside aluminium encapsulations in single storage tanks. The PCM D-mannitol is encapsulated in aluminium spherical balls. In both fixed and fluidized bed LHTS system the temperature drop of HTF decreases gradually because of the higher latent heat storage capacity of D-Mannitol [15]. In fixed bed LHTS system, the time to attain HTF temperature of 173°C from its room temperature is 150 minutes in charging process and in discharging process the time to attain its room temperature from 173°C is 330 minutes. Fluidized bed LHTS system, the time to attain HTF temperature of 172°C from its room temperature is 130 minutes in charging process and in discharging process the time to attain its room temperature from 172°C is 390 minutes.

3.4. Heat transfer effect between fixed bed and fluidized bed LHTES system

Figure 6 shows the comparison between heat absorbing (charging) and heat liberating (discharging) process of brass encapsulation in fluidized bed and fixed bed thermal storage system. Its shows that the mean value of heat energy transfer during charging in fluidized bed LHTS system has around 5.3 kW and during discharging around 3.3 kW. The rate of energy transfer in fixed bed LHTS system during charging and discharging is 3.31 kW and 1.332 kW respectively. The overall the efficiency of brass encapsulation in fluidized bed LHTS is high than fixed bed LHTS system because of the more contact time between HTF and PCM in fluidized bed LHTS system.

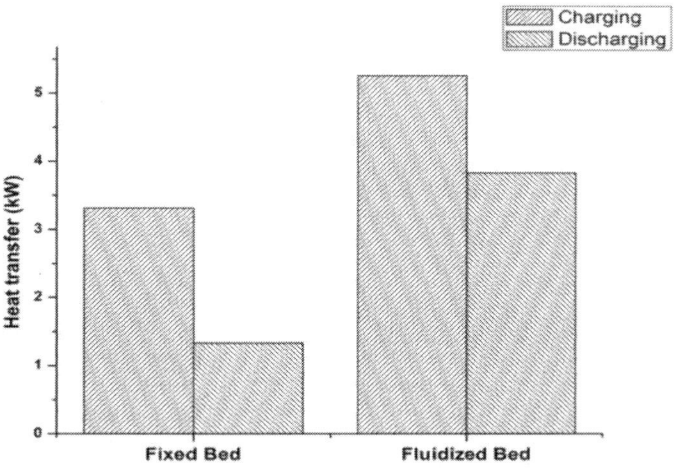

Figure 6. Heat transfer rate fixed vs. fluidized bed in brass encapsulation.

The energy exchange rate of aluminium encapsulation in fixed and fluidized bed thermal storage system is exposed the figure 7. During charging, the fluidized bed LHTS system is 2.8 kW higher than fixed bed LHTS system. During discharging process, the fixed bed LHTS 0.83 kW higher than fluidized bed LHTS system.

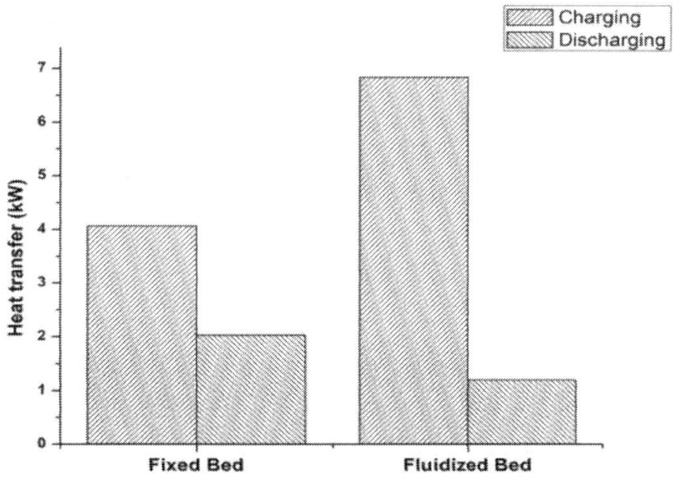

Figure 7. Heat transfer rate fixed vs. fluidized bed in aluminium encapsulation.

The mean energy transfer rate copper encapsulated PCM in fixed and fluidized bed LHTS is shown in the figure 8. The fixed bed LHTS is 3.65 kW and 2.73 kW during the process of heat absorption and liberation respectively. In fluidized bed LHTS system the heat energy transfer rate is about 5.98 kW during charging process and 2.81 kW in the case of discharging process of PCM. The results clearly show that the heat transfer rate in fluidized bed LHTS is always higher in magnitude when compared to fixed bed LHTS system. The end result of the experimentation has proved that the fluidized bed LHTS system have superior energy density than fixed bed, thus the overall effectiveness of the fluidized bed LHTS system is always high.

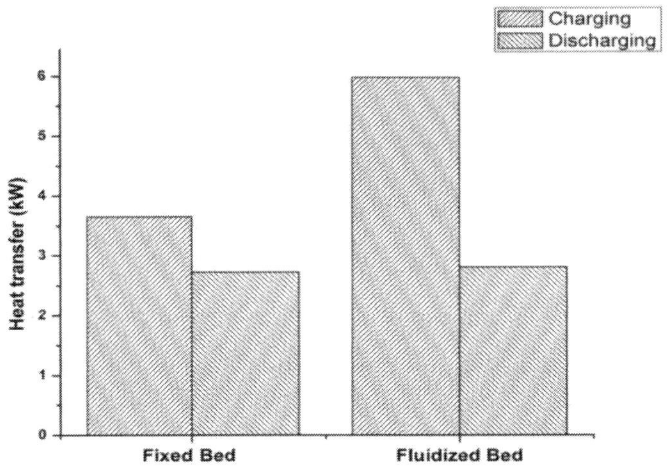

Figure 8. Heat transfer rate fixed vs. fluidized bed in copper encapsulation.

4. Conclusion

Thermal analysis of fluidized bed and fixed bed latent heat thermal storage system with the use of D-mannitol as PCM has been studied and analyzed. Comparison was made between the performance of fixed bed and fluidized bed with different encapsulation material namely copper, aluminium and brass. It is observed that from the economical point of view aluminium in fluidized bed LHTS System has highest efficiency than copper and brass. With the use of aluminium encapsulated PCM in fluidized bed LHTS system showed that the cost per kW of energy stored is around Rs. 1.8 which is 33.33 % and 22.22 % less than copper and brass respectively. The mean energy transfer rate of aluminium encapsulated PCM in both cases is about 12 % less than copper 10 % higher than brass encapsulated PCM. Thus the use of PCM in fluidized bed LHTS would give the better results of energy transfer than fixed bed LHTS system, but need to care about the HTF flow pressure drop in the storage tank.

References

[1] Nasser Mostafavinia, Saman Eghvay and Amir Hassanzadeh 2015Numerical analysis of melting of nano-enhanced phase change material (NePCM) in a cavity with different positions of two heat source-sink pairs*Indian J. Sci. Tech.***8**(S9) pp 49–61.

[2] Aceves-Saborio SS, Nakamura HH and Reistad GM 1994Optimum efficiencies and phase change temperatures in latent heat storage systems*ASME Journal of Energy Resources Technology* **116**(1) pp 79-86.

[3] Vahid Madadi, Hamid Beheshti, Touraj Tavakoli and Amir Rahimi 2015 Experimental study and first thermodynamic law analysis of a solar water heater system *J. Non-Equilibrium Thermodynamics***40**(3) pp 171–183.

[4] Selvan Bellan, Tanvir E Alam, José González-Aguilar, Manuel Romero, Muhammad M Rahman, D Yogi Goswami and Elias K Stefanakos 2015 Numerical and experimental studies on heat transfer characteristics of thermal energy storage system packed with molten salt PCM capsules*Applied Thermal Engineering***90**pp970-979.

[5] Lingkun Liu, Guruprasad Alva, Xiang Huang and Guiyin Fang 2016 Preparation, heat transfer and flow properties of microencapsulated phase change materials for thermal energy storage*Renewable and Sustainable Energy Reviews***66** pp 399-414.

[6] Govindaraj Kumaresan, Rahulram Sridhar and Ramalingom Velraj 2012 Performance studies of a solar parabolic trough collector with a thermal energy storage system *Energy***47** (1) pp 395–402.

[7] V K Jebasingh and G M Joselin Herbert 2015 Numerical simulation of elliptical absorber tube in parabolic trough collector for better heat transfer.*Indian J. Sci. Tech.* **8**(24) DOI: 10.17485/ijst/2015/v8i24/80165.

[8] Mohamed M Abdulgalil, Franc F Kosi, Mohamed H Musbah and Mirko S Komatina 2014 Effect of thermal energy storage in energy consumption required for air conditioning system in office building under the african mediterranean climate*Thermal Science***18**(1) pp S201-S212.

[9] Beemkumar N and Karthikeyan A 2016 Experimental analysis of heat transfer characteristics of solar energy *Materials today: Proceedings***3**pp 2475 – 2482.

[10] Felix Regin A, Solanki SC and Saini JS 2008 Heat transfer characteristics of thermal energy storage system using PCM capsule: A review *Renewable and Sustainable Energy Reviews***12** pp 2438–58.

[11] Medrano M, Yilmaz M O, Nogues M, Martorell I, Joan Roca and Luisa F Cabeza 2009 Experimental evaluation of commercial heat exchangers for use as PCM thermal storage systems *Applied Energy***86** 2047-2055.

[12] Nallusamy N, Sampath S and Velraj R 2007 Experimental investigation on a combined sensible and latent heat storage system integrated with constant/varying (solar) heat sources *Renewable Energy***32** pp 1206-1277.

[13] Shiina Y and Inagaki T 2005 Study on the efficiency of effective thermal conductivities on melting characteristics of latent heat storage capsules *International Journal of Heat and Mass Transfer***48**pp 373-383.

[14] Velraj R, Seeniraj R V, Hafner B, Faber C and Schwarzer K 1999 Heat transfer enhancement in a latent heat storage system*Solar Energy***65**(3) pp 171-180.

[15] Zalba B, Mann JM, Cabeza LF and Mehling H 2003 Review on thermal energy storage with phase change: materials, heat transfer analysis and applications*Applied Thermal Engineering***23**pp 251-283.

REVIEW OF PHASE CHANGE MATERIALS BASED ON ENERGY STORAGE SYSTEM WITH APPLICATIONS

R.Thamaraikannn[1], B.Kanimozhi[2], M. Anish[3], J.Jayaprabakar[4] P.Saravanan[5], A.RohanNicholas[6]

School of Mechanical Engineering, Sathyabama University, Chennai, India

E-mail: anish2010me@gmail.com

Abstract. The use of Different types of storage system using phase change materials (PCMs) is an effective way of storing energy and also to make advantages ofheating and cooling systems are installed to maintain temperatures within the well-being zone. PCMs have been extensively used in various storage systems for heat pumps, solar engineering, and thermal control applications. The use of PCM's for heating and cooling applications have been investigated during the past decade. There are large numbers of PCM's, which melt and solidify at a wide range of temperatures, making them attractive in a number of applications. This paper also outline the investigation and analysis of Phase Change materials used in Different Types of storage systems with different applications.

1. Introduction

Since the beginning of time people have beneficiated by sun. Ancient civilizations personified by the sun. Only recently suns however have we developed the ability to harness the suns awesome power .The resultant technologies have promising deduction for the future of renewable energy and sustainability Energy storage is the capture of energy produced at one time for use at later time a device that stores energy is sometimes called as accumulator. Energy comes in multiple fumes including radiation, chemical, gravitational potential, electricity, elevated temperature latent heat. Energy storage involves converts energy from forms that are difficult to store the conventional, economical storage forms. As examples energy demand can balanced day time and day time .summer heat from solar heaters can be stored interpersonally for use in kilter and cold obtained from kilter air can be provided for summer air conditioning. Storage medium include water or ice slush tanks ranging from small to massive. There are many types of storage system using phase change materials is an effective way of storing energy and also to make advantage of heating and cooling systems are installed to keep temperatures within the well being zone .PCMs have been extensively used in various storage systems. For heat pumps solar engineering and thermal control applications. The use of PCMs for heating and cooling have been investigated during the past decade .they are large number of PCMs which meld solidify at a wide range of temperatures making them beautiful in a number of applications. This paper also outline the investigation and analysis of phase change materials used in different types of storage systems with different applications .Thermal energy storage technology with PCMs with many applications those technology is very beneficial for the humans as well as the energy conversations

2. Energy Storage Systems

Thermal Energy Storage systems play an important role in the effectual functioning of different systems, and industrial waste heat advance systems. Energy can be stored and recovered as sensible heat, latent heat and also in energy storage transformations increase of any of these. In SHS system, thermal energy is stored by increasing the temperature of a solid or liquid. The SHS system uses the heat capacity and the change in the temperature of the material during the procedure of charging and discharging. The amount of stored thermal energy depend on the specific heat of the medium, the temperature alter and the amount of storage material. LHS system is settled on heat absorption or release, when heat provide to the system. When materials access certain level of temperature the additional heat energy is stored in the system without any alter in the temperature, then the stored energy in the materials experience a phase change from solid to gas or reversed. It is called as latent heat storage system. Energy Storage system, the energy is occupied and distributed while breakage or reforming molecular bonds in a fully reversed chemical reaction. In this case, the storage of heat depends on the quantity of chemical material and endothermic heat of reaction.

3.Heat Transfer Improvements Methods

Enhancement techniques are supporting tools for thermal energy storage systems. The closely enhanced design yields good results in thermal systems. The rapid development of world literature on this subject point that enhancement is now a major specialty area in thermal energy (heat transfer) Research and Development. so the industrial and domestic utilization of improvement techniques increases day by day. Heat transfer improvements techniques are needed for many latent heat thermal energy storage systems; various methods are proposed to increase the heat passing in a latent heat thermal energy storage system, such as Metallic fillers, metal matrix construction, and finned tubes were used to increase the thermal conductivity of the phase change materials as shown in figure 1.improvements related techniques are listed below. (i)Active methods of agitators, vibrators, scrapers and clay. (ii)Microencapsulated PCM. (iii) PCM contains separate high conductivity particles, Lessing rings.

(iv) PCM assorted with graphite composite material.

(v) Extensive surfaces such as fins and honeycombs.

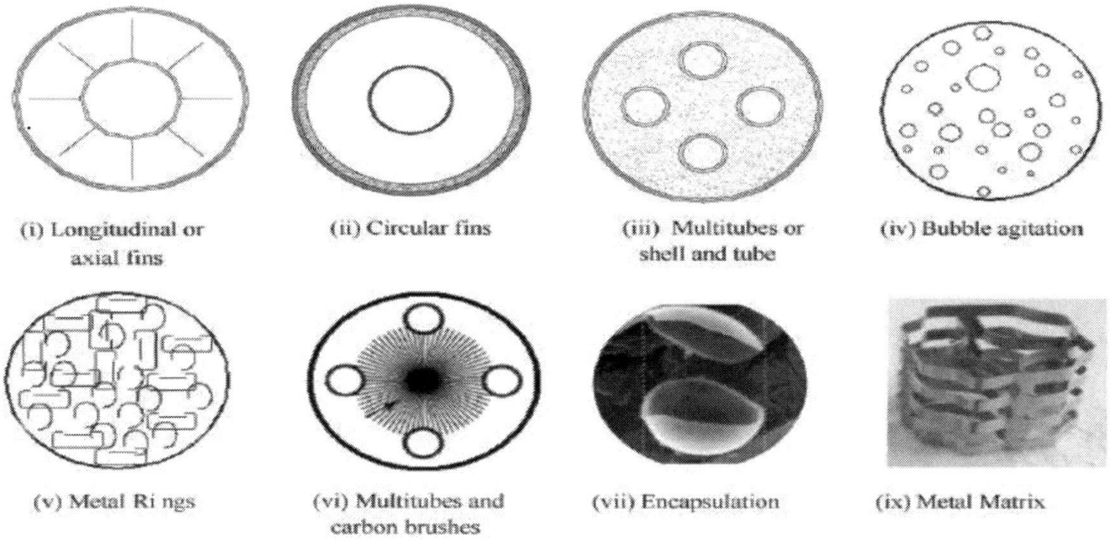

Figure.1 Different kinds of Heat Transfer Improvements fins

4. Objective:

[1]By using phase change material like paraffin and stearic acid during thermal energy storagesystemusing both sensible and latent heat storage capacity in a unit volume, while charging and discharging the solar energy is maintained at constant temp and pressure.

[2] PCMs like paraffin (aluminium is filled with paraffin) it maintain that hot water is accessible throughout the day the storage unit stores the heat in the PCM storage during the day time and supplies hot water during at night and overcast time.[3] By using he PCM (wax based storage system is improved by fixed copper, aluminium and iron spring).we can store the loss of energy during changing and discharging of solar energy.[4]His review is using the PCM the latent storage system is obtained in effective way and it has the many advantages high storage, isothermal in nature .PCM is widely used for many applications

[5] To avoid when the available energy source is irregular then a time lagging between productions is demand by providing three different aspect ratio of thermal storage tank namely 2, 1, 1/2.[6] He describe that using the pcm when the solar radiation is not available we use the sensible storage unit .when the solar radiation is available means we use the latent heat storage system.[7] The storage of solar energy with the help of thermal energy storage like solar energy storage sensible heat storage and latent heat storage with the help of PCM .[8]He describe using the solar energy the water is heated with the help of attachment of photovoltaic tank because this one reach the high efficiency when compared to others normal heater.

[9]The focus on solar power plant using the thermal energy storage technology and system to get a low economic analysis as high efficiency.[10] He describe focus on solar and power plant using the latent thermo Celine energy storage system is performed for repeated charging and discharging process and movement of process using the encapsulated PCM.[11] He describe using the PCM as salt instead of paraffin wax to increase the solidification of storage system and It has high thermal conductivity and low cost compared to the paraffin wax.[12] She describe the storage .The energy interns of temperature distribution heat transfer and energy storage rate by using multi tank storage system.[13] The solar energy storage based on the metal hybrids particularly to get the better performance, optimization and good economical feasibility.[14] He describes that energy storage system based on latent heat storage and thermo chemical heat storage mainly concentrate on limited life of methods energy source is the achievement of sustainable energy is not available at all the time the need of heat storage system obtained.[15] He describe in order to reduce operation cost of solar plant and the limitation is their low heat conductivity based on different types of thermal storage system sensible ,latent and chemical heat using PCM.[16] The simple and widely used application of solar energy include cooking, water heating power generation and agricultural drying operating the cooking on power by using different PCM(stearic acid)[17] The usage of PCM to kept the heat in the form of latent heat is increased because large amount of thermal energy stored in smaller volume. In that sodium thiosulphate ,pentahydrate used as a PCM stored in stainless steel capsule.[18] He describes the mainly concentrated on free cooling and active building cooling by the concept of thermal energy storage by using PCM as $CaCl2,6H2O$ this material is used for cooling the water during night time and circulating throughout the area.[19] He describes based on phase changing material in solar thermal energy storage system in different types Sensible heat storage ,latent heat storage and thermo chemical storage and is widely used for many applications in order to reduce the low cost medium good efficiency and stores thermal energy.[20] He describes based on PCM the indoor temperature of the fluctuation can be reduced significantly to achieve energy conservation in building with thermal comfort..

[21] The review on thermal energy storage system using phase changing material and it's widely used for many applications space heating and cooling in buildings solar application of peak energy storage on heat exchanger in improvement. It concentrates to raise in electric energy cost thermal storage

technologies.[22] To achieve thermal comfort inn buildings reducing the cooling and heating demand using PCM to obtain energy efficiency inHVAC system.[23] He describes using PCM able to charge and discharge the energy storage with a thermal power with the help of latent thermal energy storage system in order to increase the thermal conductivity of material using as pcm . which limits the power that can be extracted from the thermal energy storage [24] He explains the heat transfer co -efficent decreases with increasing phase change material as micro protected pcm suspension with water propanol mixture as the base fluid .temperature and flow rate of heat transfer in three technique the pure conduction quasi -steady and decay period [25] In order to increase the content on the pcm .its prepared by using the macro packed pcm .its prepared by using n-octedecane ,n-eicosane and n-docosane to get the thermal conductivity was more than pure pcm .and also wiedly used in aplication material in buildings [26] In order to reduce the weight and space requirement of energy storage system by using cylindrical latent heat energy storage system .based on pcmcanonic acid is used with copper fins .

[27] In order to get the satisfactory conductivity good thermal reliability and stability by using unchanged pcm in differntLauric -Palmatic -Stearic acid /expanded perlite composite[28] He describe the use of pcm in a thermal energy storage tank when blended to a heat pump system and also he mentioned to get high efficiency by using pcm tanks compared to normal tank material like salt hydride ,macro -encapsulated in plastic flat slab.[29] He describes in order to improve the speed of the charging and discharging heat process by adding the oil engine different type of paraffin wax as PCM.[30] He describes order to increase the higher heat storage capacity by using different kind of building composite PCM in different ratio and many TES application.

[31] The review on phase changing material. The phase change between 0 degree Celsius and 250 degree Celsius during heating and cooling applications by latent heat storage system and many techniques used for heat transfer.PCM salt hydrate, eutectic mixtures.

[32] By using the composite PCM to achieve the high energy charging and discharging and method to enhance the thermal conductivity by the PCM.[33] The discharging process of a latent heat thermal energy storage system with help of finned head pipes by using the high temperature PCM.[34] He describes to accrue long term productivity of LHS system by using Thermo-Physical stability of both paraffin's and salt hydrates .Its mainly focused on to improve the phase transience rate thermal conductivity latent heat storage capacity and thermo physical stability.

5.Encapsulation of Phase Change Materials

There are many merits in using an encapsulated phase change materials, which rising the heat transfer area, decrease the PCM reactivity towards the outside surroundings, and manage the changes in the storage material sizes. The phase change materials are encapsulated in clear design closed vessel with metal or plastic materials as shown in Figure.2

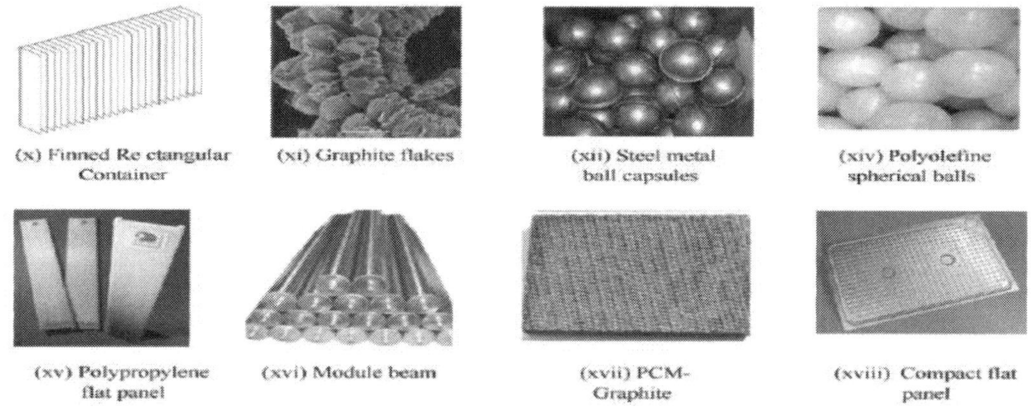

Figure.2 Heat Transfer EnhancementEncapsulation

[35] he developed a computer framework for the appraising of the temperature of solid and fluid. The

length of the packed bed contains encapsulated AL-Si PCM shots. Air was used as the heat transfer fluid (HTF) in their analyse.[36] Developed an ID porous medium framework to find the thermal characteristics of ice-water cool storage in crowded capsules for air conditioner, in which the water as PCM and alcohol as coolant for assorted porosities.[37] Analyses numerical probe on packed bed thermal framework for sensible and latent heat storage systems. The framework was compared in relation to the work factors of particle size, void fraction, particle material, flow rate alteration, working fluid inlet temperature and wall thermal losses. [38] Compare Sodium acetate try hydrate has more effectual than paraffin wax in which the storage tank comprise number of encapsulated PCM balls in which solar energy is engrossed and stored as latent heat. Large amount of solar energy can be stored in a day time and same heat can be used afterwards. It concluded Sodium acetate try hydrate has more effectual than paraffin wax and extended with thermal cooling on PCB.

Carry a transient phase change thermal storage system. The manage volume limit difference approach was used to find the equations. The EASY scheme was used to find the pressure and velocity fields of the HEAT TRANSFER FLUID. The influencing factors such as radial temperature spreading, phase change overlap position and heat accumulation have been considered. Deal with the experimental ascertaining of the thermal performance of a packed bed latent heat TES system, formed with solar flat plate aggregator, in which the TES unit contains paraffin as PCM is filled in spherical capsules. Delineate the preparation of a form-stable PCM and polyethylene paraffin compound. It contained of paraffin as a dispersed PCM and a high volume polyethylene (HDPE) as a supporting material to contribute the heat storage.

Analyse the solidification of an n-hexadecane filled in spherical enclosure. The performance of solidification process calculates for three various unchanged surface temperatures and three different initial super heats of n-hexadecane. He determined that the solidification phase propagate inside and outside the centre of the sphere. Studied four various capsules (sphere, cylinder, plate and tube) for analyse the effects of geometrical constellation. The consequence of the capsule diameter and shell thickness and the void ratio on the action of the heat storage system were also analysed. The experiment was handling by using a commercial plate heat trader as the heat storage tank. It was confirmed that the spherical capsule showed the best heat release performance among the four types of analyse capsules.

6.Thermal Energy Storage System Based on the Solar System

Thermal energy storage consists a number of technologies that store thermal energy in artificial lake for afterwards. They can be employed to equilibrium the energy demand between the day time, night time and the overcast period .Solar energy is a greater renewable energy resource, of an sporadic nature, and its effectual utilization is a part of expense and effectual storage system. No storage is utilize in solar energy system, the major part of the energy demand will be come across by theSharma et al (1999) handled experimental test on commercial grade of stearic acid, paraffin wax and acetamide . Throughout the investigate latent heat of fusion, melting temperature and specific heat of PCM have been addressed. He terminated that acetamide and paraffin wax were found to be more stable pcm than stearic acid and protracted his study with erythritol as PCM . He builds a prototype solar cooker based on an evacuated tube solar aggregator with PCM. Solar energy is stored in the PCM storage unit during daytime and is utilized for cooking in overcast period. Cooking experiments and PCM storage processes were carried out simultaneously. It was observed that noon cooking didn't affect the evening cooking, and evening cooking using PCM heat storage was found to be faster than noon cooking.

Developed a framework of open-loop passive solar water-heating system joint with sodium thisulfate pentahydrate pcm. It was determined that the water temperature at the midpoint of the storage tank reduced regularly by day untill the temperature of pcm. Heat storage action of the same solar water-heating system joined with the other salt hydrate pcm's were checked theoretically by using meteorological data and thermo physical properties of PCM with some premise, finally he proved that the PCM solar water header had a higher thermal efficiency than conventional solar water-heating system concentrated Experimental Study of Heat Transfer improvement in Latent Heat Thermal Storage System during Charging and Discharging Processes. Hybrid thermal energy storage system used for PCM , for managing simultaneously the storage of heat from solar and electric energy. This is developed and declared with experimental data. Simulations carried out for a period of 4 continuous winter months point that, with such a system, the electricity consumed for space heating is reduced above 32% Also more than 90% of the electric energy is very intense during off-peak hours. For electricity fair where time of use schemes are in effect, the return on the finance in such a thermal storage system is very desirable.

7. Conclusion

The review of this paper carried out various methods of heat transfers techniques to improve the thermal energy storage system. The paper mainly focus on PCM based thermal energy storage system, which is more adaptability and useful to the energy conservative system and wrapped current research papers in particular field.

References:

[1]Arun Prasad Raja, Basak T. and Das S.K. (2005),"Heat transfer and fluid flow in a constructal heat exchanger", Engineering in Proceedings of the Fifth International Conference on Enhanced Compact and Ultra Compact Heat Exchangers Science Engineering and Technology.

[2]Ismail K.A.R., Alves C.L.F. and Modesto M.S. (2001), "Numerical and experimental study on the solidification of PCM around a vertical axially finned isothermal cylinder", Journal of Applied Thermal Engineering, Vol. 21, pp. 53-77.

[3]Frusteri F., Leonardi V., Vasta S. and Restuccia G. (2005), "Thermal conductivity measurement of a PCM based storage system containing carbon fiber", Journal of Applied Thermal Engineering, No. 25, pp. 1623-1633.

[4] HishamEttouney, Hisham El-Dessouky and Eman Al-Kandari (2004), "Heat Transfer Characteristics during Melting and Solidification of Phase Change Energy Storage Process", Conference on Industrial Engineering and Chemical Research. No. 43, pp. 5350-5357.

[5]PiiaLamberg (2004), "Approximateanalytical model for two-phase solidification problem in a finned phase-change material storage", Journal of Applied Energy, No. 77, pp. 131-152.

[6]Ananthanarayanan V., Sahai Y., Mobley C.E. and Rapp R.A. (1987), "Modeling of fixed bed heat storage units utilizing phase- change materials", Journal of Metallurgical Trans. B, Vol. 18,pp. 339-346.

[7] Velraj R., Seeniraj R.V., Hafner B., Faber C. and Schwarzer K. (1999), "Heat transfer enhancement in a latent heat storage system", Journal of Solar Energy, Vol. 65, No. 3, pp. 171-180.

[8]Ismail K.A.R. and Stuginsky R. (1999), "A parametric study on possible fixed bed models for PCM and sensible heat storage", Journal of Applied Thermal Engineering, Vol. 19, pp. 757-788.

[9] Kanimozhi B. and Ramesh Bapu B.R. (2010), "Enhancement of Solar Water Heating System Using PCM",NationalJournal of Advanced Mechanics And Building Science,Vol. 1,No.2,pp. 48-51.

[10]Kanimozhi B. and Ramesh Bapu B.R. and Sivashanmugam M. (2009), "Thermal cooling on PCB using PCM materials", International Conference on Advanced in Mechanical and Building Science in the 3rd Millennium, ICAMP 2009, Dec. 2009, VIT, Vellore University, Vellore.

[11]Nasser Mostafavinia, SamanEghvay and Amir Hassanzadeh 2015 Numerical analysis of melting of nano-enhanced phase change material (NePCM) in a cavity with different positions of two heat source-sink pairs *Indian J. Sci. Tech.***8**(S9) pp 49–61.

[12]Aceves-Saborio SS, Nakamura HH and Reistad GM 1994 Optimum efficiencies and phase change temperatures in latent heat storage systems *ASME Journal of Energy Resources Technology* **116**(1) pp 79-86.

[13]VahidMadadi, Hamid Beheshti, TourajTavakoli and Amir Rahimi 2015 Experimental study and first thermodynamic law analysis of a solar water heater system *J. Non-Equilibrium Thermodynamics* **40** (3) pp 171–183.

[14]SelvanBellan, Tanvir E Alam, José González-Aguilar, Manuel Romero, Muhammad M Rahman, D Yogi Goswami and Elias K Stefanakos 2015 Numerical and experimental studies on heat transfer

characteristics of thermal energy storage system packed with molten salt PCM capsules *Applied Thermal Engineering* **90** pp970-979.

[15]Lingkun Liu, Guruprasad Alva, Xiang Huang and Guiyin Fang 2016 Preparation, heat transfer and flow properties of microencapsulated phase change materials for thermal energy storage *Renewable and Sustainable Energy Reviews* **66** pp 399-414.

[16]GovindarajKumaresan, Rahulram Sridhar and RamalingomVelraj 2012 Performance studies of a solar parabolic trough collector with a thermal energy storage system *Energy***47** (1) pp 395–402.

[17]V K Jebasingh and G M Joselin Herbert 2015 Numerical simulation of elliptical absorber tube in parabolic trough collector for better heat transfer. *Indian J. Sci. Tech.* **8**(24) DOI: 10.17485/ijst/2015/v8i24/80165.

[18]Mohamed M Abdulgalil, Franc F Kosi, Mohamed H Musbah and Mirko S Komatina 2014 Effect of thermal energy storage in energy consumption required for air conditioning system in office building under the africanmediterranean climate *Thermal Science***18**(1) pp S201-S212.

[19]Beemkumar N and Karthikeyan A 2016 Experimental analysis of heat transfer characteristics of solar energy *Materials today: Proceedings***3** pp 2475 – 2482.

[20]Felix Regin A, Solanki SC and Saini JS 2008 Heat transfer characteristics of thermal energy storage system using PCM capsule: A review *Renewable and Sustainable Energy Reviews***12** pp 2438–58.

[21]Medrano M, Yilmaz M O, Nogues M, Martorell I, Joan Roca and Luisa F Cabeza 2009 Experimental evaluation of commercial heat exchangers for use as PCM thermal storage systems *Applied Energy* **86** 2047-2055.

[22]Nallusamy N, Sampath S and Velraj R 2007 Experimental investigation on a combined sensible and latent heat storage system integrated with constant/varying (solar) heat sources *Renewable Energy* **32** pp 1206-1277.

[23]Shiina Y and Inagaki T 2005 Study on the efficiency of effective thermal conductivities on melting characteristics of latent heat storage capsules *International Journal of Heat and Mass Transfer***48** pp 373-383.

[24]Velraj R, Seeniraj R V, Hafner B, Faber C and Schwarzer K 1999 Heat transfer enhancement in a latent heat storage system *Solar Energy***65**(3) pp 171-180.

[25]Zalba B, Mann J M, Cabeza L F and Mehling H 2003 Review on thermal energy storage with phase change: materials, heat transfer analysis and applications *Applied Thermal Engineering***23** pp 251-283.

[26] Kanimozhi B. and Ramesh Bapu B.R. (2011), "Experimental Study of Heat Transfer Enhancement in Latent Heat Thermal Storage System during Charging and Discharging Processes", International Journal of Design and Manufacturing Technologies, Vol. 2,No. 5, pp45-49

[27] Chen S.L. and Yue J.S. (1991) "Thermal performance of cool storage in packed capsules for air-conditioning", Journal of Heat Recovery Systems and CHP, Vol. 11, No. 6, pp. 551-561. [37] UrosStritih (2003), "Heat transfer enhancement in latent heat thermal storage system for buildings", International Journal of Energy and Buildings, No. 35, pp. 1097-1104.

[22] Ye Hong Ge Xin-shi (2000), "Preparation of polyethylene paraffin compound as a form-stable solid-liquid phase change material", Conference on Solar Energy Materials and Solar Cells, No.64, pp. 37-44.

[23] Chan C.W. and Tan F.L. (2005), "Solidification inside a sphere -an experimental study", International Journal of Heat and Mass Transfer, No. 33, pp. 335-341.

[24] Jinjia Wei, Yasuo Kawaguchi, Satoshi Hirano and Hiromi Takeuchi (2005), "Study on a PCM heat storage system for rapid heat supply", Journal of Applied Thermal Engineering, No. 25, pp. 2903-2920

[25] Brousseau P. and Lacroix M. (1996), "Study of the thermal performance of a multi-layer PCM storage unit", Journal of Energy Conservation and Management, Vol. 37, No. 5, pp. 599-609.

[26] Ismail, K.A.R. and Goncalves M.M. (I999), "Thermal performance of a PCM storage unit", International Journal of Energy Conversion and Management, Vol. 40, pp. 115-138.

[27] Velraj R., Seeniraj R.V., Hafner B., Faber C. and Schwarzer K. (1999), "Heat transfer enhancement in a latent heat storage system", Journal of Solar Energy, Vol. 65, No. 3, pp. 171-180.
[28] UrosStritih (2003), "Heat transfer enhancement in latent heat thermal storage system for buildings", International Journal of Energy and Buildings, No. 35, pp. 1097-1104.
[29] Arun Prasad Raja, Basak T. and Das S. K. (2005), "Heat transfer and fluid flow in a constructal heat exchanger", Engineering in Proceedings of the Fifth International Conference on Enhanced Compact and Ultra Compact Heat Exchangers Science Engineering and Technology.
[30] Ismail K.A.R., Alves C.L.F. and Modesto M.S. (2001), "Numerical and experimental study on the solidification of PCM around a vertical axially finned isothermal cylinder", Journal of Applied Thermal Engineering, Vol. 21, pp. 53-77.
[31] Eftekhar J., Haji-Sheikh A. and Lou D.Y.S. (1984), "Heat Transfer Enhancement in a Paraffin Wax Thermal Storage System", Journal of Solar Energy Engineering, Vol. 106, No. 3, pp. 299- 306.
[32] Kanimozhi B. and Ramesh Bapu B.R. (2012), "Experimental Study of thermal energy storage in Solar System using PCM", Advanced Materials Research,No.433, pp. 1027-1032
[33] Lacroix M. and Duong (1993), "Experimental improvements of heat transfer in a latent heat thermal energy storage unit with embedded heat sources", International Journal of Energy Conversation and Management, Vol. 39, No. 8, pp. 703-716.
[34] Padmanabhan P.V. and Krishna Murthy M.V. (1986), "Outward phase change in a cylindrical annulus with axial fins on the inner tube", International Journal of Heat and Mass Transfer, No. 29, pp.1855-1866.
[35] Siegel R. (1977), "Solidification of low conductivity material containing dispersed high conductivity particles", International Journal of Heat and Mass Transfer, No. 20, pp. 1087-1089.
[36] Chow Zhong J.K. and Beam J.E. (1996), "Thermal conductivity enhancements for phase change storage media", International Journal of Heat Mass Transfer, Vol. 23, No. 1, pp. 91-100.
[37] SuatCanbazog ˇlua, Abdulmuttalip S., Ahinaslana, Ahmet Ekmekyaparb, Y. GokhanAksoya and FatihAkarsu (2005), "Thermal performance of a solar cooker based on an evacuated tube solar collector with a PCM storage unit", Journal of System Energy and Buildings, No. 37, pp. 235-242.
[38]R.J. Goldstein et al. (2006)"Heat transfer review of 2003 literature" International Journal of Heat and Mass Transfer Vol.49,pp. 451-534.

IOP Conf. Series: Materials Science and Engineering **197** (2017) 012035 doi:10.1088/1757-899X/197/1/012035

MODELLING AND OPTIMISING THE VALUE OF A HYBRID SOLAR-WIND SYSTEM

Arjun Nair[1a], Kartik Murali[1b],Anbuudayasankar.S.P[1c*] ,C.V.Arjunan[2a]

[1]Department of Mechanical Engineering, Amrita School of Engineering – Coimbatore, Amrita Vishwa Vidyapeetham University,Tamil Nadu, India. Pin – 641112

[2]Coimbatore Institute Engineering of technology, narasipuram road , Dist. Coimbatore, Narasipuram, tamil nadu 641109

[1a]arjun.nair108@gmail.com, [1b]kartik0725@gmail.com, [1c*]sp_anbu@cb.amrita.edu, [2a]arjun_nivi@yahoo.com

1. ABSTRACT

In this paper, a net present value (NPV) approach for a solar hybrid system has been presented. The system, in question aims at supporting an investor by assessing an investment in solar-wind hybrid system in a given area. The approach follow a combined process of modelling the system, with optimization of major investment-related variables to maximize the financial yield of the investment. The consideration of solar wind hybrid supply presents significant potential for cost reduction. The investment variables concern the location of solar wind plant, and its sizing. The system demand driven, meaning that its primary aim is to fully satisfy the energy demand of the customers. Therefore, the model is a practical tool in the hands of investor to assess and optimize in financial terms an investment aiming at covering real energy demand. Optimization is performed by taking various technical, logical constraints. The relation between the maximum power obtained between individual system and the hybrid system as a whole in par with the net present value of the system has been highlighted.

2. INTRODUCTION

Nowadays the price of fuels is increasing drastically mainly because of scarcity and high demand. Moreover these fuels eject out greenhouse gases which is slowly destroying the environment. Considering all these factors the people these days are largely depending on natural resources for energy for many applications like cooling, heating, electricity, etc. Solar, wind and Bio energy can be widely used these days if the environment or the place is rich in that source of natural resource. The availability of resources play a major factor for producing energy from this resources. Among all renewable energy sources solar and wind energy are more attractive to the people because these resources are available at the site at free of cost.

However, every venture exploring solar and wind energy trace the common drawback of its dependency on the unpredictable climatic and weather variations, alongside its mismatch with the demand load rate. By lowering the energy performance of the system, this shortcoming sharply decreases the lifetime of batteries. It should be noted that only an integrated exploitation of these energy resources helps keep the cost of system design low. A stand-alone solar or wind energy system is insufficient to supply continuous power citing the frequent seasonal and periodical changes (Yang et al, 2008). The drawbacks from solar and wind now a day the people are thinking towards hybrid technology i.e. combination of solar and wind. The hybrid system can get enough energy from both sources and even if the energy from one source is low at the point it will be compensated by the other. Hybrid systems has gained popularity in the past, for its application in remote systems such as radio telecommunication and satellite earth stations, or at localities inaccessible to the conventional power grids (Yang et al, 2009; Zhou et al, 2008; Diafa et al 2008). Today, upgrading the current single source systems (Solar, Hydro or Wind) into hybrid systems for grid-connection applications has been a major area of focus (Bakos and Tsagas, 2003).

Researchers have recommended many optimization procedurals such as graphical construction, probabilistic method and iterative techniques. Shah et al. (2015) contributed in the field of Performance of U.S. hybrid distributed energy systems: Solar photovoltaic, battery and combined heat and power. Until recently, the relatively high levelized cost of electricity from solar photovoltaic (PV) technology limited deployment; however, recent cost reductions, combined with various financial incentives and innovative financing techniques, have made PV fully competitive with conventional sources in many American regions. Their results show that conservatively sized systems are technically viable in any continental American climate and the details are discussed to provide guidance for both design and deployment of PV + battery + CHP hybrid systems to reduce consumer costs, while reducing energy- and electricity-related emissions.

Rahimi et al. (2015), studied the design and application of a novel hybrid sun–wind-tracking system. Their principal experiment was focused on comparison between dual-axes sun-tracking and hybrid sun–wind-tracking photovoltaic (PV) panels. Their results show that, the overall daily output energy gain was increased by 49.83% compared with that of a fixed system and an overall increase of about 7.4% in the output power was found for the hybrid sun–wind-tracking over the two-axis sun tracking system. Sinha and Chandel (2015) were studied photovoltaic–micro wind based hybrid systems for 12 locations of the Himachal Pradesh state to solve the problems in hydro power plants during extreme winter season. The results show that (i) ANN predicted data are close to measured/estimated data, (ii) The normalized solar and wind energy generation are found to range between 1034–1796 kWh/kWp/yr and 222–616.8 kWh/kWp/yr respectively for all locations and (iii) the state has good prospect of power generation from hybrid systems with major solar and minor wind components.

With the objective to minimize costs, Moussa et al. (2015), designed a hybrid solar-wind energy plant model, which included optimised variables namely, count of photovoltaic modules and wind turbines, wind turbine height and turbine rotor diameter. The results show that the hybrid plant delivered energy that combined the nature of the two complementary energy sources, and supplied energy reliably all along the year. Onar O.C et al. focused on integrating wind turbine, photovoltaic and fuel cell along with ultra-capacitor systems for grid-independent applications. Testing the dynamic characteristics of their suggested hybrid model under parameters such as wind speed, solar radiation and demand load, for a per day analysis, they found that their model displayed fantastic performance. The hybrid model of Chougule and Ravi integrated methodologies involving analytics and parameters to determine the early cost of castings. By testing their model in a unified industrial product and process designing environment, the cost determined by a product designer matched closely with that estimated by an experienced foundry engineer. Their study also shows that web-enabling of the entire system promotes collaboration between product designer, tool-maker and foundry engineer for cost reduction.

Konstantinos et al, studied the four important factors that influence the rising demand for electricity namely climatic changes, forecasted growth rates of EU Member State economies, variations in consumption rates and bringing in of latest technologies. They also did a framework for assessing power schemes for power from renewable energy sources considering the indeterministic competitive market milieu. They expected the modern deregulated electricity market to be ready to respond to this challenge. They based their expectation on their prediction of an adequate and economical supply of energy within this new model and promising new opportunities it would offer for incomings as well as present power producers. Celik (2008) proposed a battery operated model for hybrid wind-solar energy systems which included design factors such the time fraction necessary for the hybrid system to meet the load and the cost of the system. In his study he introduced new notions that integrated the autonomy and economics of the system employed in the optimisation of the technological and economical analysis. He suggested that rather than excessively raising the size of the hardware, an auxiliary source could be employed additionally within the system, resulting in increased technologically and economically optimum systems.

IOP Conf. Series: Materials Science and Engineering **197** (2017) 012035 doi:10.1088/1757-899X/197/1/012035

Dihrab and Sopian (2010) proposed a hybrid solar-wind system that was employed as the power source for the grid-connected applications in three Iraqi cities. A MATLAB simulation of the model using input parameters obtained from meteorological data of the three cities, along with the proposed sizes of the wind turbines and the photovoltaic arrays was carried out. Their results showed that their hybrid system would provide sufficient energy for villages in desert or rural areas. Kershman et al. (2008) worked on the model of a hybrid solar/wind system that operated the sea-water reverse osmosis (RO) desalination plant later, employed to supply potable water to a village on Libya's coast. In their studies, the reverse osmosis desalination plant operated using a grid-connected power supply system that combined the power from photovoltaic power generation (PV) and wind energy conversion (WEC). The results shows that the economic evaluation of the combined power system showed that the water from the Grid+WEC+RO cost more than that from the Grid+PV+RO. However, water from both the systems was expensive compared to that from the conventional fossil-operated plant.

Bakos and Tsagas (2003) analysed the technical and economic feasibility of using a grid-connected hybrid wind/solar system to meet the power demands of a typical residency in Xanthi, a Greek city, through electrical and thermal energy production. They evaluated the economics of their hybrid solar/wind system using the Life Cycle Savings (LCS) and the PayBack Period (PBP) techniques to calculate the input capital cost. Deshmukh and Deshmukh (2008) discussed about the techniques to model the constituent parts of the HRES, its design and analysis. They justified the increasing popularity for the hybrid solar/wind energy systems using the literary works focussing on improving the HRES design. Borowy and Salameh (1996) developed an analogy to optimise the storage capacity of a battery bank and a photovoltaic array considering a single hybrid solar/wind power system. In their study, the wind speed and irradiance data were compiled on an hourly basis per day for a span of 30 years and using these data, the average generated power was calculated for very same durations. Yang et al. (2008) developed simulation models for a hybrid wind/solar system which are used to calculate optimized combinations of PV module, wind turbine, and battery bank parameters for a given loss of power supply probability (LPSP). The proposed algorithm was found to deliver a good optimised sizing performance suggesting a hybrid solar/wind system to be the best solution.

Every effort to explore an efficient and economical renewable energy source requires atleast a single optimised sizing technique. The use of photovoltaic arrays, battery banks and wind turbines shoot up the investment costs and hence, such a sizing technique ensures that the hybrid system operates at an optimum investment value and power assurance. This type of optimization includes economical objectives, and it requires the assessment of the system's long-term performance in order to reach the best compromise for both reliability and cost.

In this paper, a single optimal sizing model for a stand-alone hybrid solar/wind system using battery banks is developed based on the loss of power supply probability (LPSP) and the annualized cost of system (ACS) concepts. The optimality procedure targets on finding out the configuration that yields the best compromise between the two considered objectives: LPSP and ACS. The decision variables included in the optimization process are the PV module number, wind turbine number, battery number, and also the PV module angle of slope and installation height of the wind turbine. In order for the hybrid system to operate and maintain the required power reliability at the least possible cost, the formulation of the model was carried out using an optimization technique – the genetic algorithm (GA), which is an advanced search and optimization technique; it is generally robust in finding global optimal solutions, particularly in multi-modal and multi-objective optimization problems, where the location of the global optimum is a difficult task.

3. MODELLING THE HYBRID SYSTEM COMPONENTS

Fig.1 shows a schematic representation of the proposed hybrid solar/wind power generation system. As per the design, the system includes a wind turbine, photovoltaic array, battery bank, inverter, controller and associated accessory devices and wires. In this case the system is setup such that during the day time solar panel extracts energy and directly supply to the power grid and in case excess energy it produced, it gets stored. Wind energy is also produced such that it goes to the battery and then the grid system to make sure the battery is not always in drained condition. During the night time wind energy is continuously produced like in the day time and also the stored energy in the battery is used to run the grid. To analyse the performance of the hybrid system, the overall system is modelled and analysed towards optimum cost.

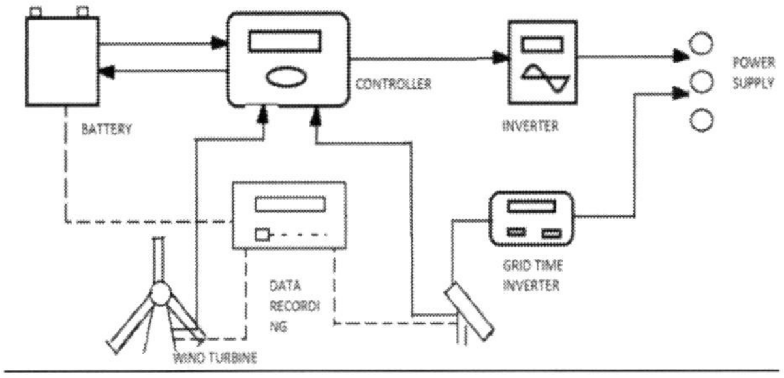

Figure 1: Block diagram of a solar / wind hybrid system

There are many times the decisions are taken while going for the grades of the accessories considering the cost and optimising the life of the products. Here the experiment set up is analysed for the following three situations

- Expensive setup using high grade instruments and materials
- Moderate cost setup using average grade instruments and materials
- Cheap setup using low grade instruments and materials

	EXPENSIVE	MODERATE	CHEAP
Initial investment	High	Average	Low
Maintenance	Low	Average	High
Life	High	Average	Low
Battery	Sealed battery	Low grade sealed battery	Lead acid battery

3.1. Supply

Hybrid solar and wind power is used to run all the household appliances. This is the why the inverter plays a very important role in this system.

IOP Conf. Series: Materials Science and Engineering **197** (2017) 012035 doi:10.1088/1757-899X/197/1/012035

Since the supply doesn't give the required output 24*7 throughout the year, hence it is necessary to use regular power supply for continuous electricity to the household appliances.

Usually there are 2 types of sources which are

- **Primary source:** Here the solar and wind energy which we extract will be the prime source of energy to run the house. Power from the electrical grid will be utilized as a backup if the battery gets drained and when battery needs to get charged again.
- **Secondary source:** Here the solar and wind energy which we extract will be a backup it the grid system supply is cut off.

Primary source has been considered for this study.

4.EQUATIONS

4.1. Optimization of variables

The important variables that describe the system are as following:

- P_{ME}: The electrical capacity of the plant
- P_S : The peak load capacity of the solar cell
- P_W : The peak load capacity of the wind turbine

4.2. Primary or Main function

To maximise the main or the primary function, the net present value (NPV) for a specific time period of time was maximised .It is done for the investment of the project lifetime. NPV is used because of its frequently appraisal criterion in plant investment and also as theory regards it superior over other parameters. This model evaluates and determines the optimal solutions for other investment parameters.

The Net present value(NPV) function to be optimized is

$$NPV = (R_{NETE} + R_{REINE})\,DF - (I_M + I_s + I_W + I_{EC} + I_C + I_{ET}) - (A_S + A_W + A_{ET})\,DF$$

It is to be noted that the maximisation function in the above statement gives the value without the taxes. All the monetary values are multiplied with a constant coefficient DF, known as the Discount Coefficient, which results in the current expression:

$DF = (1-\{1+(i-p)(1+p)\}^{-N})/(i-p)$

Where I is the interest rate, p is the inflation rate and N is the investment lifetime of the plant in a whole.

4.3. Revenues of facility

The revenues obtained by the facility are given below.R_E is the revenue obtained from sales of the net electricity supplied to the main grid:

$$R_{NETE} = C_E(1 - ne)\left(\sum\left((E_{HESt} + E_{EHEWt}) - E_{DC}\right)\right)$$

IOP Conf. Series: Materials Science and Engineering **197** (2017) 012035 doi:10.1088/1757-899X/197/1/012035

Where E_{HES} is the electricity produced from the solar cell and E_{EHEW} is the electricity produced from the wind turbine. E_{DC} is the electricity consumed by the different apparatus present during the operation. n_e is the electricity transmission losses and t = 1,....,T is the time period. E_{DC} is given as

$$E_{DC} = Ci\,(\sum E_{MSt} + E_{GTt}) + C_W(\sum E_{WT} + E_{CT} + E_{st} + E_{it})$$

Where C_i is the cost per capacity of solar energy lost and C_W is the cost lost in wind turbines. E_{MSt} is the energy lost in the solar panel in period of time and E_{GTt} is the energy lost in grid time inverter. Similarly E_{WT} is the energy lost in wind turbine during conversion, E_{CT} in the controller, E_{St} in the storage. Basically all these losses are transmission losses in different electrical equipment's.

R_{EP} is the electricity capacity available for reimbursement:

$R_{EP} = C_E P_{ME}$

Where C_E is the income from the capacity availability (₹/kW).

4.4. Expenses

Energy conversion related expenses

I_M is the base plant investment cost subtracted from the public subsidy given by government:

$$I_M = C_{MPME}(1 - S_{MB})$$

Where S_{MB} is the public subsidy of investment for the base plant and C_M is the specific investment cost (₹/kW) calculated from a known cost criterion.

A_S is the solar cell plant O&M:

$$A_S = O_S C_S P_S$$

Where O_S is the annual operation and management cost of the cell as a percentage of investment cost.

$$A_W = O_W CWP_W$$

Where O_W is the annual operation and management cost of the wind turbine as a percentage of the investment cost.

$$I_S = C_S P_S(1 - S_{MB})$$

I_S is the peak load solar cell cost minus the public subsidy.

$$I_W = C_W P_W(1 - S_{MB})$$

I_W is the peak load wind turbine cost minus the public subsidy.

I_{Et} is the investment cost of connecting the power plant to the national grid:

$$I_{ET} = (LC_{CEtv} + C_{ETf})(1 - S_{ET})$$

where LC is the total length of the transmission line (km), CETV is the changing investment cost (₹km^{-1}), CETF is the fixed connection cost and SET is the subsidy.

A_{ET} is the electricity transmission line operation and management cost as a percentage of the investment cost.

IET is the investment cost of connecting the power plant to the national grid:

$$I_{ET} = (LC_{CETV} + CETF)(1 - S_{ET})$$

where LC is the length of the transmission line (km), CETV is the variable investment cost (V km_1), CETF is the fixed connection cost (V) and SET is the subsidy.

4.5. Constraints

As with any modulation, many constraints had to be included in the mathematical formulation of the problem.

4.5.1 Energy Constraints

THE energy produced because of the solar and wind must be enough to satisfy the consumer demand otherwise there would be no use of setting up the plant and giving so much investment to it.

The constraints based on this parameter would be as the follow

$$ES + EW >= EDT \ where \ T = 1,2,3, \dots \dots, t$$

4.5.2. Legislation constraints

The legislation of Indian government says that subsidy for the particular project may be provided only if the energy produced by the plant is at least 70% or more than that of the customer demand.

4.5.3. Logical constraints

The independent variables were considered non-negative. Additionally, upper limits were placed for many of the variables.

5. CASE STUDY

The case study is done in Coimbatore which is a small city in Tamil Nadu, India. This place is selected because of uniform wind and solar energy available in this area throughout the year. Here useful inputs are used in the NPV equation and hence figures have been plotted using Mat lab to get the results. Here to account into the NPV equations few assumptions are made based on the Coimbatore environment Those technical and financial data are given below in Table 1.

Table 1. Technical Financial data

Total efficiency of plant %	55
Interest rate %	9
Inflation %	4
Investment lifetime (yr.)	15

IOP Conf. Series: Materials Science and Engineering **197** (2017) 012035 doi:10.1088/1757-899X/197/1/012035

cost per capacity of solar energy lost (₹/Kwh)	0.2
cost lost in wind turbine conversions (₹/Kwh)	0.12
income from capacity available	50000
transmission losses	20000
Specific investment cost for solar	250000
Public subsidy of investment %	40
Specific investment cost for wind turbine	600000
Annual operation and management of the cell as a percentage of investment cost	0.07
Annual operation and management of the wind turbine as a percentage of the investment cost	0.1
Investment cost of connecting the power plant to the national grid	300000
Electricity transmission line operation and management cost as a percentage of the investment cost	0.26

6. RESULTS

Figure 2 represents the change in Solar power (P_s) with net present value (NPV) for the given range of input. It shows the change of NPV value with respect to the solar power generated.

Here the different colours represent the difference in values of NPV for the given input.

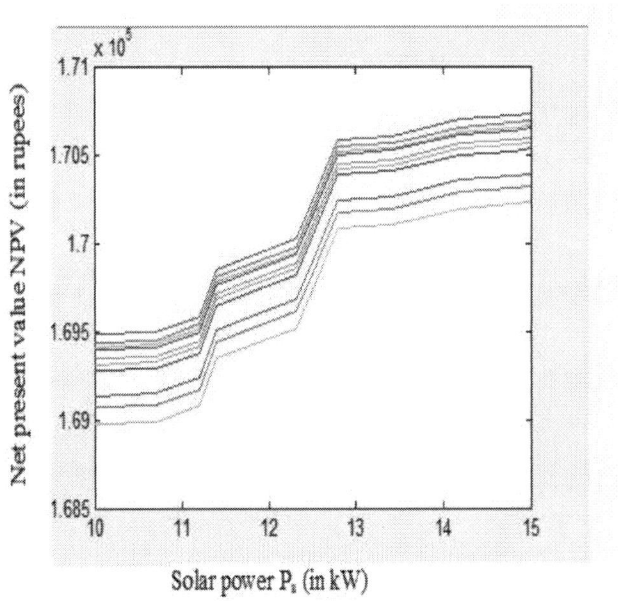

Figure 2: Solar power vs NPV

Similarly this figure 2 below represents the change in Wind power (P_w) with net present value (NPV). For the given range of input it shows how the NPV value changes for the wind power generated.

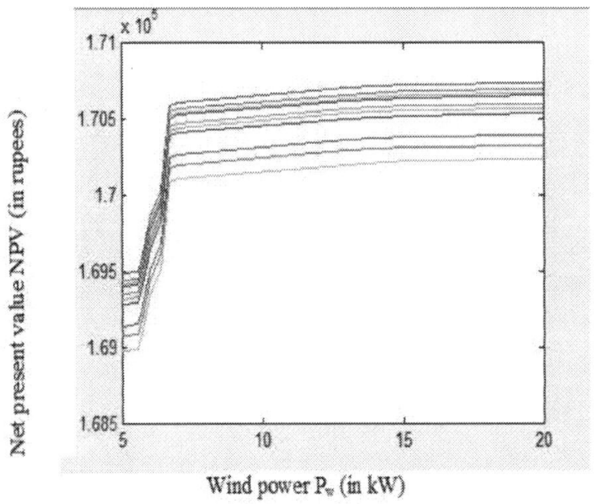

Figure 3: wind power vs NPV

Since the system is a hybrid system NPV changes with both wind and solar, so both the above figures, figure 1 and figure 2 are merged into Figure 3.

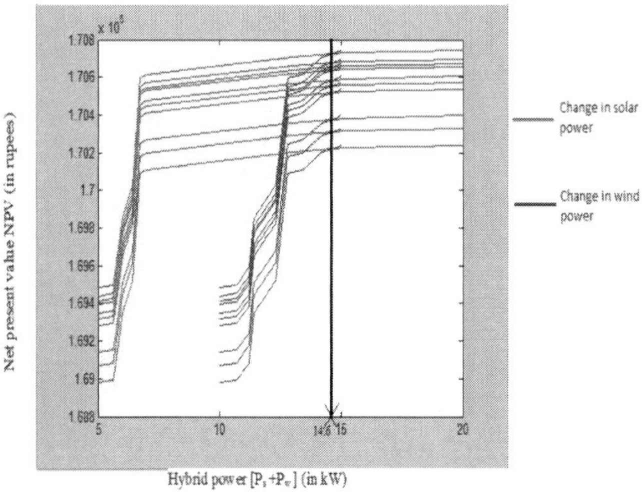

Figure 4: Hybrid power of solar and wind vs net present value

Here in figure 3 value of power both the figures merge, but there are a set of values for NPV for the optimum power. So now to get an optimum NPV value for the optimum power an need account into the maintenance, life of the components and the profit earned in the process.

For figure 1 and figure 2 the top, middle and bottom lines of the NPV which are blue, red and yellow lines respectively.

These lines represents:

Blue : Expensive setup using high grade instruments and materials

Red: Moderate cost setup using average grade instruments and materials

Yellow: Cheap setup using low grade instruments and materials

Now considering all the expenses a maintenance figure is plotted.

6.1 Maintenance figure

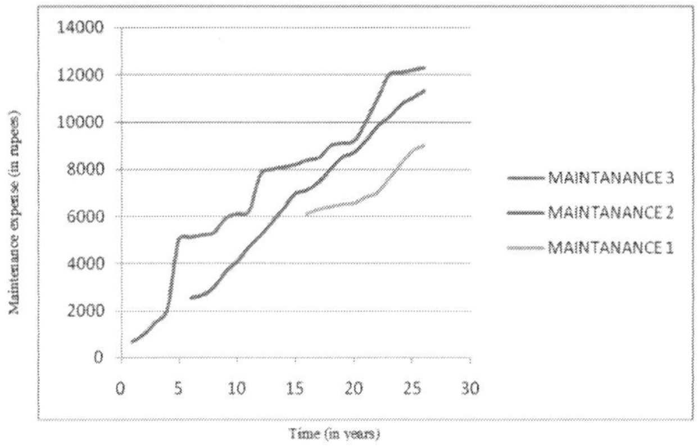

Figure 5:Time vs. maintenance cost

IOP Conf. Series: Materials Science and Engineering **197** (2017) 012035 doi:10.1088/1757-899X/197/1/012035

Here,

Maintenance 3 represents the maintenance for the cheap setup (i.e Yellow line)

Maintenance 2 represents the maintenance for the moderate setup (i.e red line)

Maintenance 1 represents the maintenance for the expensive setup (i.e blue line)

From figure 5 if cheap product is selected maintenance is high and it maintenance starts at a early period of time.

So, from the figure 5 selecting maintenance1_is optimum. Therefore the blue line has been selected

Next taking the life expectancy of all the component put together. It shows depending on the cost and quality of the product, how long it will last.

6.2 Life Expectancy

Figure 6 shows that more the cost of the product, better will be the components used and hence better will be the life time of the component.

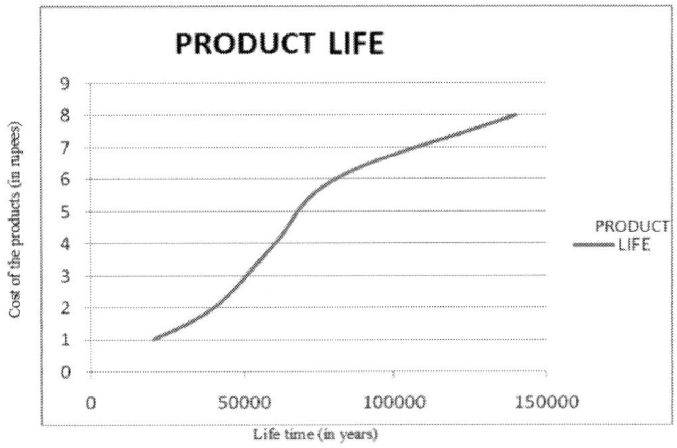

Figure 6: life time vs. cost of the products

Hence it is better to go with the blue line (expensive setup) according to Figure 6. Next is the comparison between the investment cost and profit gained in a period of 10 years using the hybrid system.

Therefore blue line is more optimum.

Table 2. Income gained over years

Initial Investment	Final Gain in 10 Years	Losses %
4	1.708	57.3
3.8	1.707	55.05
3.6	1.706	52.611
3.5	1.705	51.285
3.4	1.704	49.885
3.2	1.703	46.78
3.1	1.702	45.09
2.9	1.701	41.344
2.7	1.7	37.03
2.5	1.7	32

From the data it is found out that over the span of 10 years the profit attained is almost similar. Looking at the profit prospective, from this data it is optimum to choose the yellow line. Since, after 10 years they have got back 68% of their investment, whereas blue line have got only 42.7% of their investment. Therefore choosing yellow line is more optimum.

7. CONCLUSION

From the obtained results it can conclude that the maintenance and life expectancy of the system dictate the type of the device been chosen. In case of a low costing system it is noticed that the maintenance cost is more and the life expectancy is more. Where as in case of a system where a higher costing appliance is taken into considerations the maintenance cost is lower and the life expectancy of the product is higher. From the maintenance and life expectancy figures selecting blue line is most optimum and selecting yellow line is least optimum. Whereas when the profit attained in a span of 10 years is compared , it is profitable to go with the yellow line and least profitable to go with the blue line. Since the optimum solution for the whole system is required ,we have to take into account of the above 2 situations, hence it is better to choose the red line. The red line accounts into the profit factor as well as the maintenance and life factor. So the red line which uses the moderate cost setup and average ranged instruments and components is the most optimum. So it's better to go with the red line to attain the highest NPV and power.

The reason of going with the red line is that, in this scenario all the above mentioned parameters are in the most optimum state and this will lead to the best result.

8. REFERENCES

[1] Yang, H., Zhou, W., Lu, L., and Fang, Z 2008, Optimal Sizing Method For Stand-Alone hybrid Solar-Wind System With LPSP Technology By Using Genetic Algorithm, *Solar Energy*, Vol. **82**, pp. 354.

[2] Yang HX, Zhou W, Lou CZ 2009 Optimal design and techno-economic analysis of a hybrid solar–wind power generation system. *Appl Energy*;**86**:163–9.

[3] Zhou W, Yang HX, Fang ZH 2008, Battery behavior prediction and battery working states analysis of a hybrid solar–wind power generation system. *Renew Energy*;**33**(6):1413–23.

[4] Diafa S, Belhamelb M, Haddadic M, Louchea A 2008, Technical and economic assessment of hybrid photovoltaic/wind system with battery storage in Corsica Island. *Energy Policy*;**36**(2):743–54.

[5] Bakos GC, Tsagas NF 2003, Technoeconomic assessment of a hybrid solar/wind installation for electrical energy saving. *Energy Build* ;**35**(2):139–45.
[6] Shah, K.K., Mundada, A.S and Pearce, J.M 2015, "Performance of U.S. hybrid distributed energy systems: Solar photovoltaic, battery and combined heat and power, *Energy Conversion and Management*.
[7] Rahimi Masood, Meisam Banybayat, YaghoubTagheie, PeyvandValeh-e-Sheyda 2015,"An insight on advantage of hybrid sun–wind-tracking over sun-tracking PV system", *Energy Conversion and Management*, Pages 294–302.
[8] Sinha Sunanda and S.S. Chandel 2015. "Prospects of solar photovoltaic–micro-wind based hybrid powersystems in western Himalayan state of Himachal Pradesh in India", *Energy Conversion and Management*, Pages 1340–1351.
[9] K. Mousa, Ali Diabat, Hamzah S. Alzu'bi 2010. "Design of hybrid solar and wind power plant using optimization". *Engineering Systems Management and Its Applications*, IEEE explore, Abu Dhabhi, Pages 1-6.
[10] Onar O.C., Uzunoglu, M., and Alam, M.S 2008. "Modeling, Control And Simulation Of An Autonomous Wind Turbine/Photovoltaic/Fuel Cell/Ultra-Capacitor Hybrid Power System", *Journal Of Power Sources,* Vol. **185**, No. 2, pp. 1273-83.
[11] R. G. Chougule and B. Ravi 2005. "Casting cost estimation in an integrated product and process design environment" Vol. **3**, No.6 *IJCIM.*
[12] Celik, A.N 2003. "Techno-Economic Analysis Of Autonomous PV-Wind Hybrid Energy Systems Using Different Sizing Methods", *Energy Conversion And Management,* Vol. **44**, pp. 1951-1968.
[13] Dihrab, S.S. And Sopian, K. 2010 "Electricity generation of hybrid PV/wind systems in Iraq", *Renewable Energy*, Vol. **35**, pp. 1303-1307.
[14] G.C Bakos and N.F Tsagas 2003, Techno economic assessment of a hybrid solar/wind installation for electrical energy saving, *Energy and Buildings*, Volume **35**, Issue 2, Pages 139–145
[15] Deshmukh M.K, S.S Deshmukh 2008. "Modeling Of hybrid Renewable Energy Systems", *Renewable and Sustainable Energy Reviews,* Vol. **12**, No. 1, pp. 235-249.
[16] Borowy, B.S., and Salameh Z.M 1996. "Methodology For Optimally Sizing The Combination Of A Battery Bank And PV Array In A Wind/PV Hybrid System", *IEEE Trans. Energy Convert.,* Vol. **11**, No. 2, pp. 367-73.
[17] Konstantinos Venetsanosa, Penelope Angelopouloua, TheocharisTsoutsosb 2002, "Renewable energy sources project appraisal under uncertainty: the case of wind energy exploitation within a changing energy market environment", *Energy Policy* Greece.
[18] Yang, H., Zhou W. and Lou C 2009. "Optimal Design and Techno-economic Analysis of a Hybrid Solar-Wind Power Generation System", *Applied Energy*, vol. **86**, No. 2, pages 163-169.
[19] H.X. Yang, L. Lu, J. Burnett 2003. "Weather data and probability analysis of hybrid photovoltaic–wind power generation systems in Hong Kong", *Renewable Energy*, vol. **28**, No. 11, pages 1813-1824.
[20] Amit Kumar Yadav, S.S. Chandel 2015. "Solar energy potential assessment of western Himalayan Indian state of Himachal Pradesh using J48 algorithm of WEKA in ANN based prediction model", *Renewable Energy*.
[21] Chandel SS, Agarwal RK 2011 "Estimation of hourly solar radiation on horizontal and inclined surfaces in western Himalayas". *Smart Grid Renew Energy*,SGRE, Vol.**2** No. 1.
[22] Parita G Dalwadi, Chintan R Mehta 2012, "Feasibility Study of Solar-Wind Hybrid Power System", *International Journal of Emerging Technology and Advanced Engineering*. Vol. **2**, No. 3.
[23] Olaf Goebel, Sultan A. Kershman 2005,"Hybrid wind/PV and conventional power for desalination in Libya", *Centre for Solar Energy and Hydrogen Research* Vol. **4**, No. 3.

IOP Conf. Series: Materials Science and Engineering **197** (2017) 012036 doi:10.1088/1757-899X/197/1/012036

Performance Improvement of Energy Storage System with nano-additivesin HTF

N Beemkumar*, A Karthikeyan, Saravanakumar B, Jayaprabakar J

School of Mechanical Engineering, Sathyabama University, Chennai, India

E-mail: beem4u@gmail.com

Abstract. This paper is intended to improve the heat transfer rate of thermal energy storage system with copper oxide (CuO) as nano-additivesin heat transfer fluid (HTF) by varying encapsulation materials. The experimentation is done with different encapsulating materials like copper, brass and aluminium. The results are analysed for their thermal performance characteristics during charging and discharging processes. D-Sorbitol and therminol-66 with CuO is used as PCM and HTF respectively. A comparison was made between the different encapsulations and it was found that copper encapsulation has higher efficient, storing and recovering energy. However, its high thermal conductivity promotes larger heat losses and its cost is also high on other side. So the economical use of encapsulation material is aluminium compared to other two materials.

1. Introduction

Thermal energy storage can be defined as the storage of thermal energy of any temperature (i.e., either high or low) for temporary purposes. Thermal energy are basically stored in three different methods which include sensible heat storage, latent heat storage and some physiochemical reactions results in the thermal storage [1-4]. Sensible heat storage is the storage systems in which the heat is stored directly without the changing the phase of the material. Heat transfer fluids are used for sensible heat storage. Energy storage usually increases reliability and stabilize the energy supply which improves the performance of energy systems [5-6]. A storage system improves the performance of the power generated by levelling of load; the higher efficiency leads to energy conservation and improves the cost effectiveness.

The nanoparticles when dispersed in heat transfer fluid enhance the heat transfer in the fluids. Solid metallic particles show high thermal conductivity than non-metallic solids and metallic & non-metallic liquids. Copper particles show high thermal conductivity than Aluminium particles [7]. Nanofluids tend to deteriorate heat transfer at supercritical pressures because deposition of the nanoparticle smoothens the wall roughness and presents an additional thermal resistance [8-9]. Effective thermal conductivity of nanofluids increases with the volume fraction of nanoparticles as well as with temperature.In addition of nanoparticles gives a better enhancement with temperature for low volume fraction of particles. Many researches have been done on the improvement of heat transfer in thermal storage system but no or very minimal researches have been done to include nanoparticles in the HTF to improve the heat transfer in thermal storage system. It is decided to have CuO as a nanoparticle which is been mixed with therminol-66 oil and used as a HTF. The PCM selected is D-sorbitol which as high stability during phase transformation.

2. Experimental Setup and Methodology

The experimental set up consists of three PCM storage tanks, a storage tank, a heater and a circulating pump. The storage tank consists of both, the PCM and the HTF Each tank has a diameter of 25.6 cm and 30 cm length made of 0.6 cm thick MS plate. The storage tank is well insulated with glass wool of 10 cm thickness and pipe lines are also insulated with same material. A total quantity of 50 L of therminol-66 with an additive CuO is charged in the storage tank and also in the piping circuit, and it has been ensured that there is no leak in the oil flow path. An additional tank is also fabricated to house the heaters which will be used to provide the required heat energy. For circulating HTF gear pump is used with 0.5 HP motor. Valves are present at regular intervals to control the amount of HTF flow. The selection of PCM is depends on the required temperature range. The PCM chosen is D-sorbitol, due to its stability. The HTF is therminal-66 with nano additives of CuO. Therminol-66 is very stable and higher melting point around 390°C and CuO have higher thermal conductivity around 409w/mk. A gear pump maintains the HTF flow in the circuit, through the heater and storage tank during charging process in which the heated oil transfers heat to the PCM. The PCM is encapsulated in spherical balls of copper, aluminium and brass and the experimentation was done by changing different encapsulating material. Totally six K-Type thermocouples are used to measure the temperature at different locations with a sensitivity of approximately 41μV/°C. The experimentation has been done by two modes of operation, charging and discharging. The experimental layout and the actual PCM tank setup is shown in the figure 1 and 2 respectively.

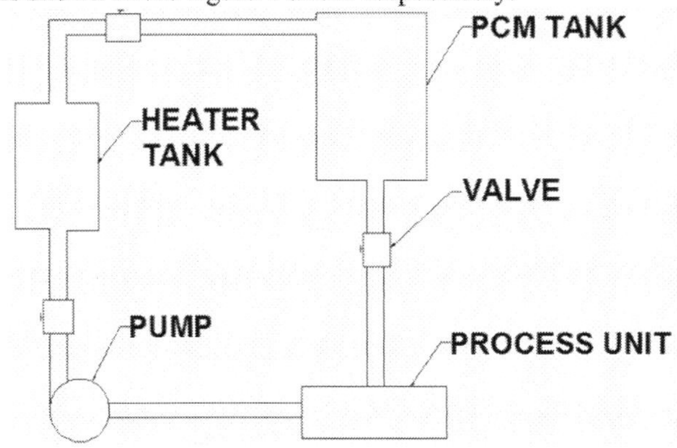

Figure 1. Experimental layout

Figure 2. PCM tank.

2.1. Charging mode

During the charging mode the HTFis circulated through the heater tank, PCM tank and process unit. The heated oil passing through the PCM tank, the PCM absorb heat energy present in the oil and then the less energy oil is made pass through the process unit (retrieval of heat energy present in the oil) then it is circulated back to the heater tank. This cyclic process is continued until PCM changes it phase from solid to liquid state. During this process the temperature oil is noted at every 10 minutes interval.

2.2. Discharging mode

During the discharging mode, the heater tank is switched is off and the oil temperature gradually decreases. The oil is made pass through the PCM tank;the PCM looses energy to the oil and the temperature of oil increases. The PCM releases heat to the HTF and changes its phase from liquid to solid. The PCM release the heat slowly. This process maintains the temperature of the fluid constant for a longer time. This process continues until the HTF reaches the temperature of 30°C (ambient temperature). This process is continued around 100 cycles. The temperature of oil observed at the interval of 10 minutes during discharge mode.

3. Results and Discussion

Figure 3 shows the charging and discharging process in the second tank. It was observed that the temperature in the second tank is slightly less than that in the first tank. The heat transferred increases initially and then varies within the specific range during charging. The heat recovered in the beginning during the discharging process varies largely. The heat stored increases sharply in the middle of the process because of a high change in temperature of the HTF during the heating process. Towards the end of the process, heat stored decreases. For the discharging process, the heat recovered decreases initially and then remains constant throughout the process. A comparison is made between heat transfer rates of three tanks. It can be interpreted that the average heat transfer rate during charging is 7.022 kW and while discharging, the heat transfer is found to be 8.051 kW. The discharging takes more time due to the slow discharge of heat from the PCM.

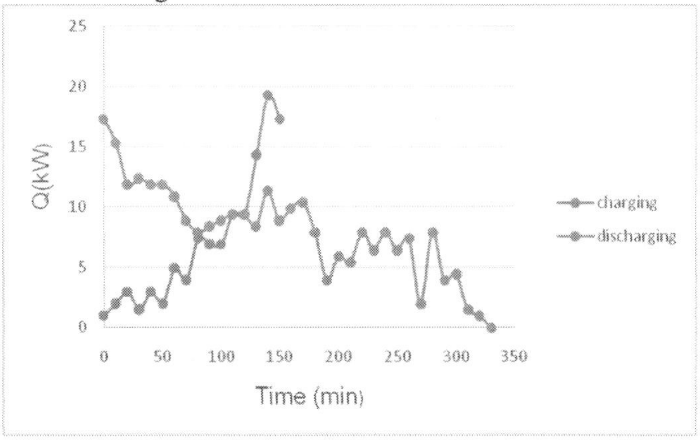

Figure 3. Charging and discharging of copper encapsulation.

From the figure 4 the charging and discharging process in the second tank was observed that the temperature in the second tank is slightly less than that in the first tank. The heat transferred increases initially and then varies within the specific range during charging. The heat recovered in the beginning during the discharging process varies largely. It can be interpreted that the average heat transfer rate during charging is around 9.562 kW and while discharging it is around 2.34 kW. Comparing charging and discharging, the heat transfer rate is very low in discharging process.

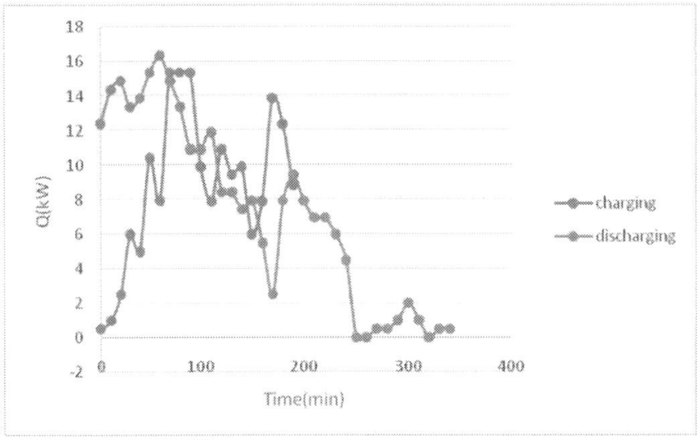

Figure 4. Charging and discharging of aluminium encapsulation.

Figure 5 shows the charging and discharging process in the second tank. It was observed that the temperature in the second tank is slightly less than that in the first tank. The heat transferred increases initially and then varies within the specific range during charging. The heat recovered in the beginning during the discharging process varies largely. The heat stored increases sharply in the middle of the process because of a high change in temperature of the HTF during the heating process. Towards the end of the process, heat stored decreases. For the discharging process, the heat recovered decreases initially and then remains constant throughout the process. The process continues for a longer period because of the slow discharge of heat from the phase changing material. A comparison is made between heat transfer rates of three tanks. It can be interpreted that the total heat transfer rate during charging for is 2.09 kW. During discharging is 11.21 kW.

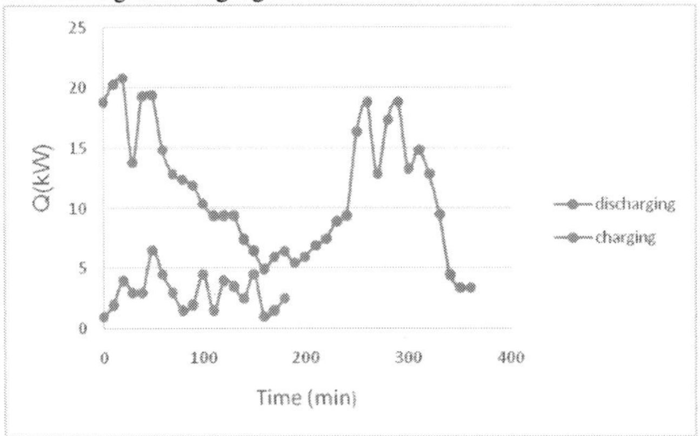

Figure 5. Charging and discharging of brass encapsulation.

4. Conclusion

A comparison of the heat stored and recovered in brass, copper, aluminium encapsulations while using therminol-66 with a nano additive of CuO as HTF. The results show that the copper encapsulated PCM balls stores the more heat energy while brass stores the least during charging mode. The same is observed during heat recovery i.e., discharging mode. But at the same time, the heat loss (difference between heat stored and recovered) is the highest in copper encapsulation followed by aluminium encapsulation and the least in brass encapsulation. This can be attributed to the high thermal conductivity of copper and a relatively low thermal conductivity for brass. Comparing three materials copper has more heat gain while charging and heat recovery while discharging. But according to economical usage the aluminium encapsulated PCM is best in all.

References

[1] Aceves-Saborio SS, Nakamura HH and Reistad GM 1994Optimum efficiencies and phase change temperatures in latent heat storage systems*ASME Journal of Energy Resources Technology* **116**(1) pp 79-86.

[2] Antony Aroul Raj and R Velraj2011Heat transfer and pressure drop studies on a PCM-heat exchanger module for free cooling applications*International Journal of Thermal Sciences***50**pp 1573-1582.

[3] Antony Farell, Brain Norton and David M Kennedy2006Corrosive effects of salt hydrated phase change materials used with copper and aluminium*Journal of Material Processing and Technology***175**pp198-205.

[4] AtillaBiyikoglu2002Optimization of a sensible heat cascade energy storage by lumped model *Energy Conversion and Management***43**pp 617–637.

[5] Felix Regin, S.C. Solanki and J.S. Saini2006Latent heat thermal energy storage using cylindrical capsule: Numerical and experimental investigations*Renewable Energy***31**pp 2025–2041.

[6] Jinjia Wei, Yasuo Kawaguchi, Satoshi Hirano and Hiromi Takeuchi2005Study on a PCM heat storage system for rapid heat supply *Journal of Applied Thermal Energy***25**pp 2903 – 2920.

[7] Horst Michels and Robert Pitz-Paal2007Cascaded latent heat storage for parabolic trough solar power plants *Solar Energy***81**pp 829–837.

[8] Govindaraj Kumaresan, Rahulram Sridhar and Ramalingom Velraj 2012 Performance studies of a solar parabolic trough collector with a thermal energy storage system *Energy***47** (1) pp 395–402.

[9] Mohamed M Abdulgalil, Franc F Kosi, Mohamed H Musbah and Mirko S Komatina 2014 Effect of thermal energy storage in energy consumption required for air conditioning system in office building under the african mediterranean climate*Thermal Science***18**(1) pp S201-S212.

[10] Beemkumar N and Karthikeyan A 2016 Experimental analysis of heat transfer characteristics of solar energy *Materials today: Proceedings***3**pp 2475 – 2482.

IOP Conf. Series: Materials Science and Engineering **197** (2017) 012037 doi:10.1088/1757-899X/197/1/012037

Recent developments in drying of food products

T.N. Valarmathi[1]*, S.Sekar[2], M. Purushothaman[3], S.D. Sekar[4], MaddelaRamaSharath Reddy[5], Kancham Reddy Naveen Kumar Reddy[6]

[1] Department of Mechanical & Production Engineering, Sathyabama University, Chennai-600 119
[2] Department of Mechanical Engineering, Velammal Engineering College, Chennai-600 066
[3] Research Scholar, Faculty of Mechanical Engineering, Sathyabama University, Chennai- 600 119
[4] Research Scholar, Department of Mechanical Engineering, R.M.K. Engineering College, Chennai-601 206
[5,6] UG Students, School of Mechanical Engineering, Sathyabama University, Chennai-600 119
*Email id: valarmathi.tn@gmail.com

Abstract. Drying is a dehydration process to preserve agricultural products for long period usage. The most common and cheapest method is open sun drying in which the products are simply laid on ground, road, mats, roof, etc. But the open sun drying has some disadvantages like dependent on good weather, contamination by dust, birds and animals consume a considerable quantity, slow drying rate and damages due to strong winds and rain. To overcome these difficulties solar dryers are developed with closed environment for drying agricultural products effectively. To obtain good quality food with reduced energy consumption, selection of appropriate drying process and proper input parameters is essential. In recent years several researchers across the world have developed new drying systems for improving the product quality, increasing the drying rate, decreasing the energy consumption, etc. Some of the new systems are fluidized bed, vibrated fluidized bed, desiccant, microwave, vacuum, freeze, infrared, intermittent, electro hydrodynamic and hybrid dryers. In this review the most recent progress in the field of drying of agricultural food products such as new methods, new products and modeling and optimization techniques has been presented. Challenges and future directions are also highlighted. The review will be useful for new researchers entering into this ever needed and ever growing field of engineering.

1. Introduction

Drying is a dehydration process used to remove the moisture present in food products by the application of heat. The heat may be supplied either by hot air or from the solar energy. Drying process is used to preserve the food products for future usage. Drying prevents the growth of bacteria and yeast formation. Drying can be achieved by using open sun drying and greenhouse drying methods. The open sun drying is presented in Fig. 1. When compared to greenhouse drying process, open sun drying process is a slow process; dried products will be of low quality due to contamination of dust particles, damages due to rain and moisture present in the air. Also there is a loss of food products due to insects, birds and animals. The quality of products obtained using greenhouse drying process is better because it is carried out in a closed environment. This closed environment is made using special cover materials like polyethylene, polycarbonate, etc. Greenhouse dryers are differentiated based on the structure of the roof such as roof even span and dome shaped dryer and based on the mode of energy transfer such as active (forced circulation)[6], and passive (natural circulation)[24]. Dryers with greenhouse drying are shown in Fig. 2. The main objective of this study is to provide a comprehensive review of literature related to the recent progress in the field of drying of agricultural food products.

Figure1. Open sun drying

Figure 2. Greenhouse solar dryers

2. Classification of drying systems

Solar dryers are classified as direct absorption type, indirect or convection type and the combination of the both. The pressure inside the drying system plays a vital role in drying of food products. Based on the operating pressure the dryers are classified into atmospheric drying, vacuum drying and freeze drying [10]. It is observed that hot air drying and vacuum drying are commonly used for drying of agricultural products. Also it is observed that hot air drying process is simple and economical[25].

Vacuum drying like microwave drying is more beneficial than hot air drying, because of short drying time, high drying rate and superior quality of dried products[26]. As yielding of microwave drying process is non-uniform it is normally carried out by combining it with other drying process to achieve better quality. Infrared drying is preferable because of uniform heating and high quality of yield. Even though the quality of dried products is good in freeze drying it is not practiced generally because it is an uneconomical process[23]. The type of dryer used is based on the need and the desired properties of drying products. Some of the new systems like desiccant [41], electro hydrodynamic and hybrid dryers [29], vibrated fluidized bed [19]&[48], fluidized bed [37]&[43], freeze [2]&[34], hot air drying [9], infrared[13,14],[16], intermittent [11], microwave [21],[49] & [53],and vacuum [44] have been utilized and investigated by researchers for drying the food products.

3. Analysis of effectiveness of drying systems

Research works have been carried out to analyze the effectiveness of various drying systems by studying the drying characteristics like moisture content, drying air temperature, air velocity, drying rate, drying time, etc. Adam Figiel [2] carried out drying of beetroot cubes using a combined convective-vacuum microwave process by freeze drying. They reported that the results such as compressive strength, antioxidant activity and rehydration have been better than the convection method and also the quality is improved with less drying time in combined method.Amin Hazervazifeh et al [5]studied the drying characteristics of apple slices and found that processing time is less in combined microwave-hot air flow method than microwave radiation and hot air drying process. And minimum energy consumption is observed in microwave radiation process. Atul Patel and Gaurav Patel[6] made CFD analysis of a forced circulation type solar dryer used conventionally for dehydrating vegetables and fruits to obtain operational augmentation in implementing the modifications in pressure and velocity values.A typical fluidized bed dryer is shown in Fig. 3.

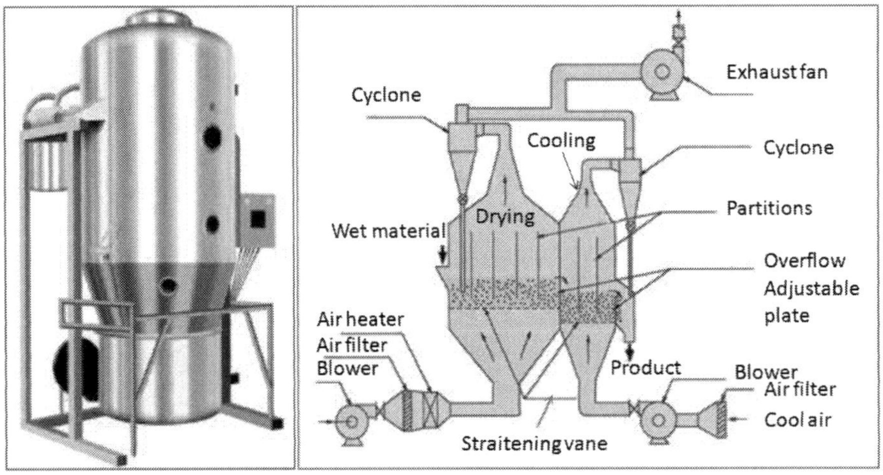

Figure 3. Fluidized bed dryer [54, 55]

Dening Jia et al [8] investigated the effect of various parameters on the drying performance of pulsed fluidized vibrated bed of biomass particles and suggested high temperature and gas flow rate for better drying.Fagunwa et al [11] developed an intermittent solar dryer for Cocoa Beans and reported that drying performance is efficient than the traditional method.Jin et al[19] studied the effect of vibration parameters of a vibrated fluid bed dryer using an optical fiber probe approach and proposed an empirical correlation to predict bed voidage.Kejing An et al[23] studied the drying characteristics of ginger rhizome using five different drying process such as hot air drying, freeze drying, infrared drying, microwave drying (MD) and intermittent microwave and convective drying(IM&CD) and found that MD and CD method is good in preserving thermo sensitive materials with less power consumption. Structures of fresh ginger observed by light microscopy (A), scanning electron micrographs (SEM) of dried gingers (B) and structural changes observed by SEM for (a-HAD), (b-IR), (c-FD), (d-MD) and (e-IM&CD) are presented in Fig. 4.

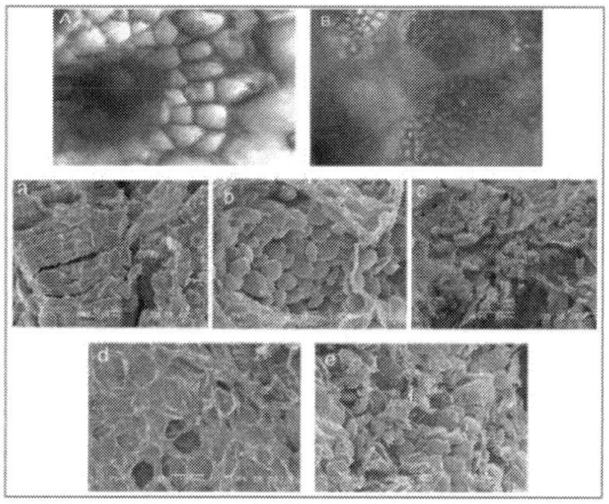

Figure 4.LM (Fresh) and SEM images of ginger (Dried) for various drying process[23]

Liliana Seremet et al [26] investigated the drying characteristics of pumpkin using hot air drying and combined drying process and concluded that the rehydration capacity of hot air drying is higher.Magdalena Zielinska and Anna Michalska [27] evaluated the drying characteristics of blueberries and found that effect of combined hot air convective drying at 90°C and microwave vacuum drying process produces better results in drying time and quality than these processes have been carried out separately. Nadine Sangster et al [32] developed a prototype for an automated cocoa drying house which is s equipped with automated roof which is automatically opened during sunlight and closed if sunlight is not there and fermenter, automatic heaters and a remote control feature for the automated components. Oleksii Parniakov et al [34] studied the effects of pulsed electric fields on vacuum freeze-drying for drying PEF treated vacuum cooled apple tissue. It is observed that electroporated tissue samples shows large pores, fast rate of moisture impregnation and large rehydration capacity hence reported that PEF treatment gives better results.Ronak Daghigh et al [39] carried out a review of solar assisted heat pump drying systems for agricultural and marine products.Ruifang Wang et al [40] carried out drying experiments using a hybrid microwave rotary drying system in soybean drying and studied the effect of drum speed, cracking ratio, etc. Ting -Jie Wang et al [45] analyzed and reported the mechanism of vibration energy transfer using wave propagation. Sensors have been used to detect the wave signals produced and calculate the pressure-wave propagation parameters. Yuting Tian et al [50] evaluated the effect of various drying methods on the qualities of the drying product. Mushroom is taken as the studying element under hot air, vacuum, microwave and microwave vacuum methods. They reported that there is a significant rise in the content of total free amino acids and the relative content of sulfur compounds of dried mushrooms in all the four methods. They observed that there is an improvement in nutrient retention and color attributes and larger amounts of taste- active amino acids are maintained and found high rehydration ratio when dried with microwave vacuum method. It is also reported that the collapse in structure also less than the other methods.

4. Modelling and optimization techniques

Modelling and optimization techniques like response surface methodology (RSM), fuzzy modelling, etc. are used to analyze the problems, in which one or more responses are influenced by many factors and to find out the quantitative relationship between the response variables and the input control variables. Also they are used to determine the significant effect of each factor on the response factor by developing mathematical models and to find out the optimal conditions. RSM technique is used by many researchers ([4], [18], [22], [33],[35], [37], [42],[44] &[51]) to optimize the drying effect of various dryers like fluidized bed dryers, on different agricultural products like Artemisia absinthium leaves, coriandrum sativum leaves, coroba slices, kefir powder, ganodermalucidium slices, green peas, olive leaves and soya bean, etc.

4.1. Response surface methodology (RSM)

Response surface methodology is a method of statistical and mathematical techniques influenced by many factors and to find out the quantitative relationship between the response variables and the input control variables. This method has been initially used for model fitting of physical experiments but later it is used for the design of experiments in process optimization. It is the process of approximating a response function, based on statistical analysis of data which has been obtained at various design points. The relationship between the control parameters and the responses is given in Equation (1) as

$$Y = f(X1, X2, \ldots Xk) + \varepsilon \qquad (1)$$

where Y is the response variable and X1, X2, ... Xk are the independent variables. The function f is called the true response function. The residual ε measures the experimental error. Coefficient of determination (R-square) is used to evaluate the model [46].

4.2. Mathematical modeling techniques of thin layer drying

Many researchers analyzed the drying characteristics of agricultural products using various mathematical models like Newton, Henderson and Pabis, Logarithmic and Weibull, Midilli et al, Page, Modified Henderson and Pabis, Two-Terms, etc. The mathematical models developed can be validated using correlation analysis ®, reduced chi-square (x^2) test and root mean square error (*RMSE*) analysis. Some of the model descriptions used by the researchers are listed in Table .1. Where *MR* is moisture ratio, M_t is moisture content at any given time (kg water kg-1 solids), M_e standing for equilibrium moisture content (kg water kg-1 solids) and M_0 representing the initial moisture content.

Table 1. Typical mathematical Models of thin layer drying

Sl.No.	Model name	Model description
1.	Approximation of diffusion	MR=a exp(-kt)+(1-a)exp(-kbt)
2.	Henderson and Pabis model	MR=a exp(-kt)
3.	Modified Henderson and Pabis	MR=a exp(-kt)+b exp(-gt)+c exp(-ht)
4.	Logarithmic	MR=a exp(-kt)+c
5.	Midilli et al	MR= aexp(-ktn)+bt
6.	Newton model	MR=exp(-kt)
7.	Page model	MR=exp(-ktn)
8.	Modified page model	MR=exp[-(kt)n]
9.	Two term model	MR=a exp(kot)+b exp(-k1t)
10.	Two term exponential	MR=a exp(-kt)+(1-a)exp(-kat)
11.	Verma et al.	MR=a exp(-kt)+(1-a)exp(-gt)
12.	Wang and Sing	$MR= 1+at+bt^2$
13.	Weibull	$MR=\exp[-(t/b)^a]$

Abano et al [1] investigated the drying characteristics like air velocity, drying time, etc. of tomato slices using RSM modeling technique and the ideal drying condition is predicted using desirability index technique.Akpinar [3] investigated the drying characteristics of mint leaves using an indirect forced convection solar dryer. The experimental values are analyzed using ten models and exergy analysis is also carried out to study the impact of parameters considered. Ali Abasi Surki et al [4] optimized the operating parameters in drying of soybean using RSM. A three-level, four-factor fractional factorial design is applied for optimization.Diamante et al [7] developed a new mathematical model for thinlayer for drying of fruits.Gokhan Gurlek et al [12] developed a new solar tunnel dryer in drying of tomato and compared twelve different mathematical models to predict the drying characteristics and reported that the two-term model gives better results.Hosain Darvishi et al [15] developed seven mathematical models to describe the characteristics of thin layer drying of pepper samples and reported that the Midilli model is more suitable for thin layer samples. They concluded that energy efficiency is increased when increase in microwave power and moisture content in microwave drying of pepper by varying the power values.
Ibrahim Doymaz[16] studied the effect of various infrared power levels on drying kinetics of carrot pomace and reported that among the twelve mathematical models developed Aghbashlo et al model is better.Ibrahim Doymaz[17] studied the effect of various infrared power levels on drying kinetics of pomegranate seeds and reported that the drying time is reduced when the power level is increased. Also they developed ten mathematical models and concluded that among the ten models The Page, Midilli et al and Weibull models have better prediction qualities than the other models. Jose Vasquez et al [20] implemented fuzzy control systems for a solar drying system integrated with thermal energy storage system and reported that the new arrangement works satisfactorily and energy is saved considerable amount than the old methods in drying of mushroom, plum and peach.Jose Vasquez et al [20] implemented fuzzy control systems for a solar drying system integrated with thermal energy storage system and reported that the new arrangement works satisfactorily and energy is saved considerable amount than the old methods in drying of mushroom, plum and peach.Lemuel M. Diamante et al [25]developed a new mathematical model for thin layer drying of fruits.Malaisamy et al [28] designed an efficient drier and implemented modeling and conventional PI controller and Fuzzy controller for maintaining temperature in the heating chambers for drying process for the efficient usage of solar energy and solar powered electrical energy for heating process. Also they compared the response with PI and concluded that Fuzzy shows better performance than PI and cardamom dryer provide better performance than copra dryer.Midilli et al [30] developed a new model to evaluate the drying characteristics for single layer drying process and also verified the effectiveness of the model with experimental data. Further they compared the new one with other available models using data obtained from literature and concluded that it is an effective model.
Minaei et al [31] analyzed the drying characteristics using 11 mathematical models in drying of pomegranate arils and reported that Midili model is best suited for vacuum drying and Page model is for microwave drying technique.Pengfei Zhao et al [36] investigated the drying characteristics and kinetics of Shengli lignite and analyzed using seven thin layer models. They concluded that among four methods of drying vibrated medium fluidized bed has been given better results and the simulation using the Midilli– Kucuk model is best suited than the other models in dewatering the lignite. Reuss et al [38] used TRNSYS simulation model for the wood drying process based on the physical properties of the drying product and concluded that the solar drying process is more advantageous.Siewkian chin et al [42] optimized the drying condition of convective hot air in drying of Ganodermalucidium slices using RSM. They concluded that drying temperature have significant effect than the other factors considered. Vega-Galvez et al [47] studied the drying characteristics of olive-waste cake using convective process with five different temperatures and tested with various mathematical models and concluded that the modified Henderson and Pabis is best suited to describe the drying curves among the other models.Zafer Erbay et al [51] optimized the operating conditions in drying of olive leaves using RSM. Zdravko Sumic et al [52] dried fresh red currants using vacuum drying and optimized the

process parameters such as temperature, pressure and drying time using RSM, regression analysis and ANOVA. They concluded that the results shows improved physic-chemical properties of lyophilized samples when compared to conventional method.

5. Conclusions

In this review the most recent progress in the field of drying of agricultural food products such as new methods, new products and modeling and optimization techniques has been presented. The quality of products obtained using greenhouse drying process is better because it is carried out in a closed environment. Hot air drying and vacuum drying are used for drying of agricultural products in which hot air drying process is simple and economical. Processing time is less in combined microwave-hot air flow method than microwave radiation and hot air drying process. In infrared drying, uniform heating and high quality of yield can be achieved. Indirect mode forced convection dryer performance is better than other methods. Solar drying system accommodated with phase change material will accelerate the drying performance of the system even during nights. Simulation models can be used to forecast and analyze the new systems developed. An improvement in nutrient retention and color attributes and larger amounts of taste- active amino acids are maintained and found high rehydration ratio when dried with microwave vacuum method. Further research work in developing hybrid drying systems and drying systems to improve energy savings are needed.

References

[1] Abano E E, Ma H and Qu W, "Optimization of Drying Conditions for Quality Dried Tomato Slices Using Response Surface Methodology", Journal of Food Processing and Preservation, Vol.38, Issue 3, (2014), pp. 996–1009.

[2] Adam Figiel, Drying kinetics and quality of beetroots dehydrated by combination of convective and vacuum-microwave methods, Journal of Food Engineering,vol.98, (2010), pp.461–470.

[3] Akpinar E.K., "Drying of mint leaves in a solar dryer and under open sun: Modelling, performance analyses", Energy Conversion and Management, vol. 51, (2010), pp.2407–2418.

[4] Ali Abasi Surki and Reza Tavakkol Afshari,N. Majnoun Hosseini and H. R. Gazor "Optimization of soya bean seed drying conditions using Response Surface methodology with respect to seed quality", Journal of Agricultural Science and Technology, Vol. 12, (2010), pp. 409-423.

[5] Amin Hazervazifeh, Ali M. Nikbakht, Parviz A. Moghaddam, "Novel hybridized drying methods for processing of apple fruit: Energy conservation approach", Energy, 103 (2016), pp.679-687.

[6] Atul Patel and Gaurav Patel, "Operational Augmentation of Forced Circulation Type Solar Dryer System Using CFD Analysis", Int. Journal of Engineering Research and Applications , vol. 4, Issue 4 (1), (2014), pp.265-268.

[7] Diamante, L. M., Ihns, R., Savage, G. P., & Leo Vanhanen, L, "A new mathematical model for thin layer drying of fruits", International Journal of Food Science and Technology, vol. 45, (2010), pp.1956–1962.

[8] Dening Jia, Oceane Cathary, Jianghong Peng, Xiaotao Bi, C.Jim Lim, Shahab Sokhansanj, Yuping Liu, Ruixu Wang and Atsushi TsuTsumi, Fluidization and drying of biomass particles in a vibrating fluidized bed with pulsed gas flow", Fuel processing Technology, vol.138, (2015), pp.471-482.

[9] Diamante L.M and Yamaguchi Y, "Response surface methodology for optimisation of hot air drying of blackcurrant concentrate infused apple cubes", International Food Research Journal, vol. 19, (2012), pp.353-362.

[10] Earle Richard L, Earle MD. Unit operations in food processing. New Zealand: Web Edition;

2004.http://www.nzifst.org.nz/unitoperations/drying1.htm.
[11] Fagunwa, A.O., Koya, O. A. and Faborode, M.O., "Development of an Intermittent Solar Dryer for Cocoa Beans", Agricultural Engineering International: the CIGR E-journal, Manuscript number 1292, Vol. XI. July, 2009.
[12] Gokhan Gurlek, Necdet Ozbalta and Ali Gungor, "Solar Tunnel Drying Characteristics and Mathematical Modelling of Tomato", Journal of Thermal Science and Technology, vol.29, (1), (2009), pp.15-23.
[13] Hebbar, H.U. and Rastogi, N.K., "Mass transfer during infrared drying of cashew kernel", J. Food Eng. Vol.47, (2001), pp.1-5.
[14] Hebbar. U. H and Ramesh. M. N, "Optimisation of Processing Conditions for Infrared Drying of Cashew Kernels with Taste", Journal of the Science of Food and Agriculture, vol.85,(2005), pp. 865-871
[15] Hosain Darvishi, Abbas Rezaie Asl, Ali Asghari, Mohsen Azadbakht, Gholamhassan Najafi, Jalal Khodaei, "Study of the Drying Kinetics of Pepper", Journal of the Saudi Society of Agricultural Sciences, Vol.13, (2014),pp.130–138.
[16] Ibrahim Doymaz, "Drying of pomegranate seeds using infrared radiation", Food Science and Biotechnology, Volume 21, Issue 5, October 2012, pp 1269–1275.
[17] Ibrahim Doymaz, "Determination of Infrared Drying Characteristics and Modelling of Drying Behaviour of Carrot Pomace",Journal of agricultural sciences, vol. 19, (2013), pp.44-53.
[18] Ilyas Atalar, Muhammet Dervisoglu, "Optimization of spray drying process parameters for kefir powder using response surface methodology", Food Science and Technology, vol.60, (2015), pp.751-757.
[19] Jin H, Zhang J and Zhang B, "The Effect of Vibration on Bed Voidage Behaviors in Fluidized Beds with Large Particles", Brazilian Journal of Chemical Engineering, Vol. 24 (03), (2007), pp. 389 – 397.
[20] Jose Vasquez, Alejandro Reyes, Andrea Mahn and Francisco Cubillos, "Experimental Evaluation of Fuzzy Control Solar Drying with Thermal Energy Storage System, Drying Technology, DOI: 10.1080/07373937.2015.1137001m, 2016.
[21] Kassem, A. S., Shokr, A. Z., El-Mahdy, A. R., Aboukarima, A. M., and Hamed, E. Y., "Comparison of drying characteristics of Thompson seedless grapes using combined microwave oven and hot air drying", Journal of the Saudi Society of Agricultural Sciences, vol.10,(2011), pp. 33–40.
[22] Karimi F, Raffiee S, Taheri Garavand A, and Karimi M, "Optimization of an Air Drying Process for Artemisia Absinthium Leaves Using Response Surface and Artificial Neural Network Models", Journal of the Taiwan Institute of Chemical Engineers, vol.43, 1, (2012), pp. 29-39., DOI10.1016/j.jtice.2011.04.005.
[23] Kejing An, Dandan Zhao, Zhengfu Wang, Jijun Wu, Yujuan Xu, Gengsheng Xiao, "Comparison of Different Drying Methods on Chinese Ginger (Zingiber officinale Roscoe): Changes in Volatiles, Chemical Profile, Antioxidant Properties, and Microstructure", Food Chemistry, 2015 (In press)
[24] Kumar A, Tiwari G N, Kumar S, Pandey M, "Role of greenhouse technology in agricultural engineering", International Journal of Agricultural Research, vol.1 (4), 3, (2006), pp.64 –72.
[25] Lemuel M. Diamante, Reiner Ihns, Geoffrey P. Savage and Leo Vanhanen, "Short communication: A new mathematical model for thin layer drying of fruits", International Journal of Food Science and Technology, Vol. 45, Issue 9, (2010), pp.1956-1962, DOI: 10.1111/j.1365-2621.2010.02345.x.
[26] Liliana Seremet (Ceclu), ElisabetaBotez, Oana-ViorelaNistor, DoinaGeorgetaAndronoiu, Gabriel-DanutMocanu, "Effect of Different Drying Methods on Moisture Ratio and Rehydration of Pumpkin Slices", FoodChemistry, 195, (2016), pp.104-109.
[27] Magdalena Zielinska, Anna Michalska, "Microwave-Assisted Drying of Blueberry (Vaccinium corymbosum L) Fruits:Drying Kinetics, Polyphenols, Anthocyanins, Antioxidant Capacity,

Colour and Texture", Food Chemistry, vol. 212, (2016), pp.671-680.

[28] Malaisamy S, A.Srinivasan A, Mohamed Rafiq M and Manimaran M, "Power Optimization and Temperature control in Solar Powered Automated Dryer Using Fuzzy Controller", International Journal of ChemTech Research, Vol.9, No.05, (2016), pp. 551-557.

[29] Metin Kiyan, Ekin Bingol, Mehmet Melikoglu and Ayhan Albostan, "Modelling and Matlab/Simulink", Energy Conversion and Management, vol.72,(2013), pp.147-155.

[30] Midilli A, Kucuk H and Yapar Z, "A New Model for Single Layer Drying", Drying Technology, vol.20 (7), (2002), pp.1503–1513.

[31] Minaei, S, Motevali, A, Ahmadi, E, and Azizi, M. H., "Mathematical Models of Drying Pomegranate Arils in Vacuum and Microwave Dryers",J. Agr. Sci. Tech. Vol. 14, (2012), pp. 311-325.

[32] Nadine Sangster, Terrence Lalla, Runul Roberts, Kristopher Meyers and Kevin Rampersad, "An automated cocoa drying house for use in a small developing country", International Journal of Engineering Inventions, Vol. 4, Issue 10 (2015),pp. 23-30

[33] Otoniel Corzo, Nelson Bracho, Alberto,Va´squez, Angel Pereira, "Optimization of a Thin Layer Drying Process for Coroba Slices", Journal of Food Engineering, vol.89, 2008, pp.372-380.

[34] Oleksii Parniakov, Olivier Bals, Nikolai Lebovka and Eugene Vorobiev, "Pulsed Electric Field Assisted Vacuum Freeze-Drying of Apple Tissue", Innovative Food Science and Emerging Technologies, vol. 35, (2016), pp. 52–57.

[35] Partheeban K, Sekar SD and Sekar Saravanan, "Optimization and Modelling Techniques of Various Dryer- A Review", International Journal of Applied Engineering Research, vol.10,50, (2015), pp.861-874.

[36] Pengfei Zhao, Liping Zhong, Ran Zhu, Yuemin Zhao, Zhenfu Luo and Xuliang Yang, "Drying characteristics and kinetics of Shengli lignite using different drying methods",Energy Conversion and Management , vol.120, (2016), pp.330–337.

[37] Rajesh R. Burande, Baburao K. Kumbhar, Prabal K Ghosh and Digvir S Jayas, " Optimization of Fluidized Bed Drying Process of Green Peas using Response Surface Methodology", Drying technology, Vol. 26, (2008), pp.920-930.

[38] Reuss M, Benkert ST, Aeberhard, Martina P, Raush G, Rentzell B V and Sogari N, "Modelling and Experimental Investigation of a Pilot Plant for Solar Wood Drying", Solar Energy, vol.59, (1997), pp.259-270.

[39] Ronak Daghigh, Mohd Hafidz Ruslan, Mohamad Yusof Sulaiman and Kamaruzzaman Sopian, "Review of solar assisted heat pump drying systems for agricultural and marine products",Renewable and Sustainable Energy Reviews, vol. 14, (2010), pp.2564–2579

[40] Ruifang Wang, Zhanyong Li and Yanhua Li, "Soybean drying characteristics in microwave rotary dryer with forced convection", Journal of Frontiers of Chemical Engineering in China, 2009, vol.3(3), 2009, pp. 289-292.

[41] Shanmugam V, Natarajan E. Experimental investigation of forced convection and desiccant integrated solar dryer. Renewable Energy, 2006;31:1239-1251.

[42] Siewkian Chin, Chung Lim Law, "Optimization of Convective Hot air drying of Ganodermalucidum Slices Using Response Surface Methodology", International Journal of Scientific and Research publications, Vol. 2, Issue 5, (2012) pp. 1-11.

[43] Sivakumar R, Ilaya Perumal A, Saravanan.R. Experimental Study of bitter guard, green peas and okra's drying characteristics in Fluidized bed dryer. IEEE, 2013; 978-1-4673-6150-7/13:371-380.

[44] Thirugnanasambandham K and Sivakumar V, "Enhancement of shelf life of coriandrum sativum leaves using vacuum drying process: Modelling and optimization", Journal of Saudi society of Agricultural sciences, vol.15(2), (2016), pp.195-201.

[45] Ting-Jie Wang, Yong Jin, Atsushi Tsutsumi, Zhanwen Wang and Zhe Cui, "Energy transfer mechanism in a vibrating fluidized bed", Chemical Engineering Journal, vol. 78, (2000), pp. 115-123.

IOP Conf. Series: Materials Science and Engineering **197** (2017) 012037 doi:10.1088/1757-899X/197/1/012037

[46] Valarmathi T N, Palanikumar K and Sekar S, "Thrust Force Studies in Drilling of Medium Density Fiberboard Panels", Advanced Materials Research, vol.622-623, (2013), pp.1285-1299/2013.
[47] Vega-Galvez. A, Miranda. M, Diaz LP, Lopez. L, Rodriguez. K and Di Scala. K., "Effective moisture diffusivity determination and mathematical modelling of the drying curves of the olive-waste cake", Bioresource Technology, vol. 101(19), (2010), pp.7265-70. doi: 10.1016/j.biortech.2010.04.040.
[48] Wang T, Jin Y, Tsutsumib A, Wang Z and Cui Z, "Energy transfer mechanism in a vibrating fluidized bed", Chemical Engineering Journal, vol.78, (2000), pp.115-123.
[49] Workneh, T.S., Raghavan, V., and Y. Gariepy (2011). Microwave assisted hot air ventilation drying of tomato slices. International Conference on Food Engineering and Biotechnology, IPCBEE (vol. 9). Singapore: IACSIT Press.
[50] Yuting Tian, Yingting Zhao, Jijun Huan, Hongliang Zeng, Baodong Zheng, "Effects of different drying methods on the product quality and volatile compounds of whole shiitake mushrooms", Food Chemistry, vol.197, (2016) pp.714-722.
[51] Zafer Erbay and Filiz Icier, Optimization of hot air drying of olive leaves using response surface methodology", Journal of Food Engineering, vol. 91, no. 4,(2009), pp. 533-541.
[52] Zdravko Sumic , Anita Vakula, Aleksandra Tepic , Jelena C Akarevic , Jasmina Vitas, and Branimir Pavlic, "Modeling and optimization of red currants vacuum drying process by response surface methodology (RSM)", Food Chemistry, vol.203, (2016), pp. 465- 475.
[53] Zhang, M., Tang, J., Mujumdar, A. S and Wang, S., "Trends in microwave related drying of fruits and vegetables", Trends in Food Science & Technology, vol.17, (2006), pp. 524–534.
[54] http://www.technoconsultancy.in/fluid-bed-dryer.htm.
[55] http://www.kurimoto.co.jp/worldwide/en/product/item/07pw/330.php.

IOP Conf. Series: Materials Science and Engineering **197** (2017) 012038 doi:10.1088/1757-899X/197/1/012038

Investigation of Sensible and Latent Heat Storage System using various HTF

N Beemkumar*, A Karthikeyan, A Manoj, J S Keerthan, Joseph Paul Stallan, Amithkishore P

School of Mechanical Engineering, Sathyabama University, Chennai, India

E-mail: beem4u@gmail.com

Abstract. The objective of the work is investigating the latent heat storage system by varying heat transfer fluid (HTF). In this experiment, the effect of using different heat transfer fluids on the combined system is studied while using a low melting phase change material (PCM) i.e., paraffin wax. The heat transfer fluids chosen are water (low boiling fluid) and Therminol-66 (High boiling fluid). A comparison is made between the heat transfers by employing both the Heat transfer fluids. In the beginning, water is made to flow as the HTF and the charging process is undertaken followed by the discharging process by utilizing the different encapsulation materials namely, copper, aluminium and brass. These processes are then repeated for therminol-66 as HTF. At the end of the experiment it was concluded that even though therminol-66 enhances the latent heat storage capacity, water offers a higher sensible heat storage capacity, making it a better HTF for low melting PCM. Similar to above said process the experiments can be conducted for high and medium range melting point PCM with variation of HTF.

1. Introduction

Energy is widely available in various forms. It is also used for various activities throughout the world. It can be converted from one form of energy to another, but it cannot be created nor destroyed according to the first law of thermodynamics. However, this energy may also get converted into some waste energy, mostly in the form of heat. This leads to some energy losses and reduction in the efficiency of energy systems. Harnessing energy when supply is limited has led to researchers conducting detailed studies with this regard.One method to overcome this problem is to store the energy when there is excess supply and use it when supply is restricted. In this paper, two heat storage units are used [1]. The first one is the heat transfer fluid that stores energy as sensible heat. The second unit is the paraffin wax which is used as PCM. It stores the energy as sensible and latent heat. The advantages of using paraffin as PCM are its high storage density, low weight, a low melting point as compared to other PCM. The PCM along with a HTF can be used as a combined sensible and latent heat system. A combined sensible and latent heat storage system overcomes the drawbacks of using either one of them separately [2]. These drawbacks include low heat transfer and low efficiency of these storage systems. Different materials have different thermal conductivities and so appropriate materials so should be chosen depending on the various applications [3]. Certain methods and techniques that can also be used to increase the amount of heat stored and recovered, thus increasing the overall efficiency of the heat storage systems [4].The study into the usage of molten salts started in the early 1980's [5]. Scientists then attributed the energy storage to applications in heating and cooling

spaces [6]. Low melting molten salts were of particular interest as a large number of fluids can be used for the heat transfer [7]. Techniques to improve the rate of heat transfer between storage systems were also studied [8]. The storage unit uses encapsulations of various materials like copper, aluminium and brass etc. This is done to increase the heat transfer between the HTF and PCM, thus increasing heat energy stored. A comparative study is carried out between water and therminol-66 as HTF. Their effect on the amount of heat stored and recovered while using a low temperature PCM is analyzed. A final comparative study of the performance of heat transfer is done for all three encapsulating material copper, brass and aluminium and the most effective encapsulation is identified.
The main purpose of encapsulation is containing the PCM, when it changes from solid state to liquid state and vice- versa. Also, it is used to increase the rate of heat storage and recovery by the use of metals of high conductivity (aluminium, brass, copper). However, one drawback of using PCM is its low sensible heat capacity which is relatively low. This disadvantage is overcome by its excellent latent heat storage capacity, which is relatively larger.

2. Experimental Setup and Methodology

The set up consists of two thermal storage tanks, a heater and a circulating pump showed in figure 1. The storage tank consists of both, the PCM and the Heat Transfer fluid. Eachcylindrical tank has a Diameter of 256 mm and 300 mm length. The storage tank is well insulated using glass wool of 0.10 m thickness and pipe lines are also insulated with same material. A total quantity of 55 L of water or therminol-66 is charged in the storage tank and the piping circuit, and it is ensured that there is no leak in the oil flow path. An additional tank is also fabricated to house the heaters which will be used to provide the required heat energy. Valves are present at regular intervals to control the amount of HTF flow. This is used to reduce build up of high pressure in the circuit, which could be dangerous.

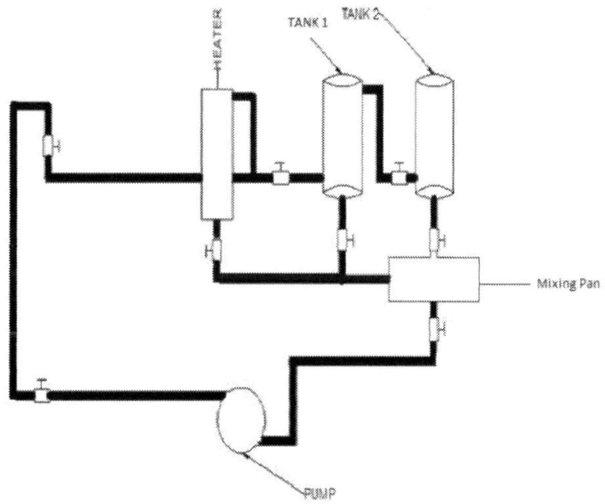

Figure 1. Experimental layout.

The selection of PCM is depends on the required temperature range. The PCM chosen here is paraffin wax, due to its stability and low melting point (50 – 60°C) that enables to use either water or therminol-66 as HTF. A gear pump maintains the HTF flow in the circuit, through the heater and storage tanks during charging process in which the heated oil transfers heat to the PCM. The PCM is encapsulated in spherical balls of copper, aluminium and brass. The copper, aluminium and brass encapsulating materials are used in this study and the experimentation was done by changing different encapsulating material. The heater is switched on along with the pump. The temperature of HTF in all the tanks and the PCM are noted after regular intervals of 10 minutes. This is the charging process. After attaining a temperature of about 85°C, the heater is switched off. Keeping the flow rate on, the above process is repeated. This is the discharging process. Hence heat is subsequently stored and recovered in these processes. The above processes are conducted for all the three type of encapsulations using water as HTF initially and then using therminol-66. The observations are noted down.

3. Results and Discussion

The results obtained from the heat transfer characteristics of the combined heat storage system using paraffin as PCM, water and therminol-66 as HTF during charging and discharging process with different encapsulation materials (i.e., copper, brass and aluminium) are discussed in detail.

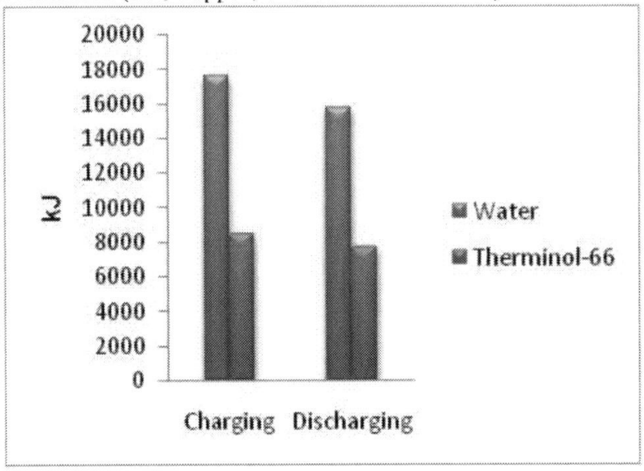

Figure 2. Charging and discharging for copper encapsulation.

Figure 2 represents the charging and discharging rates of the combined storage system while using both water and therminol-66 for copper encapsulation. A comparison is made between a high temperature heat transfer fluid (therminol-66) and a low temperature heat transfer fluid (water). It can be interpreted that the total heat stored while using water is around 18000 kJ which is almost double to that while using therminol-66 (8200 kJ). The total energy is inclusive of the sensible and latent heat energy. While discharging, the heat energy recovered is higher, while using water (17500kJ) as compared to therminol-66 (7050 kJ). Heat losses of about 50-1000 kJ are observed.

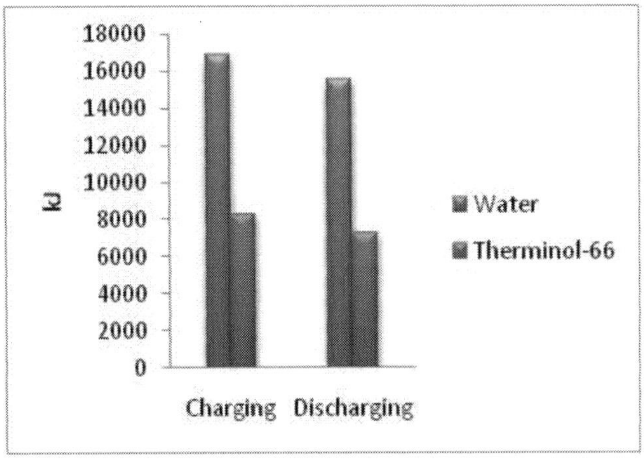

Figure 3. Charging and discharging for Aluminium encapsulation.

Figure 3 depicts the charging and discharging rates of the combined storage system while using both water and therminol-66 for aluminium encapsulation. Similar to the case for copper encapsulation, the heat stored for aluminium encapsulation is more while using water i.e., around 17000 kJ, in contrast to the heat stored using therminol-66, which is about 8050 kJ. During discharging also, water helps recover more heat (15500 kJ) as compared to therminol-66 (6100 kJ). This is due to the higher specific heat capacity of water.

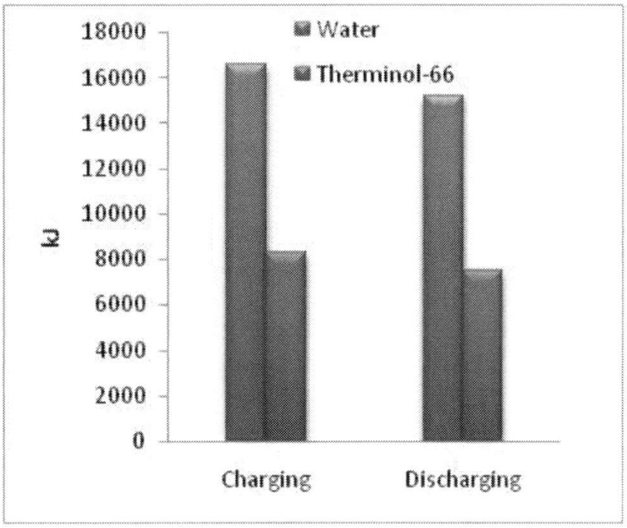

Figure 4. Charging and discharging for brass encapsulation.

For brass encapsulation from figure 4, the heat energy stored is 16565 kJ while using water while with therminol-66, it is around 8000 kJ. A similar trend is observed during discharging. The heat recovered from water (15000 kJ) is almost double of that recovered from therminol-66 (7800 kJ). This is also due to the high specific heat capacity of water and a relatively low heat capacity for therminol-66.

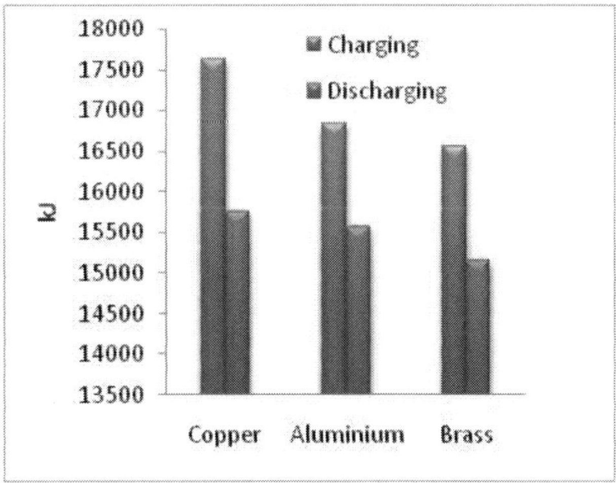

Figure 5. Comparison between different encapsulations while using water as HTF.

A comparison of the heat stored and recovered in Copper, Aluminum and Brass encapsulations while using water as HTF is shown in figure 5. From the figure, it can be easily interpreted that copper stores the most heat energy while brass stores the least. The same is observed during heat recovery. But at the same time, the heat loss (difference between heat stored and recovered) is the highest in copper encapsulation, then in aluminium encapsulation and it is the least in brass encapsulation. This can be attributed to the high thermal conductivity of copper and a relatively low thermal conductivity for brass.

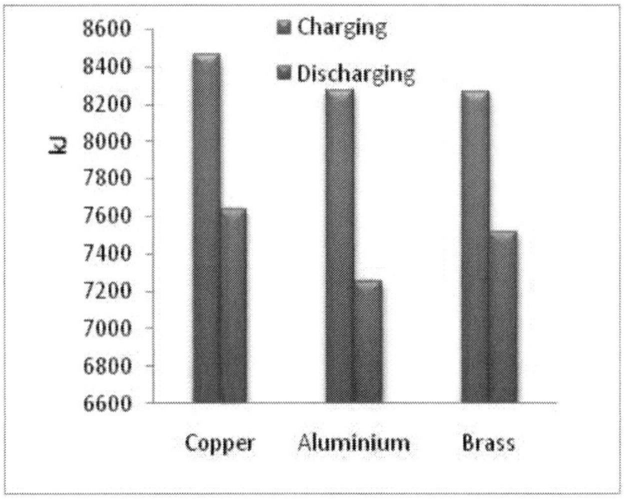

Figure 6. Comparison between different encapsulations while using therminol-66 as HTF.

Figure 6 shows a comparison between copper, brass and aluminium encapsulations while using therminol-66 as HTF.The heat stored and recovered are much lower than that while employing water. Form the comparison it can be understood again that copper is the most effective. But its high thermal conductivity leads to an increase in heat losses. The cost effectiveness is also to be considered.

4. Conclusion

The heat transfer in the combined solar thermal energy storage system was studied and analyzed. A comparative study was made between a high temperature HTF and a low temperature HTF. Their effect on a low melting PCM like paraffin was analyzed. It can thus be concluded that water, when used as the HTF stores more energy (total energy) as compared to the case where therminol-66 is used. This is due to the high specific heat capacity of water. However, the application of water as HTF is limited due to its low melting point. While heating to very high temperatures, therminol-66 can store more energy due to its high melting point (around 350 – 400°C). A comparison was also made between the different encapsulations and it was found that copper encapsulation was the most efficient, storing and recovering more energy. However, its high thermal conductivity promotes larger heat losses. Its cost is also on the higher side. Thus, water is a better heat transfer fluid when used for low melting PCM and copper is a better material for encapsulation as it improves the heat transfer rate.

References

[1] Ponshanmugakumar A, Sivashanmugam M and Stephen Jayakumar S 2014 Solar driven air conditioning system integrated with latent heat thermal energy storage *Indian Journal of Science and Technology***7**(11) pp 1798–1804.

[2] Nallusamy N, Sampath S and Velraj R 2007 Experimental investigation on a combined sensible and latent heat storage system integrated with constant/varying (solar) heat sources*Renewable Energy***32** pp 1206-1277.

[3] Vahid Agyenim Francis, Neil Hewitt, Philip Eames and Mervyn Smyth 2010 A review of materials, heat transfer and phase change problem formulation for latent heat thermal energy storage systems (LHTESS) *Renewable and Sustainable Energy Reviews***14**(2) pp 615-628.

IOP Conf. Series: Materials Science and Engineering **197** (2017) 012038 doi:10.1088/1757-899X/197/1/012038

[4] Auriemma M and Iazzetta A 2016 Numerical analysis of melting of paraffin wax with Al_2O_3, ZnO and CuO nanoparticles in rectangular enclosure*Indian Journal of Science and Technology***9**(3)DOI: 10.17485/ijst/2016/v9i4/72601.
[5] Telkes M 1980 Thermal energy storage in salt hydrates *Solar Energy Materials* **2**(4) pp 381–93.
[6] Sciacovelli A, Gagliardi F and Verda V 2015 Maximization of performance of a PCM latent heat storage system with innovative fins*Applied Energy***137** pp 707–715.
[7] Durgun E and Jeffrey C Grossman 2013 Photoswitchable molecular rings for solar-thermal energy storage *The Journal of Physical Chemistry Letters***4** (6) pp 854-860.
[8] Zalba B, Mann J M, Cabeza L F and Mehling H 2003 Review on thermal energy storage with phase change: materials, heat transfer analysis and applications*Applied Thermal Engineering***23** pp 251-283.

IOP Conf. Series: Materials Science and Engineering **197** (2017) 012039 doi:10.1088/1757-899X/197/1/012039

Analysis of Thermal Energy Storage Tank by ANSYS and Comparison with Experimental Results to Improve its Thermal Efficiency

N Beemkumar*, A Karthikeyan, Kota Shiva Keshava Reddy, Kona Rajesh, A Anderson

School of Mechanical Engineering, Sathyabama University, Chennai, India

E-mail: beem4u@gmail.com

Abstract. The discontinuous temperament of the solar power forces to consider about the energy storage. This work is to analyze the tank, amount of energy stored and its storage time. The thermal and flow analysis has been done by ANSYS with different set temperature values. The experimentation is done for various encapsulating materials with different phase change material (PCM). Findings: The results obtained from experimental work are compared with ANSYS output. The competence of the TES is calculated and further improvements are made to enhance its performance. During charging process the temperature distribution from heat transfer fluid (HTF) to PCM is maximum in copper encapsulations followed by aluminium encapsulations and brass encapsulations. The comparison shows only when the electrical power as an input source. The efficient way of captivating solar energy could be a better replacement for electrical input.

1. Introduction

A hot water storage system is used in many houses for domestic purposes to provide a reliable source of energy and this small step has given a major impact as it can be used as a substitute for many renewable resources of energy. Though there are many non-renewable source of energy, there is always a tendency to look for heated fluid to use as a source of energy [1]-[5]. This is because of the easily available solar energy. The main important feature is that it can function as a storage system in the morning during the morning by storing the energy and liberating it at off sun times when it is needed in order to run the system continuously. There is always a misconception that the liquid in the tank and its temperature is not distributed in a uniform way [6]. In real situations, the cooler, denser fluid is settled at bottom of the tank whereas the hot lighter fluid will go up to the top, provided that the water within the tank is not mixed or agitated in any way.

There is also a factor of human comfort that is there in solar energy sources as it is the cheapest source of energy [7]. One of the ways to learn the solar energy processes as well as to develop their effectiveness and performances from the process viewpoint is to pass through the dynamic modeling and numeric simulation especially which consists of strong dynamics such as huge solar tanks and collectors [8]. Recent studies have proved that usage of molten salts in the tank would increase the efficiency [9]. In recent years powerful commercial computational fluid dynamics (CFD) packages become available allowing it to deal with the complex 3D flows and mixing problems. In this way the local structure can be clearly revealed. As compared toward investigational measurements and observations, the CFD technique turns out to be a more efficient diagnosis device for thermal energy system design and analysis, a better insight at a reasonable cost, into how these systems operate. The CFD results confirmed the importance of combined effects on the performance of thermal storage tanks and showed that a properly calculated storage tank can provide improved stratification conditions. Moreover, 3D transient CFD simulations can be used as an effectual tool to optimize thermal storage tank parameters at early design stages, therefore it may add to the value of the storage tank presentation and efficiency, by optimizing the whole solar thermal energy storage system design and size [10].

The main use of Therminol-66 is the application of heat transfer fluid. There are many properties that emphasis Therminol-66 as a Heat Transfer Fluid.Therminol-66 is a high stability synthetic heat transfer fluid contribution is extensive life and very low-top up rates resulting in reduced running costs and minimal downtime for operations at downtime at 345°C. The Project aims to study the heat transfer characteristics of solar energy by comparing the results obtained in ANSYS and the results are compared with the experimental ones. In this project we are using balls of different materials i.e. copper, aluminium and brass with dissimilar PCM's such as D-Mannitol, D-Sorbitol and Paraffin wax. The heat transfer rate between Therminol-66 oil, water and PCM is studied and the results are compiled. The purpose of using different balls is to estimate which out of these balls has the maximum energy (heat) exchange rate.

2. Experimental Setup and Methodology

The experimental setup consist of three PCM tanks to store D-mannitol, D-sorbitol and paraffin wax, one heater tank for giving heat to the HTF in charging mode, one process unit to verify the harvest of the cascaded system, one oil pump, one flow meter and 12 thermocouples to sense the HTF temperature.

Figure 1. Experimental layout.

The heater and pump are turned in to on position. The temperature of HTF in all three tanks is noted after regular intervals of 10 minutes. This is the charging process. After attaining a temperature of about 250°C the heater is switched off. Keeping the flow rate on, the above process is repeated. This is the discharging process. Hence heat is subsequently stored and recovered in these processes. The above process is conducted for all the three type of encapsulations (aluminium, brass and copper) using therminol-66 as HTF. The figure 1 shows the arrangements of experimental setup, thermocouple of 10 in numbers are used to quantify the temperature of HTF at various locations. Oil circulation rate is maintained by flow meter around 26 LPM. The heat transfer rate in PCM is explained by two stages namely charging and discharging mode.

2.1. Charging mode

The HTF has been circulated through the heater source during day time. The oil passing through the heater tank absorbed heat and then sent to phase change material tanks, then it was distributed back to the heater tank. This cyclic process was continued such that PCMs are attained their corresponding melting point. The heat energy thus received by the HTF was given to the PCM encapsulations while passing through the PCMs tank. PCM received the heat from HTF during the charging mode and it gradually changed its phase from solid to liquid and stored the latent heat. This process continued until the PCMs are reached the specific melting point every 5 minutes interval the PCMs temperatures are noted using thermocouple arrangements.

2.2. Discharging mode

The circulation of HTF through the heater tank is stopped and by-passed. The temperature of the HTF gradually decreased and so the PCM inside the encapsulations released the heat outside. The oil surrounds the PCM tank gets heated up again by changing its phase from liquid to solid. This process of PCM made the HTF to maintain the temperature again for a longer time. The discharging mode enabled the thermal energy storage system to stock up the energy for a longer time that can be utilized for the evening or late night applications. This process continued until the oil temperature reaches around atmospheric condition. The time taken by the HTF to attain atmospheric condition was noted at regular intervals of 10 minutes during the Discharging mode.

The results obtained from the experimental work is compared with ANSYS output. Initially, the 3-D model of the setup is drawn with Solid Works and then imported to ANSYS window. The following inputs are given to perform the analysis.

- Input temperature- The input temperature is 90°C and it is applied on the surface of the walls of the heater.
- Time- The time mentioned here is 10 minutes which is represented in seconds (600).
- Rate of flow- The rate of flow of oil is supposed to be mentioned and it is mentioned as 26 LPM.
- Thermal coefficient of oil- The thermal coefficient of oil is 5*10-6W/m^{2}K.

The meshing is generated; divide a particular part into a number of nodes. The advantage of this meshing is that it can give an exact result. The meshed generated multi-temperature thermal energy storage system design is shown in the figure 2.

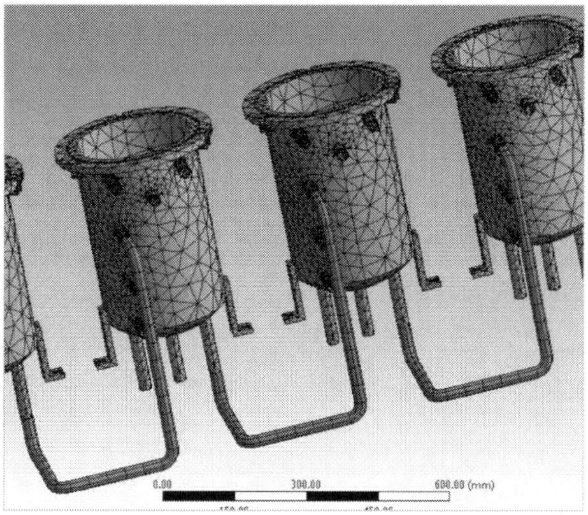

Figure 2. Mesh generated TES.

Thermal analysis is used to find out the temperature distribution and related thermal quantities in the model. The temperature distribution, sum of heat loss or gain, thermal gradients- It is the rate of change of temperature per unit in a material, thermal fluxes- it is defined as the rate of energy exchange per unit cross sectional area.

3. Results and argument

The performance of multi-temperature thermal energy storage system was analyzed with brass, aluminium, and copper encapsulations. The values obtained from the analysis are compared with experimental results.

3.1. Analysis of brass encapsulations with different PCMs

In stage 1 brass material is used to encapsulate the PCMs. The tank1 is occupied with 9 spherical brass encapsulations where D-mannitol is used as a PCM. Similarly tank 2 is occupied with 9 spherical brass encapsulations where D-sorbitol is used as a PCM. Finally tank 3 is occupied with 9 spherical brass encapsulations where paraffin wax is used as a PCM.

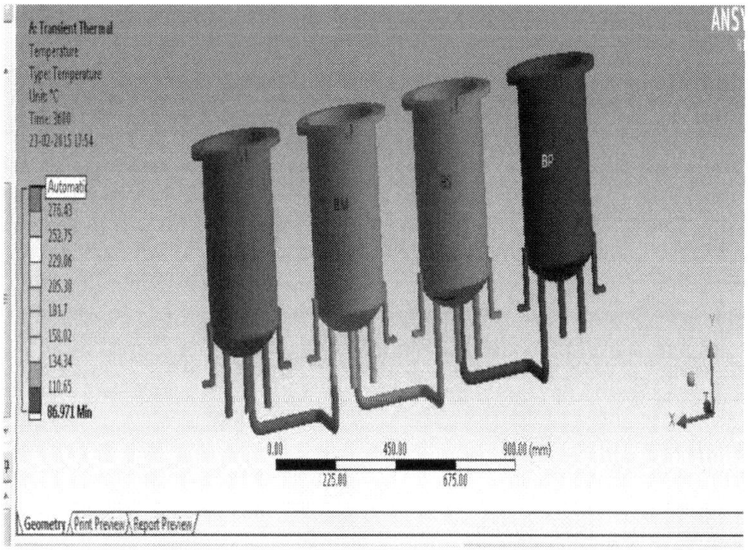

Figure 3. Temperature distribution with brass encapsulation.

The image presented in figure 3 represents the temperature distribution for brass with different encapsulations (D-Mannitol, D-Sorbitol, and Paraffin Wax). It is clearly seen that the highest temperature occurs in tank 1 which has Mannitol as a PCM in it. The temperature ranges from 120°C to 125°C (approximately). This tank is represented as BM, Brass encapsulation with D-Mannitol as PCM. The second tank which consists of D-Sorbitol has a slightly lower temperature distribution compared to that of the first. The temperature range is around 110°C to 120°C. This tank is represented as BS, Brass encapsulation with Sorbitol as PCM. The last tank consisting of Paraffin wax has the lowest range of temperature of less than 90°C. This tank is represented as BP, Brass encapsulation with Paraffin Wax as PCM.

3.2. Analysis of Aluminium encapsulation with different PCMs

In stage 2 aluminium is used as spherical encapsulation material. The tank 1 is occupied with 9 spherical aluminium encapsulations where D-mannitol is used as a PCM. Similarly tank 2 is occupied with 9 spherical aluminium encapsulations where D-sorbitol is used as a PCM. Finally tank 3 is occupied with 9 spherical aluminium encapsulations where paraffin wax is used as a PCM.

The figure 4 indicates the temperature distribution of aluminium encapsulations with different PCMs. It is found that the first tank which consists of D-Mannitol in it possesses the maximum temperature of more than 125°C which is a touch higher than brass encapsulation. This tank is represented as AM, aluminium encapsulation with D-Mannitol as PCM. The temperature distribution across the second tank having D-Sorbitol in it has an approximate temperature of 115°C or even less than that which is comparatively higher than the same encapsulation in Brass. This tank is represented as AS, aluminium encapsulation with D-Sorbitol as PCM. The last tank which consists of Paraffin wax possesses the minimum temperature of 89°C (approximately). This tank is represented as AP, aluminium encapsulation with paraffin Wax as PCM.

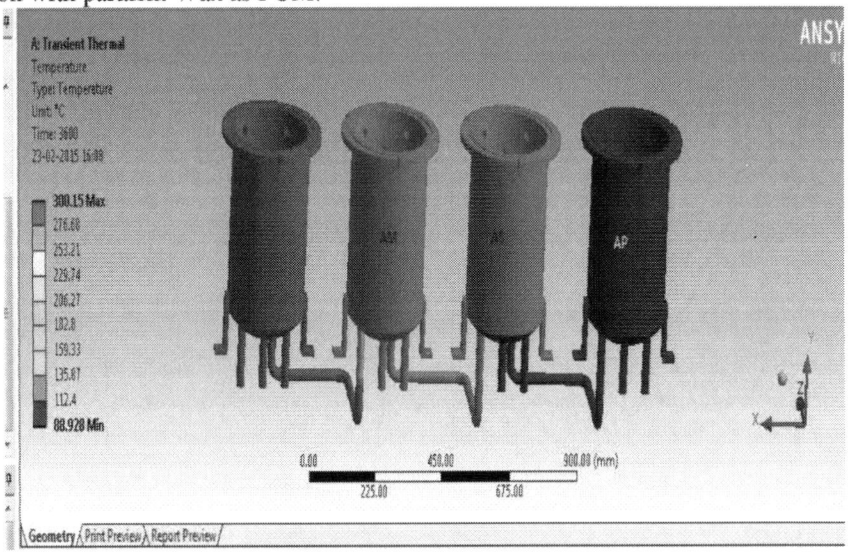

Figure 4. Temperature distribution with aluimium encapsulation.

3.3. Heat transfer effect between fixed bed and fluidized bed LHTES system

In stage 3 copper is used as spherical encapsulation material. The tank 1 is occupied with 9 spherical copper encapsulations where mannitol is used as a PCM. Similarly tank 2 is occupied with 9 spherical copper encapsulations where sorbitol is used as a PCM. Finally tank 3 is occupied with 9 spherical copper encapsulations where paraffin wax is used as a PCM.

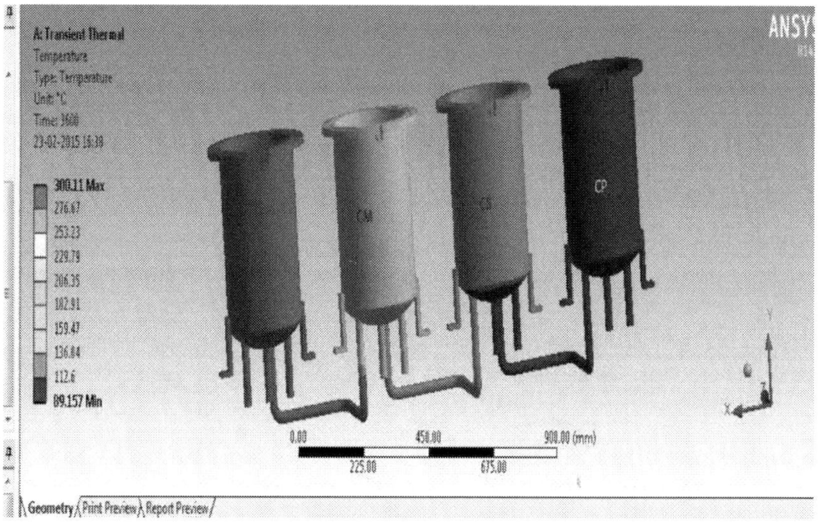

Figure 5. Temperature distribution with copper encapsulation.

The temperature distribution of copper encapsulations with different PCM is shown in the figure 5. It can be an easily concluded that copper would be the ideal material as it possesses the maximum temperature of even more than 150°C in the first tank which consists of D-Mannitol under same conditions. This tank is represented as CM, copper encapsulation with D-Mannitol as PCM. However, there is a slight drop in temperature in the second tank which possesses a temperature range from 112°C to 130°C which has D-Sorbitol in it. This tank is represented as CS, copper encapsulation with D-Sorbitol as PCM. The final tank however possesses a healthy 90°C which is still the highest compared to the other two. This tank is represented as CP, copper encapsulation with paraffin wax as PCM

3.4. Comparison between ANSYS solutions and experimental values – Brass encapsulation

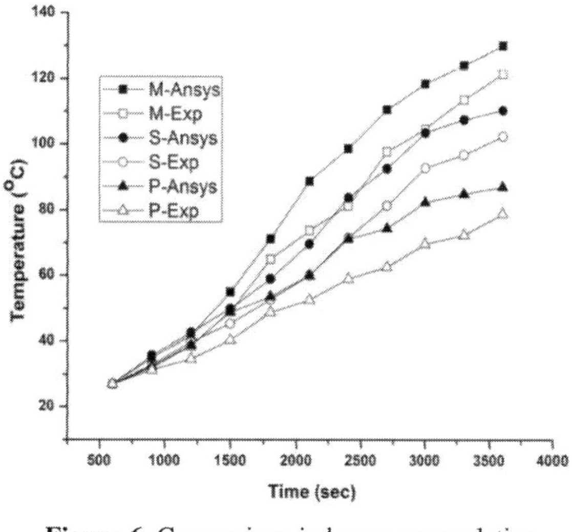

Figure 6. Comparison in brass encapsulation.

The figure 6 represents the comparison of temperature distribution when brass encapsulation filled in all tanks. M represents D-mannitol, S represents D-sorbitol, and P represents Paraffin wax. The analysis of D-Mannitol shows that the both Ansys and experimental values tend to follow the same curve with Ansys heading slightly. Although a small variation in the curve is seen exactly after 1800 seconds. The value obtained from the software which has a slightly larger value compared with experimental one. The software value after 1 hour shows it is at 130°C where as the experimental value after 1 hour is 120°C. The analysis of D-sorbitol shows that at the begining, both are started at the same temperature (27°C) which depicts the room temperature. The end temperature of Ansys software after an hour is approximately 110°C. The analysis of paraffin wax shows that the end temperature of Ansys software after an hour is approximately 85°C which is much higher than experimental ones (70°C).

3.5. Comparison between ANSYS solutions and experimental values – Aluminium encapsulation

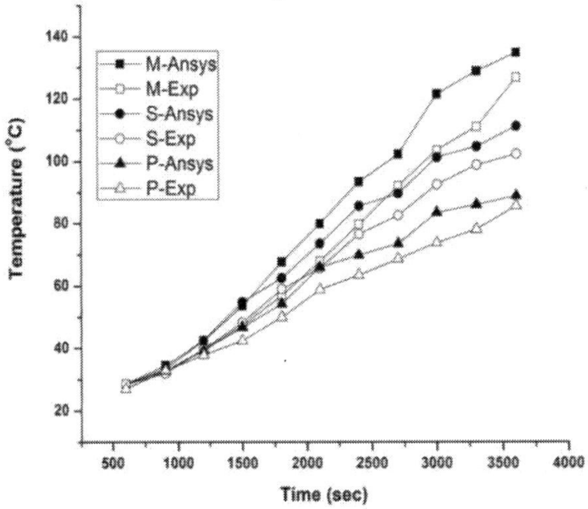

Figure 7. Comparison in alminium encapsulation.

The figure 7 represents the comparison of temperature distribution when alumium encapsulation filled in all tanks. The analysis of D-Mannitol shows that both (Analysis tool and experimental) are started at the same temperature (28°C) which depicts the room temperature. Then both follow a parallel curve where software solution has a higher value than the experimental ones. The end temperature values are approximately 135°C and 125°C in ansys solution and experimental solution respectively. The analysis of D-sorbitol shows that both curves follow the parallel curve where ansys solution has a temperature of 115°C (approximately) compared with experimental value around 102°C. The Ansys temperature is comparatively higher than the end temperature of the brass with D-Sorbitol. Similarly the paraffin wax also follows the same pattern.

3.6. Comparison between ANSYS solutions and experimental values – Copper encapsulation

The temperature distribution with the use of copper encapsulated balls is shown in the figure 8. The ansys solution temperature is comparatively higher than the end temperature of the brass encapsulation and aluminium encapsulation. The analysis of D-sorbitol shows the both experimental and ansys values are varying parallel. The analysis of paraffin wax shows that the end temperature of around 90°C which is higher than experimental values of 85°C.

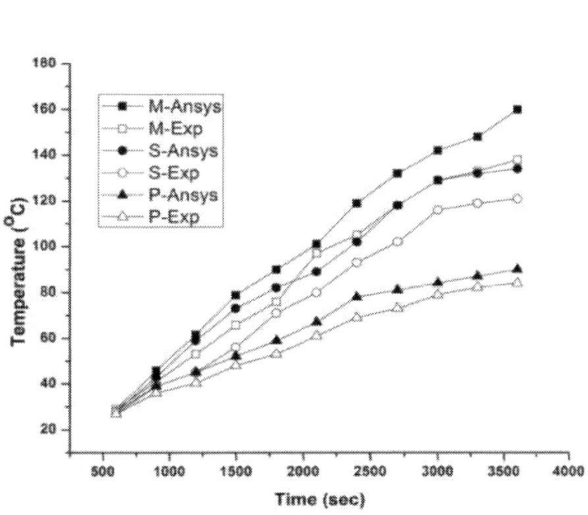

Figure 8. Comparison in copper encapsulation.

Based on the temperatures the overall system efficiency was calculated and the results are showed that the copper encapsulated D-mannitol as PCM has the highest efficiency when compared to other two PCMs namely D-sorbitol and paraffin wax.

4. Conclusion

A detailed survey has been done on the use of solar energy to meet the thermal requirement of different purposes. From detailed study designing of the apparatus according to the standard conditions and specifications are done to increase the efficiency and to reduce the heat loss. Thermal analysis is done on the model and the results were analysed and compared. The results showed that the copper encapsulation with D-Mannitol as a PCM gave the better result compared to other two (Aluminium and Brass). Similarly, D-Mannitol has the highest hidden (latent) heat of fusion compared to the other two; hence the temperature distribution of copper encapsulated with D-Mannitol as PCM reaches a maximum temperature.

References

[1] Simon Ievers and Wenxian Lin 2009 Numerical simulation of three-dimensional flow dynamics in a hot water storage tank*Applied Energy***86** (12) pp 2604-2614.

[2] Zakir Khan, Zulfiqar Khan and Abdul Ghafoor 2016 A review of performance enhancement of PCM based latent heat storage system within the context of materials, thermal stability and compatibility*Energy Conversion and Management***115** pp 132-158.

[3] A Siddique Ahmed Ghias, S Vijay Ananth, M Dev Anand and G Glan Devadhas 2016 Experimental study of thermal performance of coil in shell heat exchanger*Indian Journal of Science and Technology***9**(13): DOI: 10.17485/ijst/2016/v9i13/90571.

[4] Kamal A R Ismail and Mabruk M Abugderah 2000 Performance of a thermal storage system of the vertical tube type. Energy Conversion and Management**41**(11) pp 1165-1190.

[5] G Kumaresan, V S Vigneswaran, S Esakkimuthu and R Velraj 2016 Performance assessment of a solar domestic cooking unit integrated with thermal energy storage system *Journal of Energy Storage* **6** pp 70-79.

[6] Munoz J and Abanades A 2011 Analysis of internally finned tubes for parabolic trough design by CFD tools *Journal of Applied Energy* **88** pp 4139–49.

[7] F Zaversky, M M Rodríguez-García, J García-Barberena, M Sánchez and D Astrain 2014 Transient behavior of an active indirect two-tank thermal energy storage system during changes in operating mode – An application of an experimentally validated numerical model *Energy Procedia* **49** pp 1078-1087.

[8] Flavio Manenti and Zohreh Ravaghi-Ardebili 2013 Dynamic simulation of concentrating solar power plant and two-tanks direct thermal energy storage *Energy* **55** pp 89-97.

[9] Nasser Mostafavinia, Saman Eghvay and Amir Hassanzadeh 2015 Numerical analysis of melting of nano-enhanced phase change material (NePCM) in a cavity with different positions of two heat source-sink pairs *Indian Journal of Science and Technology* **8**(S9) pp 49–61.

[10] Wahiba Yaïci, Mohamed Ghorab Evgueniy Entchev and Skip Hayden 2013 Three-dimensional unsteady CFD simulations of a thermal storage tank performance for optimum design *Applied Thermal Engineering* **60** pp 152-163.

IOP Conf. Series: Materials Science and Engineering **197** (2017) 012040 doi:10.1088/1757-899X/197/1/012040

Charging and Discharging Processes of Thermal Energy Storage System Using Phase change materials

Dr.B.Kanimozhi[1], Kasilanka Harish[2], Bellamkonda Sai Tarun[3], PogakuSatySainath Reddy[4],Padakandla Sai Sujeeth[5]

[1]Faculty of Mechanical Engineering, Sathyabama University, Chennai-119.
[2,3,4,5] Students, Mechanical& Production Engineering Department, Sathyabama University, Chennai-119

Abstract: The objective of the study is to investigate the thermal characteristics of charging and discharge processes of fabricated thermal energy storage system using Phase change materials. Experiments were performed with phase change materials in which a storage tank have designed and developed to enhance the heat transfer rate from the solar tank to the PCM storage tank. The enhancement of heat transfer can be done by using a number of copper tubes in the fabricated storage tank. This storage tank can hold or conserve heat energy for a much longer time than the conventional water storage system. Performance evaluations of experimental results during charging and discharging processes of paraffin wax havediscussed. In which heat absorption and heat rejection have been calculated with various flow rate.
Keywords: Phase Change Materials, Solar Tank, Thermal Energy Storage System, CopperTubes,

1. INTRODUCTION

Efficient and economical technology that can be used to store large amounts of heat in a reduced volume is the subject of research for a long time. PCM plays an important role in energy conservation, which is very attractive because of its high storage density with small temperature change. It has been demonstrated that the development of a latent heat in thermal energy storages system which store heat during peak power operation and release the same during reduced power operation. Phase change material is one of the thermal storage devices. Thermal energy storage system enhanced by encapsulating with suitable PCM materials, within these surfaces heat can absorb or capture solar thermal energy through natural convection. The amount of stored heat energy depends on the specific heat of the medium, the temperature change and the amount of storage material. Latent Heat Storage (LHS) is based on the heat absorption or release when a storage material undergoes a phase change from solid to liquid or liquid to gas or vice - versa. This system provides a valuable solution for correcting the difference between the supply and demand of energy. Many phase change materials has been studied and tested for different practical uses by many scientists. This paper attempts to analyze the application of PCM in thermal storage system. The present work mainly concentrated temperature profile during charging and discharging processes in thermal energy storage system.Here some relevant literature reviews are as follows: Mohammed Mumtaz A.et.al.,[1] discussed efficient thermal energy storage system with comprehensive evaluation of suitable PCM materials in a single-effect solar absorption system affecting its performance. Guruprasad Alva, et.al., analysed [2] synthesis methods of microencapsulated phase change materials, such as physical synthesis methods like spray drying, polymerization, emulsion polymerization, interfacial polymerization, in–situ polymerization and condensation polymerization.Peizhao Lv[3] showed that the paraffin/kaolin composite with the largest particle size of kaolin (K4) has the highest thermal conductivity (0.413 W/(m K) at 20 °C) among the diverse composites. Lingkun Liu,et.al., [4] discussed enhancement techniques for phase change materials for thermal energy system in domestic application.R. Al Shannaq, and M.M. Farid[5]discuss the most important methods of encapsulation applicable to these PCMs. These methods include microencapsulation in capsules ranging in size from nano- to micro-scale, encapsulation in porous particles. Kinga Pielichowska et.al., [6] discussed flame retarding properties. The wide range of PCM applications in the construction, electronic, biomedical, textile and automotive industries is presented and future research directions are indicated.. L.C.Chow.et al [7] studied about two thermal conductivity enhancement techniques.The first technique focuses on placing encapsulated PCM of various shapes in a liquid metal medium. The second technique involves a metal / PCM composite. I. M. Bugaje [8]studied thermal response of paraffin wax contained in plastic tubes and used in latent heat storage systems was enhanced by the use of metal matrices R.Velraj et al.[9] In this paper a detailed investigation of the different heat transfer enhancement methods for the latent heat thermal storage system has been carried out, Enhancement has been done with fin configuration and by Lessing rings are used to increase the storage capacity. Ismail et al.[10] represented a model for simulation of the process of heat transfer (charging and discharging) of a latent heat storage system of packed bed of spherical capsules filled with PCM was developed and solved numerically by using a finite difference approach and moving grid technique. The numerical grid was optimized and the predicted results were compared with experimental measurements to establish the validity of the model V. Arun prasad Raja et al [11] developed numerical simulation method, he has enhanced heat transfer rate for water and air, computational fluid dynamics software (FLUENT) used. In which forced convection heat transfer in fully developed flow were compared with thermally developing laminar flow, it was found that heat transfer coefficients were higher due to small channel spacing and developing laminar flow.The author E. Assis et,al.,[12] presented the process of melting of a phase-change material (PCM) in spherical geometry has been explored experimentally and numerically. The TES unit contains paraffin as phase change material (PCM) filled in spherical capsules, which are packed in an insulated cylindrical storage tank. The water used as heat transfer fluid (HTF) to transfer heat from the solar collector to the storage tank also acts as sensible heat storage material, thermal energy storage system for the use of hot water at an average temperature of 45°C for domestic applications using combined sensible and latent heat storage concept. From the above literature a common problem in latent heat thermal storage is the poor conductivity of the phase change material. During the phase changing process, the surface heat flux decreases due to the increasing thermal resistance of the growing layer of the molten or solidified PCM. Hence, it is necessary to improve the heat transfer rate across the heating surface and phase change materials. Hence smaller diameter of copper tubes are used to improve the heat transfer rate during charging and discharging processes.

2.PHASE CHANGE MATERIALS

The Phase Change Material is the latent heat storage material. As the source temperature rises, thechemical bonds within the PCM breaks, and the material changes its phase from solid to liquid. During the charging process, the material begins to melt when the phase change temperature is reached. The temperature then stays constant until the melting process is finished. The heat that is stored during the phase change process (melting process) of the material is called the Latent Heat. It follows an isothermal behavior during the charging and discharging processes. Here, when paraffin is used as the PCM material, latent heat storage can be achieved through solid-solid, solid-liquid, solid-gas and liquid-gas phase change. However, the only phase change used for the PCM is the solid-liquid change. The PCM continues to absorb heat without a significant rise in temperature, until all the material is transformed to the liquid phase. When the ambient temperature around a liquid material falls, the PCM solidifies and releases its stored latent heat. Table 1, shows the thermo physical properties of paraffin wax

Table.1 Thermo physical Properties of paraffin

Properties		Paraffin wax
Melting point		58 °C
Latent Heat		200-210 kJ/kg
Density		920 kg /m^3
Heat capacity	Solid	2 kJ/kg·K
	Liquid	2.5 kJ/kg·K
Thermal conductivity		0.2 -1.7 W/m·K.

3. FABRICATION OF PCM STORAGE TANK

The fabrication of PCM storage tank is aconcentric cylinder with height of 80 cm in length. The inner cylinder is 35 cm and the outer cylinder is 40 cm in diameters with capacity of 45 litres. It is made with stainless steel.The tank contains numbers of copper tube each tube filled with PCM materials. The storage tank is well insulated and thermally sealed by filling the annulus with glass wool. The solar tank has 100 litres capacity; the solar collector contains 15 numbers of evacuated glass tubes. The length of evacuated glass tube is 138cm, a breadth of 100 cm, and a 4cm clearance. The height of the solar tank from the base of the collector was ensured to be 80 cm. The maximum permissible working pressure maintained in the evacuated tube collector was 0.4kg/cm^2. The outer surface of the inner glass tube is coated with a special solar selective coating, which absorbs and converts the maximum amount of solar radiation into heat. The space between the outer and inner glass tubes is evacuated and permanently sealed off, and the vacuum acts as an excellent insulator. The PCM storage tank has integratedwiththesolartank.PCM storage tank received hot water from solar tank. Solar energy has absorbed and stored in PCM storage tank as latent heat. Large quantity of solar energy can be stored in a day time and same heat can be retrieved for later use. The tank was instrumented to measure inlet and outlet water temperature at eight different levels with R20- Thermocouple arrangement. Flow meters were used to vary the flow rate.

4. EXPERIMENTAL WORKING MODEL

The experimental set-up consisting solar tank with a capacity of 100 litres and PCM storage tank with a capacity of 45 litres; the tank have instrumented to measure the HTF and PCM temperatures with a thermocouple arrangements. Flow meters were fit across the pipe line to measure the flow rates by adjusting the valve position, proper inlet and outlet pipe connections are ensured, as shown in Figure 3. Here, water is used as the Heat transfer fluid and paraffin used as phase change materials. To analyze the temperature variation of the HTF and PCM in the copper tube, the length of the copper tube is divided into four divisions or segments (levels) from the base, each division of in and out of the copper tubes is connected to thermometers with sensors. Inside the copper tube the PCM is presented, and outside the copper tube the HTF is circulated. Hence, we can read the temperature for both the HTF and the PCM at all levels. During the active phase of the sun light the heat energy can be stored in the solar tank through the evacuated glass tube collector. The Heat Transfer Fluid (HTF) from the solar tank is allowed to the PCM storage tank. Then the PCM and HTF temperature are noted every 10 minutes at four levels. During discharging process the cold water is allowed to the PCM tank with ambient temperature again the HTF and PCM temperature are noted every 10 minutes at four levels. The experiment is completed when the PCM and water temperatures are the same. The same procedure is repeated for different mass flow rates.

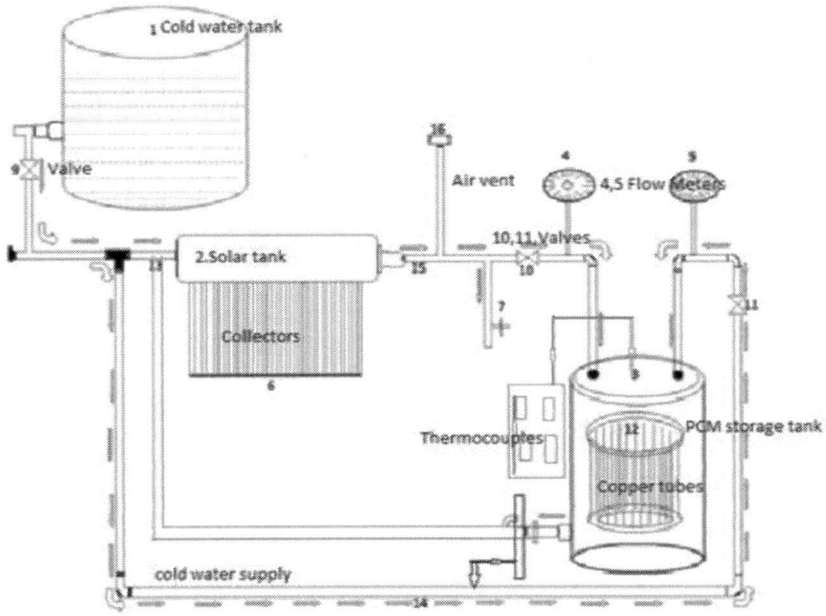

Figure 1. Experimental working model

5. RESULTS AND DISCUSSION

5.1. Charging and discharging process of paraffin wax

The charging process was carried out by applying HTF at 76^{0}C with a different flow rate of 6 kg/min, 4kg/min, and 2kg/min to the PCM storage tank. During the charging process the HTF is circulated continuously through the PCM storage tank. The Heat Transfer Fluid (HTF) exchanges its energy to copper tubes. At the beginning of the charging process, the temperature of the paraffin inside the tubes will be ambient temperature, which is lower than the melting temperature. Initially the energy is stored inside the tubes as sensible heat until the paraffin reaches its melting temperature. As the charging process proceeds, energy storage is achieved by melting the paraffin at a constant temperatureand stored as latent heat. Then the charging process is continued until the paraffin temperature reaches the value of 76^{0}Cenergy is stored as sensible heat in liquid paraffin. The discharging processes were carried out by applying the cold water at ambient temperature, which is circulated continuously through the storage tank to recover the stored heat energy. The following Figures 4 and 5 show the charging and discharging processes of paraffin wax with respect to time and temperatures.

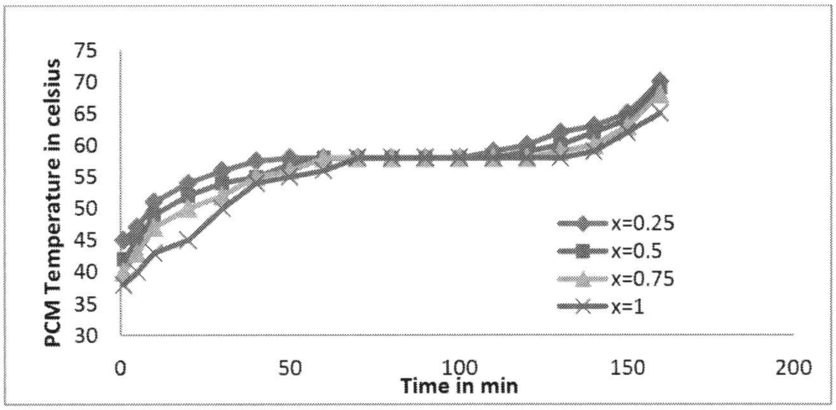

Figure 2.Charging process of paraffin wax

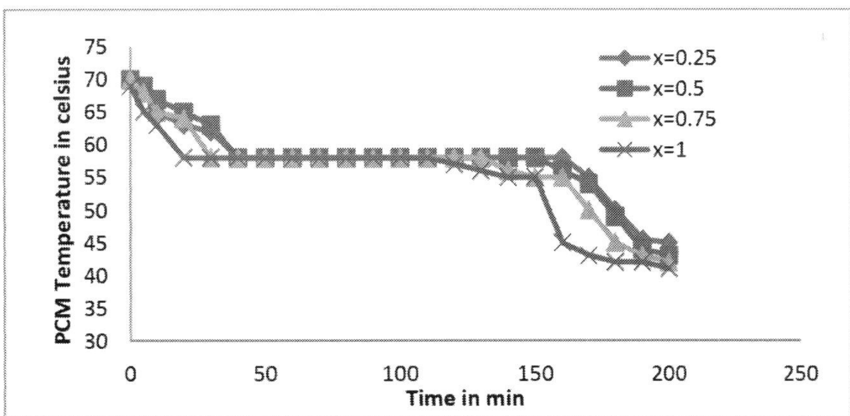

Figure 3.Discharging process of paraffin wax

The observation made from the experimental results that charging process takes 160 minutes and discharging process takes 200 minutes when the flow rate has 6kg/min. likewise for 4kg/min charging process takes 180 minutes and discharging process takes 210, and for 2kg/min the charging process takes 200 minutes and discharging process takes 220 minutes. The above results indicate that when flow rate is increase time consumption has decreases. However due to the smaller diameter of copper tube arrangement, the discharging time is reduced significantly compared with commercial thermal storage system.

6. CONCLUSION

It is concluded from the experimental results that the Phase Change Materials (PCM) plays an important role in energy conservation, which is very attractive because of its high storage density with small temperature change. Phase change material is one of the thermal energy storage materials. It has been demonstrated that the development of latent heat in thermal energy storage systems, which stores heat during peak power operation and releases the same during reduced power operations. In the fabricated storage tank heat can transfer from the HTF to the PCM as well as from the PCM to the HTF during charging and discharging processes. During the discharging process heat can be retrieved after a long interval, without losing any considerable losses. Summation of charging and discharging processes of latent heat storage material of paraffin wax has considerably higher thermal energy storage densities materials, and able to absorb or release large quantities of energy at a constant temperature by undergoing a change of phase.

REFERENCES.

[1] Mohammed Mumtaz A. Khan, R. Saidur, Fahad A. Al-Sulaiman A review for phase change materials (PCMs) in solar absorption refrigeration systems Renewable and Sustainable Energy Reviews, Volume 76, September 2017, Pages 105-137
[2] Guruprasad Alva, Yaxue Lin, Lingkun Liu, Guiyin Fang,Synthesis, characterization and applications of microencapsulated phase change materials in thermal energy storage Energy and Buildings online march 2017
[3] Peizhao Lv, Chenzhen Liu, Zhonghao RaoExperiment study on the thermal properties of paraffin/kaolin thermal energy storage form-stable phase change materials,Applied EnergyVolume 182, 15 November 2016, Pages 475–487
[4] Lingkun Liu, Di Su, Yaojie Tang, Guiyin Fang" Thermal conductivity enhancement of phase change materials for thermal energy storage: A review,Renewable and Sustainable Energy ReviewsVolume 62, September 2016, Pages 305–317
[5] R. Al Shannaq, M.M. Farid10 – Microencapsulation of phase change materials (PCMs) for thermal energy storage systems,Advances in Thermal Energy Storage Systems
[6] Kinga Pielichowska, Krzysztof Pielichowski,Phase change materials for thermal energy storage,Progress in Materials ScienceVolume 65, August 2014, Pages 67–123
[7] Chow Zhong J.K. and Beam J.E. (1996), "Thermal conductivity enhancements for phase change storage media", International Journal of Heat Mass Transfer, Vol. 23, No. 1, pp. 91-100.

[8] M. Bugaje.,Enhancing the thermal response of latent heat storage systems, international journal of energy research, vol. 21, 759ð766 (1997)
[9] Velraj R., et al.(1997), heat transfer enhancement in a latent heat storage system. Solar energy Vol.65, No.3, pp.171-180.
[10] Ismail and R. Henr_ıquez Numerical and experimental study of spherical capsules packed bed latent heat storage system Applied Thermal Engineering 22 (2002) 1705–1716.

[11] Arun Prasad Raja et al.(2005), heat transfer and fluid flow in a constructal heat exchanger. Engineering conferences International, Hoboken, NJ, USA.
[12] E. Assis et,al.,(2007) Numerical and experimental study of melting in a spherical shell International Journal of Heat and Mass Transfer 50 (2007) 1790–1804.
[13] Chow, J.K.Zhong et al. (1996), thermal conductivity enhancement for phase change storage media. Heat mass transfer, Vol.23,No.1, pp.91-100.

IOP Conf. Series: Materials Science and Engineering **197** (2017) 012041 doi:10.1088/1757-899X/197/1/012041

Experimental investigations of hybrid PV/Spiral flow thermal collector system performance using Al_2O_3/water nanofluid

R Gangadevi [1], B K Vinayagam[2], S Senthilraja [3]

[1]Assistant professor, [2]Professor, [3]Assistant professor, [1, 2, 3] Department of Mechatronics Engineering, SRM University, Chennai, India

Email: gangadevi.r@ktr.srmuniv.ac.in

Abstract. In this paper, the PV/T (Photovoltaic thermal unit) system is investigated experimentally to examine the thermal, electrical and overall efficiency by circulating Al_2O_3/water nanofluid of 1wt% and 2wt% with an optimum flow rate of 40L/H. The overall efficiency of PVT system is largely influenced by various factors such as heat due to photovoltaic action; energy radiated at the infrared wavelength of the solar spectrum, solar irradiance, mounting structure, tilt angle, wind speed direction, Ambient temperature and panel material composition. However, the major factor is considered in this study to extract the heat generated in the PV panel by using nanofluid as a coolant to increase the overall system efficiency. Therefore, the result shows that by using 2 wt% Al_2O_3/water nanofluid the electrical efficiency, thermal efficiency and overall efficiency of the PVT system enhanced by 13%, 45%, and 58% respectively compared with water.

Keywords: PVT system, electrical efficiency, thermal efficiency, Al_2O_3/water nanofluid,

1. Introduction

Hybrid PVT System absorbs the direct solar radiation and converts into electrical and thermal energy. In this system solar cell, converts sunlight into electrical energy i.e electricity, and the thermal collector captures the remaining energy and removes heat generated by the PV panel. Matin Ghadidri et al., [1] investigated the performance of PVT system using ferrofluid as a heat transfer medium. They reported that the overall system efficiency is about 76% by using ferrofluid of 3wt% as a cooling medium. Michael et al.,[2] analyzed that the copper sheet laminated PV/T system using 0.05 % copper oxide –water nanofluid .the result revealed that the thermal efficiency of 45.76%. Mohammad Sardarabadi et al., [3] investigated experimentally the effect of Silica/water on PV/T system. The result revealed that the thermal efficiency of 7.6% and 12.8% in case of 1wt% and 3wt% and the total exergy efficiency is about 24.31% at 3 wt% of Silica-water nanofluid respectively. Ying-Ying Wu et al., [4] studied the performance of photo electric hybrid system. The result showed that the unglazed PVT gives higher system performance by using nanofluid compared with water.

Wei He et al., [5] studied the performance of a thermoelectric cooling and heating system in summer and winter operation modes. The result revealed that the COP of the system reaches 1.7 and the electrical and thermal efficiency increases into 16.7% and 23.5% respectively. Khelifa et al., [6] analyzed the hybrid PV/T system using Computational Fluid Dynamics (CFD). Adnan Ibrahim et al.,

IOP Conf. Series: Materials Science and Engineering **197** (2017) 012041 doi:10.1088/1757-899X/197/1/012041

[7] presented the types of flat plate PV/T collector and its performance is evaluated by using water, air and both. Liang et al., [8] used TRNSYS simulation software for Dynamic Simulation of Hybrid Photovoltaic/Thermal Collectors. Tyagi et al., [9] presented the overview of research and development on hybrid PV/T system. Jarimi et al., [10] discussed the theoretical and experimental studies of a bi-fluid PV/T solar collector and 2D steady-state analysis is developed. The simulation result showed that the good agreement.

2. Preparation of nanofluids

The Al_2O_3 nanoparticle (size of 50nm) at low concentration of 1wt% and 2wt% are dispersed in ionized water. Figure 1 shows that the Ultrasonic probe sonicator device which operates at higher frequency and amplitude may cause heating. Therefore the nanofluid is kept in a separate container surrounded by ice cubes to prevent fluid evaporation at a higher temperature. Finally, the probe is immersed in the nanofluid to disperse the nanoparticle uniformly in the water, and the dispersion characteristics of the nanofluid are studied by a scanned electron microscope. From the SEM image Figure 2 a, b. It is clearly observed that before sonication the particles were not evenly dispersed after sonicating for 1 hour the particles are well dispersed as shown in Figure 2 c, d. The nanoparticle average diameter as measured as less than 50nm and prepared two different mass fractions of nanofluid and it was stable for 13 days without sedimentation as shown in Figure 1 a.b.

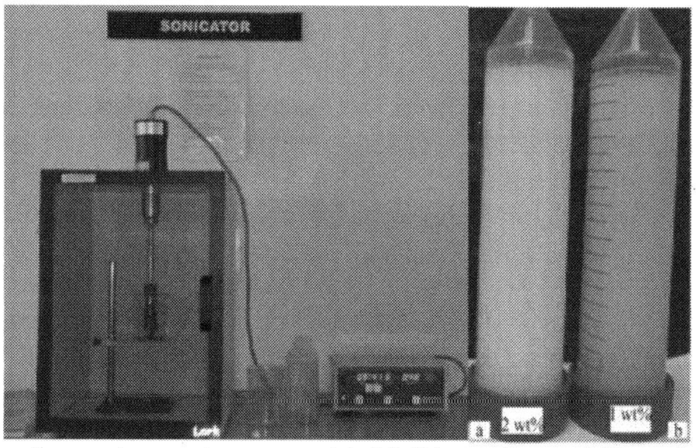

Figure 1. Ultrasonic probe sonicator Apparatus,(a),(b) Dispersion stability of prepared Al_2O_3 /water nanofluid of 2% and 1wt%

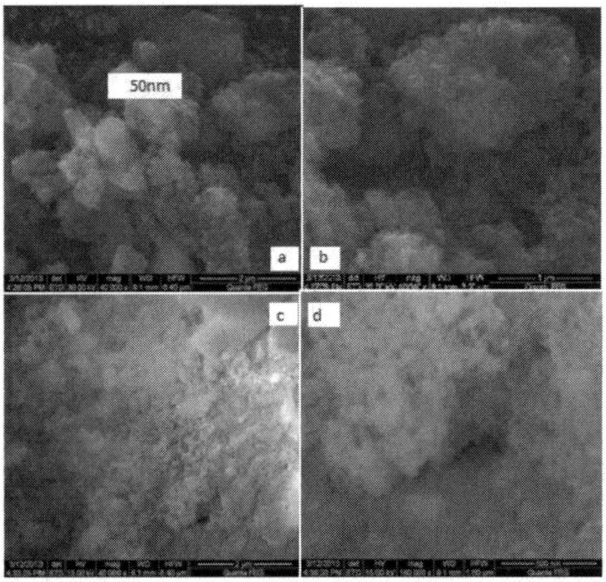

Figure 2. SEM image of Al_2O_3 /water nanofluid (a, b) 1wt% and 2 wt% before sonication (c,d) 1 wt% and 2wt% after sonication

3. Experimental setup

Figure. 3 shows the experimental setup of Spiral flow hybrid PV/T system, which consists of solar panel equipped with a spiral flow thermal collector to collect the generated heat from the PV Panel. The thermal collector made in this study is rectangular spiral type as shown in Figure 4. The PV Panel area is almost covered by the collector design to minimize the heat loss and also to improve the system electrical output. The outlet of the thermal collector is attached to a heat exchanger which is placed in city water supply tank. This transfers the heat to the cold water and then circulated back to the thermal collector. The process will continue to remove the heat generated from the PV panel. The water tank consists of two level sensor and one temperature sensor (RTD) which sense the level of water and its temperature respectively. On the other hand, two RTD is placed to measure the inlet temperature and outlet temperature of the collector for analyzing the thermal efficiency of the system. The flow meter and flow control valve are fixed and controlled to the optimum flow rate of 40LPH. The ON/OFF controller receives the signal from the temperature sensor to maintain the temperature of 60^0 in city water tank for hot water purpose. The hot water is drained continuously once the temperature reaches the set value. However, nanofluid is constantly circulated to remove the heat from the PV panel to improve the overall system performance. The characteristics of PV panel used in this experiment as shown in Table 1.

IOP Conf. Series: Materials Science and Engineering **197** (2017) 012041 doi:10.1088/1757-899X/197/1/012041

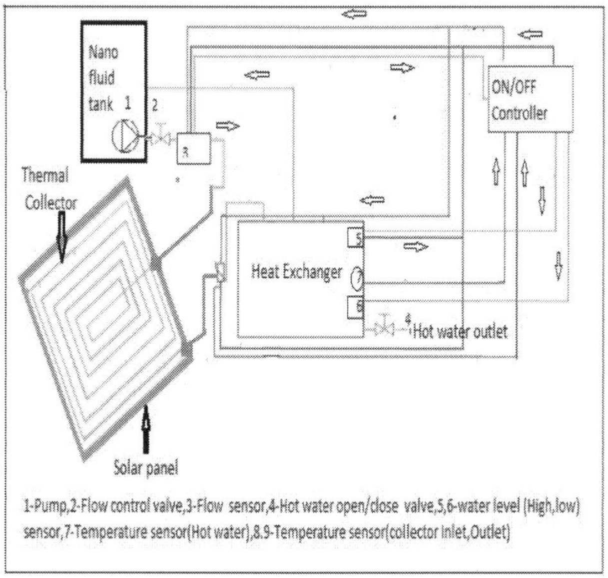

Figure 3. Schematic diagram of spiral flow hybrid PV/T system

Figure 4. Spiral flow thermal collector

Table 1: Characteristics of Poly- crystalline PV panel

Description	**Specification**
Model	Tps107s-50w-poly
Maximum power(P_{max})	50W (+-5%)
Rated voltage	17.2V
Rated current	2.91A
Open circuit voltage	21.5V
Short circuit current	3.25A
Maximum system voltage	1000V
Test conditions	Ambient 1.5, 1000w/m^2

4. Efficiency calculation of PV/T System

Solar radiation is absorbed by the PVT system, and it converts into a useful form of electrical and thermal output. The conversion efficiency of PV/T system depends on many factors which affect the overall system efficiency. The overall efficiency of the system is denoted by η_{pvt} is equal to the sum of the electrical and thermal output divided by the total solar radiation incident on the PV panel per unit area [11, 12]. The overall efficiency (η_{pvt}) of PVT system can be expressed as

$$\eta_{pvt} = \frac{\dot{E}_{Thermal} + \dot{E}_{Electrical}}{\dot{E}_{Input}} \tag{1}$$

$$\eta_{pvt} = \frac{\int_{t1}^{t2} (A_c \dot{E}_{Thermal} + A_{pv} \dot{E}_{Electrical}) dt}{A_c \int_{t1}^{t2} G dt} \tag{2}$$

$$\eta_{pvt} = \left(\dot{E}_{Thermal} + r.\dot{E}_{Electrical} \right) \tag{3}$$

Where, 'A_c' is the area of thermal collector in mm^2, 'A_{pv}' is the area of PV panel in mm^2, 'r' is the Packaging factor defined as A_c/A_{pv}, $E_{thermal}$ is the the rate of thermal output power per unit area of the collector in Watts, $E_{electrical}$ is the rate of electrical output power per unit area of the pv panel in watts 'G' is the effective solar radiation per unit area of PVT system in kW/m^2.
$E_{thermal}$ can be expressed by using Equation (2)

$$\dot{E}_{thermal} = \left(\dot{m}_f - c_{pf} (T_{f.o} - T_{f.i} \right) \tag{4}$$

Where, $\dot{m}_f$ -mass flow rate of the fluid (kg/sec) C_{pf} -specific heat of fluid (J/kg k), T_{fl}-Inlet temperature of fluid (^{0}C), T_{fo} -the outlet temperature of the fluid (^{0}C).
The electrical efficiency of PV system can be expressed by

$$\eta_{electrical} = \frac{\dot{E}_{Electrical}}{\dot{E}_{Input}} \tag{5}$$

$$\eta_{electrical} = \frac{V_{OC} * I_{SC} * FF}{A_C G} \tag{6}$$

where V_{oc}-open circuit voltage (V), I_{sc}-short circuit current (A), FF-Fill factor. FF can be defined as the quality of semiconductor cell junction and how well the solar cell can collect the carrier generated light from the sun [13].

$$FF = \frac{P_{max}}{V_{OC} * I_{SC}} \tag{7}$$

Maximum output power of PV panel is expressed by

$$P_{max} = V_m * I_m \tag{8}$$

$$\dot{E}_{el \cdot Thermal} = \frac{\dot{E}_{Electrical}}{C_f} \tag{9}$$

The PV/T system, the C_f value between 0.35 and 0.40 is reported [14]. The average value of 0.4 is applied in this study to determine the overall efficiency of the PV/T system.
Overall PV/T system efficiency can be modified from equation (3)

$$\eta_{pvt} = \left(\dot{\eta}_{Thermal} + r.\frac{\dot{\eta}_{electrical}}{0.4} \right) \tag{10}$$

The thermal efficiency is expressed as

$$\eta_{th} = \left(\frac{m\dot{C}_p \Delta T}{A_p G_T} \right) \tag{11}$$

m- mass flow rate kg/ sec, C_P-specific heat of the collector cooling medium (J/ kg ^{0}C), ΔT-Difference between fluid or air outlet temperature (^{0}C), A_p-Area covered by absorber collector (m^2) and G_T-Total solar radiation.

5. Result and Discussion

The experiment was conducted with latitude, longitude and tilt angle of 12.8, 80.0 and 32 degrees respectively. In a constant solar radiation (800W/m^2) a series of experiment was conducted by circulating pure water, Al_2O_3/water with 1wt% and 2 wt% with different flow rate to determine the optimum mass flow rate. From these experiments, the result shows that the highest system efficiency attained at 40L/H as shown in Figure 5.

5.1 Effect of temperature

The variations of PV panel temperature and collector outlet temperature for 1 wt% and 2 wt% is illustrated in Figure 6. The data was measured and collected from 9.30 to 14.30 for some selected days with that the Temperature of the PV and PV/T system was studied. The result shows that the PV panel temperature increased about 70.4 degrees. By circulating nanofluid with 1 wt% and 2 wt% the temperature decreases into 30 degrees and 32 respectively for the mass flow rate of 40L/H.

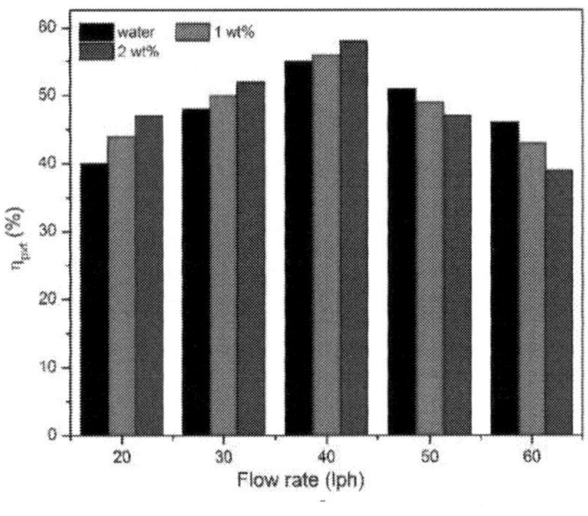

Figure. 5 Overall efficiency of PVT system with different mass flow rate

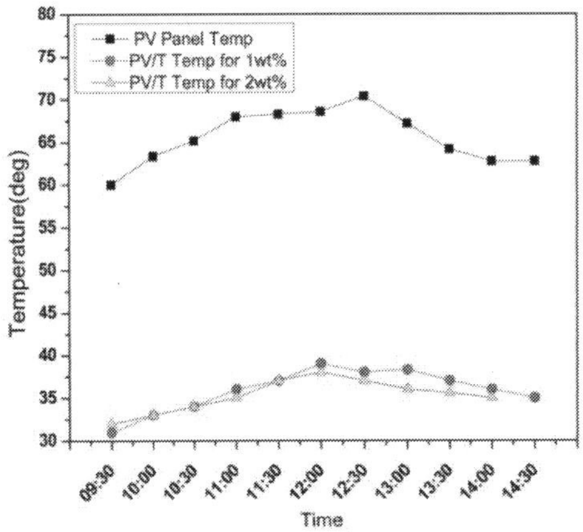

Figure 6. Variation of PV panel temperature and collector outlet temperature of a PV/T system in case of 1wt% and 2 wt% of Al_2O_3/water nanofluid.

5.2 Effect of Electrical and Thermal efficiency

The experiment result shows that the electrical efficiency of the system obtained 10% and 12% respectively by circulating water and 1wt% nanofluid. The highest electrical efficiency of 13% was obtained by circulating 2 wt% of Al_2O_3/Water nanofluid as shown in Figure 7. The more heat is extracted from the PV panel and hence increases the system electrical output.

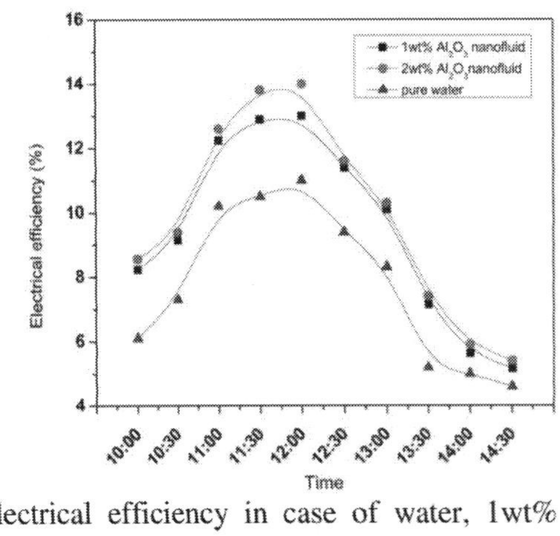

Figure 7. Variation of Electrical efficiency in case of water, 1wt% and 2 wt% of Al_2O_3/water nanofluid.

The thermal efficiency was also calculated by measuring the temperature at the inlet and outlet of the thermal collector. The experiment result shows that the highest thermal efficiency was obtained of 45% by circulating 2wt% of the nanofluid as shown in Figure 8. The thermal efficiency enhancement also depends on the thermal insulation material used to design a hybrid system. The high insulation material is the best choice to reduce the heat losses from the thermal collector to increase the overall thermal output.

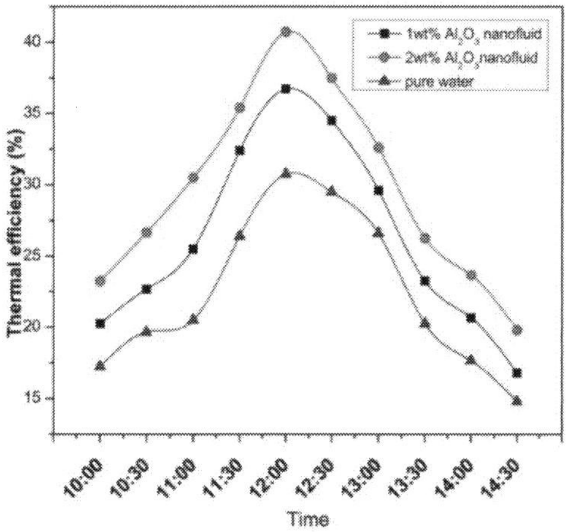

Figure 8. Variation of thermal efficiency in case of water, 1wt% and 2 wt% of Al_2O_3/water nanofluid.

6. Conclusion

In this study, the PV/T system was experimentally investigated with water, 1wt% and 2 wt% of Al_2O_3/water nanofluid. The hybrid PV/T system overall performance totally depends on the suspension sustainability of the coolant used. In this study Al_2O_3 nanoparticle is used which possess very good heat transfer characteristics when mixes with water and suspended uniformly by sonication process. The result shows that the PV panel temperature increased up to 70 degrees which may cause the reduction of PV panel life. When circulating the 2 wt% Al_2O_3/water nanofluid the temperature of the PV panel decreased into 36 degrees. The variation of electrical, thermal and overall efficiency also clearly observed. The result shows that by using 2 wt% Al_2O_3/water nanofluid the electrical efficiency, thermal efficiency and overall efficiency of the PVT system enhanced by 13%, 45%, and 58% respectively compared with water and 1wt% of Al_2O_3/water nanofluid.

References

[1]. Matin Ghadiri, Mohammad Sardarabadi , Mohammad Pasandideh-fard, Ali Jabari Moghadam, 2015, Energy Conversion and Management, Experimental investigation of a PVT system performance using nano Ferrofluids" 103 ,Pp468–476.

[2]. Jee Joe Michael , Iniyan, 2014, Solar Energy, "Performance analysis of a copper sheet laminated photovoltaic thermal collector using copper oxide – water nanofluid", 119 , Pp439–451.

[3].Mohammad Sardarabadi, Mohammad Passandideh-Fard, Saeed Zeinali Heris,2014, Energy, "Experimental investigation of the effects of silica/water nanofluid on PV/T (photovoltaic thermal units)", 66 ,Pp 264 - 272

[4]. Ying-Ying Wu, Shuang-Ying Wu, Lan Xiao,2015, Energy Conversion and Management, "Performance analysis of photovoltaic– Thermoelectric hybrid system with and without glass cover", 93 ,Pp151–159.

[5]. Wei He , JinZhi Zhou , Chi Chen , Jie Ji,2014, Energy Conversion and Management, "Experimental study and performance analysis of a thermoelectric cooling and heating system driven by a photovoltaic/thermal system in summer and winter operation modes", 84 , Pp 41–49.

[6]. Khelifa A, Touafek K, Ben Moussa H, Tabet I, Ben cheikh E hocine H, Haloui H, 2015, Energy Procedia,"Analysis of a Hybrid Solar Collector Photovoltaic Thermal (PVT)", , 74 ,Pp 835 – 843

[7]. Adnan Ibrahim , Mohd Yusof Othman, Mohd Hafidz Ruslan, Sohif Mat, Kamaruzzaman Sopian, 2011, Renewable and Sustainable Energy Reviews, "Recent advances in flat plate photovoltaic/thermal (PV/T) solar collectors". 15,Pp 352–365.
[8]. Ruobing Lianga,, Jili Zhanga, Chao Zhoua, 2015, Procedia Engineering, "Dynamic Simulation of a Novel Solar Heating System Based on Hybrid Photovoltaic/Thermal Collectors (PVT)", 121,Pp675 – 683
[9].Tyagi V V, Kaushik S.C., Tyagi S.K., 2012 Renewable and Sustainable Energy Reviews, "Advancement in solar photovoltaic/thermal (PV/T) hybrid collector technology", 16,Pp 1383– 1398
[10]. Hasila Jarimi, Mohd Nazari Abu Bakar, Mahmod Othman, Mahadzir Hj Din, 2016, Renewable Energy, "Bi-fluid photovoltaic/thermal (PV/T) solar collector: Experimental validation of a 2-D theoretical model", 85 Pp1052- 1067
[11]. Chow T T, Pei G, Fong K F, Lin Z, Chan A L S, Ji J. 2009, Applied Energy,"Energy and exergy analysis of photovoltaic thermal collector with and without glass cover", , 86,Pp 310–316.
[12]. Garg H P, Agarwal R K, 1995, Journal of Energy Conversion and Management,"Some aspects of a PV/T collector forced circulation flat plate solar water heater with solar cells", , 36 ,Pp87–99.
[13]. Chenming H, White R M, 1983, New York: McGraw-Hill,"Solar cells from basic to advanced systems" .
[14]. Kumar R, Rosen M A, 2011, Applied Thermal Engg,"Performance evaluation of a double pass PV/T solar air heater with and without fins", 31,Pp1402–1410.

IOP Conf. Series: Materials Science and Engineering **197** (2017) 012042 doi:10.1088/1757-899X/197/1/012042

DESIGN AND STATIC ANALYSIS OF AIRLESSTYRE TO REDUCE DEFORMATION

Nibin JacobMathew[1] Dillip Kumar Sahoo[2] E.MithunChakravarthy[3]

[1, 3] U.G Students, Dept. of automobile Engg Sathyabama University, Chennai
[2] Asst. professor, Dept. of automobile Engg. Sathyabama University, Chennai
Email-dillipkumarsahoo@gmail.com

Abstract: In this work a model of an air less tire is introduced with a replacement of natural rubber materials in place of synthetic rubber in tread and polyester in place of nylon in carcass. The construction and material study of various types of air less tyre is done by comparing with pneumatic tire. A brief structural study has been done on spokes of airless tyre and analyzed by ANSYS software. Analysis has been carried out on various structures like honey comb, Spokes, triangular and diamond with an applied load of 1200N. Comparison study has been carried out among various structures with different materials and it study shows that tyre with diamond structure with synthetic materials gives less deformation compared to other structure

Keywords— Tyrestructures, honeycomb, triangle, spokes,diamond and composite materials.

1. Introduction

Airless tyres or Non-pneumatic tyres area unit the tyres that aren't supported by atmospheric pressure. These tyres are known as Tweel that could be a merger of the words tyre and wheel. This is as a result of the Tweel doesn't use a conventional wheel hub assembly. The Tweel construct was initial declared by Michelin back in 2005. Its structure may be a solid inner hub mounted onto the vehicles shaft that's encircled by polyurethane spokes. This forms a pattern of wedges that facilitate to soak up the impacts of the road. These spokes look almost like those found on bicycles and plays the shock-absorbing role of the compressed gas as in an exceedingly ancient tyre. A sheer band is then stretched across the spokes that forms the environs of the tyre. It is the strain of the band and therefore the strength of the spokes that replaces the gas pressure used on ancient tyres. An airless tyre is created with differing types of spokes tension that so can enable handling varied styles of characteristics. The non-pneumatic tyres can be viewedtohave great positive implications when designed. Theinclusionoftheairlesstyresintothevehicleswillensureustheleastpossibility of blowout to occur in its performance. Addingtothe advantages that is stated forward by thenon-pneumatictyres(airless tyres), this also provides an environmentalbenefitby its usage. These tyres will never go flat and also canberetreaded, by which they never need to thrown away as inthecaseofthepneumatictyresingeneral.Hencelandfilliscutdown to a great extent contributing to theenvironmentbetterment. According to Amir Gasmi, Paul F. Joseph The tire model consists of a thin flexible annular band and spokes that connect the band to a rigid hub. The circular band is modeled exploitation recurvate beam theory that takes under consideration deformations because of bending, cutting off and circumferential extension. The impact of the spokes, which are distributed unceasingly within the model and act as linear springs, is accounted for less than in tension that introduces a nonlinear response. The quasi-static, two-dimensional analysis focuses on however the contact patch, vertical tire stiffness and rolling resistance are laid low with the stiffness properties of the band and therefore the spokes.

II. Deformation of air less tyre

The unventilated tyre (Tweel) doesn't use a conventional wheel hub assembly. A solid inner hub mounts to the shaft and is encircled by polymer spokes panoplied in a very pattern of wedges. A shear band is stretched across the spokes, forming the fringes of the tyre. On it sits the tread, the half that comes in touch with the surface of the road. The cushion shaped by the air cornered within a standard tyre is replaced by the strength of the spokes that receive the strain of the shear band. Placed on the shear band is that the tread, the half that produces contact with the surface of the road. When the Tweel is running on the road, the spokes absorb road defects identical manner atmospheric pressure will within the case of gas tyres. The versatile tread and shear bands deform briefly because the spokes bend, then quickly return to the initial form. Totally different spoke tensions may be used,PRN by the handling characteristics and lateral stiffness may vary. However, once created the Tweel's spoke tensions and lateral stiffness can't be adjusted.

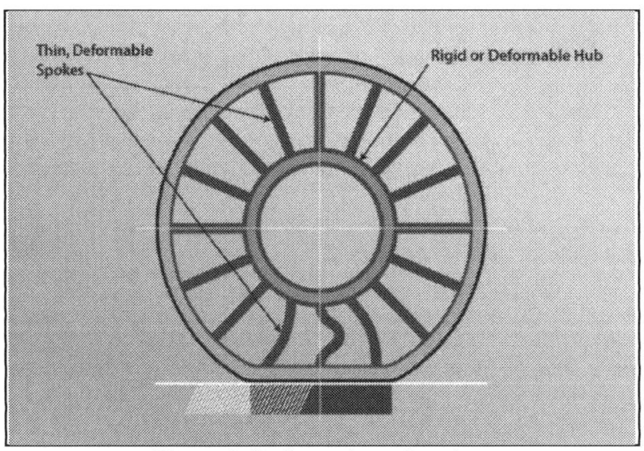

Figure1. Deformation of spokes

III Modeling

Four 3-D Air-less tire models were constructed using CREO software. Model like Honey comb, Spokes Triangular and Diamond are included for analysis. Air tire consists of tread, belt, carcass and air filled volume. The tire model refers tothe geometrical structure of Car Tyres.

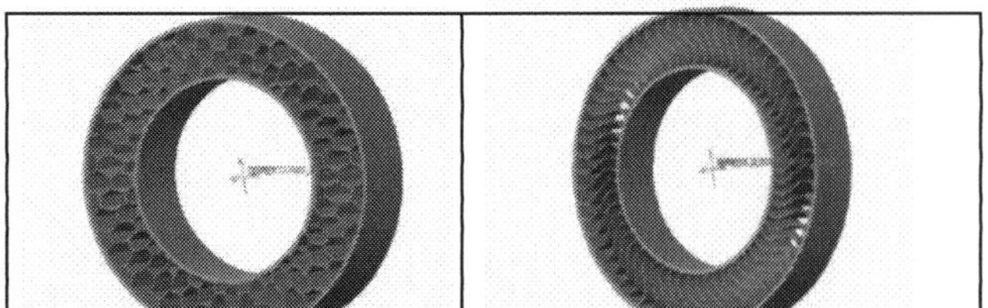

Honey comb Structure	Spoke Type Structure
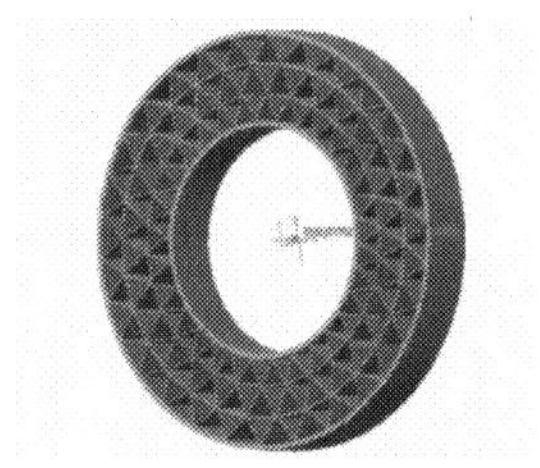 Triangular Structure	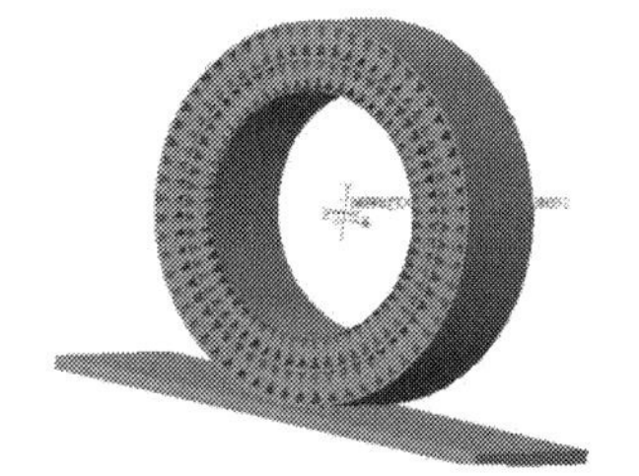 Diamond Structure

Figure 2. Various Structure of air less tyre

The honeycomb structures generally referred as the two dimensional offer the out-of-plane mechanical properties. This property of these structures makes them suitable for the thermal isolation, energy absorption, and structural protection and also as the core of the lightweight sandwich panels.
The diamond tyre structure can withstand heavy loads when applied to the tyre compared to the previous structures; This structure has a nominal total deformation brought when the force is applied on to the tyre duc to its structurc layout and solid structurc.

IV. Load and boundary condition

Air-less tire consists of layers of the inner band, outer band to create a composite structure. Steel rings also are used as reinforcing elements in Air-less tire. Tires typically face the key issues of failure thanks to heat build-up and riding comfort. so as to boost style of Air-less tire, it's necessary to be ready to predict themechanical behavior of the tire beneath applied load. Conjointly stress and strain energy distribution developed ought to be analyzed. The validation of metallic element prediction against air tire results was undertaken. The distribution of strain energy and deflection below loading was conjointly distributed victimization metallic element analysis. Vertical loading on the wheel through the appliance of a uniformly distributed edge load at the tire-rim contact region. The all tire half square measure mashed by victimization the solid Tetrahedral parts.. Tread properties square measure E=30Mpa, density=1300 kg/m3. Each bands take into account as same material properties, and also the wheel load of 1200 N is applied

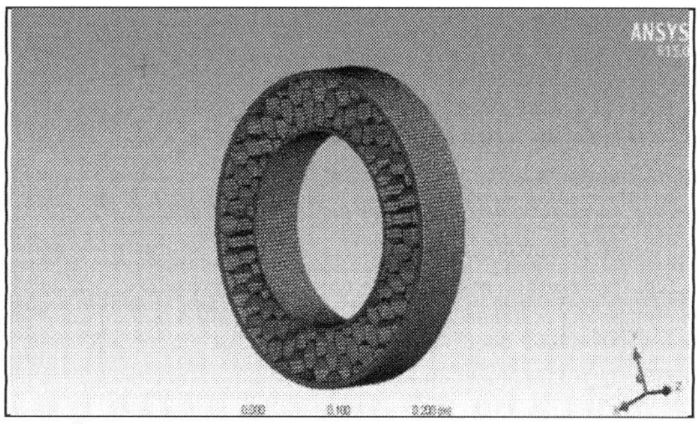

Figure 3. Honey comb Structure with Tetrahedral mesh.

IOP Conf. Series: Materials Science and Engineering **197** (2017) 012042 doi:10.1088/1757-899X/197/1/012042

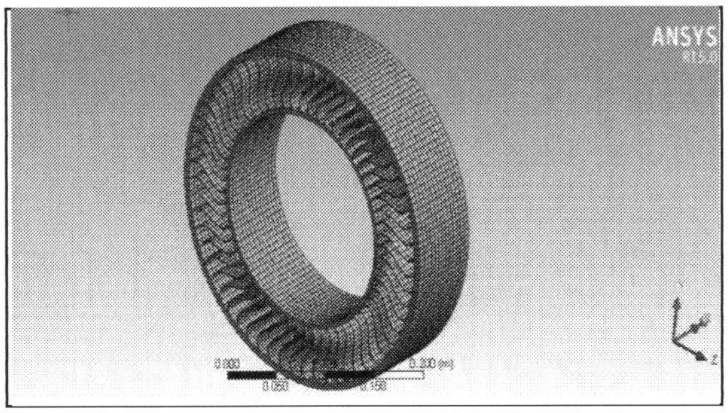

Figure 4. Spoke type Structure with Tetrahedral mesh.

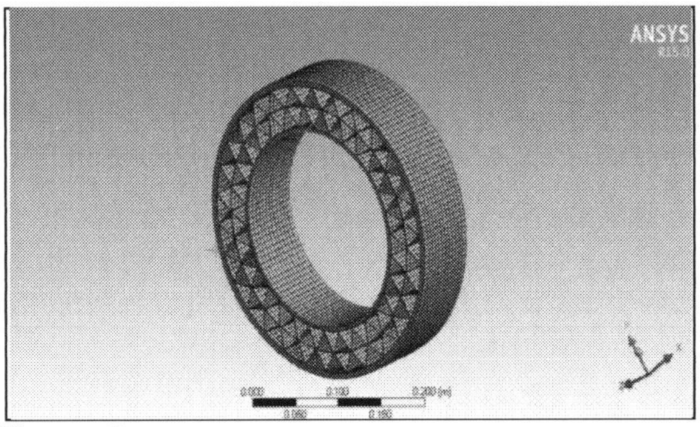

Figure 5. Triangular Structure with Tetrahedral mesh

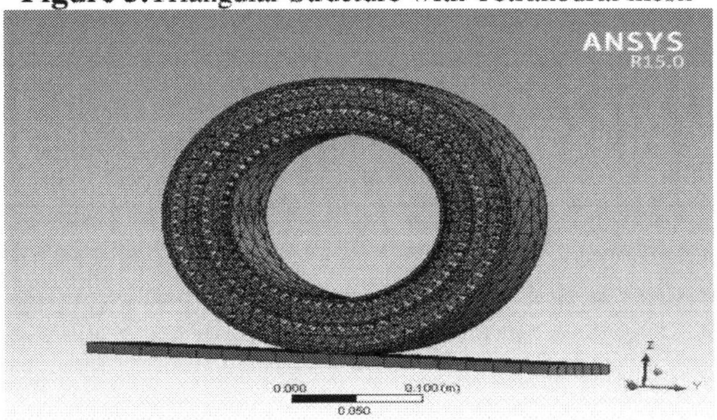

Figure 6. Diamond Structure with Tetrahedral mesh.

V. Load deflection & analysis

Model the tires by mistreatment CREO to imports ANSYS code. Road and tread were command contact stipulation not fastened. Vertical loading on the wheel through the applying of a uniformly distributed edge load at the tire-rim contact region. The deflection within the loading direction of the wheel center and therefore the displacement within the lateral direction.

(a)Honey Comb Structure

IOP Conf. Series: Materials Science and Engineering **197** (2017) 012042 doi:10.1088/1757-899X/197/1/012042

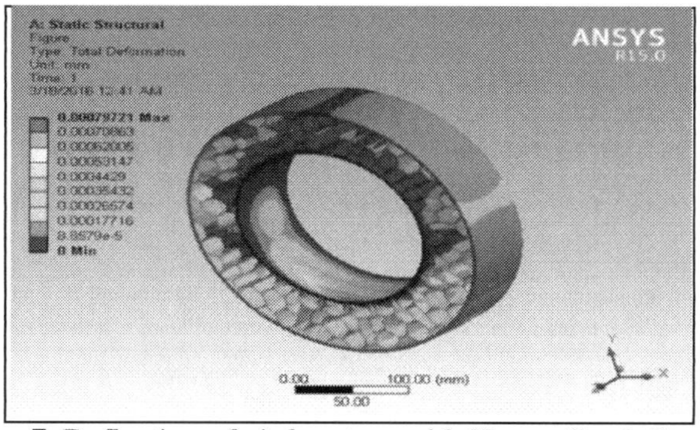

Figure 7. Deflection of air less tyre with Honey Comb Structure

The above figure explains the total deformation of the tyre when a load of 1200 N is applied, the load is acting on the centre of the axle where the deformation. of total tyre with stress and strain relationship is seen in this figure. The colour representation shows the deformation of the tyre when load is applied. The total deformation of the tyre in this type of structure is 0.00079721.

(b)Spoke type structure

Figure 8.Deflection of air less tyre with Spoke type Structure

The above figure explains the total deformation of the tyre when a load of 1200 N is applied.The normal analysis is done with the materials such as polyurethane as spokes and natural rubber as thread of the tyre and the inner layer of the tyre is used as nylon which is used in pneumatic tyres. The hub is used as aluminum which is the basic materials of a normal airless tyre.

(c) Triangular Structure

Figure 9. Deflection of air less tyre with Triangular type Structure

The above figure explains the total deformation of the tyre when a load of 1200 N is applied the load is a acting on the centre of the axle where the deformation of total tyre with stress and strain relationship is seen in this figure.

(d) Diamond Structure

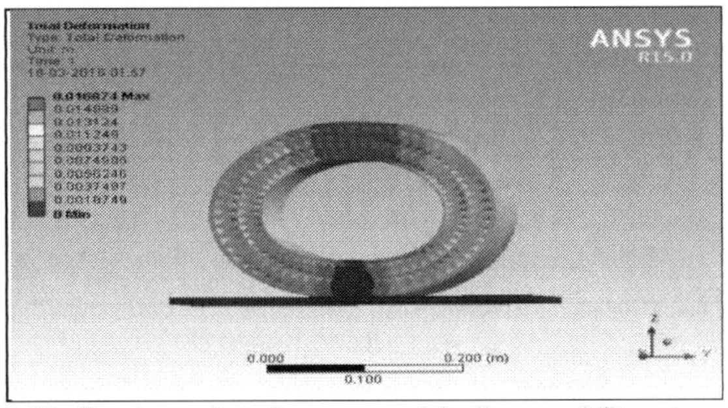

Figure 10. Deflection of air less tyre with diamond Structure

The above figure shows the total deformation of the tyre when a load of 1200 N is applied the load is acting on the centre of the axle where the deformation of total tyre with stress and strain relationship is seen in the figure.

VI. Results

The results obtained on comparison between the structures using normal and composite materials were analyzed and is stated below.

S. No	Structure	Force Applied(N)	Total Deformation (Existing Material) mm	Total Deformation (Existing Material) mm
1	Honeycomb	1200	0.00079721	1.0608e-007
2	Spokes	1200	4.6483e-7	2.8606e-007
3	Triangular	1200	8.2519e-6	1.3521e-007
4	Diamond	1200	0.016874	1.0498e-007

The performance analysis of the four various structures such as honeycomb, spokes, triangular, and diamond are given in the above table. The total deformations of the various structures are shown in which diamond structure has lesser deformation than the other three structures. Thus the diamond structure gives the high load carrying capacity.

VII. Conclusion

From the design analysis it was concluded that the Diamond tyre structure was found out to be solid, and also bears more load comparative to the other structures. The material changes brought about in the carcass and also in the tread has also contributed to the reduction the total deformation. Thus the proposed work can bear a greater amount force and at the same time exhibits a comparatively small total deformation. These types of tyres can be mainly employed for the heavy load vehicles where the load factor is a main concern.

References

[1] Anuj Suhag, Rahul Dayal.Static Analysis on Custom Polyurethane Spokes of Airless Tire. International Journal of Scientific and Research Publications, Volume 3, Issue 11, November 2013. ISSN 2250-315.
[2] Amir Gasmi, Paul F. Joseph. - Development of a two- dimensional model of a compliant non-pneumatic tire. International Journal of Solids and Structures, March 2012
[3] Balamurugan.S, Manibaalan.C, Keshore,Dr.Joshi.C.Haran.static analysis of airless Tyres.International Journal of Scientific and Research Publications, Volume 3, Issue 8, August 2013. ISSN 2250-3153.
[4] Bert Bras and Austin Cobert.-Life-Cycle Environmental Impact of Michelin Tweel Tire for Passenger Vehicles. SAE International. Published, 04/12/2011.
[5] Chih-Hsing Chu, Mu-Chi Song, Vincent C.S. Luo,(February 2005) 'Computer aided parametric design for3d tire mold production' Journal of ELSEIER.
[6] Cobert, A., Environmental Comparison of Michelin Tweel™ and Pneumatic Tire Using Life Cycle Analysis, in Mechanical Engineering. 2009, Georgia Institute of Technology: Atlanta. p. 193.
[7] Palinkas RL, Page GJ. Non-pneumatic tire with supporting and cushioning members. US patent, US 4832, 098; 1989.
[8] Periasamy.K, S.Vijayan-Design and development of air-less car tire. J.J College of Engineering & Technology, Trichy – 09. Tamil Nadu,India.International Journal of Advances in Engineering & Technology,ISSN: 22311963 Sept., 2014.
[9] Rutika Gotad, Sukanya Yadav, Aarti Dung. - Tweel Tyre Technology, International Journal of Advance Foundation and Research in Science & Engineering, 2015
[10] Zeev Bareket Paul Fancher .Representation of truck tire properties in braking and handling studies: The influence of pavement and tire conditions on frictionalcharacteristics,(December 1989).

IOP Conf. Series: Materials Science and Engineering **197** (2017) 012043 doi:10.1088/1757-899X/197/1/012043

Agricultural robot designed for seeding mechanism

Dr. K A SUNITHA G S G S Suraj[2],CH P N Sowrya[2],G Atchyut Sriram[2],D Shreyas[2] and T Srinivas [2]

Assistant Professor (Sr.G), Department of Electronics and Instrumentation Engineering,
SRM University, Chennai ,TamilNadu ,India

Students, Department of Electronics and Instrumentation Engineering,
SRM University, Chennai, TamilNadu ,India.

Abstract:

In the field of agriculture, plantation begins with ploughing the land and sowing seeds. The old traditional method plough attached to an OX and tractors needs human involvement to carry the process. The driving force behind this work is to reduce the human interference in the field of agriculture and to make it cost effective. In this work, apart of the land is taken into consideration and the robot introduced localizes the path and can navigate itself without human action. For ploughing, this robot is provided with tentacles attached with saw blades. The sowing mechanism initiates with long toothed gears actuated with motors. The complete body is divided into two parts the tail part acts as a container for seeds. The successor holds on all the electronics used for automating and actuation. The locomotion is provided with wheels covered under conveyor belts. Gears at the back of the robot rotate in equal speed with respect to each other with the saw blades. For each rotation every tooth on gear will take seeds and will drop them on field. Camera at the front end tracks the path for every fixed distance and at the minimum distance it takes the path pre-programmed.

Introduction:

Ploughing the land and sowing seeds is the root of any plantation or cultivation. Farmer sowing seeds [1] [2] manually without any machinery is the tradition way. The next modification is the introduction of tractors [3] [4]. Tractors are used to plough land and so seeds automatically. It reduced the effort to farmer. Tractors replaced the work of OX in fields. The main problem of existing technology[5] [6], it is costly and unable to be operated by the farmer. The robot discussed here can act as a better replacement of the above mentioned problems. The robot can also visualize its own path to cover entire farm. The prime function depends on distance detection by extracting pixels from the frame.

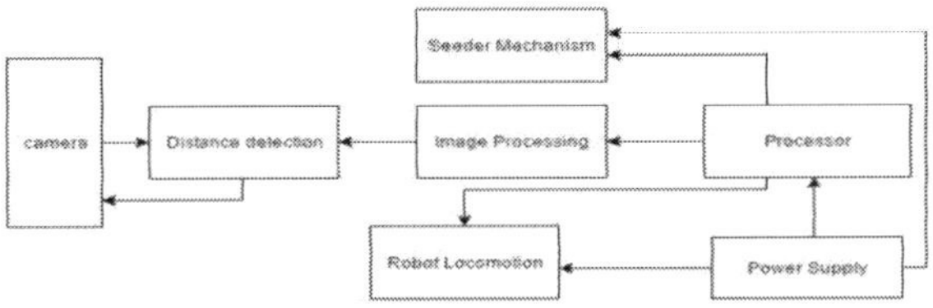

Figure 1: illustrates the entire block diagram of robot

1. Distance detection used in proposed model.

Image processing is a method which is used to convert an image into a digital form of data to perform various operations on it. The operations are performed in an image to extract the required information. In image processing, the input is given as an image or video frame to get the desired output of characteristics associated with the image. Image processing can be used for various purposes like image sharpening and restoration, image recognition, measurement of the pattern, etc. Image processing is generally done in three steps. The first step is the importing of an image. The second step is analyzing and manipulating the image which includes image enhancement, data compression, etc. The final step is the resulting image which is based on image analysis.

The sudden changes of discontinuity in an image are called edges. Edge detection technique is used to find the discontinuity in the image based on various factors like image brightness, contrast, etc. Edge detection with various masks is used for image segmentation. The various operators of Edge detection are Prewitt operator, sobel operator, Robinson compass mask, kirsch compass mask, laplacian operator. The prewitt operator is used to detect edges horizontally and vertically. Sobel operator is similar to the prewitt operator as it calculates the edges in both horizontal and vertical direction. Robinson compass masks rotate the image in all major directions to calculate the edges.

Kirsch compass mask is a derivative mask which is used to find edges in an image. Laplacian operator is a second derivate mask which is further divided into positive laplacian and negative laplacian. They both are used to find the edges in an image. Our primary concern is to detect distance from one end to the other end of field. This can be achieved by taking an image and identifying any particular object in that image. Then by taking more images one after the other and comparing the images to evaluate the distance between robot and end of the field. To determine the distance from the camera to a known Object we use triangle similarity.

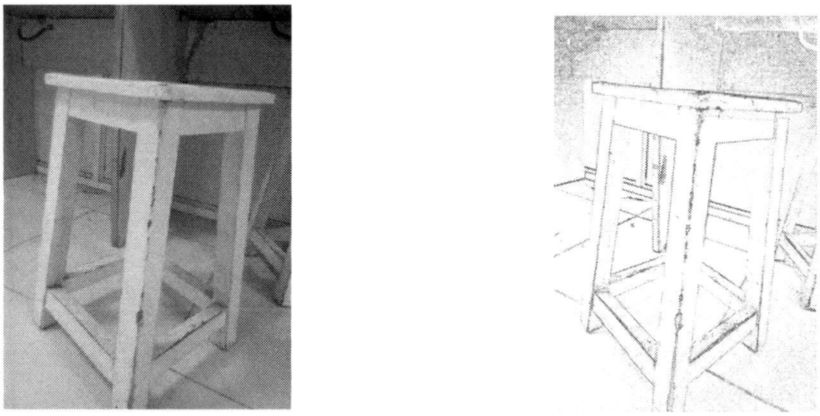

Figure 2 demonstrates the edge detection for a generalized image.

(A) Pre Image Processing **(B) Post Image Processing**

In triangle similarity we consider the number of pixels covered in one picture. The object will be placed at a distance A and of considered width X.

The number of pixels covered in this picture be K. therefore the focal length of the camera can be noted by.

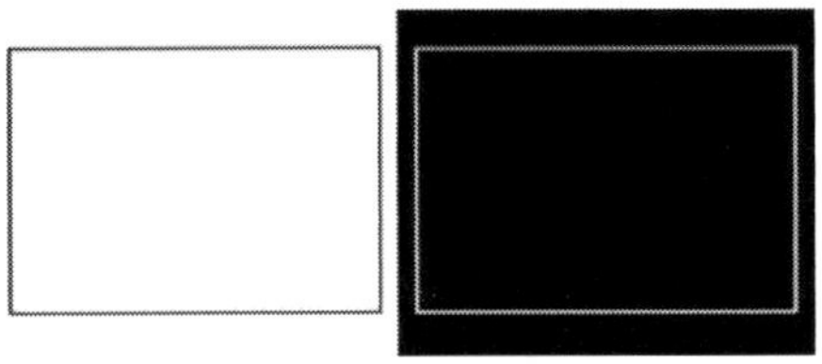

Figure 3: Square Objects Used For Edge Detection.

J = (A*K)/X.

As long as we approach the object the number of pixels covered decreases. Let the new Distance be A'. From the previous formula focal length can be determined and with change in pixel rate the new distance can be identified.

A'= (J*X)/K

The object is particularly a square shaped. It is identified easily and all the edges can be taken out. The edges of a square give good response in calculating the number of pixels linked in it. By comparing the pixel ration distance is calculated. The robot functions according to the distance calculated.

The above algorithm is coded using python. Such that the processor used in this model is able to do fast image processing using python.

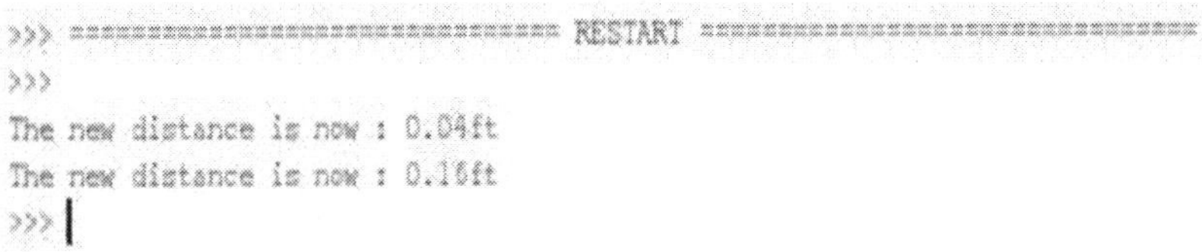

Figure 4: Results of distance detection.

2. Hardware mechanism:

This mechanism would consist of a large chassis with dimensions30cm *30cm*10cm.Themetal used for chassis is galvanized iron .We used this because it doesn't get rust easily. It is light weighted. This would help us in reducing the weight of our mechanism. There would be four rotating actuators at the bottom of chassis, which would act as wheels to the mechanism, sow used 60-rpm dc motors. We use dc motors because our model is cost efficient model; dc would help us in this area as it is available of low cost.

This motor is very much required for our mechanism as it supports us in every aspect by converting electrical energy To mechanical energy by consuming low power compared to some other motors like stepper, servo etc. The main use of this is it is easy to maintain.

Figure 5: shows the camera used in object identification and distance detection.

The torque of our motor is 5 kg's as there is an equal distribution of load on the actuators these should be ready to bear them. The wheels are covered by a conveyer belt as the movement of the mechanism is on the wet field so to bear that field and move on that we use them. Internally we would be using two rotating actuators again as there is working gear system in side which would sow the seed staking from inside the carrier. Here there is a usage of two 10rpm dc motors.

Figure 6: shows the gear mechanism used in rear part of the robot for seeding mechanism.

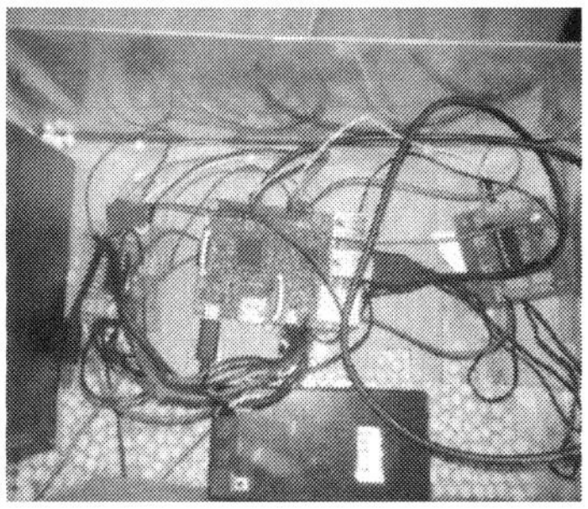

Figure 7: shows the electronics and connections for the model.

IOP Conf. Series: Materials Science and Engineering **197** (2017) 012043 doi:10.1088/1757-899X/197/1/012043

3. Results

The robot referred here is made up of galvanized ion. This adds stability and strength to the robot. Robot actuation is obtained from dc motors of 5kgtorque, speed of rotation perminuteis60.These motors are attached to wheels covered with conveyor belt. This helps in actuation and to over ride small obstacles in field. The front part of the robot is fitted with tentacles. These are made such that the tentacles get extra push from robot. At the end of tentacles saw blades are attached. Saw blades dig the land. The robot here is in particular for wet fields. Saw blades are helpful to dig which refer in ploughing the land. The rear part of the robot there are two long tooth edge ars. These help in sowing seeds. All the seeds are placed from the top of the robot and these gears are made to rotate at 10rpm. Gears are actuated with dc motors of 5kg torque. saves the image and proceeds further. After specific time it takes one more image and compares with previous image. The distance reduced is analyzed and again moves on. At a point it will change its direction of motion and the same repeats again and again. This helps to cover the entire field

The rotation of back actuators helps in pouring the seeds from the robot to the ploughed land, this can be viewed in above figure6.The complete robot is powered up with 12V dc supply. Motor driver boards are connected to all the actuators and these are controlled with raspberry pi3.The powerful 1.2GHz processor helps in controlling the whole robot easily. After the initiation of robot, it takes image and identifies the particular object by the defined parameters.

By assembling the model the robot is able to plough the land and sow seeds. The motors used for locomotion gave the best results while crossing over obstacles. The sowing mechanism has good capacity to sow al the seeds. As the processor used here is raspberry pi. It gave us fast image processing results and it bought good control of motors. The attached tentacles were able to plough the land. This robot is confined to wet land sand it can plough two rows. Seeder to sow seeds follows the same line.

4. Future Proposal

In future model, the number of gears used for seeding is increased. More number of rows can be covered easily. In the next version, dedicated processor for image detection and sowing seeds mechanism will be implemented. The powering system will be changed from12V lead acid battery to high power batteries. Use of solar energy will also be included. This makes the robot more efficient. The robot future model has a display which will be given to the farmer. It deals with the control of robot and also it transmits the video that is been focused by the robot. These changes improve the speed of the robot. This also gives better results than existing technology.

5. References

[1] Swetha S , Shreeharsha G.H''solar Operated Automatic Seed Sowing machine''..International Journal Of Advanced Agricultural Sciences And Technology 2015, Volume 4, Issue 1, pp. 67 – 71, Article ID Sci – 223 ISSN: 2320 – 026X

[2] P. Usha, V. Maheshwari, Dr.V.Nandagopal, " Design and implementatioin of seeding agricultural robot" - Journal Of Innovative Research And Solutions (Jiras) - A unit of UIIRS Print ISSN: 2320 1932 / Online ISSN – 2348 3636 Volume No. 1, Issue No.1. Page No : 138 - 143, JULY – 2015

[3] Divya C. H., Ramakrishna, H. and Praveena Gowd International Journal of Current Research Vol. 5, Issue, 03, pp.461-466, March, 2013, ''Seeding And Fertilization Using An Automated Robot''.

[4] A.O.Hannure, S.P.Kshirsagar, V.S.Kodam, O.N.Patange, "Literature Review on Automatic Seed Feeder" - International Journal Of Engineering Trends And Technology(IJETT) – Volume 36 Number 7 – June 2016

[5] Gholap Dipak Dattatraya, More Vaibhav Mhatardev, Lokhande Manojkumar Shrihari, Prof. Joshi S.G Robotic Agriculture Machine. International Journal of Innovative Research in Science, Engineering and Technology. Volume 3, Special Issue 4, April 2014

[6] Kyada, A. R, Patel, D. B .5th International & 26th All India Manufacturing Technology, Design and Research Conference (AIMTDR 2014) December 12th–14th, 2014, IIT Guwahati, Assam, India Design And Development Of Manually Operated Seed Planter Machine .

IOP Conf. Series: Materials Science and Engineering **197** (2017) 012044 doi:10.1088/1757-899X/197/1/012044

Design and Analysis of Wind Turbine Blade Hub using Aluminium Alloy AA 6061-T6

S.Ravikumar[1,*], V.Jaswanthvenkatram[2], Y.J.N.V.Sai kumar[2], S.Md.Sohaib[3]

[1,*]Assistant Professor, [2] UG students of Department of Mechanical Engineering, Sathyabama University, Chennai,Tamilnadu, India. Pin: 600 119.
[3]UG students of Department of Mechanical and Production Engineering, Sathyabama University, Chennai,Tamilnadu, India. Pin: 600 119.
[1,*]mahailakumar@gmail.com,
[2]vjvr03@gmail.com,[2]ysaikumar44@gmail.com[3]s.md.sohaib8904@gmail.com

Abstract: This work presents the design and analysis of horizontal axis wind turbine blade hub using different material. The hub is very crucial part of the wind turbine, which experience the loads from the blades and the loads were transmitted to the main shaft. At present wind turbine is more expensive and weights more than a million pounds, with the nacelle, rotor hub and blades accounting for most of the weight. In this work Spheroidal graphite cast iron GGG 40.3 is replaced by aluminium alloy 6061-T6 to enhance the casting properties and also to improve the strength-weight ratio. This transition of material leads to reduction in weight of the wind turbine. All the loads caused by wind and extreme loads on the blades are transferred to the hub. Considering the IEC 61400-1 standard for defining extreme loads on the hub the stress and deflection were calculated on the hub by using Finite element Analysis. Result obtained from ANSYS is compared and discussed with the existing design.
Key words: Cast iron GGG 40.3, Aluminium alloy 6061 T-6, IEC 61400-1 standards, Extreme loads, Finite element analysis, ANSYS.

1. Introduction

A wind turbine is a device that converts kinetic energy from the wind, also called wind energy, into mechanical energy; a process known as wind power. If the mechanical energy is used to produce electricity, the device may be called a wind turbine or wind power plant. If the mechanical energy is used to drive machinery, such as for grinding grain or pumping water, the device is called a windmill or wind pump. Similarly, it may be referred to as a wind charger when used for charging batteries. The result of over a millennium of windmill development and modern engineering, today's wind turbines are manufactured in a wide range of vertical and horizontal axis types.Fig.1 represents the image of Horizontal Axis Wind Turbine components.

Ali Muhammad et al [1] explained about Modern mega watt class wind turbines are exposed to high and complex loads. The influence of the manufacturing process on material properties is investigated includes the simulation of fatigue life and refining the existing material models lead to a more efficient utilization of the material. Arvind singh rathore et al [2] discussed an optimization model for rotor design of 750 kW horizontal axis wind turbine. In this work a blade of length 21.0 m is taken and airfoil for the blade is S809.The airfoil taken is same from root to tip. All the loads caused by wind and inertia on the blades are transferred to the hub. The stress and deflection were calculated on blades and hub by Finite element analysis method. The maximum stress in the model is less than maximum allowable stress. Tony Burton et al [3] give the information about wind resources, various theories, co-ordinates, performances related to wind turbine. It also describes about various type of geometries of turbine, different loads acting on the wind turbine and all the fundamentals of wind turbine.

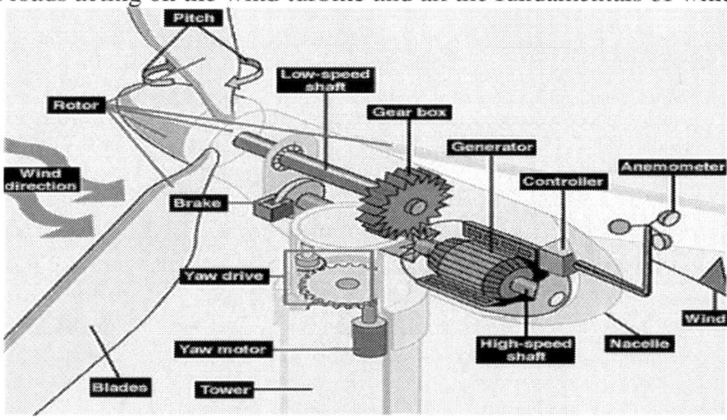

Figure.1 Components of Horizontal Axis Wind Turbine

Vinay V. Kuppast et al [6] study of effect of vibration characteristics of aluminium alloys of different compositions. The modeling and analysis is carried out using ANSYS software. Young's Modulus and the ultimate tensile strength of the 380 alloys increase with the increase in copper and silicon content. Deformation is least in case of 380 alloys and is recommended for low vibration applications.IEC-61400 [7] is an international standard published by the international Electro technical Commission regarding wind turbines. It is a set of design requirements made to ensure that wind turbines are appropriately engineered against damage from hazards within the planned lifetime. Wind classes determine which turbine is suitable for the normal wind conditions of a particular site. Turbine classes are determined by three parameters - the average wind speed, extreme 50-year gust, and turbulence. In this work aluminium alloy 6061-T6 is chosen as an alternativefor the wind turbine hub to reduce the strength-weight ratio and gradual decrease in cost of production.In ANSYS both the materials existing and proposed material are analysed under IEC 61400-1 load cases and the results of total deformation and equivalent stress is taken for comparison.

2.Experimental Set Up

The aim of the project is to reduce strength-weight ratio and decrease in cost by changing material of hub. A 3D model is designed in Pro-E and it analysed in ANSYS for both existing material and an aluminium alloy. After analyzing both the results are compared and fatigue analysis is made for aluminium alloy in order to determine whether it is capable in environmental conditions.

2.1 Mechanical properties and chemical composition of materials for study

Table 1 and 2 represents the mechanical properties and the chemical composition of the materials to study.

Table 1 Mechanical property of materials for study

Properties	Aluminium alloy 6061 T6	Cast Iron
Density	2700kg/m³	7100 Kg/m³
Tensile Strength	310Mpa	370Mpa
Yield Strength	218Mpa	250Mpa
Poisson's Ratio	0.32	0.27
Young's Modulus	69000Mpa	169000Mpa

Table 2 Chemical composition of materials for study

Aluminium alloy 6061 T6		Cast Iron	
Mg	0.937	C	3.4
Si	0.535	P	0.1
Fe	0.139	Mn	0.4
Ti	0.012	Ni	1.0
Mn	0.022	Mg	0.06
Zn	0.0983	-	-
Cr	0.022	-	-
Ni	0.005	-	-
Al	98.096	-	-

2.2Model of hub

The 3D model of the hub is modelled using the software PRO/E. PRO/E is widely used software to model the three dimensional model due to its user friendly options compared to other similar software's. The 3D model is imported to ANSYS for analysis. In order to import the model is converted into global file format namely .stp format. After converting the file into .stp format the file is imported into ANSYS. Fig. 2 (a) and (b) represents 3D model of wind turbine rotor hub and meshing of wind turbine rotor hub.

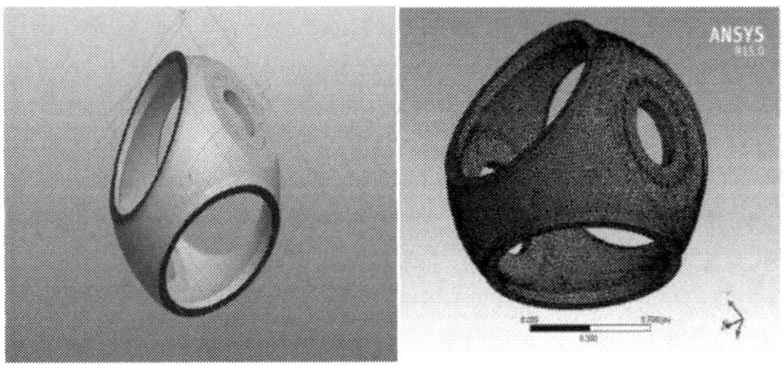

Figure. 2(a) 3D model of wind turbine rotor hub (b) Meshing of wind turbine rotor hub

IOP Conf. Series: Materials Science and Engineering **197** (2017) 012044 doi:10.1088/1757-899X/197/1/012044

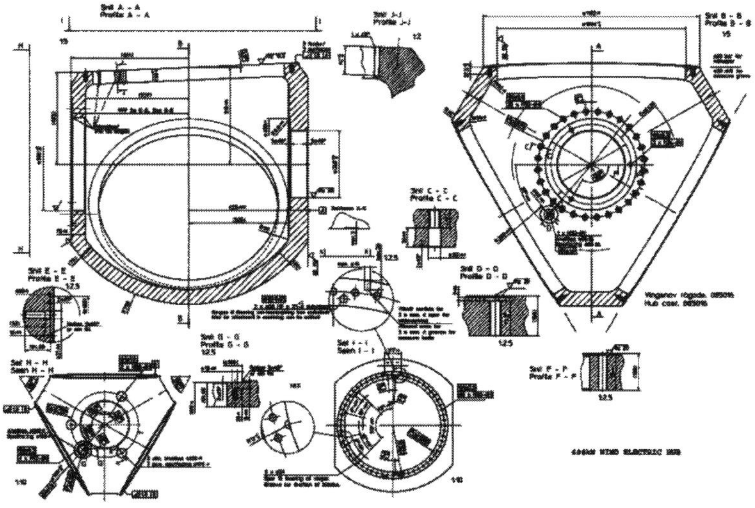

Figure. 3 2D Dimensions of 600KW Horizontal Axis Wind Turbine (SUZLON Energy)

2.3Extreme load cases on hub from IEC 61400-1

Moments and forces on the hub can then be determined from the blade forces. For a rigid rotor turbine, both flapping and lead lag moments are transmitted to hub, usually flapping is the predominant one. The blade root bending moment for each blade is $M_\beta = K_\beta * \beta$

Table 3 Extreme loads on hub from IEC 61400-1

B.M & Forces	Load Case-1	Load Case-2
301 Hub1 Mx KNm	118	-318
302 Hub1 My KNm	151	267
303 Hub1 Fx KN	25.6	31.8
304 Hub1 Fy KN	16.8	27.2
305 Hub1 Fz KN	97.9	114

2.4Analysis of materials for study

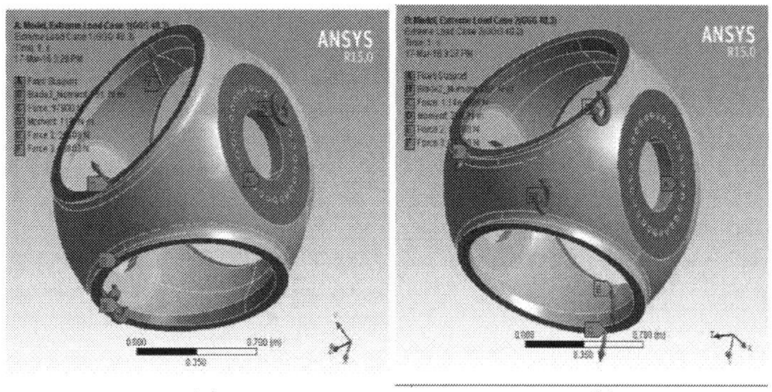

(a) (b)

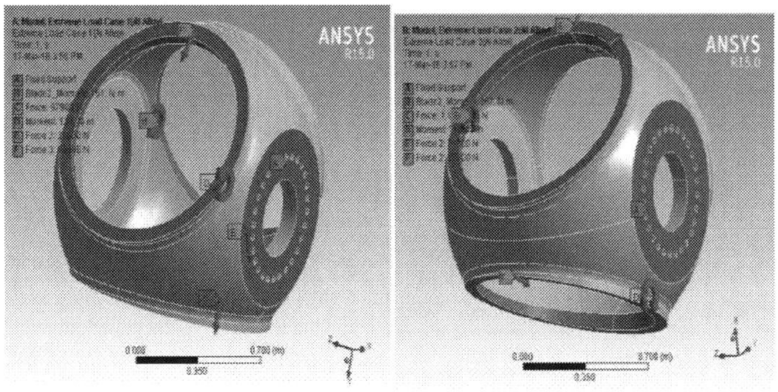

(c) (d)

Figure. 4 (a), (b) Load case 1&2 for GGG 40.3 (c), (d) Load case 1&2 for aluminium alloy materials

3.Results and Discussion

3.1Results of cast iron (GGG 40.3) for load case1and 2

GGG 40.3 is a german standard from DIN 1693 which is also called spheroidal graphite cast iron. Due to the spherical formation of the graphite, The ductile cast iron acquires special properties compared to the flake graphite; tensile strength, yield strength are improved, expansion and impact strength are increased. Deformation describes the transformation from initial to final geometry. The collective displacement of points in a body relative to an external reference frame is known as deformation. Fig 5(a) represents the total deformation change in the existing material i.e cast iron GGG 40.3 for 1 sec. The blue indicates the minimum value of 0m and red indicates the maximum value of 0.00043186m. Fig 5 (b) represents the total deformation change in the existing material i.e cast iron GGG 40.3 for 1 sec. The blue indicates the minimum value of 0m and red indicates the maximum value of 0.00053065m.The von-mises stress is often used in determining whether an isotropic and ductile metal will yield when subjected to a complex loading condition.Fig 5(c) represents the equivalent stress change in the existing material i.e cast iron GGG 40.3 for 1 sec. The blue indicates the minimum value of 29092 Pa and the red indicates the maximum value of 2.7677e7 Pa. Fig 5(d) represents the equivalent stress change in the existing material i.e cast iron GGG 40.3 for 1 sec. The blue indicates the minimum value of 27923 Pa and the red indicates the maximum value of 3.3666e7 Pa.

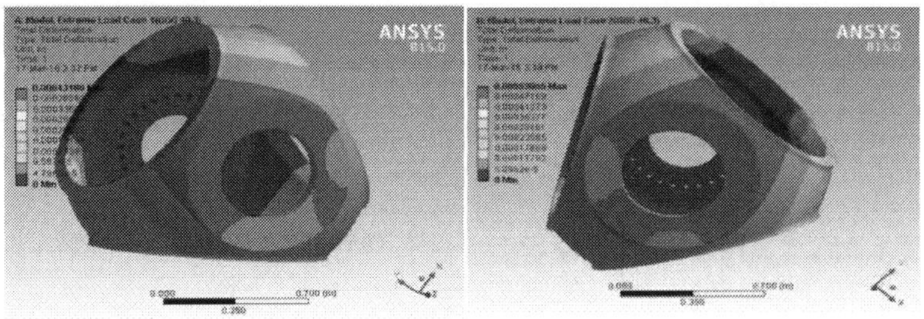

(a) (b)

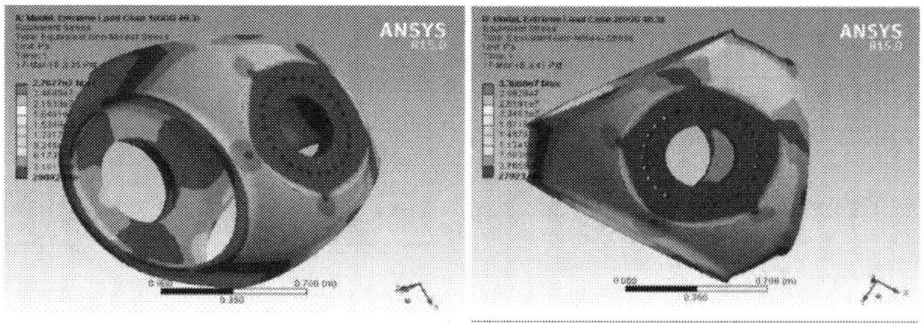

(c) (d)

Figure. 5(a), (b) Deformation for Load case 1&2 for GGG 40.3,
(c), (d) Stress for Load case 1&2 for GGG 40.3

3.2Results of Aluminium alloy for load case1and 2

Aluminium alloy6061 is a precipitation –hardened aluminium alloy, containing magnesium and silicon as its major alloying elements. T6 temper 6061 has an ultimate tensile strength of at least 3000 MPa and yield strength.Change in the shape of a body caused by the application of a force (stress). Deformation is proportional to the stress applied within the elastic limits of the material. A deformation may be caused by external loads, body forces (such as gravity or electromagnetic forces), or changes in temperature, moisture content, or chemical reactions, etc.Fig 6(a) represents the total deformation change in the existing material i.e Aluminium alloy 6061 T-6 for 1 sec. The blue indicates the minimum value of 0m and red indicates the maximum value of 0.0010323m. Fig 6(b) represents the total deformation change in the existing material i.e Aluminium alloy 6061 T-6 for 1 sec. The blue indicates the minimum value of 0m and red indicates the maximum value of 0.001268m. Equivalent stress is used to predict yielding of materials under multiaxial loading conditions using results from simple uniaxial tensile tests.Fig 6(c) represents the equivalent stress change in the existing material i.e Aluminium alloy 6061 T-6 for 1 sec. The blue indicates the minimum value of 24351 Pa and the red indicates the maximum value of 2.699e7 Pa.Fig 6(d) represents the equivalent stress change in the existing material i.e Aluminium alloy 6061 T-6 for 1 sec. The blue indicates the minimum value of 25556 Pa and the red indicates the maximum value of 3.1852e7 Pa.

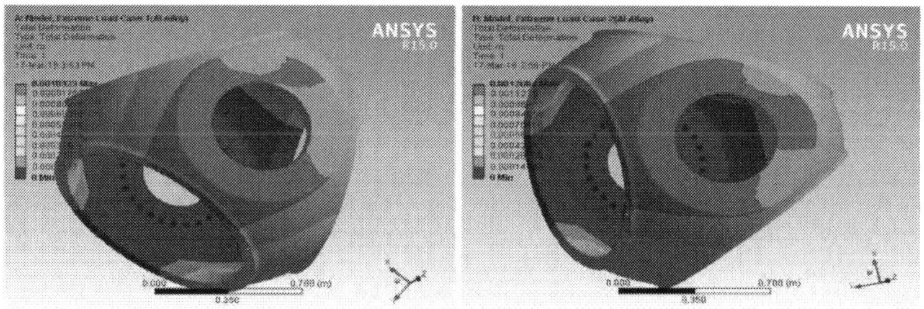

(a) (b)

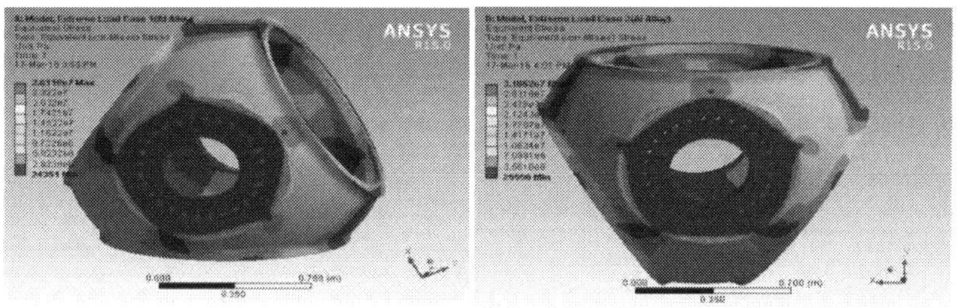

(c) (d)

Figure. 6 (a), (b) Deformation for Load case 1&2 for Aluminium alloy,
(c), (d) Stress for Load case 1&2 for Aluminium alloy

3.3Fatigue analysis of Aluminium alloy

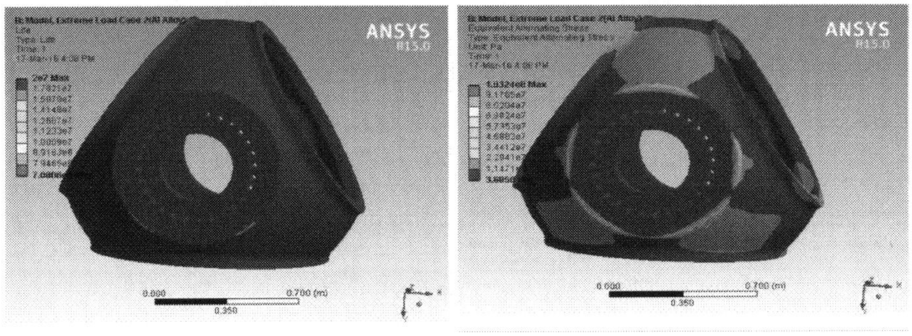

(a) (b)

Figure. 7(a) Fatigue life for Aluminium alloy hub
(b), Equivalent alternating stresses for Aluminium alloy hub

Table 4 Comparison of materials study analysis

Material	Load cases	Deformation (m)	Stress (Pa)
Cast iron	Case - 1	0.00043186 Max, 0 Min	2.7677e7 Max, 29092 Min
Cast iron	Case - 2	0.00053065 Max, 0Min	3.3666e7 Max, 27923 Min
Aluminium alloy	Case - 1	0.0010323 Max,0 Min	2.6119e7 Max, 24351 Min
Aluminium alloy	Case - 2	0.0012684 Max,0 Min	3.1852e7 Max, 25556 Min

Table 5 Fatigue analysis

Material	Safety factor	Biaxiality	Alternating stress	Life
Aluminium alloy	Max 15	Max 0.99136	Max 1.0324e8	Max 2e7
Aluminium alloy	Min 0	Min -1	Min 3.6856	Min 7.0806e6

Factor of safety (FOS), also known as (and used interchangeably with) safety factor (SF), is a term describing the capacity of a system beyond the expected loads or actual loads.N_f, as the number of stress cycles of a specified character that a specimen sustains before failure of a specified nature occurs. For some materials, notably steel and titanium, there is a theoretical value for stress amplitude below which the material will not fail for any number of cycles, called a fatigue limit, endurance limit or fatigue strength. Fig. 7 represents (a) Fatigue life for Aluminium alloy hub (b), Equivalent alternating stresses for Aluminium alloy hub. Table 4 gives the comparison of materials study analysis. Table 5 gives the results of fatigue analysis.

4.Conclusions

- Wind turbine Rotor Hub has been analyzed for two extreme load conditions as per the most important load cases according to IEC 61400-1 (wind turbine loads)
- Existing material is used for analysis (GGG40.3). Stress levels are high. Factor of safety of the analysis shows 7.6 and required is 3. This shows that the current design is much higher in safety.
- So we can use aluminum alloy for reducing material cost and total weight reduction. The stress (32MPa) and deflection (1.2mm) a value from the analysis. Further Fatigue analysis is carried out and alternating stress for the aluminum alloy is 103.2MPa which is less than the allowable yield limit(170MPa).
- The fatigue life of aluminum alloy rotor hub is safe for 20 years time of wind turbine life time. Hence using aluminum alloy rotor hub will be beneficial to the industries in cost wise.

References

[1] Ali Muhammad, Milan Ristow 2010 Fatigue Life of Wind Turbine Structural Components,NAFEMS Seminar, Simulation-based Fatigue and Lifetime Predictions , Wiesbaden, Germany.

[2] Arvind Singh Rathore, Siraj Ahmed 2011, Design and Analysis of Horizontal Axis Wind Turbine Rotor, International Journal of Engineering Science and Technology (IJEST), ISSN: 0975-5462,**3**(11)7975-7980.

[3] Fundamentals of wind energy from 'Wind Energy Hand Book'.

[4] Picture courtesy – Wikipedia and Google.

[5] Qingdao Casting Quality Industrial Co., Ltd. International Standard, UDC 669.131.7 October 1973.

[6] Vinay V Kuppast1, Vijay Kumar N Chalwa, SN, Kurbet and Aravind M Yadawad, 2014, Finite element analysis of aluminium alloys for their vibration characteristics, IJRET: International Journal of Research in Engineering and Technology, **03**(3) NCRIET-2014, Available @ http://www.ijret.org.

[7] 2014-IEC 61400-1 Edition 3.1 Design requirements, Supplied by BSB under licence from IEC for National institute of Wind Energy (NIWE) ON 09/12/2014.

Design and Analysis of Boiler Pressure Vessels based on IBR codes

[a] Balakrishnan B , [b]Kanimozhi .B
[a]Research Scholar,Department Mechanical Engineering, Sathyabama University,Chennai
[b]Faculty of Mechanical Engineering, Sathyabama University, Chennai
Email: kanihwre@gmail.com .

Abstract

Pressure vessels components are widely used in the thermal and nuclear power plants for generating steam using the philosophy of heat transfer. In Thermal power plant, Coal is burnt inside the boiler furnace for generating the heat. The amount of heat produced through the combustion of pulverized coal is used in changing the phase transfer (i.e. Water into Super-Heated Steam) in the Pressure Parts Component. Pressure vessels are designed as per the Standards and Codes of the country, where the boiler is to be installed. One of the Standards followed in designing Pressure Parts is ASME (American Society of Mechanical Engineers). The mandatory requirements of ASME code must be satisfied by the manufacturer. In our project case, A Shell/pipe which has been manufactured using ASME code has an issue during the drilling of hole. The Actual Size of the drilled holes must be, as per the drawing, but due to error, the size has been differentiate from approved design calculation (i.e. the diameter size has been exceeded). In order to rectify this error, we have included an additional reinforcement pad to the drilled and modified the design of header in accordance with the code requirements.

Key words- Boiler, IRB code, Super critical, Pressure Vessels

1. INTRODUCTION

Boiler codes define a significance of applications. A boiler is a pressure-holding container producing steam at a pressure of generally 2 bar and higher. There is, however, no universal definition. Boilers are also called steam generators. Boilers of the early eighteenth century were no more than large kettles. Since those early days boilers have made great progress in terms of size, variety, flexibility, versatility, reliability, and complexity. An attempt to classify boilers is made in the following sections.The first classification of boilers is based on what flows through the boiler tubes .Fire or smoke or flue tube or shell-type boilers in which flue gases are inside and water is outside. Water tube boilers in which water is inside and flue gases are outside. Combination (combo boiler) in which flue gas and water flow both outside and inside the tubes; the type contains an external furnace and shell-type boiler in a sequence. Fire tube boilers of smaller size are vertical and those of larger size are horizontal. The minimum capacity of a water tube boiler is usually around 5 tph. Combo boilers are generally used for solid fuel firing in lower ranges of up to 30 tph for lower pressures <25 atm

2. TYPES AND PERFORMANCE OF BOILERS

The boilers are classified according the application such as Industrial boilers (Power Boiler),Utility boilers, Marine boilers and Nuclear boilers Type of Industrial Boilers (Power Boiler) CFBC – Coal Fluidised Bed Combustion Boiler, FBC - Fluidised Bed Combustion Boiler, AFBC – Atmospheric Fluidised Bed Combustion Boiler,PCFB – Pulverised Coal Fired Boiler,Oil Fired Boiler .Super critical boiler At drum pressure of ~200 bar and SH/RH temperature of 540/565°C, limits are reached in natural circulation drum-type boilers in respect of steam pressures and temperatures with plant thermal efficiency reaching 42% on NCV and 40% on GCV. With the limits for the flue gas exit temperatures and un burnt losses reached long ago, the next improvement could come only by adopting SC conditions. It must be clarified that the SC conditions do not increase the boiler efficiency *per se*, as the efficiency is governed by stack losses and unburnt losses, neither of which is affected by the steam pressure adopted. The higher steaming conditions basically increase the overall steam cycle efficiency. There is, correspondingly, a reduction in the fuel input that also reduces the CO2 generation and sizing of auxiliaries such as the fans and pumps on per megawatt basis.

The subcritical and SC steam cycles are depicted in Figure 1.(RH is not depicted here for the sake of simplicity.) The difference is in the higher operating cycle pressure, leading to higher cycle efficiency. For the same heat loss in a condenser more work is done in a boiler and a turbine and hence the efficiency is higher. The boiler construction is different, in that there is no longer a drum.

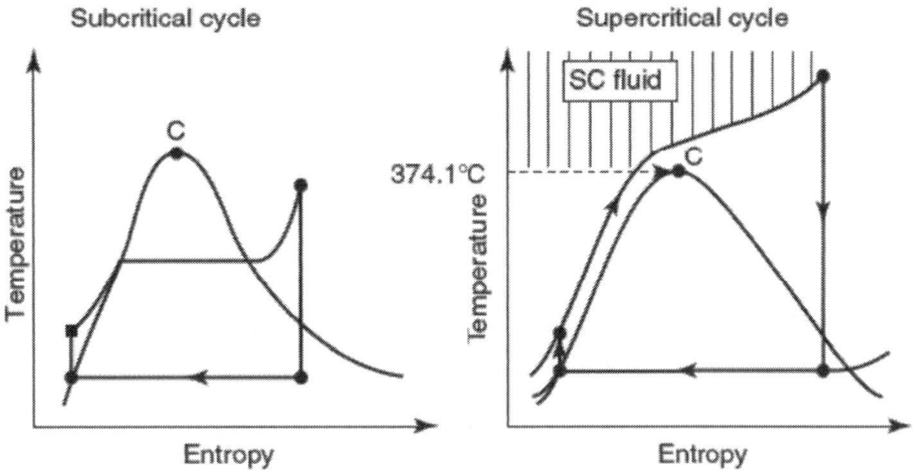

Fig 1. Sub and supercritical cycles on T-S chart

Boilers have classified according to the circulation and draft like Natural circulation or drum-type boilers ,Forced circulation boilers and Natural or balanced draft boilers Forced draft (FD) or pressurized fi red boilers, also based on Furnace Construction following are the types Two-pass boilers, One and a half-pass boilers, Single or tower-type boilers, Down-shot boilers. Boiler Component of Pressure Part such as Drum, Furnace Side,Cage Side,Vestibule Side,Primary and Final -Super Heater,Primary and Final – Reheater,Platen Super Heater,Economizer

Types of Header

Main function of the boiler header is to collect the fluid (water or steam) from the small diameter tube or the large diameter link and to redistribution, which is called "Collector". Mostly, it doesn't transfer heat and located outside of furnace or flue gas path. The header size shall be selected based on the flow, and if the flow from the tube is not surely absorbed due to small in size, pressure loss results overheating of tubes might occur. Boiler headers must be properly trapped in order to prevent carryover from the boiler from being introduced into the steam mains.

Economizer Header

The feed water is located in back-pass zone which is coming to the boiler through Economiser and the Economiser headers are located in the entrance and exit of the Economiser, fluid is water, and the material is selected based on design temperature. Mainly carbon steel material will be used.

Furnace Header

The water which is coming down through the Down comer or the Economiser Mixing Line is heated in furnace wall and furnace headers are located in the entrance and exit of the furnace wall, and inlet header is mainly located in down side, and outlet header is located in upper side.

Fluid of the inlet header is water and outlet header is both water and steam mixture, and the material is selected based on the design temperature and mainly constituted by carbon steel.

Super heater Header

The separated steam from drum or separator is sending to S/H tubes through links, so it is converted to superheated steam. Superheater headers are located in the entrance and exit of S/H tubes, and the fluid is steam. The material is selected based on the design temperature, and its low temperature part is mainly carbon steel, and high temperature part is mainly alloy steel.

Re heater Header

The steam which is entered to high pressure turbine is sending to R/H tubes, so it is converted to reheated steam. Reheater headers are located in the entrance and exit of R/H tubes, and the fluid is steam. The material is selected based on the design temperature, and its low temperature part is mainly carbon steel, and high temperature part is mainly alloy steel.

Boiler Specifications- Boilers are always specified by the four most important attributes, in addition to the types of fuel to be fired: Steam flow or evaporation, Steam outlet pressure (SOP), Steam outlet temperature (SOT),Feed water (FW) inlet temperature

Doosan Power Systems India Private Limited in Chennai has mainly involved in designing of Pressure Parts. Since 2011 Doosan Power Systems India Private Limited has adapted the technology of both Designing and Fabrication of Pressure Parts.As of Design, Checking the thickness of Header based on Design Pressure and Temperature is the main criteria followed and these Calculations are performed by Senior Engineer and Header Department pressure parts experienced Engineers. For focusing on fabrication point of view, from preparing raw material indent, purchase order sheet, receipt the inspection activities are carried by our purchase engineers and quality engineers. Finally machining sequence and welding schedule for making the header ready for dispatch has been done by supervising engineers. check malfunction/ human error while fabrication the header (i.e. Drilling of higher hole size instead the lower one), the header shall be scrap. In order to avoid the scrap of the project will provide alternative solution. Introduction of reinforcement calculation (i.e. Weld Pad Introduction) has been done and the following has been briefed in the Calculation Topics with the respective criteria's.

Design procedure and calculation

Based on the design data, thickness of header, tube and nozzle with ligament efficiency have calculated for safe design condition with normal procedure. preparing raw material indent and bill of material for the header, tube and stub for fabrication has done.The same shall be purchased and check for material testing and inspection. Then end preparation, surface finishing shall be done. After designing of header, the drill marking shall be done for stubs with proper pitch distance as per pressure parts arrangement drawing.While fabrication for a case of putting wrong drill hole diameter which is bigger than required one then the whole header may be a scarp. As per ASME code the available area (example some of header, nozzle and weld should be greater than removed area of a hole).So here we prepare design calculation and manufacturing of platen super heater header of 273 mm diameter and 18.26 mm thickness. Design modification of Header due to wrong drilling while manufacturing and inducing compensation method in nozzle location. The following table -1 show the design parameters and Fig 2 show the header pad design.

Table 1.Design parameter

Platen super heater inlet stub heater		
1.	Code of construction	ASME section 1-2013 Edition
2.	Maximum allowable Working pressure	10Mpa (g)
3.	Design temperature	450 C
4.	Hydrostatic head	Nil
5.	Other loading on head	Nil
6.	Service condition	1
7.	Hydro test pressure	15 Mpa (g)
8.	Hydro test temperature	20-40 C
9.	ASME designator	'S'
10.	Heating surface area	0 mm
11.	Test coupon requirement	Nil
12.	Corrosion allowance	0mm

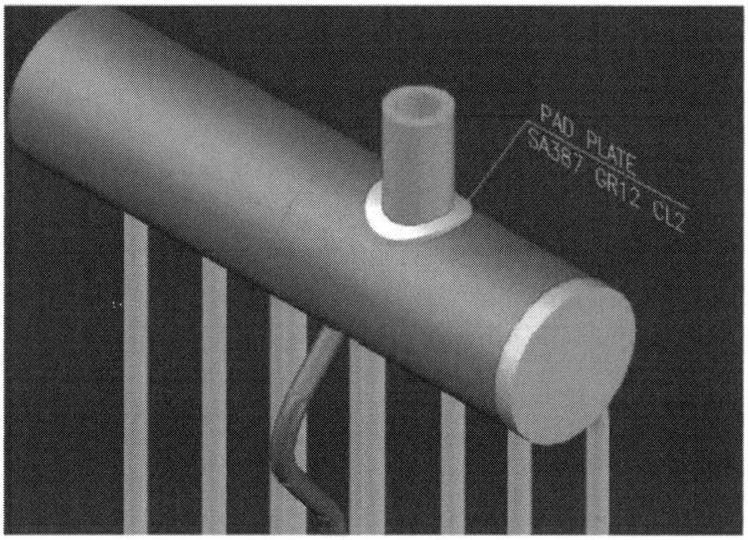

Fig.2.Header Pad design

3. DESIGN PROCEDURE AND CALCULATION

Tube thickness calculation

$$t = \frac{PDo}{2SE + P} + 0.005\ Do + e$$

$$t = \frac{10 \times 38}{2 \times 103 \times 1 + 10} + 0.005 \times 38 + 0$$

t =1.95 mm < 6.8 mm so Design is safe.

Nozzle thickness calculation

OD based calculation:

$$t = \frac{PDo}{2SE + 2yP} + C$$

ID based calculation:

$$t = \frac{PDi}{2SE - 2(1-y)\ P} + C$$

$$Di = OD - 2 \times 0.875 \times AWT$$
$$= 168.3 - 2 \times 0.875 \times 14.27$$
$$Di = 143.33\ mm$$

$$t = \frac{PDi}{2SE - 2(1-y)\ P} + C$$

$$t = \frac{10 \times 143.33}{2 \times 103 \times 1 - 2(1-0.4)10} + 0$$

t = 7.39 mm < 14.27 mm so Design issafe.

Header thickness calculation

OD based Calculation:

$$t = \frac{PDo}{2SE + 2yP} + C$$

ID based calculation:

$$T = \frac{PDi}{2SE - 2(1-y)P} + C$$

Where:

T - Minimum required thickness of the pipe.
P - Design pressure (From data sheet).
Do - Outside diameter of the pipe.
Di - Inside diameter of the pipe.
S - Allowable stress value (From ASME sec. II part D) on the basis of pipe material and design temperature.
E - 1 (For seamless pipe) otherwise efficiency of theligament.
Y- From table according to material type and temperature limit.
C - Corrosion allowance (Will be indicated in the contract Document and as per client requirement).

Equivalent Diameter of Holes

de = Do - 2 x MWT x 1.11
De - Equivalent diameter of the hole.
Do - Outside diameter of the tube.
MWT - Minimum wall thickness.
de = Do - 2 x MWT
= 38 - 2 x 6.8
de = 24.4 mm

Longitudinal ligament Efficiency

The following Fig.3 -5 shows ligament efficiency diagram.

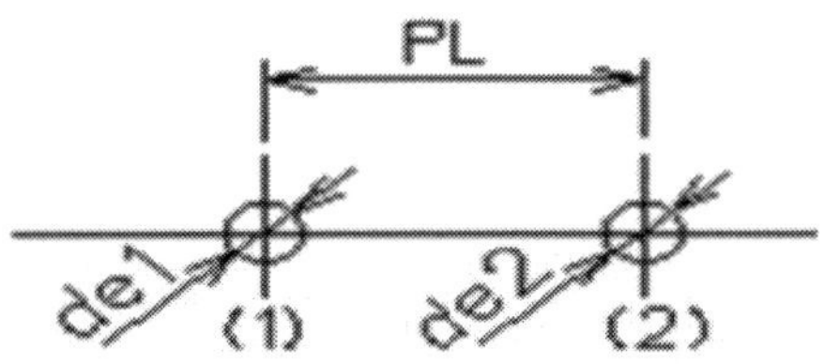

Fig.3 Longitudinal Ligament Efficiency

Formula:

$$E = \frac{PL - dm}{PL}$$

$$dm = \frac{de_1 + de_2}{2}$$

Where:

de_1 - Equivalent diameter of the $hole_1$.
de_2 - Equivalent diameter of the $hole_2$.
dm - Mean equivalent diameter of the hole.
PL - Distance between the two horizontal stubs.
E - Ligament efficiency.

Longitudinal Efficiency

$$dm = \frac{de_1 + de_2}{2}$$

$$dm = \frac{24.4+24.4}{2}$$

$$E = \frac{PL - dm}{PL}$$

$$E = \frac{172 - 24.4}{172} = 0.86$$

Circumferential Efficiency

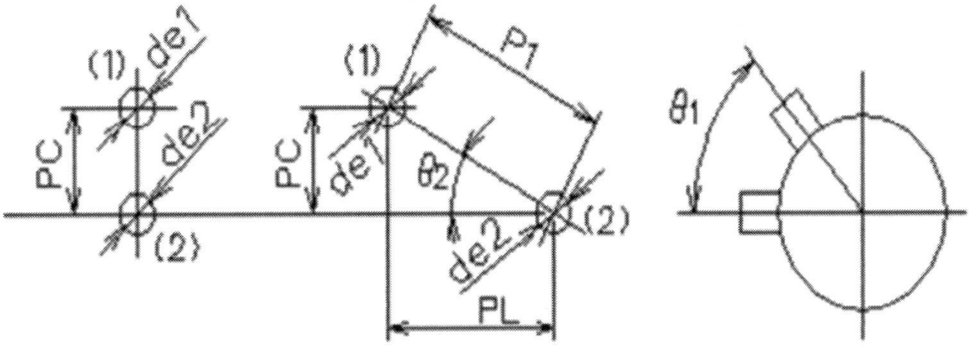

Fig .4. Circumferential Efficienc

$$Pc = \frac{¶ \text{ x } Dc \text{ x } \theta 1}{360}$$

$$= \frac{(273 \text{ x } 0.99 - 2\text{x}18.26)\ 3.14\text{x } 80}{360} = 163.11 \text{ mm}$$

$$E = \frac{2\ (Pc\text{-}de)}{Pc} = 1.70,\ \theta_2 = \tan\urcorner\ Pc/PL$$

Diagonal Efficiency:

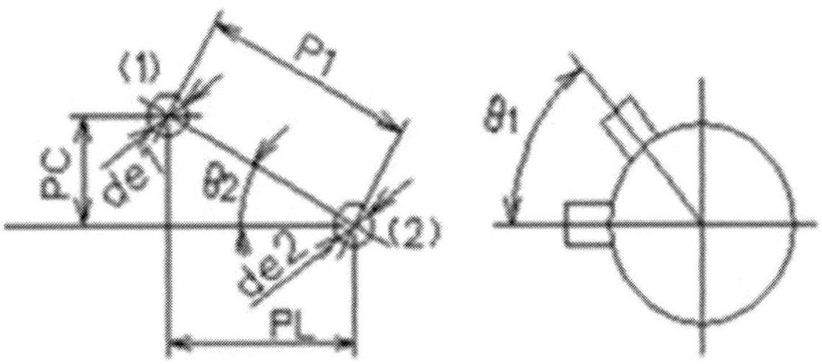

Fig .5 *Diagonal l Efficiency*

$$E = \frac{Sec^2\ \theta_2 + 1 - \left(\frac{Sec\ \theta_2}{P1\ /\ dm}\right) \text{ x } \sqrt{3} + Sec^2\ \theta_2}{0.015 + 0.005 \text{ x } Sec^2\ \theta_2} \text{ x } 100$$

Where:

θ_2 - Aligned angle between the two stubs.
de_1 - Equivalent diameter of the hole $_1$.
de_2 - Equivalent diameter of the hole $_2$.
dm - Mean equivalent diameter of the hole.
PL - Length between the two horizontal stubs.
P_1 - Length between the PL and Pc.
E - Ligament efficiency.

$$E = \frac{\mathrm{Sec}^2(44.825) + 1 - \left(\dfrac{\mathrm{Sec}\ 44.825}{121.25/24.4}\right) \times \sqrt{3} + \mathrm{Sec}^2\ 44.825}{0.015 + 0.005 \times \mathrm{Sec}^2\ 44.825} \times 100 = 0.947.$$

Headers are allowed to cut for the required length using header cutting machine using a rotating saw as belt drive. In order to make the header and saw safe (in high temperatures), the coolant oil is introduced and for smooth cutting. The required length header with the tolerance is again moved to the facing zone. For example a header of OD 273 x 18.26 thick of length 1200 length , required to cut 1050 length.

4. DRILLING AND GROOVING OF THE HOLES:

i. The header in which the holes to be drilled is sent to the drilling machine
ii. If the holes are small ranging from (2.5mm to 10mm) the holes are drilled in Radial Drilled machine
iii. The Schematic representation below shows the drilling of holes in the Radial drilling machine as shown in figures.6 &7
iv. This Radial drilling machine is not a numerical controlled one in which humans are required in order to place the drill bit in the hole location.
v. Such that hole diameter ranging up to 1 meter is drilled In Radial Boring machine is done
vi. After drilling of holes, the holes are to be faced and made grooved in order that the stub should be inserted in it.
vii. Then the fittings of nozzles , end cap and other fittings are fitted up in the header

Fig 6.Drilling bit located

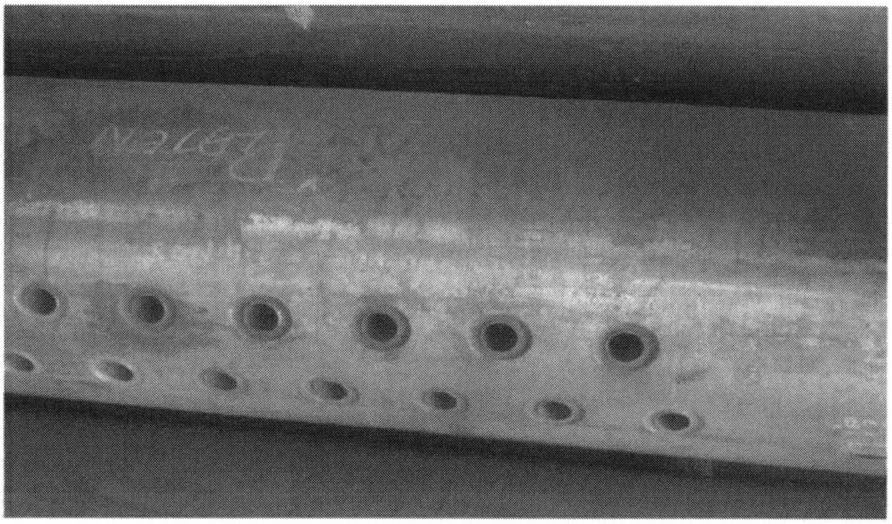

Fig.7 Finish the drilling and grooving operation

5. CONCLUSION

The design of IRB codes have done in this paper with ASME standard, which has followed with proper deign calculation and required input data hence the mandatory requirements of ASME code satisfied by the design calculation. The Actual Size of the drilled holes must be, as per the drawing, but due to error, the size has been differentiate from approved design calculation (i.e. the diameter size has been exceeded). In order to rectify this error, designed reinforcement pad has been attached.

References

1. Steam Boiler Operation by *James J.Jackson*, Prentice-Hall Inc, New Jersey, 1980.
2. Boilers by *Carl D. Shields*, McGraw Hill Book Company, U.S, 1961.
3. Industrial Heat Generation and Distribution -NIFES Training Manual Issued For CEC - India Energy Bus Project
4. Technical Papers, Boiler Congress-2000 Seminar, 11 & 12 January 2000
5. Industrial Boilers by *David Gunn and Robert Horton*, Longman Scientific & Technical, New York
6. Steam Generation, Distribution and Utilisation by TERI, GTZ and EMC
7. Efficient Operation of Boilers by National Productivity Council
8. Linden, A.J.,1 949, Proceedings of Institution of Mechanical Engineers, Vol.160, P.233.
9. Structural Research and Analysis Corporation, California, une 1994, COSMOS/'4 Users Guide Version 1.71
10. Indian Standard Code of Practice for General Construction in Steel. Second Revision (1S:800-1984)

IOP Conf. Series: Materials Science and Engineering **197** (2017) 012046 doi:10.1088/1757-899X/197/1/012046

FLEXURAL BEHAVIOR OF ALUMINUM HONEYCOMB CORE SANDWICH STRUCTURE

VIDYASAGAR MATTA[1], J SURESH KUMAR[2], DUDDU VENKATARAVITEJA[3],GUGGULLA BHARATH KUMAR REDDY[4]

[1]Assistanat Professor,[2]Research Fellow,[3]Student,[4]Student
Faculty of Mechanical Engineering,Sathyabama University, Chennai-600119, Tamilnadu India.
* E-mail address:[1]vsr255@gmail.com,[3]teja0085@gmail.com

Abstract----This project is concerned with the fabrication and flexural testing of aluminium honey comb sandwich structure which is a special case of composite materials that is fabricated by attaching two thin but stiff skins to a light weight but thick core. The core material is normally low density material but its high thickness provide the sandwich composite with high bonding stiffness.Honeycomb core are classified into two types based on the materials and structures. Hexagonal shape has a unique properties i.e has more bonding strength and less formation time based on the cell size and sheet thickness. Sandwich structure exhibit different properties such as high load bearing capacity at low weight and has excellent thermal insulation. By considering the above properties it has tendency to minimize the structural problem. So honey comb sandwich structure is choosed. The core structure has a different applications such as aircraft, ship interiors, construction industries. As there is no proper research on strength characteristics of sandwich structure. So, we use light weight material to desire the strength. There are different parameters involved in this structure i.e cell size, sheet thickness and core height.In this project we considered 3 level of comparison among the 3 different parameters cell size of 4, 6 and 8 mm, sheet thickness of 0.3, 0.5 and 0.7 mm, and core height of 20,25 and 30 mm. In order to reduce the number of experiment we use taguchi design of experiment, and we select the L8 orthogonal array is the best array for this type of situation, which clearly identifies the parameters by independent of material weight to support this we add the minitab software, to identify the main effective plots and regression equation which involves the individual response and corresponding parameters.Aluminium material is used for the fabrication of Honeycomb sandwich structure among the various grades of aluminium we consider the AL6061 which is light weight material and has more strength. By the power press used as forming method we fabricate the honey comb core and stacking the sheets with adhesive as epoxy resin or laser beam welding and sandwich structure will form with two face sheets.Then the specimen is taken to be tested to know the flexural behaviour by the flexural test as 3 point and 4 pont bend test. After testing of two different tests then we get the force vs displacement curve by this we can know the maximum force and by loading configurations and its displacement or deflection then we can calculate flexural stiffness and core shear modulus by the variation of three parameters .Our ultimate aim is to achieve maximum strength by minimum weight.

1. Introduction

A sandwich-structured composite is a special case of composite materials that is fabricated by attaching two thin but stiff and skins to a light weight ,thick core as shown in figure 1. The core material is normally low strength material, the higher thickness provides a sandwich composite with high bending stiffness with overall low density.

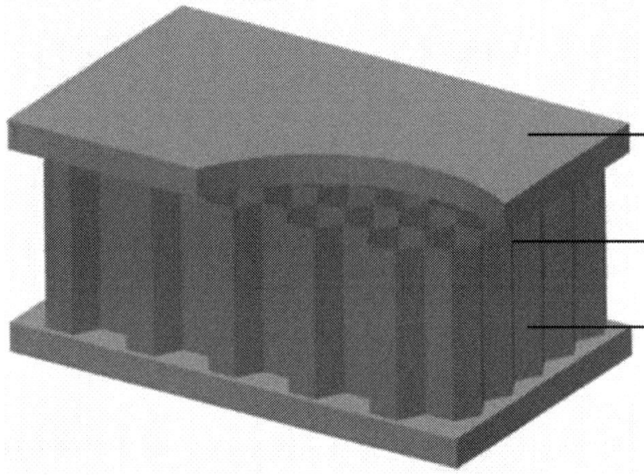

Figure 1. Honeycomb Core Sandwich Structure

Honeycomb sandwich materials are being widely used in weight sensitive and damping structures where high flexuralrigidity is required in many fields mainly in the automobile industry. Honeycomb core sandwich panel is formed by adhering two high strength-rigidity thin-face sheets with a low-density, honeycomb core possessing highstrength and stiffness,by varying the core , thickness and material of the face sheet, Which is possible to obtain various properties and desired performance, particularly in high strength-to-weight ratio. Sandwich construction is widely used because of its ability to provide high bending stiffness coupled with lightweight. Sandwich panels are made of two stiff, strong skins separated by a light weight core.

The objectives of the study is to **e**xamine the influence of aluminium honey comb sandwich structure having core of different sheet thickness, cell size and core height on the static flexural stiffness of sandwich panel with face sheets and to investigate the extent of improvement in the flexural performance characteristics of the aluminum honeycomb sandwich panel.

2. Methodology

Taguchi's **Design of Experiment** is a well ordered and effective way to find the contribution of structural parameters. This method utilizes the orthogonal array of combinations to reduce the number of experiments. Flexural stiffness and core shear modulus in honeycomb core sandwich structure is usually higher in the out-of plane loading because of the high load carrying capacity compared to the in plane loading. As the Flexural stiffness and core shear modulus highly depends on the material properties and structural parameters and hence there is a need to find the contribution. To identify the most influencing structure parameters three different parameter has chosen such as cell size, cell wall thickness and core height. Three variations of the parameters or factors are representative of their low, medium and higher

IOP Conf. Series: Materials Science and Engineering **197** (2017) 012046 doi:10.1088/1757-899X/197/1/012046

levels respectively assuming that the factor effects are linear when the factor changes from one level to the other one. Three factors at three levels results in nine combinations and hence experimental trials are listed in table

Table 1. Flexural stiffness and core shear modulus of Specimen

Trial #	Factors			Flexural stiffness (Mpa)	Core shear modulus(Mpa)
	Cell size(mm)	Core sheet thickness (mm)	Core height(mm)		
1	4	0.3	20	0.53	0.2921
2	4	0.5	25	2	0.837
3	4	0.7	30	6.46	2.215
4	6	0.3	25	0.78	0.355
5	6	0.5	30	2.19	0.9085
6	6	0.7	20	2.73	1.6497
7	8	0.3	30	6.99	2.444
8	8	0.5	20	0.58	0.2876
9	8	0.7	25	4.2	1.69

3. Experiments and Specimens

To Evaluate static flexural stiffness of sandwich panels with different cell wall thickness, cell size and core height of the core by 3-point and 4- point Bend Tests. Flexural tests as per ASTM C393 were carried out to determine sandwich structure bending stiffness (D) and Core Shear Modulus (G) on the Sandwich panels. Three-point and four-point bend tests were conducted at the method of each loading case Flexural testing with central loading has been carried out in accordance with ASTM C 393 using indigenously designed and developed Digital Flexural Test System . The movable carriage of the machine is brought by the loading bar and load is applied gradually at the cross head displacement ratio of 2 mm/min and Plot of load versus deflection is recorded. The two methods were most usually used for the determination of flexural properties of laminates are the three-.point and four-point tests. The states of stress in specimens subjected to three-point or four-point bending tests are somewhat different and may lead to differences in the results.

Sandwich structures subject to three-point bending test are to analyze the local buckling modes of the upper face sheet and cores. TPBT involves the combined load of the bending moment associated with the plastic buckling of compressed upper face sheet, transverse shear between face sheet and core and collapse of compressed cores. The loading configuration was shown in figure 2

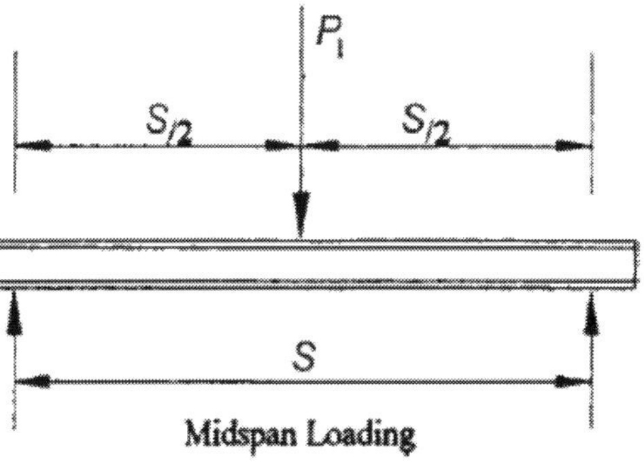

Figure 2 Three Pointbend test and loading configuration

Four point bend test is used effectively to measure the bond, strength in four-point bending a uniform maximum moment and an area of maximum tension at the bottom of the specimen is achieved, it provides values for the modulus of elasticity in bending, flexural stress, strain and the flexural stress-strain response of the material. This test is very similar to the three points bending flexural test. The major difference is that the addition of a 4th bearing brings a maximum larger area of the beam to the maximum stress, as opposed to only the material right under the central bearing.

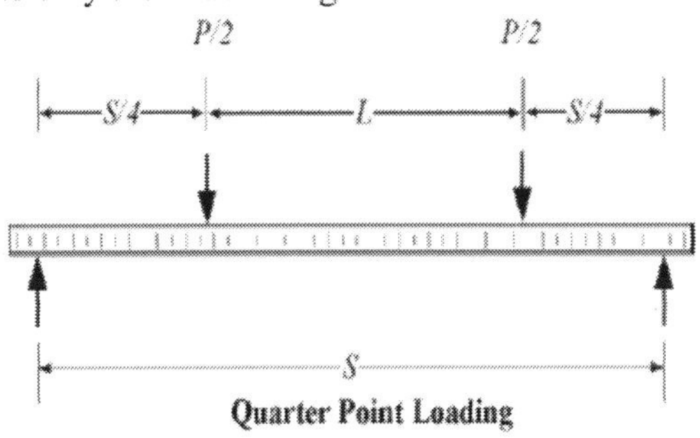

Figure 3 Four Point Loading Configuration

One 3-Point Mid-Span Loading Configuration and One 4-Point Quarter-Span Loading Configuration— The solution for this case (L1 = 0, L2 = S2/2) then calculate the flexural stiffness, shear rigidity, and core shear modulus for each selected value of P1

***D* = flexural stiffness, N-mm2 [lb-in.2],**

$$D = \frac{P_1 S_1^3 \left(1 - \frac{11 S_2^2}{8 S_1^2}\right)}{48 \Delta_1 \left(1 - \frac{2 P_1 S_1 \Delta_2}{P_2 S_2 \Delta_1}\right)}$$

U = transverse shear rigidity, N [lb]

$$U = \frac{P_1 S_1 \left(\frac{8 S_1^2}{11 S_2^2} - 1\right)}{4 \Delta_1 \left\{\left(\frac{16 P_1 S_1^3 \Delta_2}{11 P_2 S_2^3 \Delta_1}\right) - 1\right\}}$$

IOP Conf. Series: Materials Science and Engineering **197** (2017) 012046 doi:10.1088/1757-899X/197/1/012046

$$\Delta = \frac{P(2S^3 - 3SL^2 + L^3)}{96D} + \frac{P(S-L)}{4U}$$

G = core shear modulus, MPa [psi],

$G = \frac{U(d-2t)}{(d-t)^2 b}$

Where,
D = beam mid-span deflection, mm [in.],
P = total applied force, N [lbf],
d = sandwich thickness, mm [in.],
b = sandwich width, mm [in.],
t = facing thickness, mm [in.] ,
S = support span length, mm [in.],
L = load span length, mm [in.]

3.1.Materials and specimens

Preparation of honeycomb core sandwich structure and face sheets according to ASTM standards of length 200mm and width 50mm and it is fabricated by using power press method. In all, nine specimens were prepared with different cell size, sheet thickness core height with an insert face sheet. Use of appropriate test methods as per standards and guidelines in order to evaluate the flexural properties like stiffness and core shear modulus by the digital flexural system we get the maximum load carrying capacity for clear understanding of static behavior of Honeycomb sandwich structure.

Commercially available of aluminum of different thickness 0.3, 0.5 and 0.7 mm is used for making the honeycomb sandwich structure the material was cut into correct size and built to a length of 120cm and width of 10 cm by using shearing machine.

Press brake is used to forming the hexagonal structure by aluminum sheet. A press brake is a special type of machine press which bends the sheet metal into required shape and size. A good example of the type of work a press brake can do is the back plate of a computer case. The die press sheets were trimmed according to the ASTM standards of length of 200mm and width of 50mm through band saw.

Adhesive bonding, are currently being used for joining aluminum alloy facing material and honeycomb core. An adhesive is the mixture of resin and hardener mixed in equal ratio and weight was then applied. Adhesive joints for the sandwich construction is having advantages such as fewer pieces, lower weight, good load distribution and lower cost. Epoxy resin is the best adhesive for the metallic materials to stack one over another.

4. RESULTS AND DISCUSSION

The Flexural stiffness (D) is the resistance of a member against bending deformation. It is a function of elastic modulus E, the area moment of inertia I of the beam cross-section about the axis of interest, length of the beam and beam boundary condition. Bending stiffness of a beam can be analytically derived from the equation of beam deflection when it is applied by a force. Bending stiffness in beams is also known as Flexural rigidity.

The shear modulus is used for the deformation which takes place when a force is applied parallel to one face of the object while the opposite face is held by another equal force. That's why the shear modulus is sometimes called the modulus of rigidity. In first approximation there is no change in volume in this deformation.

Load vs. deflection curve for three point bend test were shown in the figure

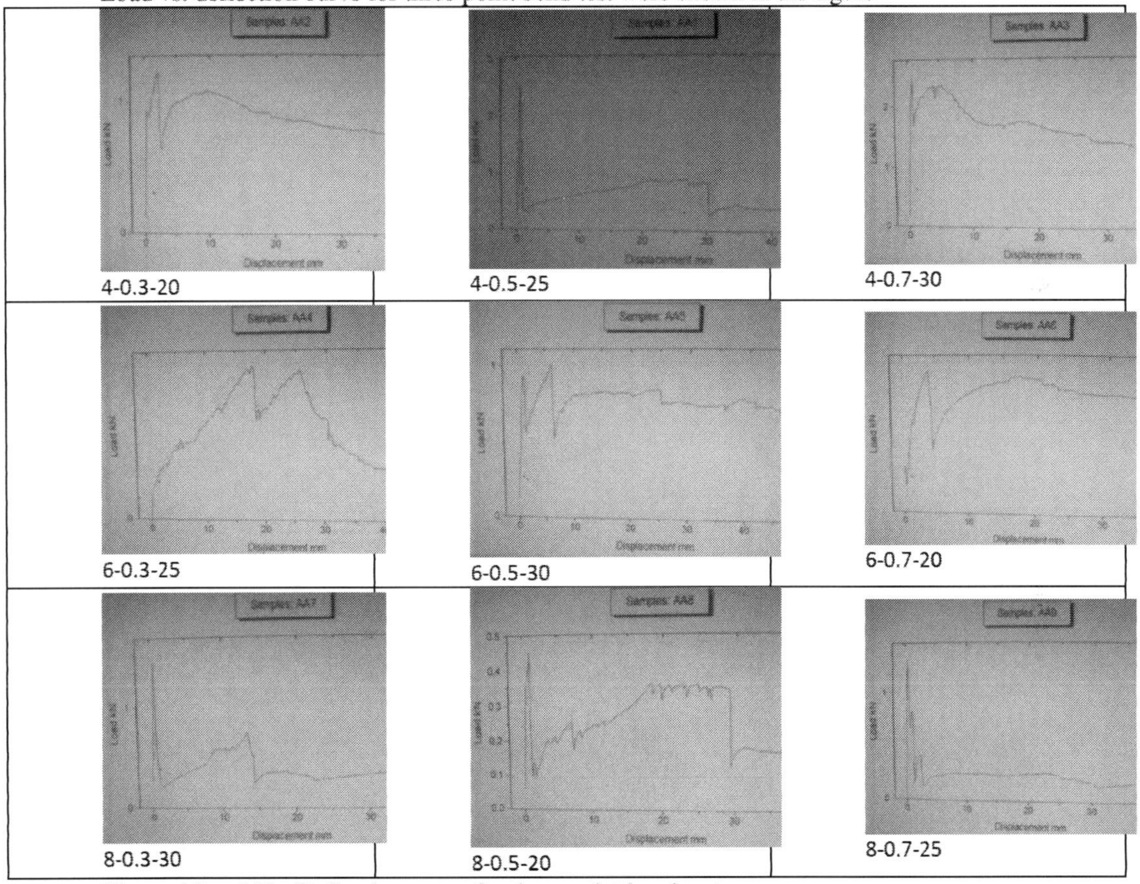

Figure4. Load Vs. Deflection curve for three point bend test

From the above graphs the maximum load and its corresponding deflection of specimen in three point bend test is tabulated in table 2.

Table2. Details of load and deflection of three point bend specimen.

Experiment	Deflection (mm)	Yield load (kN)	Young's Modulus (Mpa)
1	0.8045	2.777	7901

IOP Conf. Series: Materials Science and Engineering **197** (2017) 012046 doi:10.1088/1757-899X/197/1/012046

2	2.397	1.22	2269
3	0.7618	2.512	2233
4	17.83	0.7497	6573
5	6.268	1.007	1061
6	4.026	0.7941	396.4
7	0.7061	1.513	1717
8	0.9906	0.456	1662
9	0.4074	1.319	3603

Load vs deflection curve for three point bend test were shown in the figure

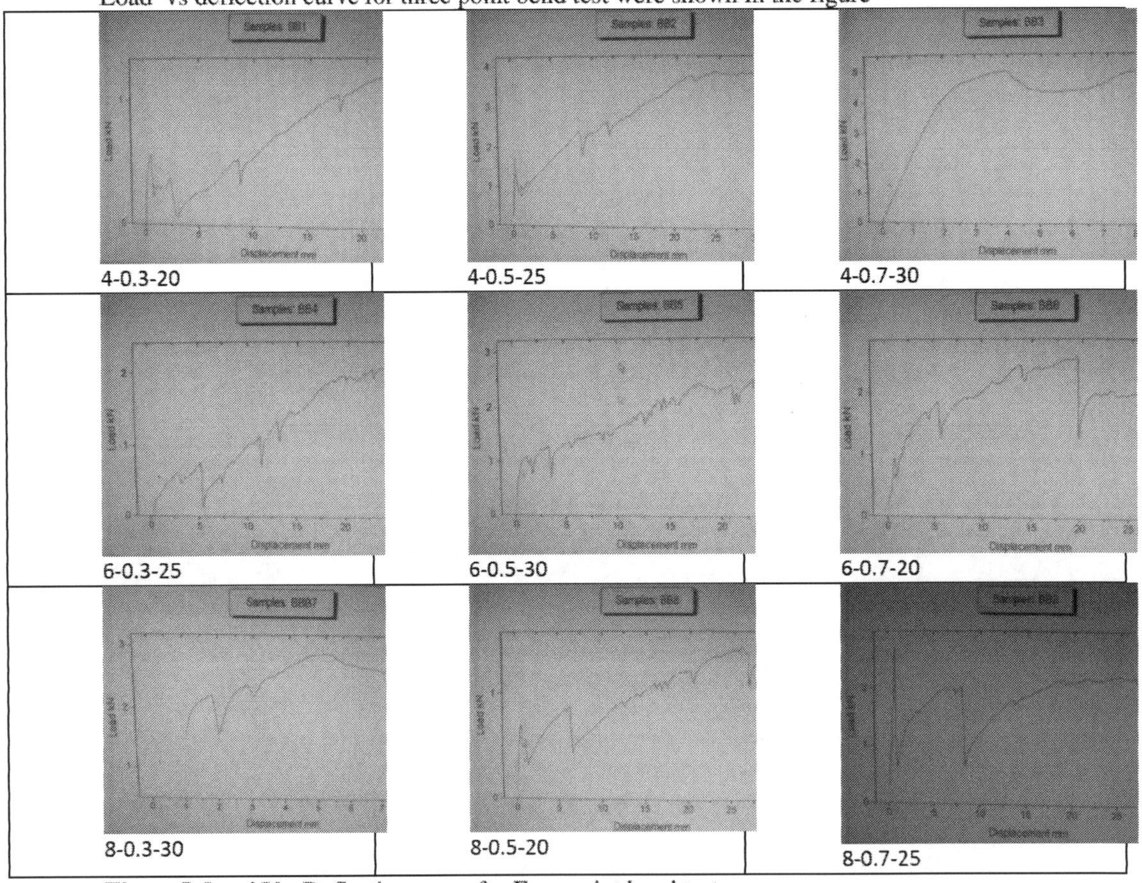

Figure5. Load Vs. Deflection curve for Four point bend test

From the four point bend test,the maximum load and its deflection is tabulated in the below table.

Table 3. Load vs Deflection of Four point bend Test

Experiment	**Deflection (mm)**	**Yield load (kN)**	**Youngs Modulus**
1	22.24	1.203	59.23
2	24.39	3.838	189.2
3	8.593	5	905.4
4	26.62	2.196	57.68
5	27.42	2.911	42.26
6	19.52	2.561	245
7	5.016	2.88	486.9
8	25.76	1.426	67.84
9	0.7586	2.689	2506

In the main effects plot, the deviation from horizontal line indicates the process parameter has more influence on response variables. Cell size and cell wall thickness the lower level has average effect and medium level has poor effect and higher level has very good effect on core shear modulus and flexural stiffness. Whereas core height effect is moderate in the beginning and for higher value the properties got increased more.

IOP Conf. Series: Materials Science and Engineering **197** (2017) 012046 doi:10.1088/1757-899X/197/1/012046

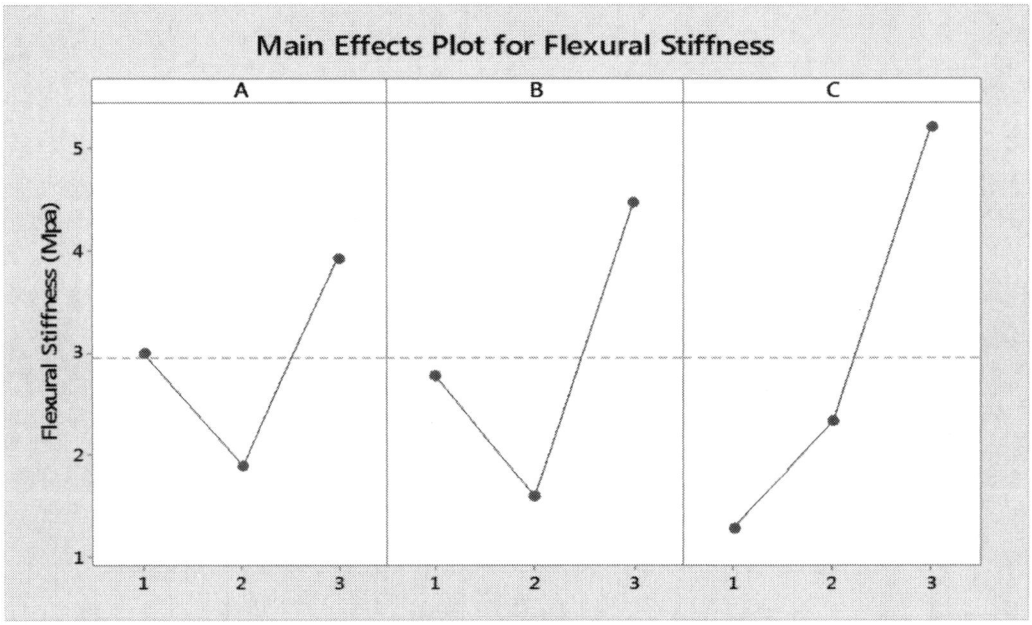

Figure6. Main Effects Plot for Flexural Stiffness

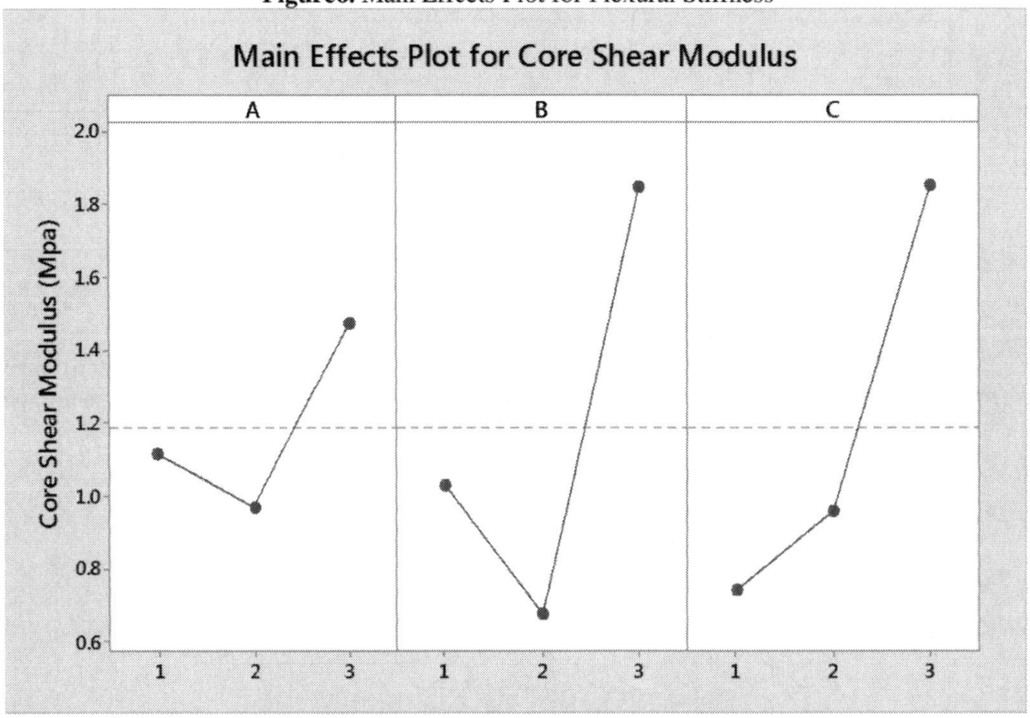

Figure7. Main Effects Plot for Core Shear Modulus

The interaction plots among all the parameter were shown in figure. There exists a strong interaction among all the parameters.

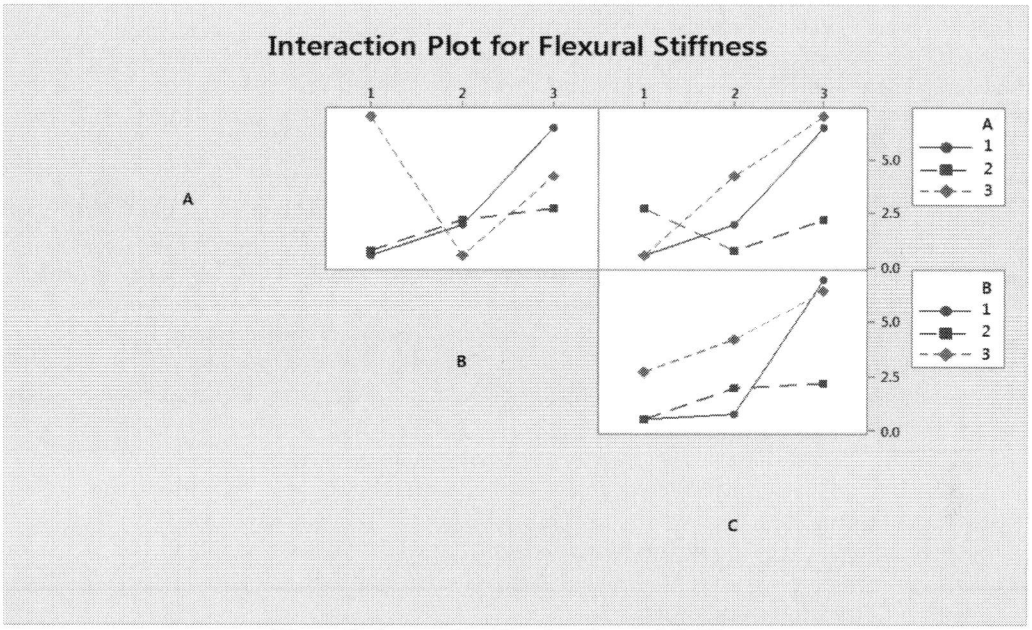

Figure8. Interaction Plot for Flexural Stiffness

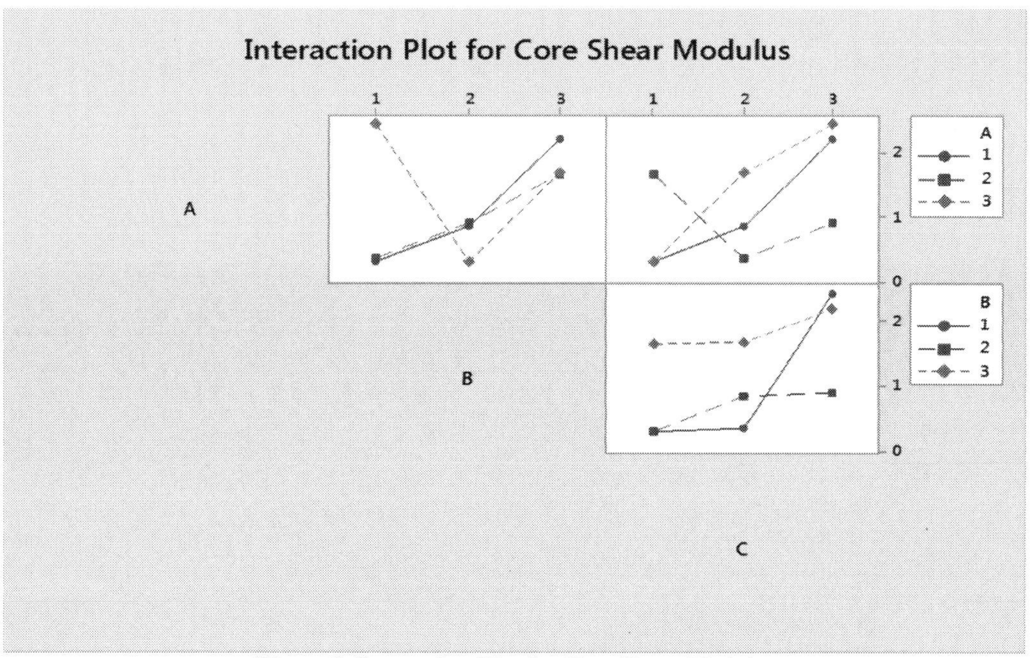

Figure9. Interaction Plot for Core Shear Modulus

IOP Conf. Series: Materials Science and Engineering **197** (2017) 012046 doi:10.1088/1757-899X/197/1/012046

5. CONCLUSION

In this finding, Taguchi's design of experiment with orthogonal array was used to optimize the response characteristics of aluminum honeycomb core sandwich structure. The order of importance of the influencing factor is as follows: core height, cell wall thickness and cell size respectively.

References

[1]. Aktay.L, A.F. Johnson, B. Kroplin, Engineering Fracture Mechanics 75 (2008) 2616–2630

[2]. Allen.H, Analysis and Design of Structural Sandwich Panels. PergamonPress,Oxford, 1969.

[3]. ASTM C274-99. ASTM standard terminology of structural sandwich constructions, West Conshohocken, PA: ASTM International; 1999.

[4]. ASM International Metals Park. 1987. Engineered Materials Handbook,

[5]. Belouettar S, Abbadi A, Azari Z, Belouettar R, Freres P. Experimental investigation of static and fatigue behaviour of composites honeycomb materials using four point bending tests. Compos Struct, 2009;87(3):265–73.

[6]. Frostig Y, M. Baruch, Bending of sandwich beams with transversely flexiblecore, American Institute of Aeronautics and Astronautics, Journal 28 (3) (1990) 523–531.

[8]. Gdoutos E.E, I.M. Daniel, K.A. Wang, J.L. Abot, Nonlinear behavior of Composite sandwich beams in three-point bending, Experimental Mechanics 41 (2) (2001) 182–189

[9]. Gibson L.H, M.F. Ashby, Cellular Solid – Structure and Properties, second ed.,Cambridge University Press, Cambridge, 1997.

IOP Conf. Series: Materials Science and Engineering **197** (2017) 012047 doi:10.1088/1757-899X/197/1/012047

Design Intent for CAD Modeling Features Using Boolean Operations

[1]Sonawane Chandrakant R, [2]Ghadge Sujit

[1]Mechanical Engineering Department, Symbiosis Institute of Technology, Symbiosis International University, Pune, Maharashtra, India.
[2] PG Student, Pune, Maharashtra, India.

Corresponding author's e-mail address : chandrakant.sonawane@sitpune.edu.in

Abstract: The objective of this paper is to add one more enhancement to design intent by adding a rule to find the intersection edges created between Boolean features. Design Intent is a core module in CAD software which is used for smart design of referencing elements to create required features. In general, the particular design intent will form a particular rule which can be utilized for specified purpose. In this paper, a design intent rule for Edge Blend feature is designed. The rule is also implemented and integrated with CAD software. The major contributions of this paper is to create a new intent design rule which picks the edges of the feature Present design intent rule is intended to pick the intersection edges of the feature, in doing so; the intent will avoid referencing to topology over referencing to hierarchy objects for greater reliability

1. Introduction

Today's Computer Aided Design (CAD) software's are designed to improve the productivity of the user. In the interest of robust update, body in white workflow avoids topology referencing. Update feature replaces work flow in CAD software and are very robust. This success of replace workflow relies on referencing hierarchy objects. Input of topology (an edge or a face) is less reliable as compared to the input of a hierarchy object (a body or a feature). While modelling, the user generally has the intent to select bodies, faces, curves and points. Many of the commercially available software like PTC PRO-E, Siemens-NX, and Dassault Systems-CATIA have their own mechanism to select the object in graphics window. CAD has been evolving by incorporating various ways of collecting the entities by finding new ways of smart selections. Many researchers' were working on smart selection of an entity resulting various rules have been added to empower the selections from graphic window and part navigator.

P. Hachenberger et al [1] developed algorithm for the geometric operations on "Selective Nef Complexes"(SNC). Their algorithm is based on Boolean operations where partitions of three dimensional entities are carried out by planes and whole domain is divided into the cells. In order to optimize the performance they implemented the two methods of finding topological intersections viz. kd-Tree method and fast box-intersection algorithm method. J. M. Smith et al [2] worked on B-rep models which contains polyhedral boundaries and presented an algorithm which typically utilizes Boolean operations. Their algorithm is so robust that for given Boolean with valid connectivity the

algorithm always generates a result. The algorithm is also finds and produce results with fewer numerical errors associated with geometries or arithmetic or even with input data.
Gang Mei et al [3] developed approach for triangulated faces. Both open and closed surfaces had been considered. Their Boolean operation based algorithm identifies the surfaces/edges formed due to intersection, union and subtraction features. Yu Peng et al [4] developed an algorithm for planar polygon constructed with Boolean operations. The simplex theory based mathematical model is developed and applied in the algorithm to indentify the required intersecting surfaces. Cha-Soo Jun et al [5] utilized the topological transition points over the interaction of curve. They pre-processes the curve and surfaces to locate these topological transition point. Hence their algorithm can compute the interactions points created due to the parametric regular surface with a set of available parallel planes.
Ouyang Ying-xiu et al [6] developed an algorithm which also based on finding the topological transition points. Their algorithm utilizes the available topological relations to find first topological transition point/loop. Then these previously computed loop/points utilized to compute new subsequent loop occurring on the intersecting geometries. Kyu-Yeul Lee et al [7] presented an algorithm based on tracing principle. For surface boundaries identification, they utilized trace-terminating condition along the surface-surface intersection curves. A. Borkowski et al [8] presented mathematical approach for finding intersection curves. Their work mainly focuses on locating terrain edges for an available three-dimensional entity. Pinghai Yang et al [9] developed algorithm based on the moving least-squares (MLS) surface. They utilized MLS to obtain point-sampled geometry from the underlying surface representation. They also derived a closed form formula which helps to evaluate the curvature of planar surface/curves.
Laurent Buse et al [10] developed the generalized matrix based representations of the parameterized surfaces where the intersection curve of two topological surfaces as the zero set of a matrix determinant is obtained. In this way, a compact and efficient representation of intersection curves is obtained by allowing reduction of a few geometric operations on curves to matrix operations using linear algebra. Manuel Ventura et al [11] developed methodology to compute the intersection formed due to the complex curves. They consider the application of modeling the ship-Hull which involves interaction of many complex curves and surfaces. The major focus is to study the surfaces form due to the combinations of non uniform rational B-Splines (NURBS). Other algorithms, methodology and/or mathematical analysis work for computing intersection curves and computing the two rational parametric surfaces can also be found in [12, 13, 14].
From above discussion it is clear that researchers' are continuously trying to modify the Boolean operations to make the task of finding or locating the intersection edges more generic, scalable, robust, and of reduced computational efforts. Many algorithms for Boolean operations on polygons were based on the use of bitmaps and have many drawbacks like large memory usage. Researchers described various methods for Boolean operations on solids/planes/surfaces with specified boundary representation. Their main interest is to reduce an available three dimensional (3D) geometrical problem to two dimensional (2D) one using reconstruction or other suitable ways which ultimately makes the algorithm more robust. But such algorithm could process objects with planar surfaces.
This paper aims to add a rule to find the intersection edges created between Boolean features resulting in enhancement to CAD design. The key task in implementing this new rule is to evaluate the intersection edges, which should ultimately correspond to the optimized solution. The proposed new rule is capable to evaluate intersection surface/edge of the Boolean bodies even if it has one or more intersecting bodies. The scope is limited to the evaluation of intersection of the Boolean like features

2. Basics of Selection Intent

In CAD software, there are various features to carry out modelling operations. Every feature takes some input and returns desired output. When user launches a particular feature command, a user Interface (UI) dialog pops up asking for various inputs. As far as topological inputs are concerned,

user may have intent to specify point, curve, face, or body. There are number of ways to specifying the inputs. The term referencing points to the way in which input is specified.
Topological referencing as shown in figure 1 is the method of input specification, in which user directly specifies the basic topology required for that feature. For example, "Measure Face" command needs a face. If user specifies it with a single click on individual face, this becomes a case of topological referencing. Every feature asking for topological input is enabled with this type of referencing. In Hierarchical referencing, it is not required that user specifies input of the same level of Topology as required by the command. The user actually specifies a topologically higher entity, and the required entity is evaluated from the input.

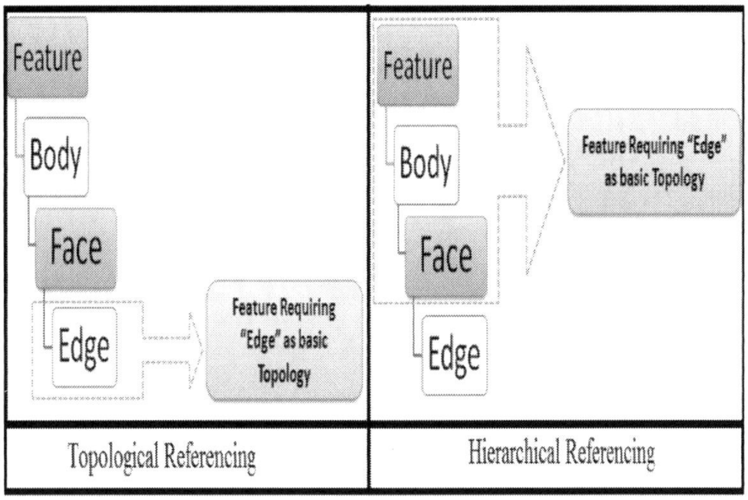

Figure 1. Topological referencing and Hierarchical referencing.

For example, let's assume user wants to evaluate of perimeter of a face using "Measure Length" command. Measure Length command actually works on curves. So, basic topology here is a curve. Individually specifying every edge of the face is example of topological referencing, while if user sets "Face Edges" as a rule and specifies the edges by selecting the face directly, it is the example of hierarchical referencing.

3. Computational Algorithm

Figure 2 shows the algorithm for select Feature/Edge. The flow chart is self explanatory. The filter selection does validation before collection of edges and then create rule for selection. Figure 3 shows the workflow diagram for computer code generated.

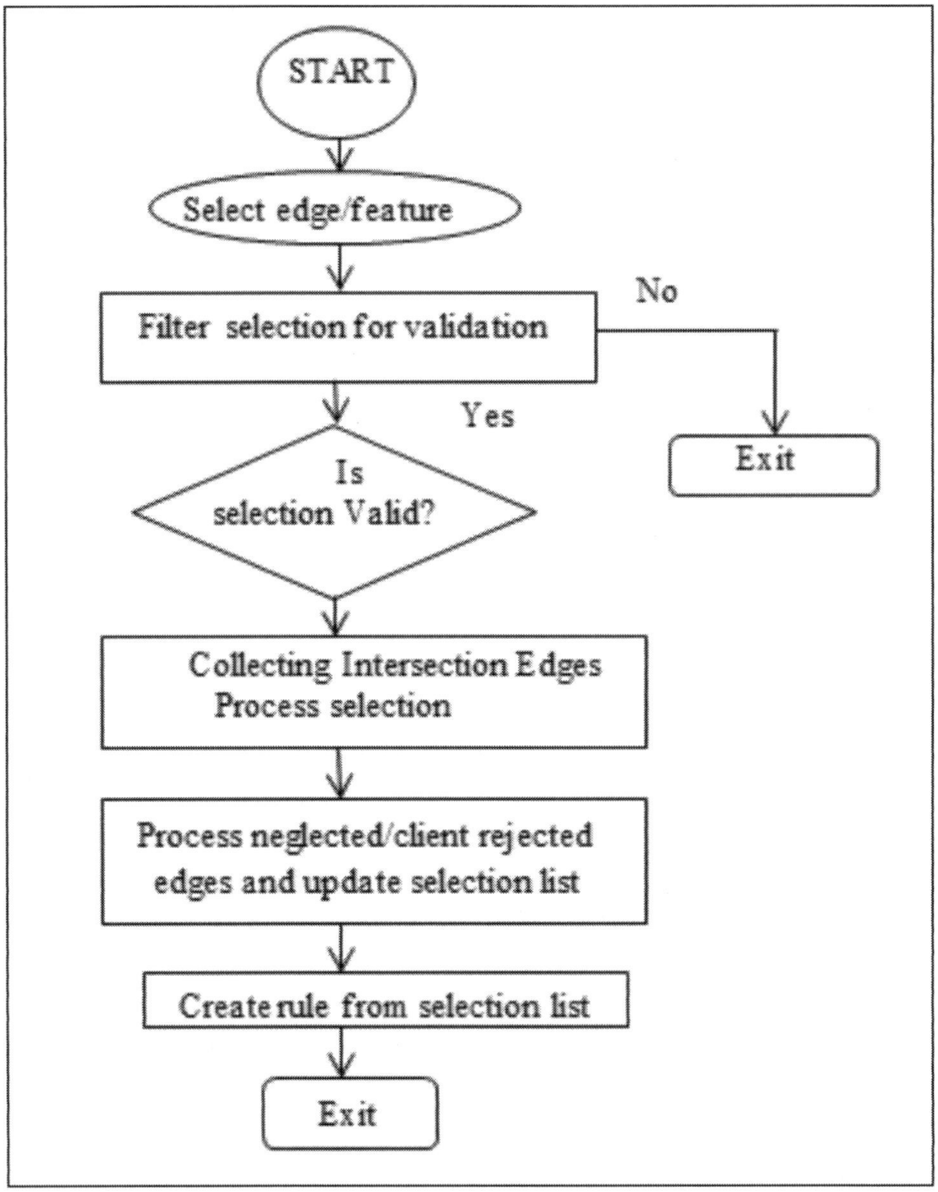

Figure 2. Activity Flowchart.

IOP Conf. Series: Materials Science and Engineering **197** (2017) 012047 doi:10.1088/1757-899X/197/1/012047

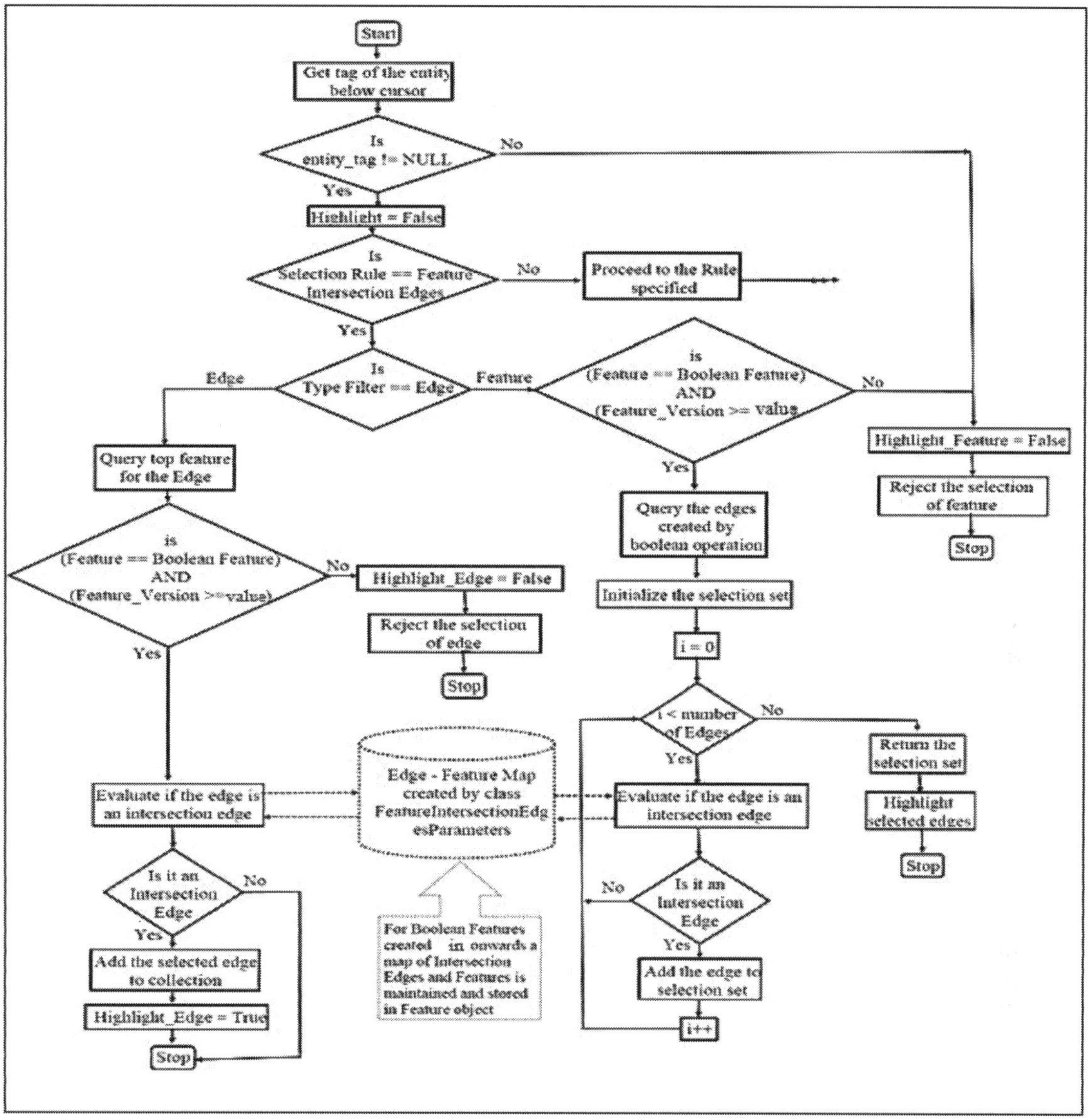

Figure 3. Workflow diagram.

4. Time Saving Analysis

As the new rule assists to grab the intersection edges at very ease it is good to realize how much time saving is achieved with the new enhancement. Here, we tried to apply edge blend to various complex parts using conversional single selection rule and "Feature Intersection Edges" rule.

Table 1 shows that how Feature Intersection Edges are saving the time to apply blend at intersection of Boolean features like Unit, Subtract and Intersect. From Table 1 it can be observed that about 80% - 95% time saving is achieved using new selection intent rule.

Table 1: Time saving analysis

(*I - Total Blend Length (mm); II - No. of clicks Required with single selection; III - Time required with single Selection (sec); IV - No. of clicks Required with Feature Intersection Edge Rule; V - Time required with Feature Intersection Edge Rule; VI - Percentage Time Saving)*

Sr No	Image of the Part	I	II	III	IV	V	VI
1.	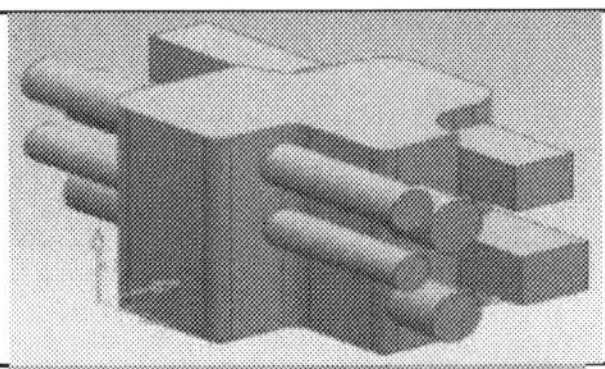	562	24	30	1	2	93.33
2.	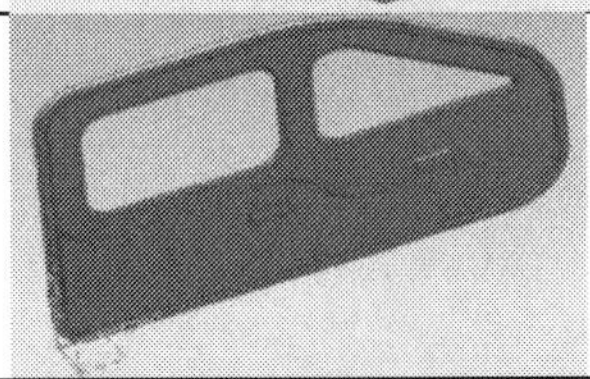	1745	34	270	1	30	88.89
3.	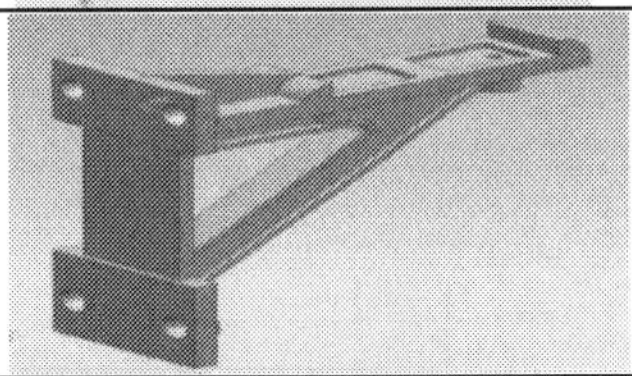	550	89	240	1	45	81.25
4.		845	157	789	1	94	88.09
5.	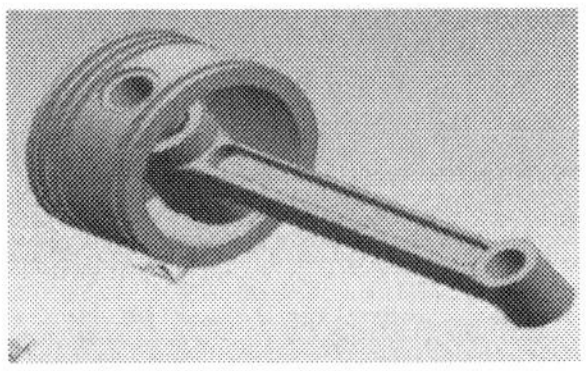	200	78	450	1	40	91.11
6.		2457	540	2480	1	120	95.16
7.		5480	480	1170	1	185	84.19

5. Conclusion

The New rule for selection intent is developed and applied successfully. It is evident from the testing results that new rule has enhanced CAD's Design Intent capability to incorporate the idea of hierarchical referencing. The new Design Intent rule leaves behind the caveat of not implementing for interpart edges.
There is scope to give thought for implementing the rule for interpart collections too. Also, new rule can be used for picking the edges for other modelling features. The scope was limited to Boolean like features and future enhancement is possible so that intersection edges can be picked for non-Boolean like features.
The current rule enhanced the usability of edge blend tool of CAD software. Design time of the designer is reduced and the accuracy of work will be increased.

Acknowledgment

Thanks to Siemen, Pune for their valuable interaction and suggestions contributed in the research work of the new rule of Selection Intent.

References

[1] Peter Hachenberger, Lutz Kettner, Kurt Mehlhorn, Boolean operations on 3D selective Nef complexes: Data structure, algorithms, optimized implementation and experiments, Computational Geometry, 2007, Vol. 38, issue1-2, 64–99.
[2] J. M. Smith and N.A. Dodgson, A topologically robust algorithm for Boolean operations on polyhedral shapes using approximate arithmetic, Computer Aided Design, 2007, Vol. 39, issue 2, 149-163.
[3] Gang Mei and John C. Tipper, Simple and Robust Boolean operations for Triangulated Surfaces, arXiv preprint arXiv:1308.4434.
[4] Yu Peng, Jun-Hai Yong, Wei-Ming Dong, Hui Zhang, Jia-Guang Sun, A new algorithm for Boolean operations on general polygons, Computers and Graphics, 2005, Vol. 29, issue 1, 57-70.
[5] Cha-Soo Jun, Dong-Soo Kim, Deok-Soo Kim, Hyun-Chan Lee, Tien-Chein Chang, Surface slicing algorithm based on topology transion, Computer Aided Design, 2001, Vol. 33, issue 11, 825-838.
[6] Ouyang Ying-xiu, Tang Min, Lin Jun-cheng, Dong Jin-xiang, Intersections of two offset parametric surfaces based on topological analysis, Journal of Zhejiang University Science, 2004, Vol. 5(03), 259-268.
[7] Kyu-Yeul Lee, Doo-Yeoun Cho, Tar-Wan Kim, A Tracing Algorithm for Surface-Surface Intersections on Surface Boundaries, Journal of Computer Science and Technology, 2002, Vol. 17, No 6, 843-850.
[8] A. Borkowski, W. Keller, Global and local methods for tracking the intersection curve between two surfaces, Journal of Geodesy, 2005, Vol. 79, issue 1, 1-10.
[9] Pinghai Yang, Xiaoping Qian, Direct Boolean intersection between acquired and designed geometry, Computer Aided Design, 2009, Vol. 41, issue 2, 81-94.
[10] Laurent Busé, ThangLuu Ba, The surface/surface intersection problem by means of matrix based representations, Computer Aided Geometric Design, 2012, Vol. 29, issue 8, 579-598.
[11] Manuel Ventura and C. Guedes Soares, Surface intersection in geometric modeling of ship hulls, Journal of Marin Science and Technology, 2012, Vol. 17, issue 1, 114-124.
[12] V. B. Sunil, Rupal Agarwal, S. S. Pande, An approach to recognize interacting features from B-rep CAD models of prismatic machined parts using a hybrid (graph and rule based)

technique, Computers in Industry, 2010, Vol. 61, issue 7, 686-701.
[13] Tobias Dipper, Xun Xu, Peter Klemm, Defining, recognizing and representing feature interactions in a feature-based data model, Robotics and Computer-Integrated Manufacturing, 2011, Vol. 27, issue 1, 101-114.
[14] Li-Yong Shen, Jin-San Cheng, XiaohongJia, Homeomorphic approximation of the intersection curve of two rational surfaces, Computer Aided Geometric Design, 2012, Vol. 29, issue 8, 613-625.

IOP Conf. Series: Materials Science and Engineering **197** (2017) 012048 doi:10.1088/1757-899X/197/1/012048

Design and Analysis of Drive Shaft using Kevlar/Epoxy and Glass/Epoxy as a Composite Material

P Karthikeyan[1], RGobinath[2], L Ajith Kumar[3], D Xavier Jenish[4]

[1,2]Assistant Professor, Sathyabama University
[3,4]U.G Student, Sathyabama University
Email: kksivakodi@gmail.com

Abstract. In automobile industry drive shaft is one of the most important components to transmit power form the engine to rear wheel through the differential gear. Generally steel drive shaft is used in automobile industry, nowadays they are more interested to replace steel drive shaft with that of composite drive shaft. The overall objective of this paper is to analyze the composite drive shaft using to find out the best replacement for conventional steel drive shaft. The uses of advanced composite materials such as Kevlar, Graphite, Carbon and Glass with proper resins ware resulted in remarkable achievements in automobile industry because of its greater specific strength and specific modulus, improved fatigue and corrosion resistances and reduction in energy requirements due to reduction in weight as compared to steel shaft. This paper is to presents, the modeling and analysis of drive shaft using Kevlar/Epoxy and Glass/Epoxy as a composite material and to find best replacement for conventional steel drive shafts with an Kevlar/epoxy or Glass/Epoxy resin composite drive shaft. Modeling is done using CATIA software and Analysis is carried out by using ANSYS 10.0 software for easy understanding.The composite drive shaft reduces the weight by 81.67 % for Kevlar/Epoxy and 72.66% for Glass/Epoxy when compared with conventional steel drive shaft.

1. Introduction

A driveshaft in automobile is used to transmit power from engine to wheels. Driveshaft must be operated by constantly changing angles between the transmission and axle. A high quality steel is a common material used for constructing drive shaft. The bending natural frequency of the beam is directly proportional to the square of the length and proportional to the square root of specific modulus. Because of that the steel drive shafts are manufactured by two pieces to increase the bending natural frequency. The two pieces of drive shaft with three universal joints, a bracket and a center supporting bearing, which increases the total weight of a vehicle and decreases the power transmission due to its inertial mass(5).Replacing conventional metallic structures with composite structures has many advantages due to its higher specific strength and stiffness of composite materials(3). Also its weight is lesser than steel and aluminum with similar strength. Composite drive shaft is manufactured by single piece it will eliminate the assemblies of three universal joints, a bracket and a center supporting bearing of two piece steel drive shaft. Also, composite drive shaft will have lower modulus of elasticity which results in decrease in stress and best shock absorber when torque is more[6].The advanced composite materials such as Kevlar, graphite, carbon and glass with proper resins ware resulted in notable achievements in automobile industry(4). Because of their elastic propertiesit will increase the torque they can carry as well as withstand the rotational speed. This project provides the

IOP Conf. Series: Materials Science and Engineering **197** (2017) 012048 doi:10.1088/1757-899X/197/1/012048

design and analysis of Kevlar/Epoxy and Glass/Epoxy as a composite material for the drive shaft and by comparing their result with conventional steel drive shaft to find the best replacement for it.

2. Material Selection

2.1 Materials for Steel drive shaft

Steel (SM45C) is a material mostly used for conventional drive shaft. The steel material which are used for conventional drive shaft must satisfy the specifications such as capacity of torque transmission, bending natural frequency and capability of buckling torque. The properties of the material steel (SM45C) are shown in Table 1.

2.2 Composite Materials for drive shaft

A composite material is a material made by combining two or more constituent materials with different physical or chemical properties and to produce a material with uniqueness different from the individual components. The individual components remain separate and dissimilar within the completed structure. In practice high performance composites are formulated using a bulk material(the matrix) and fiber or fabric reinforcement to increase the strength and stiffness of the matrix.Reinforcements may be in the form of fibers, particles or flakes which gives the length, orientation, shape and material.The matrix factors gives thecompressive strength, transverse modulus, shear modulus and strength, co-efficient of thermal expansion and thermal resistance.

Glass Fiber

These are made by blending quarry products like sand, kaolin, limestone, colemanite at 1,600°C to get liquid glass then it is passed through micro-fine bushings and at the same timeit is cooled to produce glass fiber filaments from 5-24μm in diameter. By varying the quarry products, different types of glass can be produced. E-glass is the most common reinforcing fiber used in polymer matrix composites. because of its good tensile and compressive strength, stiffness and poor resistance. It is available in the form of strand, yarns and Rovings. Strand is a compactly associated bundle of filaments, yarns are closely associated bundle of twisted filaments and rovings are loosely associated bundle of untwisted filaments.

Table1 Mechanical properties of Steel (SM45C)

Properties	Symbols	Units	Steel
Young's modulus	E	GPa	207
Shear modulus	G	GPa	80
Poisson's Ratio	V	---	0.3
Density	ρ	Kg/m^3	7600
Yield strength	Sy	MPa	370
Shear strength	Ss	MPa	---

Kevlar Fiber

Kevlar fiber is developed by Stephanie Kwolek at DuPont in 1965, it was first used as a replacement for steel in racing tyres because of its high strength. Later it is used as a fiber or as a ingredient of composite materials in ropes or fabric sheets. Kevlar fiber has a tensile strength of about 3.620×10^6 KPa, and relative density of 1.44.

Kevlar's structure consists of relatively rigid molecules which will perform a sheet like structures fairly like silk protein. By measuring high tensile strength to weight ratio of Kevlar with steel, Kevlar is 5 times stronger than steel because of that it has many applications on bicycle tires, racing body armour and to make modern drumheads due to its high impact strength

Table 2. Material properties of composite materials

Properties	Symbols	Kevlar/Epoxy	Glass/Epoxy	Units
Young's modulus X direction	E	95.71e9	5e10	Pa
Young's modulus Y direction	E	10.45e10	1.2e10	Pa
Young's modulus Z direction	E	10.45e10	1.2e10	Pa
Major Poisson's Ratio XY	V	0.34	0.3	---
Major Poisson's Ratio YZ	V	0.37	0.3	---
Major Poisson's Ratio ZX	V	0.34	0.3	---
Shear modulus XY	G	25.08e9	5.6e9	Pa
Shear modulus YZ	G	25.08e9	5.6e9	Pa
Shear modulus ZX	G	25.08e9	5.6e9	Pa
Density	ρ	1402	2100	Kg/m^3

3. Material Fabrication and Testing

The specimens were prepared with the Kevlar/Epoxy and Glass/Epoxy laminates according to the ASTM standard. The specimens were undergoing for tensile testing, flexure testing and impact testing by Universal testing machine and Impact testing machine.

3.1 Laminate Materials

In this laminate, Kevlar fiber and Glass fiber were used as a reinforcement and matrix Epoxy. Resin, adhesive LY556 and Hardener HY951 were used at the ratio of 10:1.Epoxy polymer materials physically consist of two necessary components, a resin and a hardener. Initially it will be in the form of liquid and by chemical reaction it will be converted into the solid. These type of polymers are mechanically strong and more adhesive during the conversion of liquid to solid.The resin and hardener are prepared with the right quantity relative to one another to bring out the chemical reaction and these are kept in a separate container in liquid form.

3.2 Laminate Methodology

The material fabrication usually involves mixing or saturating the reinforcement and then allowing the matrix to bind together into a rigid structure with the help of chemical reaction. The operation is usually done in an open mold by Hand Lay-up method. Resins are concentrated by hand into fibers which are in the form of woven, stitch or bonded fabrics. This is usually done by rollers or brushes, for forcing resin into the fabrics by means of rotating rollers and a tub of resin. Then the laminates are left to cure under normal atmospheric conditions.

Step 1 - The surface of the mold is cleaned and a release agent is applied. The release agent will be in the form of liquid or solid and the improper cleaning of mold will affect the function of release agent.

Step2 - The primary layer of fiber glass structure is made with same resin is laid up on the mold surface. This layer protects the laminate from surface irregularities and surface abrasion. A peel ply is placed on the top of the primary layer, it will be removed finally.

Step 3 - Prepare the epoxy resin by mixing the adhesive (LY556) and hardener (HY951) in the ratio 10:1 and mixing it continuously for lamination.

Step 4 - The first peel ply is placed upon the mold and subsequent plies are placed one upon another. Press each layer after applying resin to get better lamination. Pressure will be given by a roller or other small hand tool, it will also remove vacuum between the layers. It is important that the reinforcement material have sufficient tack so that it sticks slightly to the layers of peel plies.

Step 5 - Make more number of layers to get the lamination of 3mm thickness. Finally place the primary fabric glass fiber, it will cover both the lamination surface. Apply uniform load along the surface of the lamination of 4 to 5 kg about 48 hours.
Step 6 - The reinforcement plies are cut according to design specifications(150mmX25mm). They can be cut by hand using hand shears or a steel knife.
Step 7 - Sealant tape is placed around the entire periphery of the lay-up. The vacuum bag is cut to size and placed over the lay-up.

3.3 Mechanical Testing

Tensile Testing

Tensile test is one of the fundamental mechanical test, commonly used to select material under quality control and reaction of material under different types of force. A specimen is prepared in a square section depending on the standard gauge length of the machine. Both end of the specimen was holded by the cross heads and they are firmly gripped during testing. Force will be applied on the specimen by driven cross head until it fractures, during this process applied force and elongated length is measured to find the tensile strength.

Flexure Testing

Flexure testing is used to find the bending strength of the material and modulus of the rupture. Flexure test is done on the same tensile testing machine, here the specimen is placed on the bench wise and force will be applied on the center of the specimen until it gets fractured. The figure 1 shows specimen testing.

(i) (ii) (iii)

Figure 1.(i)Tensile Testing Machine, (ii) Flexure Testing Machine and (iii) Impact TestingMachine

Impact Testing

The impact test is used find the toughness of the material it is also known as the V-notch test, determines the amount of energy absorbed by a material during fracture. The impact testing machine has a pendulum of known length and mass, then the pendulum of mass is made to impact a notched specimen. This absorbed energy is a measure of a toughness of the given materials and acts as a tool to study temperature dependent ductile brittle conversion.

4. Modeling and Analysis of Drive Shaft

4.1 Modeling of Drive Shaft

The 3-D modeling of composite drive shaft is done by using CATIA and analyzed using ANSYS. A comparisons has been made between composite and steel drive shaft with respect to weight, cost and strength.The figure 2 shows the 3-D modeling of drive shaft.

IOP Conf. Series: Materials Science and Engineering **197** (2017) 012048 doi:10.1088/1757-899X/197/1/012048

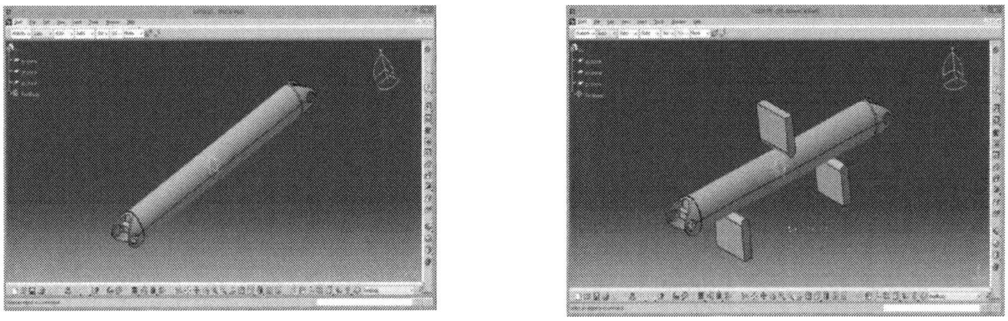

Figure 2.3D Modeling of Drive shaft in CATIA

4.2 Analysis of Drive Shaft

Analysis is carried out by using ANSYS 10.0 software, the following figures shows total deformation and Equivalent Stresses on Tensile test, Flexure test and Impact test of steel, Kevlar/Epoxy and Glass/Epoxy drive shafts under the application of 750N load. In that red zone indicates the maximum deflection area and blue zone indicates the minimum deflection area, which is shown by probe.

Total Deformation on Tensile Testing -The figure 3 shows the analysis of total deformation on tensile testing for Steel, Kevlar/Epoxy and Glass/Epoxy Drive Shafts Respectively.

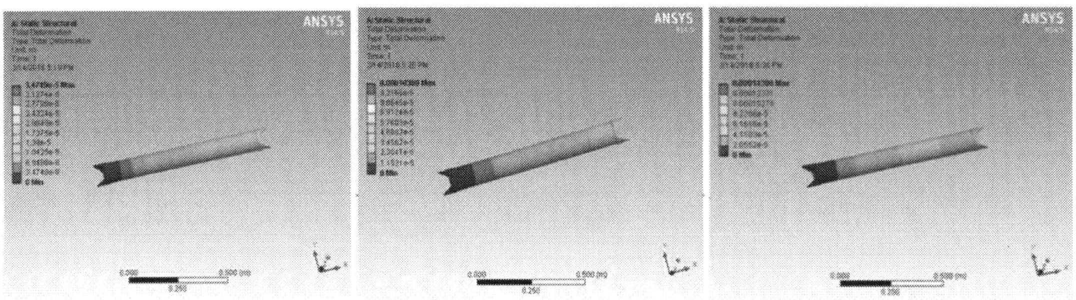

(i) (ii) (iii)

Figure 3. Total Deformation on Tensile Testing of (i) Steel, (ii) Kevlar/Epoxy (iii) Glass/Epoxy Drive Shafts Respectively.

Equivalent Stress on Tensile Testing-The figure 4 shows the analysis of equivalent stress on tensile testing for Steel, Kevlar/Epoxy and Glass/Epoxy Drive Shafts Respectively.

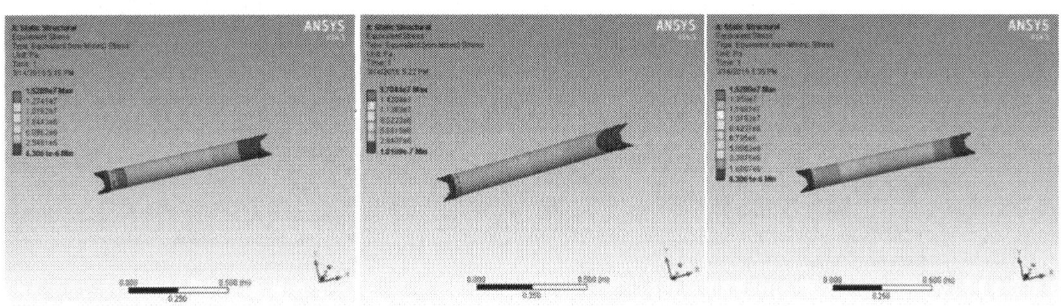

(i) (ii) (iii)

Figure4. Von-Misses Stress on Tensile Testing of (i) Steel, (ii) Kevlar/Epoxy (iii) Glass/Epoxy Drive Shafts Respectively.

IOP Conf. Series: Materials Science and Engineering **197** (2017) 012048 doi:10.1088/1757-899X/197/1/012048

Total Deformation on Flexure Testing-The figure 5 shows the analysis of total deformation on flexure testing for Steel, Kevlar/Epoxy and Glass/Epoxy Drive Shafts Respectively.

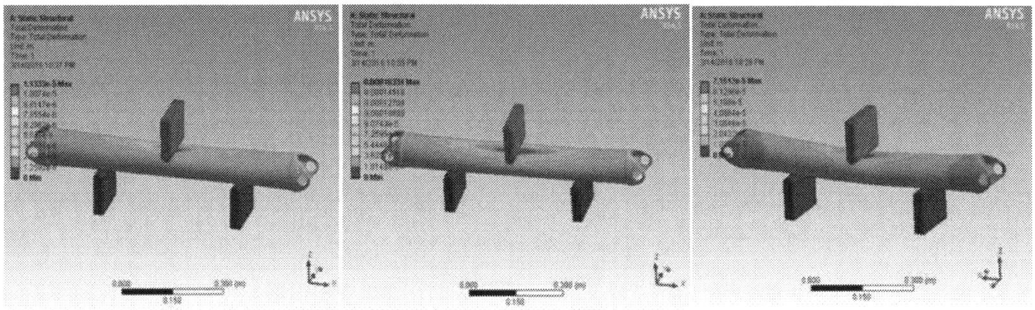

(i) (ii) (iii)

Figure 5. Total Deformation on Flexure Testing of (i) Steel, (ii) Kevlar/Epoxy (iii) Glass/Epoxy Drive Shafts Respectively.

Total Deformation on Impact Test-The figure 6 shows the analysis of total deformation on impact test for Steel, Kevlar/Epoxy and Glass/Epoxy Drive Shafts Respectively.

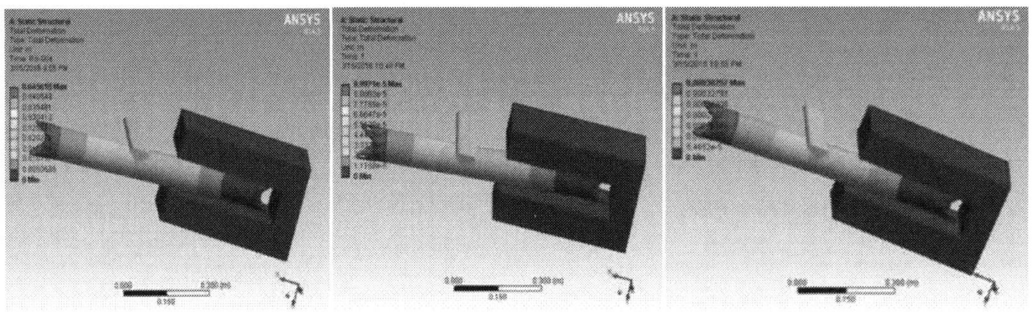

(i) (ii) (iii)

Figure 6. Total Deformation on Impact Testing of (i) Steel, (ii) Kevlar/Epoxy (iii) Glass/Epoxy Drive Shafts Respectively.

Equivalent Stress on Impact Testing-The figure 7 shows the analysis of equivalent stress on impact test for Steel, Kevlar/Epoxy and Glass/Epoxy Drive Shafts Respectively.

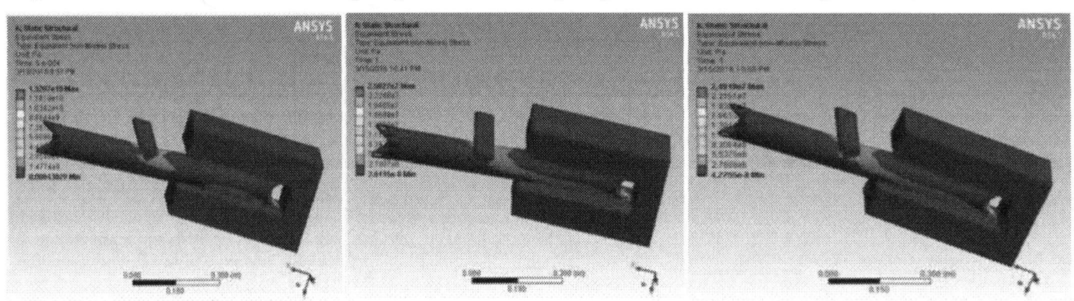

(i) (ii) (iii)

Figure 7. Von-misses stress on Impact Testing of (i) Steel, (ii) Kevlar/Epoxy (iii) Glass/Epoxy Drive Shafts Respectively.

5. Resultsand Discussions

The figure8 shows the deformation of Steel, Kevlar/Epoxy and Glass/Epoxy with respect to load. For tensile testing and Flexure testing, the deformation of steel is less when compare to Glass/Epoxy and Kevlar/Epoxy. Glass/Epoxyis having higher deformation then the Kevlar/Epoxy. But the deformation of Steel is higher than Glass/Epoxy and Kevlar/Epoxy under impact load. Kevlar/Epoxy and Glass/Epoxy are having more deformation under tensile and flexure load hence it replaces the two pieces of steel drive shaft into single shaft with better transmission of power. Because of their elasticity, they decrease the stress and act as a best shock absorber when the torque is more. Also Kevlar/Epoxy and Glass/Epoxy having less deformation under impact load so that it can absorb more energy because of its toughness.

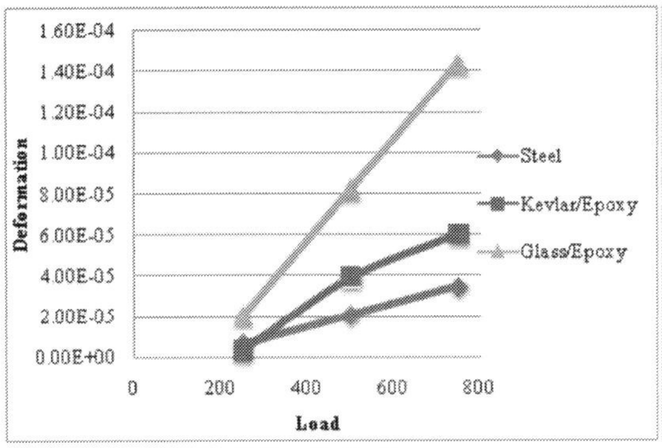

(i) Tensile Testing

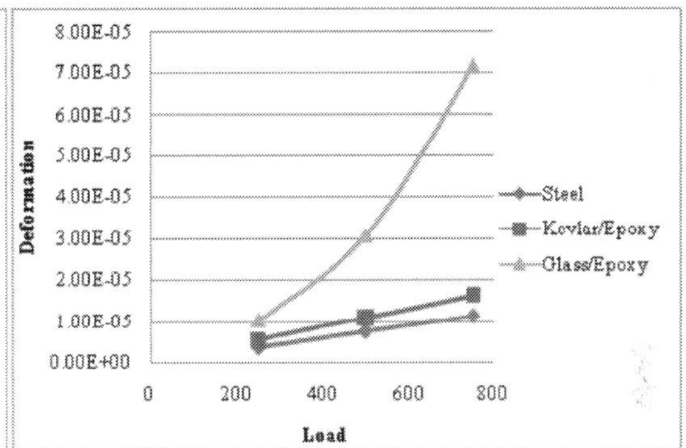

(ii) Flexure Testing

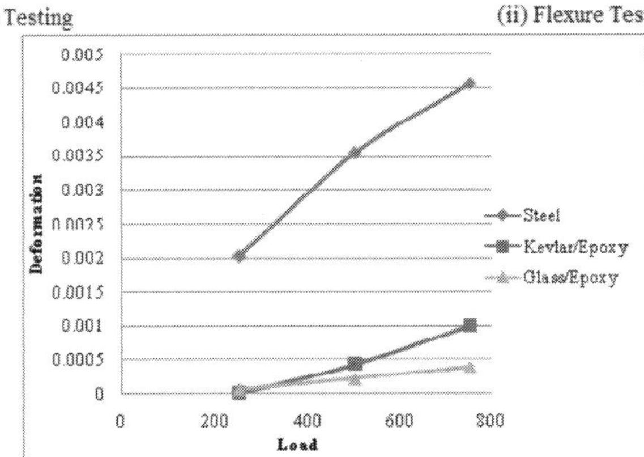

(iii) Impact Testing

Figure 8. Graphs Load Vs Deformation

Table 3. Results of Weight Optimization

Parameters	Steel	Kevlar/Epoxy	Glass/Epoxy
d_o (mm)	100	100	100
d_i(mm)	90	90	90
L(mm)	1250	1250	1250
Weight(Kg)	14.30	2.62	3.91
Weight saving (%)	-	81.67	72.66
Tensile Strength(%)	-	72.5	69.7

IOP Conf. Series: Materials Science and Engineering **197** (2017) 012048 doi:10.1088/1757-899X/197/1/012048

From table 3 it is observed that the composite drive shaft is lighter and more economical than the conventional steel drive shaft with similar design specifications. It is observed that the weight reduction of drive shaft is achieved up to 81.67% in case of Kevlar/Epoxy and 72.6 in case of Glass/Epoxy than steel.

6. Conclusion

This project made an attempt tomake a better replacement for a conventional steel drive shaft by a composite material of Kevlar/Epoxy and Glass/Epoxy. It can be observed that the Kevlar/Epoxy and Glass/Epoxy are having more deformation under tensile and flexure load hence it replaces the two pieces of steel drive shaft into single shaft with better transmission of power. Because of their elasticity, they decrease the stress and act as a best shock absorber when the torque is more. Also Kevlar/Epoxy and Glass/Epoxy having less deformation under impact load so that it can absorb more energy because of its toughness. The composite drive shaft reduces the weight by 81.67 % for Kevlar/Epoxy and 72.66% when compared with conventional steel drive shaft. The introduction of Kevlar/Epoxy and Glass/Epoxy composite materials will reduce the weight of the drive shaft without any drop in torque transmission capacity and stiffness.

References

[1] Ashwin Russelle, N. Naganambi 2014 *Elsevier Procedia Materials Science* 5 (2014) 204 – 211

[2] Branislav Duleba , Ludmila Dulebova, Emil Spisak 2014 *Elsevier Procedia Materials Science* 96(2014)70-74

[3] Chaitanya G Rothe, A.S.Bombatkar 2015 *International Journal of Innovative and Emerging Research in Engineering* } vol 2 p-ISSN 2394 - 5494

[4] Harshal Bankar, Viraj Shinde and P.Baskar 2013 *IOSR Journal of Mechanical and Civil Engineering* Vol.10,No. 1,pp. 39-46

[5] Kumar Rompicharla.R.P and Co. 2012 *International Journal of Engineering Research and Technology* Vol.1,Issue7,ISSN:2278-0181

[6] Narayana V.L., Mr. D. Mojeswararao and Mr. M.N.V.R.L.Kumar 2012 *International Journal of Innovative and Emerging Research in Engineering* Vol. 1 Issue 6

[7] Parshuram D, Sunil Mangsetty 2013 *The International Journal of Engineering And Science* Vol 35 pp 1--6 Volume 2, Issue 01, Pg.No. 160-17, ISSN: 2319 – 1813

[8] Srinivasa Moorthy.R, Yonas Mitiku and K. Sridhar 2013 *American Journal of Engineering Research* e-ISSN-2320-0847 p-ISSN-2320-0936, Vol 02, Issue 10, pp-173-179

IOP Conf. Series: Materials Science and Engineering **197** (2017) 012049 doi:10.1088/1757-899X/197/1/012049

Finite element analysis of chip formation usingale method

V.Jayaprakash[1]

[1]Asst Professor, School of Mechanical Engineering, Sathyabama University, India
E-mail: meerajayaprakashh@gmail.com

Abstract. In recent times, many studies made in FEM on plain isotropic metal plate formulation. The stress analysis plays the significant role in the stability of structural safety and system. The stress and distortion estimation is very helpful for designing and manufacturing product well. Usually the residual stress and plastic strain determine the fatigue life of structure, it also plays the significant role in designing and choosing material. When the load magnitude increases the crack starts to form, decreasing the work load and the residual stress reduces the damage of the metal. The manufacturing process is a key parameter in process and forming the part of any system. However, machining operation involves complex thing like hot development, material property and other estimates based on transition of the plastic strain and residual stress. The reduction of residual stress plays the complexity role in the finite element study. This paper deals with the manufacturing process with less residual stress and strain. The results shows that, by applying the ALE method in machining we can reduce the load on the work piece hence the life type of the work piece can be increased. We also investigate the cutting tool wear and there efficiency since it is a essential machine member in fabrication technology. ABAQUS platform used to solve the machining operation

1. Introduction

The chip formation on the isotropic metal is very difficult since the property of the isotropic material behaves adversely to the removal of material. This paper deals about the continuity of the chip formation due to the residual stress and the plastic strain. We have used certain methods based on the references in metal removal process and there crucial benefits of specimen:

The ALE method was applied since,

(1) It calculates the short dynamic performance time of the analysis and to find the continual event of processin large model.

(2) It can be used in the exceptional analysis by complex contact. The contact algorithm compared with certain conditions which contains the relative thing. Therefore it is clear that forming a metal removal in complex contact is very difficult.

(3) It uses 11/11 algorithm to send, the large deformation of theoretical model which possibly carries on the big rotation and in the deformation.

(4) It can be used to carry out a heat insulation of stress analysis, if the inelastic dissipation estimate causes in the heat of material.

(5) It considers automatic or semi systems to clear about dynamics procedure. These characteristics to say ideally made for analyzing in the residual stress of metal removal

The part body is meshed into the element name: S4R: A 4-node doubly curved thin or thick shell, reduced integration, hourglass control, finite membrane strains and the element shape: quad.

IOP Conf. Series: Materials Science and Engineering **197** (2017) 012049 doi:10.1088/1757-899X/197/1/012049

Adiabatic analysis typically used for imitating the high-speed processes of manufacture of involvement much inelastic density and the acceptable heating.
Description of discretization of the specimen has a specification of 66250 nodes and 239825 elements

1.1 Prediction of tool wears and tool life
The conventional way to classify tool wear for cutting operation in the many industrial sectors used to perform cutting test analysis under various types of cutting conditions and then analyzes the tool wear data using optimization and other techniques.
The disadvantage of using this approach, we can't get a good amount of accuracy in calculation of tool wear. The modelling procedure requires a large number of experimental test which is time and cost consuming. The finite element analysis successfully implemented in olden days to simulate various cutting operations. We used the Abacus software pyton coding to calibrate the required values.

1.2 Wear Mechanism and tool wear models
We have taken two types of model to study this chapter
1. Empirical tool life models
2. Tool wear rate models

2. Results and Discussion

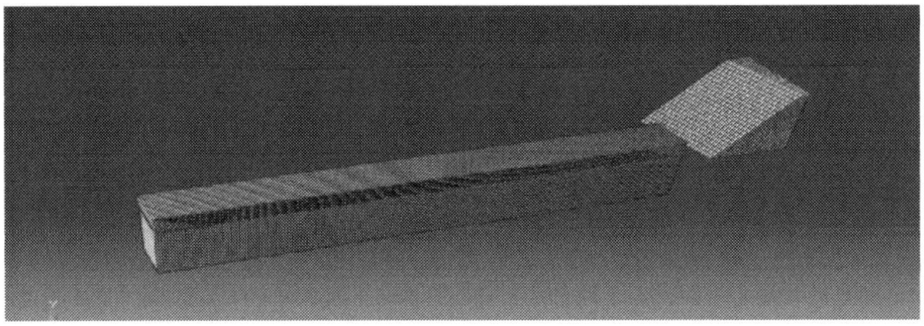

Figure 1. Meshed chip model

The figure is the complete setup of the system; the work piece is meshed fine to visualize the formation of chip very uniform without breaking.

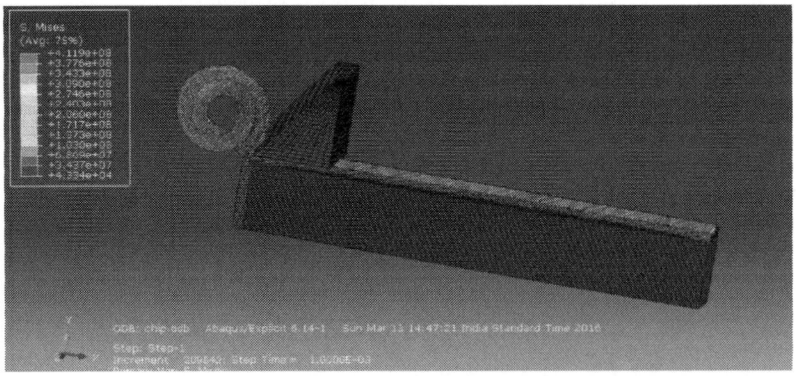

Figure 2. Von misses stress of plain carbon steel

IOP Conf. Series: Materials Science and Engineering **197** (2017) 012049 doi:10.1088/1757-899X/197/1/012049

When the tool touches the specimen the specimen starts to react for the deformation. That deformation is negligible but it can affect the material property.
It can be avoided when the impact velocity of the tool is reduced.

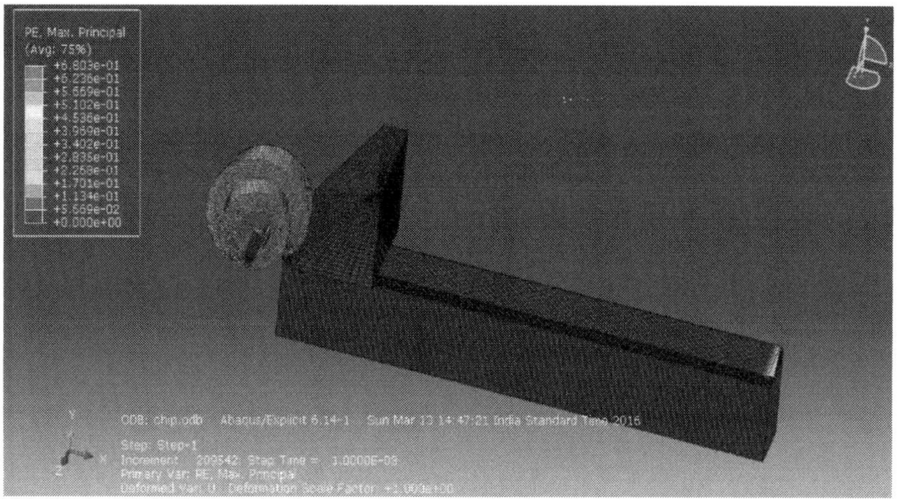

Figure 3. Plastic strain of plain carbon steel

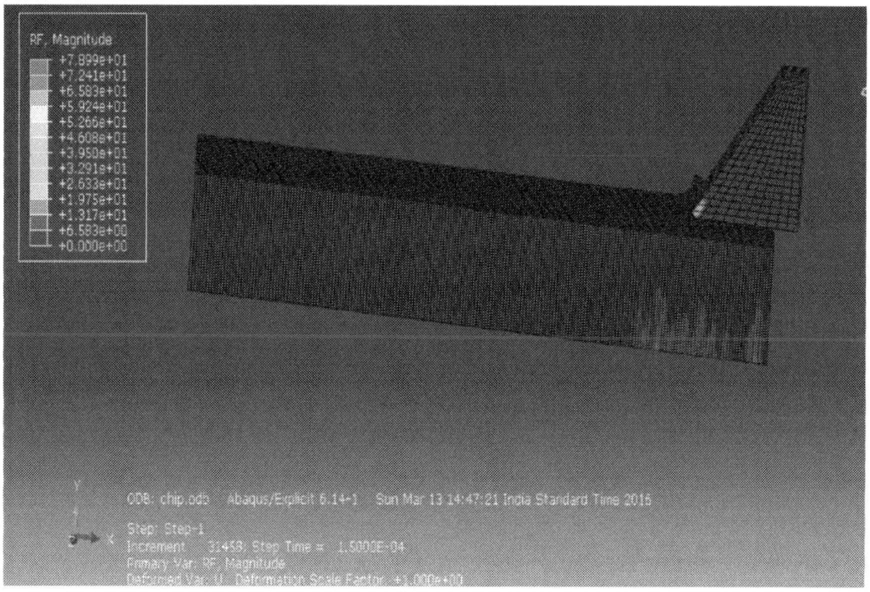

Figure 4. Reaction force of plain carbon steel

The reaction force of the specimen is zero, its shows that the magnitude of the force acting is very negligible range and the plastic strain after the chip formation shows the dissipation of the temperature is very good.

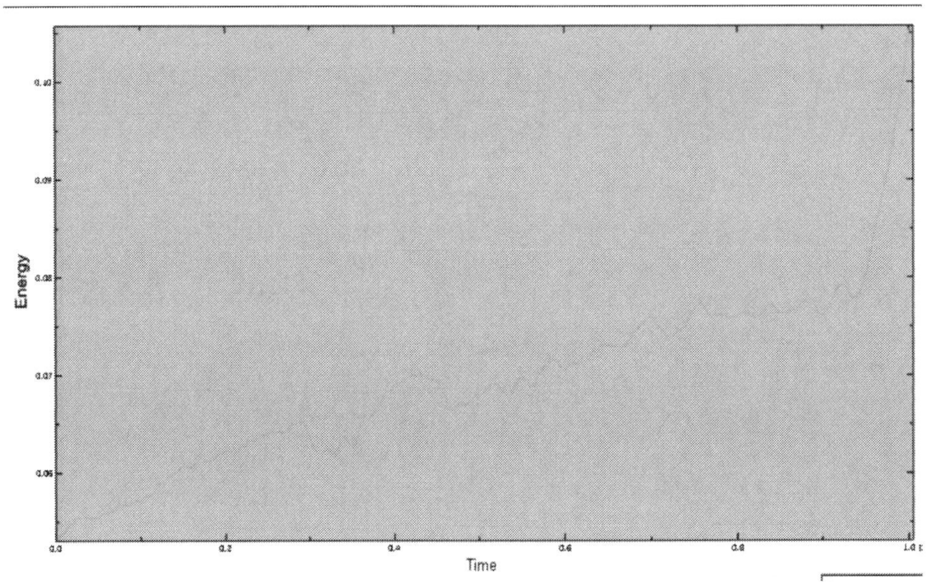

Figure 5. Kinetic energy of plain carbon steel

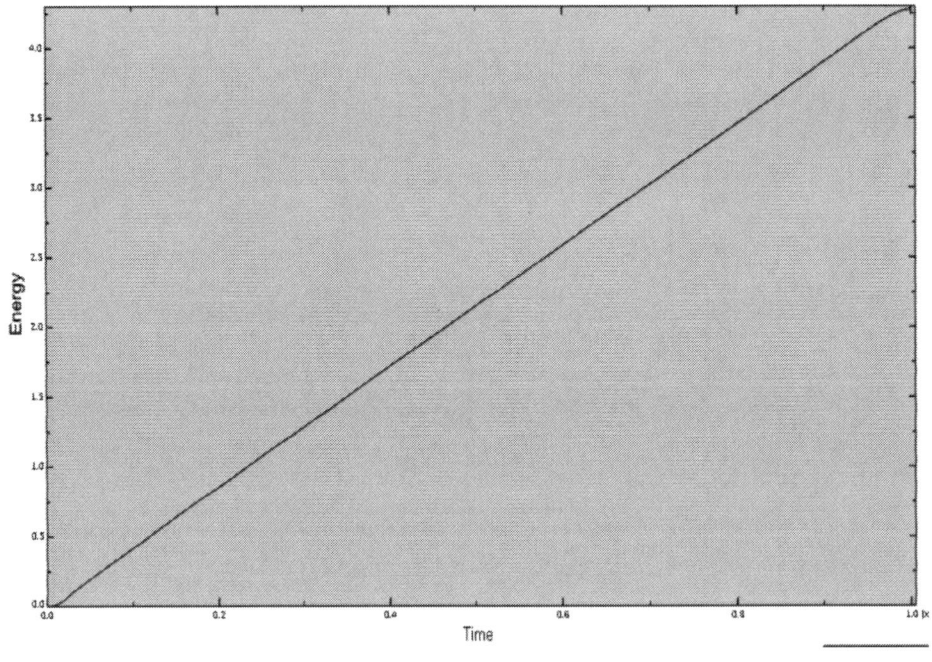

Figure 6. Plastic dissipation rate of plain carbon steel

The kinetic energy and the plastic dissipation of the carbon steel is rapidly increasing with respect to the formation of chip.

The chip formation of the specimen is very good since the kinetic energy is not dropping to the acceptable limit.

IOP Conf. Series: Materials Science and Engineering **197** (2017) 012049 doi:10.1088/1757-899X/197/1/012049

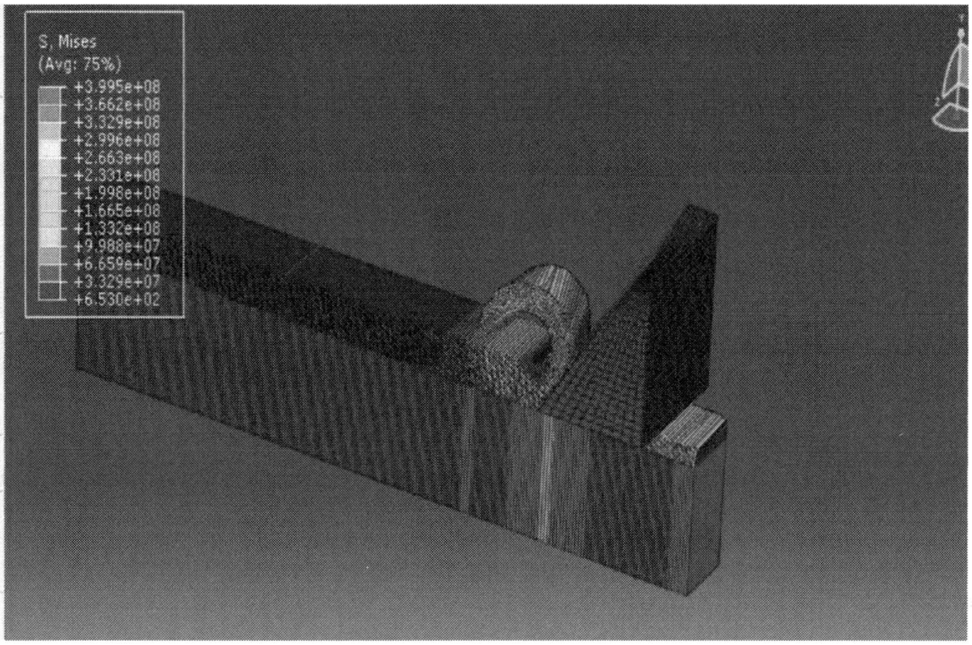

Figure 7. Von misses stress of mild steel

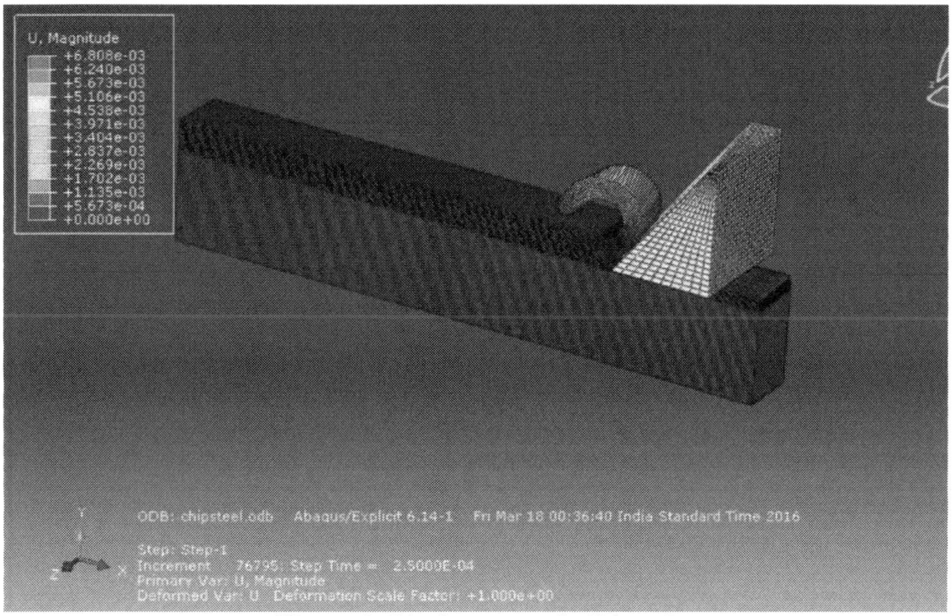

Figure 8. Displacement of mild steel

The diaplcement and the plastic strain magnitude for the mild steel is higher than the carbon steel.

Table 1. Stress comparison table

Material	Stress N/m^2	Plastic strain	Displacement (mm)
1.143	0.286	0.593	0.286
1.143	0.067	0.796	0.067
1.143	0.077	0.782	0.286

Table 2. Tool wear rate

	Plain carbon steel		Mild steel	
Parameter	HSS	Carbide tool	HSS	Carbide tool
Tool wear rate in mm	0.03	0.02	0.0034	0.015

3. Conclusion

The chip formation parameters of both mild steel and carbon steel predict that the machining of the carbon steel takes more loads during machining. The result concludes that load and the tool wear rate is directly proportional to the formation of the chip. The healthy chip gives better output finish and less wastage of the metals.

- The ALE method shows the wear rate with clarity to find the life span of the tool.
- The Johnson cook damage declares that the tool wear rate of the carbon steel is high when compared to the mild steel.

References

[1] G.Suttera, G.Lista 2013 Very high speed cutting of Ti–6Al–4V titanium alloy – change in morphology and mechanism of chip formation, International Journal of Machine Tools and Manufacture.

[2] Dinesh Setti et.al 2015 Performance evaluation of Ti–6Al–4V grinding using chip formation and coefficient of friction under the influence of nanofluids,International Journal of Machine Tools and Manufacture.

[3] S. Manigandan 2015 Computational Investigation of High Velocity Ballistic Impact Test on Kevlar 149, Applied Mechanics and Materials.

[4] A.Molinari,X.Soldani,M.H.Miguélez 2013 Adiabatic shear banding and scaling laws in chip formation with application to cutting of Ti–6Al–4V,Journal of the Mechanics and Physics of Solids.

[5] Chengyong Wang et.al2014 Research on the Chip Formation Mechanism during the high-speed milling of hardened steel, International Journal of Machine Tools and Manufacture.

[6] P. Gunasekar 2015 Torsional Extrusion Processing of Titanium Alloy Ti 6Al-6V-2SN under High Feeding Rates, Applied Mechanics and Materials.

[7] Manigandan 2015 Computational Investigation of High Velocity Ballistic Impact Test on Kevlar 149. In Applied Mechanics and Materials.

[8] Gunasekar, P., Manigandan, S., Anderson, D., & Devipriya, J 2017 Evaluation of Fe-Epoxy metal nanocomposite in glass fiber and Kevlar. International Journal of Ambient Energy.

IOP Conf. Series: Materials Science and Engineering **197** (2017) 012050 doi:10.1088/1757-899X/197/1/012050

Design and Analysis of High Pressure Hydraulic Filter for Marine Application

Toshin Momin[1], RP Chandrasekar[2], S Balasubramanian[3], Dr. AM Junaid Basha[4]

[1]Senior Research Fellow, DRDO, CVRDE, Chennai
[2] Scientist 'C', DRDO, CVRDE, Chennai
[3] Technical Officer 'C', DRDO, CVRDE, Chennai
[4] Scientist 'G' Additional Director, DRDO, CVRDE, Chennai

E-mail: Momin.toshin@gmail.com

Abstract. Filter is a critical component in ahydraulic system for maintaining the cleanliness of the fluid to required class level. InMarine applications very high reliable filter is required to operate continuously in saline environment. Design anddevelopment of high pressure hydraulic filter for Marine application is a challenging task. The design involves selection of special materialsandstringent qualification tests as per International standards. The present paper describes various stages of design and development of high pressure hydraulic filter for Marine application.

1. Introduction

Hydraulic system of Marine vessel is used for various utility operation viz. sailing of vessel andoperation ofcontrols. Thesystem oil cleanliness has to be maintainedby filters to achieveprolong operationallife of valves and actuators. Filter to meet the system requirement has to be designedwith utmost care to meet the specification requirement. Also, filter has to be designed to cater for high reliability insevere environment conditions, like high salinity, high and low temperature cycle, vibration, acceleration and shock.The filters used in marine vessels are subjected to shock up to 30g. The design of the components for Marineinvolves special material, process planning, quality control and stringent quality acceptance test procedure to International Marine standards.Most of the hydraulic filters in modern marine vessel uses disposable filter elements made up of high strength glass fiber media.The filter rating of such glass fiber media filters varies from 5 to 25 microns. The glass fiber media has better contamination control and good service life over the conventional reusable stainless steel mesh type media [1]. The detail design and analysis of the selected high pressure hydraulic filter for Marine application are described and results of FEM analysis are presented.

2. High pressure hydraulic filter

Atypical Marinefilteras shown in Figure 1 considered for the present work is hydraulic pressure line filter installed between the pump and other critical components of the system. Three such filters are used for cruise, banal and alert lines. Pressure relief valve is provided before the filter to control the pressure of the system. The filteris made of grey cast iron and coated with special coating to prevent corrosion from marine environment.

The filter assembly shown in Figure 1 consists of filter head with in and out ports.The filter is integrated with four mounting holes at the top of head.Itis designed with a handle type shutoff valve which disconnects the element with the fluid line to make it possible for safe removal of element, without spilling the oil.The filterhead is also equipped with bypass valve, non-return valve (NRV), differential pressure indicator (DPI), bleed of valve and commuter valve. DPI is having a thermal lockout arrangement, which prevents faulty indication in case of thickening of fluid in sub-zero temperature. Drain plug is provided at the bottom of bowl. The specifications of filter are given in table 1.

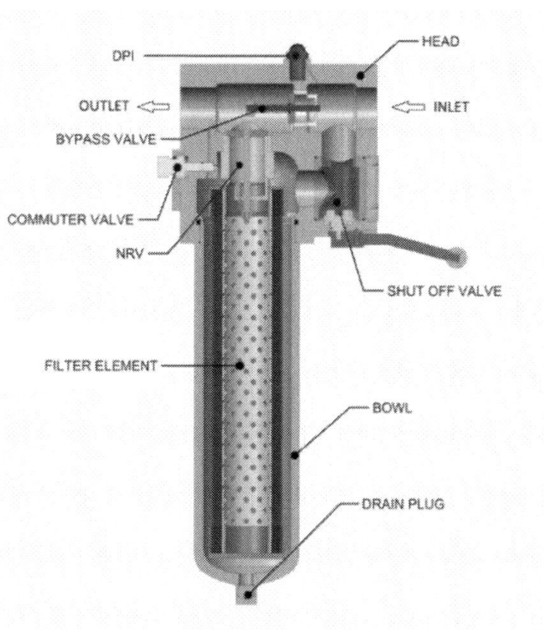

Figure 1. Filter assembly

Table 1. Specifications of filter

Filtration rating	β10(C)≥1000
Fluid flow rate	50 lpm
Working pressure	205 bar
Operating Temperature	- 5°C to 100°C
System fluid	FHARI SHELL 2841
Clean element pressure drop @ rated flow	≤ 0.3 bar
Assembly unit pressure drop @ rated flow	1.5 bar
Element differential collapse pressure	20 bar
Housing proof pressure	307 bar
DPI activation pressure	2 bar
Bypass valve cracking pressure	4 bar
Direction of flow	OUT to IN

3. Design of high pressure filter

Filter design is a two-step process, first the filter element is designed to meet the specification requirement, then housing is designed which accommodates the element to withstand the specified pressure ratings with minimum pressure drop at the rated flow.

3.1 Filter element design

The filter element serves to trap contaminants. The filtration depends on the properties of the fluid being filtered i.e. viscosity, density and the concentration of the solids in suspension. The filter element is made out of glass fiber media of requisite filter rating and is supported with rayon paper and stainless steel course mesh on both sides as shown in Figure2. Layers of media are pleated and wound over a perforated stainless steel centre tube.The pleated filter media along with centre tube, top and bottom covers are joined carefully with special adhesive. Finally the fabrication integrity of filter element is checked. The filter element assembly is shown in Figure3.

While designing a filter element some basic parameters like element pressure drop, critical pressure for centre tube collapse and critical pressure for centre tube buckling are calculated.

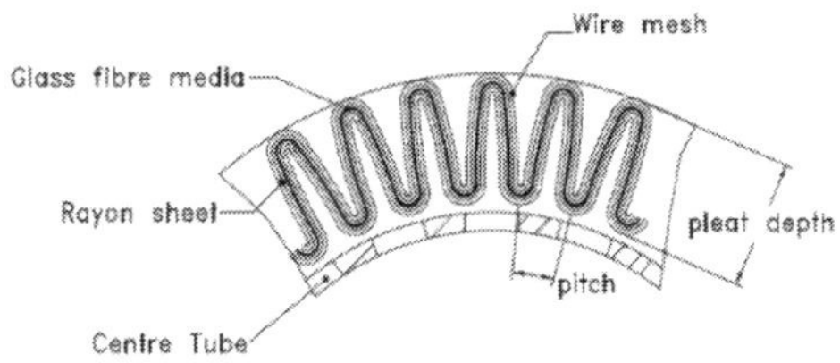

Figure 2. Pleated media

3.2. Estimation of clean element pressure drop

An important characteristic of the filter media is their resistance to the flow of fluid. This resistance is a basic property of the filter material and is related to the physical structure of the filter element. The pressure drop generated due to fluid passing through the filter element can be obtained from Darcy's equation as givenin equation 1 [2].

$$\Delta P = \frac{Q \times \mu \times t}{A \times k \times \varepsilon} \quad \cdots\cdots (1)$$

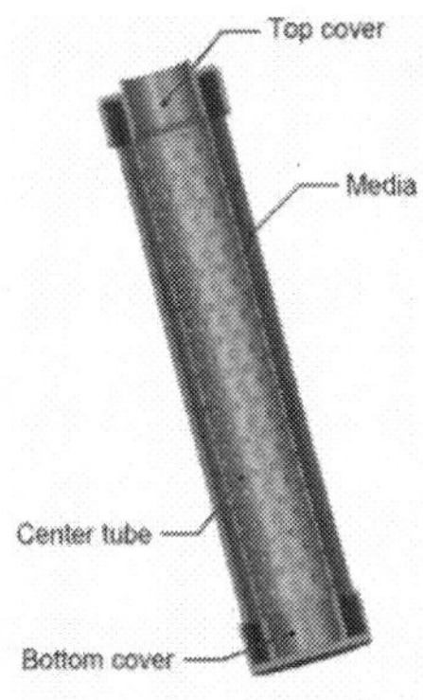

Figure 3. Element assembly

Where:

Δp = pressure drop across the porous medium (N/mm^2)
Q = fluid flow rate (mm^3/s)
μ = dynamic viscosity of the fluid (Ns/mm^2)
t = thickness of the filter material (mm)
A = filtering area (mm^2)
k = permeability constant for the filter material (mm^2)
= $d_2 / 32$
d = pore size (mm)
ε = porosity

Pleated structure is used for the filter to provide large surface area in small envelope and thereby keeping housing size to a minimum. Total filtrationarea can be calculated as,

$$\text{Filtering area } A = h \times l \quad \cdots\cdots (2)$$

Where

h = height of element (mm)
l = developed length of element (mm)
= $2 \times W \times N$
N = no. of pleats
W = pleat depth (mm)

Number of pleats (N) in filter element is obtained byconsidering pitch of 1.5 to 2mm at inner diameter of element and is given as

$$N = \frac{\pi \times D_2}{\text{Pitch}} \quad \cdots\cdots (3)$$

Filter element pleat depth (W) is given as

$$W = \frac{D_1 - D_2}{2} \quad \cdots\cdots (4)$$

where:

D_1= outer diameter (mm)
D_2 = inner diameter (mm)

As shown in Figure2, there are five sheets in filter element viz. two sheets of wire mesh, twosheets of rayon fabric and one sheet of glass fiber media. Therefore total pressure drop across filter element will be sum of pressure drop across all five sheets.
For the specified filter element the total pressure drop across all sheets is estimated as 0.2bar which is less than the specified pressure drop of 0.3 bar.

3.3. Critical load for buckling of inner tube due to compressive load

During the extreme operating conditions the differential pressure rises very high across the filter element. This will cause buckling of the inner tube of the filter element resulting in failure due to collapse. Therefore it is essential to ensure the material selected and structural strength of the centre tube along with welded joints of top and bottom covers to withstand the high compressive load. The critical compressive load for the buckling of the centre tube is estimated from Johnson's equation as given below [3].

$$P_{cr} = f_y \times A\left[1 - f_y\left(\frac{h}{k}\right)^2 \times \frac{1}{4\pi^2 \times E}\right] \quad \cdots\cdots (5)$$

Where

$$\text{Equivalent Area, (A)} = 0.5 \times \frac{\pi}{4}\{D^2 - d^2\}$$

$$\text{Moment of Inertia, (I)} = \pi\left[\frac{D^4 - d^4}{64}\right]$$

$$\text{Least radius of gyration, (k)} = \sqrt{I/A}$$

Load at Collapse Pressure (P),

$$P = 0.5 \times \frac{\pi}{4}\left[{D_1}^2 - {D_2}^2\right] \times \text{Collapse pressure} \quad \cdots\cdots (6)$$

IOP Conf. Series: Materials Science and Engineering **197** (2017) 012050 doi:10.1088/1757-899X/197/1/012050

Where:

Pcr = Critical compressive Load (N)
P = Load at collapse pressure (N)
D = Outside diameter of tube (mm)
d = Inside diameter of tube (mm)
h = Length of the tube (mm)
D_1 = Outside diameter of element (mm)
D_2 = Inside diameter of element (mm)
p = Differential Collapse pressure (N/mm^2)
fy = Yield strength (N/mm^2)
E = Young's modulus (N/mm^2)

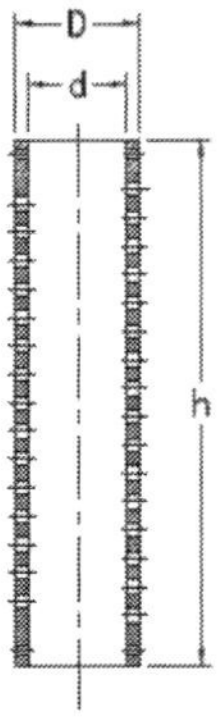

Figure 4. Centre tube

For the specified filter, the critical load for buckling of inner tube due to compressive load is obtained as 44500 N, which is more than the load of 2900 N at collapse pressure. Hence the centre tube will not collapse under collapse pressure load due to its strength.

3.4. Estimation of critical radial pressure for buckling of inner tube

The filter element is subjected to high radial pressures which will result in failure of the centre tube along with top and bottom cover. Therefore it is important to estimate the critical radial pressure that the filter element could withstand based on the specified collapse pressure rating. The critical radial pressure should be more than the specified collapse pressure rating required for the filter. It is given byequation 7 in the following [3].

$$P_{cr} = \frac{\pi^2 \times D \times P_1}{l^2 \times a} \qquad \text{...... (7)}$$

Where,

$$D = \frac{E \times h^3}{12 \times (1 - \mu^2)}$$

$$n_1 = \frac{n \times l}{\pi \times a}$$

$$Z = \frac{l^2 \times [1 - \mu^2]^{0.5}}{a \times h}$$

$$P_1 = \frac{(1 + {n_1}^2)^2}{{n_1}^2} + \frac{12 \times Z^2}{\pi^4 \times {n_1}^2 \times (1 + {n_1}^2)^2}$$

Where

h = Thickness of tube, (mm)
a = Inside radius of tube (mm)
l = Length of tube (mm)
E = Young's modulus (N/mm^2)
μ = Poison's ratio
n = No. of modes

IOP Conf. Series: Materials Science and Engineering **197** (2017) 012050 doi:10.1088/1757-899X/197/1/012050

For the specified filter, the critical radial pressure is obtained as 696barwhich is more than specified collapse pressure of 20bar, hence the centre tube will not collapse.

4. FEM analysis of centre tube

In order to study the pattern of stresses experienced by the centre tube under the specified operating conditions, an FEM analysis has been carried out using ABAQUS software. The material of stainless steel to AISI: 316L has been selected with the mechanical properties as shown in table 2.The FEM analysis is discussed in the following.

Centre tube has been modeled as eight node shell element for a sector of 45° such that one period of the hole pattern is included in the model. Fixed boundary condition is applied on both end nodes where it is held at top and bottom covers and the pressure load is applied appropriately. A static analysis has been carried out and the maximum stress in the centre tube is computed as 97.5 N/mm2 for the working pressure of 205bar. The stress pattern of the filter element is shown in the Figure 5

Table 2. SS316L Material properties

Density	7900 kg/m^3
Young`s modulus	200000 N/mm^2
Poisson Ratio	0.3
UTS	540 N/mm^2
(0.2%) Proof strength	400 N/mm^2
Endurance limit	270 N/mm^2

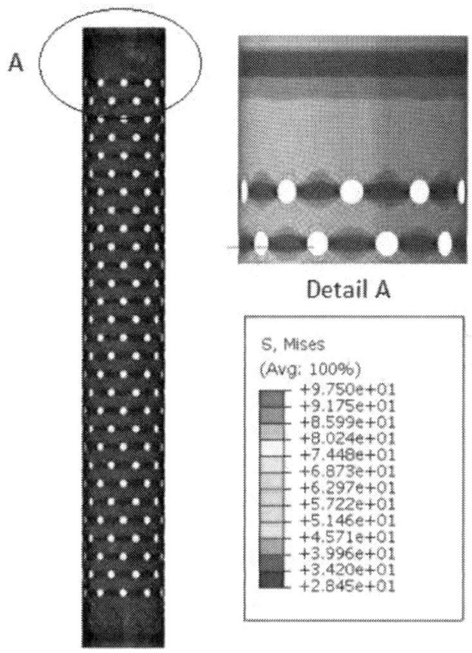

Figure 5. FEM analysis of centre tube

4.1 Filter housing design

The filter housing consists of a filter head and filter bowl. They are coated with protective coating to prevent rusting from saline and humid environment of sea. The filter head provides housing for non-return valve and shutoff valve mechanism.The handle type shutoff valve is made up of ball and seat arrangement and is designed to serve leak proof joint upto maximum pressure of 1.5 times the operating pressure. The ball is made from stainless steel and seat is made up of soft teflon material to acquire perfect seating and leak proof arrangement.The filter bowl is used to house the filter element, and is provided with drain port.

4.2Filter bowl

The filter bowl is coupled with filter head with help of threads. During the replacement of filter element, the bowl is withdrawn from the filter head and the new element is positioned with bowl and reassembled.

The filter head and bowl are designed to meet the following pressure requirements:

Normal working pressure : 205 bar

Proof pressure	:307 bar
Burst pressure	:410 bar

In order to study the pattern of stresses experienced by the bowl under the specified operating conditions, an FEM analysis has been carried out. Fixedboundary condition is applied on bowl top portion where it is mounted. The maximum stress distribution on bowl of the filter is observed to be 239.3 N/mm^2 as shown in Figure6.

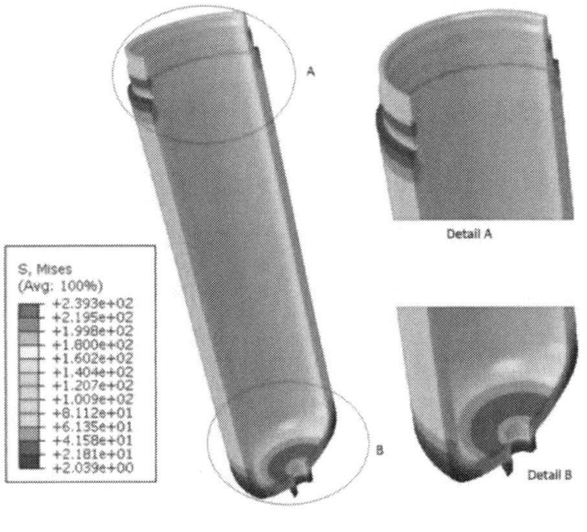

Figure 6. FEM Analysis of filter Bowl

4.3 Filter bypass valve design

The bypass valve as shown in Figure7, limits the differential pressure across the element and can provide uninterrupted flow when element is clogged. It is mounted in the filter head between the inlet and outlet port.

Bypass Valve serves to:

1) Prevent collapse of the element
2) Ensures uninterrupted fluid supply to system till clogged element is replaced
3) Ensures that the differential pressure across the filter unit is within the specification

The poppet type bypass valve allows unidirectional flow only.

The design requirement of bypass valve is to maintain a pressure difference of not more than 4bar between inlet and outlet ports. This set pressure is known as cracking pressure of the valve. If the pressure difference between inlet and outlet increases beyond set pressure of 4 bar the force generated is enough to compress the spring thus unseating the poppet from retainer and allowing the fluid to pass.

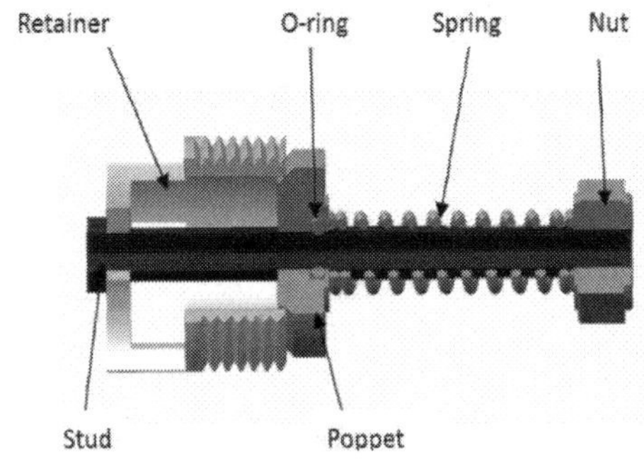

Figure 7.Bypass valve

The dynamic friction acting between the seal and the stud is

IOP Conf. Series: Materials Science and Engineering **197** (2017) 012050 doi:10.1088/1757-899X/197/1/012050

also being considered in the design calculation, since it contributes significantly in resisting to the movement of poppet. The frictional force acting on O-ring seal can be estimated from the following equation given below

$$f = \mu F \qquad \cdots\cdots(8)$$

Where,

μ = Co-efficient of friction between seal and stud

Compressive or radial force (F) on seal is given by,

$$F = \pi \times d_o \times D_o \times E\left[1.25\left(\frac{x}{d_o}\right)^{1.5} + 50\left(\frac{x}{d_o}\right)^{6}\right] \qquad \cdots\cdots (9)$$

Where

d_o = O-ring wire diameter (mm)
D_o = O-ring diameter(mm)
E = Young's Modulus of elasticity (mm)
x = Seal deformation (mm)

Spring deflection per coil is given by,

$$\delta = \frac{8 \times W \times D^3}{G \times d^4} \qquad \cdots\cdots (10)$$

Where,

D = spring coil mean diameter (mm)
d = spring wire diameter (mm)
G = Modulus of rigidity (N/mm^2)
W = Axial Load on spring (when solid), (N)

$$= \frac{\tau \times \pi \times d^3}{8 \times D}$$

τ = Allowable shear stress in spring material (N/mm^2)

$$= \frac{\text{Tensile Strength} \times 0.5}{\text{Factor of safety} \times \text{Wahl's factor (k)}}$$

k = Wahl's stress correcting factor

$$= \frac{4C - 1}{4C - 4} + \frac{0.615}{C}$$

C = spring index

= D/d

IOP Conf. Series: Materials Science and Engineering **197** (2017) 012050 doi:10.1088/1757-899X/197/1/012050

Force balance on poppet is done as shown,

Differential pressure force = spring pre-compression force + frictional force

$$\Delta p \text{ x Area} = \text{stiffness} \times \text{precompression} + f \qquad \cdots\cdots (11)$$

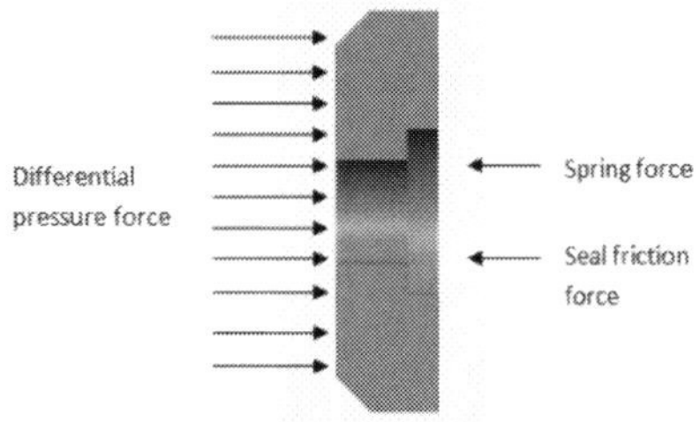

Figure 8. Forces acting on poppet

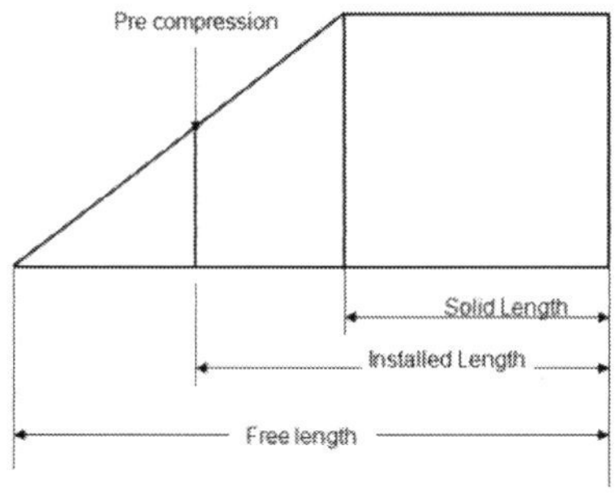

Figure 9. Spring load characteristics

For the specified filter, spring pre-compression required for accurate opening of valve at differential cracking pressure of 4bar has been computed as 7 mm. Other dimensions of spring are selected from standard design data book.

5. Conclusion

The high pressure glass fiber media cleanable filter required for hydraulic system of the Marine hydraulic system has been designed with respect to pressure drop, critical radial pressure and compressive load factors. An FEM analysis has also been carried out to study the level of stresses experienced under collapse pressure. The critical radial pressure estimated is well within the collapse pressure rating of the filter. The stresses observed due to the maximum working pressure are well within the yield strength of the filter element. Hence the design of filter has been accepted for further manufacturing and qualification testing for obtaining Navy certification.

REFRENCES

[1] Parker filtration's Handbook of Hydraulic filtration- Parker Hannifin, Hydraulic Filter Division, Europe.
[2] Fluid mechanics and hydraulics by Dr. Jagadish Lal
[3] Design of filter element, M/s.Pal filter, U.K.

IOP Conf. Series: Materials Science and Engineering **197** (2017) 012051 doi:10.1088/1757-899X/197/1/012051

Experimental Analysis of Exhaust Manifold with Ceramic Coating for Reduction of Heat Dissipation

Saravanan J[1], Valarmathi T N[*2], Rajdeep Nathc[3], Prasanth Kumar[4]

[1]PG student, Department of Mechanical & Production Engineering, Sathyabama University, Chennai.
[2]Professor, Department of Mechanical & Production Engineering, Sathyabama University, Chennai.
[3,4] UG students , Department of Mechanical & Production Engineering, Sathyabama University, Chennai.

*Email id: valarmathi.tn@gmail.com

Abstract. Exhaust manifold plays an important role in the exhaust system, the manifold delivers the waste toxic gases to a safe distance and it is used to reduce the sound pollution and air pollution. Exhaust manifold suffers with lot of thermal stress, due to this blow holes occurs in the surface of the exhaust manifold and also more noise is developed. The waste toxic gases from the multiple cylinders are collected into a single pipe by the exhaust manifold. The waste toxic gases can damage the material of the manifold. In this study, to prevent the damage zirconia powder has been coated in the inner surface and alumina (60%) combined with titania (40%) has been used for coating the outer surface of the exhaust manifold. After coating experiments have been performed using a multiple-cylinder four stroke stationary petrol engine. The test results of hardness, emission, corrosion and temperature of the coated and uncoated manifolds have been compared. The result shows that the performance is improved and also emission is reduced in the coated exhaust manifold.

1. Introduction

The exhaust manifold is used in a multiple cylinder IC engine which collects the exhaust gases from the cylinders into one pipe. During working it dissipates heat and emits harmful undesirable substances like noxious or toxic. Many research works like better change in design, coating, etc. are carried out to reduce these effects to increase the performance of the engine.

Azadiet al [1]performeda review of thermal barrier coating applications in diesel engines to select a proper type and to find coating effects. The coating system has effects on the fuel consumption, the power and the combustion efficiency, pollution contents and the fatigue lifetime of engine components. Durga Prasad et al [2] designed an exhaust manifold with different coatings and analyzed with modelling and finite element methodology techniques. Gopaal and Kumara Varma [3] used finite element analysis technique to reduce the design difficulties of exhaust manifold and concluded that the design made by them is safe and effective. Kandylas and Stamatelos [4] used mathematical modeling and CFD software to analyze the heat transfer in the exhaust manifold. Mahesh S Vasavan and Jotaniya [5] presented a detailed review on heat transfer analysis in automotive exhaust system.Mouna Challa and Imran Quraishi[6]explained the selection of advanced material using CES software for exhaust manifold based on the benefits and also suggested new manufacturing methods like deformation process, investment casting process, etc. for the exhaust manifold using new materials.Navaneethakrishnan and Vasudevan[7] iinvestigated the performance, emission characteristics of a single cylinder four stroke direct injection diesel engine operating on biodiesel. After comparisontheyrevealed that blends of TME, DEE with diesel up to 20% and 20% by volume provide better engine performance (BSFC and BTE) increased up to 3-4% and the exhaust emission is decreased dramatically. Prabhat Kumar Jhaet al [8] carried out experimental studies to reduce the exhaust gas temperature and gas emissions of a Spark Ignition (SI) engine by injecting water in the intake manifold gives better results. Serkan Saridag and Onjen Tak [9] investigated the mechanical properties of zirconia in the scientific literature and revealed that zirconia clearly measures up to any other equivalent manufactured material.

2. Materials and methods

2.1. Specimen preparation for coating

The waste exhaust gases corrode the manifold material due to this some blow holes will occur and noise pollution will increase. In order to avoid these type of failure, in this work some ceramic powders are used on both inner surface and outer surface of the cast iron exhaust manifold. The inner surface of the cast iron exhaust manifold is coated by zirconia powder to arrest the temperature in inside itself this will reduce the heat dissipation. The outer surface of the cast iron exhaust manifold is coated by alumina and titania in the combination range of 60% - 40%. The coated and uncoated manifold which is used for this work is shown in Figure 1 (a) and (b) respectively. The coating is performed using plasma spray technique. The coating for the cast iron exhaust manifold is done by using plasma spray coating technique for a thickness of 250 μm.

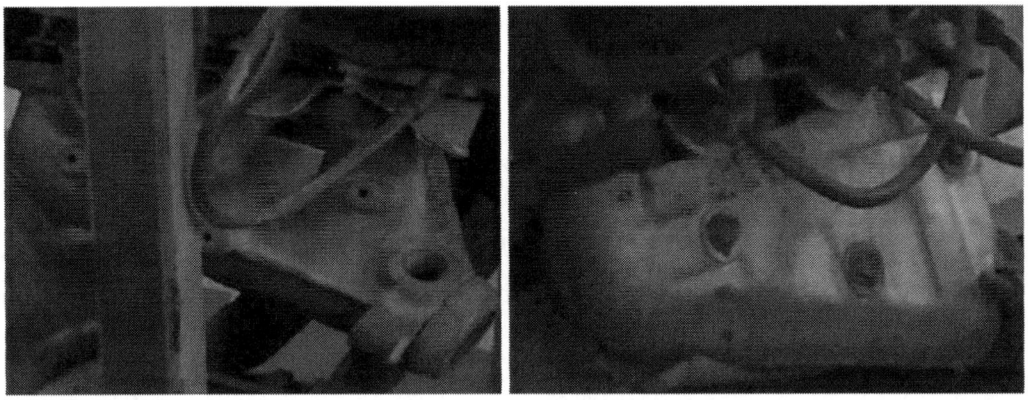

Figure 1.(a) Coated (b) Uncoated exhaust manifold

2.2. Experimental setup

A multiple-cylinder four stroke stationary petrol engine has been used to conduct the experiments. The experimental setup is shown in Figure 2.Infrared thermometer testing machine is used to analyse the temperature of the manifold. And the readings of temperature have been taken. Emission test has been carried out using automotive emission analyzer testing machine. In this experiment the loads used are varying load which varies from 15 to 45N loads. After the load is applied the engine reaches a steady condition at a particular point, during this particular point the readings such as time taken for 20cc fuel consumption, outer surface temperature, inner surface temperature, HC, CO and NOx are taken.

Figure 2. Experimental setup

3. Results and Discussion

3.1. Microstructure studies

Properties of the coated specimen using plasma spray process were analyzed by using SEM and EDX. The coating thickness achieved is in the order of 125μ. Optimum compositional and structural conditions of the coating component have been analyzed. The micro graph of the fracture surface of the uncoated and coated cast iron specimen samples is shown in Figure 3. The structure exhibited particles of both materials were deformed on impact during the plasma spray and melted on the specimen surface is shown in the Figure 3(c). From the microstructure images it is observed that the grain particles size is less for coated cast iron specimen when compared with a cast iron specimen without coated.

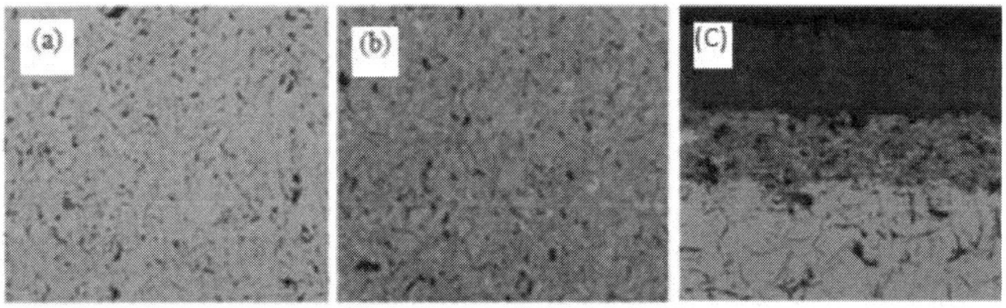

Figure 3. Microstructure of (a) Uncoated (a) polished (b) etchedand (c) Coated cast iron specimens

3.2 Hardness test

In this work specimens have been prepared to test the hardness of both coated and uncoated cast iron material. Vickers hardness test has been carried out for testing the hardness of the specimens shown in Figure 4. The results are presented in Table 1. From the results it is observed that coated cast iron specimen is giving the best results.

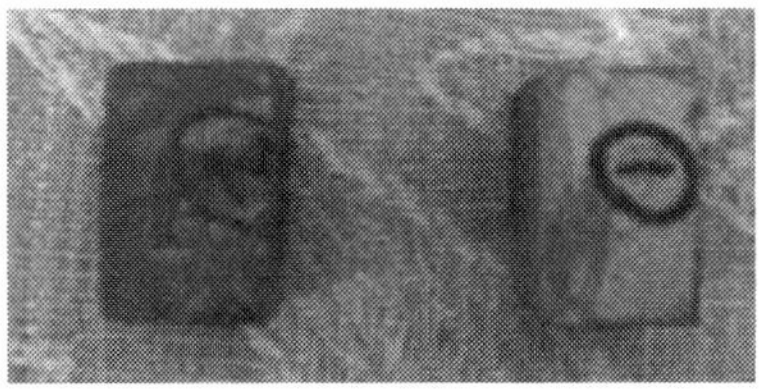

Figure 4. Coated specimens used for hardness test

Table 1 Results for the hardness test

Trail No.	Hardness values	
	Coated	Uncoated
1	89	84
2	90	85
3	89	84

3.3 Corrosion test

In this project salt spray test is used for testing the corrosion of a specimen. The salt spray test is a standardized and popular corrosion test method, used to check corrosion resistance of materials and surface coatings. Usually the materials to be tested are metallic and finished with a surface coating which is intended to provide a degree of corrosion protection to the underlying metal.The results are presented in Table 2. Input parameters considered for salt spray process is listed below

- Chamber temperature: 34.5 - 35.5 °C
- pH value : 6.65 - 6.85
- Salt solution collected : 1.0 – 1.5 ml/hr
- Concentration of solution : 5% Nacl
- Air pressure : 14 – 18 Psi
- Component loading in chamber position: 30 ° angle.

Table 2. Results of corrosion test

Time	**Rust formation**	
	Uncoated	**Coated**
After 12 Hrs	White rust	No rust

3.4 Emissions test

Automotive emission analyzer is used to measure the gas emission density. It is a five-gas analyzer, which can detect five gases viz. CO, HC, CO_2, O_2 and NOx emitted during the test. It also gives air - fuel ratio and air surplus rate. It gives accurate and instantaneous results especially when compared to the similar commercially available equipment. The emission testing machine is shown in the Figure 5.

Figure 5. Emission indicator set up

3.5 Effect of temperature

The performance and characteristics of the coated and uncoated cast iron exhaust manifold has been performed in the multi- cylinder engine. The outer surface temperature in the manifold is measured by the infrared thermometer testing machine. The inner surface temperature is measured by the thermocouple. The surface temperature obtained for both outer and inner coated and uncoated cast iron exhaust manifold at various load conditions have been compared and presented in Figure 6. It is observed that the use of coated surface reduces the temperature developed when compared to the use of uncoated surface. It is observed that the use of coated surface reduces the temperature developed when compared to the use of uncoated surface. From the temperature study it is asserted that the zirconia coated exhaust manifold arrest the temperature when compared to uncoated exhaust manifold. So this results show that heat dissipation is decreased in the coated exhaust manifold.

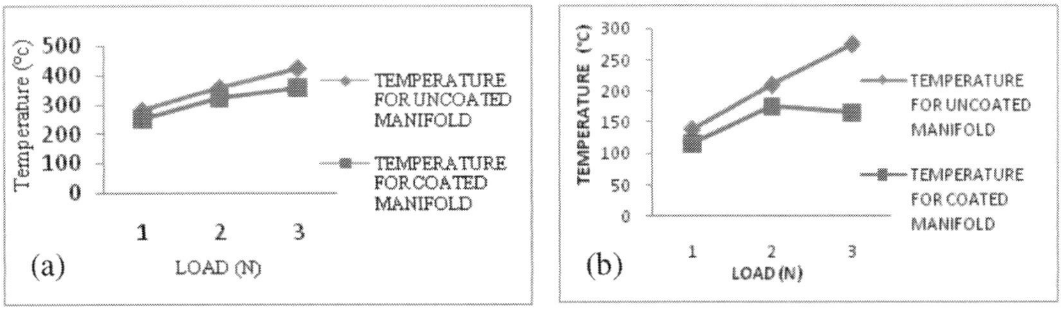

Figure 6. Comparison of (a) outer and (b) inner surface temperature

3.6 Effect of emission of NOx and CO

The comparison of nitrogen oxide emission from the waste exhaust gases for both coated and uncoated cast iron exhaust manifold is presented in Figure 7(a). From Figure it is observed that the emission of NOx is less in coated manifold specimen. The comparison of carbon monoxide emission from the waste exhaust gases for both coated and uncoated cast iron exhaust manifold is presented in Figure 7(b). From Figure it is observed that the emission of CO is less in coated manifold specimen.

IOP Conf. Series: Materials Science and Engineering **197** (2017) 012051 doi:10.1088/1757-899X/197/1/012051

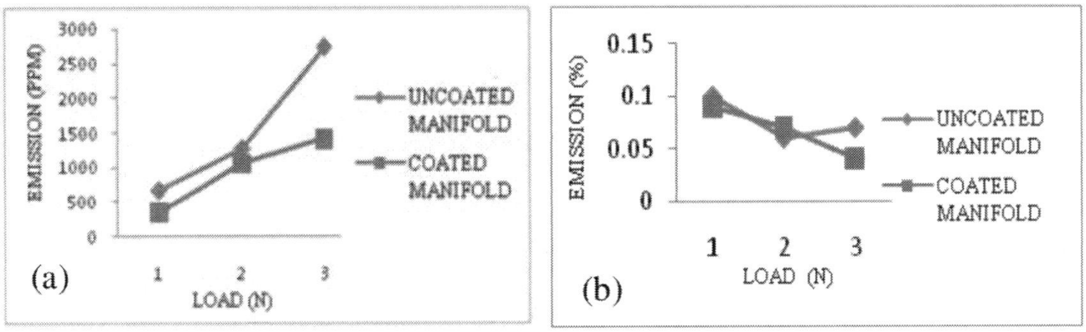

Figure 7. Comparison of (a) NOX and (b) CO emission for both coated and uncoated manifold

4. Conclusion

Based on the experimental results it is concluded that

- The surface temperature, carbon monoxide emission and nitrogen oxide emission is minimized by coating the exhaust manifold at rated load.
- Hardness is increased in the coated manifold.
- No rust formation in the coated manifold.

Finally it is concluded that the coated cast iron gives better performance than the uncoated cast iron exhaust manifold.

References

[1] Azadi M, Baloo M, Farrahi G. H, Mirsalim S. M. A review of thermal barrier coating effects on diesel engine performance and components lifetime. International Journal of Automotive Engineering, 2013 March; 3, 306 – 317.

[2] Durga Prasad K V V and Sanmala Rajasekhar, Mohan Krishna K, Hari Narayana Rao J. Thermal analysis of exhaust manifold coating with high temperature resistant materials. International Journal of Science Engineering and Advance Technology. 2015 November; 3(11), 1207- 1212.

[3] Gopaal, Kumara Varma M M M. Exhaust manifold design – FEA approach. International Journal of Engineering Trends and Technology. 2014 November; 17, 485 - 489,.

[4] Kandylas I P, Stamatelos A M. Engine exhaust system design based on heat transfer. Computation Energy Conversion & Management, 1998 December; 10, 1058 - 1072.

[5] Mahesh S Vasavan and Jotaniya P V. A review on heat transfer analysis in automotive exhaust system. International Journal of Innovative Research in Science, Engineering and Technology. 2015 February; 4 (2), 558 - 561.

[6] Mouna Challa and Imran Quraishi. Exhaust manifold. International Journal of Engineering Research & Technology. 2012 December; 1(10), 1- 6.

[7] Navaneethakrishnan P and Vasudevan D. Experimental study on performance and exhaust emission characteristics of a C.I. engine fuelled with tri compound oxygenated diesel fuel blends. Indian Journal of Science and Technology. 2015 January; 8(1), 96–102.

[8] Prabhat Kumar Jha, K. Karunamurthy, Joy Das, Rahul Malik. Study of exhaust gas temperature of S I engine using water, injection. Indian Journal of Science and Technology, 2016 August, 9(35),Doino:10.17485/ijst/2016/v9i35/92817.

[9] Serkan Saridag and Onjen Tak. Basic properties and types of zirconia: An overview", World Journal of Stomatol, 2013 August 20; 2(3), 40 – 47.

IOP Conf. Series: Materials Science and Engineering **197** (2017) 012052 doi:10.1088/1757-899X/197/1/012052

Fabrication of SiC particulate reinforced polyester matrix composite and investigation

R. Selvam[1], S. Ravi[2], R.Raja[3]

[1] Research scholar, Department of Mechanical Engineering, Bharath Institute of Higher Education and Research, Chennai, India.
[2] Professor and Head, Department of Mechanical Engineering, Sriram College of Engineering, Chennai, India.
[3] Professor and Head, Department of Mechanical Engineering, Bharath Institute of Higher Education and Research, Chennai, India.
E-mail: selvam_r7 @yahoo.com

Abstract. Polymer composite provokes a new alternative material to engineering and domestic application. Polymeric nano composite have been intensively investigated due to the performance improvement when a small amount of nano sized particulates are added to matrix. The distinguished properties of SiC particulates influence to make a polymeric composite. This composite material has many application such as mechanical, automobile, marine, appliances and packaging. The composite material is fabricated in deferent weight ratio and it is characterized to understand the mechanical behavior, which was studied by various testing method under external load.

Keywords: dispersed, polyester, biodegradable

1. Introduction

Polyester contributes significantly to the composite as the liaison agent, which has led to bind the particulates closer. Thermosetting plastics, such as polyester resins are widely used as matrix material which distributes the stresses uniformly in all direction and it resists shocks and vibration . Polyester resin found to be the most widely used resin system, particularly in the marine industry. Thermosetting plastic systems consist of liquid mixtures of low molar mass reactants, such as monomers, which upon polymerization from highly cross linked network polymers [1, 2, 10].

The particulate filler dispersed in composite exhibits improved properties. Polymer composites fabricated so have made them possible to find major application as high strength low weight materials.

In recent years nonmetals are used as reinforcement in polymer composite, SiC particles are prospective reinforcing material for various composites. This study concerned about evaluation of mechanical properties such as tensile, flexural, compressive and impact strength for different weight ratio of SiC particulate and polyester composite. This reveals that the PMC have a potential to replace conventional material in many application.

1.1 General character of SiC and Polyester

1.1.1 Biodegradability of PMC

The synthetic polyester are among the most promising materials of environmental friendly biodegradable polymer composite. They are adding natural inorganic material like SiC.

Aliphatic polyester degrades the polyester by using bacteria. Recently Lee et al [1] reported the biodegradation of aliphatic polyester based composites under compost.

1.1.2 Properties of SiC

Silicon carbide was the hardest synthetic material. It has Mohs hardness rating of 9. In addition to hardness, it has fracture characteristic that make them extremely useful in abrasive wheel. Its high thermal conductivity, together with high temperature strength, low thermal expansion and resistance to chemical reaction, makes silicon carbide is valuable in the manufacture of high temperature bricks and other refractory. It is also classed as a semiconductor, having an electrical conductivity between that of metals and insulating materials [6,7]

1.1.3 Properties of Polyester

The density of polymer is between 1.22-1.38 g/cm^3. It is able to form desired shape. It can be wet or dry and still remain strong and it dries very quickly. It has resistance to stretching and wrinkling. More over polyester offers resistance to shrinking, abrasion, colorless and transparent. The other hand, when burned it gives off a strong odour. The molten residue can cause many harmful burns when it comes in contact with human skin. Polyester melts at a temperature of 249 – 288 °C [5,6,7].

2 Fabrication of SiC particulate reinforced polyester matrix composite

- Preparation of SiC particles
- Preparation of polyester resin
- Fabrication of PMC

2.1 Preparation of SiC particles

Silicon carbide is produced from quarts by carbo thermal reaction, see figure 1. A mixture of pure quarts (SiO_2) sand and carbon in the form of finely ground coke is built up around a carbon conductor within a refractory kiln electrical resistance type furnace. There SiC is produced by high temperature around 2700°C for 40 hours [7,8]

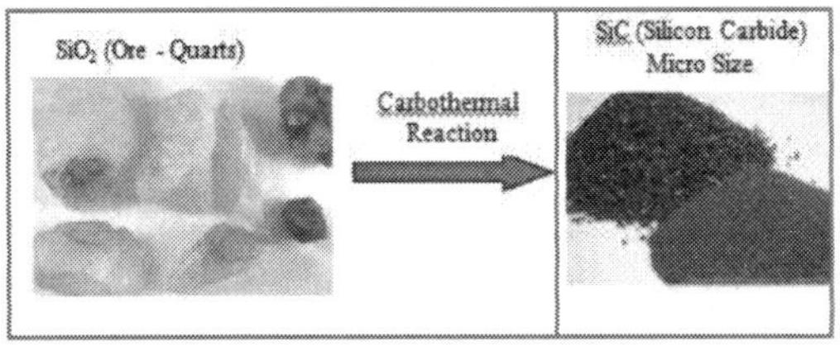

Figure 1 Conversion of SiC from quartz

2.2 Preparation of polyester

Polyester resin is prepared by condensation polymerization method. Malice acid and ethylene glycol are used to make the resin, which are diacid and dialcohol. Malice acid ($C_4H_6O_3$)

is an organic compound made from living organism. Polymerization relies on the reaction between malice acid and ethylene glycol, since the water is released during formation of polymer chain has been produced, refer the figure 2.

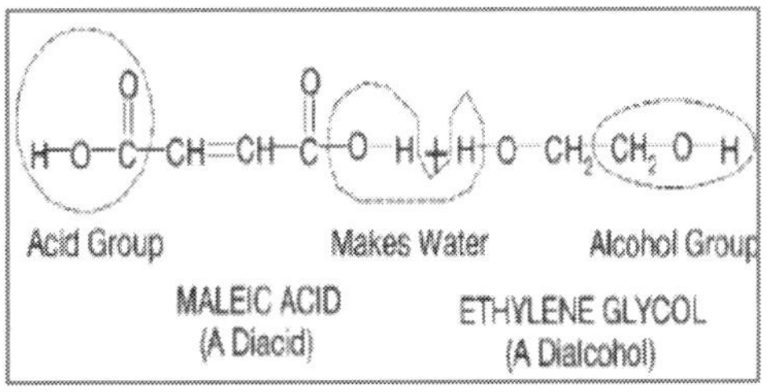

Figure 2 Formation of Polyester

The curing agent methyl e ethyle keton peroxide and rthe accelerator copalt napthanate are procured and measured its Ph values using the Ph meter, whose values obtained are 4.23 for resin, 7.35 for MEKP and 6.33 for copalt napthanate. Refere the figure 2.1 amd 2.2

Figure 2.1 Resin, Accelerator, Catalyst and Releasing agent

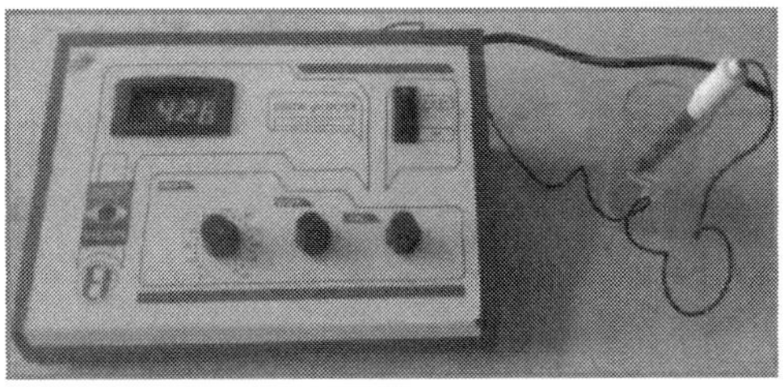

Figure 2.2 Checking pH value

2.3 Fabrication of PMC

The PMC is prepared for different weight ratio of SiC particulates and resin. Specimen 1 is pure polyester, specimen 2 has 30% of SiC and specimen 3 has 40% of SiC refer table 1.

Table 1 Quantity of SiC and Polyester as 30% and 40% composite

Sample No	Mixing Of Macro Size SiC& Polymer Matrix %	Weight Of Macro Size SiC (gm)	Weight Of Polymer Matrix (gm)	Curing Time hours
1	30:70	1.05	2.45	4
2	40:60	1.40	2.10	4

The composite is fabricated by in situ process and shaped into the required form as per ASTM standard for further testing. The dimension of the mould is shown in the table 2 and 3.

First the releasing agent and non binding polymer is coated on to the mould it will make the finished and cured part release easily from the mould. The gel coat polyvinyl alcohol is sprayed on to the mould and then the mixture of polyester resin and SiC is poured into the mould. Hand roller is employed to spread the material and dead weight (10 kg) is used to compact the composite and get homogeneity and flat surface. Curing time given to the product is around 4 hours. The specimen are lapped and kept in a closed container for further testing [6, 10].

In the beginning at the interfacial region, properties of the particle differ from those of the matrix. But after the passing of sometime the properties of both particulate and matrix become same. Most critical processing parameters are dispersion of clumps or agglomeration of particulates which create defect sides that will initiate failure & limit the efficiency of particulate to carry load [5].

2.3.1 *Mould Preparation*

For the sample preparation the first and foremost step is the preparation of the mould which has the dimension as per ASTM where the composite to be prepared.

Table 2. ASTM Standard for Specimen Preparation

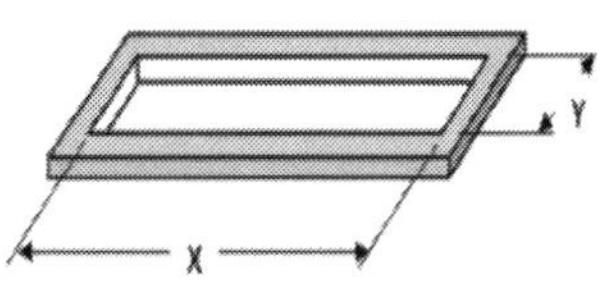

S.No	Name of Test	ASTM STD	Specimen size		
			X (Length)	Y (Width)	t (Thickness
1	Tensile test	D638-03	250	25	3
2	Compressive test	D695	Refer Table 4.2		
3	Flexural test	D790	154	13	3
4	Impact test	D256	64	12.7	3

IOP Conf. Series: Materials Science and Engineering **197** (2017) 012052 doi:10.1088/1757-899X/197/1/012052

Table 3 Compressive test Specimen size as per ASTM D695

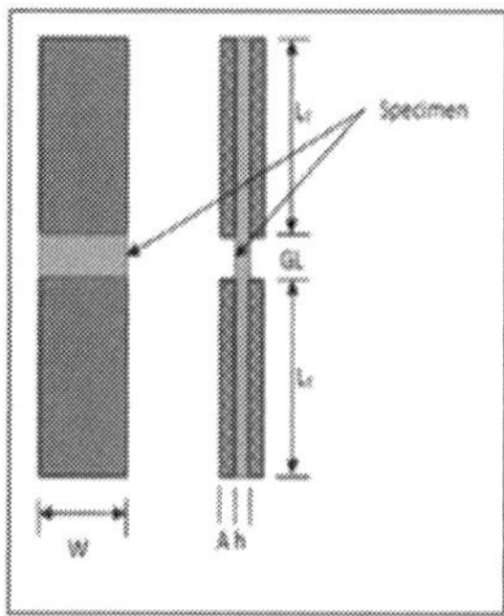

Specimen width (W)	10 mm
End tab length (L_T)	35 mm
Gauge length (GL)	5 mm
Specimen thickness (h)	2 mm
End tab thickness (A)	2 mm

3 Testing and Investigation

After the composite specimen made to finish as standard was tested under various condition. The specimens for the mechanical testing are shown in table 4.

Table 4 Specimen for testing as per ASTM standard

S.No	Specimen photograph	Name test
1	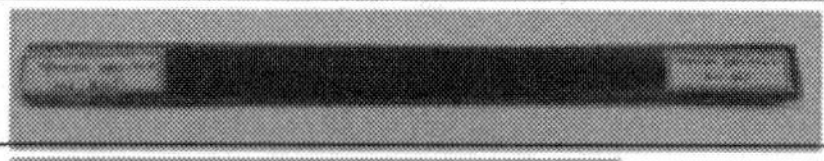	Tensile test
2		Flexural test
3		Compression test
4		Impact test

3.1 Tensile Test

The tensile test specimen as per ASTM D368 was tested in UTN. The stress strain curve was plotted for the determination of ultimate tensile strength [3,4,6]

From the test, the tensile strength of pure polyester is 22.89 N/mm^2, 30% SIC and polyester composite is 12.41 N/mm^2 and 40 % SiC is 16.41 N/mm^2. From this report interaction between particulates and the matrix is less under tensile load by comparing tensile strength of pure polyester refer the figure 3.1

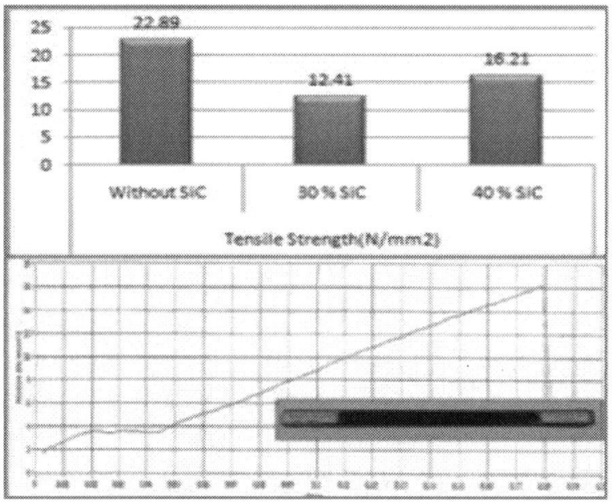

Figure 3.1 Comparison of tensile strength with test chart

3.2 *Compression Test*

Compression test is carried out in universal testing machine. The graph is plotted between stress and strain by using the software. Figure 3.2 shows stress and strain diagram for the composite of 30 % SiC which has compressive strength of 557.72 N/mm^2 and it is seen that the compressive strength of composite without SiC and 40 % SiC are 476.47 N/mm^2 and 675.91 N/ [3,4,6].

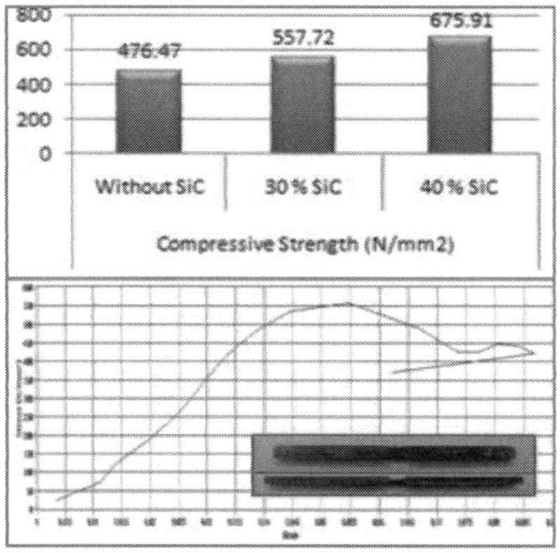

Figure 3.2 Comparison of compressive strength with test chart

IOP Conf. Series: Materials Science and Engineering **197** (2017) 012052 doi:10.1088/1757-899X/197/1/012052

3.3 Flexural Test

Flexural test were conducted using 3-point bending method according to ASTM D7264. Flexural test was performed to study the behavior and ability of material under bending load.

The beams were suitably instrumented for measuring deflections at several locations including the midspan deflection with dial gauges and LVDTs. The flexural strength of pure polyester is 1.865 N/mm^2, 30% of SiC is 1.509 N/mm2 and 40% of Sic is 1.304 N/mm^2

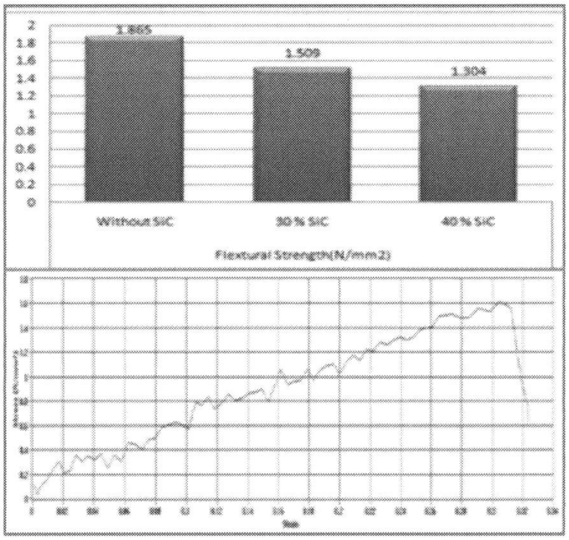

Figure 3.3 Comparison of Flexural strength with test chart

3.4 Impact Test

Impact test was performed by using Chorpy test according to the ASTM D256. This test was used to check and determine whether the material meets specific impact strength. Here the impact strength was observed 2J for all the three specimens.

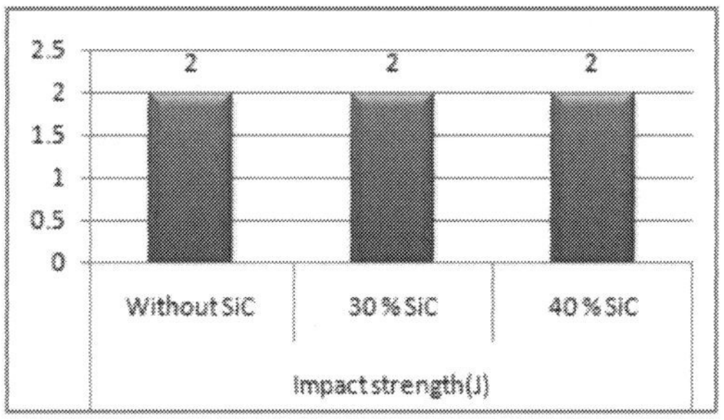

Figure 3.4 Comparison of Impact strength

4 Conclusion

Interfacial region in beginning at the points of particulate and the matrix properties differ. Polymer Composite having interface size as small as in nano scale the tensile and flexural strength could be increased.

Most critical processing parameters having the ability to disperse, agglomerations of particulates which lead failure & limit the efficiency of composite to carry load.

These Materials have good compressive strength, wear resistance. Hence we can use this material for piston, impeller etc. SiC Polymer composite is one of the alternate materials for the engineering, industrial application. The polymer composite material will replace the demand for other materials like aluminum, steel etc which are going to vanish in the near future as per the statistical survey.

Based on the results obtained from this study, it is concluded that 40% of SiC and 60 % of polyester is suitable for light weight and compressive load applications; also it is very good for shock absorption application. Composite properties further improved by particulate size and shape, which enhances adhesion between reinforcement and matrix.

5 References

[1] Adlan Akram Mohamad Mazuki, Hazizan Md Akil, Sahnizam Safiee, Zainal Arifin Mohd Ishak, "Degradation of dynamic mechanical properties of pultruded kenaf fibre reinforced composites after immersion in various solutions" (2010).

[2] G.Velmurugan,S.P.Venkatesan,P.V.Prakash,N.Sathishkumar,N.VijayakumarMechanical Testing of Hybrid Composite Material (Sisal and Coir).International journal of scientific and research publication, (2014).

[3] R.Selvam, S.Ravi Extraction and Synthesis of SiC Nano Particulate to Enhance Matrix Composites, International Journal of Research in Advent Technology.3,1-10 (2015)

[4] Siwei Huang, Ling Zhou, Mei-Chun Li, Qinglin Wu, Yoichi Kojima and Dingguo Zhou, "Preparation and Properties of ElectrospunPoly(Vinyl Pyrrolidone) / Cellulose Nanocrystal / Silver Nanoparticle Composite Fibers Materials,9, 523 (2016)

[5] Guillermo Bastos, Faustino Patiño-Barbeito, Faustino Patiño-Cambeiro and Julia Armesto "Nano-Inclusions Applied in Cement-Matrix Composites" Materials, 9, 1015,(2016)

[6] Rahul Mangire, Malur N. Srinivasan "Mechanical Behavior of Glass Fiber Reinforced Polymer Pultruded Composite Gratings" *Modern Mechanical Engineering*, 3, 142-146

[7] Aru.n.P.S,S.Sathish,C.Narendhar Fabrication of Polystyrene Composite Reinforced with Silicon Carbide Nanoparticles, International Journal of Current Engineering and Technology.(2014)

[8] Kalpit P.Kaurase Polymer Matrix Nano Composite Materials-A Critical Review(Sep-2014).International Journal of Engineering Research&Technology(IJERT).ISSN:2278 0181

[9] R.A.Andrievski-Synthesis,structure and properties of nanosized silicon carbide – adv.mater.sci.22, 1-20, (2009)

[10] V. Ramasamy and G. Suresh , Mineral Characterization and Crystalline Nature of Quartz in Ponnaiyar River Sediments, Tamilnadu, India. American-Eurasian Journal of Scientific Research 4 (2): 103-107, (2009)

IOP Conf. Series: Materials Science and Engineering **197** (2017) 012053 doi:10.1088/1757-899X/197/1/012053

Effect of Steel slag on Wear Characterization of Aluminium Composite Using Taguchi Technique

K S Sridhar Raja[1,*], V K Bupesh Raja[2], Aaro Rinold.F[3], Abhilash.L.S[4]

[1,]Department of Mechanical & Production Engineering, Sathyabama University, Chennai, India.

[2]Professor, Department of Automobile Engineering, Sathyabama University, Chennai, India.

[3,4] Under Graduate Students, Department of Mechanical & Production Engineering, Sathyabama University, Chennai.

[*]Email:sridhar2raja@gmail.com

Abstract: A wear behavior of steel slag reinforced aluminium A356 composite was fabricated by stir casting technique. Here an attempt has been made by reinforcing steel slag particle of size (1-5micron) in the aluminium matrix alloy fabricated through stir casting process. Dry sliding wear behavior of the composite was studied by pin on disc method. The experiment was conducted through Taguchi technique. A L_{16} orthogonal array was developed using ANOVA and the regression analysis was determined to find the optimum parameter. The parameters such as applied load, sliding speed, percentage reinforcement, and the distance travelled are considered for wear test. By considering the' smaller the best' the dry sliding wear was analyzed. Based on analysis of variance (ANOVA) the single-to-noise ratio are used to investigate the wear rate parameters. By studying the SEM analysis the worn out parts are studied.

1. Introduction

Metal matrix composite has found many applications in the areas of automobiles, aircrafts, and marine. MMCs have been used commercially in reinforced piston and aluminium crank case which strengthen its cylinder surface and brake pads. It has high thermal and corrosion resistance and improves fatigue strength at higher temperatures. Aluminium is the most popular metal matrix composite that are attractive due to its low density high thermal and electrical conductivity good wear resistance and high damping capacity [1-3]. In general aluminium matrix are reinforced with SiC, Al_2O_3, mica, but very rare research have been in SiO_2, BN, B_4C.Since boron carbide and boron nitride are hardest material and is most widely used in armors, nuclear reactor etc. can be used as the reinforcing medium [4]. The aluminium composites are currently used in automobile industries for fabrication of brakes, pistons in engines[5-8]The steel slag is the by product that has been produced during the separation of molten metal from impurities in induction furnace. A large amount of slag was dumped as a waste material which will affect the land debris. Since the properties of steel slag are similar to silica an attempt has been done by considering these as the reinforcing material. HuLin Li et al, in his research on PTFE copper composite was prepared by compression molding. The result analysed that sliding speed influence the wear rate of the composite[9]. M.Walczak et al., discussed the tribological test on AlSi9Mg with 20% of SiC and they concluded that the wear rate gets reduced [10]. Shaoyang Zhang and Fuping Wang in their research investigate the wear properties of aluminium with Sic particle and it was observed that friction coefficient decreases as the load and the speed increases[11]. Hemanth Kumar.T.R.et.al., investigated the aluminium 2618 reinforced with TiO_2 . By using Taguchi technique a mathematical model was developed and their result delivers that when the sliding speed and the reinforcement percentage increases the wear rate reduces gradually[12].

2 Experimental Details

2.1. Materials selection

Aluminium AlSi6Mn alloy was used matrix and its chemical composition given in Table 1. Its average tensile strength is 135Mpa and average hardness was 107HV. The steel slag which was the byproduct obtained from melting of metals from induction furnace was used as the reinforcing material which has the following chemical composition given in Table 2. By varying the weight percentage of steel slag (3%, 6%, 9%, 12%) the composite was fabricated by stir casting technology. In order to achieve good mechanical properties and good interfacial bonding between the matrix and the reinforcing particle stir casting technique was adopted to fabricate the specimen. The wear specimen of size 30mm x 8mm x8mm was machined as per ASTM G99 standard for testing.

Table 1. Chemical composition of A356 alloy

Si	*Fe*	*Cu*	*Mn*	*Mg*	*Ni*	*Zn*	*Ti*	*Al*
6.58	*0.16*	*0.06*	*0.06*	*0.57*	*0.01*	*0.01*	*0.14*	*balance*

Table 2. Chemical composition of steel slag

Fe	*Mg*	*Mn*	*Si*	*Ca*	*Cr*	*Ti*	*Al*
10	*0.32*	*15.66*	*57.01*	*2.45*	*0.58*	*0.63*	*13.35*

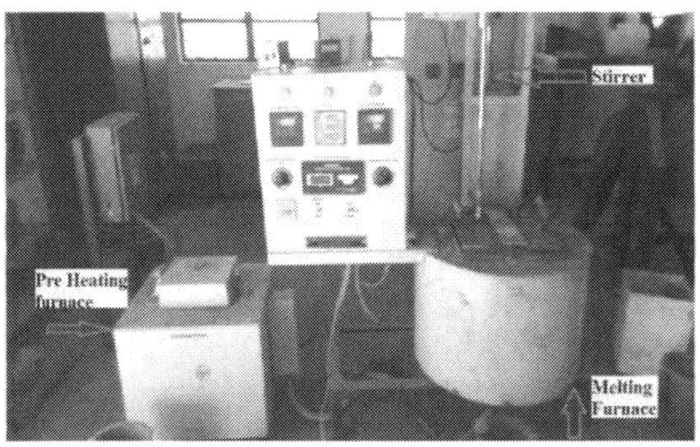

Figure 1. Experimental Setup of stir casting

2.2 Wear Study

The wear characteristic was analyzed by using dry sliding wear test as shown in fig.2 By using pin on disc method the wear behavior of the composite was studied. The composite material was taken as the pin which was pressed against the rotating EN31 carbon steel disc. Different parameters like applied load (N), Sliding speed(S), weight percentage of slag (%), distance travelled (D) are identified as the influencing parameter for wear rate. Based on Taguchi technique L_{16} orthogonal array was developed and the experiment were conducted accordingly. The objective of 'Smaller the best' was considered and the wear response was studied by signal to noise ratio. The process parameters are shown in Table.3. By using logarithmic transformation the S/N ratio were analyzed.

Figure 2. Pin on Disc wear setup

2.3 Design of Experiments (DOE)

Design of Experiments (DOE) is one of the most powerful and important statistical technique that was used to optimize the parameters. By considering four levels and four parameters an orthogonal array was developed. In the present study the wear parameters chosen are applied load (N), sliding speed (m/s), % reinforcement (%) and the distance travelled are considered. According to L_{16} orthogonal array was developed and the experiments were conducted [14]. The wear rates are tabulated with respect to orthogonal array. The response to be studied is the wear rate with the objective of 'smaller the best'. To determine the finest process design the experimental outcome uses signal to noise ratio to transform the output. The process parameters are shown in Table.3. By using logarithmic transformation the SN ratio were analyzed.

$$\text{S/N ratio } (\eta) \qquad = -10\log_{10}\left(\frac{1}{n}\sum_{i=1}^{n} y_{ij}^{2}\right)$$

Table.3 Process parameter for wear test

	Levell	Level 2	Level 3	Level 4
Load (N)	20	40	60	80
Sliding speed(m/s)	0.5	1.0	2.5	2
% reinforcement (%)	3	6	9	12
Distance (m)	500	1000	1500	2000

3. Result and Discussions

The experimental results were calculated and its response are calculated by signal to noise ratio using equation (1) and are listed in Table.4. The signal to noise ratio are predicted based on the experimental values , and the optimum parameters resulting in wear rate are shown in fig.3. The fig.3 clearly shows that the speed plays major influence. In table.4 the higher the value of S/N ratio lower the wear rate are identified at 60N load and at 2m/s of speed ,at 6% reinforcement the lower wear rate occurs.

Table.4. L_{16} Orthogonal array of Taguchi method for wear rate

Expt. No.	Load (N)	Speed (m/s)	Percentage Reinforcement (%)	Distance (m)	Wear rate (mm^3/m)	S/N Ratio (wear)db
1	20	0.5	3	500	0.0011664	50.1437
2	20	1	6	1000	0.0029403	51.3035
3	20	1.5	9	1500	0.0016468	62.1849
4	20	2	12	2000	0.0016468	58.6630
5	40	0.5	6	1500	0.0014409	50.6322
6	40	1	3	2000	0.0050483	55.6672
7	40	1.5	12	500	0.0020121	55.6672
8	40	2	9	1000	0.0024145	56.8273
9	60	0.5	9	2000	0.0006036	45.9371
10	60	1	12	1500	0.0065120	53.9270
11	60	1.5	3	1000	0.0032560	52.3435
12	60	2	6	500	0.0016280	64.3850
13	80	0.5	12	1000	0.0016280	43.7257
14	80	1	9	500	0.0011664	49.7463
15	80	1.5	6	2000	0.0029403	55.7669
16	80	2	3	1500	0.0016468	55.7669

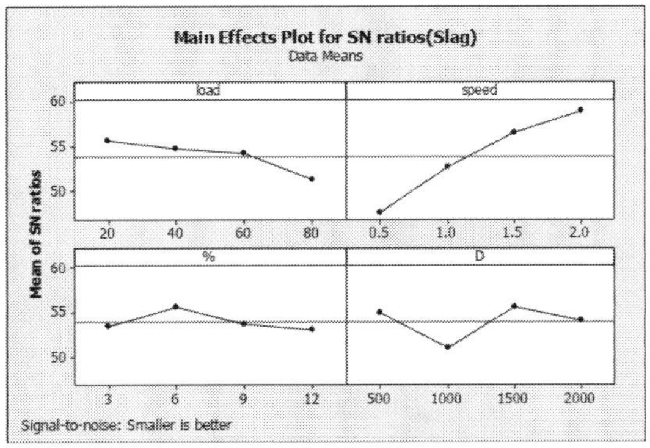

Figure.3 Main effect plot for means-wear rate

Table 5. Analysis of Variance for wear rate (mm^3/m)

Source	DF	SS	MS	F	P
REGRESSION	4	1.023e-07	2.559e-08	11.50	0.001
RESIDUAL ERROR	11	2.448e-08	2.225e-09		
TOTAL	15	1.2688e-07			

The response values are tabulated in Table 5, and it shows that the speed plays major role in wear rate. The ANOVA table shown the speed has major influence on the wear of the composite.

3.1 *Multiple linear regression models*

A multi linear regression model is developed using statistical software MINITAB R16.It shows the relationship between independent/predictor variable and a response variable by fitting a linear equation to the observed data. The regression equation for wear rate is

Wear rate = 0.00341+0.000028L-0.00238S+0.000117%-0.00000D
R-Sqr (adj) = 73.7%

Table 6 Response table for wear rate

Level	Load	Speed	% reinforcement	Distance
1	55.57	45.47	53.48	54.99
2	53.57	52.66	54.39	51.05
3	53.14	56.49	52.66	54.50
4	51.25	58.91	53.00	53.00
Delta	4.32	13.44	1.73	3.94
Rank	2	1	4	3

3.2 Confirmation Experiment

Based on the Taguchi's technique the optimum parameter was selected and the confirmation test was compared with the regression model. The experimental values are compared with the regression model which was developed by mathematical model.

Table 7. Result of confirmation experiment

Level	Load (N)	Sliding speed (S)	Weight percentage of steel slag (%)	Distance (D)	Predicted wear rate	Experimental wear rate
13	80	0.5	12	1000	0.0016280	0.001841

4. Conclusion

The steel slag reinforced aluminium composite was fabricated and the dry sliding wear characteristic was analyzed and optimized y Taguchi Technique. From the study the following conclusions was obtained

- Optimum wear rate was obtained.
- The wear rate was dominated by sliding speed followed by applied load, distance travelled and the percentage reinforcement of the steel slag.
- ANONA test concluded that sliding speed increases the wear rate of the composite material decreased significantly.

References

[1] Sharma S C, Sathish B M, Girish B M, Kamath R, and Asanuma H 1998 *Wear* . 31 183.
[2] Sharma S. C, Girish B M, Kamath R, and Sathish B M 1999 *Material Science* 315 298.
[3] Davis L W, and Bradstreet S W 1970 *Wear* 113 234.
[4] Rajesh S, Rajakarunakaran S, and Sudhakara Pandian R 2012 Materials Physics and Mechanics 15 150.
[5] Mahdavi S, and Akhlaghi F 2011 Tribology Letters 44 (1) 1.
[6] Mindivan H, Kayali E S, and Cimenoglu H 2008 Wear 265 (5–6) 645.
[7] Mishra A, Sheokand R, and Srivastava R K 2012 International Journal of Scientific and Research Publications 2 (10) 1.
[8] Rudnik D, and Sobczak J , Wyd. Instytutu Transportu Samochodowego, Warsaw, 2001(in Polish).
[9] HuLin Li, ZhongWei Yin , Dan Jiang, LiYong Jin, and YuQing Cui 2015 Wear 328-329 17.
[10] Walczak M, Pieniak D,and Zwierzchowski M 2 0 1 5 Archives of civil and mechanical engineering 1 5 1 1 6.
[11] Shaoyang Zhang, and Fuping 2007 Wang Journal of Materials Processing Technology 182 122.
[13] Hemanth Kumar T R, Samy R P, and Chandrashekar T K 2013 Indian Journal of engineering & Materials Science 120 329.
[14] Sridhar Raja K S, and Bupesh Raja V K 2015 International Journal of Applied Engineering Research 10(8) 6360.

IOP Conf. Series: Materials Science and Engineering **197** (2017) 012054 doi:10.1088/1757-899X/197/1/012054

Computational analysis of frp composite under different temperature gradient

P.Gunasekar[1] **S.Manigandan**[2]

[1,2] Asst professor, Department of Mechanical Engineering, Sathyabama University, India

Abstract. Composite material strength depends on the stiffness of fiber and the resin which is used for reinforcement. The strength of the laminate can be increased by applying good manufacturing practices. The strength is directly depending on the property of resin. The property of the any compound subjected to changed when they exposed to the temperature. This paper investigates the strength of laminate when they subjected to different temperature gradient of resin while manufacturing. The resin is preheated before adding hardener with them. These types of laminate reinforced with resin at different levels of temperature 20c, 40c, and 60c. These different temperature resin are used for reinforcement and the specimen tested. The comparative results are made to find how the stiffness of laminate changes with respect to the thermal property of resin. The results are helpful to obtain high strength laminate.

1. Introduction

Recent years, quick advance in construction materials skill have enabled civil engineers to attain impressive gains in the economy and safety. The earliest FRP materials used glass fibers rooted in polymeric resins that were made available at World War II. The combination of high-strength, stiffness structural fibers with low cost and lightweight. The environmentally challenging polymers resulted in composite materials with mechanical properties and durability superior than metals. Fiber materials such as boron, carbon, and aramid, were used in space exploration and air due to its higher strength, higher stiffness, and lower density. Stable cross-sectioned FRP composite are produced for use in the construction industry for building bridge and other superstructures. Recent research on pultruded FRP shapes is very high.

The natural fiber composites are promising as reasonable alternatives to glass-reinforced composites. Natural fiber composites such as hemp fiber-epoxy and china reed fiber-PP are predominantly striking in automotive applications because of lower cost density. The natural fibers conventionally have been used to fill and reinforce thermosets; have attracted better concentration due to super recyclability behavior. Natural fiber composites are also claim to tender environmental reward such as lower greenhouse gas emissions, lowest pollutant emission and enhanced energy recovery.

1.1 FRP on Polyester Resin

Polyester resin reinforcement with natural fibers like coconut fibers are exhibits higher tensile strength and light weight, it can easily deform to high strain values due to its high flexibility stamina. The reference paper predicts that sisal-glass composite were undergone high compression and poor tensile under dynamic loading. It also pragmatic that the tensile properties are superior compare to the sisal-

glass matrix. Fiber reinforced composites create average impact to temperatures when the thermal source involved. Form [1][2] as the temperature increases, the impact strength of the composites decrease steadily [6-8]. The bagasse fiber has been reinforced with polyester resin in addition to its mechanical properties were distinguished. It is observed that tensile property is increasing with increase in proportion of polyester.
Experimental Details

1.2 Materials

Raw materials used in this experiment is:

- Glass Fiber and Jute -Polyester
- Accelerator
- Catalyst

Jute Fiber: Jute is a lengthy fiber, spongy in nature it can be spin into coarse and strong threads. It is produced from genus corchorus plant. Jute is one of the finest and lowest cost natural fibers, and it's only to cotton in amount bent and mixture of uses. Jute fibers are contains mostly of the plant materials such as cellulose.
Glass Fiber - Polyester: Glass Fiber Reinforced Polyester is reinforced polymer completed of a plastic matrix reinforced by glass fiber. Fiber glass is a light weight, strong, and robust.
The composite laminate specimen's are geared up using the hand layup technique and the specimen is investigation in ASTM standards. The simplest technique of manufacturing adapted to laying down bidirectional, a liquid thermosetting resin is worked into the reinforcement by hand with a brush or roller. The process is frequent number of times alike to the number of layers required for the final composite. Resin and catalyst agents are pre-mixed and usually designed to harden at room temperature. The major advantage of this process is its great flexibility and cheap. The prepared specimen is tested for the mechanical strength.

2. Results and discussion

For this research we have opted, four specimen of same property.
Specimen 1 Glass fiber with polyester resin at room temperature
Specimen 2 Glass fiber with polyester resin preheat 28 C
Specimen 3 Glass fiber with polyester resin preheat 32 C
Specimen 4 Glass fiber with polyester resin preheat 35 C
The specimens are tested and the results are compared with each other to find the best optimized process for making the laminate.

Table 1. Stress comparison table

Material	S-I	S-II	S-III	S-IV
Ultimate/ Break Load	8	9.75	11.7	12.0
Disp at Fmax(mm)	13	14.2	17.04	19.5
Max.Disp(mm)	15	16.9	20.28	22.5
Area mm²	63.7	63.7	63.7	63.7
Ult.Stress(Mpa)	218	233	279.6	327.0
Yield stress (Mpa)	125	135	162	187.5
YS/UTS Ratio	0.29	0.35	0.42	0.4
Elongation	0.16	0.21	0.252	0.2

Table 2. Stress comparison table

Material	S-I	S-II	S-III	S-IV
Ult/Break Load	0.4	0.45	0.5175	0.52
Disp at Fmax(mm)	2.5	2.8	3.22	3.25
Max.Disp(mm)	3.5	3.9	4.485	4.55
Area	63	63	72.45	81.9
Ult.Stress	9	12	13.8	11.7

Table 3. Stress comparison table

Material	S-I	S-II	S-III	S-IV
Ultimate Stress (Mpa)	218	233	279.6	327.0
Expt.Ultimate Stress (Mpa)	220	233	249	258.0
Maximum Displacement	15	16.9	20.28	22.5
Expt.MaxDisp	16.9	16.9	15.3	16.3
Expt.Ultimate Stress (Mpa)	7.5	11	13	12.5
Ultimate Stress (Mpa)	9	12	13.8	11.7
Maximum Displacement	3.5	3.9	4.485	4.55

Expt.MaxDisp	3.5	3.9	10.2	14

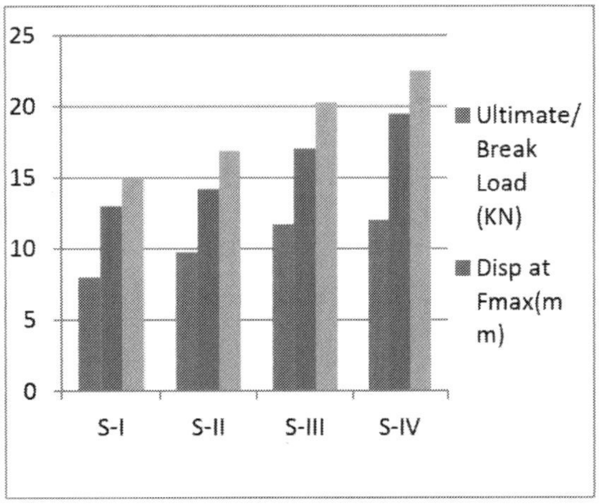

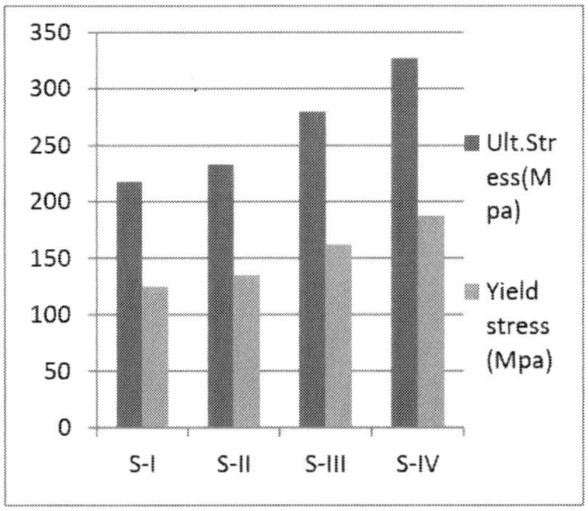

Figure 1. Ultimate stress vs Displacement

Figure 2. Stress vs Yield stress

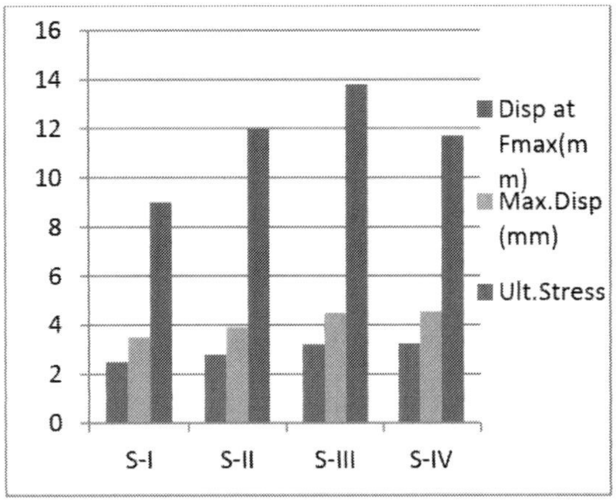

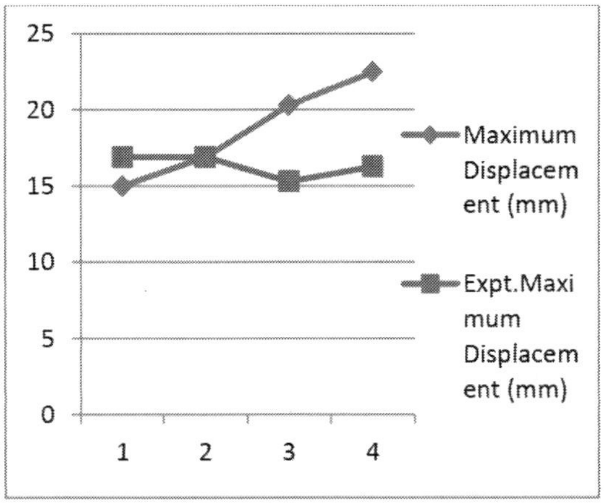

Figure 3. Stress vs strain

Figure 4. Strain validation

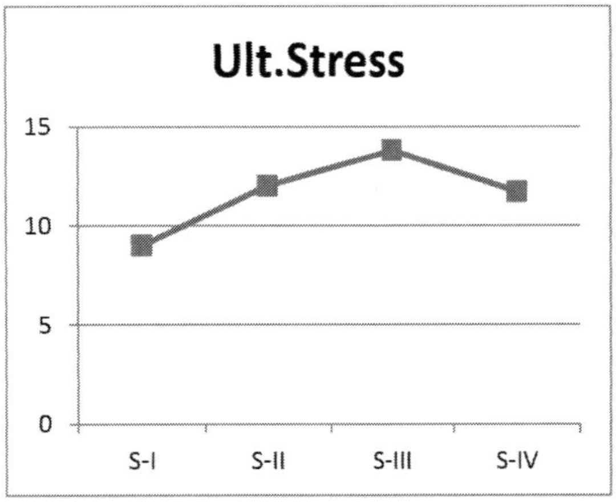

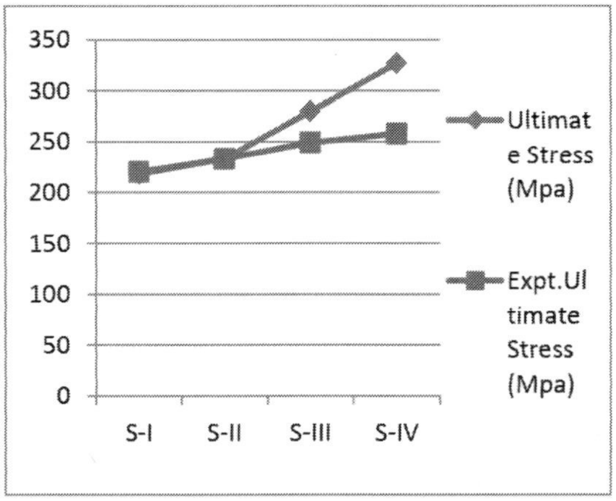

Figure 5. Ultimate stress plot

Figure 6. Validation tensile ultimate stress

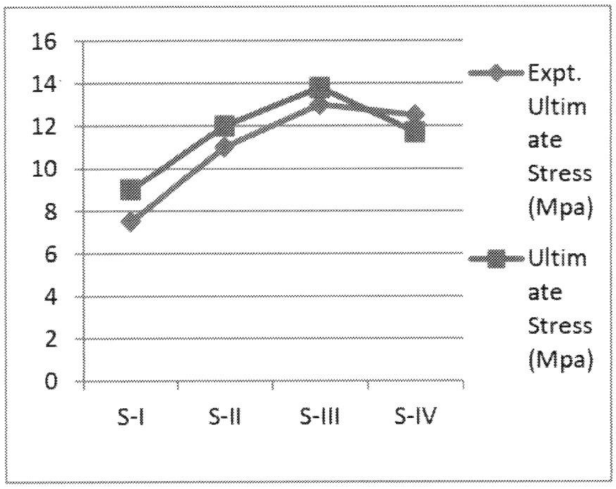

Figure 7. Validation of compression stress

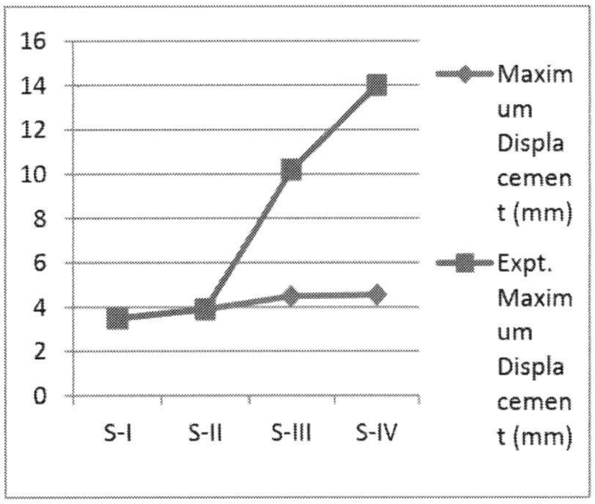

Figure 8. Validation of compression strain

3. Conclusion

All the four types of specimen tested for the mechanical behaviour and their results are plotted above. The comparative data's and plot shows that the behaviour of the laminate is very good and superior as the temperature of the resin changes.

The results predict that as the temperature of the pre mixing increases the strength of the fiber is increased significantly.
It also concluded that the strength of fiber is depending on the resin temperature.

References

[1] Gunasekar, P., Manigandan, S., Anderson, D., &Devipriya, J 2017 Evaluation of Fe-Epoxy metal nanocompopsite in glass fiber and Kevlar. International Journal of Ambient Energy.

[2] Puck, A., H. Schürmann 1988 Failure analysis of FRP laminates by means of physically based phenomenological models, Composites Science and Technology.

[3] S. Manigandan 2015 Computational Investigation of High Velocity Ballistic Impact Test on Kevlar 149, Applied Mechanics and Materials.

[4] Sriramula, Srinivas, Marios K. Chryssanthopoulos 2009 Quantification of uncertainty modelling in stochastic analysis of FRP composites, Composites Part A: Applied Science and Manufacturing.

[5] Karbhari, V. M 2003 Durability gap analysis for fiber-reinforced polymer composites in civil infrastructure, Journal of composites for construction.

[6] Rahimi, Hamid, Allan Hutchinson 2001 Concrete beams strengthened with externally bonded FRP plates, Journal of composites for construction.

[7] S. Manigandan 2015 Determination of Fracture Behavior under Biaxial Loading of Kevlar 149, Applied Mechanics and Materials

IOP Conf. Series: Materials Science and Engineering **197** (2017) 012055 doi:10.1088/1757-899X/197/1/012055

Evaluation of Thrust force in Drilling Woven roving Glass fibre reinforced Aluminium Sandwich laminates with TiAlN coated drill using Taguchi analysis

G. Ramya Devi[1], **K. Palanikumar**[2]

[1]Research Scholar, Sathyabama University, Chennai, India Department of Mechanical Engineering,St. Peter's College of Engineering and Technology, Chennai, India
[2]Department of Mechanical Engineering,Sri Sai Ram Institute of Technology,Chennai-600 044, India
Email id: gramyadevi@gmail.com palanikumar_k@yahoo.com

Abstract: TiAlN is a high-performance coating which outshines in coarse and hard-to-machine materials like cast iron, aluminium alloys, tool steels, and nickel alloys. This paper presents the prediction and evaluation of thrust force and Torque in drilling of Woven roving Glass Fibre Reinforced Plastic and Aluminium sandwich laminate. The Prediction is based on Taguchi method. The experimental results specify that the feed rate and the drill diameter are the most significant factors affecting the thrust force, while the feed rate and spindle speed contribute the most to the surface roughness. In this study, the objective was to establish a correlation between the feed rate, spindle speed and drill diameter with the induced thrust force and Torque in drilling sandwich laminate.

Keywords: Thrust force measurement; Sandwich Laminate; Fibre metal laminates; Drilling; Taguchi analysis

1. INTRODUCTION

Fibre metal laminates are well known for their fatigue resistance and damage tolerance properties. After its successful utilization in Fuselage of Airbus A380, it attracted many researchers world-wide to study its characterisation in various application fields. Drilling is a primary machining operation for assembly of these fibre metal laminates for structural applications. Reducing thrust force and torque reduces the delamination. Therefore it is more essential to reduce thrust force and torque to improve the quality of hole produced. Palanikumar analysed the models of comparable nature made of carbon fiber composites where the size of the delamination region has been revealed to be associated to the thrust force developed during the drilling process and the value of critical thrust force is also observed[1]. Mohan et al diagnosed that speed and drill diameters are the more significant parameters on thrust force than the specimen thickness and the feed rate. Highest speed on the lowest specimen with the smallest drill size and lowest feed rate are the best combination that produces lowest thrust force [2]. There have been many research developments are going on for optimizing the cutting parameters to get the better productivity in drilling process, For these reasons there have been research developments with the objective of optimizing the cutting conditions to obtain a better productivity in drilling process. Ghani et al. used a same approach using Taguchi's method in end milling to optimize the cutting parameters. Thrust force of Core drill made of diamond is experimentally investigated in drilling carbon fiber reinforced plastic by Tsao [4], the result shows that the large grit size of diamond, minimum thickness of core drill, lower speed and feed produced minimum thrust force. Wu [5] employed the multifaceted drills to minimize thrust force induced in drilling operation. Mathew et al investigated the thrust force and torque developed while trepanning tool is used for drilling glass fibre reinforced plastic [6]. Julian et al studied the impact of workpiece contituents and cutting speed on the forces induced in drilling CFRP made of three different combinations of resins and two types of woven carbon fibre fabrics [7]. Davim et al investigated the impact of cemented carbide drill geometry, cutting speed and penetration rate on the thrust force in drilling glass fibre reinforced plastic. The result proved that spur geometry of cemented carbide drill exhibited less thrust force compared with 118° point angle. From ANOVA data analysis it is proved that cutting pressure gets reduced with increasing penetration speed and it becomes highest significant factor among the other factors considered to affect both the cutting pressure and thrust force developed during drilling [8]. Basavarajappaa et al discussed the effect of spindle speed, feed rate on feed force developed in drilling hybrid metal matrix composites, the composite is made by stir casting, The effect of these cutting parameters on burr height and surface finish is also studied, the Taguchi analysis is employed to analyse the datas and it is revealed that the output is greatly influenced by the feed rate that the speed [9] Many adverse damages like delamination, fibre pull outs produced during drilling reduce the fatigue strength of composite laminates thereby reducing its durability.[10] Davim et al. [11,12,13], Sardinas et al. [14] observed that cutting speed has more effect on delamination during conventional drilling of composite laminates. All the structures in metallic materials respond in a same manner under machining loads and conditions, whereas composite laminates having anisotropic structures behave in a different manner at different locations of its structure based on the application of load. This variation may lead to defects like delamination. The thrust force produced during drilling is one of main indexes to define machinability of composite laminates due to the fact that it directly affects the quality of drilled holes, especially delamination [15]. There is a critical value for thrust force which is believed to exist below which no delamination occurs [16]. In this paper, the sandwich laminate of 5 layers is made with alternating layers of aluminium and glass fibre mat and its drilling characteristics are studied and analysed using Taguchi method.

2. METHODS AND MATERIALS

The metal used for the current study is Aluminium 1050 which is a popular grade of aluminium for all-purpose sheet metal work where moderate strength is required. Alloy 1050 has excellent corrosion resistance, high ductility and highly reflective finish. WRM mat of GFRP is selected for its excellent flexural strength. The sandwich laminate is fabricated by Hand-Layup process. The Aluminium sheet of grade AA 1050 H14 of 1.2mm thickness is used in this study, for making the sandwich laminate with Glass fiber of Woven roving mat. The sandwich laminate consists of 5 layers in which the Aluminium forms the top, middle and bottom layers. Glass fibre mat are placed in between the aluminium sheets. The Aluminium sheets are uniformly roughened on both sides using a very fine emery paper. Woven Roving mats (WRM) made of E-Glass fibre are also cut to the specimen shape. The Epoxy resin (LY 556) and hardener (Aradur HY951) are mixed in the ratio of 10:1 by weight respectively. The mould on which the layers are to be placed is coated with wax. Aluminium sheet is placed on the wax coating, resin mixture is applied over the roughened Aluminium surface. Woven Roving mat is placed over the coating. The above steps are repeated for next three layers. Now the mould is closed by a wax coated acrylic sheet. Weights are placed over it to get a uniform pressure while curing. Excess resin oozed out is wiped. The hybrid laminate is taken out of the mould and epoxy paste sticking on the sides is cleaned. The fabricated Sandwich laminate is cut to a size of 100mmX50mm for so that it can be fit into the fixture provided experimental setup. The thrust force is measured using the Kistler dynamometer which is placed just below the fixture. The prepared specimen for the measuring the thrust force is shown in figure 1.

IOP Conf. Series: Materials Science and Engineering **197** (2017) 012055 doi:10.1088/1757-899X/197/1/012055

The Drill bit selected for this study is Titanium Aluminum Nitride (TiAlN) coated drill of 6mm, 8mm and 10mm diameters, which is shown in Figure 2. TiAlN coating is applied where heat resistance and extra hardness is required for machining abrasive materials. Aluminium oxide layer formed gives better tool life. This coating is mainly selected where no coolant or little coolant is being used for machining; TiAlN has a high surface hardness with a coefficient of friction less than Titanium Nitride (TiN). TiAlN is a high-performance coating which outperforms in drilling materials like cast iron, aluminum alloys, tool steels, and nickel alloys.

Figure 1. GFRP/Al Sandwich Laminate Specimen

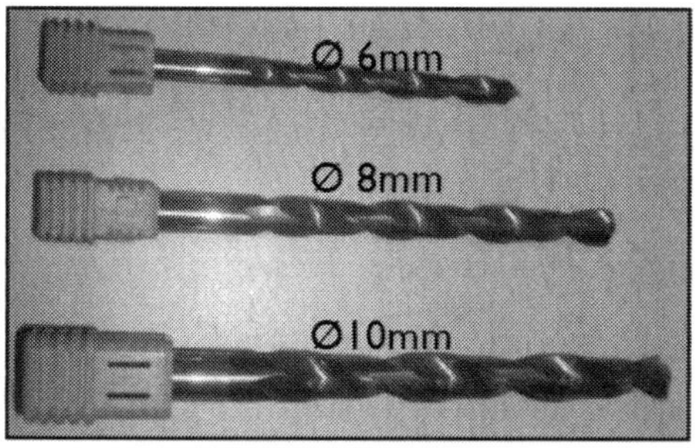

Figure 2. TiAlN Coated drills

3. EXPERIMENTAL SETUP

Drilling process is carried out in a SIEMENS CNC Vertical Machining Center using TiAlN coated drill of 3 different diameters 6mm, 8mm and 10mm. The machining parameters such as speed and feed are varied and their ranges are shown in table 1. The thrust forces during drilling were measured using a Kistler type piezoelectric dynamometer and the signal was transmitted and converted by a Kistler type controller unit and recorded on a personal computer. Each test was repeated twice and the average is taken for discussion. The experimental setup is shown in the figure 3.

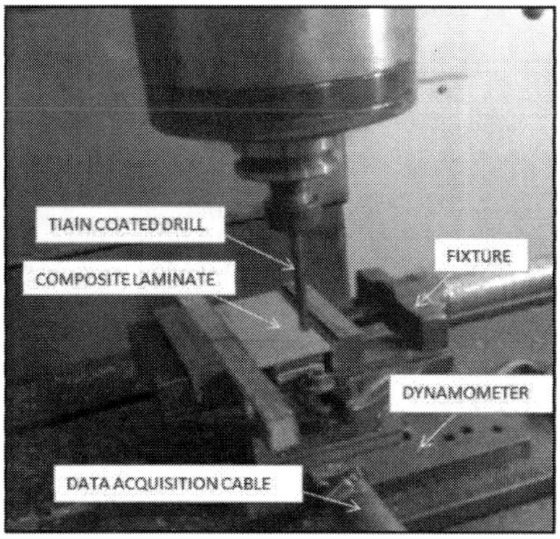

Figure 3. Experimental Setup

Input Parameters for Drilling Operation		
A:Diameter (mm)	B:Feed (mm/min)	C:Speed (rpm)
6	100	1000
8	200	2000
10	300	3000

Table 1. Drilling input parameters

After the drilling operation is done for each set of input variables, the chips are removed and the workpiece is removed from the fixture. The thrust force and torque developed during the drilled are analysed.

4. RESULTS AND DISCUSSION

Drilling of sandwich panels is different than drilling of monolithic materials as drill passes through alternate fibre and metal layers which have different properties. The variation in the properties of the constituents of composite materials makes difficult in understanding the mechanism of the material removal rate. During drilling of sandwich laminates the uneven load distribution between the fibre and metal layers leads to series of fractures. The thrust force and torque developed in the drilling operation influence much on the size of the delamination. There is also a critical value for the thrust force below which damage does not occur. Drilling induced damages are mainly due to these forces. Torque is associated to the feed rate, higher the value of feed rate higher will be the torque and high torque may also reduce the tool life.

Response Surface Methodology is used to analyse the effect of drilling parameters on thrust force and torque. Design Expert software is used to check the significant factors. Orthogonal array of L_{27} is used to design the experiments. The fig. 4 shows clearly the ANOVA table for the Thrust Force, it denotes the model is significant. The parameters A and C are the most significant on thrust force. Value of 'Prob>F' less than 0.05 denotes the factors are significant which is shown in the fig. 4 In this case, A, B, C, A^2, C^2 and AC are significant model terms. Values greater than 0.1 indicates the non-significant factors, i.e., the parameters, Drill Diameter, Spindle Speed and Feed rate and the interaction between Diameter and Feed rate affects the thrust force developed.

Response: Thrust Force

ANOVA for Response Surface Quadratic Model

Analysis of variance table [Partial sum of squares]

Source	Sum of Squares	DF	Mean Square	F Value	Prob > F	
Model	7.745E+005	9	86056.18	124.92	< 0.0001	significant
A	1.790E+005	1	1.790E+005	259.78	< 0.0001	
B	11643.38	1	11643.38	16.90	0.0007	
C	4.987E+005	1	4.987E+005	723.97	< 0.0001	
A^2	14880.24	1	14880.24	21.60	0.0002	
B^2	12.91	1	12.91	0.019	0.8927	
C^2	9922.67	1	9922.67	14.40	0.0014	
AB	679.51	1	679.51	0.99	0.3346	
AC	59671.20	1	59671.20	86.62	< 0.0001	
BC	0.053	1	0.053	7.742E-005	0.9931	
Residual	11711.07	17	688.89			
Cor Total	7.862E+005	26				

The Model F-value of 124.92 implies the model is significant. There is only a 0.01% chance that a "Model F-Value" this large could occur due to noise.

Figure 4. ANOVA generated by the Design Expert software for the response 'Thrust Force'

IOP Conf. Series: Materials Science and Engineering **197** (2017) 012055 doi:10.1088/1757-899X/197/1/012055

Response: Torque

ANOVA for Response Surface Quadratic Model

Analysis of variance table [Partial sum of squares]

Source		Sum of Squares	DF	Mean Square	F Value	Prob > F	
Model		11398.74	9	1266.53	647.64	< 0.0001	significant
	A	*1498.78*	*1*	*1498.78*	*766.40*	*< 0.0001*	
	B	*62.16*	*1*	*62.16*	*31.79*	*< 0.0001*	
	C	*9688.32*	*1*	*9688.32*	*4954.12*	*< 0.0001*	
	A^2	*17.90*	*1*	*17.90*	*9.15*	*0.0076*	
	B^2	*0.29*	*1*	*0.29*	*0.15*	*0.7040*	
	C^2	*75.71*	*1*	*75.71*	*38.71*	*< 0.0001*	
	AB	*0.40*	*1*	*0.40*	*0.20*	*0.6569*	
	AC	*55.17*	*1*	*55.17*	*28.21*	*< 0.0001*	
	BC	*4.408E-003*	*1*	*4.408E-003*	*2.254E-003*	*0.9627*	
Residual		33.25	17	1.96			
Cor Total		11431.98	26				

The Model F-value of 647.64 implies the model is significant. There is only a 0.01% chance that a "Model F-Value" this large could occur due to noise.

Figure 5. ANOVA generated by the Design Expert software for the response 'Torque'

The fig. 5 shows clearly the ANOVA table for the Torque, the effects are almost same as that of thrust force, the parameters drill dia and feed are more significant on Torque. Value of 'Prob>F' less than 0.05 denotes the same in the table of fig 5, the interaction between Diameter and Feed rate also affects the toque.

The surface plot can help to visualize the response surface. These are used to promote the required response values and working conditions. The surface plots displays how a thrust force and torque relates to two factors based on model equation. The surface plot shows only two variables, the third variable is kept as constant.

The 3D surface plot generated by the Design Expert software showing the interactions between the parameters affecting Thrust Force is shown in the fig. 6 which denotes that the higher the feed rate and diameter the higher will be the thrust force value, where on the other hand higher the spindle speed lower will be the thrust force. Fig 6 [a] shows the interaction between feed rate and diameter affecting thrust force, fig 6[b] shows the interaction between feed and spindle speed and fig 6[c] shows the interaction between spindle speed and diameter.

The three dimensional surface plot shown in fig. 7 is the interaction between parameters that is affecting torque. It is clearly visible that torque is most affected by feed rate and it increases with the increase in feed rate and drill diameter, whereas it decreases slightly with the increase in speed. Therefore feed rate and drill diameter have high influence on the thrust force and torque. . Fig 7 [a] shows the interaction between feed rate and diameter affecting thrust force, fig 7[b] shows the interaction between feed and spindle speed and fig 7[c] shows the interaction between spindle speed and diameter.

IOP Conf. Series: Materials Science and Engineering **197** (2017) 012055 doi:10.1088/1757-899X/197/1/012055

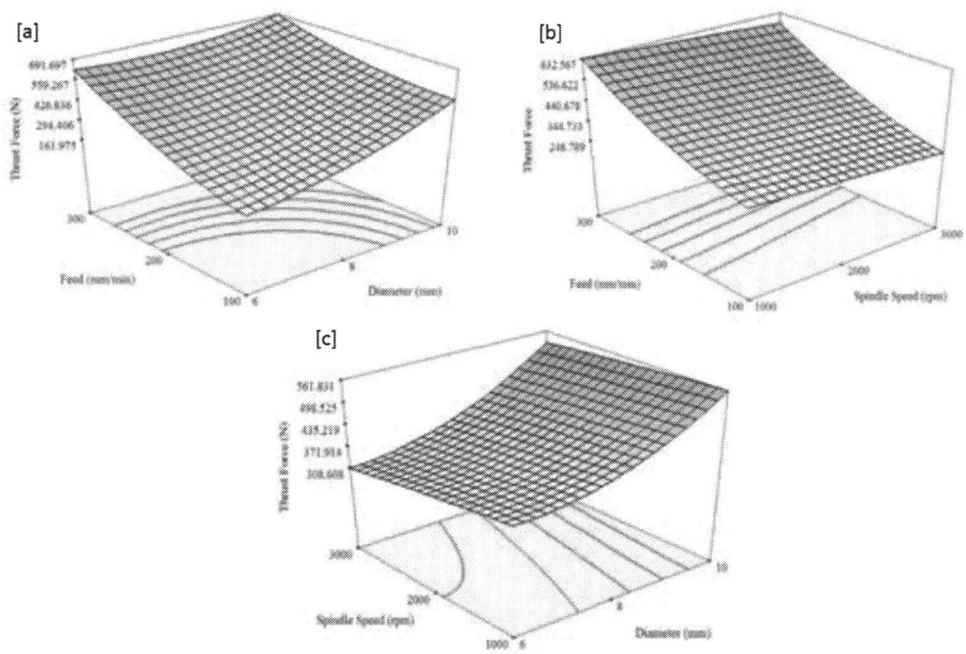

Figure 6. 3D surface plot between the parameters affecting Thrust Force
[a] Interaction between feed rate and diameter [b] Interaction between feed and spindle speed
[c] Interaction between spindle speed and diameter

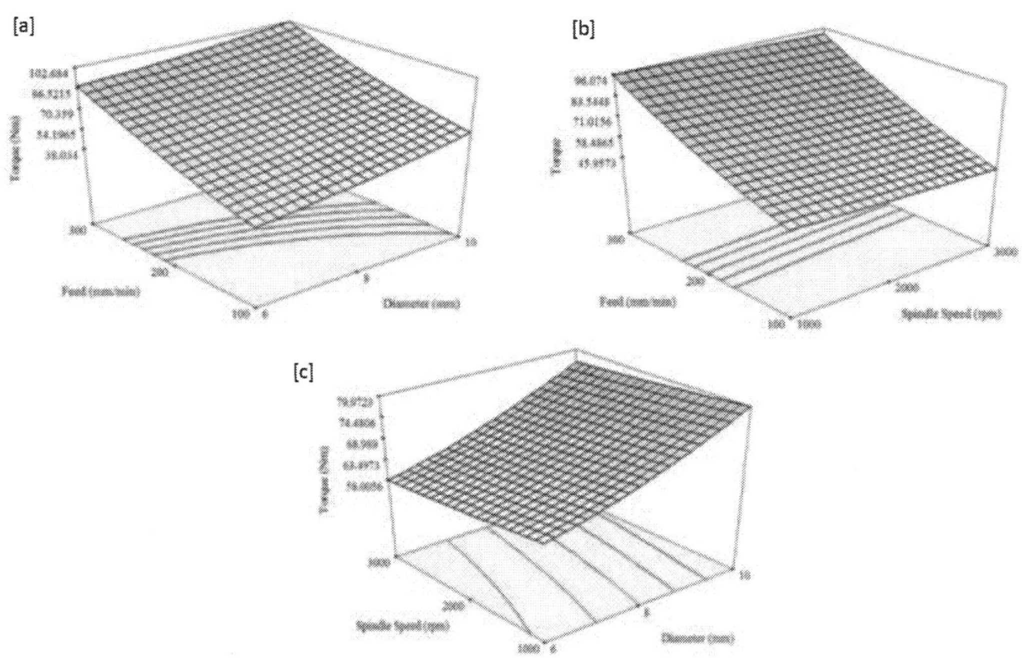

Figure 7. 3D surface plot between the parameters affecting Torque
[a] Interaction between feed rate and diameter [b] Interaction between feed and spindle speed
[c] Interaction between spindle speed and diameter

Scanning Electron microscopy of drilled surface is shown in fig.8, the surface roughness of the drilled portion is clearly visible in fig.8[a], it shows the fibres which are uncut during drilling operation, the broken resin particles are also spread over the surface wrecks the surface finish of the hole produced, higher the spindle speed, the finer the surface finish, this is due to the reason that at higher speed the thrust force produced is less. Fig. 8[b] shows traces of the initiation of delamination of the layers, but the surface finish of the hole produced at low feed is comparatively better compared with surface produced at high feed.

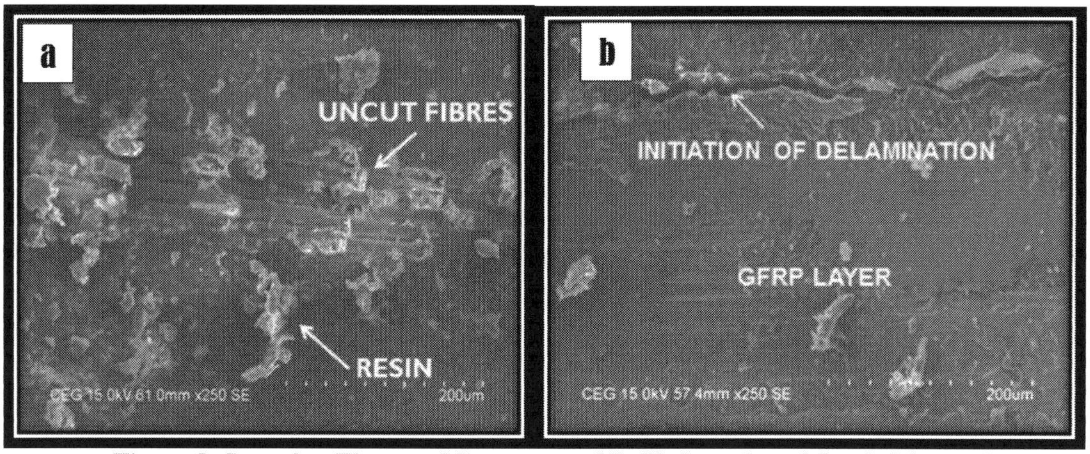

Figure 8. Scanning Electron Microscopy of Drilled portion of Sandwich panel [a] Surface produced at high feed and low speed [b] Surface produced at low feed and high speed

5. CONCLUSION

An experimental and Taguchi based analysis for thrust force and torque associated with various input parameters in drilling of sandwich laminate is presented in this paper. The following conclusions are drawn from the above analysis:

- 10mm drill offers the highest drilling thrust force compared with other drills.
- The two major cutting parameters affecting thrust force is the diameter and feed rate.
- The thrust force increases with increase in diameter and feed rate.
- Within the selected cutting range, the combination of high diameter (6 mm), low feed rate (100 mm/min) and high speed (3000 rpm) would be the best experimental conditions when drilling laminates.
- Scanned Electron Microscopy images show that uncut fibres occurs in hole produced at high feed rate and low spindle speed.
- 6mm diameter hole surface quality is better than the 10mm diameter hole surface due to the fact that the thrust force experienced is more, which indicates that the thrust force has more effect on the hole surface.

REFERENCES

[1] K. Palanikumar 2011 Experimental investigation and optimisation in drilling of GFRP composites *Measurement* 44 2138–2148.

[2] N.S. Mohan, A. Ramachandra and S.M. Kulkarni 2005 Influence of process parameters on cutting force and torque during drilling of glass–fiber polyester reinforced composites *Composite Structures* 71 Issues 3–4 407–413

[3] Ghani JA, Choudhury IA and Hassan HH. 2004 Application of Taguchi method in the optimization of end milling parameters *Journal of Material Processing Technology* 145 84–92.

[4] Tsao CC, Hocheng H. 2007 Parametric study on thrust force of core drill *J. Mater. Process. Technol.* 192 37-40.

[5] S.M. Wu, J.M. Shen and L.H. Chen 1982 Multifacet drills *Proceedings of the14th National SAMPE Technology Conference* pp. 456–463.

[6] J. Mathew, N. Ramakrishnan and N.K. Naik Investigations into the effect of geometry of a trepanning tool on thrust and torque during drilling of GFRP composites *J. Mater. Process. Technol.* 91 1–11.

[7] Julián Luis Merino-Pérez, Raphaël Royer, Eleanor Merson, Aiden Lockwood, Sabino Ayvar-Soberanis and Matthew B. Marshall 2016 Influence of workpiece constituents and cutting speed on the cutting forces developed in the conventional drilling of CFRP composites *Composite Structures* 140 621–629.

[8] Davim JP, Reis P and António CC 2004 Experimental study of drilling glass fiber reinforced plastics (GFRP) manufactured by hand lay-up. *Compos Science Technology* 64 289–97.

[9] S. Basavarajappaa, G. Chandramohanb and J. Paulo Davim 2008 Some studies on drilling of hybrid metal matrix composites based on Taguchi techniques *J. Mater. Process. Technol.*196 332–338.

[10] Mishra R, Malik J, Singh I and Davim JP 2010 Neural network approach for estimate the residual tensile strength after drilling in uni-directional glass fiber reinforced plastic laminates. *Material Design* 31 2790–5.

[11] Davim JP, Reis P and Antonio CC 2004 Experimental study of drilling glass fiber reinforced plastics (GFRP) manufactured by hand lay-up *Composite Science Technology* 64 289–97.

[12] Davim JP and Reis P 2003 Drilling carbon fiber reinforced plastics manufactured by autoclave – experimental and statistical study. *Material Design* 24 315–24.

[13] Davim JP ans Reis Pedro 2003 Study of delamination in drilling carbon fiber reinforced plastic (CFRP) using design experiments *Composite Structures* 59 481–7.

[14] Sardinas RQ, Reis P and Davim JP 2006 Multi-objective optimization of cutting parameters for drilling laminate composite materials by using genetic algorithms *Composite Science Technology* 66 3083–88.

[15] Khashaba UA, El-Sobaty IA, Selmy AI and Megahed AA 2010 Machinability analysis in drilling woven GFR/epoxy composites: part I – effect of machining parameters *Composites* Part A 41 391–400.

[16] Hocheng H and Tsao CC. 2005 The path towards delamination-free drilling of composite materials *J Mater Process Technol* 167 251–64.

Measurement and analysis of thrust force in drilling sisal-glass fiber reinforced polymer composites

M. Ramesh[1,*] and A. Gopinath[2]

[1]Department of Mechanical Engineering, KPR Institute of Engineering and Technology, Coimbatore - 641407, Tamil Nadu, India.
[2]Department of Mechanical Engineering, Sri Sai Ram Engineering College, Chennai - 600044, Tamil Nadu, India.

E mail: mramesh97@gmail.com[1], gopinathraph@gmail.com[2]

Abstract. Drilling of composite materials is difficult when compared to the conventional materials because of its in-homogeneous nature. The force developed during drilling play a major role in the surface quality of the hole and minimizing the damages around the surface. This paper focuses the effect of drilling parameters on thrust force in drilling of sisal-glass fiber reinforced polymer composite laminates. The quadratic response models are developed by using response surface methodology (RSM) to predict the influence of cutting parameters on thrust force. The adequacy of the models is checked by using the analysis of variance (ANOVA). A scanning electron microscope (SEM) analysis is carried out to analyze the quality of the drilled surface. From the results, it is found that, the feed rate is the most influencing parameter followed by spindle speed and the drill diameter is the least influencing parameter on the thrust force.
Keywords: Drilling; Thrust force; Response surface methodology; ANOVA; Sisal-glass fiber composites.

1. Introduction

The composite materials are only an alternative for traditional materials due to its properties such as high strength-to-weight ratio, good wear resistance and damage tolerance, lower thermal expansion and lightweight, provided they can be machined to any shapes and sizes at an acceptable cost and quality [1-6]. Some of the factors which affect the drilling process are spindle speed, feed rate, drill geometry and properties work material. With the advent of composite materials and their wide use in structural applications, it has become necessary to drill holes into the laminates, to facilitate bolting or riveting to the main load bearing structure. According to statistics, more than 60% of the scrapped parts during aircraft assembly are caused by unqualified holes [7, 8]. Drilling of composite materials can destruct of fiber continuity and cause damages such as surface delamination, fiber/resin pull out, matrix thermal degradation etc [9-11]. Some of the challenges faced during drilling include dimensional variation, high temperature distribution, disintegration of material etc [12]. Machining of composite materials has started during 1970's and concentrating on drilling of conventional fiber reinforced composites [13]. Then later stages the naturally available plant fibers as alternate reinforcement in the polymer matrix composites due to their inherent superior properties which make them highly competitive against conventional materials for any structural applications[6]. The selection of the reinforcement, fiber orientation, tool geometry and material, and cutting conditions are very important, while machining composite materials [14-16].

In order to overcome the difficulties during machining an experiment has been carried out by using cemented carbide end mill and the results are compared with glass fiber reinforced polymer composites (GFRPs) [17]. From the experiment, it is found that the delamination and surface roughness of the natural fiber reinforced composites are better than the GFRPs. The effect of machining speed and feed rate on delamination in drilling of glass-hemp fiber reinforced composites with different fiber volume fractions was studied [18]. From the study, it is found that the damage around the hole is predominant at higher feed rates and the holes are in elliptical shape along the direction of fibers. Optimization of cutting parameters in drilling of hemp fiber reinforced composites are carried out by Dillibabu et al. [19] and found that the feed rate and cutting speed have largest contribution on the delamination. The hybrid composites from natural fibers instead of orthopedics alloys for internal fixation of fractured bones on human body have been developed by Chandramohan and Marimuthu [20]. From the study it is found that the composites made from natural fibers are best materials, when compared with composites made from orthopedic materials. Drilling of composites is carried out by using HSS-Co twist drill, multi twist drill and brad & spur drill and reported that the brad & spur drill produces less damage and thrust force when compared to the other two drills [21]. In the present experimental study, sisal-glass fibers reinforced composite laminates have been prepared and the influence of various cutting parameters, such as spindle speed, feed rate and tool diameter on thrust force is examined in the drilling of composites. The modeling of drilling parameters is carried out using RSM and the adequacy of the models has been checked by using ANOVA.

2. Experiment and measurement

2.1. Experimental setup

The machining operation is carried out on an auto feed drilling machine, with a variable speed (Model: A/30/GT, Make: CKP Industries), attached with the multi component piezo electric dynamometer (Model: 9257B, Make: Kistler). The drilling set up is presented in Figure 1. The orthogonal components of the cutting forces are measured, using a multi component piezoelectric dynamometer. The drilling operation was performed using a carbide drill with three different feed rates, spindle speeds and drill diameters.

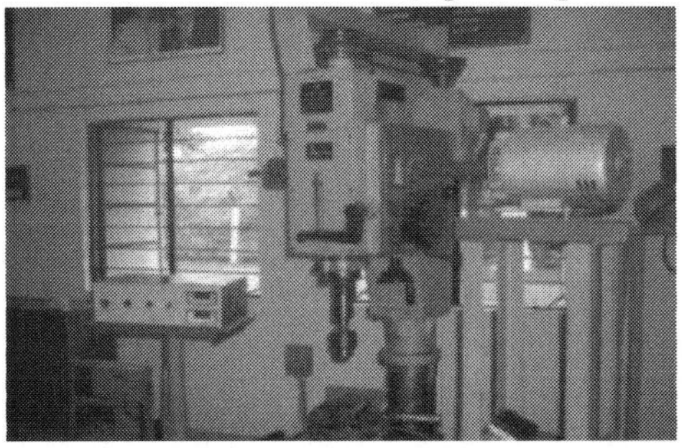

Figure 1 Drilling set up

2.2. Experimental results

The drilling of the composites is carried out, using carbide drills. The experimental results of the thrust force during drilling is presented in Table 1.

Table 1. Experimental results of thrust force

Trial No.	Levels of factors			Thrust force, F_Z(N)
	Speed, V (rpm)	*Feed rate, f (mm/rev)*	*Tool dia, d (mm)*	
1	1000	0.04	6	136.32

2	1000	0.04	9	162.92
3	1000	0.04	12	212.4
4	1000	0.06	6	196.79
5	1000	0.06	9	196.2
6	1000	0.06	12	230.21
7	1000	0.08	6	233
8	1000	0.08	9	326.02
9	1000	0.08	12	295.58
10	2000	0.04	6	127.4
11	2000	0.04	9	158.75
12	2000	0.04	12	230.03
13	2000	0.06	6	169.05
14	2000	0.06	9	200.21
15	2000	0.06	12	232.21
16	2000	0.08	6	230.35
17	2000	0.08	9	237.12
18	2000	0.08	12	256.67
19	3000	0.04	6	122.54
20	3000	0.04	9	136.14
21	3000	0.04	12	147.54
22	3000	0.06	6	132.48
23	3000	0.06	9	137.33
24	3000	0.06	12	176.47
25	3000	0.08	6	114.85
26	3000	0.08	9	187.87
27	3000	0.08	12	213.46

2.3. Observations on the drilling of sisal-glass fiber composites

It is known that the variables in the drilling process, such as spindle speed, feed rate, and drill diameter, have great influence on the thrust force and the quality of the holes drilled. The typical thrust force developed during the drilling of sisal-glass fiber composites is presented in Figure 2. The figure shows that the thrust force is minimum at the entry of the drill and the maximum thrust force is observed at the middle of the hole. When the drill penetrates the composite laminate, the thrust force increases, and it decreases when the drill bit exits out on the other side of the laminate.

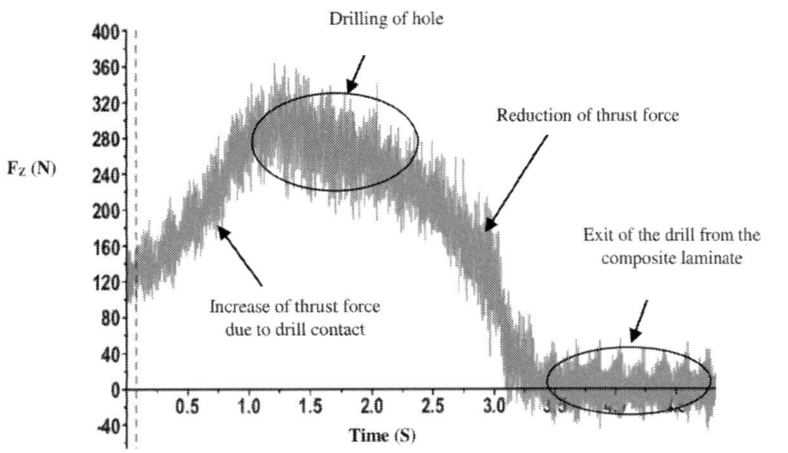

Figure 2 Typical thrust force observed in the drilling of sisal-glass fiber composites

2.4. Response surface regression model for sisal-glass fiber composites

The machining parameters considered in this experiment are the spindle speed (V) in rpm, feed rate (f) in mm/rev, and drill diameter (d) in mm. The quadratic response models for thrust force are developed for the observed experimental results of sisal-glass fiber composites. The regression model developed for the thrust force (F_Z) for the composites is presented in Eq. 1.

$$F_Z = (54.52963)+(0.097219*V)+(862.50000*f)+(14.50185*d)-(0.97000*V*f)-(1.25000E\text{-}004*V*d)-(22.50000*f*d)-(1.80944E\text{-}005*V^2)+(25972.22222*f^2)-(0.16975*d^2) \quad (1)$$

The summary statistics of the results of measured and the RSM predicted values for thrust force are presented in Table 2.

Table 2. Results of experimental and RSM predicted values for thrust force

Standard order	Experimental thrust force (N)	RSM predicted thrust force (N)
1	136.30	144.4
2	162.90	210.3
3	212.40	190.1
4	196.80	204.3
5	196.20	244.9
6	230.20	230.4
7	233.00	253.5
8	326.00	260.2
9	295.60	350.1
10	127.40	132.5
11	158.80	164.3
12	230.00	242.3
13	169.10	177
14	200.20	204.9
15	232.20	244.6
16	230.40	242.6
17	237.10	262
18	256.70	254.3
19	122.50	132.4
20	136.10	140.3
21	147.50	155.6
22	132.50	136.8
23	137.30	146.5
24	176.50	192.2
25	114.90	121.4
26	187.90	192.5
27	213.50	230.4

2.5. Analysis of thrust force using ANOVA

The ANOVA results obtained for the thrust force during drilling of sisal-glass fiber composites are presented in Table 3. From the table it can be observed that the R^2 values are 90.24%, which shows that high correlation exists between the developed model and the experimental results. Further, it can be asserted that the developed models agree well with the experimental results, and hence, the developed RSM models can be used for predicting the thrust force in the drilling of these composites. From the ANOVA tables it is further observed, that all the input parameters are significant terms, which influence the thrust force in the drilling of sisal-glass fiber composites. Among all three parameters, the feed rate is the most influential one, as its F-value is higher when compared to the other parameters. The next contributing factor is the spindle speed followed by the drill diameter.

Table 3. ANOVA for thrust force in the drilling of sisal-glass fiber composites

Source	Sum of squares	DoF	Mean square	F value	Prob>F
Model	68563.62	9	7618.18	17.46	< 0.0001
V	21403.81	1	21403.81	49.07	< 0.0001
f	24264.58	1	24288.08	55.68	< 0.0001
d	15705.83	1	15705.83	36.01	< 0.0001
V^*f	4516.32	1	4516.32	10.35	0.0051
V^*d	1.69	1	1.69	3.869E-003	0.9511
f^*d	21.87	1	21.87	0.050	0.8255
V^2	1964.45	1	1964.45	4.50	0.0488
f^2	647.57	1	647.57	1.48	0.2397
d^2	14.00	1	14.00	0.032	0.8599
Error	7415.36	17	436.20		
Total	75978.98	26			
			R-Squared=90.24%		

3. Results and discussion

3.1. Effect of the machining parameters on thrust force

The influence of the cutting speed on thrust force in the drilling of sisal-glass fiber composites is presented in Figure 3. From the figure it can be observed that the thrust force decreases when the speed increases, that is the thrust force at 3000 rpm is lower than the 2000 and 1000 rpm. The influence of the feed rate on thrust force is depicted in Figure 3 (b). The feed rate plays a vital role on the thrust force, and the experimental results showed that the thrust force is increased by increasing the feed rate. The reason is that while increasing the feed rate the load on the drill bit increases, which in turn, increases the thrust force. Figure 3(c) shows the variation of the thrust force with respect to the drill diameter. The drill diameter also shows almost the same trend as feed rate, in which the thrust force increases with the increase in the drill diameter. The thrust force is the minimum at a lower diameter, and maximum at a higher diameter of the drill, because the higher diameter drill increases the contact area between the tool and the work piece, and also the subsequent increase in load. The optimum cutting conditions for machining of composites is higher cutting speed, lower feed and dept of cut [22].

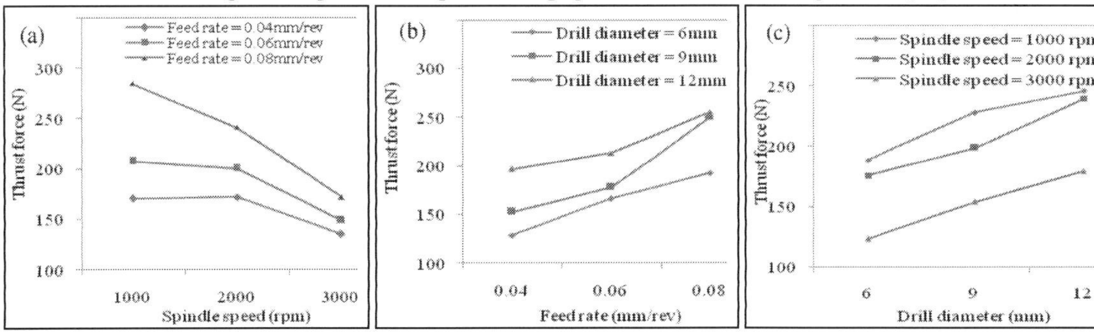

Figure 3 Influence of (a) spindle speed, (b) feed rate and (c) drill diameter on the thrust force in the drilling of sisal-glass fiber composites

3.2. Scanning electron microcopy (SEM) analysis

Figure 4 shows the SEM micrographs of the surface of the drilled holes, observed in drilling of sisal-glass fiber composites using carbide drills, under various cutting conditions. The surfaces of the microstructure show the smooth edges, and some damages like fiber pull out, fiber dislocation, uncut fibers and uneven distribution of fibers etc. These are the common problems in the drilling of any material, including composites. Figure 4(a) is the SEM image of the sisal-glass fiber composite laminate subjected to drilling at the low feed rate of 0.04 mm/rev and at the low spindle speed of 1000 rpm. From the micrograph the cutting edges of the fibers are clearly visible. The SEM image of the composite laminate subjected to drilling at the medium feed rate of

0.06mm/rev and at the medium spindle speed of 2000rpm is presented in Figure 4(b). The fibers at the inner surface of the drilled hole are clearly seen from the image.

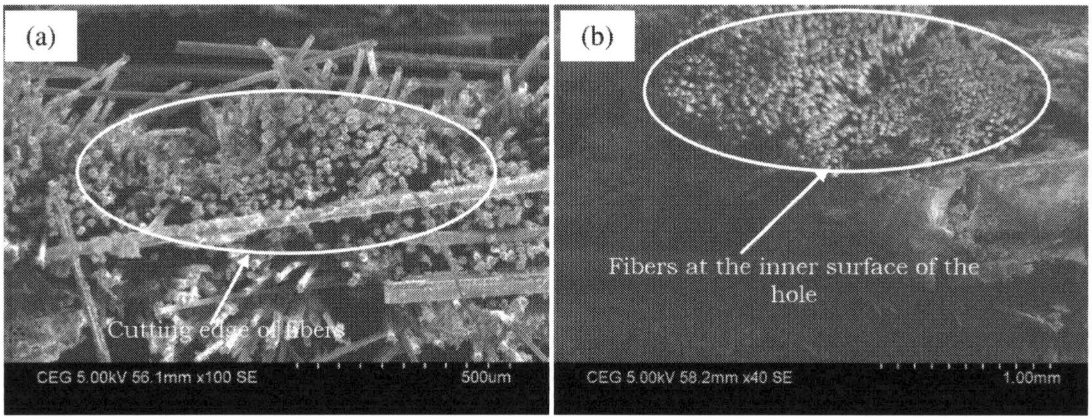

Figure 4 SEM images of the inner surfaces of the drilled holes in the drilling of hybrid composites

3.3. Comparison of experimental results with RSM model

The comparative results for thrust force between the experimental results and the models developed using RSM, are presented in Figure 5. The figure clearly shows that there is prediction potentials exist between the experimental results and the developed model; hence, this model can be effectively used for the prediction of thrust force in drilling of hybrid composites.

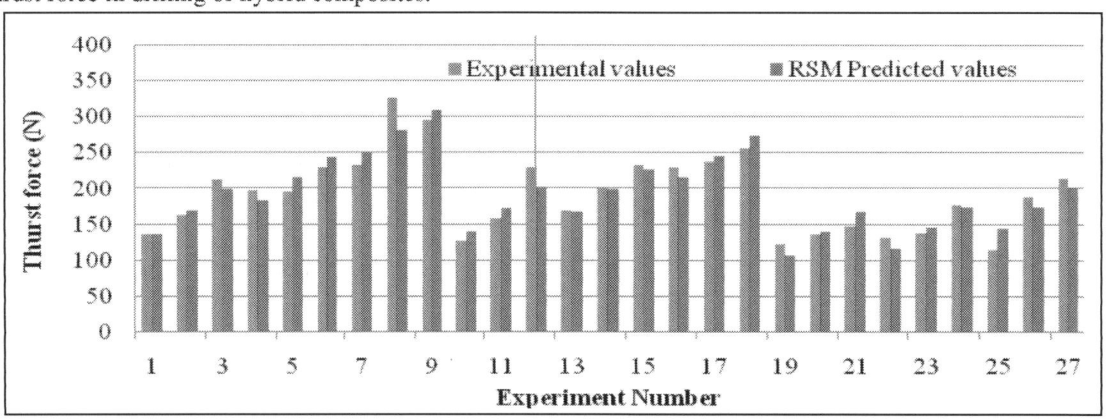

Figure 5 Comparison plot for thrust force in the drilling of sisal-glass fiber composites

4. Conclusion

The drilling experiments are planned and conducted to predict the influence of cutting parameters on thrust force in drilling of sisal-glass fiber composite laminates by using carbide drills. Based on the experimental results the following conclusions have been arrived.

- The maximum thrust force absorbed by the sisal-glass fiber composite laminates is 326 N.
- The thrust force increases with an increase in the feed rate and drill diameter, and decreases when the spindle speed increases for all cutting combinations.
- The thrust force is highly influenced by the feed rate followed by the spindle speed. The drill diameter has the least influence on the thrust force.
- The coefficient of correlation for the models is nearly equal to 1, and hence, these models can have very good prediction potentials.
- From the SEM analysis, the interfacial relationship between the fiber and the matrix, fiber debonding during drilling and the internal cracks of the drilled surfaces are observed.
- The results indicated that the high spindle speed, low feed rate and drill diameter are preferred for drilling of sisal-glass fiber reinforced hybrid composites.

5. References

[1]. Songmene V and Balazinski M 1999 Machinability of graphitic metal matrix composites as a function of reinforcing particles. *CIRP Ann. Manuf. Technol.* **48** 77.

[2]. Ramesh M, Sudharsan P and Palanikumar K 2015 Processing and mechanical property evaluation of flax-glass fiber reinforced polymer composites. *Appl. Mech. Mater.* **766-767** 144

[3]. Bhoopathi R, et al. 2015 Experimental investigation on mechanical properties of hemp-banana-glass fiber reinforced composites. *Appl. Mech. Mater.* **766-767** 167

[4]. Ramesh M, Palanikumar K and Reddy K H 2013 Mechanical property evaluation of sisal-jute-glass fiber reinforced polyester composites. *Compos. Part B Eng.* **48** 1

[5]. Ramesh M, Palanikumar K and Reddy K H 2013 Comparative evaluation on properties of hybrid glass fiber-sisal/jute reinforced epoxy composites. *Proc. Eng.* **51** 745

[6]. Dandekar C R and Shin Y C 2012 Modeling of machining of composite materials: a review. *Int. J. Mach. Tool. Manuf.* **57** 102

[7]. Khashaba U 2004 Delamination in drilling GFR-thermoset composites. *Compos Struct.* **63** 313

[8]. Liu DF, Tang YJ, Cong WL. A review of mechanical drilling for composite laminates. Composite Structures. 2012; 94: 1265–79.

[9]. Davim J P, Reis P and Antonio C C 2004 Experimental study of drilling glass fiber reinforced plastics (GFRP) manufactured by hand lay-up. *Compos. Sci. Technol.* **64(2)** 289.

[10]. Zitoune R, et al. 2011 Behaviour of composite plates with drilled and moulded hole under tensile load. *Compos. Struct.* **93** 2384

[11]. Wern C W, Ramulu W and Shukla A 1996 Investigation of stresses in the orthogonal cutting of fiber-reinforced plastics. *Exp. Mech.* **36(30)** 33

[12]. Kolhar A S 2007 Masters Thesis, Wichita State University, USA.

[13]. Doran J H and Hanley F 1972 Machining of boron/epoxy composites. *Met. Eng. Qly.* **12(3)** 38.

[14]. Singh I and Bhatnagar N 2006 Drilling of unidirectional glass fiber reinforced plastic (UDGFRP) composite laminates. *Int. J. Adv. Manuf. Technol.* **27** 870

[15]. Yusup N, Zain A M and Hasim S Z M 2012 Evolutionary techniques in optimizing machining parameters: Review and recent applications (2007-2011). *Exp. Sys. Appl.* **39** 9909

[16]. Dandekar C R and Shin Y C 2008 Multiphase finite element modeling of machining unidirectional composites: Prediction of debonding and fiber damage. *J. Manuf. Sci. Eng.* **130(5)** doi: 10.1115/1.2976146

[17]. Dillibabu G, Sivajibabu K and Gowd UM 2013 Effect of machining parameters on milled natural fiber reinforced plastic composites. *J. Adv. Mech. Eng.* **1** 1

[18]. Naveen P N E, Yasaswi M and Prasad R V 2012. Experimental investigation of drilling parameters on composite materials. *IOSR J. Mech. Civil Eng.* **2** 30.

[19]. Dillibabu G, Sivajibabu K and Gowd UM 2013 Optimization of machining parameters in drilling hemp fiber reinforced composites to maximize tensile strength using design experiments. *Ind. J. Eng. Mater. Sci.* **20** 385.

[20]. Chandramohan D and Marimuthu K 2010 Thrust force and torque in drilling the natural fiber reinforced polymer composite materials and evaluation of delamination factor for bone graft substitutes-A work of fiction approach. *Int. J. Eng. Sci. Technol.* 2(10) 6437.

[21]. Jayabal S and Natarajan U 2010 Optimization of thrust force, torque, and tool wear in drilling of coir fiber-reinforced composites using Nelder–Mead and genetic algorithm methods. *Int. J. Adv. Manuf. Technol.* **51** 371

[22]. Ramesh M, Elvin R P, Palanikumar K and Reddy K H 2011 Surface roughness optimization of machining parameters in machining of composite materials. *Int. J. Appl. Res. Mech. Eng.* **1** 26

QUALITY OF WORK LIFE – A STUDY

Dr.Mily Velayudhan T.K[1], Yameni M.D[2]

[1]Research Supervisor, Sathyabama University, Chennai.- 600119, Tamil Nadu, India

[2]AssistantProfessor,Loyola institute of technology chennai, tamil nadu

mily.rajeev@gmail.com , yamenidhanapal70@gmail.com

Abstract: The main objective or the purpose of this research is to investigate and identify the significance of work environment towards the performance and also to study the effectiveness of the QWL in the organization.**Methods/Analysis:**In order to meet the stated objectives a structured questionnaire was framed and data was collected using convenience sampling from 123 employees of the steel manufacturing organization in Chennai, and to study the significant association chi-square was used by the researcher.**Findings**:QWL of the employees of this steel company can be improved by conducting some more training classes for the employees who are falling in the category of more than 3 to 4 years of experience and >4 years of experience which would boost their self confidence and help them attain their level of satisfaction. Similarly the organization can give some more security to the employees falling in the category of 41 and above so that they feel quite secure in the hand of organization and they can give their paramount performance.**Novelty /Improvement:**This empirical article on **Quality of Work life – A Study's** structured questionnairecan be applied as an Employee opinion Survey taken in once in 6 months on knowing the quality of work life. By doing this surveyorganizations can get to know the quality of work life of the employees and take necessary steps to improve theQWL among all the Employees. It also helps the employers to know that their employees who are working in their organization are happily working leading to good QWL which will boost up their performance to come happily daily to their work place.Key Words: Human behavior, Quality of work Life, Employee performance

1. INTRODUCTION

Human behaviour is the concern for all. Parents are really worried about the behaviour of their children, teachers complain about the lack of interest in the students, and the employers are vinegary about the lack of devoir from their employees. The enigma remains still moot why people behave as they do. When the reputed firm's highly paid employees go on strike in some other form of clash the typical reaction of the organization is at chagrin. The assumption is that when an employee is well remunerated for his time and labour he is suppose to be stimulated to carryout responsibilities entrusted to him or her and he should be happy with his job but the traditional methods of motivating the employees have become ineffective. Technological developments can be taken as a key factor, even though advanced technology is increasing the efficiency but it has impact on work-place environment. The new form of work organization must stand for an optimal balance between technical demands of job and social needs of the people performing it. This mounting interest in the new form has been accompanied by the detonation of the term QWL.QWL will have direct and not direct association with the economic and social welfare of large portion of population which lies beyond the domain of Industry. Quality of work life is significant to all the organizational inputs which leads to employees' satisfaction and influence organizational effectiveness[1]. The importance of quality of work life is to extend jobs and working conditions that are outstanding for employees as well as economic strength of the organization. It refers to the level of motivation, satisfaction, commitment, involvement of an employees and appraising for the bestperformance of the employees as per Organizational objectives an individual experience with respect to their work. Improved QWL naturally helps to improve the employee's and world also improves the performance of the Industry or enterprises[2].

India must go beyond China in steel production these are the words of Prime Minister Narendra Modi he related steel production to India's defence alertness and said that India cannot afford to lag behind China in this field. Modi made a strong point for his "**Make in India**" initiative. When I talk of 'Make in India', I am not ready to lag behind anyone." This could be achieved by improving the Quality of work life of an organisation which will help the employees to work efficiently for the upliftment of themselves as well as the organisation and the upliftment of the nation as a whole. India is likely to stand as the world's second biggest producer of crude steel in the next 10 years, from third position to moving up, as its capacity is estimated to increase to about 300 MT by 2025 – 2% of Indias GDP will be contributed by India's Steel sector and employees over 6,00,000 people and if these people are retained in their job it would definitely lead organizations into miles and bounds and this could be achieved by good QWL in the

IOP Conf. Series: Materials Science and Engineering **197** (2017) 012057 doi:10.1088/1757-899X/197/1/012057

organisation.Though QWL originated nearly thirty years ago 'the researchers' interest in this field is still noticeable[13,14]. Usually, management encouraged employees to separate between work and family[15]. This situation tells that work and individual life functions independently. Thus there is no conflict (Titmuss 1968). Modern social and demographics changes view that individual will function very effectively only when the personal life and working life are balanced. This footing gives rise to the study on QWL. (Sarina Muhamad Noor & Mohamad Adli Abdullah,2012)The Quality of Work Life (QWL) has made up increasing interest and importance in all the countries of the World. It is very important in the background of commitment to work, motivation and job performance. It also is a means to facilitate the glee of human needs and goal achievement. Work life of course means everything including their environment both internal and external either in office, factory or field-working. What is expected froma a worker? What are the conditions of the work place? What is the compensation in the form of wages that the worker gets? What are the incentives offered to him? How about his satisfaction with the work environment and the compensation? These are the questions are to be tackled by the Researchers in their study in quality of work life. Quality of work Life is completely referred to as recovering the the human factor and working life. It mostly refers to friendly or unfriendly of a job environment for the people involved in it. Quality of Work Life generally try to insist employees the feelings of security, equity, pride, autonomy ,ownership ,democracy, responsibility and flexibility. Today's educated workforce expects more than just pay from their work. Dynamic work environment always demands equal importance to both human need and technology, where the individual perspective play a very important role in democratization of work relations and humanization of work atmosphere. The holistic approach can contribute to high employee perception of QWL in an organization (Anbarasan and Mehta, 2009)[11,12].

2. Theoretical Foundation

Employees' performance is behaviour; the theoretical substructure of this research can best be elucidated using employee behaviour theories found in the field of organizational behaviour. Organizational behaviour addresses employee's behaviour at the individual, group, and organizational levels as totally different one. Employee behaviour is influenced by factors such as leadership ,motivation, job satisfaction, attitudes, , personality, stress and group dynamics (Luthans, 2005). Though, this study makes use of some of these factors including job satisfaction, personality factors, competence etc as the prime variables. The study was backed by expectancy theory, the Big Five personality theory, job characteristic model, and the different models of employee competence. One of the organizational behaviour theories which accounts for performance is that of the expectancy theory. In addition, there are a number of studies over the past several years that have demonstrated that among many theories expectancy theory has basically been useful for the prediction of job performance (Galbraith and Cummings, 1967). QWL includes adequate and fair compensation, work conditions, opportunity for development and growth, involvement and recognition, workload, job security, social integration, work life balance. This was established by Kurt Lewin's Field theory of employee's behaviour which shows the relationship between performance and QWL . According to this theory, individual behaviour is influenced by how one perceives and reacts to the surroundings provided by the organization in their study by (Kohler & Mathieu, 1993). In this research, the quality of work life environment that is what it means. The theory goes on to state that the association between performance and effort is mediated by individual abilities and features (which include personality and competence) and role perception. Such personality characteristics includes control. Lawler (1982) in his studies postulates, the more a person is oriented toward internal control, the additional to that he will feel that his performance will lead to required outcomes and vice versa. Competencies are also important for performance; thus, it refers to the ability to display a particular performance-relevant behaviour which is portrayed by accurate understanding of what the job requirements are (Lawler and Porter, 1968).

Bagtasos in his study said that the development of QWL can be etched way back to the earlier management. Frederick Taylor and Elton Mayo in their theories suggested that human is important in determining firms' performance. While both have different ways of looking at human aspects, with Taylor being more quantitative, while Mayo focused more on human as well as environment, their ideas on appreciating human as a person is valid. The current development on QWL is further enhanced by the Maslows need hierarchy theory and, Herzberg's Two factor theory, McClleland three needs theory and Alderfer') and spill over effect (Sirgy Efraty, Siegel & Lee, 2001). These theories says that the basic needs of individual facade from time to time and there are some spill over effect to their personal life and family. The evolution of QWL began in late 1960s on emphasizing the human dimensions of work that was focused on the quality of the association between the worker and their environment. Quality of work life is an idea of behavioural scientist, and the term was first made known by Davis at the Forty-Third American Assembly on the changing World of work at Columbia university's Arden House. The select participants assembled there concluded in their final remarks that "improving the place, nature of work ,the organisation, lead to better work performance and a better quality of life in the society[3,4,5].

IOP Conf. Series: Materials Science and Engineering **197** (2017) 012057 doi:10.1088/1757-899X/197/1/012057

Cohen and Rosenthal (1980) in their study in QWL had an intentionally designed effort to bring out increased labour management, and jointly solve the problem of improving organizational performance and employee's satisfaction by the co-operation. **Greenhaus & Beutell (1985)** in their study in detail studied about the work-life conflict which arised from the incompatible emotional and behavioural demands of work and non-work roles, such that contribution in one role is made more hard by participation in the other. This conflict can result in greater stress at home or at work, poorer health, higher turnover and absenteeism, reduced job satisfaction. **Hochschild (1989), Kelley and Voydanoff (1985), Hochschild (1989)Thompson & Walker (1989)** in their study exposed that working women face well-documented conflicts due to their continuing role as primary caretakers for their homes, children, or elderly parents and also it is women's greater responsibility to take care of children and other family members and they experience more obstacles than men resulting to common household problems. **Greenhaus and Beutell (1985)** and Greenhauset al. **(1989)** has examined the conflict of between family and work, **Goodstein (1994)** and **Ingram and Simons (1995)** in their article did a study which said that an institutional perspective on organizations responses to work-family related issues. In addition**, Campbell, Campbell and Kennard (1994)** also have studied the effects of family restraint on the work commitment and job performance of women. **Hyman and Summers (2004)** in their study have classified seven major problems which are linked with current issues related to work-life balance these are deviation of adoption across different organisation as well as sectors, lack of making of policies at organizational level, restricted employee's power over the opening and functioning of policies , policies are primarily to meet business needs rather than those of employees. **Chan and Einstein (1990)** in their study explored that QWL reflects as a concern for people's experience at work, their relationship with other people, their work settings and their effectiveness on the job. **Higgins and Duxbury (1992)** did articulate in their study that work conflict is a greater source of work-family conflict whereas personal or family lives, interfere with work are linked for fewer hours but work that interferes with life matters as revealed by the researcher **Reynolds (2005). Clark and Kirchmeyer, (2000)** Their study said that there is a negative effect of work imbalance on quality of life and established that the damaging effect is due to critical levels of work-to family conflict & stress. If there is a balance of engagement in work & family roles which is expected to be associated with that of the individual well-being because such balance offcourse reduces work-family conflicts. **David Lewis et al (2001)** did a indepth study on the extrinsic and intrinsic attributes of quality of work life. The objective of the research was to test even if extrinsic or intrinsic or previous traits test predict satisfaction with QWL in health care. The findings shows that, supervisor style, commitment and discretion, all play a role in determining QWL. Female employees were less satisfied with these traits than male. **Linda K. Johnsrud (2006)** did her research on Quality of faculty work life in the University of Hawaii to describe the changes in QWL from 1998 to now. The objective of the study was to find out the current level of satisfaction on QWL. Variables were used in Relations with the community service, faculty relation, salary department chair, campus service and demographic factors were used in this study,and the result showed that salary was one of the main variable for satisfaction of the employees from year 1998 to 2006. Faculty relations and community services was the most positive elements in in that University for work life and other finding was campuses' faculty were generally more satisfied than other employees of that university. **J. Gnanayudam & AjanthaDharmasiri (2008)** in their article concentrated on the Influence of quality of work life on organizational commitment by doing an investigation on unsatisfactory level of commitment among workers in medium and large organizations in the clothing industry in Sri Lanka. The result showed that QWL has a positively significant relation with the commitment variable and moderator effect of HRDC variable. **W.N. thalang et al. (2010)** studied on quality of work life indicators as a csr of electrical and electronics private Organizations in Thailand. Objective of the research was find out the quality of Work Life Indicators as a Corporate Social Responsibility The result showed that QWL gives indication of a effective CSR, developing a good Quality of Work Life (QWL) is crucial. **Ayesha T. (2012)** has evaluated the quality of work life of the faculty members of private universities in Bangladesh with the objective was to investigate into the factors affecting the overall perception of QWL. It is found that high satisfaction is found amont the female with regards to QWL when compared to male. Teaching experience of teachers with less than one year experience is more positive in their thought about their QWL and its related dimensions compared to experienced teacher. AnandPawar (2013) has done a totally differentstudy on QWL and job satisfaction of employees in VTPS and found out that the level of satisfaction among the employees with regard to various job related aspects, also study was done to in depth to measure the QWL and job satisfaction of employees, which included major factors in their study and they included the following variable too they are (i) wages and salaries; (ii) rewards system (iii) a safe and healthy environment; (iv) working conditions; (v) interpersonal relations and (vi) superiors are considerate and helpful as their variables of their study. The result showed that there is dissatisfaction in the interpersonal relations between the cader-wise QWL and no proper grievance handling procedure were adopted among the employees which affect the job satisfaction[6,7,8]. **S.Khodadadi et al (2014)** in their article investigated the QWL dimensions effect on the employees' job satisfaction. The results of the study showed that the salary and benefits' and its policies have a significantly positive effect on Shuhstar's Shohola Hospital employees' job satisfaction.

Quality of work life is crucial for organizations to persistently attract and retain employees which was found by (Akder 2006) in their study and also it has become significant in the last two decagon due to the increasing demands of today's business environment and family structure. Lau (2000) defined QWL as the favourable conditions and environments of a workplace that support and sustain employees' satisfaction by providing them with job security and reward. QWL encompasses various facets such as working conditions, working time, mode of wages payment, health hazards, and management behaviour throughout the process of responding to the needs of the employees Therefore, QWL involves some financial and non-financial benefits, as well as management behaviour towards workers.

Islam and Siengthai (2009) did an indept study on some of the key elements of QWL like employee involvement ,job security, better reward system, employee benefits, and organizational performance. QWL can also be defined as a feeling that employees have in approach towards their jobs, colleagues, and the organization (Heskett et al., 1994). Thus, if employees have good opinion towards their jobs, peers, and the organization, it signifies that they are happy doing their work; and consequently, the QWL is good. Lawler (1982) Quality of work life is defined by as one of the most important attribute of the employee perceptions for their physical and mental happiness at work and these perceptions could be for or against. Thus, it compass working conditions, working time, mode of wages payment, and health related issues. Therefore, quality of work life integrates both financial and non-financial benefits and management behaviour towards workers[17,18,19,20].

Quality of work life gives heed to various qualities of work life factors as proposed by Walton (1975), Mishra (1996), and Ellis and Pompli (2002) in their study. In Walton's (1975) article he discussed that there are eight major conceptual categories that sling to QWL, namely: adequate and fair compensation; safe and healthy working conditions; immediate opportunity to establish human capacities; opportunity for continued growth and security; social integration in the work place, work and total life space, and social relevance of work life. (M. Swapna & S. Gomathi, 2013) proposed nearly six variables measuring Quality of Work Life which included job related factors and also employees growth related factors like career development and training & development. Thus the variables used to measure QWL in this study are: 1. job satisfaction 2. Working condition 3. Well-being 4. Home work interface / Work life balance 5. Career anticipation and compensation 6. Training and Development. The researcher has used all these variables while framing the questionnaire for collecting the data[16]. There were 21 questions of close ended and multiple choice questions asked in the questionnaire on these variables. Convenience sampling was adopted for the current study[9,10].

3. Research Methodology

The main objective of this study is to investigate and identify the significance of work environment towards the performance and also to study the effectiveness of the QWL in the organization. In order to meet the stated objectives a structured questionnaire was framed and data was collected using convenience sampling from 123 employees of the steel manufacturing organization in Chennai, and to study the significant association chi-square was used by the researcher.

4. ANALYSIS AND FINDINGS

- **Null Hypothesis (Ho):** There is no association between Age and Satisfaction level of health and safety.
- **Alternate Hypothesis (H1):** There is association between Age and Satisfaction level of health and safety.

Table No: 1 Showing the Chi- Square test for Association between Age and satisfaction on health and safety conditions (No. of respondents and the row percentages)

Age	Satisfaction level of health and safety conditions					Total	Chi-square Value	P-value
	Highly Satisfied	Satisfied	Neutral	Dissatisfied	Highly Dissatisfied			
< 20	14(35%)	5(12.5%)	6(15%)	7(17.5%)	8(20%)	40		
21-30	7(23.3%)	10(33.3	8(26.6	3(10%)	2(6.6%)	30		

		%)	%)					
31-40	19(44.18 %)	13(30.2 %)	4(9.3%)	5(11.6%)	2(4.6%)	43	20.470[a]	.000*
41-50	3(37.5%)	2(25)	1(12.5 %)	2(25%)	0(0%)	8		
> 50	1(50%)	0(0%)	0(0%)	1(50%)	0(0%)	2		
Total	44	30	19	18	12	123		

- **Source: Primary data Note:* - Indicates Significant at 5% level.**

<u>Table :1 Figure:1</u>

The Table:1 analyses the relationship between Age and satisfaction of the employees on health and safety measures of the organisation. It is inferred that out of 123 respondents the respondents falling under the age group of < 20 and 21 – 30 and 31-40 are highly satisfied with 35%,33.3%,and 44.18% on the health and safety conditions of the organisation when compared to employees falling in the category of 41-50 and >50. The chi square value is 20.470. The calculated P value is less than 0.05. Hence the null hypothesis is rejected at significant level of 5%, and we accept the alternative hypothesis and it says there is association between age and satisfaction of employees on health and safety measures in the organisation, it is also clear that people who are falling in the category of less than 40 years of age are either satisfied or highly satisfied with the health and safety conditions of the company than the employees falling under the category of 41 and above. Hence from the table it is inferred that there is association between Age and satisfaction of health and safety conditions in the organisation.

- **Null Hypothesis (Ho):** There is no association between Training programme and No. of years of experience.
- **Alternate Hypothesis (H1):** There is no association between Training programme and No. of years of experience.

Table No: 2 Showing the Chi- Square test for Association between Training Programme and No of years of experience . (No. of respondents and the row percentages)

	Training programme		Chi-square	P-

No. of years of experience						Total	Value	value
	Highly satisfied	Satisfied	Neutral	Dissatisfied	Highly Dissatisfied			
< 1yr	4(26.6%)	2(13.3)	8(53.3%)	1(6.6%)	0(0%)	15	40.886[a]	.000*
1-2 yrs	23(41%)	1(1.7%)	13(23.2%)	17(30.3%)	2(3.5%)	56		
2-3 yrs	2(8.3%)	4(16.6%)	14(58.3%)	3(12.5%)	1(4.1%)	24		
3 -4 yrs	6(33.3%)	4(22.2%)	3(23%)	3(16.6%)	2(11.1%)	18		
>4 yrs	4(40%)	4(40%)	0(0%)	2(20%)	0(0%)	10		
Total	39	15	38	26	5	123		

Source: Primary data Note:* - Indicates Significant at 5% level

The Table: 2 analyses the relationship between years of experience and satisfaction of training programmes.It is inferred that out of 123 respondents 41% of the respondents with less than 2 years of experience feel that they are highly satisfied with training programmes conducted by organisation whereas 58.3% of the respondents falling in the category of 2-3 years experience feels that the training programme conducted in their organisation is neutral and 11.1% of the respondents falling in the category of 3-4 years feels that they are not satisfied with the training programmes conducted in their organisation.The chi square value is 40.886. The calculated P value is less than 0.05. Hence the null hypothesis is rejected at significant level of 5%, and we accept the alternative hypothesis and it says there is association between years of experience and satisfaction of training programmes conducted in the organisation. Hence from the table it is inferred that years of experience and satisfaction of training programmes has significant association on the performance of the employees.

5. Suggestion& Conclusion

In the past certain values were attributed with work. Work was considered as worship and people had sincerity and commitment to work. But now a day's employee will not believe in such values of work. He works for his salary, he works if the conditions of work are helpful and affable and terms of employment are favourable to him. As such, the work norms have been changing from time to time. Employees at the grass root level experience a sense of frustration because of low level of wages, poor working conditions, not favourable terms of employment, inhuman treatment by their superiors and the like whereas managerial workforce feel frustrated because of unfriendliness over their conditions of employment, role conflicts, interpersonal conflicts, lack of freedom in work, absence of challenging work, job pressures etc. **Hickman** Life presented six strategies in this context for improving quality of work life.

They are:

- Development of Careers and career paths
- Work design
- Organization Reward System
- Design and Maintenance of Group and inter group relationship
- Managerial Practice and
- Internal and external strategies for change

Improved performance leads to improved quality of work life. Moreover, result revealed that quality of work life toward workers development like training of the employees , worker's union, participation in decision making variables, management should come forward to meet worker's demand that they have positive impact on firm performance. The over all, performance of an organisation depends completely on the performance of its people, in spite of of the organisation's size, purpose or other characteristics.

Based on the discussion of the literature review, prior studies have established the relationship between QWL and performance. Quality of work life programs should be linked with such affective outcomes such as increased job satisfaction, improved employee performance to the extent that they develop employee participation, and involvement and responsibility. Studies have also shown that there is a association between personality and job satisfaction and that there are many different personality factors which is correlated with job satisfaction. Again, broad research has proven that job contentment does not occur in aloofness, as it is dependent on organizational variables such as structure, size, pay, working conditions, and leadership, which create the organizational climate in the study of (Schneider & Snyder, 1975; Kerego & Mthupha, 1997; Boeyens, 1985). The Quality of Work Life mean to develop enhance and utilize human resource effectively, to improve Quality of products, productivity , services and reduce cost of production per unit of output and to satisfy self esteem of the workers psychological needs,their participation, and recognition in job, etc.,

QWL of the employees of this steel company can be improved by conducting some more training classes for the employees who are falling in the category of more than 3 to 4 years of experience and >4 years of experience which would boost their self confidence and help them attain their level of satisfaction. Similarly the organization can give some more security to the employees falling in the category of 41 and above so that they feel quite secure in the hand of organization and they can give their paramount performance. We must understand that half of our daily life is spent at work places and work life has become an integral part of our total life. Making work place happier has to be not only mandatory part of HR role but it is to be carried out by the HR functionaries with the same passion, spirit, enthusiasm, commitment and energy so management should make sure that all the employees working in their organization are happily working leading to good QWL which will boost up their performance to come happily daily to their work place.

6. References

1. Bagtasos, M. R. Quality of work life: A review of literature. DLSU Business & Economics Review ,2011,Vol. 20.(2),Pp. 1-8.
2. C.P Garg et al "Quality of work life an overview", International journal of physical and social science,2012 ,Vol:2,issue :3,Pp:231-242.
3. Dr.Sorav sadri & Prof.Conrad Goves "Sustainable quality of work life and job satisfaction" Journal of Economic Development environment and people, 2013,Vol.2,Issue:4 PP: 25-29.
4. Dr. Zabin Ullah Hashemi and Mansour Yari kam rani "Investigation of the relationship between QWL and customer relationship management in the police force of one of the cities in the west of Iran". Advances in Enviornment Biology, 2014, 8(21) pp: 869-874.
5. Dr. A Jayakumar & K.Kalaiselvi "Financial services and management research: International journal of Marketing, 2012, Vol: 1, issue10, Pp: 140-151.
6. Florence Muindi & Peter k'Oobonyo "Quality of work, life, personality,Job satisfaction,competence, and job performance a critical review of literature" ,European Scientific journal , 2015, (11-26) pp:223-226.

7. Fattah Nazem et.al. Modeling social capital in banks based on qulality of work life", Advances in Enviornment Biology,2014, Pp: 1371.
8. H Ramesh Babu & M Ramesh"Relationship between job satisfaction and quietly of work life of employees in service sector", Economic and Business, 2013,Vol:13 issue: 2 pp: 58-71.
9. Josiah Roman Aketch, & Odhiambo Odera et.al "Effects of Quality of Work Life on Job Performance: Theoretical Perspectives and Literature Review",Journal of Social Sciences 2012,vol.4(5):Pp 383-388.
10. Harish.K1 , Subashini K"Quality of work life in Indian industries –A case study" International Journal of Innovative Research in Science, Engineering and Technology, 2014,Vol:3,Issue :10,Pp 16799-16804
11. Preethi Vijaimadhavan& Dr.D.Venkatarama Raju "An empirical study on relationship among quality of work life and its factors" IOSR-JBM journal of business management , 2013,Vol:12 Issue:3.Pp:20-28.
12. Radha Yadav &Ashu Khanna "Literature Review on the Quality of work life and their dimensions" ISOR Journal of Humanities & Social Sciences (IOSR JHSS), 2014, Vol. (19) issue: 9, Pp:71:80.
13. Euhee Shin and Heekyung kim "Factors Influencing health – related quality of life among male workers in Korea" Indian Journal of Science & Technology.2015,Vol.8.Pp236-246.
14. Hye-Young Kim and Young -Kwan Yang "Relation Research on Job Demand-Control-Support, Work-Leisure Conflict, Quality of Life in Tourism Employee's" 2016 Indian Journal of Science and Technology, Vol 9(26),Pp:2-6

IOP Conf. Series: Materials Science and Engineering **197** (2017) 012057 doi:10.1088/1757-899X/197/1/012057

Website References

15. Steel Industry in India: http://www.ibef.org/industry/steel.aspx#sthash.kz5Ld3Hg.dpuf Date accessed:02/12/2015.

16. Indian Economy:http://www.ibef.org/economy/indian-economy-overview#sthash.ab96i1ve.dpuf Date accessed: 02/12/2015.

17. Times of India: http://timesofindia.indiatimes.com/business/india-business/India-must-surpass-China-in-steel-production-Modi/articleshow/46778245.cms Date accessed:02/12/2015.

18.Quality Work Life among Factory Workers in Malaysia: http://ac.els-cdn.com/S1877042812004569/1-s2.0-S1877042812004569-main.pdf?_tid=8e977788-bf65-11e5-aa67-00000aacb360&acdnat=1453288041_a6711e914bbf03fa72462483a9211d7eDate accessed: 02/12/2015.

19. QUALITY OF WORK LIFE AN OVERVIEW: http://www.ijmra.us/project%20doc/IJPSS_MARCH2012/IJMRA-PSS867.pdf Date accessed 02/12/2015.

20.QUALITY OF WORK LIFE-AN OVERVIEW: http://indianresearchjournals.com/pdf/IJMFSMR/2012/October/12.pdf Date accessed: 01/12/2015.

21. Quality of work life and organizational performance: Empirical evidence from Dhaka Export Processing Zone: http://www.ilo.org/legacy/english/protection/travail/pdf/rdwpaper37a.pdf Date accessed: 01/12/2015.

IOP Conf. Series: Materials Science and Engineering **197** (2017) 012058 doi:10.1088/1757-899X/197/1/012058

EVOLUTION AND RECENT TRENDS IN FRICTION DRILLING TECHNIQUE AND THE APPLICATION OF THERMOGRAPHY

Mathew Alphonse[1], V.K.Bupesh Raja[2], K.Logesh[3], N.MuruguNachippan[4]

[1]Research Scholar, Department of Mechanical Engineering, Sathyabama University, Chennai, Tamil Nadu, India
[2]Professor & Head, Department of Automobile Engineering, Sathyabama University, Chennai, Tamil Nadu, India
[3]Assistant Professor, Department of Mechanical Engineering,Vel Tech University, Chennai, Tamil Nadu, India
[4]Assistant Professor, Department of Automobile Engineering, Vel Tech University, Chennai, Tamil Nadu, India

Email:[1]gnmathew@gmail.com,[2]bupeshvk@gmail.com,[3]klogesh7@gmail.com, [4]murugunachippan@gmail.com

Abstract: Friction drilling is a new trend of hole making process in which the frictional force between the tool and work material is used for hole making, by thrusting of the tool into the work material without chips formation as like conventional drilling.During friction drilling ductility of the material increases with heat, which causes the material to extrude on the front and backsides of the hole, which forms the boss and bush respectively on the work piece, so it provides the required projection for threading. The damage to the surface occurring during the hole making, takes place at high temperature, which could be monitored using thermography. By using this technique damages occur during manufacturing can be monitored and minimized.Due to high temperature during drilling various phenomena like thermal phase transition, plastic flow of material and extrusion may occur, which directly influences mechanical properties and may cause plastic material from the hole surface.Friction drilling is used for hole making process of aerospace, automotive, commercial and industrial products.
***Keywords*:**Friction Drilling, Extrusion, Thermography, boss, bush.

1. INTRODUCTION

In conventional drilling the main defects and damages are at the plate entrance and exit in the hole wall. These type of errors happen mainly due to anisotropy of the material[1]. Recent technological development over conventional drilling is friction drilling. Friction drilling is a new trend of hole making without chips, with maximum of 12mm thickness and the friction drilling needs high force for hole making[2]. Rotating a conical tool at high speed causes work material to soften and the tool to penetrate to penetrate the work

IOP Conf. Series: Materials Science and Engineering **197** (2017) 012058 doi:10.1088/1757-899X/197/1/012058

piece,due to the heat generated by friction and thus making a hole on the work piece. While heat increases ductility of material extrudes itself on front and back side of the hole[3].

The extruded work material forms boss and bush on the work piece, where bushing is formed 3 times the size of work material. The advantages of friction drilling is bushing increases depth for the threading, its clean and chip less hole making process[4].Major concern during machining of composites is delaminating, mainly at plate entrance and exit[5]. Delamination factors are characterized by the damages caused at the entrance and exit. Major role for reducing damages at the entrance and exit is by high speed cutting or by minimizing point angle and low feed rate[6]. Several investigation have studied the damages caused during drilling by using thermograph, these thermography technique can help in avoiding manufacturing related damages[2].

Several investigators have used different materials for research, on hole making through conventional friction drilling. Different work piece material used are carbon/epoxy, woven composite, glass epoxy laminate, carbon fiber reinforced plastic etc. Tools with coated and non-coated have been studied and it is observed that coating on the tool plays a major role in reducing damages[7]. Quality of drilled hole can be improvedby using optimization techniques and suitable parameters and tool material can be established[8, 9]

2. EXPERIMENTAL SETUP

Friction drilling experimental are generally carried out by using a three axis computer numerical controlled vertical machining center. Typical setup of friction drilling has a work piece held rigidly on top of a piezo electric drilling dynamometer. The fig 1 shows the difference stages of friction drilling process, for making hole.

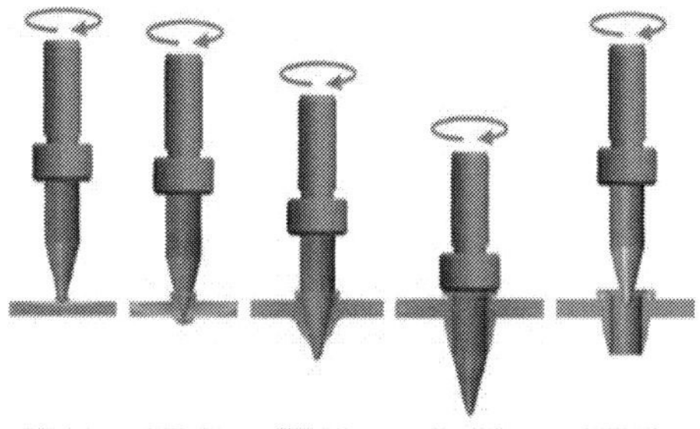

Figure 1 Stages of tool position infriction drilling[9]

In stage 1 the work piece and tool comes in contact, in stage 2 the conical tool is penetrate the work piece due to the heat which developed by the frictional force. In this stage thrust force on the tool is at its peak and extrusion of material can be also seen, due to the ductility of the work material[3]. In stage 3 bushing can be identified due to the extrusion of material in the exit side of the hole. In stage 4 back extrusion is formed on the work piece at the entry side of the hole[3]. Schematic diagram of CNC machining centreexperimental set up for friction drilling setup was shown in fig 2[8].

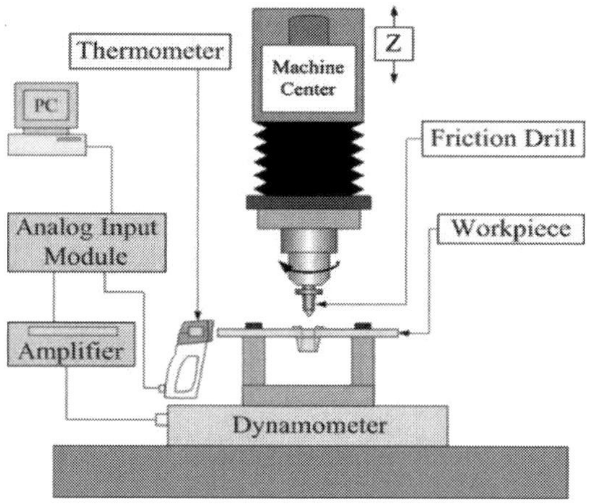

Figure 2 Experimental setup of Friction Drilling[8]

During the friction drilling process the work piece is held on the dynamometer. Further to a thermometer/ sensor is fixed on the work piece while machining, for measuring temperature in drilling zone[8].

3. THRUST FORCE AND TORQUE

In [9]experimentally investigated ALSI 1020 steel, observed that peak thrust force of 700N occurs at a tool travel of 2.5mm from the starting contact with the work piece as shown in fig 3 at the initial position. Further torque rose to 1.5 N-m from 0.5 N-m with increase in torque, thrust force decrease to 300N.

IOP Conf. Series: Materials Science and Engineering **197** (2017) 012058 doi:10.1088/1757-899X/197/1/012058

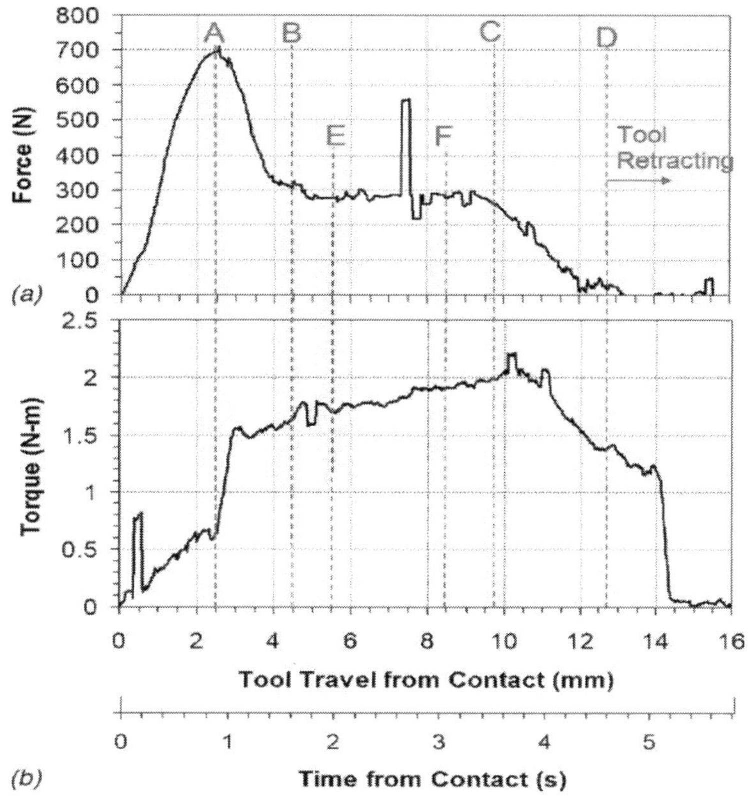

Figure 3. Torqueand Thrust Force in friction drilling[9]

In that position torque will be increased to 1.7 N-m. Suddenly after third position torque and thrust decreases, owing to this discoloring rings are found inside and outside of hole[9]

In [3]conducted experiment with spindle speed of 3000rpm and tool feed rate of 2.53 mm/s, 4.23mm/s and 5.93 mm/s, and measured tool thrust force and torque was measured using thermocouple with coefficient of friction of 0.7 at feed rate 5.95mm/s , 4.23 mm/s, 2.54mm/s, its observed that peek thrust force and torque are about 800N, 750N and 700N respectively as shown in fig 4. For the above paper its observed that during machining process, temp which is developed was about 400° C, 350° C, 270° C at 2.74mm, 3.48mm, 5.11mm respectively.

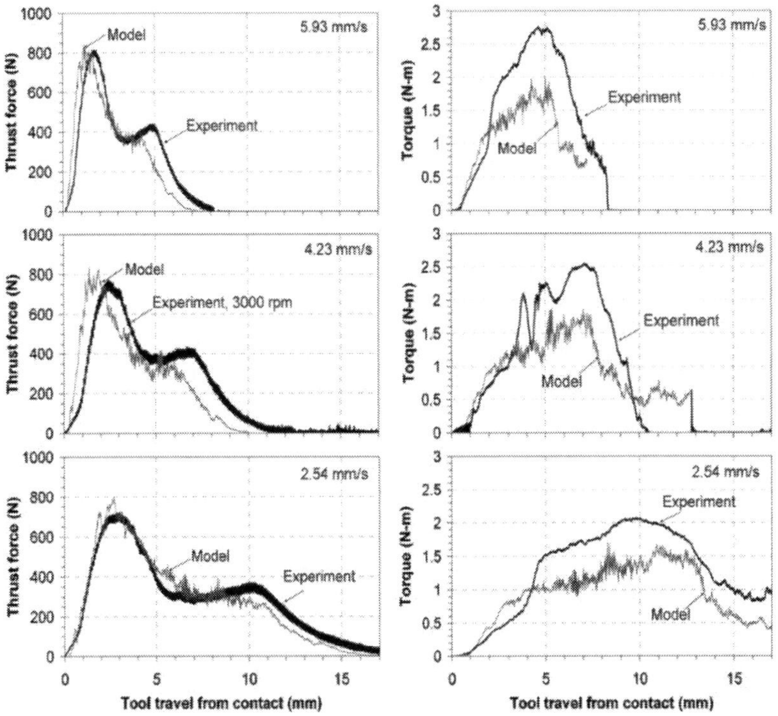

Fig 4 Comparison of the model predicted trust force and torque versus experiment in friction drilling (3000 rpm spindle speed and 5.93 mm/s, 4.23 mm/s and 2.54 mm/s) for 0.7 coefficient of friction[3]

For the prediction of thrust force and torque in friction drilling process, it's purely based on pressure and contact area between tool and work piece.Boopathiet al[4] concluded in there paper that at constant feed rate, thrust force decreases linearly. This implies heat energy produced will be greater at higher speed and thus concluded that high tool speed, thrust decreases during penetration of hole. The investigators observed that for Aluminum peak thrust force is 1512N, for brass Peak thrust force 1798N, and Stainless steel peak thrust force 2745N for penetrating a hole.

4. MATERIAL SELECTION

A New generation material which is applied in aerospace, automotive etc was sandwich material, which is made by thin sheets and polymer/resin[11]. Sandwich material was prepared by using roll-joining process with the help of skin material and polypropylene, these will give good load bearing capacity[13].The tensile modulus and tensile strength of Glass fiber reinforced epoxy based composites increases with the tensile strength[10, 11].

In[2] conducted friction drilling analysis with two different materials glass epoxy TSE-2 laminate and sandwich laminate,the author concluded that heat generated is more when compared with sandwich composite. While machining with Bosh drilling machine, HSS drill bit are used. R piquet et al[15]used thin carbon/epoxy.Drill bit made up of k20 rated micro grain tungsten carbide, but two different tool geometry i.e op1 and op2 both having 3 cutting edges, maximum cutting edge is 59°, minimum cutting edge is 0°, clearance angle is 0°, rake angle for both op1 and op2 in 0°, in op1 no chamfer angle is available, author concluded that if feed rate is decreased quality can be increased.

In [7] investigated parameters of nano coated drilling on CFRP/ aluminum sandwich material. Tools used are tungsten carbide drill with nano coating and non nano coated tool. Author observed that during drilling on material aluminum and composites thrust force is reduced with the help of non-coated tool, while surface roughness is minimum for sandwich material. Scott f miller et al[16] studied microstructure alteration associated with friction drilling of Titanium, Aluminum, Steel it was observed that microstructure is affected by heat produced during friction drilling process. It was observed that while friction heat and temperature is also increased, results in material elongation and bushing.

Sanjay Rowat et al[30] investigated characteristics of High Speed drilling of woven composites and found that, chisel edge of drill cause hard graphite fiber fracture inside the soft epoxy material, this happened based on fracture mechanism, thrust force increases with low feed rate and medium speed. Jose Mathew et al[17], reviewed carbon fiber reinforced plastic composite material and concluded difference in thermal properties of the carbon fiber matrix of material in composites and the difficulties in maintaining good quality cuts to maintain high quality better focusing behavior of pulsed Nd: YAG laser and high beam intensity gives a smaller thrust load during cutting. This creates minimum defects concluded that damages are obtained by repetition rate, pulse duration, cutting speed and beam energy.

5. TOOL SELECTION

In Friction drilling drill bit is made up of cylindrical rod made up of Tungsten Carbide with grain size super finish. A diamond grinder is used for grinding into a conical shape for friction drilling tool as shown in fig 5 and fig 6. Fig 5 shows Friction Contact Area Ratio 100% Fig 6 shows FCAR of 50%. ie, friction contact area ratio in this drill four friction contact area shown in fig .FCAR was defined in friction contact area/ circumference area as shown in fig 5.

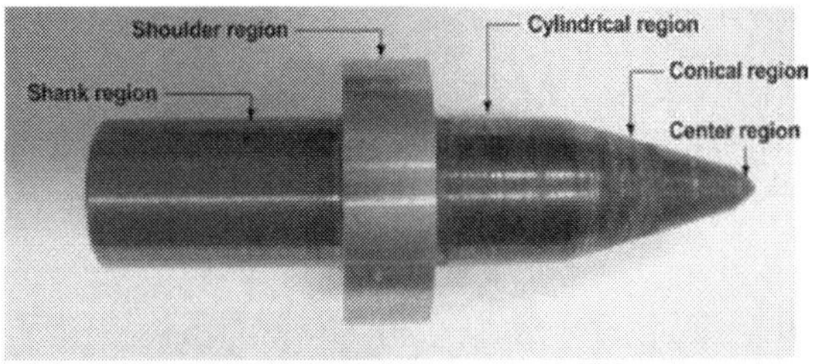

Figure 5 FCAR of 100% [8]

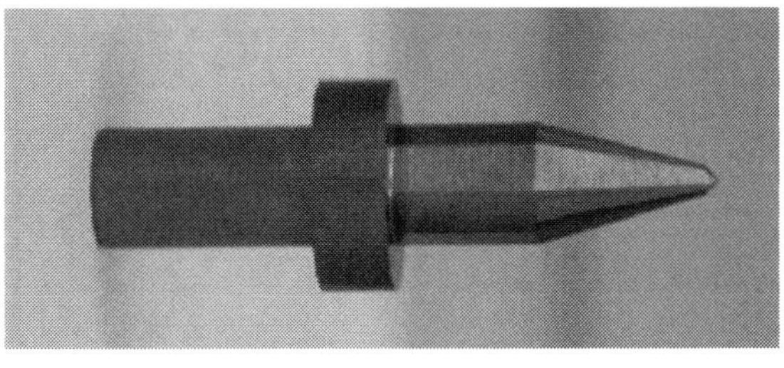

Figure 6 FCAR of 50% [8]

M Boopathi et al [4] investigated the use of tungsten carbide tool for making hole, they performed experiments on brass, aluminum and stainless steel, and concluded thatbrinell hardness for each metal

Aluminum, Brass, Stainless steel is 122,208,477 respectively and incase of aluminum high adhesion of work piece and material transfer on microstructure. Very less material transfer on brass and no evidence of melting of brass, no evidence for work material transfer and melting for stainless steel.

Redouanezitoune et al [7] experimentally investigated the use of Nano coated cutting tool and withoutNano coated tool, two types of tungsten carbide tool were used for the study. The author concluded that thrust force for coated drill is (10-15)% lesser compared with uncoated drill while machining on a composite plate. Similarly thrust force for coated drill is 50% lesser compared with uncoated drill while machining on aaluminum plate.

R Piquet et al [15] investigated specific cutting tool is shown in fig 7 op1 and op2 tool is made up of double fluted twist drill uses micrograin tungsten carbide than carbides which are tougher. Tool is made with tough and wear resistant in tool op1 chamfer angle with zero clearance angle in op2 tool also chamfer on cutting edge.

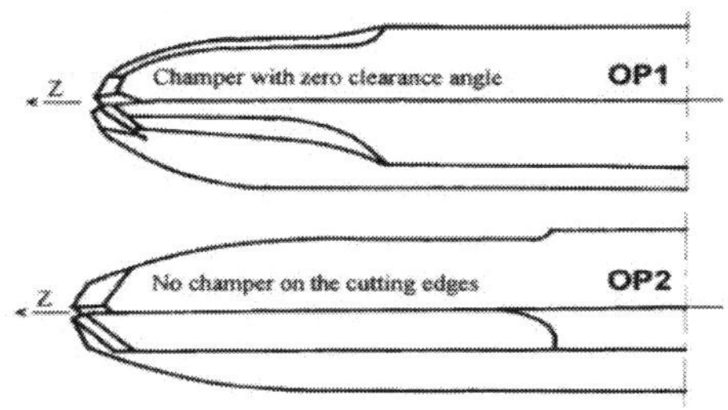

Figure 7 Specific tool geometry for op1 and op2 tool [15]

Geometric comparison between f2 and op1, working section shown in fig 8. Author concluded that op1 tool has less damage than f2 drill tool, at plate entrance and exit while experiment done on carbon/epoxy plate. Op1 too has excellent result. Experiment carried out at constant feed rate of 0.5 to 62.5 micron.

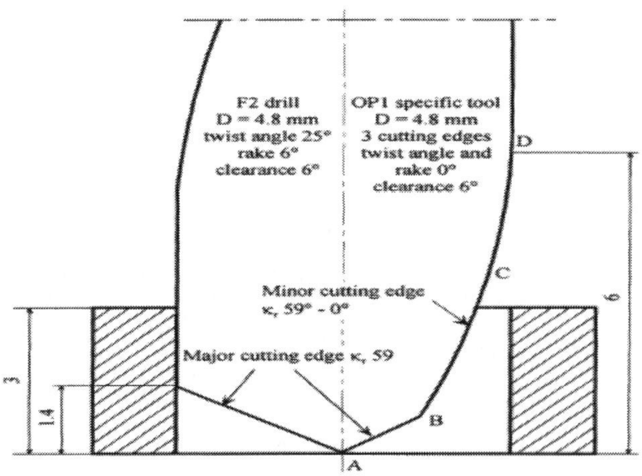

Figure 8 Geometric comparison between f2 and op1 [15]

6. DELAMINATION

Delamination is the main problem during drilling process in composite material. Due to deforming forces and temperature. Delamination occursnon optimal due to process parameters[2]. Delamination mainly occurs at both entrance and exit. Mechanism which is responsible for delamination is push out at exit and peel up at entrance as shown in fig 9

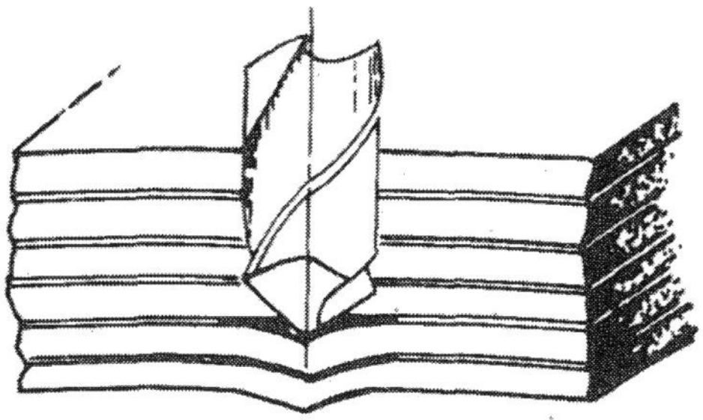

Figure 9 Push out delamination at exit [5]

Where push out happens when drill bit when it approaches the end, uncut thickness become smaller and the deformation resistance decreases, so at this point delamination occurs due to the load exceeds the inter laminate bond strength. Peel up at entrance happens due to the downward acting trust force as shown in fig 10[5].

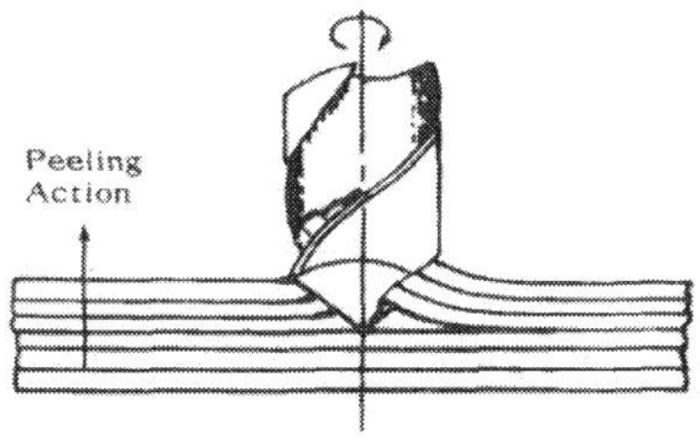

Figure 10 Peel up at delamination entrance [5]

For getting higher quality holes without delamination the use of special drill and feed method [27]. Schema of the visualization of delamination is shown in fig 11[18].

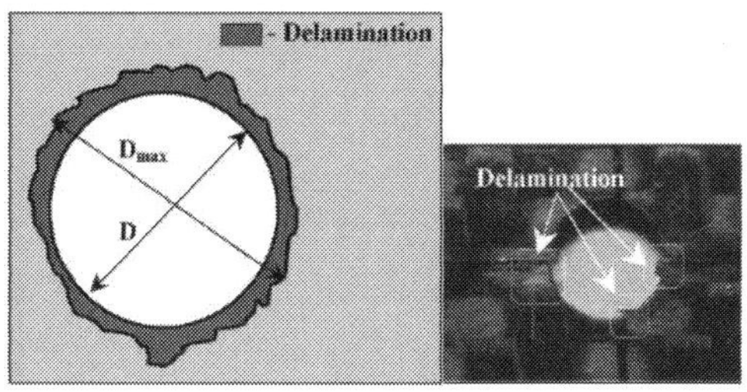

Figure 11 Schema of the visualization of the delamination [18]

From the study it is concluded that delamination at exit is lesser when compared with entrance damages. For higher cutting speed and higher feed delamination increases with cutting parameters in composite plate. While comparing Brad n Spur drill with straight shank drill delamination is less in Brad n Spur drill. Major focus for reducing delamination was cutting velocity [18]. The layered component have a good agreement with model prediction and used for minimize delamination defects [19].

Delamination factors can be studied by evaluating cutting speed, feed rate, and point angle. More over by low feed rate and point angle damages can be reduced [6]. For machining high quality drilled hole following areas should be analyzedhole diameter, peel up delamination, circularity, push out delamination. For mainly these parameters major role is played by spindle speed and feed rate [20]. Major problem for delamination caused to the geometry, with decrease of feed rate delamination size also decreased [21]. U In[22] experimentally investigated delamination on different materials like cross winding/polyester, continues winding with filler/polyester, woven/ polyester, chopped /polyester and woven /epoxy composite. Machining done on all the materials, the author observed that push out delamination is lower in both in chopped/composite and woven composite with the reduction of thrust force [22].

Industries are in greater importance to thermo mechanical tool loading because hole geometry and surface integrity are affected greatly[23].In [24] developed a new method for reducing delamination using Artificial Neural Network(ANN) method as shown in fig 12.

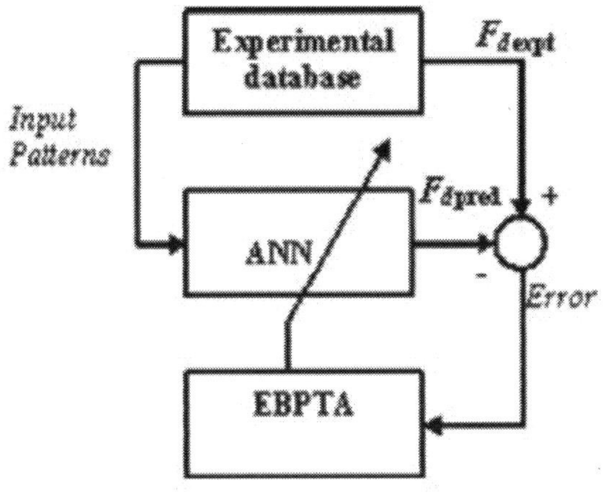

Figure 12 Methodology of ANN modeling[24]

Concept of friction/ heating is adopted both in friction drilling and Flow Drill Screw driving process (FDS). In normal friction drilling process tool penetrates and forms a extrusion, so surface area of treading is also formed. So again it need high torque for tightening and clamping. FDS used thread tool as shown in fig 13.

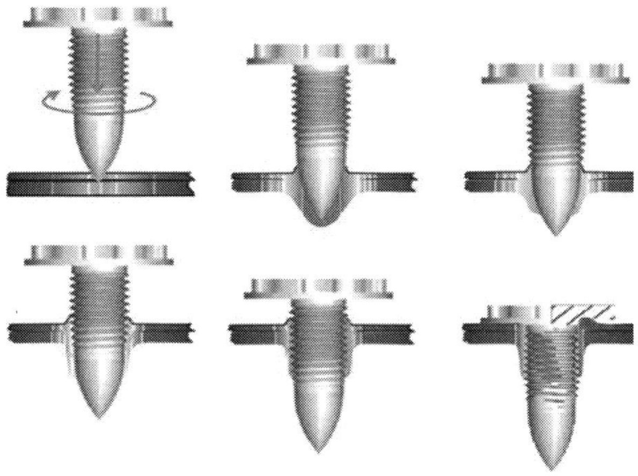

Figure 13 Six steps FDS process [14]

Steps involved are same as of Friction drilling, but its limited to light weight [14].During FDS process heat generated follows columbic friction law, which depends on following parameters coefficient of friction between tool and work material and velocity reduction of sliding contact.

7. SURFACE ROUGHNESS

Micro fiber fracture, fiber pull outs and matrix cracking are the main problem during machining process this may affect surface roughness. By low feed rate and reduction of thrust force surface roughness can be improved, with low feed and high speed temperature increases the low coefficient of thermal conductivity and low transition of plastics [26].

8. FAILURE MECHANISM

Mechanical behaviour of sandwich composite structure or any other material can be experimentally studied by using tensile testing and 3 point bending test. Failure behaviour of tested structure obtained from stress- strain curve were under various loading condition. Shear failure and elastics and plastic zone failure were identified using above testing method, a new model was be developed from obtained results [31].

9. THERMOGRAPHY

Mechanical damages of the structure are monitored by using thermography technique. The study of the critical temperature analyzed by the heat during drilling process and can be used to predict the mechanical damages and hence avoided by applying suitable techniques like selection of appropriate drill bit, cooling of drillbitetc[2]. While friction drilling process measuring work material deformation and temperature is difficult. FEM technique also can be adopted for this process as shown in fig 14, shows deformation in meshed form, plastic strain, temperature and von-mises stress at feed rate 4.23mm/s, speed 3000rpm, coefficient of friction 0.7etc[3].

IOP Conf. Series: Materials Science and Engineering **197** (2017) 012058 doi:10.1088/1757-899X/197/1/012058

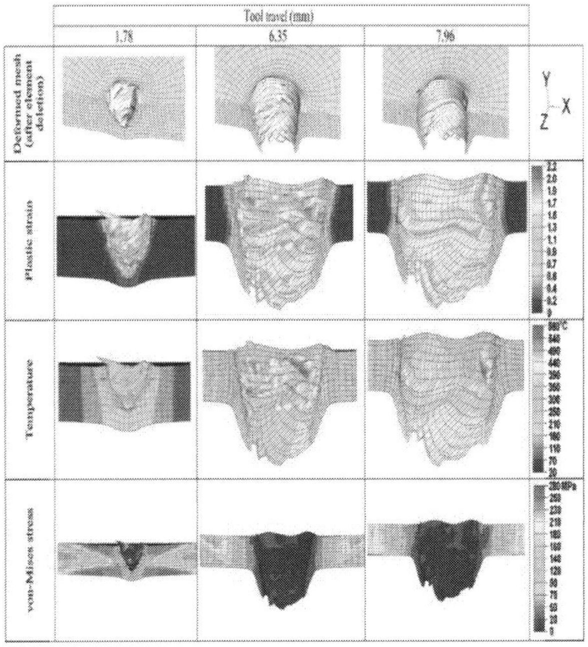

Figure 14 Deformed mesh and plastic strain, temperature and von mises stress [3]

While doing friction drilling process temperature in work piece and tool is high,for measuring temperature Infrared(IR) camera is used for the thermography study. The range of IR camera is about 3 to 5 microns. Minimum detecting temperature is about 250°C and max of about 1000°C and it varies with IR camera. Fig 15 shows picture of thermography.

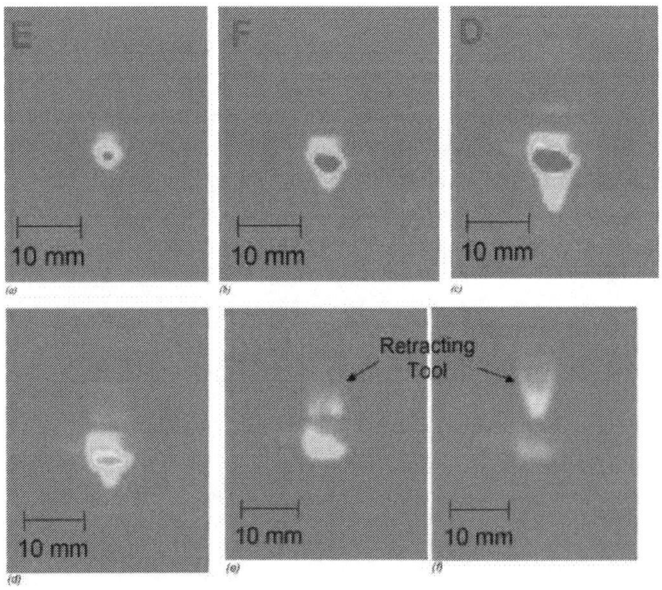

Figure 15 Infrared Camera image of friction drilling [9]

Work piece used in the study is AISI1020 cold rolled carbon steel plate and db mate of wc in co matrix. In fig 15(e) and 15 (f) tool is retracted, high intensity with bright spot on 15(c) in fig 16 stress max temp and min temp which is noted by thermography technique [9].

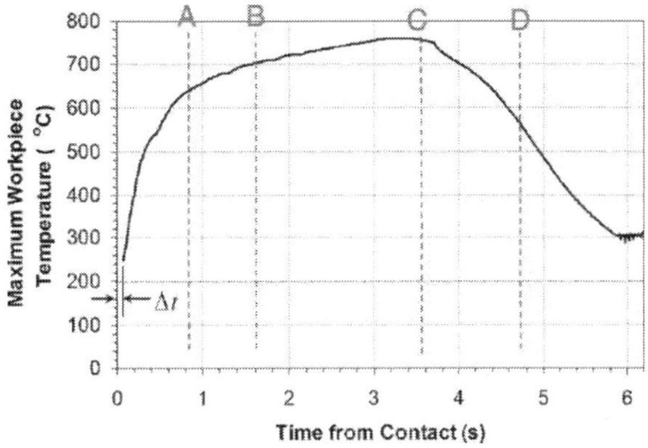

Figure 16 Maximum and Minimum work piece temperature detected by infrared camera [9]

How thermo grams are important are studied as experimental process in the paper image processing and data analysis [28]. In Scanning Electron Microscope(SEM) images are shown fig 17, where thermal damage of the carbon fiber composite material is visibility seen. These damages are happened due to the raise in temperature during machining process.

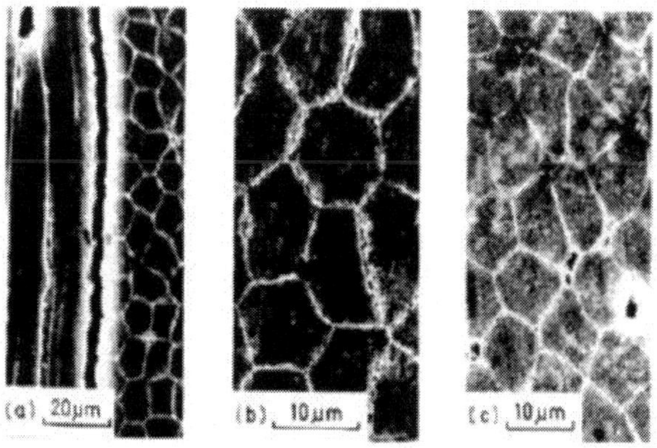

Figure 17 SEM image carbon fiber composite material [29]

10. OPTIMIZATION TECHNIQUE

Optimization technique is a powerful tool for reducing mechanical damages by optimizing the process parameters. After conducting different number of experiment trials, these trials were examined by Analysis of Variance (ANNOVA). For thermal friction drilling process using optimization techniques are carried out. Design of experiments is adopted by taguchi method and analysis of variance by Annova[8].

11. CONCLUTION

Friction drilling is a non –traditional process for holemaking, this technique is the most recent trend in hole making which currently applied in all major mechanical industries. Friction drilling is used in all sectors like Automobile, Aeronautical etc. Thermography is the new technique for identifying machining damages these damages are happened due to the influence of high temperature. A Boss and Bush is formed on front and backside of the work piece, due to the increase in ductility of the work material increases. This literature review reveals that the sandwich material is suitable for friction drilling process, because sandwich material have advantages likevery low weight, high stiffness and durability. Friction drilling drill bit is made up of cylindrical bar made up of tungsten carbide with Nano coating FCAR of 100% and 50%. By reducing low feed rate, point angle and proper cutting speed delamination can be reduced.

REFERENCES

[1]. R Piquet, B Ferret, F Lachaud, P Swider,2000 Experimental analysis of drilling damage in thin carbon/epoxy plate usingspecial drills*Composites part A applied science and manufacturing,***31**P. 1107 - 1115.

[2]. PrzemysławSitek, AndrzejKatunin, Analysis of drilling processof composite structures – Part I: Evaluation of thermal condition*Modelowanieinżynierskienr***55**ISSN 1896-771X, P 88-94.

[3]. Scott F. Miller, Albert J. Shih, 2007Thermo-Mechanical FiniteElement Modeling of the FrictionDrilling Process *Journal of Manufacturing Science and Engineering***Vol. 129**, P 531-538.

[4]. M.Boopathi, S.Shankar, S.Manikandakumar, R.Ramesh, 2013 Experimental Investigation of Friction Drilling on Brass, *Aluminumand Stainless Steel",Procedia Engineering***64** P. 1219 – 1226.

[5]. Ho-Cheng, C. K. H. Dharan, 1990 Delamination During Drilling inComposite Laminates *Journal of Engineering for Industry* **Vol. 112** P 236 – 239.

[6]. V.N. Gaitonde, S.R. Karnik, J. Campos Rubio, A. EstevesCorreia, A.M. Abra, J. Paulo Davim,2008Analysis of parametric influence on delamination inhigh-speed drilling of carbon fiber reinforcedplastic composites*Journal of materials processing technology***203**, P. 431–438.

[7]. RedouaneZitoune,VijayanKrishnaraj , BelkacemSofianeAlmabouacif , Francis Collombet ,Michal Sima , Alain Jolin, 2012Influence of machining parameters and new nano-coated tool on drilling performance of CFRP/Aluminium sandwich*Composites: Part B***43** P. 1480–1488.

[8]. Wei-Liang Ku , Ching-Lien Hung , Shin-Min Lee , Han-Ming Chow, (2011) Optimization in thermal friction drilling for SUS 304stainless steel *International Journal Advanced Manufacturing Technology* **53**: P. 935–944.

[9]. Scott F. Miller,RuiLi,HsinWang,Albert J. Shih, 2006Experimental and NumericalAnalysis of the Friction Drilling Process*Journal of Manufacturing Science and Engineering* **Vol. 128**, P 802 -810

[10]. Sivasaravanan S, V K Bupesh Raja, Manikandan, 2014 Impact characterization of Epoxy LY556/E-Glass fibre/ nano clay hybrid nano composite material*Procedia Engineeing***97**P. 968-974.

[11]. K Logesh, V K Bupesh Raja, R Subramanian, A Raju Kumar Sharma, 2015A Review on characteristics of sandwich material and its enhanced application *Journal of Applied Engineering research***Vol 10** No 84, P. 258-261.

[12]. K Logesh, V K Bupesh Raja, 2015Formability analysis for enhancing forming parameters in AA8011/PP/AA1100 sandwich material*International Journal Advanced Manufacturing Technology.*

[13]. K Logesh, V K Bupesh Raja,Investigation of Mechanical properties of AA8011/PP/AA1100 sandwich material*International journal of ChemTech research* **Vol 8**, No 3, P. 1749-1752.

[14]. J pauloDavim, 2001 A note on the determination of optimal cutting conditions for surface finish obtained in turning using design of experiment *Journal of materials processing technology* **116** P. 305-308.
[15]. R. Piquet, B. Ferret, F. Lachaud, P. Swider,2000 Experimental analysis of drilling damage in thin carbon/epoxy plate usingspecial drills *Composites: Part A* **31**, P.1107–1115.
[16].Scott F Miller, peter J Blau, Albert J Shih,2005Microstructure Alterations Associated with friction Drilling of Steel, Aluminum and Titanium*Journal of Materials Engineering and Performance* **Vol 14(5)**,P. -647-653.
[17]. Jose Mathew , G.L. Goswami , N. Ramakrishnan , N.K. Naik, 1999 Parametric studies on pulsed Nd:YAG laser cutting of carbon fiber reinforced plastic composites *Journal of Materials Processing Technology* **89-90.**
[18]. J. Paulo Davim, Pedro Reis,2003 Drilling carbon fiber reinforced plastics manufactured by autoclave- experimental and statistical study *Materials and Design* **24**, P.315–324.
[19].D.K. Shanmugam, T. Nguyen, J. Wang, 2008 A study of delamination on graphite/epoxycomposites in abrasive waterjet machining*Composites: Part A* **39** P.923–929.
[20]. VijayanKrishnaraj , A. Prabukarthi , ArunRamanathan , N. Elanghovan , M. Senthil Kumar ,RedouaneZitoune , J.P. Davim 2012 Optimization of machining parameters at high speed drilling of carbon fiber reinforced plastic (CFRP) laminates*Composites: Part B* **43**, P.1791–1799.
[21]. E. Kilickap,2010 Optimization of cutting parameters on delamination based on Taguchi methodduring drilling of GFRP composite *Expert Systems with Applications* **37** P. 6116–6122.
[22]. U.A. Khashaba, 2004 Delamination in drilling GFR-thermoset composites *Composite Structures* **63** P. 313–327.
[23]. J. C. Outeiro, P. Lenoir, A. Bosselut, Thermo-mechanical effects in drilling using metal working fluids and cryogenic cooling and their impact in tool performance, *SPRINGER,* P.1-12.
[24].S.R. Karnik a, V.N. Gaitonde b, J. Campos Rubio c, A. EstevesCorreia d, A.M. Abrão c, J. Paulo Davim, 2008 Delamination analysis in high speed drilling of carbon fiber reinforcedplastics (CFRP) using artificial neural network model*Materialsand Design* **29** P. 1768–1776.
[25]. Jamie D. Skovron, R. Rohan Prasad, DurulUlutan, LaineMears,DuaneDetwiler, Daniel Paolin,,Boris Baeumler,,Laurence Claus, 2015Effect of Thermal Assistance on the Joint Quality of Al6063-T5A During Flow Drill Screwdriving*Journal of Manufacturing Science and Engineering* **Vol. 137** / 051019, P 1-8.
[26]. El-Sonbaty, U.A. Khashaba, T. Machaly, 2004 Factors affecting the machinability of GFR/epoxy composites*Composite Structures* **63** P. 329–338
[27]. AndrzejKatunin ,Małgorzata John , KamilJoszko ,Anita Kajzer,Characterization of quasi-static behavior of honeycomb core sandwich structures",*Modelowanieinżynierskienr53*,ISSN 1896-771X , P. 78-84.
[28]. C. Ibarra-Castanedo, D. Gonz_alez, M. Klein, M. Pilla, S. Vallerand, X. Maldague,2004 Infrared image processing and data analysis" , *Infrared Physics & Technology* **46** P. 75–83.
[29]. W. S. Lau , W. B. Lee , S. Q. Pang, Pulsed Nd:YAG Laser Cutting of Carbon Fiber Composite Materials", Department of Manufacturing Engineering, Hong Kong Polytechnic/Hong Kong, P. 178-182.
[30]. Sanjay Rawat , HelmiAttia, 2009 Characterization of the dry high speed drilling process of wovencomposites using Machinability Maps approach *CIRP Annals - Manufacturing Technology* **58** P. 105–108.
[31]. Kyu Yeol Park, Jin Ho Choi , Dai Gil Lee, 1995Delamination –free and High efiiciency drilling of carbon fiber reinforced plastics*Journal of composite materials*, **Vol.29**, No.15, P.1988-2002.

IOP Conf. Series: Materials Science and Engineering **197** (2017) 012059 doi:10.1088/1757-899X/197/1/012059

Evaluation and Selection of Best Priority Sequencing Rule in Job Shop Scheduling using Hybrid MCDM Technique

Kalla Kiran Kumar[1,5], Dega Nagaraju[2], S Gayathri[3] and S Narayanan[4]

[1]B.Tech Mechanical Engineering, VIT University, Vellore, Tamil Nadu, India
[2,4]School of Mechanical Engineering, VIT University, Vellore, Tamil Nadu, India
[3]Research Scholar, VIT University, Vellore, Tamil Nadu, India.

[5]Email: kirankumar.kalla7@gmail.com

Abstract. Priority Sequencing Rules provide the guidance for the order in which the jobs are to be processed at a workstation. The application of different priority rules in job shop scheduling gives different order of scheduling. More experimentation needs to be conducted before a final choice is made to know the best priority sequencing rule. Hence, a comprehensive method of selecting the right choice is essential in managerial decision making perspective. This paper considers seven different priority sequencing rules in job shop scheduling. For evaluation and selection of the best priority sequencing rule, a set of eight criteria are considered. The aim of this work is to demonstrate the methodology of evaluating and selecting the best priority sequencing rule by using hybrid multi criteria decision making technique (MCDM), i.e., analytical hierarchy process (AHP) with technique for order preference by similarity to ideal solution (TOPSIS). The criteria weights are calculated by using AHP whereas the relative closeness values of all priority sequencing rules are computed based on TOPSIS with the help of data acquired from the shop floor of a manufacturing firm. Finally, from the findings of this work, the priority sequencing rules are ranked from most important to least important. The comprehensive methodology presented in this paper is very much essential for the management of a workstation to choose the best priority sequencing rule among the available alternatives for processing the jobs with maximum benefit.

1. Introduction

Job sequencing is determining the order in which jobs are processed at one or more workstations. Generally, a variety of tasks are carried out at each workstation. When workstations are densely loaded, the order of processing is very crucial in terms of costs associated with jobs waiting for processing and the cost of idle time at the workstations. If schedules are not precisely planned to avoid job traffic jam, waiting lines will develop. Such situations become complicated and lay pressure on management to establish scheduling procedures to process the workload efficiently.

In this line of direction, there are a number of priority sequencing rules or heuristics that are used to select the order in which the jobs will be processed at a workstation. The decision about which job to be processed next is made with priority sequencing rules whenever the workstation becomes available for further processing. Employing priority sequencing rules is worthwhile as the last minute information on operating conditions is also integrated into schedule as it evolves. The effectiveness of any given sequence generated by a priority sequencing rule is judged in terms of performance measures. Some of the performance measures are mean job completion time, average number of jobs in the system, mean job tardiness, utilization etc. Selecting the best priority sequencing rule to

IOP Conf. Series: Materials Science and Engineering **197** (2017) 012059 doi:10.1088/1757-899X/197/1/012059

determine the sequence of processing the jobs at a workstation is a challenging problem. Out of the usable priority sequencing rules, no choice is clearly the best for a management of a workstation to decide. Hence, a thorough method of selecting the right choice is essential in managerial decision making perspective. The methodology which is adopted should be able to structure complex problems well by considering multiple criteria explicitly so that it leads to more informed and improved decisions. The methods which fulfil the above said quality are multi criteria decision making (MCDM) techniques.

MCDM techniques consider and solve decision and planning problems comprising multiple criteria. Their objective is to support decision-makers in taking the best decision. The analytic hierarchy process (AHP) is a well-structured MCDM technique in this line of direction. The AHP method is a decision-making approach developed by Saaty [1,2,3]. Since the AHP approach is based on judgements form a decision maker having real time knowledge and experience, it agrees well with the behaviour of the decision maker. Rangone developed an analytic hierarchy process framework for measuring and comparing the overall performance of various manufacturing divisions based on non-financial and financial performance criteria [4]. Kamal adopted the analytic hierarchy process (AHP) for superior decision making in project management i.e. for selection of best contractor to execute a project [5]. Author presented group decision making using the AHP and used Expert Choice professional software for implementing the AHP. Liang developed AHP-based model for evaluation of project continuation or termination which is based on benchmarking method [6]. Author illustrated the formulated approach using research and development (R&D) case study in Taiwan. Tahriri formulated AHP based supplier selection model and utilized it in a steel manufacturing company in Malaysia [7]. Amanda et al. opted the analytic hierarchy process (AHP) with ratings for prioritization of research and development projects in a Brazilian aerospace institution [8]. Dalalah et al.adopted the analytic hierarchy process (AHP) for selection of cranes [9]. Authors considered three crane types as alternatives and used Expert Choice software to perform experimental assessments. Subramanian and Ramanathan reviewed the literature to find the significant research gaps that exist in the application of the AHP method and prepared a comprehensive listing of AHP applications [10]. Sivakumar et al. adopted analytic hierarchy process (AHP) and Taguchi loss functions for evaluation and selection of relevant vendor for the production process in the green mining industries based on economic, environmental benefits and risk factors [11]. Luthra et al. investigated the barriers in the supply chain to implement sustainable consumption and production (SCP) and evaluated fifteen barriers using Analytic hierarchy process (AHP) to determine their relative importance [12].

In recent times, the concept of competitive benchmarking is adopted in establishing performance measures and setting goals to measure the present product performance against that of best companies [13,14]. The competitive benchmarking led to the evolution of the technique for order preference by similarity to ideal solution (TOPSIS) to solve MCDM complications. The uncommon feature of TOPSIS technique is that the chosen alternative has the closest geometric distance from the positive ideal solution (PIS) and the farthest geometric distance from the negative ideal solution (NIS). Aizhen adopted TOPSIS method for thorough and extensive assessment of environmental quality [15]. Li-juan highlighted the significance of supplier selection in supply chain management (SCM) and formulated a model for supplier selection using TOPSIS technique [16]. Wang and Hsu employed TOPSIS technique to evaluate the business operation performance of ten listing companies in the Taiwan stock market to guide investors in selection of target stock shares and investment financial programs [17]. Tong adopted TOPSIS method to derive the overall performance index (OPI) for multiple responses and to determine the optimal factor/level combination with the maximum OPI value [18]. Huang et al. proposed a combined entropy weight and TOPSIS method for selection of suitable information system [19]. Vijay and Sankarpresented a logical procedure using TOPSIS method for evaluation of CNC machines in terms of specifications and cost [20]. Li et al. adopted entropy weight and TOPSIS method for evaluation of safety conditions of four coal mines in order to improve the safety levels and ensure safe production of coal mines [21]. Monjezi et al. adopted TOPSIS as a MCDM technique to investigate the blasting operation in a lime stone mine and to select the most relevant blasting

pattern[22]. Lin et al. presented an integrated approach adopting analytic hierarchy process (AHP) and the technique for order preference by similarity to ideal solution (TOPSIS) to facilitate designers in identifying customer requirements and design characteristics [23]. Bhutia and Phiponadopted combined AHP and TOPSIS hybrid methodology for supplier selection in a supply chain cycle [24]. They have calculated the priority weights for each criterion based on Analytic Hierarchy Process (AHP) and then applied these weights to the TOPSIS method to rank suppliers. In this paper, an integrated methodology is demonstrated for evaluating and selecting the best priority sequencing rule by using hybrid multi criteria decision making technique (MCDM), i.e., analytical hierarchy process (AHP) coupled with technique for order preference by similarity to ideal solution (TOPSIS).
The remainder of this paper is classified as follows. The proposed model, AHP methodology and TOPSIS methodology are discussed in section 2. Then in section 3, the study area job shop sequencing, priority sequencing rules (alternatives) and performance measures (evaluation criteria) are explained. In section 4, numerical illustration of the proposed methods and results are presented. And lastly section 5 concludes the paper.

2. The Proposed Model

The overall procedure of the present study is shown in Figure 1. After determining the study area i.e. job shop sequencing, different priority sequencing rules are identified. Then for evaluation and selection of the best priority sequencing rule, the criteria are determined. According to these criteria, the required data for the entire procedure is collected from the related workstation. After building the evaluation criteria hierarchy, the criteria weights are determined by applying the AHP method. The performances of the alternatives corresponding to the evaluation criteria are performed and tabulated. Finally, MCDM technique TOPSIS is employed to obtain the final ranking results. The major steps of the entire procedure are described in detail in the following subsections.

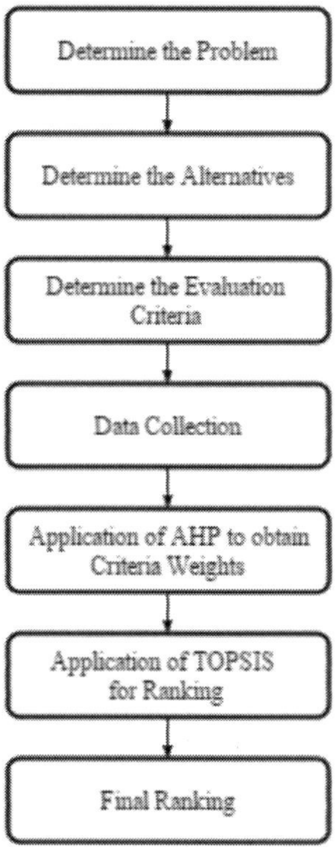

Figure 1. The Overall Procedure.

2.1. Analytic Hierarchy Process (AHP) Methodology

The Analytic Hierarchy Process (AHP) is a multi-criteria decision making technique that allows qualitative and quantitative, financial and non-financial measures to be considered and trade-offs among them to be addressed. The AHP technique integrates different measures into a single global score for ranking the alternatives. It is mainly based on pair-wise comparison judgements [5]. The AHP methodology is described in seven steps.

Step 1: Formulation of relative comparison decision matrix (A)

$$A_{MxM} = \left[b_{ij}\right] = \begin{bmatrix} 1 & b_{12} & b_{13} & \cdots & \cdots & b_{1M} \\ \frac{1}{b_{12}} & 1 & b_{23} & \cdots & \cdots & b_{2M} \\ \frac{1}{b_{13}} & \frac{1}{b_{23}} & 1 & \cdots & \cdots & b_{3M} \\ \vdots & \vdots & \vdots & \vdots & \vdots & \vdots \\ \vdots & \vdots & \vdots & \vdots & \vdots & \vdots \\ \frac{1}{b_{1M}} & \frac{1}{b_{2M}} & \cdots & \cdots & \cdots & 1 \end{bmatrix}$$

where $i, j = 1,2,...,M$.

Let $C_1, C_2,...,C_M$ denote the set of criteria, while b_{ij} represents a quantified judgement on a pair of criteria, C_i and C_j. Saaty [1,2,3] constituted a measurement scale for pair-wise comparison.

Step 2: Computation of normalized decision matrix. Each set of column values is added and then each value is divided by its respective column sum value.

Step 3: Calculation of priority weights. The mean of rows of normalized decision matrix is determined and the priority weights of the decision-maker's criteria are obtained. A set of M numerical weights $w_1, w_2,...., w_M$ are determined.

Step 4: Determination of the vector Aw^T, where A is the pairwise comparison matrix and w^T is the M -dimensional column weight vector.

Step 5: Computation of $\frac{1}{M}\sum_{i=1}^{i=M}\frac{i^{th} entry\, in\, AW^T}{i^{th} entry\, in\, W^T}$

Step 6: Calculation of the Consistency Index (CI)

$$CI = \frac{(Step\,5\,Result) - M}{M - 1}$$

Step 7: Comparison of Consistency Index (CI) to the Random Index (RI) for the proper value of M. The value of the Random Index is foundfrom Table 1.

Table 1. The Random Index Table.

Matrix Order	1	2	3	4	5	6	7	8	9
RI	0	0	0.58	0.9	1.12	1.124	1.32	1.41	1.46

CR (consistency ratio) isthe final ratio to be determined. If consistency ratio is lesser than 0.1, the pairwise comparison matrix is consistent and the attained weights can be used. The CR is computed as follows: $CR = \frac{CI}{RI}$.

2.2. TOPSIS Methodology

Step 1: Determine a decision matrix for the ranking. The matrix structure is expressed as follows:

$$M = \begin{array}{c} \\ a_1 \\ a_2 \\ \vdots \\ a_i \\ \vdots \\ a_J \end{array} \begin{array}{c} \begin{matrix} F_1 & F_2 & \cdots & F_j & \cdots & F_n \end{matrix} \\ \begin{bmatrix} f_{11} & f_{12} & \cdots & f_{1j} & \cdots & f_{1n} \\ f_{21} & f_{22} & \cdots & f_{2j} & \cdots & f_{2n} \\ \vdots & \vdots & \cdots & \vdots & \cdots & \vdots \\ f_{i1} & f_{i2} & \cdots & f_{ij} & \cdots & f_{in} \\ \vdots & \vdots & \cdots & \vdots & \cdots & \vdots \\ f_{J1} & f_{J2} & \cdots & f_{Jj} & \cdots & f_{Jn} \end{bmatrix} \end{array}$$

Where a_j represents the alternatives, $j = 1,2,3,...,J$; F_i denotes i_{th} attribute or criterion, $i = 1,2,3,...,n$ and f_{ij} is a crisp value specifying the performance rating of each alternative a_j against each criterion F_i.

Step 2: Determine the normalized decision matrix R(=[r_{ij}]). The normalized value r_{ij} is computed as:

$$r_{ij} = \frac{fij}{\sqrt{\sum_{j=1}^{J} f_{ij}^2}} \quad j = 1,2,3,...,J;\ i = 1,2,3,...,n.$$

This step converts various dimensional attributes into non-dimensional attributes that allows comparisons across criteria.

Step 3: Determine the weighted normalized decision matrix. It is obtained by multiplying the normalized decision matrix by its associated weights. The weighted normalized value v_{ij} is determined as:

$$v_{ij} = w_i \times r_{ij} \quad j = 1,2,3,....,J;\ i = 1,2,3,...,n.$$

Where w_i denotes the weight of the i_{th} criterion or attribute and $\sum_{i=1}^{n} w_i = 1$.

Step 4: Determine the positive-ideal solution (PIS) and negative-ideal solution (NIS).

$$A^* = \left\{\tilde{v}_1^*, \tilde{v}_2^*, ..., \tilde{v}_i^*\right\} = \left\{\left(\max_j v_{ij} \,|\, i \in I'\right), \left(\min_j v_{ij} \,|\, i \in I''\right)\right\}$$

$$i = 1,2,3,...,n;\ j = 1,2,3,....,J.$$

$$A^- = \left\{\tilde{v}_1^-, \tilde{v}_2^-, ..., \tilde{v}_i^-\right\} = \left\{\left(\min_j v_{ij} \,|\, i \in I'\right), \left(\max_j v_{ij} \,|\, i \in I''\right)\right\}$$

$$i = 1,2,3,...,n;\ j = 1,2,3,....,J.$$

Where I' is associated with the benefit criteria and I'' is associated with the cost criteria.

Step 5: Determine the separation measures, adopting the n-dimensional Euclidean distance. The separation of each alternative from the PIS (D_j^*) is calculated as follows:

$$D_j^* = \sqrt{\sum_{i=1}^{n} \left(v_{ij} - v_i^*\right)^2} \quad j = 1,2,3,....,J.$$

Likewise, the separation of each alternative from the NIS (D_j^-) is calculated as follows:

$$D_j^- = \sqrt{\sum_{i=1}^{n} \left(v_{ij} - v_i^-\right)^2} \quad j = 1,2,3,....,J.$$

Step 6: Determine the relative closeness to the ideal solution. The relative closeness of the alternative a_j is obtained as:

$$CC_j^* = \frac{D_j^-}{D_j^* + D_j^-} \quad j = 1, 2, 3, \ldots, J.$$

Where the CC_j^* index value varies between 0 and 1. The higher the index value implies the better the performance of the alternatives.
Step 7: Rank the preference order.

3. Job Shop Sequencing

Several jobs need to be carried out at one or more workstations. Generally, different types of tasks are performed at each workstation. If schedules are not cautiously planned to avoid blockage, waiting lines will develop. Such complex situations put pressure on administration to work out scheduling procedures to manage the workload effectively. One way to develop schedules in job shops is by adopting priority sequencing rules, which allows the schedule for a workstation to evolve over a period of time. The selection of job to be processed is made with priority sequencing rules whenever the workstation is ready for further processing.

Priority sequencing rules are simple heuristics adopted to select the sequence in which the jobs are to be processed. The most common are mentioned below.

- First come, first served (FCFS): Jobs are processed in the order in which they arrive at the workstation.
- Last come, first served (LCFS): Jobs are processed in the order of last to first in which they arrive at the workstation.
- Shortest processing time (SPT): Jobs are processed in order of the processing time required at the workstation, with the job requiring the least processing time at the workstation scheduled first.
- Longest processing time (LPT): Jobs are processed in order of the processing time required at the workstation, with the job requiring the longest processing time at the workstation scheduled first.
- Earliest due date (EDD): Jobs are processed in the order in which they are due for delivery to the customer.
- Critical ratio (CR): Jobs are processed in order of increasing critical ratio (the ratio of time required by work left to be done to time left to do the work)
- SLACK - Jobs are processed in order of increasing slack time (time until due date minus remaining time to process)

The above mentioned priority sequencing rules are generally based on the following assumptions.

- The set of jobs is known; no new jobs arrive after processing begins; and no jobs are cancelled.
- Setup time is deterministic.
- Setup time is independent of processing sequence.
- Processing time is deterministic.
- There will be no interruptions while processing such as machine breakdowns, accidents etc.

Priority sequencing rules try to minimize completion time, number of jobs in the system, and job lateness, while maximizing facility utilization. The effectiveness of any given sequence is judged through performance measures. The criteria (performance measures) for evaluating job sequencing rules are listed below.

C_1-Average number of jobs in the system
C_2-Utilization
C_3-Average job tardiness
C_4-Total flow time
C_5-Maximum job tardiness
C_6-Total tardiness
C_7-Average job completion time

IOP Conf. Series: Materials Science and Engineering **197** (2017) 012059 doi:10.1088/1757-899X/197/1/012059

C_8-Makespan/Total processing time

Here only second criterion is a benefit criterion and the others are cost criteria. From the available priority sequencing rules, no choice is clearly the best. Hence, a comprehensive method of selecting the right choice is required. For evaluating and selecting the best priority sequencing rule, a hybrid multi criteria decision making technique (MCDM), i.e., analytical hierarchy process (AHP) with technique for order preference by similarity to ideal solution (TOPSIS) is adopted and demonstrated in the following sections.

4. Numerical Illustration

Our application is related to selecting the right choice from the available priority sequencing rules for a workstation located in Vellore. The management desires to find the best rule from the seven alternatives. A committee of decision makers is formed and the required data for the entire procedure is collected from the workstation.

4.1. Analytic Hierarchy Process (AHP) Methodology

In this section, AHP technique is applied to calculate the weights of the criteria. To determine the relative weights, decision makers are asked to make pairwise comparisons using a suitable scale shown in Table 2. Then, a comprehensive pair-wise comparison matrix is built by unifying their grades. The final pair-wise comparison matrix is built as in Table 3. As per the procedure stated in section 2.1 the priority weights are calculated and tabulated in Table 3. From Table 3 as the CR (Consistency Ratio) value is less than 0.1, the judgements are acceptable.

Table 2. The Fundamental Scale for Pairwise Comparisons.

Intensity of Importance	Definition
1	Equal importance
3	Moderate Importance
5	Strong importance
7	Very strong importance
9	Extreme importance

[a]Intensities of 2,4,6 and 8 are used to express intermediate values. Intensities of 1.1,1.2,1.3, etc. are used for elements that are very close in importance.

Table 3. Pairwise Comparison Matrix and Priority Weights.

	C_1	C_2	C_3	C_4	C_5	C_6	C_7	C_8	Priority Weights	Consistency Index	Random Index
C_1	1.000	2.000	0.500	2.500	4.000	3.000	0.400	1.500	0.1396	0.0276	1.41
C_2	0.500	1.000	0.333	2.000	3.500	3.000	0.286	0.667	0.0954		
C_3	2.000	3.000	1.000	3.500	4.500	4.000	0.500	2.500	0.2114		
C_4	0.400	0.500	0.286	1.000	2.000	1.500	0.250	0.400	0.0594		
C_5	0.250	0.286	0.222	0.500	1.000	0.667	0.167	0.222	0.0344		
C_6	0.333	0.333	0.250	0.667	1.500	1.000	0.200	0.286	0.0446		
C_7	2.500	3.500	2.000	4.000	6.000	5.000	1.000	3.000	0.2927		
C_8	0.667	1.500	0.400	2.500	4.500	3.500	0.333	1.000	0.1225		
Consistency Ratio										0.0196	

4.2. Application of TOPSIS Method

The weights of the criteria are calculated by AHP till now, and then these values will be used in TOPSIS. Processing time (including setup times) and due dates for six jobs waiting to get processed at the workstation are collected and tabulated in Table 4.

Table 4. Data collected from Workstation.

Job	Processing Time (days)	Due Date (days)
A	2	7
B	8	16
C	4	4
D	10	17
E	5	15
F	12	18

The performance measures (Criteria) with respect to each of the priority sequencing rules (alternatives) are calculated and tabulated in Table 5. Thus Table 5 is considered as the pairwise comparison decision matrix of alternatives with respect to each criterion. The normalized decision matrix is formed as in Table 6. After the weighted normalized decision matrix is formed, positive ideal solution (PIS) and negative ideal solution (NIS) are determined as in the following:

$$A^{*} = [0.0412, 0.0432, 0.0439, 0.0176, 0.011, 0.0093, 0.0865, 0.0463]$$

$$A^{-} = [0.0685, 0.0261, 0.1248, 0.0291, 0.0167, 0.0263, 0.1434, 0.0463]$$

Then the separation of each alternative from PIS and NIS are calculated, i.e. the D_j^* and the D_j^- of the seven alternatives are calculated by using formulae given in section 2.2 and tabulated in Table 7.

Table 5. Performance Measures calculation with respect to Priority Sequencing Rules.

	Average no. of jobs in the system	Utilization (in %)	Average job tardiness (days)	Total flow time (days)	Maximum job tardiness (days)	Total tardiness (days)	Average job Completion Time (days)	Total Processing time /Makespan (days)
FCFS	2.93	34.17	9	120	23	54	20	41
LCFS	4.07	24.55	16	167	34	96	27.83	41
SPT	2.63	37.96	6.67	108	23	40	18	41
LPT	4.37	22.91	18	179	35	108	29.83	41
EDD	2.68	37.27	6.33	110	23	38	18.33	41
CR	3.24	30.83	9.67	133	24	58	22.17	41
SLACK	3.24	30.83	9.5	133	26	57	22.17	41

Table 6. Normalized Decision Matrix.

	C_1	C_2	C_3	C_4	C_5	C_6	C_7	C_8
FCFS	0.3289	0.4076	0.2952	0.3285	0.3183	0.2952	0.3285	0.3780
LCFS	0.4569	0.2929	0.5249	0.4571	0.4706	0.5249	0.4571	0.3780
SPT	0.2953	0.4529	0.2188	0.2956	0.3183	0.2187	0.2956	0.3780
LPT	0.4906	0.2733	0.5905	0.4900	0.4844	0.5905	0.4899	0.3780
EDD	0.3009	0.4446	0.2076	0.3011	0.3183	0.2078	0.3010	0.3780
CR	0.3637	0.3678	0.3172	0.3640	0.3322	0.3171	0.3641	0.3780
SLACK	0.3637	0.3678	0.3116	0.3640	0.3599	0.3116	0.3641	0.3780

4.3. Results and Discussion

Finally, the closeness coefficients of seven alternatives are calculated by this formula: $CC_j^* = \frac{D_j^-}{D_j^* + D_j^-}$ $j = 1,2,3,....,J$. These values are tabulated in Table 7. According to the closeness coefficients of seven alternatives, the ranking order of the seven alternatives is determined as EDD> SPT> FCFS> SLACK> CR> LCFS> LPT. In the chosen case study EDD is determined as the most appropriate priority sequencing rule for processing the jobs in the workstation. In other words, EDD rule is closer to the PIS and farther from the NIS.

Table 7. Computations of D_j^*, D_j^- and CC_j^*.

	D_j^*	D_j^-	CC_j^*	Ranking order
FCFS	0.02225	0.08424	0.7910	3
LCFS	0.08827	0.01795	0.1690	6
SPT	0.00240	0.10431	0.9775	2
LPT	0.10620	0.00006	0.0006	7
EDD	0.00199	0.10497	0.9814	1
CR	0.03371	0.07291	0.6838	5
SLACK	0.03291	0.07383	0.6917	4

5. Conclusion

The objective of this study is to demonstrate the methodology of evaluating and selecting the best priority sequencing rule for processing jobs at a workstation by using hybrid multi criteria decision making technique (MCDM). This paper used analytical hierarchy process (AHP) to derive criteria weights based on pairwise comparison and adopted technique for order preference by similarity to ideal solution (TOPSIS) for evaluating and selecting the best priority sequencing rule. Then the numerical illustration (case study) of eight main criteria and seven alternatives is taken into consideration and the methodology is applied to it. The results guide the management of the workstation to choose the best priority sequencing rule among the available alternatives for processing the jobs with maximum benefit. For further research, the usage of fuzzy sets to describe uncertainties in different factors can be incorporated and hybrid MCDM techniques can be adopted under fuzzy environment.

References

[1] Saaty TL 1990 *European journal of operational research.* **48**9-26.
[2] Saaty TL 1990 *Decision making for leaders: the analytic hierarchy process for decisions in a complex world.*RWS publications.
[3] Saaty TL and Kearns KP 2014 *Analytical planning: The organization of system.* Elsevier.
[4] Rangone A 1996*International Journal of Operations & Production Management.***16**104-19.
[5] Al-Harbi KM2001 *International journal of project management.* **19**19-27.
[6] Liang WY 2003*Benchmarking: An International Journal.***10**445-56.
[7] Tahriri F, Osman MR, Ali A, Yusuff RM and Esfandiary A 2008*Journal of Industrial Engineering and Management.* **1**54-76.
[8] da Silva AC, Belderrain MC and Pantoja FC 2011*Journal of Aerospace Technology and Management.* **2**339-48.
[9] Dalalah D, Al-Oqla F and Hayajneh M 2010*Jordan Journal of Mechanical and Industrial Engineering.* **4**567-78.
[10] Subramanian N and Ramanathan R 2012 *International Journal of Production Economics.***138**215-41.
[11] Sivakumar R, Kannan D and Murugesan P 2015*Resources Policy.* **46**64-75.
[12] Luthra S, Mangla SK, Xu L and Diabat A 2016*International Journal of Production Economics.*

[13] Camp RC Benchmarking: the search for industry best practices that lead to superior performance 1989 *Milwaukee, Wis.: Quality Press; Quality Resources*.

[14] Zairi M and Leonard P 1994 *Practical Benchmarking: The Complete Guide.* Springer Science & Business Media.

[15] Aizhen LC 1999 *[J]. JOURNAL OF GEOLOGICAL HAZARDS AND ENVIRONMENT PRESERVATION.* **2**.

[16] Li-juan MA 2002 *[J]. Industrial Engineering and Management*. **6**008.

[17] Wang TC and Hsu JC Evaluation of the business operation performance of the listing companies by applying TOPSIS method 2004 *InSystems, Man and Cybernetics, 2004 IEEE International Conference.* **2** 1286-1291. IEEE.

[18] Tong LI, Wang CH, Chen CC and Chen CT 2004 *European Journal of Operational Research.* **156**433-44.

[19] Huang J Combining entropy weight and TOPSIS method for information system selection 2008*In2008 IEEE Conference on Cybernetics and Intelligent Systems*. 1281-1284. IEEE.

[20] Athawale VM and Chakraborty S A TOPSIS method-based approach to machine tool selection 2010 *InInternational conference on Industrial engineering and operations management*.

[21] Li X, Wang K, Liu L, Xin J, Yang H and Gao C Application of the entropy weight and TOPSIS method in safety evaluation of coal mines 2011 *Procedia Engineering*. **26**2085-91.

[22] Monjezi M, Dehghani H, Singh TN, Sayadi AR and Gholinejad A 2012*Arabian journal of geosciences*. **5**95-101.

[23] Lin MC, Wang CC, Chen MS and Chang CA 2008*Computers in industry*. **59**17-31.

[24] Bhutia PW and Phipon R 2012*IOSR Journal of Engineering*. **2**43-50.

IOP Conf. Series: Materials Science and Engineering **197** (2017) 012060 doi:10.1088/1757-899X/197/1/012060

Evaluation and selection of material handling equipment in iron and steel industry using analytic hierarchy process

Sajja Varun[1*], Raj Harshita[2], Sesha Pramod[3] and Dega Nagaraju[4]

[1*,2,3]B.Tech, Production & Industrial Engineering, School of Mechanical Engineering, VIT University, Vellore-632014, Tamil Nadu, India,
[4]Department of Manufacturing Engineering, School of Mechanical Engineering, VIT University, Vellore-632014, Tamil Nadu, India.

[*]**E mail**: svarun_2013@vit.ac.in

1. Abstract: The paper presents the analytic hierarchy process (AHP) as a potential decision making method for use in the selection of the most suitable material handling (MH) system in an iron and steel industry. In this study, AHP is used in assessing the various material transportation systems employed in a steel manufacturing industry and to decide the best equipment to be used.Information on the use of AHP in evaluating MH equipment is provided and an AHP model is proposed to guide the management of an iron and steel Industry, i.e., JSW Steel Ltd. Most important factors while selecting material transportation equipment and their relative influence on the objective of decision-making model are found. A total of seven decision criteria and five different alternatives are considered for this purpose. Each alternative is evaluated in terms of the decision criteria and the relative importance (or weight) of each criterion is estimated. From the obtained pairwise comparison matrices, the best alternative is chosen.This paper provides a good insight into a decision-making model to guide managers for assessing the various material transportation equipment that are commonly employed in a steel manufacturing plant.

2. Introduction

The Analytic Hierarchy Process(AHP) is a multi-criteria decision approach[1].The AHP has attracted the interest of many researchers mainly due to the nice mathematical properties of the method and the fact that the required input data are rather easy to obtain.Instead of endorsing a "right" choice, the AHP helps decision makers find one that best suits their goal and their understanding of the issue. It provides a comprehensive and rational framework for structuring a decision problem, for representing, measuring and quantifying its elements, for relating those elements to overall objectives, and for assessing alternative solutions.

It uses a multi-level hierarchical structure of objectives, criteria, sub criteria and alternatives.The appropriate data is derived by using a set of pairwise correlations.These comparisons are used to obtain the weights of significance of the choice criteria and the relative performance measures of the alternatives in terms of each individual decision

IOP Conf. Series: Materials Science and Engineering **197** (2017) 012060 doi:10.1088/1757-899X/197/1/012060

criteria.If the comparisons are not perfectly consistent,then it provides a mechanism for improving consistency.

Typical applications where AHP has been used are in:

- Prioritizing factors and requirements that impact software development and productivity,
- Choosing among several strategies for improving safety features in motor vehicles,
- Estimating cost and scheduling options for material requirements planning(MRP),
- Evaluating the quality of research or investment proposals.

Some of the industrial engineering applications of the AHP include its use in integrated manufacturing[2], in the evaluation of technology investment decisions[3], in flexible manufacturing systems[4], layout design[5], and also in other engineering problems[6]. A number of criticisms have been launched at AHP over the years.It is said that in order to elicit the weights of the criteria by means of a ratio scale, the method asks decision-makers meaningless questions[7].It is pointed out that this method can suffer from rank reversal[8,9].

The objective of this paper is to introduce the application of AHP in the evaluation and selection of Material Handling Equipment in an Iron and Steel industry.The various factors used for evaluation are installation cost,maintenance and operation cost,speed of movement,capacity,operator requirement, installation space required and ease of operation.Using these criteria,pairwise comparison matrices are developed to select the best material handling equipment from the available alternatives.Pairwise comparisons are used to determine the relative importance of each alternative in terms of each criterion.These comparisons are quantified by using a scale.Such a scale is a one-to-one mapping between the set of discrete linguistic choices available to the decision maker and a discrete set of numbers which represent the importance,or weight,of the previous choices.The scale proposed by Saaty is depicted in Table 1.

Table 1.Scale of Relative Importance (according to Saaty(1980))

Intensity of Importance	**Definition**	**Explanation**
1	Equal importance	Two activities contribute Equally to the objective.
3	Weak importance of One over another	Experience and judgment Slightly favor one activity Over another
5	Essential or strong importance	Experience and judgment Strongly favor one activity Over another.
7	Demonstrated importance	An activity is strongly Favored and its dominance Demonstrated in practice
9	Absolute importance	The evidence favoring one Activity over another is of Highest possible importance
2,4,6,8	Intermediate values between two adjacent judgments.	When compromise is needed.

IOP Conf. Series: Materials Science and Engineering **197** (2017) 012060 doi:10.1088/1757-899X/197/1/012060

Material handling equipment selection is an important function in the design of a material handling system, and thus, a crucial step for facilities planning[10]. Using proper material handling equipment can enhance the production process, provide effective utilization of manpower, increase production, and improve system flexibility. The importance of material handling equipment selection cannot be overlooked. However, with the wide range of material handling equipment available today, determination of the best equipment alternative for a given production scenario is not an easy task.This problem can be solved by using the AHP technique

3. Methodology

Five alternative material handling systems,i.e,conveyor,industrial truck, automated guided vehicle(AGV),rail and crane need to be evaluated in terms of the six decision criteria:installation cost,operation and maintenance cost,speed of movement, volume capacity, operator requirement, space occupied and ease of operation.This can be done by following the standard procedure of AHP as described below:

1.Developing the pair-wise comparison matrix for each criteria.
2.Synthesizing the pair-wise comparison matrix.
3.Obtaining the random consistency index (RCI) from table
4.Calculating the priority vector for each criterion.
5.Calculating λ_{max} using the consistency index value CI
6.Calculating the consistency ratio CR
7.Checking the consistency of the pair-wise comparison matrix to check whether the decision-maker's comparisons are consistent or not.

By following the AHP procedure described above,the hierarchy of the problem can be developed as shown in Fig 1.

Fig 1.Hierarchy of the MH Equipment problem

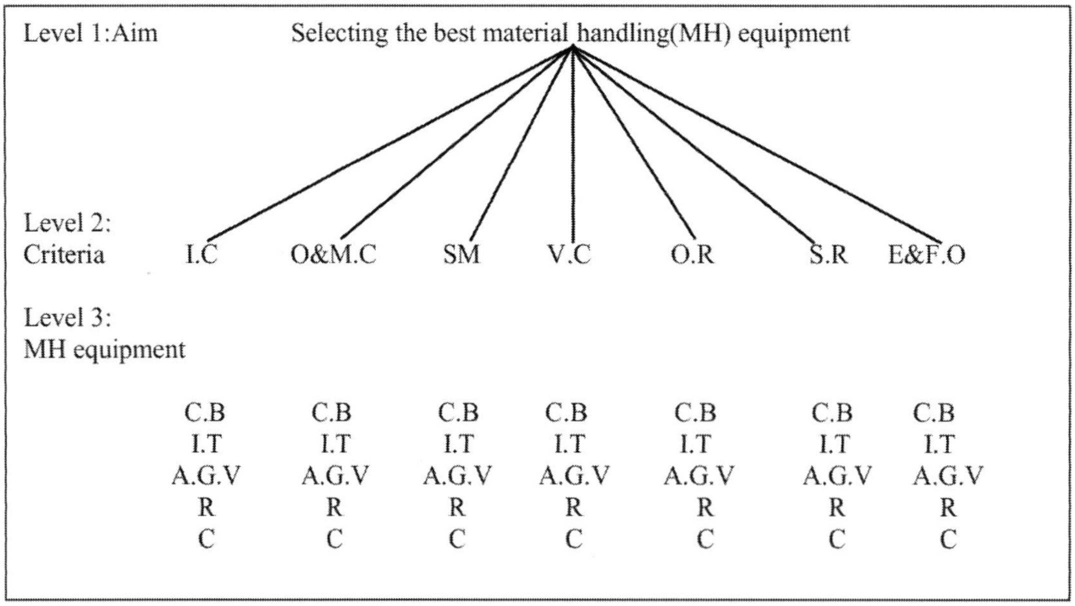

I.C=Installation Cost
S.M=Speed of Movement
O.R=Operator Requirement
E.O=Ease and Flexibility of Operation
O&M.C=Operation and Maintenance Cost
V.C=Volume Capacity
S.R=Space Requirement

R=Rail
C.B=Conveyor Belt
A.G.V=Automated Guided Vehicle
I.T=Industrial Truck
C=Crane

Table 2. RCI values for different values of n:

n	1	2	3	4	5	6	7	8	9
RCI	0	0	0.58	0.90	1.12	1.24	1.32	1.41	1.45

Firstly,considering the installation cost of each material handling equipment,the pairwise comparison matrix is developed as shown below in Table 3:

Table 3. Pair-wise comparison matrix for installation cost:

I.C	CB	IT	AGV	R	CR
CB	1.0000	0.3333	5.0000	0.5000	3.0000
IT	3.0000	1.0000	7.0000	2.0000	6.0000
AGV	0.2000	0.1429	1.0000	0.2000	0.3333
R	2.0000	0.5000	5.0000	1.0000	4.0000
CR	0.3300	0.1667	3.0000	0.2500	1.0000

Synthesizing the pair-wise comparison matrix is performed by dividing each element of the matrix by its column total. The priority vector in Table 4 can be obtained by finding the row averages.For example,the priority of conveyor with respect to the criterion 'installation cost' can be obtained by dividing the sum of the rows (0.1531+0.1555+0.2380+0.1265+0.2093) by the number of equipment (columns),i.e, 5,in order to obtain the value 0.1765.

The priority vector for installation cost, indicated in Table 4, is given below.

$$\begin{pmatrix} 0.1765 \\ 0.4369 \\ 0.0438 \\ 0.2620 \\ 0.0808 \end{pmatrix}$$

Table 4. Synthesized matrix for installation cost

I.C	C	IT	AGV	R	OC		Priority Vector
C	0.1531	0.1556	0.2381	0.1266	0.2093		0.1765
IT	0.4594	0.4667	0.3333	0.5063	0.4186		0.4369
AGV	0.0306	0.0667	0.0476	0.0506	0.0233		0.0438
R	0.3063	0.2333	0.2381	0.2532	0.2791		0.2620
OC	0.0505	0.0778	0.1429	0.0633	0.0698		0.0808

IOP Conf. Series: Materials Science and Engineering **197** (2017) 012060 doi:10.1088/1757-899X/197/1/012060

Now, estimating the consistency ratio is as follows:

$$0.1765\begin{pmatrix}1\\3\\0.2\\2\\0.33\end{pmatrix} + 0.4369\begin{pmatrix}0.33\\1\\0.14\\0.5\\0.16\end{pmatrix} + 0.0438\begin{pmatrix}5\\7\\1\\5\\3\end{pmatrix} + 0.2620\begin{pmatrix}0.5\\2\\0.2\\1\\0.25\end{pmatrix} +0.0808\begin{pmatrix}3\\6\\0.33\\4\\1\end{pmatrix}$$

$$= \begin{pmatrix}0.9144\\2.2818\\0.2208\\1.3756\\0.4086\end{pmatrix}$$

Dividing all the elements of the weighted sum matrices by their respective priority vector element, we obtain:

$\frac{0.9144}{0.1765} = 5.1802$, $\frac{2.2818}{0.4369} = 5.2231$,

$\frac{0.2208}{0.0438} = 5.0461$, $\frac{1.3756}{0.2620} = 5.2509$

$\frac{0.4086}{0.0808} = 5.0551$

To obtain λ_{max}, we compute the average of these values, i.e:

$\lambda_{max} =$

$$\frac{5.1802 + 5.2231 + 5.0461 + 5.2509 + 5.0551}{5}$$

$=5.1511$.

Now, we compute the consistency index, CI, as follows:

$$CI = \frac{\lambda_{max} - n}{n-1} = \frac{5.1511 - 5}{5 - 1} = 0.0337$$

Selecting the suitable value of random consistency ratio, RI, for a matrix size of 5 using Table 2 , we obtain RI = 1.12. We then compute the consistency ratio, CR, as follows:

$$CR = \frac{CI}{RI} = \frac{0.0337}{1.12} = 0.0245.$$

As the value of CR is less than 0.1,the judgments are acceptable.The same procedure can be followed for all the decision alternatives.In addition to this,we also use the same pair-wise comparison procedure to set priorities for all the seven criteria in terms of significance of each in contributing to the overall objective.Table 11 shows the pair-wise comparison matrix and priority vector for the seven criteria.

For the criteria of Maintenance and Operation cost, priority vector is in Table 5,

Table 5.Pair-wise comparison matrix for operation cost

0. C	C	IT	AGV	R	OC	Prioriy Vector
C	1.0000	6.0000	3.0000	7.0000	2.0000	0.4397
IT	0.1667	1.0000	0.3333	1.0000	0.2500	0.0626
AGV	0.3333	3.0000	1.0000	3.0000	0.3333	0.1501
R	0.1429	1.0000	0.3333	1.0000	0.2000	0.0577
OC	0.5000	4.0000	3.0000	5.0000	1.0000	0.2899

λ_{max}=5.0811,CI= 0.0203, CR = 0.0181 < 0.1(Accept)

Table 6.Pair-wise comparison matrix for speed of movement

S.M	C	IT	AGV	R	OC	Priority Vector
C	1.0000	0.3333	7.0000	3.0000	5.0000	0.2642
IT	3.0000	1.0000	9.0000	5.0000	7.0000	0.5077
AGV	0.1429	0.1111	1.0000	0.2500	0.5000	0.0385
R	0.3333	0.2000	4.0000	1.0000	3.0000	0.1287
OC	0.2000	0.1429	2.0000	0.3333	1.0000	0.0610

λ_{max} =5.1629,CI =0.0407,CR = 0.0364 < 0.1(Accept)

Considering the capacity in Table 7, i.e ,the volume of material different equipment can hold,

Table 7. Pair-wise comparison matrix for capacity

Ca	C	IT	AGV	R	OC	Priority Vector
C	1.0000	0.5000	5.0000	0.2500	2.0000	0.1468
IT	2.0000	1.0000	6.0000	0.3333	4.0000	0.2429
AGV	0.2000	0.1667	1.0000	0.1250	0.2500	0.0375
R	4.0000	3.0000	8.0000	1.0000	5.0000	0.4790
OC	0.5000	0.2500	4.0000	0.2000	1.0000	0.0938

λ_{max}=5.2045,CI =0.0511, CR = 0.0457 < 0.1(Accept)

Table 8. Pair-wise comparison matrix for operator requirement

O.R	C	IT	AGV	R	OC	Priority Vector
C	1.0000	4.0000	0.3333	7.0000	3.0000	0.2754
IT	0.2500	1.0000	0.2000	2.0000	0.3333	0.0739
AGV	3.0000	5.0000	1.0000	7.0000	4.0000	0.4672
R	0.1429	0.5000	0.1429	1.0000	0.3333	0.0460
OC	0.3333	3.0000	0.2500	3.0000	1.0000	0.1376

λ_{max}=5.2063,CI = 0.0516, CR = 0.0461 <0.1(Accept)

Taking into account the space required for different equipment, as shown in Table 9,

Table 9. Pair-wise comparison matrix for space occupied

S.R	C	IT	AGV	R	OC	Priority Vector
C	1.0000	0.3333	0.2000	4.0000	2.0000	0.1324
IT	3.0000	1.0000	0.3333	5.0000	3.0000	0.2430
AGV	5.0000	3.0000	1.0000	7.0000	5.0000	0.4931
R	0.2500	0.2000	0.1429	1.0000	0.5000	0.0478
OC	0.5000	0.3333	0.2000	2.0000	1.0000	0.0837

λ_{max} =5.1758, CI =0.0439, CR =0.0392 < 0.1(Accept)

Considering the criteria of ease and flexibility of operation,as shown in Table 10,

Table 10. Pair-wise comparison matrix for ease of operation

E.O	C	IT	AGV	R	OC	Priority Vector
C	1.0000	0.2500	0.2000	0.5000	2.0000	0.0856
IT	4.0000	1.0000	0.5000	3.0000	5.0000	0.2967
AGV	5.0000	2.0000	1.0000	4.0000	5.0000	0.4343
R	2.0000	0.3333	0.2500	1.0000	2.0000	0.1218
OC	0.5000	0.2000	0.2000	0.5000	1.0000	0.0616

λ_{max} =5.1095, CI =0.0273, CR =0.0244 < 0.1(Accept)

Priority vector for comparison matrix of decision criterion is shown in Table 11,

Table 11. Pair-wise comparison matrix of various criterion:

	IC	O&M.C	S.M	V.C	O.R	S.R	E.O		Priority Vector
IC	1.0000	3.0000	5.0000	2.0000	9.0000	6.0000	7.0000		0.3584
O&M.C	0.3333	1.0000	3.0000	0.5000	5.0000	3.0000	4.0000		0.1571
S.M	0.2000	0.3333	1.0000	0.3333	5.0000	3.0000	4.0000		0.1082
V.C	0.5000	2.0000	3.0000	1.0000	7.0000	5.0000	6.0000		0.2388
O.R	0.1111	0.2000	0.2000	0.1429	1.0000	0.3333	0.5000		0.0272
S.R	0.1667	0.3333	0.3333	0.2000	3.0000	1.0000	3.0000		0.0646
E.O	0.1429	0.2500	0.2500	0.1667	2.0000	0.3333	1.0000		0.0387

λ_{max} = 7.3623, CI = 0.0603, CR = 0.0457 < 0.1(Accept)

The overall priority for various alternatives is computed as shown in Table 12.

Table 12. Priority matrix for MH EQUIPMENT selection:

	I.C (0.3584)	O&M.C (0.1571)	S.M (0.1082)	V.C (0.2388)	O.R (0.0272)	S.R (0.064)	E.O (0.0387)	Overall priority vector
C	0.1765	0.4397	0.2642	0.1468	0.2754	0.1324	0.0856	0.2814
IT	0.4369	0.0626	0.5077	0.2429	0.0739	0.2430	0.2967	0.2981
AGV	0.0438	0.1501	0.0385	0.0375	0.4672	0.4931	0.4343	0.1383
R	0.2620	0.0577	0.1287	0.4790	0.0460	0.0478	0.1218	0.2400
OC	0.0808	0.2899	0.0610	0.0938	0.1376	0.0837	0.0616	0.1146

The overall priority for various alternatives is computed as shown in Table 12.

For ex,the overall priority of conveyor is given as = 0.3584(0.1765) + 0.1571(0.4397) + 0.1082(0.2642) + 0.2388(0.1468) + 0.0272(0.2754) + 0.064(0.1324) + 0.0387(0.0856) = 0.2814.

4. Conclusion:

The selection of Material Handling equipment involves complex decision making problems that require discerning abilities and methods to generate sound decisions. This paper has presented the AHP as a decision-making tool that allows the consideration of multiple criteria. The actual process of conducting this analysis has helped us prioritize the criteria in a manner that otherwise might not have been possible. Bounded rationality and limited cognitive processes make it impossible for the decision maker to consider all the factors involved in a complex decision making activity. Without decision support methodologies such as AHP, managers might base their decisions on only a subset of important criteria while not understanding their relative importance and interactions. As several criteria are involved in this problems, AHP is considered superior to other decision making approaches. Material transportation equipment selection is an important factor in the design of MH systems, and thus turns out to be a crucial facet in facilities planning.

When applying the AHP method in the field of MH equipment associated with an Iron and Steel Industry i.e, JSW Steel Ltd. , the most suitable alternative was found to be an Industrial Truck. According to the selected criteria, it was also found that a Conveyor System is the next best alternative and an Overhead Crane was evaluated to the the least suitable option.

Even though we have examined an Iron and Steel Industry, this approach can be applied to any manufacturing firm where a best alternative has to chosen against multiple criteria.

IOP Conf. Series: Materials Science and Engineering **197** (2017) 012060 doi:10.1088/1757-899X/197/1/012060

References:

[1] Saaty.T.L, The Analytic Hierarchy Process, McGraw-Hill, *RWS Publications*, Pittsburgh, PA,1990.

[2] Putrus.P, Accounting for Intangibles in Integrated Manufacturing (nonfinancial justification based on Analytic Hierarchy Process).

[3] Boucher T.O. and E.L. McStravic, "Multi-attribute Evaluation Within a Present Value Framework and its Relation to the Analytic Hierarchy Process," *The Engineering Economist*. 1991, 37, 55-71.

[4] Wabalickis R.N, "Justification of FMS With the Analytic Hierarchy Process," *Journal of Manufacturing Systems*, 1988, 17, 175-182.

[5] Cambron K.E. and Evans G.W, "Layout Design Using the Analytic Hierarchy Process," *Computers & Industrial Engineering*. 1991, 20, 221-229.

[6] Wang, L., and T. Raz, "Analytic Hierarchy Process Based on Data Flow Problem," *Computers & Industrial Engineering*. 20, 355-365, 1991.

[7] Watson SR, Freeling ANS, Assessing attribute weights, *Omega* 1982, 10(6).

[8] Belton V, Gear T, The legitimacy of rank reversal- a comment, *Omega* 1958, 13(3):143-4.

[9] Dyer JS, Remarks on the analytic hierarchy process, *Management Science*, 1990, 3:249-58.

[10] James M. Apple. Plant Layout and Material Handling.3rd edn.The Ronald Press Co,2007.

IOP Conf. Series: Materials Science and Engineering **197** (2017) 012061 doi:10.1088/1757-899X/197/1/012061

Evaluation and Selection of Predicaments in Pharmaceutical Supply Chain using AHP under Fuzzy Environment

E.V.S. Nikhil[1*], V. Sai Ram[2], V. Charan Yadav[3], Kalla Kiran Kumar[4] and Dega Nagaraju[5]

[1,2,3,4]B.Tech Mechanical Engineering, VIT University, Vellore-632014, Tamil Nadu, India
[5]School of Mechanical Engineering, VIT University, Vellore-632014, Tamil Nadu, India

Email: nikhilepuri18@gmail.com

Abstract. Pharmaceutical sector plays an important role in the medical and health system. Due to the globalization of the business, increasing demand and supply for drugs, growing regulatory requirements, all stages of the pharmaceutical supply chain (SC) are facing numerous predicaments. The traditional way of selection and evaluation of these predicaments is customarily done using technical information. This approach lacks the ability to project the burning issue that to be addressed first. Hence, a computing method of selecting the crucial issue from the existing issues is essential in a pharmaceutical supply chain. This paper considers seven different predicaments as criteria and five sub-criteria under each main predicament of a pharmaceutical supply chain. The intention of this project is to manifest the process of assessing and selecting the issue that to be addressed first by using multi-criteria decision making technique (MCDM), i.e., fuzzy analytical hierarchy process (FAHP). The criteria and sub-criteria weights are calculated and priority assessment of the predicaments is done by using FAHP. Finally, from the findings of this work, the predicaments are ranked from most important to least important. This gives information to the decision maker (DM) to solve the issue that is affecting the SC the most with respect to the others.

1. Introduction

Pharmaceuticals are the most significant elements of present day scenario. For the Product to reach the customer, the product has to be subjected to many processes such as manufacturing, packing and transporting from the industries. This sequence of processes is known as supply chain (SC). In order to deliver the product in a good quality and in fail safe condition to the customer, one has to check the SC and study the components in detail. In this sequence of processes, the industry faces a lot of predicaments to achieve high productivity.

These pharmaceuticals are the people's life and death matter, so manufacturers are corroborating quality and reliability in each and every part of the SC. If a healthy SC is to be obtained, the factors which are affecting the SC should be taken care with proper guidance. Many authors addressed the pharmaceutical SC's. Mousazadeh et al. [1] developed a bi-objective mixed integer linear programming (BOMILP) for the design of pharmaceutical SC, explaining strategic issues like opening of an industry and the distribution centres. Weraikat et al. [2] investigated the pharmaceutical reverse SC using Lagrangian relaxation method. A bonus sharing technique is also proposed based on each investment. Lemmens et al. [3] reviewed the SC network model to find the relevancy of the network design to the issues pertaining to it. They also addressed the scepticism in the reviewed literature and embraced the variability in lead time and demand. Dua Weraikat [4] explored the role of providing the incentives to customers and to improve sustainability for a real pharmaceutical reverse supply chain (RSC). A technique is proposed to share the RCV's saving among the producer and the 3PL companies. Sometimes the factors which are affecting the SC might be skimmed off when the demand is high. So these problems are to be listed according to the effect they make on the SC. To achieve these requirements, MCDM techniques are taken into consideration. From these set of techniques in MCDM, the AHP method is one. Karanik [5] worked on how to choose best alternative by using Analytic hierarchical process (AHP). A new way is developed to reconstruct the inconsistent matrix. Luthra et al.

[6] investigated the barriers in the SC to implement sustainable consumption and production (SCP) and found 15 hindrances using AHP. Sivakumar et al. [7] adopted AHP technique based on pairwise comparisons and designed a framework to perform sensitivity analysis which is a measure to alleviate subjectivity of the judgments. Before making a common decision the Consensus reaching models are mainly used to make decision making. Dong and Cooper [8] discussed the peer-to-peer dynamic adaptive consensus and within the group, they adopted AHP A Markov chain method to determine the decision makers 'weights of importance with respect to other group members. Dağdeviren et al. [9] proceeded with the AHP and the Technique for Order Performance by Similarity to Ideal Solution (TOPSIS) for selecting the best and optimal weapon in Defence systems. To resolve the problems that were arising frequently in the present information and to do reduce the bias for the fuzziness in human judgment and preference, the fuzzy set theory is introduced to treat ill-defined MCDM problems. Tavana et al. [10] identified admissible criteria and sub-criteria by using SWOT analysis and then Intuitionistic FAHP is embraced to find the weights for criteria and sub-criteria in Reverse Logistics outsourcing.

Wang and Chin [11] explained the drawback of FAHP process, fuzzy preference programming (FPP) and then proposed a new technique: Logarithmic fuzzy preference programming (LFPP) for FAHP priority derivation. Authors concluded by showing that LFPP methodology will produce unique optimal priority vector for any fuzzy pairwise comparison matrix. Gao et al [12] assessed the tunnel fire risk assessment of subway in hardware facilities by combining the fuzzy consistent matrix and AHP. Kusumawardani and Agintiara [13] investigated the use of the FAHP-TOPSIS method to the difficulty of human resource selection. Russo and Camanho [14] unfolded a systematic review of the literature on the real cases that applied AHP to evaluate how the criteria are being defined and measured. Authors pointed that the three main AHP methodology functions are: structuring complexity, measurement, and synthesis. Kong and Liu [15] evaluated the success on E-Commerce and found the factors of success in E-Commerce by applying FAHP process. Mikhailov and Tsvetinov [16] addressed uncertainty and imprecision in the service process. They applied Analytic Hierarchy Process to prioritize the best alternatives and offers for decision makers. Liu et al. [17] adopted improved fuzzy AHP process to make a decision on the wind power integration schemes. Various MCDM (Multiple Criteria Decision Making) techniques are used in the literature which includes several algorithms such as AHP, Fuzzy AHP, TOPSIS, and Fuzzy TOPSIS etc.

In the above-mentioned techniques, Fuzzy AHP is used in the present study to evaluate the issue that is most affecting the pharmaceutical supply chain.

The remnants of the paper are organized as follows. Evaluation Criteria, Methodology, and Fuzzy AHP process are explained in section 3. Then in section 4 Numerical Illustration is carried out and the results are explained. Finally, section 5 concludes the paper.

2. Constructing the Evaluation Criteria

In this section, we describe the required evaluation criteria used for predicaments selection. The necessary data were determined and Table 1 presents the final set of criteria for the evaluation of the predicaments.

Table 1. The Evaluation Criteria

Notation	Main Criteria	Sub-criteria	Notation
C_1	**Order management**	Accurately promising dates based on fulfillment planning lead times/estimates.	M1
		Responding to changing customer order delivery expectations.	M2
		Managing different rules and order management process for each customer.	M3
		Combating the rising costs related to the fulfillment.	M4
		Number of staff required to manage order entry process	M5
C_2	**Ware house management**	Inaccurate quantities.	M6
		Capacity	M7
		Damage	M8
		Product identification	M9
		Lack of training	M10
C_3	**Shortage avoidance**	Lack of advanced warning systems.	M11
		Supply/demand issues.	M12
		Manufacturing difficulties.	M13
		Shortage of raw materials.	M14
		Business/economic issues.	M15
C_4	**Temperature control**	Monitoring the high-risk products	M16

		Failures of the refrigerators.	M17
		Failures of temperature monitors.	M18
		Transportation with inbuilt coolers and refrigerators.	M19
		wrong calibration of thermometers	M20
C_5	**Shipment**	High cost	M21
		Optimal fleet management	M22
		Risk management	M23
		Leasing of vehicles	M24
		Availability of labor	M25
C_6	**Expiration**	Running of old stocks/high supply than demand.	M26
		Wastage of the products	M27
		Storage conditions.	M28
		Package conditions.	M29
		Transport conditions.	M30
C_7	**Inventory management**	Unqualified managers.	M31
		Forecast management	M32
		Lack of communication.	M33
		Having too many stocks keeping units(SKU)	M34
		Wrong denoting of the stockpiles	M35

3. Methodology

The focus is to evaluate the major issues affecting the SC and to pick the most important issue which is to be addressed first. For doing this, an example has been taken with seven main criteria and five sub-criteria under each and this will be evaluated using AHP method under Fuzzy environment to approach the evaluation. After the criteria are evaluated, the sub-criteria is addressed using the approach as mentioned above. Finally, the main issue governing the SC will be revealed.

3.1 The Fuzzy AHP Methodology

Step1. Construction of fuzzy pair-wise comparison matrix:

The fuzzy judgment matrix $A=\{a_{ij}\}$ of n criteria using pair-wise comparison is made by the use of TFNs as follows:

$$A=\begin{bmatrix} 1 & a_{12} & \cdots & a_{1n} \\ a_{21} & 1 & \cdots & a_{2n} \\ \cdots & \cdots & \cdots & \cdots \\ a_{n1} & a_{n2} & \cdots & 1 \end{bmatrix}$$

Where a_{ij} is a fuzzy triangular number.

Step2.Calculating the value of Fuzzy Synthetic Extent

The value of fuzzy synthetic extent S with respect to the i_{th} criterion is computed based on the aggregated pair-wise comparison matrix, $A=\{a_{ij}\}$, as follows

$$S_i=\sum_{j=1}^{n} a_{ij}\otimes\left[\sum_{i=1}^{n}\sum_{j=1}^{m} a_{ij}\right]^{-1}$$

Where

$$\sum_{j=1}^{m} a_{ij}=\left(\sum_{j=1}^{m} l_j, \sum_{j=1}^{m} m_j, \sum_{j=1}^{m} u_j\right) \text{ and } \sum_{i=1}^{n}\sum_{j=1}^{m} a_{ij}=\left(\sum_{i=1}^{n} l_j, \sum_{i=1}^{n} m_j, \sum_{i=1}^{n} u_j\right)$$

Step3. Approximation of fuzzy priorities

The relative preferences of one criterion when compared to others i.e. the degree of possibility are computed using Chang's method as expressed below on the basis of these fuzzy synthetic values

IOP Conf. Series: Materials Science and Engineering **197** (2017) 012061 doi:10.1088/1757-899X/197/1/012061

$$V(S_i \geq S_j) = \begin{cases} 1, & if\ m_i \geq m_j \\ \dfrac{(u_i - l_j)}{(u_i - m_i) + (m_j - l_j)}, & if\ l_j \leq u_i \\ 0, & others \end{cases},$$

where $i, j = 1, \ldots\ldots, n; j \neq i$

The relative preferences for a TFN S_i to be higher than the number of n TFNs can be expressed as $V(S_i \geq S_1, S_2, S_3, \ldots. S_k) = \min(S_i \geq S_1, S_i \geq S_2, \ldots., S_i \geq S_k) = w(S_i)$ where $k \neq i$ .Each value of w (S_i) represents the relative preference of one criterion over others or weight which is a non-fuzzy number.

Step4. Determination of Normalized Weights

The normalized weights W(S_i) will be formed in terms of a weights vector as follows:

$W = (w(S_1), w(S_2), \ldots.. w(S_n))^T$

Step5. Establish final global weights

4. Numerical Illustration

In the following section, fuzzy AHP method is proposed for selecting the burning predicament of the pharmaceutical supply chain that to be addressed first. A pair-wise comparison matrix is built by three decision makers using fuzzy linguistic variables shown in Table 2.

The synthetic fuzzy extent value for main criteria is calculated as

S1= (7.03, 9.93, 17) ⊗ (49.754, 64.6740, 95.6480)
= (0.0735, 0.1535, 0.3417)

S2= (9.2, 11.486, 15.38) ⊗ (49.754,64.6740,95.6480)
= (0.0962, 0.1776, 0.3091)

S3 = (7.65, 10.3, 16.6) ⊗ (49.754, 64.6740, 95.6480)
= (0.08, 0.1593, 0.3336)

S4= (9.288, 11.278, 14.668) ⊗ (49.754, 64.6740, 95.6480)
= (0.0971, 0.1744, 0.2948)

S5= (2.6, 3.2, 3.8) ⊗ (49.754,64.6740,95.6480)
= (0.0272, 0.0495, 0.0764)

S6= (5.63, 7.8, 13.3) ⊗ (49.754, 64.6740, 95.6480)
= (0.0589, 0.1206, 0.2673)

S7= (8.356, 10.68, 14.9) ⊗ (49.754, 64.6740, 95.6480)
= (0.0874, 0.1651, 0.2995)

Table 2. Linguistic Variables

Linguistic scale of importance	Triangular Fuzzy Numbers
Equal	(0,0,0)
Extremely low advantage	(0,0.1,0.2)
Very low advantage	(0.1,0.2,0.3)
Medium low advantage	(0.3,0.4,0.5)
Medium advantage	(0.4,0.5,0.6)
Medium high advantage	(0.5,0.6,0.7)
High advantage	(0.6,0.7,0.8)
Very high advantage	(0.7,0.8,0.9)
Extremely high advantage	(0.8,0.9,1)

Table 3. The Fuzzy Evaluation Matrix for Main Criteria

Criteria no.	1	2	3	4	5	6	7
1	(0,0,0)	(0.3,0.4,0.5)	(0.1,0.2,0.3)	(0.6,0.7,0.8)	(2.5,3.33,5)	(3.33,5,10)	(0.2,0.3,0.4)
2	(2,2.5,3.33)	(0,0,0)	(2,2.5,3.33)	(0.2,0.3,0.4)	(1.25,1.428,1.66)	(1.25,1.428,1.66)	(2.5,3.33,5)
3	(3.33,5,10)	(0.3,0.4,0.5)	(0,0,0)	(0.5,0.6,0.7)	(1.66,2,2.5)	(1.66,2,2.5)	(0.2,0.3,0.4)
4	(1.25,1.428,1.66)	(2.5,3.33,5)	(1.428,1.66,2)	(0,0,0)	(1,1.11,1.25)	(1.11,1.25,1.428)	(2,2.5,3.33)
5	(0.2,0.3,0.4)	(0.6,0.7,0.8)	(0.4,0.5,0.6)	(0.8,0.9,1)	(0,0,0)	(0.1,0.2,0.3)	(0.5,0.6,0.7)
6	(0.1,0.2,0.3)	(0.6,0.7,0.8)	(0.4,0.5,0.6)	(0.7,0.8,0.9)	(3.33,5,10)	(0,0,0)	(0.5,0.6,0.7)
7	(2.5,3.33,5)	(0.2,0.3,0.4)	(2.5,3.33,5)	(0.3,0.4,0.5)	(1.428,1.66,2)	(1.428,1.66,2)	(0,0,0)

Table 4: The Sub Criteria fuzzy evaluation matrix with respect to C_1

Criteria	M1	M2	M3	M4	M5	Row total	Normalised Weights
M1	(0,0,0)	(0.3,0.4,0.5)	(0.5,0.6,0.7)	(0.6,0.7,0.8)	(0.8,0.9,1)	(2.2,2.6,3)	0
M2	(2,2.5,3.33)	(0,0,0)	(0.4,0.5,0.6)	(0.6,0.7,0.8)	(0.7,0.8,0.9)	(3.7,4.5,5.67)	0.169298
M3	(1.42,1.66,2)	(1.66,2,2.5)	(0,0,0)	(0.4,0.5,0.6)	(0.5,0.6,0.7)	(3.98,4.76,5.8)	0.184426
M4	(1.25,1.42,1.66)	(1.25,1.42,1.66)	(1.66,2,2.5)	(0,0,0)	(0.2,0.3,0.4)	(4.36,5.14,9.67)	0.280859
M5	(1,1.11,1.25)	(1.11,1.25,1.42)	(1.42,1.66,2)	(2.5,3.33,5)	(0,0,0)	(6.03,7.35,9.67)	0.365417

These values are compared as per the above mentioned methodology and the V values are obtained:

$V(SC_1 \geq SC_2)$= 0.8337 $V(SC_1 \geq SC_3)$= 0.9619
$V(SC_1 \geq SC_4)$= 0.8512 $V(SC_1 \geq SC_5)$=1
$V(SC_1 \geq SC_6)$=1 $V(SC_1 \geq SC_7)$= 0.9202

$V(SC_2 \geq SC_1)$=1 $V(SC_2 \geq SC_3)$= 1
$V(SC_2 \geq SC_4)$=1 $V(SC_2 \geq SC_5)$=1
$V(SC_2 \geq SC_6)$=1 $V(SC_2 \geq SC_7)$=1

$V(SC_3 \geq SC_1)$=1 $V(SC_3 \geq SC_2)$= 0.8619
$V(SC_3 \geq SC_4)$= 0.882 $V(SC_3 \geq SC_5)$=1
$V(SC_3 \geq SC_6)$=1 $V(SC_3 \geq SC_7)$= 0.9559

$V(SC_4 \geq SC_1)$=1 $V(SC_4 \geq SC_2)$= 0.9634
$V(SC_4 \geq SC_3)$=1 $V(SC_4 \geq SC_5)$=1
$V(SC_4 \geq SC_6)$=1 $V(SC_4 \geq SC_7)$=1

$V(SC_5 \geq SC_1)$=1 $V(SC_5 \geq SC_2)$=1
$V(SC_5 \geq SC_3)$= 0.425 $V(SC_5 \geq SC_4)$= 0.225
$V(SC_5 \geq SC_6)$= 0.1256 $V(SC_5 \geq SC_7)$= 0.196

$V(SC_6 \geq SC_1)$= 0.7464 $V(SC_6 \geq SC_2)$= 0.5266
$V(SC_6 \geq SC_3)$= 0.6933 $V(SC_6 \geq SC_4)$= 0.5356
$V(SC_6 \geq SC_5)$= 1 $V(SC_6 \geq SC_7)$= 0.6319

$V(SC_7 \geq SC_1)$=1 $V(SC_7 \geq SC_2)$= 0.8761
$V(SC_7 \geq SC_3)$=1 $V(SC_7 \geq SC_4)$= 0.9037
$V(SC_7 \geq SC_5)$=1 $V(SC_7 \geq SC_6)$=1

Then priority weights are calculated as per the methodology
$d'(C_1)$ =min (0.8337,0.9619,0.8512,1,1,0.9202) = 0.8337
$d'(C_2)$ =min (1,1,1,1,1,1) = 1
$d'(C_3)$ =min (1,0.8619,0.882,1,1,0.9559) = 0.8619
$d'(C_4)$ =min (1,0.9634,1,1,1,1) = 0.9634
d'(C5) =min (1,1,0.425,0.225,0.1256,0.196) = 0.1256
$d'(C_6)$ = min (0.7464,0.5266,0.6933,0.5356,1,0.6319) = 0.5266
$d'(C_7)$ = min (1,0.8761,1,0.9037,1,1) = 0.8761
Priority weights form W'= (0.8337, 1, 0.8619, 0.9634, 0.1256, 0.5266, 0.8761)
These weights are to be normalized and the new weights are projected as

W' = (0.160719, 0.192779, 0.166156, 0.185723, 0.024213, 0.101517, 0.168893)
After the calculation of the priority weights of the main criteria, we go for the calculation of the priority weights of the sub criteria or alternatives w.r.t each main criteria are determined.After the evaluation of main criteria we go for the sub-criteria with same methodology:Considering order management first, the evaluation of the sub-criteria is shown below. The normalized weights are calculated using fuzzy AHP. Normalized Weights for all the sub-criteria with respect to their corresponding main criteria is tabulated in Table 5.

Table 5. Normalized weights of sub-criteria

Under C_1	Under C_2	Under C_3	Under C_4	Under C_5	Under C_6	Under C_7
0	0	0	0.008467	0.008723	0.012159	0
0.169298	0.255825	0.274003	0.28828	0.078	0.16426	0.070715
0.184426	0.25494	0.269865	0.342783	0.3225	0.256991	0.233963
0.280859	0.260356	0.193144	0.15744	0	0.270039	0.311208
0.365417	0.228879	0.262988	0.20303	0.5869	0.296551	0.384113

Now from all these normalized weights we calculate the overall global weights.

Table 6. Overall Global Weights

Weights of main criteria	Sub-criteria weights						Overall weights
0.160719	w.r.t C1	0	0.169298	0.1844	0.2808	0.365417	0.160705
0.192779	w.r.t C2	0	0.255825	0.25494	0.2603	0.228879	0.192768
0.166156	w.r.t C3	0	0.274003	0.269865	0.193144	0.262988	0.166156
0.185723	w.r.t C4	0.0084	0.28828	0.342783	0.15744	0.203	0.185705
0.024213	w.r.t C5	0.008723	0.078	0.3225	0	0.5869	0.04119
0.101517	w.r.t C6	0.012159	0.16426	0.256991	0.270039	0.296551	0.101517
0.168893	w.r.t C7	0	0.070715	0.233963	0.311208	0.384113	0.168893
						Total	1.000000

From Table 6. From the overall weights calculated, the sequence of the criteria that we can inference is
C2 > C4 > C7 > C3 > C1 > C6 > C5.
In this manner, the quality engineer can approach the issues of the pharmaceutical supply chain according to the above-obtained order.

5. Conclusion

In this paper, the evaluation and selection of predicaments that are majorly affecting the SC of pharmaceutical industries are elevated. The fuzzy AHP methodology is implemented in order to know the order in which the issues affecting the pharmaceutical supply chain are to be addressed. This approach projects the burning issue that to be addressed first. Then the numerical illustration of seven main criteria and five sub-criteria under each main criteria is taken into consideration and the methodology is applied to it. From the above illustration, it is clear that the main governing issue for SC's in pharmaceutical industries is WAREHOUSE MANAGEMENT. Here in Fuzzy AHP the usage of criterion weights and the sub-criteria weights makes the problem more accurate and reliable. This helps the decision maker (DM) to find and correct the issue that affects the SC the most than the rest. By approaching the Ware House Management predicament first the entire SC can be more profitable than addressing other issues first.

IOP Conf. Series: Materials Science and Engineering **197** (2017) 012061 doi:10.1088/1757-899X/197/1/012061

References

[1] Mousazadeh, Mohammad, S. Ali Torabi, and B. Zahiri. "A robust possibilistic programming approach for pharmaceutical supply chain network design." *Computers & Chemical Engineering* 82 (2015): 115-128.

[2] Weraikat, Dua, Masoumeh Kazemi Zanjani, and Nadia Lehoux. "Coordinating a green reverse supply chain in pharmaceutical sector by negotiation." *Computers & Industrial Engineering* 93 (2016): 67-77.

[3] Lemmens, Stef, et al. "A review of integrated supply chain network design models: Key issues for vaccine supply chains." *Chemical Engineering Research and Design* 109 (2016): 366-384.

[4] Weraikat, Dua, Masoumeh Kazemi Zanjani, and Nadia Lehoux. "Two-echelon pharmaceutical reverse supply chain coordination with customers incentives." *International Journal of Production Economics* 176 (2016): 41-52.

[5] Karanik, Marcelo, et al. "Reconstruction methods for AHP pairwise matrices: How reliable are they?." *Applied Mathematics and Computation* 279 (2016): 103-124.

[6] Luthra, Sunil, et al. "Using AHP to evaluate barriers in adopting sustainable consumption and production initiatives in a supply chain." *International Journal of Production Economics* 181 (2016): 342-349.

[7] Sivakumar, Ramakrishnan, Devika Kannan, and Palzha Murugesan. "Green vendor evaluation and selection using AHP and Taguchi loss functions in production outsourcing in mining industry." *Resources Policy* 46 (2015): 64-75.

[8] Dong, Qingxing, and Orrin Cooper. "A peer-to-peer dynamic adaptive consensus reaching model for the group AHP decision making." *European Journal of Operational Research* 250.2 (2016): 521-530.

[9] Dağdeviren, Metin, Serkan Yavuz, and Nevzat Kılınç. "Weapon selection using the AHP and TOPSIS methods under fuzzy environment." *Expert Systems with Applications* 36.4 (2009): 8143-8151.

[10] Tavana, Madjid, et al. "An integrated intuitionistic fuzzy AHP and SWOT method for outsourcing reverse logistics." *Applied Soft Computing* 40 (2016): 544-557.

[11] Wang, Ying-Ming, and Kwai-Sang Chin. "Fuzzy analytic hierarchy process: A logarithmic fuzzy preference programming methodology." *International Journal of Approximate Reasoning* 52.4 (2011): 541-553.

[12] Gao, Jun-peng, et al. "Application of the model based on fuzzy consistent matrix and AHP in the assessment of fire risk of subway tunnel." *Procedia engineering* 71 (2014): 591-596.

[13] Kusumawardani, Renny Pradina, and Mayangsekar Agintiara. "Application of Fuzzy AHP-TOPSIS Method for Decision Making in Human Resource Manager Selection Process." *Procedia Computer Science* 72 (2015): 638-646.

[14] de FSM Russo, Rosaria, and Roberto Camanho. "Criteria in AHP: a systematic review of literature." *Procedia Computer Science* 55 (2015): 1123-1132.

[15] Kong, Feng, and Hongyan Liu. "Applying fuzzy analytic hierarchy process to evaluate success factors of e-commerce." *International Journal of Information and Systems Sciences* 1.3-4 (2005): 406-412.

[16] Mikhailov, Ludmil, and Petco Tsvetinov. "Evaluation of services using a fuzzy analytic hierarchy process." *Applied Soft Computing* 5.1 (2004): 23-33.

[17] Liu, Siyuan, et al. "A comprehensive decision-making method for wind power integration projects based on improved fuzzy AHP." *Energy Procedia* 14 (2012): 937-942.

IOP Conf. Series: Materials Science and Engineering **197** (2017) 012062 doi:10.1088/1757-899X/197/1/012062

Reduction of Defects in Al-6061 Friction Stir Welding and Verified by Radiography

D Kumaravel[1], Dr V K Bupesh Raja[2], Chakravarthy Potnuru[3], Navakanth Polina[4]

[1-4]Sathyabama University, Chennai, India

E-mail: kevinkumaravel@gmail.com

Abstract. Friction Stir Welding is a new innovating process of joining of two work pieces. It is an relatively a new joining process and highly useful in welding method, which can produce high strength weld without using any toxic materials like electrodes. In this method, weld is obtained by frictional produced between shoulder and work piece [1, 2]. Main parameters which are to be considered for FSW are spindle speed and feed rate. By providing suitable parameter during welding defects will not be occurring. Also, FSW is an eco-friendly process because there is no fumes production and no filler material. To get high quality of weld, then high heat should be generated. In this paper, Al-6061 material is welded by H-13 tool with different parameters and quality of weld is examined by using a non destructive testing method called Radiography.

1. Introduction

Friction Stir Welding is a process of joining two surfaces using a third body tool. Friction Stir welding process was invented and experimentally proved by The Welding Institute (TWI) UK in December 1991. TWI filed successfully patents in Europe, USA, Japan and Australia. Since, FSW invention has received worldwide new attention. FSW has been proved to be an effective process for welding aluminium, brass, copper and also for other metals which are having low melting temperature materials. The latest study in FSW has been targeted to at expanding the usefulness of this procedure in the metals having high melting temperature materials also like carbon and stainless steels. Major parameters which affect the weld quality are spindle speed and tool feed rate [3]. These parameters are responsible for heat generation and intermixing of work pieces. Internal defects produced due to improper welding are not verified by normal method. These defects can be detected by a special method called radiography. Radiography is one of the non-destructive techniques which can find defects without creating any damage. It is a technique used to view internal structure of a material [4]. Aim of this paper is to prove that by proper spindle speed and feed rate. Thus, the defect can be reduced.

Figure 1. Friction stir welding

2. Material

2.1 Material

Al-6061 has good strength, good weldability and excellent corrosion resistance to atmospheric conditions and salt. Used for marine fittings, electrical fittings, valves and cycle parts. Known for its high tensile strength and intensive texture. It has very good corrosion resistance and very good weldability even though decreased strength in the weld area . Hence it not preferred for very complex cross sections. Al-6061 is generally utilized as a part of designing structures and segments where light weight or erosion resistance is required.

Table 1. Physical property of Al- 6061

Physical Property	Value
Density	2.7 gm / cm^3
Melting Point	Approximately 580^{0}C
Modulus of Elasticity	70 – 80 GPa
Poisson's Ratio	0.33

2.2.Microstructure of Al-6061

The microstructure shows the grains of Mg_2Si in aluminium solid solution. The sample is in the longitudinal direction as it is evident from the parallel grain flow. The grains size of the particles is ranging from 10 – 15 microns. Some direction of forming and its grain size is ranging from 30 to 40 microns [5, 6].

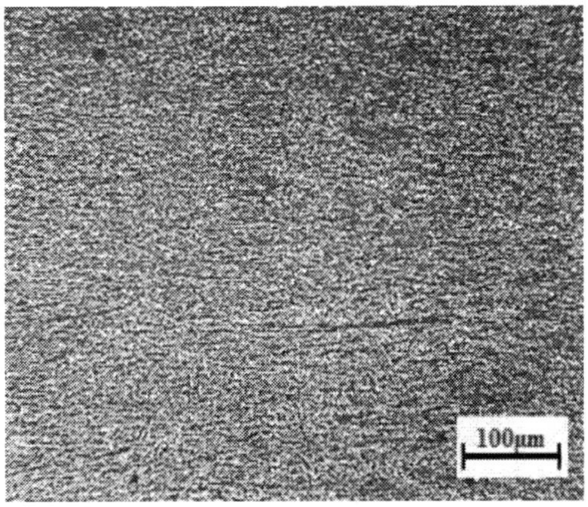

Figure 2. Microstructure of Al-6061

3.Experimentation

Friction Stir weld between two Al-6061 material with the dimensions of 300 X 100 X 3 mm were successfully produced in vertical milling machine. To weld these aluminium work pieces, H-13 tool with cylindrical probe with diameter 3 mm and with shoulder diameter of 10 mm was used. Friction Stir welds were produced by varying the rotational speed between 1000 and 2000 rpm and feed rate of 15 and 20 mm/min. Tool speed 1000, 1400 and 2000 rpm were employed while 15 and 20 mm/min were the feed rates. These feed rates were considered as representing the low high setting feed rates respectively [7]. To get high hardness for a tool, it is subjected to nitriding before weld. After weld,

IOP Conf. Series: Materials Science and Engineering **197** (2017) 012062 doi:10.1088/1757-899X/197/1/012062

the material was inspected by X-ray radiography. Different weld parameters were used in this method is given in table 2.

Table 2. Weld Parameter

Work piece	Speed	Feed
1	1000	15
2	1000	20
3	1400	15
4	1400	20
5	2000	15
6	2000	20

4. Testing

4.1 Visual inspection

In order to reduce the defects on friction stir welding, we need to select the correct parameter. If the spindle speed increases, heat generated between tool and work piece get increased. This operation is done, in order to prove that quality of weld get increased if we increase the speed value and decrease feed value. During first process, spindle is made to rotate at 1000 rpm and tool is made to move at 15 mm/min. Since the tool speed is less, heat created between the tool and work piece is less. Therefore, due to in sufficient heat, the work piece could not obtain good weld. From the table, it is shown that friction stir welding at 1000 rpm and feed of 15 mm/min produced less quality weld, due to in sufficient heat produced. At 1400 rpm and feed 15 mm/min, tool speed is little high and heat created between the tool and work piece will get increase. As compare with before, tool speed increased and tool feed reduced. Hence, heat generation increased and quality of weld improved further. During welding of 2000 rpm and feed 15 mm/min, the tool and work piece is subjected to high friction to obtain good quality weld.

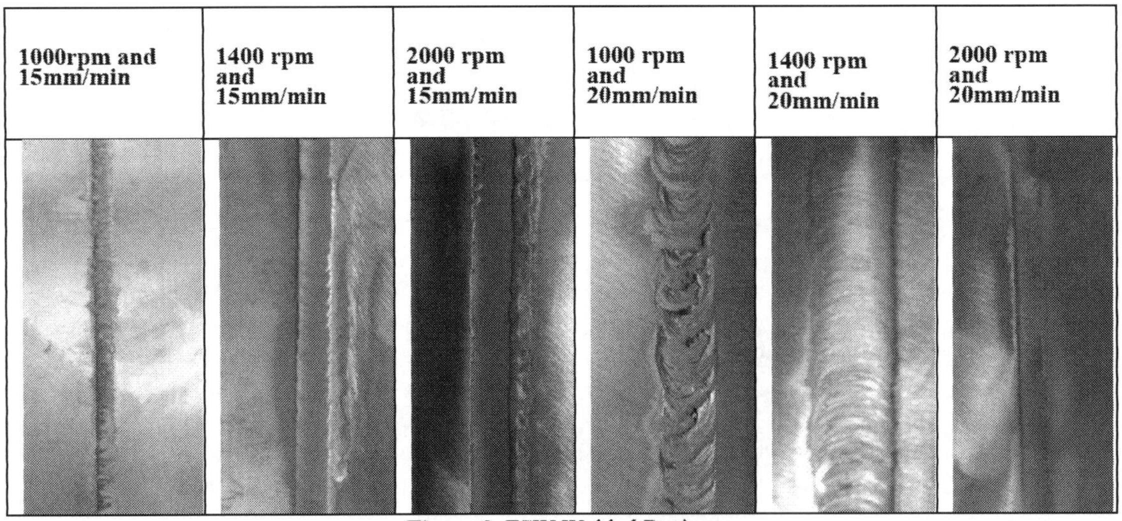

Figure 3. FSW Welded Region

IOP Conf. Series: Materials Science and Engineering **197** (2017) 012062 doi:10.1088/1757-899X/197/1/012062

4.2 Radiography inspection

Radiography is an imaging technique uses electromagnetic radiation other than visible light especially X-rays to view the internal structure of a non-uniformly composed and opaque object. To obtain the radiography image of the weld ,X-ray is projected toward the object. Internal defects which are not predicted by visual inspection can be obtained by using radiography method [9].

1400 rpm and feed 20 mm/min	2000 rpm and feed 15 mm/min	2000 rpm and feed 20 mm/min

5. Result

It is apparent that increasing the feed rate effectively increase the amounts of defects formed during welding, it can be attributed to the fact that less heat is generated at high feed rate which in turn is not enough to consolidate the weld with exception of the weld produced at 1000 rpm and 20 mm/min which can be interpreted that the high rotational speed at 2000 rpm compensated for the high feed rate of 20 mm/min and hence assisted in consolidating the weld produced at this setting. At 2000 rpm and 15 mm/min has been found to be the optimum tool rotational speed for defect free weld for Al-6061 work piece of 3 mm thickness.

6. Conclusion

Non Destructive testing techniques through X-ray radiography were successfully conducted on welds produced at various parameter combinations. With this testing result of various work piece it is found that, if the spindle speed increases more heat will produce on the work piece, which make the weld with good quality. Though, high speed available on the tool, increasing feed rate will compensate the heat generation. Hence, by selecting correct speed and feed rate we can reduce the defect produced during friction stir welding.

IOP Conf. Series: Materials Science and Engineering **197** (2017) 012062 doi:10.1088/1757-899X/197/1/012062

7. Reference

[1] Mishra R S, Friction stir welding and processing, *Material science and engineering*, (2005) 1-78.

[2] Starink M J, Deschamps A, The strength of friction stir welded and friction stir processed aluminium alloys, Scripta materialia (2008) 337-382.

[3] Lombard H, Hattingh D G, Optimizing FSW process parameter to minimize defects and maximize fatigue life in 5083-H321 aluminium alloy, *Engineering fracture mechanics*, (2008) 341-354.

[4] Sergei Yu.Tarasov, Valery E.Rubtsov, Radiography detection of defect in friction stir welding on aluminium alloy.

[5] Jian-Qing Su, Tracy Nelson W, Microsructure evolution during FSW of high strength aluminium alloys, *Material science and engineering* (2005) 277-286.

[6] Afrin N, Chen D L, Microstructure and tensile properties of friction stir welded AZ31B magnesium alloy, *Material science and engineering*, (2008)179-186.

[7] Sato Y S, Arkom P, Effect of microstructure on properties of friction stir welded inconel alloy 600, *Material science and engineering*, (2008) 250-258.

[8] Esther.T, Andrian.C.S, Non-destructive testing of dissimilar friction stir welds, World congress on engineering (2012) 1-5.

[9] Saravannan T, Lahiri B, Non destructive evaluation of friction stir welded joints by X-ray radiography and infra red thermography, *Precidia Engineering*, (2014) 469-475.

[10] Won-Bae Lee, Chang –Young Lee, Microstructural investigation of friction stir welded pure titanium, *Material science*, (2005) 3315-3318.

IOP Conf. Series: Materials Science and Engineering **197** (2017) 012063 doi:10.1088/1757-899X/197/1/012063

EXPERIMENTAL INVESTIGATION AND TAGUCHI OPTIMISATION OF DRILLING PROPERTIES ON TEAK WOOD REINFORCED EPOXY RESIN

J.Lilly Mercy[1*], Mohamed Shaqir Tanvir[2], K.Swaroopkanth[3]

[1,2,3]Department of Mechanical and Production Engineering, Sathyabama University

[1*]Email: lillymercy.j@gmail.comPh: +91-9940319604

Abstract. The drilling properties of teak reinforced epoxy resin composite are explored in this work. The thrust force and temperature during the drilling process was found and optimised. Nine holes were drilled in accordance with L_9 orthogonal array on Medium Density Fibre board and Teak wood reinforced epoxy composite board and the thrust force and temperature induced during drilling is measured. Drilling experiments were conducted using CNC Vertical drilling machine and the thrust force was measured using dynamometer and temperature using infra-red thermometer. The experiments were conducted with varying levels of spindle speed and feed rate and optimised using Taguchi optimisation. It was observed that higher thrust and temperature were observed while drilling teak wood composite due to the high mechanical strength of teak wood. The hard and brittle properties of the resin seemed to be more pronounced in the composite. The experimental results were optimised to find the best combination of input parameters for reduced thrust and temperature. When speed increases, thrust force decreases and temperature increases. When feed increases, thrust force increases and temperature decreases. Experimental findings encouragesto use teak wood reinforced epoxy resin as a substitute for the traditionally used Medium Density Fibre Board. The percentage of mixing of teak dust can be increased with various resin combinations to arrive at the best suitable combination for obtaining optimal mechanical properties.

***Keywords*:**Wood Composite, Taguchi Optimisation, Drilling properties, Thrust force, Temperature.

1. Introduction

Wood composites find its applications in interior decoration industry which involves furniture, kitchen cabinets, office tables etc. During the assembly process of joining the panels together for these applications, drilling is an unavoidable machining operation. Often drilled holes exhibit poor surface finish, delamination of outer layer and improper hole diameter which results in proper fitting of bolts and screws during assembly. Due to stress developed in the misaligned joints the drilled holes enlarges leading to unbalanced assembly of wooden parts. The problems encountered in the quality of drilled holes can be reduced to a large extent by controlling drilling parameters.Feed rate and thrust force were found to be the critical parameters contributing to the delamination of the composite. The influence of cutting parameters on the torque and cutting forces during drilling of GFRP laminate were

analysed and it was found that the drill size as well as the speed of the drill bit had major influence on the thrust force. A number of experiments were conducted on the composite by varying the parameters and the results were then optimised by Taguchi's design of experiment method. Composites consisting wood fibers generally include added teak wood, maple wood, oak wood, cane strawand bagasse. Teak wood was added to manufacture the wood plastic which not only rendered a natural wood like appearance but also imparted the sturdiness of the teak. Medium density fiber board is the commonly engineered wood board in the present times.
Therefore this experiment aims at the study of characteristics of teak wood specimen and comparison to that of a medium density fiber board, procured from the market.

2. Materials and Methods

2.1 Fabrication of teak wood composite

Teak dust and chips were procured from the local saw mill. It is mixed in the ratio of 1:5 with LY556 epoxy resin at room temperature manufactured and supplied by Araldite. Mechanical stirrer is used to mixed wood and epoxy composites till the resin forms a uniform viscous liquid. The hardener HY951 is then added to the resin at the ratio of 1:10. Aluminium moulding die 300x300mm is used where the resin mixed with teak dust is poured. This mould was kept in hot press moulding machine at a pressure is 1MPA and temperature was maintained at 60° C. Figure 1 shows the fabricated Teak wood composite plate and Figure 2 shows the MDF board purchased.

Figure 1. Fabricated Teak wood composite plate

Figure 2. Purchased Medium density fibre board

2.2 Input parameters

The drilling parameters spindle speed and the feed rate of the drill is included in this study. Previous studies explain the dominant impact of these 2 parameters during the drilling process. The factors and levels taken for the study are shown in Table1.

Table1. Experimental factors and levels

Factors	**Levels**		
	1	**2**	**3**
Speed (rpm)	600	852	1260
Feed (mm/rev)	0.05	0.07	0.09

The drilling experiment was conducted using a CNC drilling machine having spindle speed range of 60 to 6000 rpm and a table of size 710*400mm was manufactured and supplied by Bharat FrietzWemer. The feed rate was set at three levels in arithmetic progression starting from 0.05rpm and having a common difference of 0.02rpm and the recommended speed corresponding for each level was taken for experimentation. The CNC drilling machine used for the experiment is shown below in figure 3.

Figure3. CNC drilling machine used for the study

2.3 Measured Output responses
The output responses measured are Thrust force and Temperature during the drilling process. Thrust represents the force at which the drill bit forces through the work piece. Hard and tough materials exhibit higher resistance to the cutting action and hence the thrust increases. However, thrust force depends on the nature of drill bit, the angle of cutting lip etc. The major 2 factors of the drilling process, the speed and feed rate of the drill on the Thrust force evolved is assessed in this study. The temperature generated during the drilling process is measured using an infrared thermometer, which measures the temperature through infra-red radiations. The Thrust force and Temperature needs to be low to produce a neat drilling process.

2.4 Design of Experiments
Conventional method of conducting experiments by varying one parameter at a time leads to many experiments which are costly and time consuming. Hence Taguchi Design of experiments was considered to reduce the number of experiments, but at the same time not lose the accuracy in getting reliable conclusions[11]. In the present study considering 2 factors and 3 levels, Taguchi's L_9 orthogonal array was chosen and hence 9 drilling experiments were conducted varying the speed and feed rate of the drill.

2.5 Experimental results

The results obtained from the drilling experiments by drilling the fabricated teak wood composite and the purchased MDF board are shown in Table 2. The MDF board after drilling is shown in Figure 4 and the Teak wood board after drilling is shown in Figure 5.

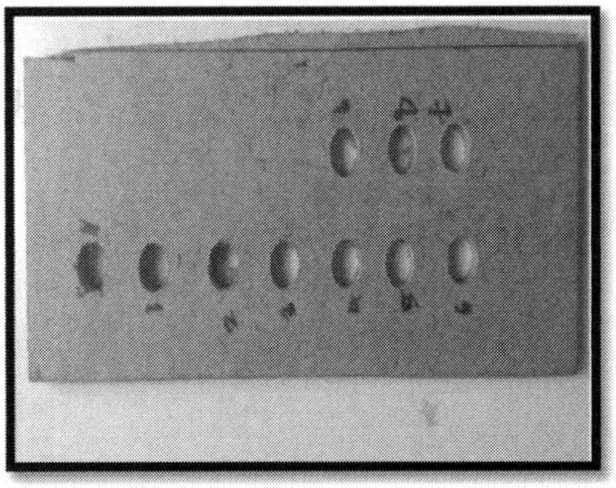

Figure 4. Drilled MDF wood board

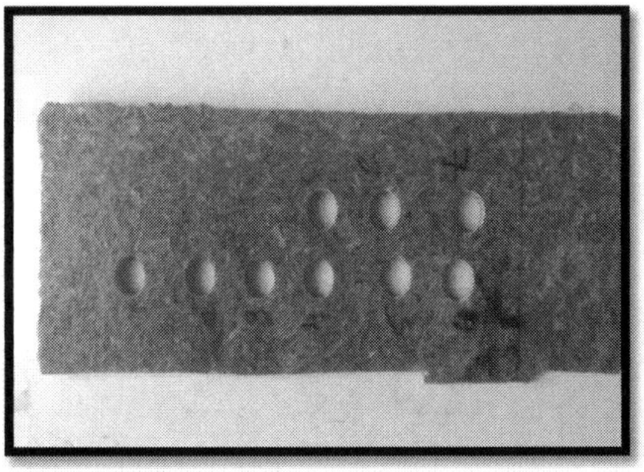

Figure 5. Drilled Teak wood board

Table2. Experimental results on drilling MDF and Teak wood composite

Experiment No	Input Parameters		Output Responses			
			Teak wood Composite		MDF wood board	
	Speed(rpm)	Feed(mm/rev)	Thrust Force (N)	Temperature (C)	Thrust Force (N)	Temperature (C)
1	600	0.05	69.55	32.35	31.86	33.16
2	600	0.07	78.4	31.14	47.42	32.37
3	600	0.09	100.4	30.79	56.76	31.54
4	852	0.05	58.87	33.21	29.02	33.71
5	852	0.07	61.48	32.47	37.26	33.09
6	852	0.09	72.4	31.72	38.86	32.27

7	1260	0.05	38.22	33.83	23.38	34.04
8	1260	0.07	53.56	33.10	26.26	33.59
9	1260	0.09	61.16	32.67	32.55	32.62

3. Taguchi Optimisation of Output responses

3.1 Response table

The responses for the individual input parameters were analysed for its effect on the thrust force and temperature. Table 3 shows the response table for each level of speed and feed rate on the output responses.

Table3. Response Table for Thrust Force and Temperature

Levels	Thrust Force				Temperature			
	Teak wood board		MDF wood board		Teak wood board		MDF wood board	
	Speed	Feed	Speed	Feed	Speed	Feed	Speed	Feed
1	82.7	55.54	45.34	28.08	31.42	33.3	32.35	33.63
2	64.25	64.48	35.04	36.98	32.46	32.23	33.02	33.01
3	50.98	77.98	27.39	42.72	33.2	31.72	33.41	32.14

3.2 Discussion

From the response table shown in table 3, the following conclusions can be drawn. When the speed increases, the Thrust force decreases by level and when the feed increases the thrust force also increases. This is because higher speed makes the cutting lips to pass through frequently over the periphery of the hole, making the cutting process smooth and easier[12]. Increase in thrust force due to increase in feed rate is due to the fast progress of drill bit into the work piece within a short span of time[13]. The thrust force value of teak wood board is high compared to MDF board. We can infer that teak wood board is stronger than MDF wood board offering high resistance during drilling. In case of temperature, when the speed increases, the temperature also increases whereas when the feed increases the temperature decreases. The increase in temperature due to increase in speed can be attributed to the fact that the friction between the cutting surfaces is more as the frequency of the lips touching the wood surface is more. More feed facilitates deeper penetration of the drill bit into the workpiece, providing more space for heat dissipation. The temperature generated during drilling of teak wood board is almost the same when compared to MDF.

4. Conclusion

The teak wood composite comprising of teak wood powder and epoxy resin could be successfully fabricated. The comparative study of the drilling properties of MDF wood board and Teak wood board gives the inference that Teak wood board produces more Thrust Force during drilling when compared to MDF, making it less suitable for hole drilling applications. However, in the view of the strength of the wood composite, Teak wood board seems to offer better resistance to cutting process. The strength of the teak wood powder and the epoxy resin seems to pronounce the nature of the composite.

References

[1] J.PauloDavim, V.C.Clemente, Sergio Silva, Drilling investigation of MDF, Journal of Materials Processing Technology, 2008, 203(1-3), 537-541.

[2] S.Prakash, K.Palanikumar, Modeling for prediction of surface roughness in drilling MDF panels using Response Surface Methodology, Journal of Composite Materials, 2011, 45(16), 1639-1646.

[3] S.Prakash, K.Palanikumar, N.Manoharan, Optimisation of delamination factor in drilling medium-density fibre boards(MDF) using desirability-based approach, The International journal of Advanced manufacturing Technology, 2009, 45(3), 370-381.

[4] J.PauloDavim, V.N.Gaitonde, S.R.Karnik, An investigative study of delamination in drilling of medium density fibreboard(MDF) using response surface models, The International journal of Advanced manufacturing Technology, 2008, 37(1),49-57.

[5] S.Jain, D.C.H.Yang, Effects of feed rate and Chisel edge on delamination in Composites drilling, Journal of Engineering for Industry, 1993,115(4), 398-405.

[6] N.S.Mohan, A.Ramachandra, S.M.Kulkarni, Influence of process parameters on cutting force and torque during drilling of glass fiber polyester reinforced composites, Composite Structures, 2005,71(3-4), 407-413.

[7] Patil YP, Gajre B, Dusane D, Chavan S, Mishra S, Effect of maleic anhydride treatment on steam and water absorption of wood polymer composites prepared from wheat straw, cane bagasse, and teak wood sawdust using Novolac as matrix. Journal of Applied Polymer Science, 2000, 77(13), 2963–2967.

[8] KaminiSewda, S.N.Maiti, Dynamic Mechanical Properties of high density Polyethylene and teak wood flour composites, Polymer Bulletin, 2013,70(10), 2657-2674.

[9] S.Prakash, J.Lilly Mercy, K.Palanikumar, S.Ramesh, M.I.Rizwan Jamal, A.James Michael, Experimental Studies on Surface roughness in drilling MDF composite panels using Taguchi and Regression Analysis Method, Journal of Applied Sciences, 2012, 12(10), 978-984.

[10] A.Krishnamoorthy, J.Lilly Mercy, K.S.M.Vineeth, Manoj Kumar Salugu, Delamination Analysis of Carbon Fibre reinforced Plastic (CFRP) composite plates by Thermographic technique, Materials Today Proceedings, 2015, 2, 3132-3139.

[11] A.Velayutham and Krishnamurthy, Effect of point geometry and their influence on thrust and delamination in drilling of polymeric composites, Journal of Materials Processing Technology, 2007, 185, 204-209.

IOP Conf. Series: Materials Science and Engineering **197** (2017) 012064 doi:10.1088/1757-899X/197/1/012064

Process improvement by cycle time reduction through Lean Methodology

R Siva[1*], Mahamed naveed khan patan[1], Mane lakshmi pavan kumar[1], M Purusothaman[1], SAntony pitchai[1]and Y Jegathish[1]

[1]Department of Mechanical Engineering,Satyabhama University, Chennai 600119, Tamil nadu, India.

* E-mail :siva.mech@sathyabamauniversity.ac.in

Abstract - In present world, every customer needs their products to get on time with good quality. Presently every industry is striving to satisfy their customer requirements. An aviation concern trying to accomplish continuous improvement in all its projects. In this project the maintenance service for the customer is analyzed. The maintenance part service is split up into four levels. Out of it, three levels are done in service shops and the fourth level falls under customer's privilege to change the parts in their aircraft engines at their location. An enhancement for electronics initial provisioning (eIP) is done for fourth level. Customers request service shops to get their requirements through Recommended Spare Parts List (RSPL) by eIP. To complete this RSPL for one customer, it takes 61.5 hours as a cycle time which is very high. By mapping current state VSM and takt time, future state improvement can be done in order to reduce cycle time using Lean tools such as Poke-Yoke, Jidoka, 5S, Muda etc.,

1. Introduction

Lean manufacturing is a technique, philosophy using to eliminate the hidden wastes that present in the companies.it is the most useful quality tool in manufacturing sector. In this paper the lean implementation is made in service sector for an aviation service sector. The Process Electronics Initial Provisioning (eIP) that generates Recommended Spare Parts List (RSPL) is taken for implementation. Initial Provisioning is the process of recommending, selecting and procuring equipment, parts, and material to the range and depth necessary for support of a customer's fleet operation for an initial period. The quantitative problems identified for the process RSPL at Service industry are as follows. Delivery time is 61.5 hours which is more than a week. High waiting time. Manual validation. Error due to over utilization of man power and proper work standard procedure is not prepared for the process. Visual management is not there. High processing time due to time stamp is not captured in informatica confirmation mail in successful upload of baseline. Errors in the process due to Increase in manual intervention, No Engine model validation while uploading baseline. The objectives are to reduce the non-value activities in the generation process by using proper lean tools. To reduce the Dock to Dock time to meet the demand. To reduce the generation cost of the RSPL. To utilize the man power efficiently. To prepare work standard procedure for the processes. To improve the accuracy of the RSPL reporting. To analyze root cause for the errors.

2. Process flow of E – Initial provisioning

The generation of recommended spare parts list flow chart is plotted in the figure 1. First the Initial Provisioning request will be received from external customers for particular enginemodel. The

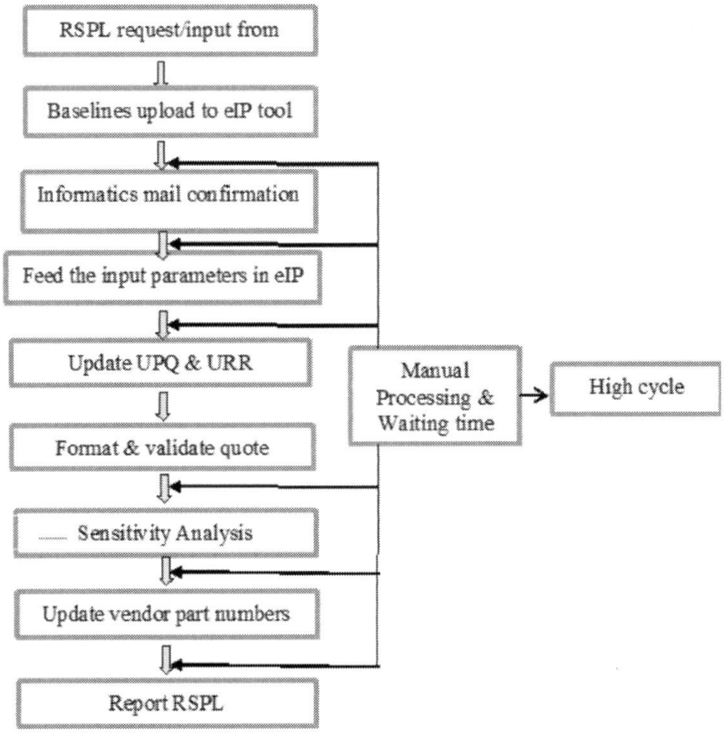

Figure 1. RSPL Process flowchart

respective engine model's baseline technical data is manually updated as per the inputs received from the Product Support Engineering and is uploaded to eIP system. Then the informatics runs in the back end and it inserts the new records to the eIP system records by replacing the old records. After 2 or 3 hours eIP system will generate a confirmation mail on the update of new records. Then the input parameters fed to the eIP system to request RSPL. The next process is updating Unit Pack Quantity (UPQ) and Unscheduled Removal Rate (URR) and thereafter, a recommended spare part list output is formatted and quote indicator is validated. Then the sensitivity analysis is done based on the requirement and recommended by quantity is validated. The respective vendor part numbers updated against the part numbers. Eventually, recommended spare parts list is reported in the required format to the customers. Initial Provisioning Commercial Spares, it contact with a new customer, or an existing customer who has purchased a new product/model. It is an initial opportunity to make a favorable first impression with the customer that could result in securing long-term material orders/agreements. A RSPL is the product of Initial Provisioning, it is the list of parts that recommend to customers to purchase at the point of entry into service. A RSPL includes many data attribute fields including the cost of the parts as well as the recommended buy quantity RBQ of each part. eIP has two major components, RSPL generation process and RBQ generation process. The RSPL generation process dictates which parts appear on the RSPL and the RBQ generation uses the various data attributes (take from both Initial provisioning commercial and Tech Data) to determine the quantity of the recommended parts the customer should purchase. eBaseline will move the data attribute management aspect of the RSPL generation process out of eIP/Tech Database and move it into the electronic Bill of Material (eBOM) tool. The proposed methodology for lean implementation for the process Recommended Spare Parts List is as follows. a) Data collection b) Select the electronics initial provisioning tool c) Identify the process to implement Lean d) Observe the process flow e) Plot the current state Value Stream Mapping f) Identify Value added and Non value added activities g) Analyze the eIP functional problem, which arises in the current state map h) Data analysis and development of solution stage and Implementation i) Plot the future state Value Stream Mapping J) Identify the roadmap for further improvement.

3. Literature survey

V. Crute et al., (2003) [7] in this paper the key drivers for Lean has been discussed in aerospace and examined the assumption that cross-sector transfer has been examined by assuming difficulties. Individual plant context with sector specific factors difficulties were explained in the case study. It has been concluded that lean implementation is not difficult when compared to automobile industries. The key drivers specified for lean manufacturing are a) plant specific b) best performing determination should differ in different departments. c) Concentrate on product than the functional. d) New strategies has to follow instead of following past manufacturing cases. e) Should have culture supports and learning through experimentation to get good response. f) Extra resources can opt to complete the job instead changing the layouts. E. Andres Lopeza (2015) [3] this paper gives a practical guideline to implement lean service by considering waste concepts, values and some inherent characteristics. A conceptual model has been created with two cycles named conceptual cycle and practical cycle. In this, various lean techniques were analyzed and categorized into the two cycles. It gives how to begin a lean implementation in a service sector by understanding of service, inherent aspects and close knowledge of customer value. Finally it gives a practical guideline for the introduction of lean philosophy in service organizations. P.P. Wang (2014) [6] an ontology based product service configuration approach has developed and discussed in this paper. It deals with three categories namely value attributes of product service, the modular system of product service, the proposed approach to product–service configuration. The above three categories were demonstrated by an example of configuring service packages of civil aircraft materials. Mass customization is discussed with product service and that gave high satisfaction to customer. To provide customization-oriented menu, the product–service modularity method is proposed with ontology-based configuration method. G. F. Barbosa (2014) [4] In this paper a frame work has been prepared by using the lean concepts and compared with the specific application of projects case study in automated production. It gives an information about the automation benefits for engineers. This paper explains the interconnection between customer satisfaction and manufacturing strategies favored by the lean manufacturing rules. Brandon Stump (2012) [1] this paper deals about integration of flexible, quick response manufacturing and theory of constraints in mass customization (MC). A boat case study has been discussed and their pre and post implementations were compared. Low level and high level mass customization are compared. For high level MC the flexible, quick response manufacturing are suggested as an appropriate choices to improve. Mohammad Ali Maasouman (2016) [5] this paper deals lean manufacturing effectiveness over input and outputs measures. Two manufacturing cells data were collected through questionnaires, historical data (both quantitative and qualitative) and analyzed as a case study. Scoring system were given for non-conformance and the performance were analyzed to show the lean effectiveness. Finally a visual maturity model with methodology was prepared to implement the lean manufacturing gradually at shop floor. Chee-Cheng Chen (2007) [2] in this paper a performance measurement system called an integrated dynamic performance measurement system (IDPMS) has been created to meet requirements for today's manufacturing environment. Its integrating three areas named company management, process improvement, and the factory shop floor. This performance measurement system introduced in planning and establishing a manufacturing performance to delight customers. The proposed method is analyzed with two product groups

(Desktop power and Telecom. power) and proved it could help firms in selecting and rewarding the best manufacturing teams. The above literature survey discussion clearly shows the lean implementation for a company is results in the reduction of lead time. By the tool Kanban and 5S the inventories between the processes are get reduced. So for this project the lean implementation is taken to reduce the lead time and the tools also used to reduce the supporting activities for lead time.

4. Current state value stream map for the process RSPL

The right area to implement the Lean is discussed and selected with the help of the company executives,

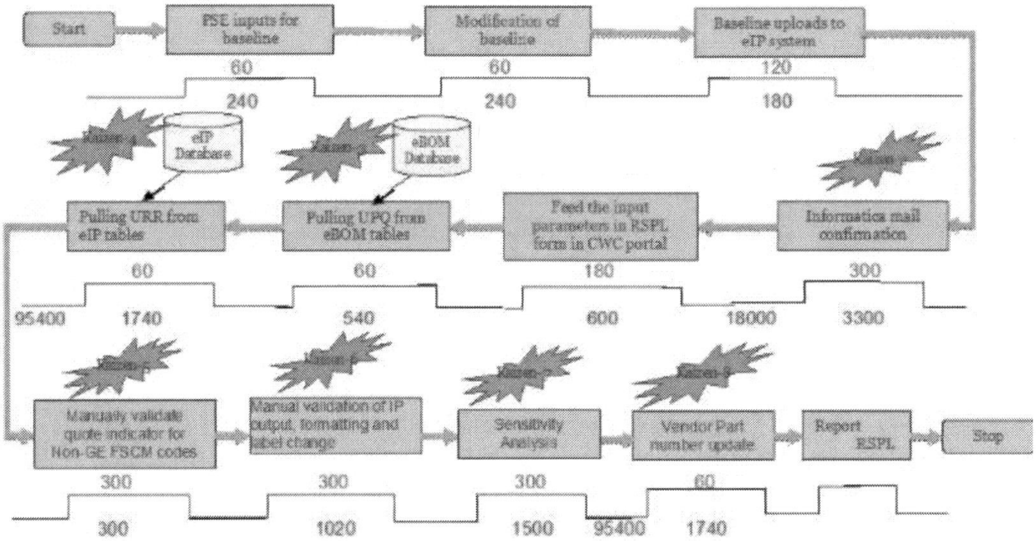

Figure 2. Current state VSM

those who are involved in the eIP process. Next the typical process is selected from the Electronics – Initial Provisioning, which is identified by as improvement areas. To create a current state map, a detailed discussion was conducted with the team members and manager.The Process Recommended Spare Parts List is selected and the number of RSPL required is 5 RSPL/month. The Cycle time is 61.5 hours. Working hours per shift/day is 8 hours (1800 hours/Year).lunch break is 30 min. Number of working days per week is 5 days/week.Takt time is calculated by the ratio of available working time per shift (in sec) with the customer demand per shift. By calculation Takt time = 540000/5=108000s (30 hours), from the current state VSMshown in figure 2. The following times were observed. Value added time = 900s, Non-value added time = 219600s, Total processing time = 12600 s, Total cycle time = 221400s.

5. Improvements

To reduce the bottleneck operations and the non-value added activities, some of the lean tools are selected. First the Visual management tool is introduced by doing Proper display system for process flow. Do and don't do pictures are placed in every process to avoid the error process. 5S tool has been implemented and reduced e-waste. Poke yoke is introduced by designing error detection and prevention into RSPL generation process with the goal of achieving error free IP runs. It is difficult and expensive to find all defects through the inspection methods, and correcting it with solutions. Designed eIP system to a partially automate instead of full automate in the generation process to reduce the expense and to automatically stop when errors are detected. After Jidoka, processors can frequently monitor multiple baselines (reducing labor costs) and many quality issues can be detected immediately. A strategy (Kaizen) is introduced to make continuous improvements by initiating all the employees to work together. By this kaizen, the cross functional work will be initiated and it could eliminates the wastes continuously in manufacturing process. Another common approach is to ask 5 - why tool. By asking why each time, the solution for the problem will be discovered easily. It helps to eliminate the problem truly by applying corrective action to the root cause of the problem. The cause and effect diagram for the high cycle time to generate RSPL as shown in the figure 3. Man: Anyone who is involved with the

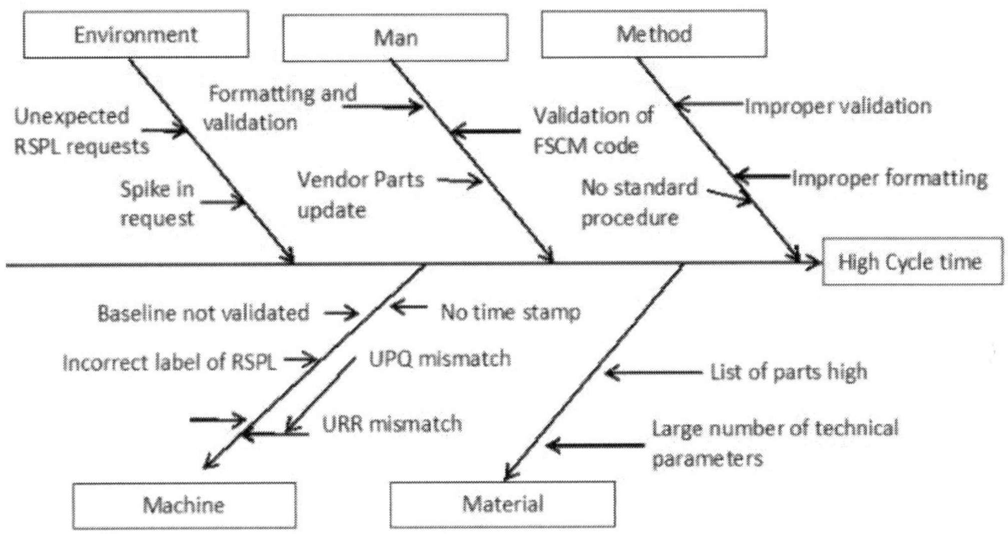

Figure 3. Cause and effect (Fishbone) diagram

process. Methods: How the process is performed and the requirements for doing it. Machines: Any type of equipment required to accomplish the job. Materials: Raw materials, pens, paper, etc. used to produce the product. Measurements: Data generated from the process that are used to ensure its quality. Environment: The natural conditions, such as, temperature, and culture in which the process operates.

6. Future state value stream map

The implementation is carried out with the future state value stream map in the electronics – initial provisioning process for the generation of recommended spare parts list. After thesuccessful implementation, the non-value added activities which identified from the current state value stream

map

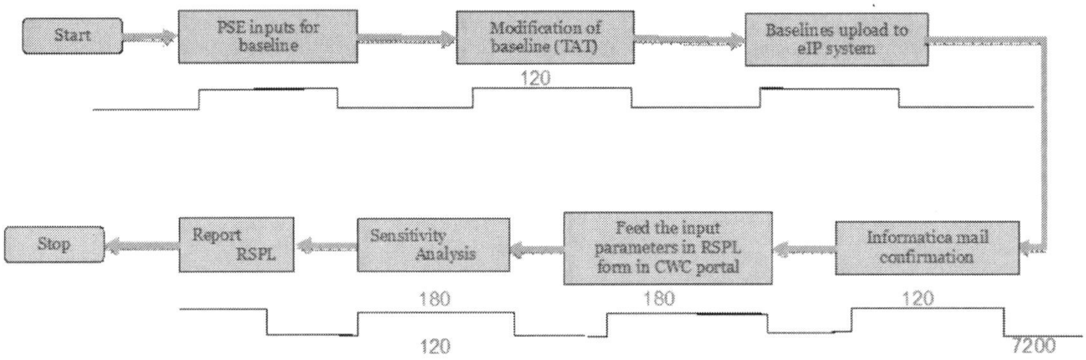

Figure 4. Future state value stream map

is reduced. The lead time, value added time, non-value added cycle time and added processing time observed from the current state value stream mapping are 3690 min, 30 min, 3660 min and 210 min. Mistake proofing and automation identified in the Future state value stream mapping. The bottleneck (high processing and waiting time) process is reduced to 132 min to meet the takt time 256 min. The planned RSPL generation time has to be 90% to 95% of the takt time. So the processing and waiting time are reduced to 90% of the takt time to run a smooth flow process. The future state VSM is shown in figure 4.

7. Results and discussion

The general parameters comparison is shown in the table 1. By the results the cycle time, error, non-value added time, generation cost are reduced. Proper work standard procedure is introduced in the process for generating RSPL. This activity reduced the waiting time in the generation process.

Table 1. Results of Lean implementation

Study Parameters	Before implementation	After implementation
Total processing time (s)	12600	720
Total non-value addes time (s)	219600	7320
Total Value added Time (s)	900	600
VA/NVA ratio	0.82	8.37
Bottleneck operation : Manual processing and Waiting time		
Cycle time (s)	221400	7920

8. Conclusion

The lean tools are successfully introduced for service industry to improve their quality of service.The following points are shown as conclusions from the result and discussion.It is concluded and shows cycle time reduction is achieved by incorporating the lean tools in the Aviation industry. The methodology carried out in lean implementation could serve as a guideline for implementing lean concepts.The project carries genuine advantage of focusing on eliminating all kinds of non-value added activities in the service industry. It focuses the cycle time reduction by eliminating nonvalue-added time through VSM to identify improvement points and 5S, Jidoka, Muda, Visual management, Poka yoke, Cause and effect, Work standard procedure to eliminate inefficiencies.The empirical results drawn from the project implementation serve to demonstrate that an operative decision has helped to improve the lean parameters, in particular to reduce the cycle time and increase the value added time and shows the transformation of a former service organization into a better lean organization that has set a lowest cycle time.

9. Scope of future work

Automating the Sfile and Vfile generation processes. Construct the RSPL and RBQ calculation procedure spare part class wise to understand the Electronics – Initial Provisioning tool functionally. The waiting time is reduced in the future state with the help of automation tool. But still, there is a scope in informatica confirmation mail generation. This waiting time waste will be greatly reduced if the organization introduces informatica to run in every 30 minutes.

References

[1] Brandon Stump and Fazleena Badurdeen 2012 Integrating lean and other strategies for mass customization manufacturing: a case study*J Intell Manuf***23**109–124.
[2] Chee-Cheng Chen and Wen-Ying Cheng 2007 Customer-focused and product-line-based manufacturing performance measurement*Int J Adv Manuf Technol***34** 1236–1245.
[3] E.Andres Lopeza, I.Gonzalez Requenaa, A.Sanz Loberaa 2015 Lean Service: Reassessment of Lean Manufacturing for Service Activities*Procedia engineering***132** 23 – 30.
[4] G.F.Barbosa, J.Carvalho, E.V.G.Filho 2014 A proper framework for design of aircraft production system based on lean manufacturing principles focusing to automated processes*Int J Adv ManufTechnol***72** 1257–1273.
[5] Mohammad Ali Maasouman and Kudret Demirli 2016 Development of a lean maturity model for operational level planning, *Int J Adv ManufTechnol***83**1171–1188.
[6] P.P.Wang, X.G.Ming, Z.Y.Wu, M.K.Zheng, Z.T.Xu. 2014 Research on industrial product–service configuration driven by value demands based on ontology modeling, *Computers in Industry* **65** 247–257.
[7] V.Crute, Y.Ward, S.Brown, A.Graves 2003. Implementing Lean in aerospace—challenging the assumptions and understanding the challenges, *Technovation***23**917–928.

IOP Conf. Series: Materials Science and Engineering **197** (2017) 012065 doi:10.1088/1757-899X/197/1/012065

Plasma Nitriding of CP Titanium Grade-2 and Ti-6Al-4V Grade-5

J.R.Deepak[1]*,V.K.Bupesh Raja[2], J.Senthil Kumar[3], Subin Thomas[4], Thomas Raju Vithaiyathil[5]

School of Mechanical Engineering, Sathyabama University Chennai- 600119

deepakcape@gmail.com

Abstract. Titanium metal is considered to be asset material due to its high tribological properties. Since these tribological properties like hardness, roughness, wear resistance etc. are influenced by the surface properties of the material, so obviously any changes in the surface of the material has direct impact on the tribological properties too. Nitriding is a heat-treating process that diffuses nitrogen into the surface of a metal to create a case hardened surface. The main objective is that to implement the plasma nitriding process to both CP Titanium grade-2 and Ti-6Al-4V grade-5 and to observe the improvements in the tribological properties with respect to the parent materials.

Key words: Plasma Nitriding, Ti-6Al-4V, CP Grade Titanium Alloy

1. Introduction

In this research work we are emphasizing on applying the concept of plasma nitriding on commercially pure Grade-2 Titanium and Grade-5 Titanium and to compare the change in the tribological properties of the raw samples of the titanium metals and the nitrided samples of the same. The plasma nitriding process is done under two different time durations, one being for 10 hours and other being for 24 hours. The coupon size of the metals, over which the nitriding is done, are dimensioned 40x25x3 mm. Metals are cut in EDM machine for minimum wastage and good surface edge finish of the work pieces. These two metals, one being pure and other alloyed, have excellent hardness and roughness properties. Out of all the grades of titanium, these two grades are mostly being used for industrial purposes. Some other grades are also used for fields like dentistry, chemical industries etc. Also it is noted that the wear resistance of the alloyed metal is up to the level. The main properties of the metal that are to be tested are hardness, roughness, wear resistance and change in the microstructure.

It was also inferred from the previous studies and articles regarding the surface enhancement of titanium and its alloys. A noted one is that of Bekir.S.Yilbas et al./ Surface and Coatings Technology 80 (1996) 287-292, where specimens of alloyed titanium were nitrided in H_2-N_2 (1:8 ratio) plasma and the wear property was discussed. Also noted contributions were made by S. Ganesh Sundara Raman, Department of Metallurgical and Materials, Engineering, IIT Madras, Tribology International 43 (2010) 152–160 regarding the same.

2. Experimental work

The samples of dimensions 40x25x3 mm are diffused with nitrogen by plasma nitriding process. The plasma nitriding of CP titanium Grade-2 and Ti-6Al-4V Grade-5 will be subjecting to nitriding durations of 10 hours and 24 hours. The work piece has been cut into the required size (40mm*25mm*5mm), all the edges of the work piece must be cleaned and polished to ensure that no impurities are present on the metal surface. The cleaned work pieces are loaded into the nitriding chamber. Once the chamber has been sealed and all the air within the chamber removed, thereby creating a vacuum. This ensures that there are no other residual gasses within the chamber. Once a stable environment is established, nitrogen gas is passed into the chamber via an opening as shown in figure. The nitrogen gas is highly active and contains ionized molecules formed due to intense electric fields. This gas is generated around the work piece. The ionized molecules bombard onto the surface of the titanium work pieces, thereby creating molecular bonds with the surface of the metal. This is made to go on for a certain period of time. Based on this time the thickness of the nitrided coating varies. The first batch of work pieces is removed after a time period of 10 hours and the second batch after 24 hours.

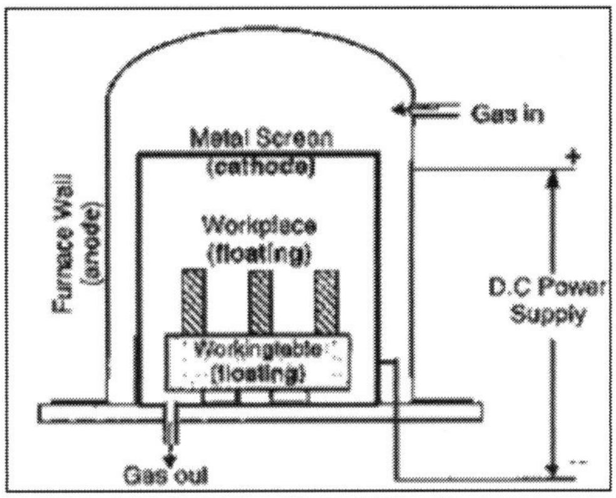

Fig 1: Plasma Nitriding Setup

3. Results And Discussion

The Tests conducted during this research are as follows:

3.1Hardness Test:

Table 1: Vickers Hardness of Plasma Nitrided (10 Hours) Samples

CP Titanium Grade-2	*TI-6AL-4V Grade-5*
178.1	366.3
173.2	374.7
187.8	380.8
194.0	408.8
205.1	385.9

Table-2: Vickers Hardness of Plasma Nitrided (24 Hours) Samples

CP Titanium Grade-2	*TI-6AL-4V Grade-5*
159.1	339.7
161.2	381.5
159.8	377.8
175.4	380.0
174.5	362.4

Comparing the hardness values of raw sample and nitrided sample, we could see a significant change in the same. When comparing the harness value of grade-2 titanium, the raw sample has a hardness value of 152 (Vickers scale) while the mean hardness value of plasma nitrided samples for 10 hours and 24 hours are 166 and 187.64 (Vickers scale) respectively. When coming to grade-5 titanium, the raw sample has a hardness value of 170 (Vickers scale) while the mean hardness value of plasma nitrided samples for 10 hours and 24 hours are 368 and 383 respectively.

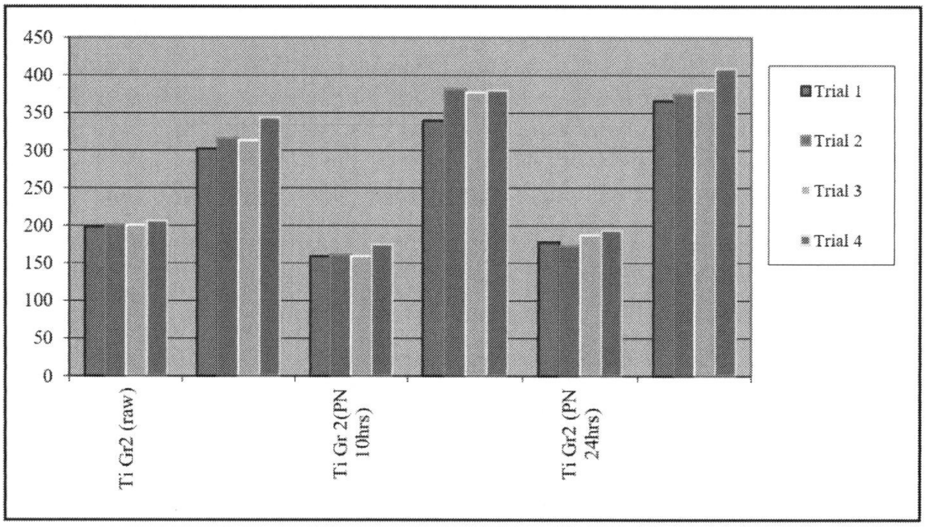

Fig 2: Vickers Hardness Comparison Graph

Though we could see a vast change in the hardness of raw sample and the nitrided samples for grade-5 titanium, we would generally consider the optimum rise of the physical property. The reason for the significant rise in the hardness values for nitrided sample of grade-5 titanium is due to the coating thickness. Excessive coating thickness may sometimes have negative impact on other tribological properties. So the best result is observed in plasma nitrided samples of grade-2 titanium.

3.2 Roughness Test:

Comparing the roughness values of raw sample and nitrided sample, we could also see a mild change in the same. The following interpretation has been observed:

Table 3: Roughness Values of Plasma Nitrided (24 Hours)

Sample (Nitrided)	**Lay 1**		**Lay 2**	
	R_a μm	R_z μm	R_a μm	R_z μm
CP Titanium grade-2*	7.26	60.7	4.28	46.6
Ti-6Al-4V grade-5*	7.23	56.2	7.41	68.1
CP Titanium grade-2**	4.59	53.5	3.30	45.6
Ti-6Al-4V grade-5**	3.96	59.7	5.08	58.9

Beginning with the lay 1, we could see that plasma nitrided (both10 hours and 24 hours) grade-2 samples show more roughness when compared to the roughness of raw sample of grade-2 titanium, raw sample of grade-5 titanium and plasma nitrided samples (both 10 hours and 24 hours) of grade-5 titanium.

IOP Conf. Series: Materials Science and Engineering **197** (2017) 012065 doi:10.1088/1757-899X/197/1/012065

In lay 2, we see that plasma nitrided (both 10 hours and 24 hours) grade-5 samples showing more roughness than raw samples of grade-5 titanium, raw sample of grade-2 titanium and plasma nitrided samples (both 10 hours and 24 hours) of grade-2 titanium.

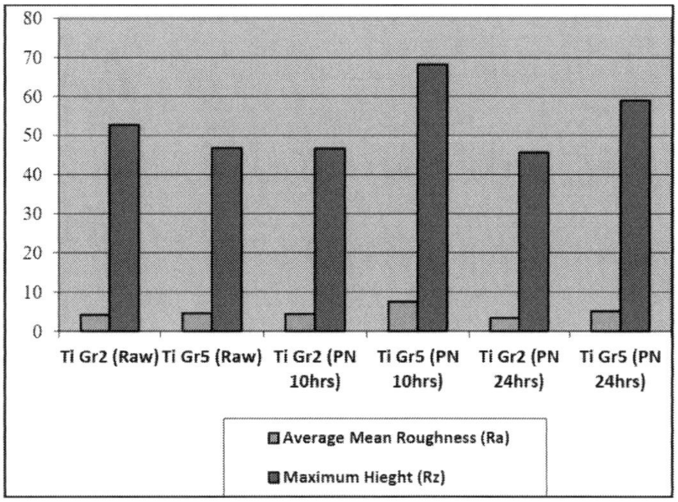

Fig 3: Roughness Comparison Graph

3.3 Wear Test:

The wear for raw sample of grade-5 titanium is recorded to be around 53.1 microns for 236.24 seconds and the wear for plasma nitrided sample (24 hours) of grade-5 titanium is noted to be around 21 microns for 237 seconds.

Table 4: Wear Test Values

Time interval (sec')	Wear (microns)	Frictional force (N)
0.98	-1.37	0.01
50.973	-11.70	3.38
100.965	-1.70	3.11
150.957	4.41	3.07
200.948	9.79	3.43
236.237	13.95	3.36

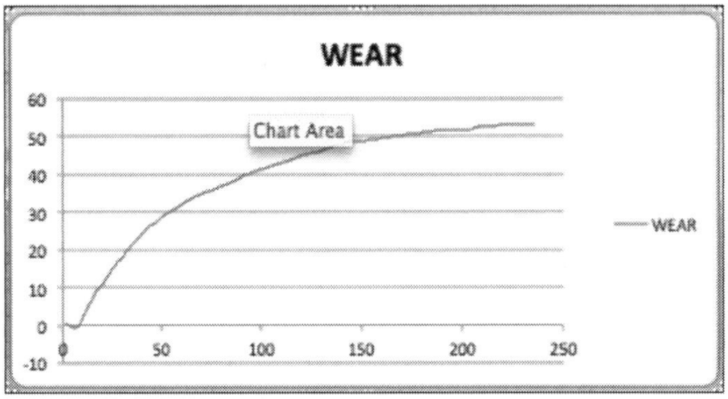

Fig 4: Wear Comparison Graph (Time interval in X axis)

Fig 5: Wear Test Setup

This giving us the clear picture that plasma nitrided sample has excellent wear resistance. Also the point to be noted is that the load condition given for the raw sample was 5 N while the load condition for nitrided sample was between 10 – 15 N.

4. Conclusion

After the thorough discussions of the result, we summarize the following under each testing as follows. By analyzing and logical reasoning, we arrived in a conclusion that plasma nitrided samples, both 10 hours and 24 hours, seem to have increased in their hardness property when compared to the raw samples. Both plasma nitrided samples (10 and 24 hours) of grade-2 titanium and grade-5 titanium show significant increase in the roughness values with respect to the lay of roughness tested, we prefer lay-1, hence concluding that plasma nitrided grade-2 samples have best surface roughness property. Since wear testing was done only on the alloyed samples, it was noted that there was a significant increase in wear resistant for 24 hours plasma nitrided sample of grade-5 titanium when compared to the raw sample. In this research work we have concluded that plasma nitriding is the best method of nitriding technique in use and this is the best suited method for affecting the structures of the metal and thereby increasing the tribological properties of the CP Grade-2 Titanium and Ti-6Al-4V Grade-5.We have shown the immense increase in the hardness of the metal. This increase in hardness would widen the area of usage of titanium. Since titanium is generally highly corrosion resistant, by nitriding it will further increase its corrosion properties. The best nitriding technique is plasma which gives us the highest hardness values. The microstructure images are studied and are classified based on alpha beta components. The alpha and beta particles are all uniformly distributed. This research work carried out by us would further lead to future aspects like what are all the possibilities of using both the grades of titanium. This leaves us with the conclusion that plasma nitriding is the best suited method in the improvement of the tribological properties like hardness, roughness, wear resistance etc. with the study of its microstructures for evidence collection, there by widening its application specially in boiler industries, marine field and aerospace industries.

References

[1] A. Raveh (1993), Title of the paper, Mechanism of r.f Plasma nitriding of Ti-6Al-4V alloy, Materials Science and Engineering, Vol No A167, Pg No 155-164.
[2] C. Anandan (2008), Title of the paper, Micro Raman study of the effect of oxide layer on nitriding of Ti–6Al–4V, Applied Surface Science, Vol No 254, Pg no 2783–2789.
[3] B.S. Yilbas (2000), Title of the paper, Laser treatment of Ti-6Al-4V alloy prior to plasma nitriding, Journal of Materials Processing Technology, Vol No 103, Pg no 304-309.
[4] Bekir S. Yllbas, (1996), Title of the paper, Plasma nitriding of Ti-6Al-4V to improve some tribological properties, Surface and coatings, Vol No 80, Pg No 287-292.
[5] D. Nolan (2006), Title of the paper, Sliding wear of titanium nitride thin films deposited on Ti-6Al-4V, Surface & Coatings Technology, Vol No 200, Pg No 5698–5705.
[6] E. Koyuncu (2009), Title of the paper, Investigation of surface properties of high temperature nitrided titanium alloys, Journal of achievements in material and manufacturing technology, Vol No 37, Pg No 434-441.
[7] G. Straffelini (2004), Title of the paper, lubricated rolling–sliding behavior of ion nitrided and untreated Ti–6Al–4V, WEAR, Vol No 256, Pg No 221-456.
[8] V. Fouquet (2004), Title of the paper, Plasma assisted nitridation of Ti-6Al-4V, Applied Surface Science, Vol No 221, Pg No 248-258.
[9] P. Budzynski (2006), Title of the paper, Surface modification of Ti–6Al–4V alloy by nitrogen ion implantation, WEAR, Vol No 261, Pg No 1271–1276.
[10] Mahfujur Rahman (2006), Title of the paper, Saddle free fast atom beam source□A new low pressure plasma nitriding for Ti-6Al-4V, Thin Solid Films, Vol No 515, Pg No 129 – 134.
[11] N. KashaevT (2005), Title of the paper, Assessment of the application potential of the intensified glow discharge for plasma nitriding of Ti-6Al-4V, Surface & Coatings Technology, Vol No 200, Pg No 502-506.
[12] N. Tsuji (2008), Title of the paper, evaluation of surface modified Ti-6Al-4V alloy by combination of plasma carburizing and deep rolling, Materials Science and Engineering, Vol No A 488, Pg No 139-145.
[13] S. Taktak (2004), Title of the paper, Diffusion kinematics of explosively treated and plasma nitrided Ti-6Al-4V alloy, Vol No 75, Pg No 247–259.

IOP Conf. Series: Materials Science and Engineering **197** (2017) 012066 doi:10.1088/1757-899X/197/1/012066

SALT BATH NITRIDING OF CP TITANIUM GRADE-2 AND TI-6AL-4V GRADE-5

J. R. Deepak[1], V.K. Bupesh Raja[2] K.Arun Kumar[3], Varun Radhakrishnan.H[4], Sonu Skariah Thomas[5],

School of Mechanical Engineering, Sathyabama University Chennai- 600119

Email id : deepakcape@gmail.com

Abstract. Titanium is known for its strong affinity towards nitrogen. The metal forms a nitride form case of high hardness when the salt bath nitriding is carried out. The selection of Grade 2 Titanium and Grade 5 Ti6Al4V Titanium alloy, stands with the fact that CP Grade 2 Titanium is the most formable and corrosion resistant amongst the pure grades of Titanium and grade 5 alloy is biocompatible and also has excellent tribological properties. This research work attempts to solve the problem of galling by comparing the morphology of the nitride case produced in Commercially Pure Grade 2 Titanium and the nitride formations produced in the Titanium Ti6Al4V alloy through Salt bath nitriding for a time span of 24 hours. Salt Bath Nitriding imparts unique improvements in Roughness, Hardness and Wear resistance of the samples thereby widening the applications of the material.

Index Terms—SurfaceEngineering Salt Bath Nitriding, Ti-6Al-4V, EDM

1. Introduction

Although many researchers have studied about the nitriding of common ferrous metals, the proper mechanism regarding the nitridingof Titaniumand its alloys has not been completely understood. This paper presents the effect of different Salt Bath Nitriding parameters on the structure of nitrided zone of Commercially Pure Titanium (Grade 2) and

Grade 5 (Ti6Al4V) alloy and their influence on mechanical properties on the samples. The salt bath nitriding process introduces nitrogen into the surface of metal parts to improve tribological properties and performance. Nitriding is performed at sub critical treatment temperatures, typically at 1075 F (580°C) in a molten salt bath. Process Time can bereduced if the thickness is compromised. Liquid nitriding is considered as the benchmark for uniformity, consistency andflexibility. It provides the best corrosion resistance and short process time. The nitride surface layer is well integrated with bulk; hence it is not susceptible to flaking or peeling. Liquid nitriding achieves results by means of chemical reaction (at temperatures lower than conventional heat treatment) that leaves a hard nitride compound layer on the surface of the component. The liquid nitriding process is carried out for duration of 24 hours and is post processed by imparting the process of Air quenching and Water Quenching on distinct samples and the final values of Hardness, Roughness and Wear are noted. These test results are then compared with the results of the raw samples of both the grades of Titanium and the usefulness of introducing a nitride layer on the surface of grade 2 and grade 5 Titanium is understood by comparing the results before and after nitriding.

2. Experimental Setup

Liquid nitriding is a process in which anhydrous ammonia is introduced into cyanide - cyanate bath. The salt bath process uses the principle of the decomposition of cyanide to cyanate and the liberation of nitrogen within the salt for diffusion into the steel surface. The work piece has been cut into the required size (40mm*25mm*5mm), all the edges of the work piece must be cleaned and polished to ensure that no impurities are present on the metal surface. The cleaned work pieces are loaded into the nitriding chamber. Maintenance of dimensional and geometric stability during liquid nitriding is enhanced by hardening of parts prior to nitride treatment. Four plates each of Grade 2 and Grade 5 are obtained for the experiment. The apparatus used for the experiment is an enclosed chamber or a bath known as the Salt Bath. Inside the chamber salts of ammonia or cyanide are supplied prior to the Nitriding process. The specimens are then held on a fixture which is placed inside the molten salt bath containing the nitriding salts. The individual components should not be in contact with each other during the process. The overall process temperature is maintained in the range of (560 – 580)°C for a time period of 24 hours. The hardness and depth of the diffusion zone will determine fatigue strength. In addition to case depth and hardness also the obtained compressive residual stress state in the case is of major importance.

Factors affecting Nitriding are as follows:

- Quantity of parts per batch.
- Process Temperature, typically (500-580) degree celcius.
- Complexity of the components in terms of blind holes which adds the time involved in cleaning the component after the process.
- Ancillary steps like cleaning, polishing, impregnation etc.

3. Results and Discussion

The Nitriding of the 2 different grades of Titanium in Salt bath apparatus for 24 hours led to introduction of a surface coating of lustrous silver/black colour on the metal. This is the nitrided layer which acts as a surface coating for the metal and the values of Hardness, Roughness and Wear of the nitrided samples of CP Titanium grade 2 and grade 5 alloyed Titanium were noted down and comparison was made with the raw samples.

3.1. Hardness Test

The Vickers method is based on an optical measurement system. The Micro hardness test procedure, ASTM E-384, specifies a range of light loads using a diamond indenter to make an indentation which is measured and converted to a hardness value. The following observations were made post the hardness test

Table1. Vickers Hardness Values of Liquid Nitrided (24 Hours) Ti Grade-2

L.N+W.Q	L.N+A.Q	H+L.N+W.Q
187.2	186.3	370.2
189.2	220.4	369.5

250.2	215.8	371.4
266.9	221.2	372.2

Table2. Vickers Hardness Values of Liquid Nitrided (24 Hours) Ti6Al4V Grade-5

L.N+W.Q	L.N+A.Q	H+L.N+W.Q
349.3	368.6	367.3
351.1	368.9	369.7
352.3	368.9	370.2
353.1	370.6	370.7

It is found that the Average value of hardness has considerably increased in case of Liquid Nitrided samples of CP Grade 2 Titanium and Grade 5 Titanium alloy. ThePre Hardening of both grades of titanium increased the Hardness values to a notable degree and Water Quenching post Nitriding in Salt Bath furnace increased the Hardness of the component. This is of major importance in applications like manufacturing of fasteners, automotive components like nuts, bolts and also in environments where the metal is subjectedto continuous stress. Maximum hardness value was observed to be 372.2 in case of CP Titanium grade 2, Hardened, Liquid Nitrided and Water Quenched sample compared to 202 in case of raw sample of CP Titanium. This clearly shows the advancement in hardness after nitriding. Also when the same process was carried out in grade 5, high hardness is achieved

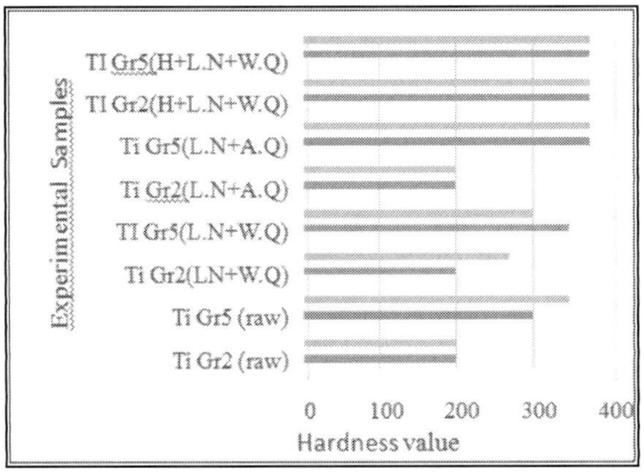

Figure 1. Comparison of Micro Hardness values of Ti & Ti6Al4V Raw and Liquid Nitrided (Water & Air Quenched)

3.2. Roughness Test

Roughness is a component of surface texture. It is quantified by the vertical deviation of real surface from its ideal form. If these deviations are large, the surface is rough. Roughness test results displayed almost no major deviation in the surface of the samples but due to the presence of nitrided layer, a slight increase in roughness was observed while interpreting the Arithmetic mean value. This might be due to the fact that the components which were prone to liquid media for a time period of 24 hours increased theroughness incertain samples. This can be avoided by reducing the nitriding process time and carrying out the process in a more controlled environment.

Table3:Roughness Value of Liquid NitridedTi Grade-2&Ti6Al4V Grade-5

Sample	Layer1		Layer2	
Roughnessvalue	R_a µm	R_z µm	R_a µm	R_z µm
Grade2Raw	7.28	55.2	4.12	52.7
Grade5Raw	3.65	48.8	4.53	46.8
Grade2(L.N+A.Q)	4.92	49.6	4.33	61.3
Grade2(H+L.N+W.Q)	2.91	31.0	3.35	50.9
Grade2(L.N+W.Q)	4.27	48.4	7.77	62.0
Grade5(L.N+A.Q)	6.92	46.6	3.63	52.5
Grade5(H+L.N+W.Q)	8.94	46.2	8.24	51.2
Grade5(L.N+W.Q)	4.72	49.7	4.62	49.7

Note:$R_a$referstoArithmeticaveragevalueofRoughness
$R_z$referstoMeanRoughnessdepthvalue

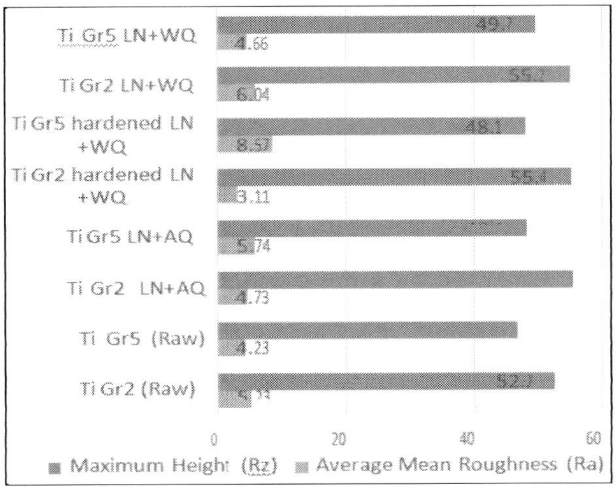

Figure 2. Comparison of Roughness Comparison Graph of Liquid NitridedTi Grade-2&Ti6Al4V Grade-5

3.3 Wear Test

The wear test was carried out in Ducom TR-20LE-PHM-250. Pin on disc rig to test the wear resistance of Titanium grade 5 alloy liquid nitrided for 24 hours, water quenched as the sample showed deviations from other samples. A pin on disc tribometer consists of a stationary "pin", which is Titanium metal in this case, under an applied load in contact with a rotating disc. The disc in which the pin is placed is made of hardened high Carbon, high Chromium Steel, 60HRC hardness which has excellent wear resistance. The pin surface is slided against the surface of the rotated disc. The time and sliding distance determine the amount of wear in the surface of the pin. A test load of 5 N was introduced in case of raw sample and a test load of 10 N was introduced on the nitrided sample to infer the rate of wear resistance in nitrided sample. The load is also directly proportional to the wear caused in the surface of the pin. This is the Wear and friction test rig. The Wear, Time, Sliding diameter, coefficient of friction are the various test parameters. The presence of nitride layer in the sample improved the wear resistance of the metal. The final value of amount of wear in the nitrided sample was observed to be 23.56 micrometers whereas the raw sample showed a value of 53 micometers during the wear test for half the load. Liquid nitriding is therefore suggested for Titanium parts and components in industries where the metal is exposed to high wear and friction applications along with high corrosion resistance.

TABLE-4:WEARTESTVALUES

Time(seconds)	Wear (micrometers)	Frictional Force(N)
55.14	5.0	3.58
85.52	9.70	3.29
130.72	15.63	3.31
190.42	21.28	3.20
236.64	23.56	3.76

Post test examination led to some interesting observations and results. The Grade 5 raw sample showed very high wear resistance values. This stands with the fact that Titanium is a wear resistant metal and Grade 5 which is widely used for manufacture of Aerospace components and medical implants, is amongst the most wear resistant grades available. The amount of wear is determined by measuring appropriate linear dimensions of both specimens before and after the test, or by weighing both specimens before and after the test.

4. Conclusions

The test results and the observations give a conclusion that the liquid nitriding process improves the tribological properties of the titanium grade 2 and Ti-6Al-4V alloy. Grade 2 Titanium which is usually not used in harsh saline and chemical environment due to its pure state has been benefited by nitriding in Salt Bath media. This is a very important fact as nitriding Commercially pure Titanium makes it equally competent with alloyed Ti grade 5 and it can be used in more useful and new areas which was once restricted and did not make its place into. The compound layer is responsible for the improved wear properties observed in Ti-6Al-4V alloy. In comparison with the roughness values, it is studied that Ti Grade 5 6Al-4V alloy liquid nitrided for 24 hours plus Water Quenched displayed excellent wear resistance and roughness characteristics. The wear can be further reduced by subsequent treatment by a dry film lubricant coating on one member of the couple. We have hereby presented the effect of introducing a nitride coating on the surface of Grade 2 CP Titanium and Grade 5 Ti-6Al-4V alloy in a salt bath media and the usefulness of using this age told technique of surface treatment. Vigorous research work is being carried out around the globe for improving the surface properties of Titanium, especially Grade 5 Titanium which is the work house industry metal and it is concluded that liquid nitriding is the most effective nitride process for improving tribological properties of the metal with minimum distortion among other advantages like short process time and the most cost-effective method of nitriding. The Trademark Arker process Salt Bath nitriding is amongst the most industrially accepted processes with almost no environmental impact and hazards, is effective in nitriding Titanium grade 2 and grade 5 and usage of the metal in more useful, high-end applications such as in Supersonic jet engines and parts of high performance super cars to an advanced level. Thus we conclude the research work by standing with the fact that further research work and experiments regarding the tribological properties of Titanium and by carrying out surface engineering techniques on the metal will open ways to further beneficial and cost-effective exploitation of Titanium.

References

[1] Anandan. C, Rajam. K.S , MicroRaman study of the effect of oxide layer on nitriding of Ti–6Al–4V Vol. 254, Applied Surface Science,2008.
[2] Budzynsk. P, Surface modification of Ti–6Al–4V alloy by nitrogen ion implantation Vol. 261 ,Wear, 2006.
[3] Cuthill.J.R, W.D. Hayas, R.E. Seebold, Nitriding Phenomena in Titanium and the 6Al-4V Titanium alloy Vol. 64.A,No.1, Journal of Research of the National Bureau of Standards, A. Physics and Chemistry, 1959
[4] Frank Czerwinski, Thermochemical treatment of metals Vol. 80, Intechopen,2012.
[5] George Vander Voort, Metallographic Preparation of Titanium and its Alloys Vol. 3, Issue 3 , Beulher,2015
[6] Gotman.I, E.Y Gutmanas, Department of Materials Science and Engineering, Technion, Israel Institute of Technology, Titanium nitride- based coatings on implantable medical devices, Advanced biomaterials and devices in medicine, Haifa, Israel 2014.
[7] Guilherme AS, Henrique GE, Zavanelli RA, Mesquita MF, Surface roughness and fatigue performance of commercially pure titanium and Ti-6Al-4V alloy after different polishing protocols,2005

[8] Kenneth G. Budinski, Tribological properties of titanium alloys Vol .151, Wear, 1991.
[9] Koyuncu.E, F.kahraman,O. karedeniz, Investigation of surface properties of high temperature nitrided titanium alloys Vol 37, Journal of achievements in material and manufacturing engineering, 2009.
[10] Li Yuan-hui, Luo De-fu, Wu Shao-xu, School of Material Science and Engineering, Jiangsu University, Zhenjiang Jiangsu, QPQ Salt Bath Nitriding and Corrosion Resistance Vol. 118, Solid state phenomena, China,2006.
[11] Mittemeijer.E.J, Max Planck Institute for Intelligent Systems, ASM Handbook, Fundamentals of Nitriding and Nitrocarburizing Vol. 4A, Steel Heat Treating Fundamentals and Processes, 2013
[12] Veiga.C, J.P. Davim,A.J.R. Loureiro, Department of Mechanical Engineering, Coimbra Institute of Engineering, Rua Pedro Nunes, Quinta da Nora, Properties and applications of Titanium Alloys: A brief review, Reviews on Advanced Material Science, Coimbra , Portugal 2012.
[13] Weisenburger JN, Hovendick SM, Garvin KL,Haider H, How durable are titanium nitride coatings on total hip replacements, J Bone joint surg,2011.
[14] Zhecheva A, Sha W,Malinow S,Long A, Enhancing the microstructure and properties of titanium alloys through nitriding and other surface engineering methods. Surf coat technol, 2005.

Experimental Investigation of Mechanical Properties of Welded Corten Steel A588 Grade Plate Using ER70S - 6 Filler Material for Construction Application

J.R.Deepak[1]*, V.K.Bupesh Raja[2], Mittapalli Janardhan Guptha[3] Palaparthi Hari Durga Prasad[4], V.Sriram [5]
School of Mechanical Engineering, Sathyabama University Chennai- 600119

deepakcape@gmail.com

Abstract. ASTM A588 Grade A steel plate is a high strength, low alloy structural steel with 0.19 % of carbon content. When exposed to the atmosphere, A588 Grade A is suitable for construction in the bare (paint – free) condition. The main problems are lack of fusion, lack of penetration and corrosion on heat affected zone. In this research work Corten ASTM A588 Grade steel of 3mm thickness is electroplated with copper and then both raw and copper electroplated are welded by GMAW welding process with ER70S-6 as a filler material. The welded ASTM A588 is cut according to ASTM size for further testing of mechanical properties. Considering its welding strength after the process of electroplating, this research clearly states the metal can be utilized for better results in any given field. Here both the tensile and hardness are higher in copper electroplated welded when compare to raw welded.

Key words: Corten A588 Grade, ER70S – 6, Welding, Electroplating

1.Introduction:

Corten steel is also called as weathering steel which belongs to a steel alloy. Basically it is a high strength low alloy steel were it is used as a paint free material. It also forms rust like top layer called patina with good corrosion stability. The metallurgical properties varies from weathering steel and ordinary structural steel is the addition of chromium, copper and nickel alloying elements by which weathering steel has high resistance to corrosion. It is available in different standards like ASTM A242, ASTM A588 and ASTM A606. In this work corten A588 is used as base metal.

1.1. Base metal Corten A588:

HSLA (High-strength low-alloy) steels are mainly used for their higher strength and corrosion resistance. It has atmospheric corrosion resistance than carbon steels. Using shear cutting process the base metals are cut into 300*150*3 (all dimensions are in mm)

Table 1. Chemical composition of Corten ASTM A588 Grade Steel

Elements	C	Si	Mn	P	S	Cr	Ni	Cu
Composition %	0.120	0.298	0.391	0.087	0.013	0.546	0.218	0.302

1.2. Filler material (ER70S-6):

Abbreviation for ER70S-6:

ER: An electrode or filler rod

70: A minimum of 70000 pounds of tensile strength per square inch of weld.

S: Solid wire

6: The amount of deoxidizing agent, cleaning agent and copper coating on the electrode.

It is a mild steel solid wire used for good strength, proper welds and problem free in high temperature, high speed etc. it is basically used for the gas mixture of 100% co2 and 80% co2 & 20% Ar.

Table 2. Chemical composition of ER70S-6

Elements	C	Si	Mo	S	Cr	Ni	Cu
Composition, %	0.1	0.8	0.12	0.03	0.15	0.13	0.5

2. Problems definition

Lack of fusion, lack of penetration, cracks on heat affected zone and corrosion in heat affected zone are the main problems in welded corten A588 Grade steel plates

3. Experimental work

3.1. Electroplating of Corten A588 Grade steel plate coated with copper:

In this research work the raw milled corten steel A588 plates is electroplated with copper. Firstly, degreasing is done for removal of dirt and dust and pickling is followed for removing the rust. Pre copper coating is done and then rinsed in water followed by drying process.

Figure 1. Raw milled Corten A588 Grade steel Plate

IOP Conf. Series: Materials Science and Engineering **197** (2017) 012067 doi:10.1088/1757-899X/197/1/012067

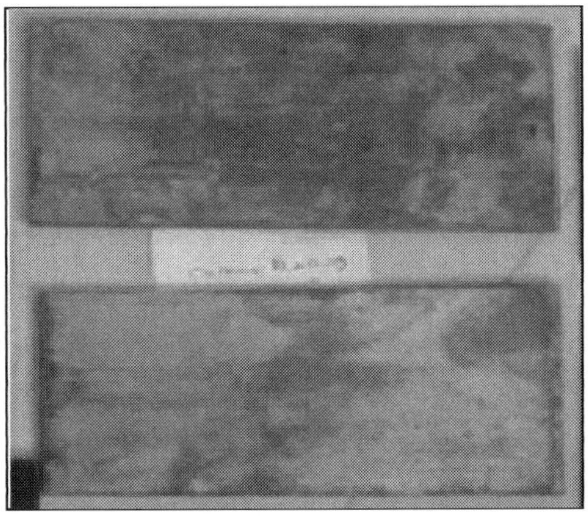

Figure 2. Corten A588 Grade steel Plate Electroplated with Copper

3.2. Welding Process:

GMAW (Gas Metal Arc Welding) process is also called as MIG Welding (Metal Inert Gas welding). This helps to produce good weld due to the smaller thickness of filler material. Gas used for welding is 80% C02 and 20% Ar.

3.3. EDM Cutting Process:

After the welding is done these are cut in ASTM E8/E8-M-11 standards for tensile and hardness testing using EDM cutting process.

4. Results

4.1. Tensile test:

The tensile testing is carried out in Micro Tensile testing setup as shown in figure which is used to find the capacity of the material with varying loads.

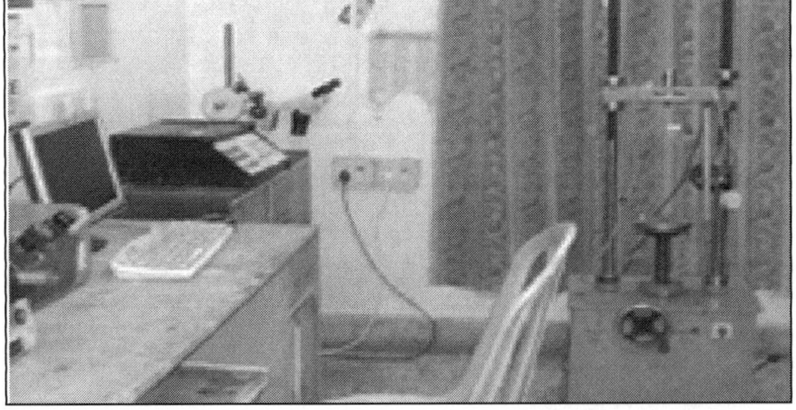

Figure 3. Tensile Testing machine Setup

Table 3. Tensile Testing Machine specification

Machine Name	**Tensile Testing Machine**
Testing load range	Max 5 Ton Load Condition (HVN(1 kg))
Make	Associated Scientific Engineering Works, New Delhi
Digital Encoder make	Auto Instruments - Kholapur
Gear rotation speed (for gradual loading)	1.25. 1.5 & 2.5 mm /min
Software details	FIE make India

The tensile testing Specimen as per ASTM E8/E8-M-11 is cut by Wire cut EDM process

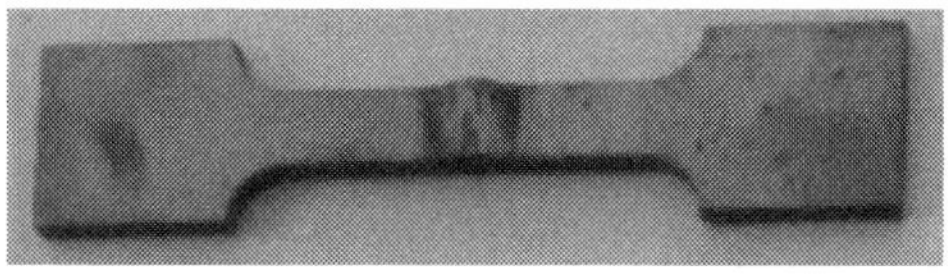

Figure 4. Tensile Testing Specimen as per ASTM E8/E8-M-11

4.2. Tensile results:

Table 4. Tensile Testing results

Description	Welded Raw milled Corten A588 Grade steel Plate	Welded Corten A588 Grade steel Plate Electroplated with Copper
Ultimate Breaking Load (KN)	9.480	12.635
Fracture Location	Base	Base
Displacement at FMAX (mm)	10.100	11.300
Max Displacement (mm)	12.900	14.000
Area (mm^2)	18.000	24.000
Ultimate Stress (KN/mm^2)	0.527	0.526
Elongation %	17.143	25.714
Yield stress (KN/mm^2)	0.309	0.422
YS / UTS Ratio	0.588	0.801

IOP Conf. Series: Materials Science and Engineering **197** (2017) 012067 doi:10.1088/1757-899X/197/1/012067

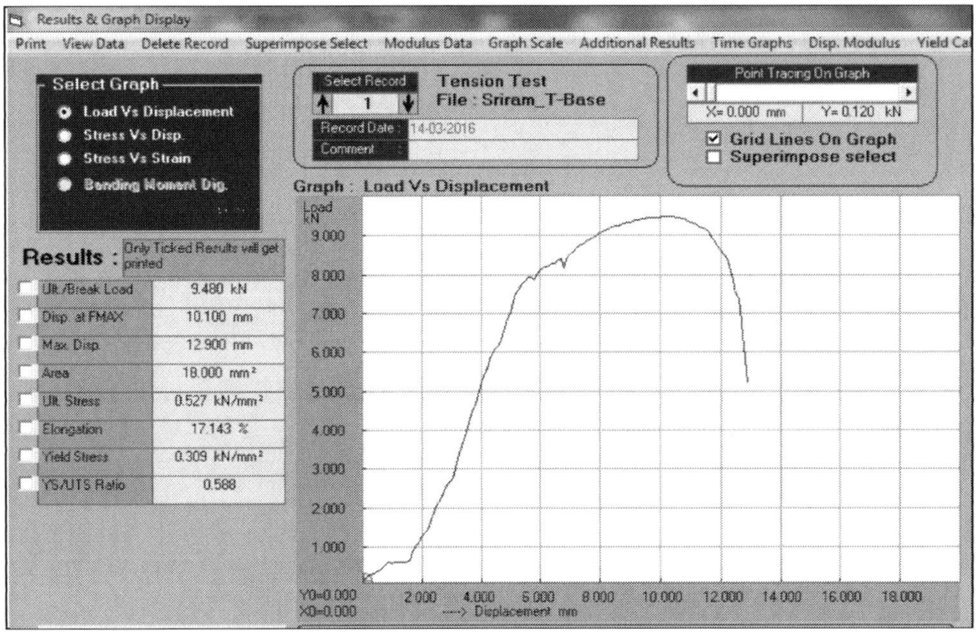

Figure 5. Tensile graph (Load Vs Displacement) of Welded Raw milled Corten A588 Grade steel Plate

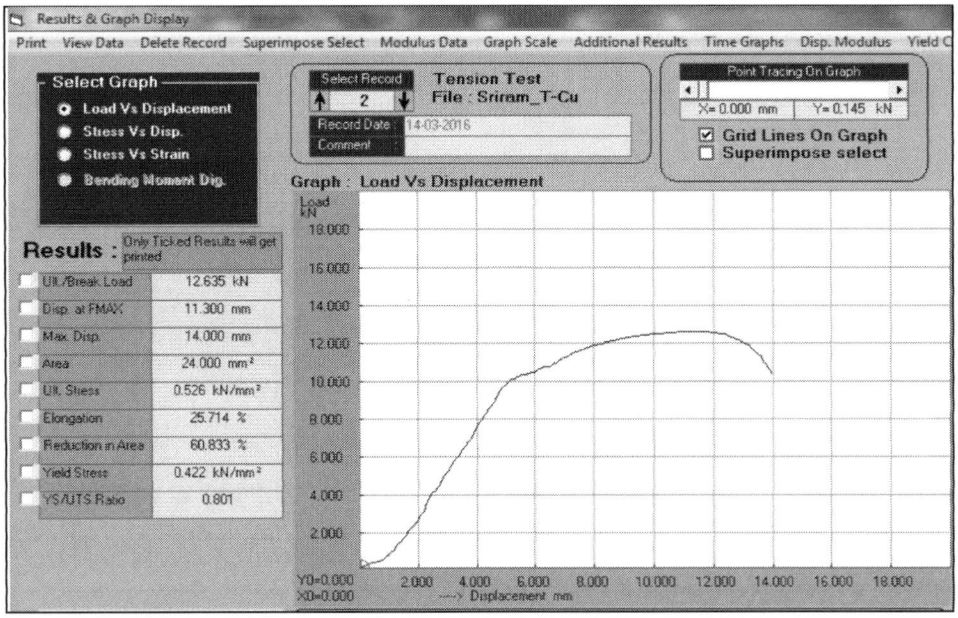

Figure 6. Tensile graph (Load Vs Displacement) of Welded Corten A588 Grade steel Plate Electroplated with Copper

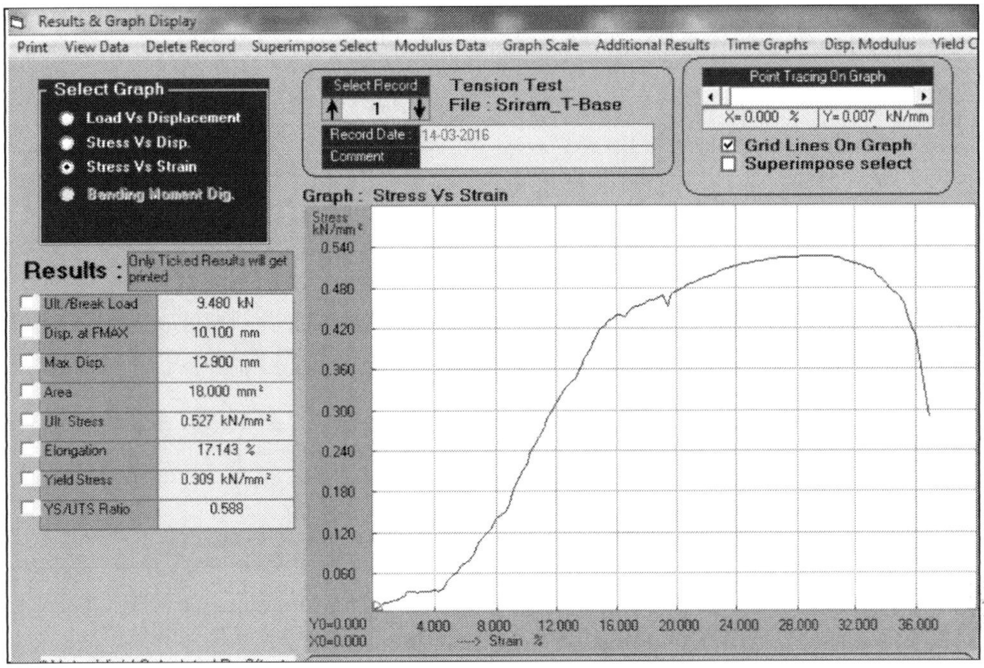

Figure 7. Tensile graph (StressVs Strain) of Welded Raw milled Corten A588 Grade steel Plate

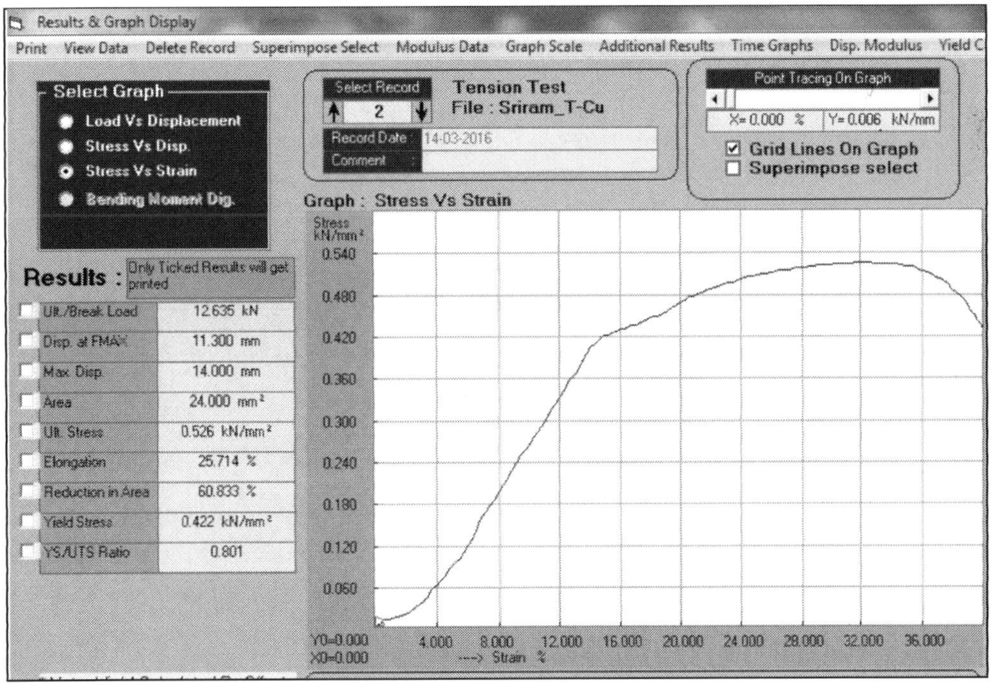

Figure 8. Tensile graph (StressVs Strain) of Welded Corten A588 Grade steel Plate Electroplated with Copper

IOP Conf. Series: Materials Science and Engineering **197** (2017) 012067 doi:10.1088/1757-899X/197/1/012067

4.3. Hardness test:

For metals it is to test resistance of permanent deformation. It is also related to the plastic flow stress of materials.

Table 5. Micro Hardness Testing Results (HVN)

Samples	Base Material (BM)	Heat Affected Zone (HAZ)	Weld Zone(WZ)	Heat Affected Zone (HAZ)	Base Material (BM)
Welded Raw milled Corten A588 Grade steel Plate	185.85	209.0	216.6	209.6	180.6
Welded Corten A588 Grade steel Plate Electroplated with Copper	188.6	218.1	251.8	246.7	182.2

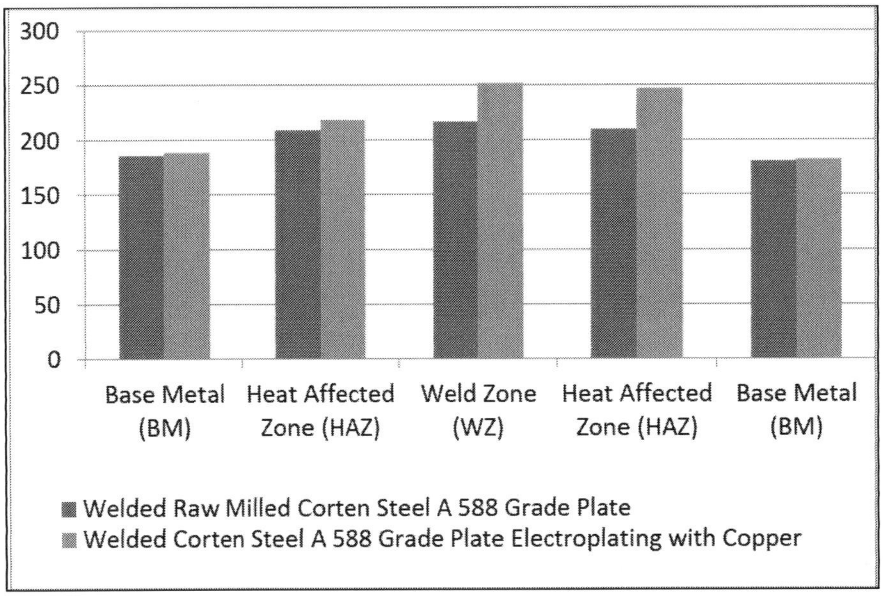

Figure 9. Micro hardness Bar Chart of Welded Corten A588 Grade steel Plate Electroplated with Copper

5. Conclusion

- Copper Electroplated plate has attained an additional elastic property and ductility with a maximum displacement of 14.00 (mm).
- Tensile Test Comparison (Load Vs Displacement) the Copper electroplated metal plates has improved tensile stress compared to the base metal.
- Tensile strength of Copper Electroplated has higher Ultimate Breaking Load 12.635 (KN).
- These changes occur due to electroplating of copper to the parent metal due to which additive stress has been created so breaking period of metal will increase results in increase in tensile properties.

- Electroplating of Copper in the parent metal has resulted in good improvement in the mechanical properties as a result of this weld and heat affected zone does not break due to ductility.
- The welded zone has the highest hardness values. Copper Electroplated plate has the highest hardness value of 251 (HVN)

References:

[1] ASTM C101-04:2004 Guide for Estimating the Atmospheric Corrosion Resistance of Low-Alloy Steels. ASTM International, USA.
[2] EN 10025-5:2004 Hot rolled products of structural steels – Part 5: Technical delivery conditions for structural steels with improved atmospheric corrosion resistance. CEN.
[3] EN 1993-1-1:2005 Eurocode 3: Design of steel structures – Part 1-1: General rules and rules for buildings. CEN.
[4] EN 1990:2002 Eurocode: Basis for structural design. CEN, 2004.
[5] EN 10025-1:2004 Hot rolled products of structural steels – Part 1: General technical delivery conditions. CEN.
[6] EN 1993-2:2006 Eurocode 3: Design of steel structures – Part 2: Steel bridges. CEN.
[7] EN 1993-2:2006 Eurocode 3: Design of steel structures – Part 2: Steel bridges. CEN.
[8] Corten weather steel: technical data, archived from the orginal on 13 January 2010, retrieved 13 January 2010.
[9] Elizabeth A. Harris (august 27, 2012). " Constructing a façade both rugged and rusty" . nytimes.com. Retrieved 10 September 2012.
[10] Harris , Elizabeth (27 august 2012). " Constructing a façade both rugged and rusty". Nytimes. Retrieved 27 September 2015.
[11] Manual of steel construction, 8th edition second revised printing. Chicago : American institute of steel construction.

IOP Conf. Series: Materials Science and Engineering **197** (2017) 012068 doi:10.1088/1757-899X/197/1/012068

Experimental Investigation Nano Particles Influence in NPMEDM to Machine Inconel 800 with Electrolyte Copper Electrode

K Karunakaran[1] and M Chandrasekaran[2]
[1]Research Scholar and Assistant Professor, Department of Mechanical Engineering, VELS University, Chennai, Tamil Nadu, India.
[2]Professor, Department of Mechanical Engineering, VELS University, Chennai, Tamil Nadu, India.

E-mail: ksivasakthi1966@gmail.com

Abstract— The recent technology of machining hard materials is Powder mix dielectric electrical Discharge Machining (PMEDM). This research investigates nano sized (about 5Nm) powders influence in machining Inconel 800 nickel based super alloy. This work is motivated for a practical need for a manufacturing industry, which processes various kinds of jobs of Inconel 800 material. The conventional EDM machining also considered for investigation for the measure of Nano powders performances. The aluminum, silicon and multi walled Carbon Nano tubes powders were considered in this investigation along with pulse on time, pulse of time and input current to analyze and optimize the responses of Material Removal Rate, Tool Wear Rate and surface roughness. The Taguchi general Full Factorial Design was used to design the experiments. The most advance equipments employed in conducting experiments and measuring equipments to improve the accuracy of the result. The MWCNT powder mix was out performs than other powders which reduce 22% to 50% of the tool wear rate, gives the surface roughness reduction from 29.62% to 41.64% and improved MRR 42.91% to 53.51% than conventional EDM.

1. Introduction

The most modern advanced material removal process which is useful to enhance machining performance and efficiency is Powder Mixed dielectric fluid in Electrical Discharge Machining (PMEDM). It enlarges the machining efficiency and surface finish by faster sparking and increase thermal conductivity within discharge arise results faster erosion from the work piece's surface. Jeswani (1981) investigated the effect of the addition of Graphite powder in Kerosene and used as dielectric in EDM. The authors highlighted that generally the mix of Nano powder in electrical discharge machining for increasing material removal rate (MRR) and Tool wear rate (TWR). And in specific the addition of graphite Nono powders in kerosene for using as the dielectric medium (4g/l) brought improvement of 60% material removal rate and 15% of Tool wear rate in 10μm inter Electrode Gap and 30% reduction of breakdown voltage in the 50μm inter Electrode Gap. Kuldeep et al., (2011) considered chromium power suspension in dielectric fluid. Prabhu and Vinayagam (2011) investigated single wall CNT suspension in dielectric for machining AISI D2 tool steel in EDM and observed that good thermal conductivity, reduced white layer formation and absorb heal as well. The Prihandana et al. (2011) reported that suspension of nano sized powder of SiC and Al2O3 in dielectric fluid in micro electric discharge machining, improved the surface finish by generating 14% to 24% of average surface roughness as well as free from micro cracks and it is also mentioned that nano-graphite powder mixed dielectric produced a high sparking gap size.

2. Research GAP

Some research works were reported on EDM in machining performance improvements on such difficult-to-cut materials. Kuppan et al. (2008)] investigated EDM deep hole drilling on Inconel 718. The process variables like peak current, pulse on-time, duty factor and electrode speed were considered for their investigation as factors. In which the authors developed mathematical models for the responses of MRR and depth averaged surface roughness (DASR) and used response surface methodology. Rajyalakshmi and Venkata Ramaiah, (2013) investigations were for the performance measures like MRR, SF, and SG in Wire EDM on Inconel 825 material they used Taguchi grey relational analysis for optimizing process parameters. MuthuKumar et al. (2010) investigation was in WEDM process to optimize parameters by Gray–Taguchi Method for machining Incoloy800 super alloy with multiple performance characteristics such as MRR, surface roughness and Kerf. Mao-yong LIN et al. (2013) analyzed parameter optimization on micro milling EDM for the machining of Inconel 718. The another attempt was made by Muthu Kumar et al. (2010) on WEDM process to optimize parameters for improving responses of MRR, SR and Ker width for machining Titanium alloy.

The authors recently developed a Mathematical Model for Radial Overcut on Electrical Discharge Machining for Incoloy 800. And they used Response Surface Methodology (Muthu Kumar et al. (2014)). Rathi and Mane (2014) studied the Effect of Powder Mixed dielectric in EDM of Inconel 718 machining with powders like Silicon carbide, Aluminum oxide, Graphite. With other parameters of Current, Pulse on time (μs) and Duty cycle to optimize the responses of MRR and TWR. Prajapati and Haresh (2015), investigated SiO2 PMEDM in machining En-8 steel for optimize surface roughness. The Peak current, pulse on time and concentration of powder were considered for factors to optimize the response of surface finish. The concentration levels varied from 0 to 8 g/l but the dielectric without SiO2 powder mix gives minimum SR as 1.9 μm and roughness increases with increase of powder concentration. Bhiksha et al (2015) investigated effects of Graphite powder concentration in machining Ti-6Al-4V. The Graphite powder concentration varied from 4.5 to 13.5. Up to 9 g/l powder concentration the surface finish improved then it decrease of concentration. Satpal and Kalra (2014) investigation was Tungsten Powder in Dielectric in machining EN 24 steel, and reported that by increasing Powder Concentration from 0 to 4 g/l Powder Concentration, mean of the means of MRR is increased by 90.78%. Syed et al (2015) used Aluminum Powder concentration variation fro 0 g/l to 2g/l with kerosene in the machining of Al/SiC composites and found that in lower peak current (2A), the MRR increases with the increase of Al powder concentration in the dielectric fluid up to a certain limit. Kuldeep et al (2011) used Chromium Powder Suspended Dielectric to machine EN-8 steel, the range of concentration was 2 to 6 g/l. And concluded that MRR shows an increasing trend of increase in powder concentration.

The trend shows that MRR will increase further with further increase in concentration and the TWR increase with a lower range of powder concentration but then decrease. The Shriram et al. (2014) analyzed the effect of PMEDM in machining Tungsten Carbide of different grades W20, W30 and W40 with powders of Al, Gr and SiC in the concentration of 4, 8 and 12g/l respectively powder separately. The authors reported that the maximum MRR gained at 8 g/l of SiC powder with Flushing pressure 1.5 Kg/cm2. Tzeng and Lee [11] investigated the effect of Al, Cr, Cu and SiC powders on PMEDM in machining SKD-11 and reported that the properties of powders like concentration, density, size, thermal conductivity and electrical resistivity are significantly influenced in the machining performance and in particularly at a fixed concentration, the smallest size of the particle led to highest MRR. Hence, in this research, we used Nono sized particles as a very fine powder for and about to 5 mm in size were used, hence the process is Nano Powder Mix Electric Discharge Machining

(NPMEDM). That is this paper deals about NPMEDM, in which about 5Nm sized particle powders of multi wall CNT mixed in kerosene dielectric, SiC Nano Powder Mixed in kerosene dielectric fluid, and TiC Nano Powder Mixed in kerosene dielectric fluid along with kerosene and servotherm oil dielectric environments considered with other process parameters with three level of Peak Current, Pulse on Time, and Pulse off time to optimize the responses of MRR, TWR and SR in machining Inconel 800 work material with electrolyte copper electrode. With the best literature available till now, no such work was reported in any literature.

3. Research Design

3.1: Need For the Research

Inconel or incony series are Nickel based super alloys which are widely used in high temperature and high pressure applications such as, nuclear reactors, gas turbines, high temperature chemical vessels and electric power generation equipment (Hewidy et al. (2005)). The Incony 800 has high thermal strength, highly abrasive carbide particles high hardness, low thermal diffusivity. But Inconel 800 has high tendency to welding with tool material and forming built–up edge, which extremely difficult-to-machine (Narutaki et al. (1993)). Hence it is challenging to machine Incony 800 in Traditional machining processes. Alternatively the non-traditional machining methods like The Electro Chemical Machining, Abrasive water jet machining, Laser beam Machining, Electrical Discharge Machining (EDM) are attractive alternative offers for such hard materials. So in this study the Incony 800 material machining process is preferred.

3.2: Aluminum Nano particles

The most and widely used powders in PMEDM are Si and Al. The mixing of Aluminum powder in dielectric and machining with copper electrode is suggested for machining W300 Die Steel (Syed and Kuppan (2012)), for AISI D3 die steel (Jamadar and kavade (2014)), for H13 Steel (Gurtej et al. (2012), for D2 Die steel (Gurule and Nandurkar (2012), for AISI H -11 material (Mir et al. (2012), for and Hastelloy (Paramjit et al. 2012) . for AISI 1045 steel (Shivam Goyal, Rakesh Kumar Singh 2014) and in particularly mixing with kerosene in recommended by (Gurule and Nandurkar (2012), (Mir et al. (2012) and Chow et al. (2000). Wong et al. (1998) suggested Al powder at concentration of 2 g/l for obtaining mirror finish in SKH-15 Work piece and Syed et al. (2013) suggested 4 g/l aluminum powder in water. Here the Al powder mixed with kerosene and stirred continuously about 8 to 9 hours in the Magnetic stirrer for reducing the size of Particles as very fine and about to 5 Nm, before they mix in the actual dielectric.

3.3: Silicon Nono Particles

Banker (2014) state that the abrasive powders like Sic and alumina in mixed with dielectric improves the MRR. Mohri et al. investigated the effects of Si powder addition and reduced surface roughness about 2μm. Pecas and Henriques (2003) investigated and reported that the positive influence of 2 g/l concentrations of the silicon powder towards the reduction of the operating time required to achieve a specific surface quality in machining AISIH13 mould steel. Sukhvir and Jujhar (2014) used silicon powder in the dielectric fluid of the EDM ranging from 0.4g/l to 2 g/l and reported that the optimal MRR and TWR obtained with Powder Concentration =1.2 g/l in machining ASTM A-105 Steel. Kansal et al. (2007) studied the effect of silicon powder mixing into the dielectric fluid of EDM on machining suggested at high concentration of 4 g/l silicon powder in AISI D2 die steel to achieve high MRR .Similarly the silicon powder in Kerosene and machining with copper electrode was suggested by Nimo et al (2014) for EN 8 steel, Razak (2014) for Stavax material, Soumakant Padhee et al. (2012) for EN 31 Steel, Here also the before mixing of Si powder in dielectric fluid in the actual dielectric, it was stirred continuously about 8 to 9 hours in the Magnetic stirrer for reducing the size of Particles as very fine and about to 5 Nm with kerosene sample.

3.4: Multi-Wall Carbon Nano Tubes (MNCNTs)

CNT based nano fluids is of special interests to researchers because of its novel properties like efficient thermal conductors (Thermal conductivity 3,000 W/mK in axial direction and small values in a radial direction), high strength to weight ratio, amazing strength (Young's modulus of 1TPa), exclusive current carrying capacity (Conductivity of CNTs is 109 A/cm2) electrical properties, excellent field emitter and has a high aspect ratio. The CNTs are fullerene-related to structures that consist of either a grapheme cylinder or a number of concentric cylinders (Darji et al. 2014). The authors investigated MWCNTs NPMEDM for EN-31 Steel, with 0.5 g/l mixed with kerosene and reported that the MRR was improved averagely 19% and TWR was decreased averagely 8.51% with respect to the input parameter. Hence multi wall CNT stirred continuously with kerosene about 8 to 9 hours to reducing the maximum size of about 5 Nm before they added in to dielectric of the EDM.

3.5: Nano powder Concentration

Yan and Chen (1994) reported that the powder particles contribution in the reduction of surface cracks and improve surface finish (lowest surface roughness), for obtaining homogeneity of white layer formation and maintain a correct balance of the discharge energy density and the discharge rate were observed for a powder concentration within the range of 2 to 5 g/l. So here 2g/l is preferred for both Aluminum and silicon powder mix with kerosene. The amount of kerosene added in size reduction process were included total mixing ratio. In case of MWCNT, Mai and Hocheng (2011) warned that high concentrations of CNT powder in the gap often made the EDM process unstable. So this study considered the same 0.5g/l with kerosene as dielectric and 2g/l for aluminum and silicon Nono powder with kerosene to machine inconel 800 nickel based super alloy.

4. Experimental Design

This research is motivated for an industrial requirement. The manufacturer deals various kinds of jobs of inconel 800. The practical issues like, statistical approximation and formula based parameter computations often difficult to fix the parameter for machining. Sometimes the calculated parameters couldn't produce the required finish due to statistical approximation or labor mistake of calculating parameters. And the manufacture demanded that a parameter choice table (like paint shade card for selecting the suitable/desired colour) to fix factored for desired responses and for their regular use. Hence the research is designed completely experimental. Taguchi general full factorial design was preferred for the experimental design.

The Taguchi General full factorial is used to design the experiments and the same was used for analyzing the influencing of factors considered for producing optimal responses. The factors were taken in the rage they used for processing the order. The factors with their levels and responses considered were furnished in the Table 1. The factors of Input Current, Pulse on Time, pulse off time and the nano-powders and their concentrations for mix with kerosene to use as dielectric fluid were preferred based on the literature review to analyze the response like tool wear rate, Material removal rate and surface roughness. The objective is investigating the influence of nano-particle (nono-powder) suspension on dielectric fluid in machining hard nickel based super alloy inconel 800 work piece. The experimental design is tabulated in Table-2a to Table -2l. Total 108 experiments were obtained from the MINITAB release 16 software, for three levels of parameters in pulse on time, pulse off time, Current and 4 levels of dielectric medium (3x3x3x4 =108).

Table 1. Factors and Responses

Factor	Level 1	Level 2	Level 3	Level 4
Peak Current (A)	5	10	15	-
Pulse on time (µs)	6	7	8	-
Pulse off time (µs)	4	5	6	-
Dielectric fluid (kerosene mixing with)	EDM oil *	Al (2 g/l)	Si(2 g/l)	MWCNT (0.5 g/l)
Responses	MRR	TWR	SR	-

*No Nano Powder mixed case (conventional practices)

IOP Conf. Series: Materials Science and Engineering **197** (2017) 012068 doi:10.1088/1757-899X/197/1/012068

TABLE I. FACTORS AND RESPONSES

Run Order	Std Order	Current (A)	Pulse On (μs)	Pulse Off (μs)	Nano Powder (g/l)	MRR (g/Sec)	TWR (g/Sec)	Ra (μm)
1	44	10	6	4	MWCNTs (0.5)	0.27587	0.00050	1.01
2	94	15	7	5	Al (2)	0.54882	0.00329	2.02
3	59	10	7	5	Si (2)	0.58576	0.00215	1.79
4	43	10	6	4	Si (2)	0.21343	0.00058	1.07
5	7	5	6	4	Si (2)	0.11894	0.00040	1.01
6	71	10	8	5	Si (2)	1.03866	0.00411	2.09
7	92	15	7	4	MWCNTs (0.5)	0.82496	0.00253	1.13
8	6	5	6	4	Al (2)	0.09830	0.00040	1.22
9	51	10	7	3	Si (2)	0.47209	0.00199	1.13

TABLE II. FACTORS AND RESPONSES

Run Order	Std Order	Current (A)	Pulse On (μs)	Pulse Off (μs)	Nano Powder (g/l)	MRR (g/min)	TWR (g/min)	Ra (μm)
10	52	10	7	3	MWCNTs (0.5)	0.60346	0.00178	1.00
11	101	15	8	4	No NP (0)	0.67268	0.00705	2.05
12	32	5	8	4	MWCNTs (0.5)	0.94256	0.00313	1.16
13	64	10	8	3	MWCNTs (0.5)	1.07264	0.00411	1.14
14	17	5	7	4	No NP (0)	0.30753	0.00174	1.65
15	19	5	7	4	Si (2)	0.45024	0.00153	1.12
16	22	5	7	5	Al (2)	0.37984	0.00188	1.78
17	50	10	7	3	Al (2)	0.40252	0.00218	1.29
18	25	5	8	3	No NP (0)	0.50210	0.00422	1.61

TABLE III. FACTORS AND RESPONSES

Run Order	Std Order	Current	Pulse On (μs)	Pulse Off (μs)	Nano Powder (g/l)	MRR (g/min)	TWR (g/min)	Ra (μm)
19	21	**5**	7	5	No NP (0)	0.30573	0.00225	1.93
20	13	5	7	3	No NP (0)	0.30676	0.00168	1.45
21	81	15	6	5	No NP (0)	0.25684	0.00147	1.86
22	62	10	8	3	Al (2)	0.71494	0.00536	1.46
23	61	10	8	3	No NP (0)	0.61234	0.00587	1.68
24	49	10	7	3	No NP (0)	0.34833	0.00248	1.49
25	18	5	7	4	Al (2)	0.37210	0.00161	1.45
26	16	5	7	3	MWCNTs (0.5)	0.54673	0.00124	0.98
27	1	5	6	3	No NP (0)	0.07219	0.00037	1.31

TABLE IV. FACTORS AND RESPONSES

Run Order	Std Order	Current (A)	Pulse On (μs)	Pulse Off (μs)	Nano Powder (g/l)	MRR (g/min)	TWR (g/min)	Ra (μm)
28	99	15	8	3	Si (2)	0.94047	0.00631	1.45
29	65	10	8	4	No NP (0)	0.60530	0.00557	2.03
30	58	10	7	5	Al (2)	0.46888	0.00239	1.93
31	34	5	8	5	Al (2)	0.69590	0.00413	2.12
32	66	10	8	4	Al (2)	0.73120	0.00524	1.79
33	41	10	6	4	No NP (0)	0.15471	0.00656	1.49
34	3	5	6	3	Si (2)	0.10233	0.00031	0.99
35	63	10	8	3	Si (2)	0.85807	0.00489	1.27
36	40	10	6	3	MWCNTs (0.5)	0.23132	0.00040	0.94

TABLE V. FACTORS AND RESPONSES

Run Order	Std Order	Current (A)	Pulse On (μs)	Pulse Off (μs)	Nano Powder (g/l)	MRR (g/min)	TWR (g/min)	Ra (μm)
37	96	15	7	5	MWCNTs (0.5)	0.93914	0.00248	1.82
38	107	15	8	5	Si (2)	1.01378	0.00630	2.18
39	48	10	6	5	MWCNTs (0.5)	0.34454	0.00049	1.11
40	53	10	7	4	No NP (0)	0.34230	0.00249	1.78
41	88	15	7	3	MWCNTs (0.5)	0.74389	0.00245	1.04
42	72	10	8	5	MWCNTs (0.5)	1.43593	0.00361	2.03
43	30	5	8	4	Al (2)	0.61350	0.00384	1.69
44	108	15	8	5	MWCNTs (0.5)	1.39543	0.00564	2.11
45	90	15	7	4	Al (2)	0.52210	0.00321	1.64

TABLE VI. FACTORS AND RESPONSES

Run Order	Std Order	Current (A)	Pulse On	Pulse Off	Nano Powder (g/l)	MRR (g/min)	TWR (g/min)	Ra (μm)
46	15	5	7	3	Si (2)	0.43523	0.00138	1.09
47	46	10	6	5	Al (2)	0.20078	0.00065	1.63
48	83	15	6	5	Si (2)	0.39860	0.00117	1.15
49	82	15	6	5	Al (2)	0.31891	0.00125	1.72
50	37	10	6	3	No NP (0)	0.12975	0.00057	1.40
51	29	5	8	4	No NP (0)	0.50725	0.00407	1.92
52	27	5	8	3	Si (2)	0.71903	0.00349	1.21
53	28	5	8	3	MWCNTs (0.5)	0.91896	0.00287	1.08
54	10	5	6	5	Al (2)	0.11180	0.00047	1.50

TABLE VII. FACTORS AND RESPONSES

Run Order	Std Order	Current (A)	Pulse On (μs)	Pulse Off (μs)	Nano Powder (g/l)	MRR (g/min)	TWR (g/min)	Ra (μm)
55	78	15	6	4	Al (2)	0.26650	0.00122	1.38
56	38	10	6	3	Al (2)	0.15427	0.00053	1.21
57	14	5	7	3	Al (2)	0.36433	0.00152	1.26
58	89	15	7	4	No NP (0)	0.43249	0.00342	1.87
59	57	10	7	5	No NP (0)	0.37756	0.00280	2.09
60	73	15	6	3	No NP (0)	0.18745	0.00117	1.43
61	5	5	6	4	No NP (0)	0.08239	0.00042	1.39
62	54	10	7	4	Al (2)	0.41140	0.00232	1.56
63	68	10	8	4	MWCNTs (0.5)	1.14130	0.00428	1.22

TABLE VIII. FACTORS AND RESPONSES

Run Order	Std Order	Current (A)	Pulse On (μs)	Pulse Off (μs)	Nano Powder (g/l)	MRR (g/min)	TWR (g/min)	Ra (μm)
64	55	10	7	4	Si (2)	0.49449	0.00230	1.14
65	12	5	6	5	MWCNTs (0.5)	0.19290	0.00036	1.02
66	100	15	8	3	MWCNTs (0.5)	1.18295	0.00522	1.30
67	97	15	8	3	No NP (0)	0.41304	0.00346	1.55
68	33	5	8	5	No NP (0)	0.56255	0.00486	2.29
69	39	10	6	3	Si (2)	0.18438	0.00048	1.05
70	47	10	6	5	Si (2)	0.25278	0.00056	1.15
71	106	15	8	5	Al (2)	0.92407	0.00735	2.35
72	69	10	8	5	No NP (0)	0.67892	0.00570	2.45

TABLE IX. FACTORS AND RESPONSES

Run Order	Std Order	Current	Pulse On	Pulse Off	Nano Powder (g/l)	MRR (g/min)	TWR (g/min)	Ra (µm)
73	9	5	6	5	No NP (0)	0.08968	0.00055	1.63
74	11	5	6	5	Si (2)	0.14936	0.00040	1.05
75	4	5	6	3	MWCNTs (0.5)	0.12835	0.00025	0.88
76	95	15	7	5	Si (2)	0.68974	0.00292	1.87
77	86	15	7	3	Al (2)	0.49270	0.00313	1.34
78	60	10	7	5	MWCNTs (0.5)	0.80297	0.00181	1.72
79	84	15	6	5	MWCNTs (0.5)	0.54649	0.00095	1.09
80	103	15	8	4	Si (2)	0.95988	0.00637	1.47
81	20	5	7	4	MWCNTs (0.5)	0.58082	0.00136	1.05

TABLE X. FACTORS AND RESPONSES

Run Order	Std Order	Current	Pulse On	Pulse Off	Nano Powder (g/l)	MRR (g/min)	TWR (g/min)	Ra (µm)
82	76	15	6	3	MWCNTs (0.5)	0.33273	0.00075	0.96
83	79	15	6	4	Si (2)	0.33247	0.00126	1.09
84	8	5	6	4	MWCNTs (0.5)	0.15437	0.00035	0.97
85	23	5	7	5	Si (2)	0.47466	0.00161	1.65
86	45	10	6	5	No NP (0)	0.16180	0.00763	1.76
87	87	15	7	3	Si (2)	0.58643	0.00286	1.17
88	31	5	8	4	Si (2)	0.72383	0.00373	1.23
89	24	5	7	5	MWCNTs (0.5)	0.64989	0.00142	1.59
90	26	5	8	3	Al (2)	0.59742	0.00382	1.39

TABLE XI. FACTORS AND RESPONSES

Run Order	**Std Order**	**Current**	**Pulse On**	**Pulse Off**	**Nano Powder (g/l)**	**MRR (g/min)**	**TWR (g/min)**	**Ra (µm)**
91	42	10	6	4	Al (2)	0.17630	0.00062	1.31
92	102	15	8	4	Al (2)	0.81330	0.00663	1.81
93	104	15	8	4	MWCNTs (0.5)	1.23825	0.00535	1.39
94	67	10	8	4	Si (2)	0.89846	0.00496	1.29
95	105	15	8	5	No NP (0)	0.74297	0.00859	2.54
96	75	15	6	3	Si (2)	0.26374	0.00091	1.07
97	93	15	7	5	No NP (0)	0.44281	0.00388	2.19
98	70	10	8	5	Al (2)	0.83280	0.00479	2.26
99	35	5	8	5	Si (2)	0.87644	0.00364	1.96

TABLE XII. FACTORS AND RESPONSES

Run Order	**Std Order**	**Current**	**Pulse On**	**Pulse Off**	**Nano Powder (g/l)**	**MRR (g/min)**	**TWR (g/min)**	**Ra (µm)**
100	85	15	7	3	No NP (0)	0.67632	0.00754	1.92
101	91	15	7	4	Si (2)	0.63184	0.00314	1.19
102	74	15	6	3	Al (2)	0.22187	0.00100	1.24
103	2	5	6	3	Al (2)	0.08590	0.00033	1.14
104	56	10	7	4	MWCNTs (0.5)	0.64253	0.00187	1.08
105	36	5	8	5	MWCNTs (0.5)	1.19067	0.01133	1.90
106	80	15	6	4	MWCNTs (0.5)	0.42598	0.00099	1.03
107	98	15	8	3	Al (2)	0.79329	0.00680	1.67
108	77	15	6	4	No NP (0)	0.23205	0.00130	1.57

5. Experimental Method

The EDM set-up employed in this experimental study is Electronica Machine Tools make Xpert 1model die sinking type CNC EDM machine. Kerosene is a dielectric fluid in this machine because of its property of very low viscosity and it gets flushed away easily. The fixed time of 5 minutes per experimental was permitted. The dielectric mediums are considered separately in the conduct of experiments. The Taylor Hobson makes, Surtronic3+ branded contact type profile-meter was employed in surface roughness measurement with 0.8mm cut length and sampling number is three. The laboratory balance of semi micro with an accuracy of 0.00001g, was employed in measuring weight loss tool and work material in before and after every run to compute MRR and TWR. The observations of Experiments servotherm oil mixed with Kerosene dielectric fluid environment. It is conventional practices to machine Inconel 800 jobs by the manufacturers in other words, no powder mixed in dielectric fluid or conventional EDM and not PMEDM environment. In the case of Nano Powder mixed EDM, the environments of Aluminum nano-powder mixed with Kerosene dielectric fluid, silicon nano-powder mixed with the Kerosene dielectric fluid environment and MWCNTs Nono-powder mixed with the Kerosene dielectric fluid environment was separately considered for experimentation due to practical constraints of random experiments those environments as in order of expernet run. The observations were recorded as per Taguchi general Full Factorial Design of Experiments in Table 2*a* to Table 2*l*.

6. Results and Discussion

6.1: Material Removal Rate

The material removal rate performance by nano particle suspended EDM machining on Inconel 800 nickel based super alloy material is graphically illustrated in the Figure 1 to Figure 9. In which A indicates No Nano Poweder case, B indicates Si Nano Poweder case, C indicates Al Nano Poweder case and D indicates MWCNTs Nano Poweder case. Figure 1 to Figure 3 are the MRR performance of Nano Powders in NPMEDM at Current 5A, 10A and 15A respectively, when the Pulse off time is 3 μs, the Figure 4 to Figure 6 are for the Pulse off time 4 μs and Figure 7 to Figure 9 are for the Pulse off time is 5μs. on other words the Figure 1, Figure 4 and Figure 7 are the MRR at Current 5A, the the Figure 2, Figure 5 and Figure 8 are the MRR when the input Current 10A, and the the Figure 3, Figure 6 and Figure 9 are the MRR when the Peack current 15A. It is observed that when the input current increases the MRR increases, when the pulse on time increases the MRR increases. At high pulse on time the NPMEDM performance is more significant than lower pulse on time. When the pulse off time increases, the MRR increases. Hence all these parameters influence the MRR performance of NPMEDM. The NPMEDM performance with MWCNT is most significant and MRR is high at high plus off time, high pulse on time and High input current. The low MRR observed at no nano powder mixed in dielectric environment machining.

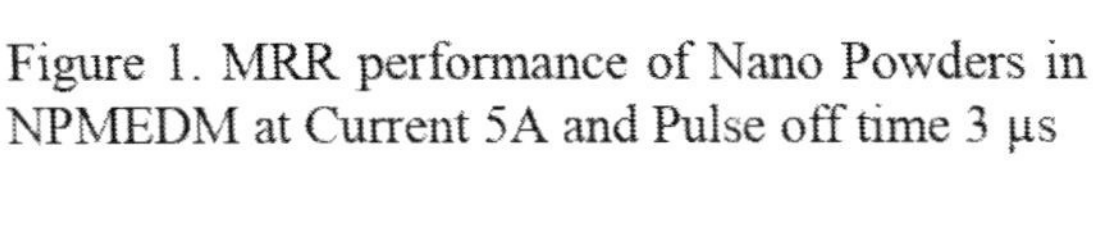

Figure 1. MRR performance of Nano Powders in NPMEDM at Current 5A and Pulse off time 3 μs

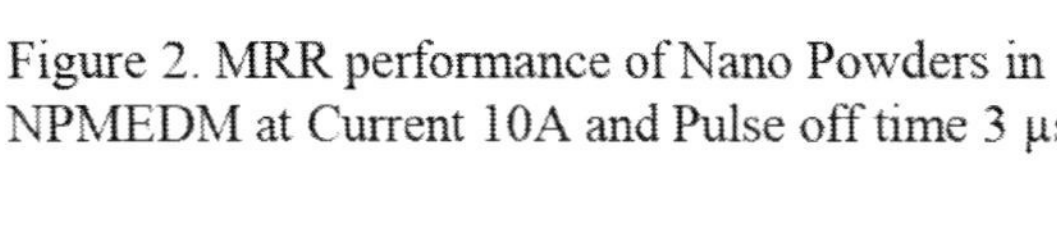

Figure 2. MRR performance of Nano Powders in NPMEDM at Current 10A and Pulse off time 3 μs

IOP Conf. Series: Materials Science and Engineering **197** (2017) 012068 doi:10.1088/1757-899X/197/1/012068

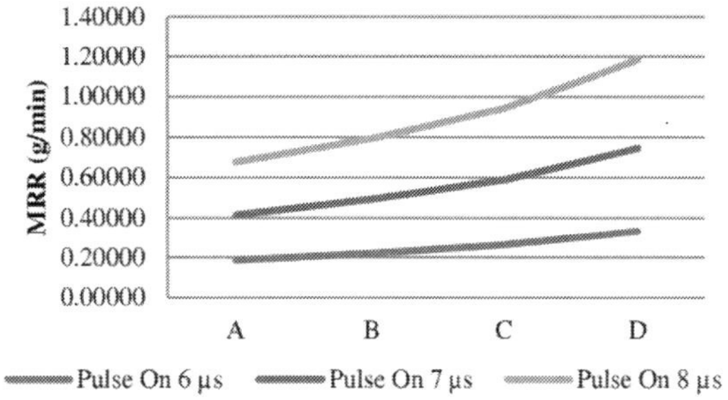

Figure 3. MRR performance of Nano Powders in NPMEDM at Current 15A and Pulse off time 3 µs

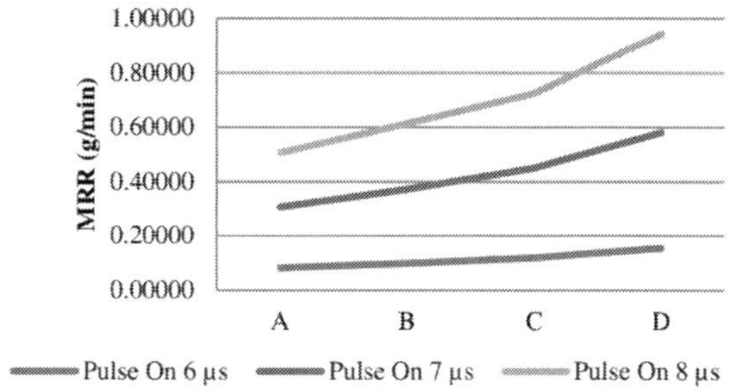

Figure 4. MRR performance of Nano Powders in NPMEDM at Current 5A and Pulse off time 4 µs

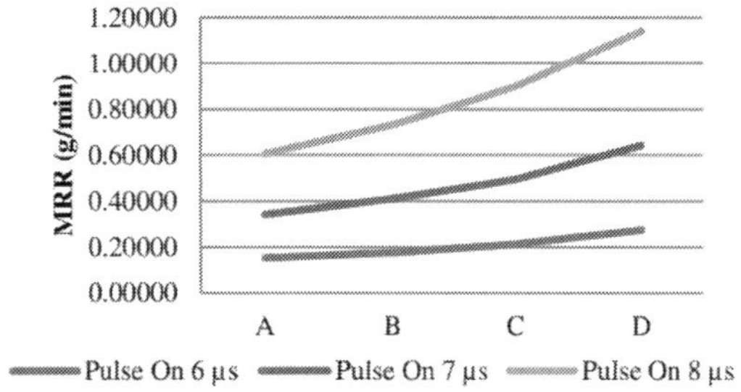

Figure 5. MRR performance of Nano Powders in NPMEDM at Current 10A and Pulse off time 4 µs

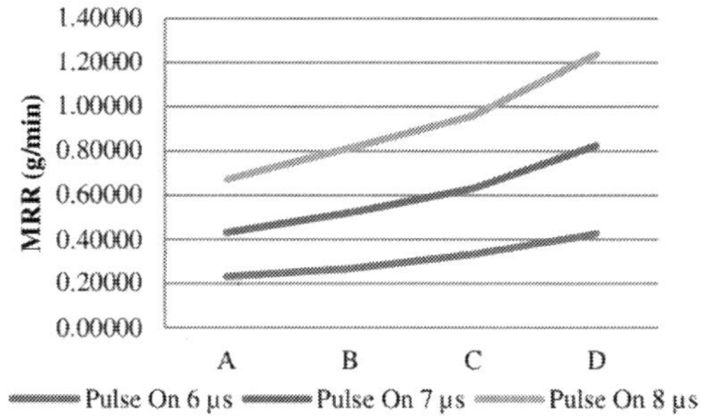

Figure 6. MRR performance of Nano Powders in NPMEDM at Current 15A and Pulse off time 4 µs

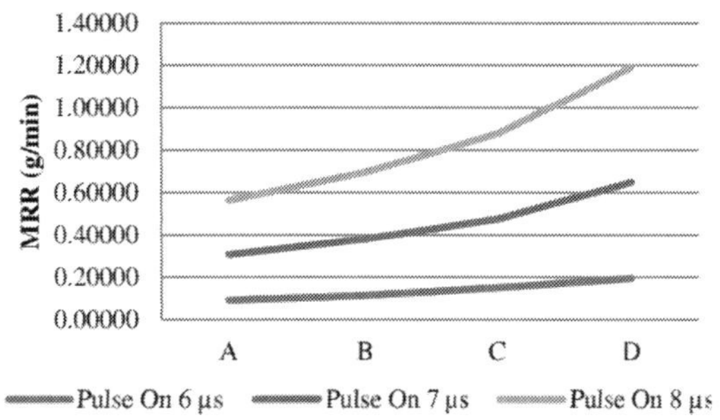

Figure 7. MRR performance of Nano Powders in NPMEDM at Current 5A and Pulse off time 5 µs

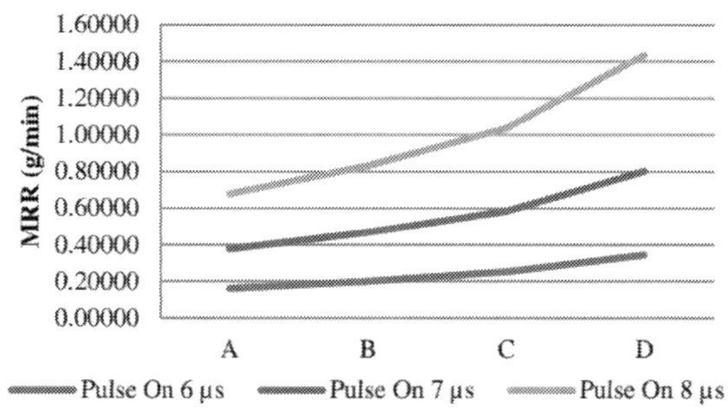

Figure 8. MRR performance of Nano Powders in NPMEDM at Current 10A and Pulse off time 5 µs

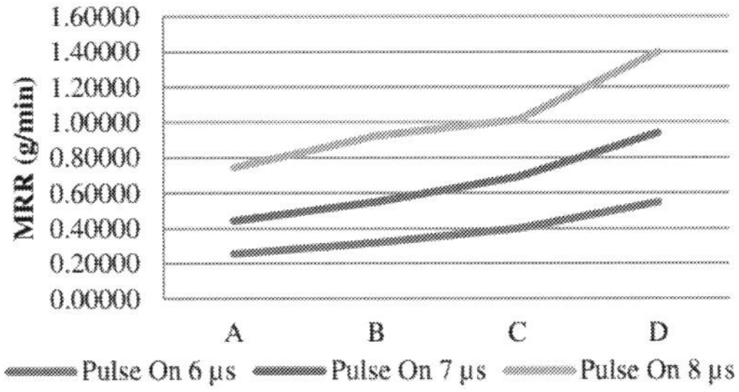

Figure 9. MRR performance of Nano Powders in NPMEDM at Current 15A and Pulse off time 5 µs

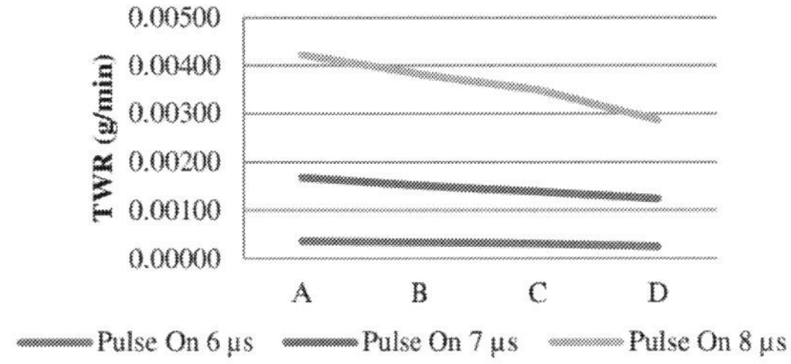

Figure 10. TWR performance of Nano Powders in NPMEDM at Current 5A and Pulse off time 3 µs

6.2: Tool Wear Rate

The influence of the factors in reduction of tool wear was depicted in the figure 10 to Figure 18. From These graphs in the Firure 10 to Figure 12 TWR performance given by Nano Powders in NPMEDM at Current 5A, 10A and 15A respectively, when the Pulse off time is 3 µs, the the Firure 13 to Figure 15 is for Pulse off time 4 µs and the the Firure 16 to Figure 18 for the Pulse off time is 5µs. The Figure 10, Figure 13 and Figure 16 are the TWR when peak Current 5A, the Figure 11, Figure 14 and Figure 17 are the TWR when peak Current 10A, and the The Figure 12, Figure 15 and Figure 18 are the TWR when peak Current Current 15A. The tool wear rate is high at no nano powder mixed in dielectric environment machining. The addition of nano powder decreases the tool wear. The high tool wear reduction by nano fluids was observed at higher pulse on time as well as high pulse off time. The input current influence on tool wear rate. At higher the current the tool wear also higher.

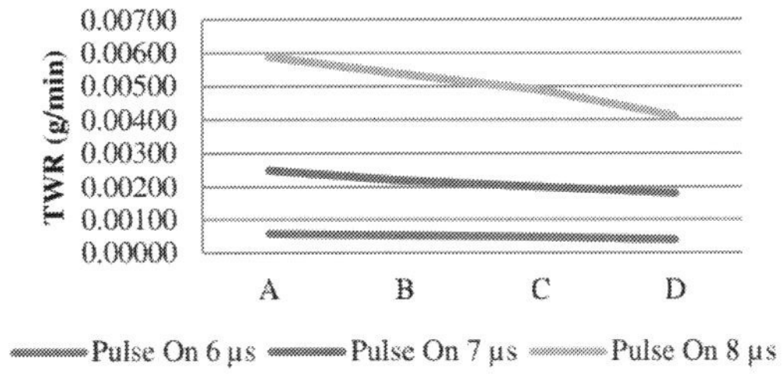

Figure 11. TWR performance of Nano Powders in NPMEDM at Current 10A and Pulse off time 3 µs

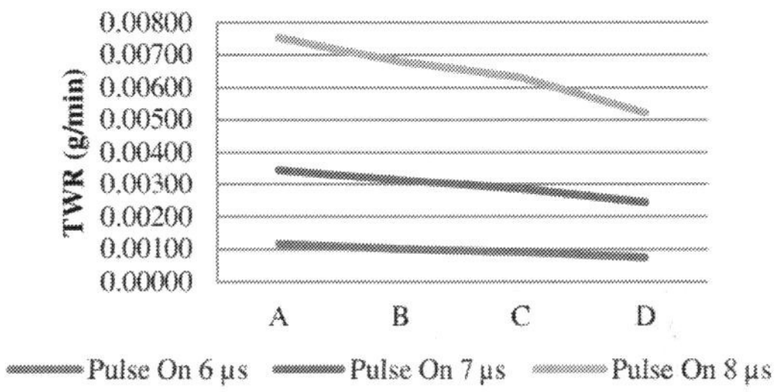

Figure 12. Performance of Nano Powders in NPMEDM at Current 15A and Pulse off time 3 µs

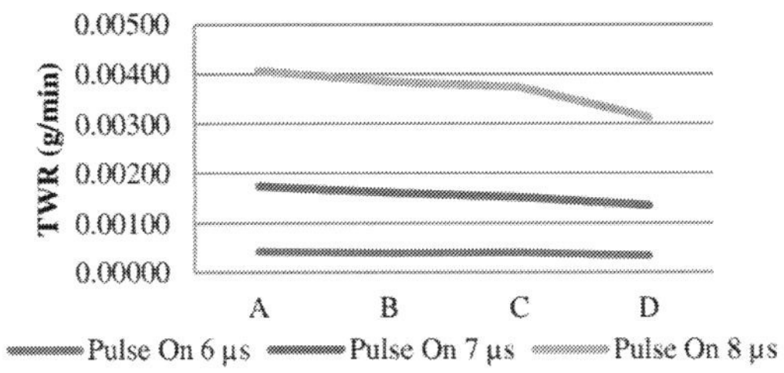

Figure 13. TWR performance of Nano Powders in NPMEDM at Current 5A and Pulse off time 4 µs

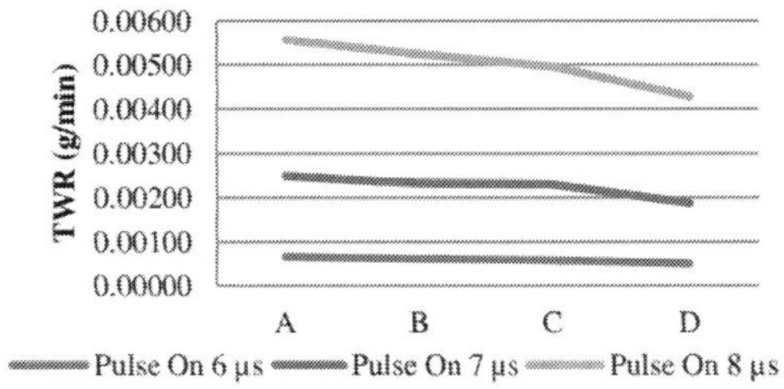

Figure 14. TWR performance of Nano Powders in NPMEDM at Current 10A and Pulse off time 4 µs

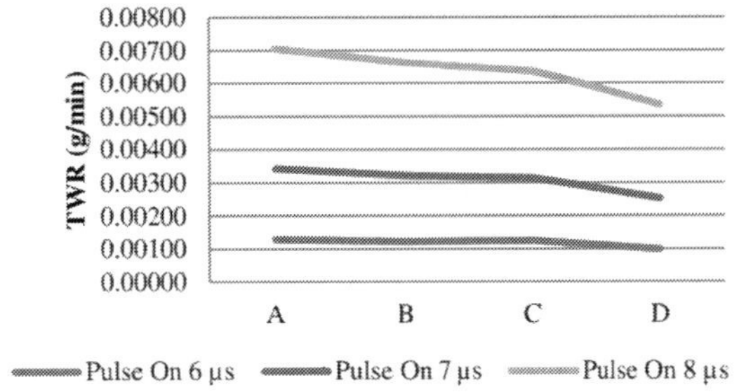

Figure 15. TWR performance of Nano Powders in NPMEDM at Current 15A and Pulse off time 4 µs

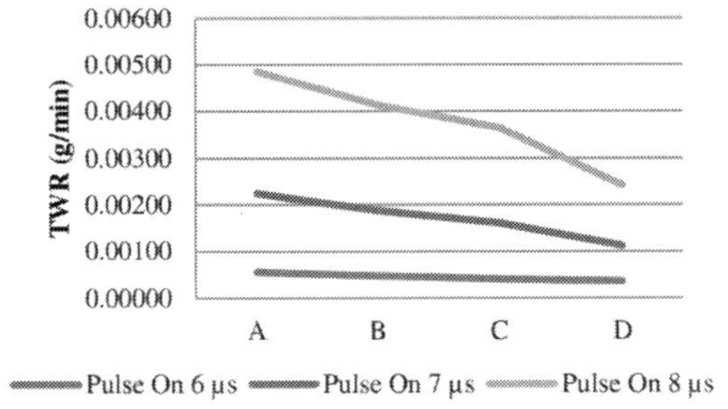

Figure 16. TWR performance of Nano Powders in NPMEDM at Current 5A and Pulse off time 5 µs

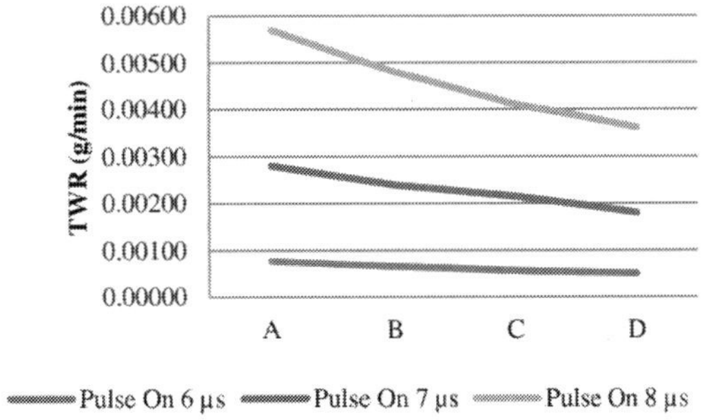

Figure 17. TWR performance of Nano Powders in NPMEDM at Current 10A and Pulse off time 5 µs

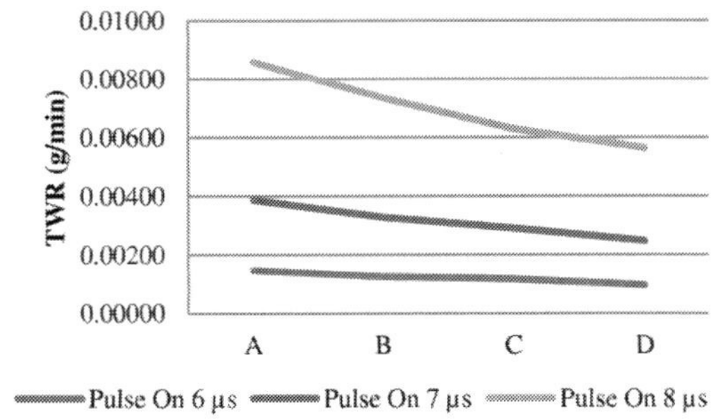

Figure 18. TWR performance of Nano Powders in NPMEDM at Current 15A and Pulse off time 5 µs

6.3: Surface Roughness

The surface roughness is one of the measures of performance of surface quality. As above way the surface roughness performances were depicted from Figure 19 to Figure 27. And it is evident form those graphs that surface roughness reduction in NPMEDM is highly significant. The lower surface roughness was observed at 6 µs pulse on time. The higher the surface roughness was observed at higher pulse on time. Like above all the factors influence in roughness of machined work piece. The roughness reduction is steep when adding nano sized abrasive powders (Al and Si) in dielectric fluid but best MWCNT as nano-powder in NPMEDM.

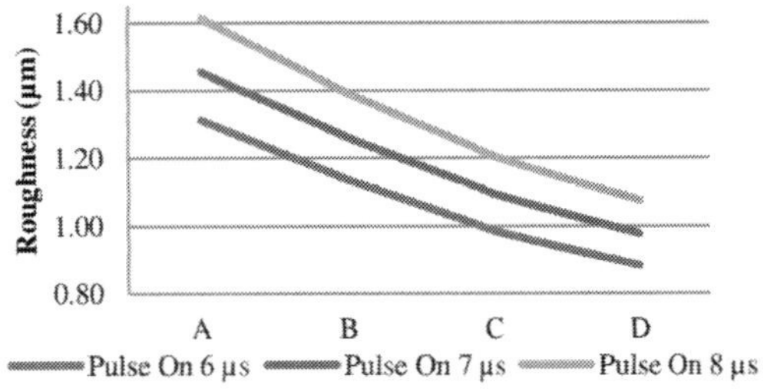

Figure 19. Surface Roughness performance of Nano Powders in NPMEDM at Current 5A and Pulse off time 3 µs

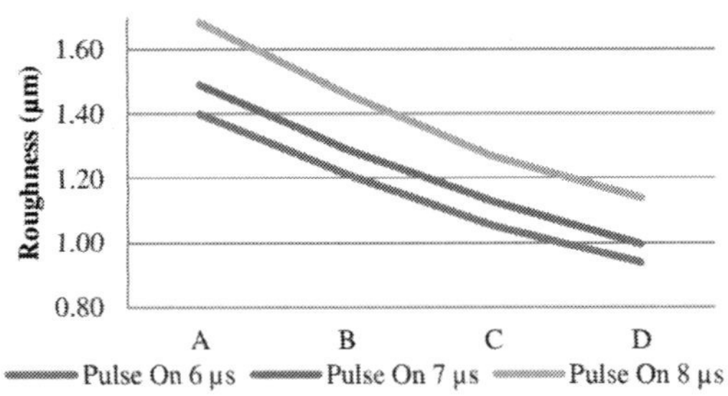

Figure 20. Surface Roughness performance of Nano Powders in NPMEDM at Current 10A and Pulse off time 3 µs

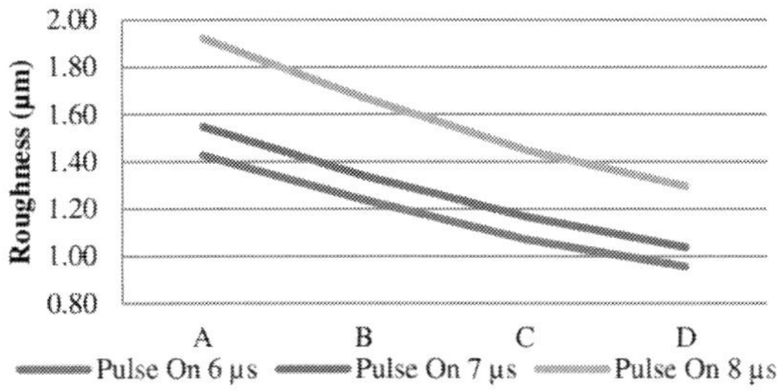

Figure 21. Surface Roughness performance of Nano Powders in NPMEDM at Current 15A and Pulse off time 3 µs

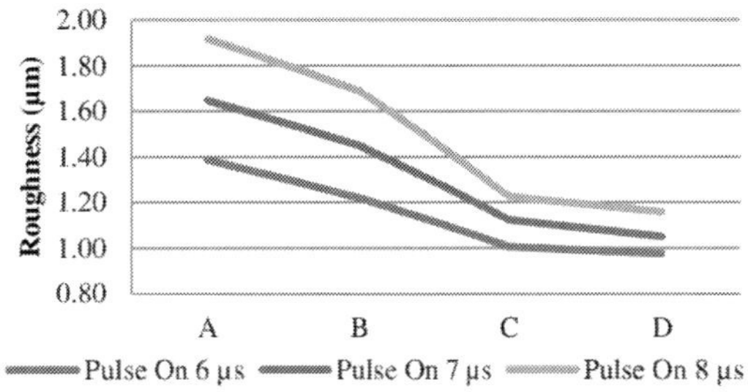

Figure 22. Surface Roughness performance of Nano Powders in NPMEDM at Current 5A and Pulse off time 4 µs

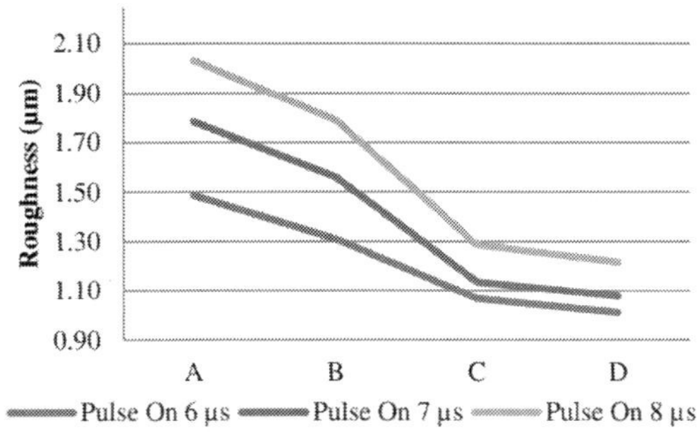

Figure 23. Surface Roughness performance of Nano Powders in NPMEDM at Current 10A and Pulse off time 4 µs

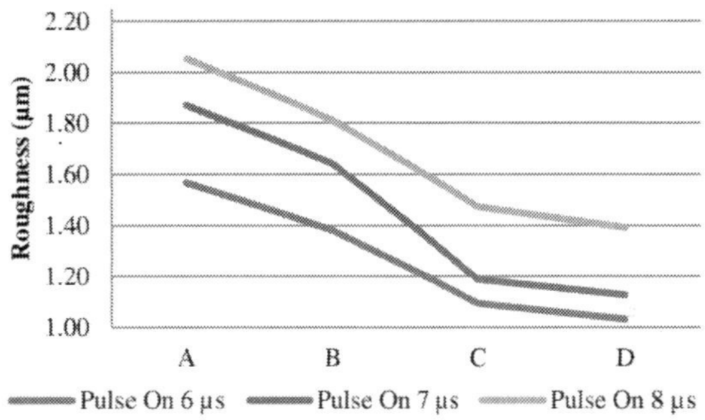

Figure 24. Surface Roughness performance of Nano Powders in NPMEDM at Current 15A and Pulse off time 4 µs

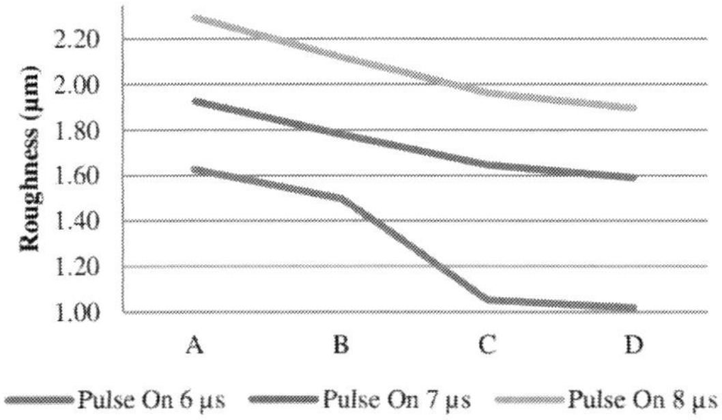

Figure 25. Surface Roughness performance of Nano Powders in NPMEDM at Current 5A and Pulse off time 5 µs

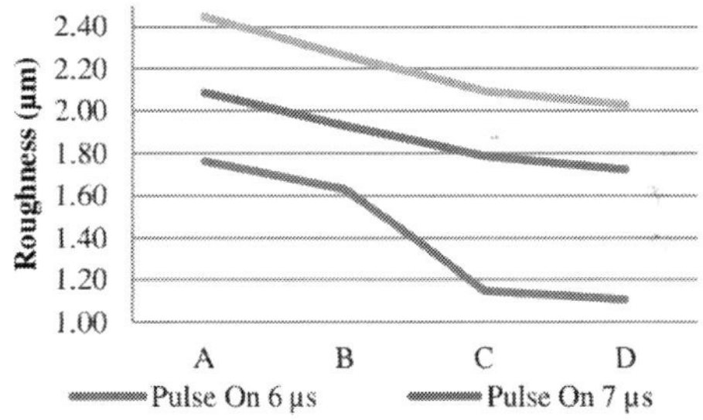

Figure 26. Surface Roughness performance of Nano Powders in NPMEDM at Current 10A and Pulse off time 5 µs

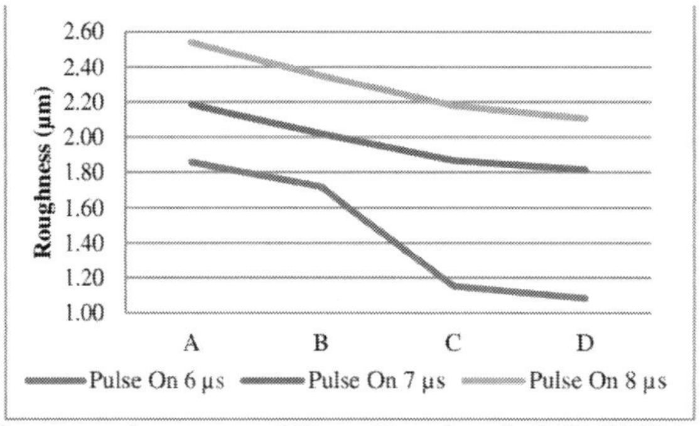

Figure 27. Surface Roughness performance of Nano Powders in NPMEDM at Current 15A and Pulse off time 5 µs

7. Conclusion

The NPMDEM machining on inconel 800 material experimentally investigated and analyzed well. The NPMEDM performances were compared with conventional EDM with equal priority. The result shows that the NPMEDM outperforms in terms of higher material removal rate, minimum tool wear and minimal surface roughness in machined work piece. In particularly MWCNTs powder mixed NPMEDM environment. All the factors which considered for investigation were contributing for the response of MRR, TWR and SR. The observation table can be used for parameter setting to obtain desired result by the manufacturer like shade card for pain selection. It is easy and convenient to indentify suitable parameters for desired outcome of jobs. Here some in specific conclusions

The Input current significantly influence in MRR. The pulse on time and pulse off time influence in responses were clearly depicted in figure 1 to figure 3.

The minimum material removal rate (0.04721g/min) was observed in silicon powder mixed NPMEDM environment at pulse off time 3 µs, pulse on time 7µs and the input current of 10A and the highest material removal rate 1.435926 g/min observed in at pulse off time 5 µs, pulse on time 8µs 10A current at MWCNT mixed NPMEDM in machining inconel 800 work piece.

In the case of roughness of machinated surface, the minimum roughness of 0.88 µm was observed in MWCNTs mixed NPMEDM at current 5A, pulse on time 6µs and pulse off time 3µs. the highest roughness of 2.54 µm conventional EDM of kerosene mixed with EDM oil in dielectric fluid environment in the parameters of 15A current, pulse on time 8µs and pulse off time 5µs. in case of NPMEDM 2.35 µm roughness in aluminum nano powder mix case at 15A current, pulse on time 8µs and pulse off time 5µs.

The maximum tool wear 0.00859 g/min was observed in conventional EDM at 15A current, pulse on time 8µs and pulse off time 5µs. In case of NPMEDM, the MWCNTs mixed NPMEDM environment produces 0.00753 g/min of tool wear at 15A current, pulse on time 6µs and pulse off time 3µs. but the same NPMEDM environment produced minimum tool wear of .00025 g/min at 5A current, pulse on time 6µs and pulse off time 3µs

References

[1] O.Kuldeep, R. K. Garg, K. K. Singh, “Experimental Investigation and Modeling of PMEDM Process with Chromium Powder Suspended Dielectric,” Int. J. Appl. Sci. Engg. 9 (2) (2011), pp. 65-81.

[2] T. Prabhu, B.K. Vinayagam6, “Analysis of surface characteristics of AISI D2 toolsteel material using electric discharge machining process with single-wall carbon nano-tubes,” Int. J. Machining and Machinability of Materials. (10) (2011), pp. 99-119.

[3] Gunawan Setia Prihandana, Muslim Mahardika, M. Hamdi, Y. S. Wong, Kimiyuki Mitsui, “Accuracy improvement in nanographite powder-suspended dielectric fluid for micro-

electrical discharge machining processes," Int. J Adv Manuf. Technol. (56) (2011), pp. 143-149.

[4] R.Boopathi and S.Sundaram, 'Influence of Process Parameters for Electrical Discharge Machine Using Nano Particle and Brass Electrode', International Journal of Research in Advent Technology, Vol.2, No.11, (2014), pp.45-49.

[5] M. L. Jeswani, 1981. Effect of the addition of Graphite powder to Kerosene Used as the dielectric fluid in Electrical Discharge Machining., Elsevier Sequoia S.A., Lausanne 133-139 .

[6] Hyun-Seok TAK, Chang-Seung H, Characteristic evaluation of Al2O3/CNTs hybrid materials for micro-electrical discharge machining. elsevier (2011)s28-s38.

[7] Yashesh Darji, Pankaj Koradiya and Jigesh R.Shah , Experiment investigation of EDM parameter MRR and TWRwith Multi Wall Carbon Nano Tubes, International Journal of Mechanical Engineering and Technology , Volume 5, Issue 7, July (2014), pp. 184-192

[8] V. Muthukumar,, N. Rajesh, R. Venkatasamy, A. Sureshbabu , N. Senthilkumar, Mathematical Modeling for Radial Overcut on Electrical Discharge Machining of Incoloy 800 by Response Surface Methodology Procedia Materials Science 6 (2014) 1674 – 1682

[9] C. Mai & Hong Hocheng" Advantages of carbon nanotubes in electrical discharge machining Int J Adv Manuf Technol DOI 10.1007/s00170-011-3476-2 (2011).

[10] MuthuKumar, V., SureshBabu, A., Venkatasamy, R., Raajenthiren, M., 2010. Optimization of the WEDM parameters on machining Incoloy800 Super alloy with multiple quality characteristics, International Journal of Engineering Science and Technology 2, 1538-1547.

[11] MuthuKumar,V., SureshBabu, A., Venkatasamy, R., Raajenthiren, M., 2010. Multi response optimization of machining parameters of Wire-EDM using grey relational analysis in the Taguchi Method. International Journal of Applied Engineering Research 5, 2325-2338.

[12] Kuppan, P., Rajadurai, A., Narayanan, S., 2008. Influence of EDM process parameters in deep hole drilling of Inconel 718, Int J Adv Manuf Technol 38, 74–84.

[13] Rajyalakshmi, G., Venkata Ramaiah, P., 2013. Multiple process parameter optimization of wire electrical discharge machining on Inconel 825 using Taguchi grey relational analysis. Int J Adv Manuf Technol 69, 1249–1262.

[14] Lin, M.Y., Tsao, C.C., Hsu, C.Y., Chiou, A.H., Huang, P.C., Lin, Y.C., 2013. Optimization of micro milling electrical discharge machining of Inconel 718 by Grey-Taguchi method. Trans. Nonferrous Met. Soc. China 23, 661−666.

[15] Mahendra G. Rathi, Deepak V. Mane Study on Effect of Powder Mixed dielectric in EDM of Inconel 718, International Journal of Scientific and Research Publications, Volume 4, Issue 11, November 2014pp1-7

[16] Sukhvir Singh and Jujhar Singh, Optimization of Powder Mixed Electric Discharge Machining using ASTM A-105 Steel by Response Surface Methodology, International Journal of Enhanced Research in Science Technology & Engineering, Vol. 3 Issue 8, August-2014, pp: (291-303).

[17] R.A.Prajapati and Haresh Patel, Experimental Investigation of Powder mix EDM using Silicon Dioxide as a Powder additive for Surface Roughness, International Journal of Advance Engineering and Research Development Volume 2, Issue 1, January -2015 pp. 28-34

[18] Bhiksha Gugulothu, D. Hanuma ntha Rao and G. Krishna Mohana Rao, Optimization of EDM process parameters and Graphite powder concentration on electrical Discharge machining of Ti-6Al-4V alloy using Taguchi method, International Journal of Advances In Production And Mechanical Engineering, Volume-1,Issue-5,2015.

[19] Satpal Singh and C.S. Kalra Experimental Study of PMEDM on EN 24 Steel with Tungsten Powder in Dielectric, International Journal on Emerging Technologies 5(1): 153-160(2014)

[20] Khalid Hussain Syed, G. P. Anuraag, Gangarapu Hemanth and Shaik Abdul Subahan, Powder-Mixed EDM Machining of Aluminium-Silicon Carbide Composites.

[21] Indian Journal of Science and Technology, Vol 8(S2), 133–137, January 2015.

[22] Shriram Y. Kaldhone, Mukund V. Kavade and Umashankar Rawat, Effect of Powder Mixed Dielectric on Performance Measures of EDM for Tungsten Carbide, International Journal of Innovative Research in Advanced Engineering Volume 1 Issue 10 (2014) pp. 106-111

[23] Kuldeep Ojha, R. K. Garg and K. K. Singh, Experimental Investigation and Modelling of PMEDM Process with Chromium Powder Suspended Dielectric, International Journal of Applied Science and Engineering 2011. 9, 2: 65-81

[24] Y.F. Tzeng, C.Y. lee, ―Effects of powder characteristics on EDM efficiency‖, Intl. J. Adv. Manuf. Tech, 17, 2001, 586-592.

[25] K.S. Banker , S. P. Parmar and B. C. Parekh Review to Performance Improvement of Die Sinking EDM Using Powder Mixed Dielectric Fluid International Journal of Research in Modern Engineering and Emerging Technology, Vol. 1, Issue: 3, April – 2013 pp. 57-62.

[26] Wong, Y.S., Lim, L.C., Rahuman, I., and Tee, W.M, "Near-Mirror-Finish Phenomenon in EDM using Powder-Mixed Dielectric", Journal Of Materials Processing Technology, Vol. 79, 30-40,1998.

[27] Tzeng, Y.F and Lee, C.Y, "Effects of Powder Characteristics on Electrodischarge Machining Efficiency", International Journal of Advance Manufacturing Technology, Vol. 17, 586-592 ,2001.

[28] Zhao, W.S., Meng, Q.G. and Wang, Z.L, "The Application of Research on Powder Mixed EDM in Rough Machining", Journal of Materials Processing Technology, Vol 129, 30-33,2002.

[29] Pecas, P. and Henriques, E, "Influence of Silicon Powder-Mixed Dielectric on Conventional Electrical Discharge Machining", International journal of machine tool and manufacturing, Vol. 43, 1465-1471, 2003.

[30] Kansal, H.K., Singh, Sehijpal and Kumar, P, "Effect of Silicon Powder Mixed EDM on Machining Rate of AISI D2 Die Steel", Journal of Manufacturing Processes,Vol. 9, 13-22,2007.

[31] Chow, H.M., Yang, L.D., Lin, C.T. and Chen, Y.F, "The Use of SiC Powder in Water as Dielectric for Micro-Slit EDM Machining", Journal of Materials Processing Technology, Vol. 195, 160-170,2008.

[32] Kansal, H.K., Singh, Sehijpal and Kumar, P, "Numerical Simulation of Powder Mixed Electric Discharge Machining (PMEDM) using Finite Element Method" Mathematical and Computer Modeling, Vol. 47, 1217-1237,2008.

[33] Kung, K.Y., Horng, J.T. and Chiang, K.T, "Material Removal Rate and Electrode Wear Ratio Study on the Powder Mixed Electrical Discharge Machining of Cobalt - Bonded Tungsten Carbide", International Journal of Advanced Manufacturing Technology, Vol. 40, 95-104,2009.

[34] M.Y. Ali, N. Atiqah and Erniyati (2011), " Silicon Carbide Powder Mixed Micro Electro Discharge Milling of Titanium Alloy", International Journal of Mechanical and Materials Engineering (IJMME), Vol.6 (2011), No.3, 338-342,2011.

[35] Jabbaripour, B., Sadeghi, M.H., Shabgard, M.R. and Faraji, H, "Investigating Surface Roughness, Material Removal Rate and Corrosion Resistance in PMEDM of γ-TiAl Intermetallic", Journal of Manufacturing Processes, Vol. 15, 56-68,2013.

[36] M. A. Razak, A. M. Abdul-Rani, and A. M. Nanimina, "Improving EDM Efficiency with Silicon Carbide Powder Mixed Dielectric Fluid", International Journal of Materials, Mechanics and Manufacturing, Vol. 3, No. 1, 2014.

[37] Mahendra G. Rathi, Deepak V. Mane, " Study on Effect of Powder Mixed dielectric in EDM of Inconel 718", International Journal of Scientific and Research Publications, Volume 4, Issue 11, November 2014 1 ISSN 2250-3153,2014.

[38] N Mohri, N Saito, T Takawashi, K Kobayashi, Mirror-like Finishing by EDM, Proceedings of the 25th International Symposium on Machine Tool Design and Research, UK, 1985, 329–336.

IOP Conf. Series: Materials Science and Engineering **197** (2017) 012069 doi:10.1088/1757-899X/197/1/012069

Investigation on the Effect of Friction Welding Parameters on Impact Strength in Dissimilar Joints

C.Shanjeevi, J.Jeswin Arputhabalan, Rohan Dutta and Pradeep

Department of Mechanical and Production Engineering
Sathyabama University, Chennai

shanjeevii@gmail.com

Abstract. Due to increase in demand of joining dissimilar material combinations in applications such as cryogenic fluids, power generation industries and reactor cooling systems, the solid state welding is most suitable in the current scenario. In this study, the experiment is undergone to evaluate the impact strength on the friction welded joint of copper and AISI 430 ferritic stainless steel material in the weld interface. The experiment is conducted by varying the input parameters such as friction pressure, upset pressure, burn-off length and rotational speed using Taguchi's L9 orthogonal array. The welded joints were examined under scanning electron microscopy (SEM) and type of failure mode is discussed.

1. Introduction

Friction welding process is increasing rapidly in dissimilar material combinations with defect free joint and also with less time consumption. Joining of copper and ferritic stainless steel is widely used nowadays for many industrial applications such as power generation industries, refrigeration cabinets, food processing, electrical cabinets, and automotive industries, chemical and nuclear industries. Friction welding is a solid state joining process by continuous rubbing action which produces heat between the circular interfaces to make plastic state. When sufficient energy is developed, the rotation stops and upset or compressive force is given to consolidate the weld and form a solid state bond. Friction pressure, upset pressure, burn-off length and rotational speed are the most important parameters to be used in the friction welding process.

Tomoyuki et al was examined to investigate the feasibility of the friction welding of the ferritic and martensitic steel combinations [1]. Paulraj Sathiya et al investigated on similar material of ferritic stainless steel by friction welding. An experimental study was focused on the effect of welding parameters on multiple performance characteristics by friction welding process. Experiments conducted based on Taguchi method showed that the joints exhibited high quality weld for the optimized process parameters [2]. A study was focussed on dissimilar material of austenitic stainless steel and low carbon using friction stir welding process. It was found that due to hot deformation in the welded zone, the austenite region produces small austenite grains is then transformed to fine ferrite and pearlite structure when reaches to room temperature [3].

Sammaiah et al experimentally investigated on the friction welding of 6063 aluminium alloy with AISI 304 austenitic stainless steel to determine the correlation between the microstructure and the joint strength. Few researchers have worked with copper combinations as well. Seong-Yeon Kim et al determined the factors affecting the joint performance of copper to titanium material and found that upset pressure plays a major role. It was stated that the tensile strength of dissimilar joint increased with increasing upset pressure [5].

A combination of sintered powder metallurgical steel to wrought copper material was made by friction welding and found more deformation on copper side due to high thermal conductivity [6]. Mumin sahin discussed the mechanical and metallurgical variations at weld interface of austenitic stainless steel and copper material with friction welding [7]. Subsequently the study was also focused on mechanical and metallurgical properties of austenitic stainless steel and copper material including surface roughness in the weld interface. The tensile strength in welded joint shows higher than copper material [8]. Radoslaw et al investigated the study of mechanical properties and microstructure of friction welded joint of ductile iron with stainless steel and studied the phase transformation and fracture morphology during friction welding [9].

Madhusudhan et al studied the microstructure and mechanical properties of similar and dissimilar welds of austenitic stainless steel, ferritic stainless steel and duplex stainless steel by using friction welding process. The impact strength and notch tensile strength of electron beam and friction weldments are studied and it shows lower toughness in both the weldments than parent metals [10]. Subhash et al studied the friction welding of austenitic stainless steel and low alloy steel related to the microstructure and mechanical properties. Hardness in the weld interface is harder than the parent metals. Notch tensile strength and impact toughness increases with increase in rotational speed and decreases constantly [11].

Yokoyama et al evaluated the tensile strength and absorbed energy under high and low loading rates. The tensile strength and energy of the friction welded joint increase with a high loading rate and fracture position of the joints varied with respect to the different loading rate [12]. Jian et al studied the friction stir welded 18Cr-2Mo ferriti stainless steel thick plate and the microstructure and mechanical properties are evaluated by showing fine grained equiaxed ferrite in the welded joint [13]. Based on the literature findings, the joining of dissimilar materials are studied widely based on strength and metallurgical aspects of ferritic stainless steel with other combinations and good amount of literature are available in friction welding process. However the dissimilar joint of ferritic stainless steel and copper material are very limited. In this study, the effect of impact strength and its fracture behaviours are discussed.

2. Experimental Procedure

Dissimilar materials of copper and ferritic stainless steel with cylindrical rods of 25 mm diameter were purchased. The chemical composition of the dissimilar material is given in Table 1 and 2. The friction welded samples are machined to a diameter of 24 mm and 75 mm in length. The surface is well polished and cleaned by using acetone. Experiments are conducted based on Taguchi's L9 orthogonal array with three levels and four factors is shown in Table 3. The joints made by friction welding are shown in Figure 1.

Table 1. Chemical composition of Ferritic stainless steel

Element	C	Si	Mn	P	S	Ni	Cr	Fe
%	0.076	0.551	0.689	0.023	0.016	0.19	16.52	81.58

Table 2. Chemical composition of Copper

Element	Cu	Fe
%	99.99	<0.01

Table 3. Experimental Factors and their Levels

Parameters	Level 1	Level 2	Level 3
Friction pressure (MPa)	22	33	43
Upset pressure (MPa)	65	87	108
Burn-off length (mm)	1	2	3
Rotational speed (rpm)	500	1000	1500

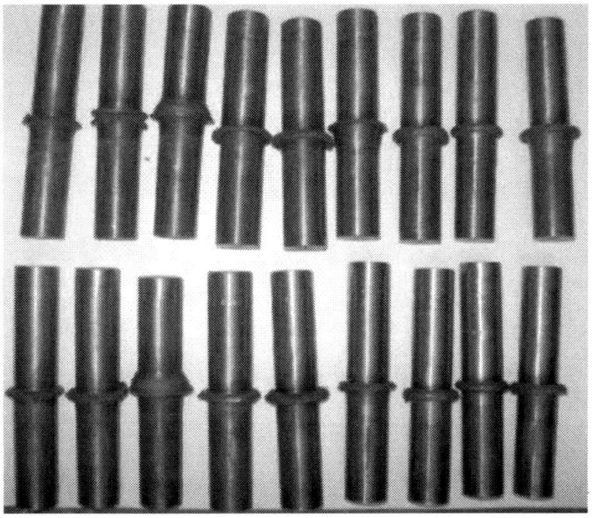

Figure 1. Friction welded joints

Table 4. Experimental Results

S.No	Friction Pressure (MPa)	Upset Pressure (MPa)	Burn-off length (mm)	Rotational Speed (rpm)	Impact Strength (J/cm^2)
1	22	65	1	500	42
2	22	87	2	1000	38
3	22	108	3	1500	34
4	33	65	2	1500	36
5	33	87	3	500	38
6	33	108	1	1000	36
7	43	65	3	1000	14
8	43	87	1	1500	46
9	43	108	2	500	44

3. Results and Discussion

3.1. Macroscopic analysis

At a macroscopic visual inspection, friction joint is characterized by the presence of plastically deformed flash material, extruded in both parallel and normal directions in respect to the force-motion plane, as shown in Figure 1. The flash length was comparable along the two extrusion directions. Flash material was expelled even from the corners, which is considered to be a necessary condition for obtaining sound joint.

3.2. Mechanical Properties

The impact test is conducted as per ASTM standard and impact values of the welded joint are listed in Table 4. On the basis of the impact strength of dissimilar friction-welded joints, the fractures occurred in the weld interface is shown in Figure 2. The experimental values of impact toughness values were ranged from 14 to 46 J/cm^2. As upset pressure decreased, the toughness decreases results in lower value of 14 J/cm^2. When the upset pressure increases, the joint was evidently superior to that of lowering upset pressure and results in higher value of 46 J/cm^2.

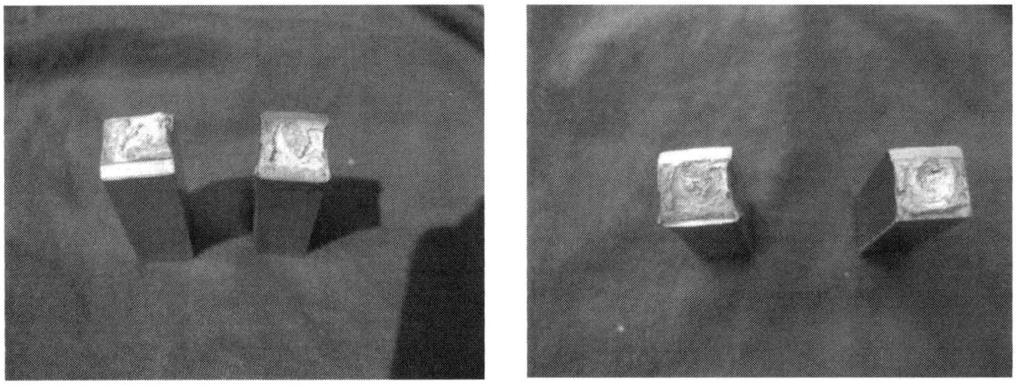

Figure 2. Impact fractured sample on ferritic stainless steel and copper

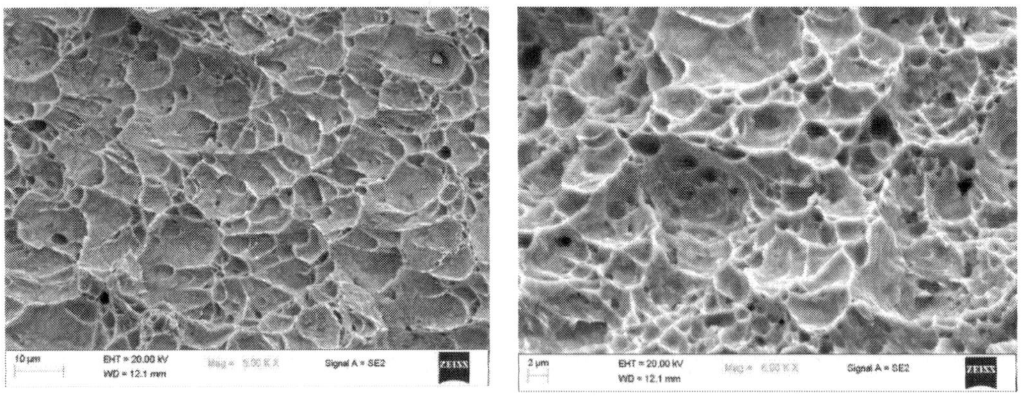

Figure 3. SEM analysis of impact fractured sample on ferritic stainless steel and copper

This is resulting due to fine grain formation with higher degree of working at the weld interface. Due to temperature rise at the weld interface and presence of intermetallic layers, alloying elements accumulate resulting in poor weld strength. The friction welded joint with a high absorbed energy shows more deformation in the copper material, whereas the friction welded joint with a low absorbed energy fractures in the weld interface without showing any clear plastic deformation. Since, ductile type of fracture appeared in copper side, SEM result shows with dimple features as observed in Figure 3.

4. Conclusion

The dissimilar material combination of ferritic stainless steel and copper material are friction welded and its toughness of the weld interface was evaluated. Based on the result analysis, the following conclusions were obtained.

- Friction welding has been successfully joined with ferritic stainless steel and copper material. The impact toughness values obtained on joints were varied with one another.
- The quality and the strength of the bond produced are also varied. Higher the burn-off length with low upset pressure decreases in impact toughness of friction-welded joint whereas with low burn-off and high upset pressure results in increase in toughness of welded joint.
- The lowest impact toughness obtained in friction welded joint was 14 J/cm2 and highest impact toughness shows with 46 J/cm2.
- Due to temperature rise at the weld interface and presence of intermetallic layers, accumulation of alloying elements results with poor impact toughness of the welded joint. Fracture analysis was made in the impact tested sample with different magnifications result in ductile mode of fracture with dimple formation.

References

[1] Tomoyuki Uwaba, Shigeharu Ukai, Tatsuyoshi Nakai, Masayuki Fujiwara, "Properties of friction welds between 9Cr-ODS martensitic and ferritic–martensitic steels", Journal of Nuclear Materials, 367–370, 1213–1217, 2007.

[2] Paulraj Sathiya, S.Aravindan, A.Nooral Haq, "Optimization for friction welding parameters with multiple performance characteristics", Int.J.Mech Mater Des, 3:309-318K.Y, 2006.

[3] M. Jafarzadegan, A. Abdollah-zadeh, A.H. Feng, T. Saeid, J. Shen, H. Assadi, "Microstructure and mechanical properties of a dissimilar friction stir weld between austenitic stainless steel and low carbon steel", Journal of Materials Science & Technology, Volume 29, Issue 4, Pages 367-372, April 2013.

[4] P. Sammaiah, Arjula Suresh, G. R. N. Tagore, "Mechanical properties of friction welded 6063 aluminum alloy and austenitic stainless steel", J Mater Sci, 45:5512–552, DOI 10.1007/s10853-010-4609-y, 2010.

[5] Seong-yeon kim, Seung-boo jung, Chang-chae shur, Yun-mo yeon, Dae-up kim, "Mechanical properties of copper to titanium joined by friction welding", Journal of Materials Science, 38, 1281 – 1287, 2003.

[6] Jayabharath.K, Ashfaq.M, Venugopal.P, Achar.D.R.G. Investigations on the continuous drive friction welding of sintered powder metallurgical (P/M) steel and wrought copper parts. Materials Science and Engineering A, 454-455: 114-123, 2007.

[7] Mumin Sahin, "Joining of stainless steel and copper materials with friction welding", Industrial Lubrication and Tribology, 61/6: 319–324, 2009.

IOP Conf. Series: Materials Science and Engineering **197** (2017) 012069 doi:10.1088/1757-899X/197/1/012069

[8] Shanjeevi.C, Satish Kumar.S, Sathiya.P, "Evaluation of Mechanical and Metallurgical properties of dissimilar materials by friction welding", Procedia Engineering ,64 1514-1523, 2013.
[9] Radosław Winiczenko, Mieczysław Kaczorowski, "Friction welding of ductile iron with stainless steel", Journal of Materials Processing Technology, Volume 213, Issue 3, Pages 453-462, March 2013.
[10] G. Madhusudan Reddy & K. Srinivasa Rao, "Microstructure and mechanical properties of similar and dissimilar stainless steel electron beam and friction welds", Int J Adv Manuf Technol, 45:875–888, 2009.
[11] G Subhash Chander , G Madhusudhan Reddy , A Venugopal Rao, " Influence of rotational speed on microstructure and mechanical properties of dissimilar metal AISI 304 - AISI 4140 continuous drive friction welds", Journal of Iron and Steel Research, International, 19(10): 64-73, 2012.
[12] T.Yokoyama, K.Ogawa, "Impact tensile properties of 6061 aluminium alloy to SUS 304 stainless steel friction-welded butt joints", Welding International, 17 (7) 514–523, 2003.
[13] Jian Han, Huijun Li, Zhixiong Zhu, Frank Barbaro, Laizhu Jiang, Haigang Xu, Li Ma, "Microstructure and mechanical properties of friction stir welded 18Cr–2Mo ferritic stainless steel thick plate", Materials & Design, Volume 63, Pages 238-246, November 2014.
[14] Sathiya.P, Aravindan.S, Noorul Haq.A, "Effect of friction welding parameters on mechanical and metallurgical properties of ferritic stainless steel", Int.J.Adv.Manuf Technology, 31: 1076-1082, 2007.
[15] G.Taguchi, "Taguchi on Robust Technology Development Methods", ASME press, New York, pp.1-40, 1993.
[16] G.Taguchi and S.Konishi, "Taguchi methods, orthogonal arrays and linear graphs. In: Tool for Quality Engineering", American Supplier Institute, pp.35-40, 1987.

IOP Conf. Series: Materials Science and Engineering **197** (2017) 012070 doi:10.1088/1757-899X/197/1/012070

Optimization of Drilling Parameters for Reducing the Burr Height in Machining the Silicon Carbide Particle (SiCp) Coated with Multi Wall Carbon Nano Tubes (MWCNT) Reinforced in Aluminum Alloy (A 356) Using Meta Modeling Approach

[a]Sangeetha.M., [b]Prakash.S.

[a]Research Scholar, Department of Mechanical Engineering, Sathyabama University,Chennai, Tamil Nadu.
[b]Professor and Dean, School of Mechanical Engineering, Sathyabama University, Chennai, Tamil Nadu.

Email id: [a]inba_sangee@yahoo.com, [b]prakash_s1969@yahoo.com.

Abstract –This paper explains the optimization of drilling parameters using meta modeling approach to reduce the burr height while machining Silicon Carbide Particle (SiCp) coated with Multi Wall Carbon Nano Tubes (MWCNT) and reinforced in aluminum alloy (A 356). The specimen is prepared by the combination of sonication and stir casting processes. The volume fraction of MWCNT used is 1.5% and the volume fraction of SiCp is 10%. The combination of input parameters for drilling the holes is designed using Taguchi experimental design technique. The input parameters chosen for drilling operations are spindle speed, feed rate and drill diameter. The ranges of input parameters are listed in Table 1. The tools used for drilling operation are made up of solid carbide drill bit. Meta model is a mathematical and statistical model whose second-order model can be fitted by factorial design. The optimization model can be improved significantly by the second-order model compared to the first-order model. Twenty - seven holes are drilled using vertical machining center in the prepared specimen (A 356/MWCNT coated SiCp). Desirability function shows the optimized values of input parameters to obtain minimum burr height. Meta modeling approach is used to design a model using input parameters and output response burr height. The residuals plot shows the predicted values are closer to the measured values. This plot explains that the Meta model is adequately used to predict the burr height. The optimized values of input parameters for obtaining minimum burr height are the combination of high speed, low feed and low drill diameter. The minimum value of burr height observed in this experiment is 0.002mm and it is obtained in the optimized combination of N_3, f1 and d_1.

Keywords: Burr Height; Desirability Function; Math Modelling; Taguchi Design;

1. Introduction

Burr is a small protrusion and it is produced during any machining process. Burr is formed due to the plastic deformation or tearing of materials. In drilling the burr formed around the hole edge which results in improper seating of fasteners for assembly purposes. Burr is removed by any of the finishing operation such as deburring and edge finishing. This finishing operation is about 30 % of total cost of material produced. The presence of burr in moving components results in unwanted friction and heat. Burr is reduced by the optimized values of input parameters [1].

T.Rajmohan and T.Palanikumar performed a drilling operation on hybrid metal matrix composites (A356/SiCp/Mica) using solid carbide tool. They discussed the use of Response Surface Methodology to minimize the burr height. They analyzed the optimized input parameters to obtain reduced burr height of 0.143mm. The optimized input values are low speed of 1000rpm, low feed of 50mm/min and high percentage (15%) of reinforcement [1].

Domingo and et al. conducted a drilling operation on thermoplastic polymeric material and polyether-ether-ketone reinforced with 30% glass fibers. The input values for machining chosen are drill bit speeds of 6000, 7000 and 8000 rpm, feed rates of 300, 400 and 500 mm/min and drill types of high-speed steel, tungsten carbide and diamond-tipped tungsten carbide. The author proved that the consumption of energy is reduced and the surface quality is improved at the drill bit speed of 7000 rpm and the feed rate of 400 mm/min by using tungsten carbide drill [2].

Jayaram and Mahesh Kumar performed a machining operation on AA6063 T6 alloy, the input parameters taken are spindle speed, feed and depth of cut and analyzed the inputs using Grey Relational Analysis(GRA). In the analysis, it is proved that the feed (57.365%) and depth of cut (25%) are majorly depends on the effect of output [3].

L.H. Manjunathan and P.Dinesh, fabricated a composite by mixing Multi Wall Carbon Nano Tubes with the base alloy in order to increase the strength of base metals [4].

Saravanakumarand P. Sasikumar conducted a drilling operation on hybrid aluminium composites using TiN coated solid carbide drill and analyzed the effect of drilling parameters on the burr height. They confirmed that the minimum burr height is obtained at the spindle speed of 3000rpm, feed of 50mm/rev and graphite percentage of 6. The output response burr height is greatly influenced by percentage of graphite [5].

Eder Silva Costa and et.al tested the micro alloyed steel and recorded the burr height. They proved that the burr growth depends on the tool wear. It is confirmed that the burr increase with drill wear and it grows after 64% of drill life at the cutting velocity of 45m/min. and the burr grows after 82%of drill life at the velocity of 60m/min [6].

Shanti Parkash and et.al conducted a drilling operation on Al-fly ash composites and verified the burr height. They presented the optimized input parameters to reduce the burr height. The optimized values of input parameters are low spindle speed of 60m/min, medium feed of 0.07mm/rev and low point angle of 100° [7].

Nibir Saha and et.al performed a turning operation on an aluminium alloy and studied the burr formation using profile projector. They proved that burr height increased with the increase in input parameters. Burr height increased with feed values from 0.6mm/rev to 0.8mm/rev and it is decreased with the feed from 0.2 to 0.4 mm/rev. The minimum value of burr height is obtained at the optimized input values such as the depth of cut = 1mm, feed = 0.22mm/rev and cutting velocity = 235.5m/min [8].

Ravinder Kumar Chauhan used Taguchi method to obtain minimum burr height in drilling stainless steel AISI202 and obtained the optimized input parameters namely speed of 45m/min, point angle of 135° and feed of 0.12mm/rev [9]

G.Vijaya Kumar and P.Venkataramaiah focused on minimizing burr height in drilling aluminium metal matrix composites using Desirable Fuzzy approach. They proved that the spindle speed, tool material and reinforcement material are greatly influenced the burr height [10].

IOP Conf. Series: Materials Science and Engineering **197** (2017) 012070 doi:10.1088/1757-899X/197/1/012070

From the above literature reviewed it is clear that in all experiments a composite material is fabricated and its burr height is minimized by analyzing the input parameters. The novelty of this paper is the treated ceramics. The ceramics are not reinforced directly in the base metal it is treated with multi wall carbon nano tubes and then it is reinforced.

2. Materials and methods

2.1 Materials

Base metal chosen for this experiment is A 356.Aluminium Cast Alloy (A 356). It has some unique properties such as excellent cast ability, weldability, good in pressure tightness, toughness and it has good resistance to corrosion, and it performs well in in the conditions of high temperature and pressure, making it ideal for various high-wear applications. A 356 has a specific application in the fields of automobile as cast wheel and in aerospace as inner turbo frame.Other applications are aircraft fittings, transmission cases, wheels, bridge parts, pump parts and control parts.

Silicon Carbide Particle (SiCp) has good thermal and chemical stability during the process of synthesis and under severe service conditions. Multi wall carbon nano tube (MWCNT) forms good interfacial bonding between the reinforcement and matrix and it reduced the surface roughness and tool wear while machining.

2.2 Methods

2.2.1 Sonication Process

Sonication process is the vibration process in which the silicon carbide particle is coated with multi wall carbon nano tubes in the presence of poly vinyl acetate as a sticking material. Various percentage of carbon nano tubes are coated on the silicon carbide particles such as 0.5%, 1%, 1.5% and 2%. The resultant product is verified under scanning electron microscope.

2.2.2 Casting Process

In casting process, the molten aluminium is mixed with coated silicon carbide particle by semisolid stir casting process.

In this experiment two muffle furnace are used. In one furnace the aluminium is melted to its melting point of700°C and the melted aluminium mixed with preheated ceramics at 605°C for five minutes. Then it is poured in to the preheated die to get a required shape.

2.2.3 D*rilling Process*

The fabricated specimen is subjected to drilling with input parameters spindle speed, feed rate and drill diameter. The spindle speed values ranges from 600 to 1860 rpm. The feed rate values ranges from 25 to 75 mm/min and the drill diameter ranges from 4 to 12mm.The drilling operation is carried out in Vertical Machining Centre. Figure. 1explains the experimental set up for drilling process and measuring of output response Burr Height.

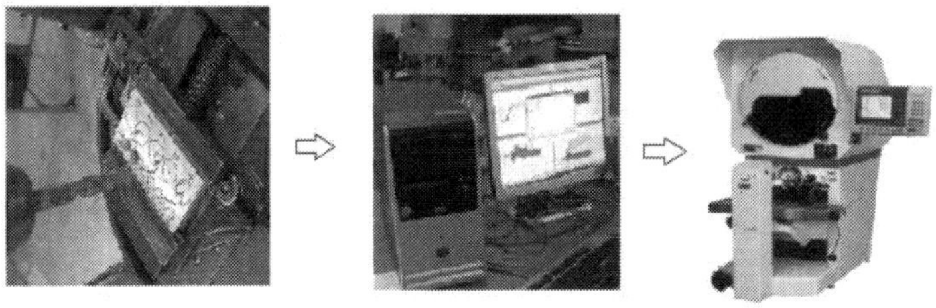

Figure1. Experimental set up for measuring Burr Height

2.2.3 Measurement of Burr height
In this experiment the hybrid metal matrix composite is drilled. Due to machining operation, some protrusion of material formed along the boundary. At high temperature, the tool material subjected to wear and cracks in workpiece which leads to burr. Burr height is the buildup edges due to rise in temperature and it results in poor surface quality. Burr height is measured using Profile Projector. The optimized combination of input parameters is obtained using Taguchi technique.

Table 1. Input parameters for drilling and their levels

Parameters	Levels		
	1	2	3
Spindle Speed, N (rpm)	600	1260	1860
Feed, f (mm/min)	25	50	75
Drill diameter, d (mm)	4	8	12

2.2.4 *Modeling of Burr Height using Response Surface Methodology*

Response Surface Methodology is an essential technique to find out the relationship between the process parameters. It is used to develop a mathematical relationship between the output response, Burr height and input parameters speed, feed and drill diameter.
Figure. 2 shows the normal probability plot of experiment values and predicted values. The points are closer to the lines and the deviations are less. This figure explains the good correlation between the experimental and model values.

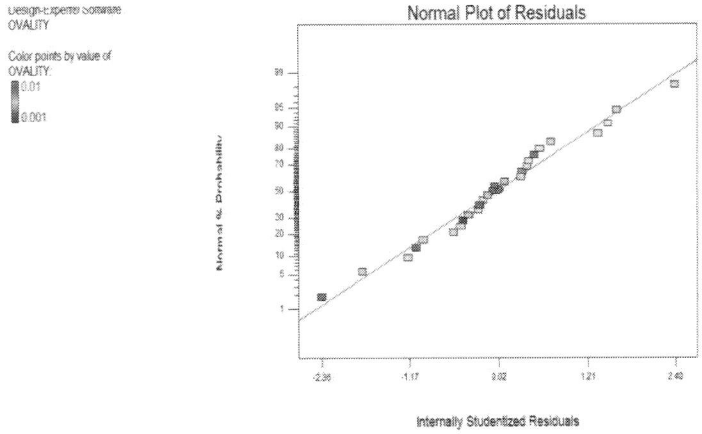

Figure 2. Normal probability Plot

2.2.5 Optimization of Burr height using Response Surface Methodology

Desirability based optimization is chosen in this experiment and it is mostly adaptable for multi performance objective. Desirability ranges between zero to one at the goal. For more than one output responses these goals are combined to form one desirability function.

3. Results anddiscussions

3.1 Results of Sonication

Results of sonication shows the equal distribution of carbon nano tubes over the ceramics. Figure. 3 shows the micro structure of multi wall carbon nano tubes, silicon carbide particles and resultant product of sonication.

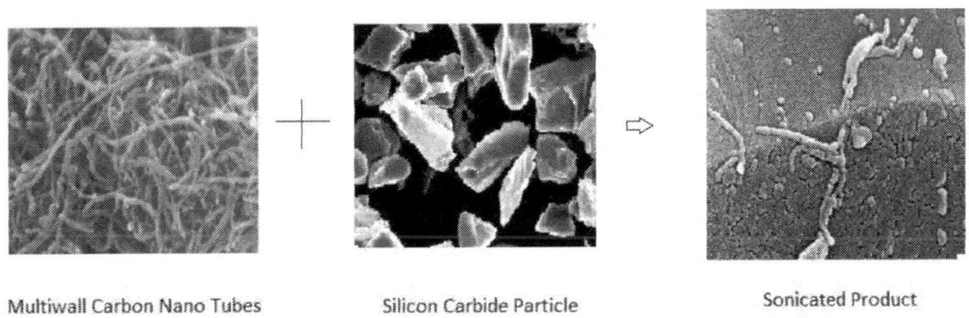

Figure 3. Mixing of multi wall carbon nano tube with silicon carbide particle and the resultant product of sonication

3.2 Results of Stir Casting

Results of stir casting shows the presence of carbon, silicon carbide particle and aluminium. Figure. 4 shows the micro structure of coated silicon carbide particle and aluminium alloy and the result of stir casting.

IOP Conf. Series: Materials Science and Engineering **197** (2017) 012070 doi:10.1088/1757-899X/197/1/012070

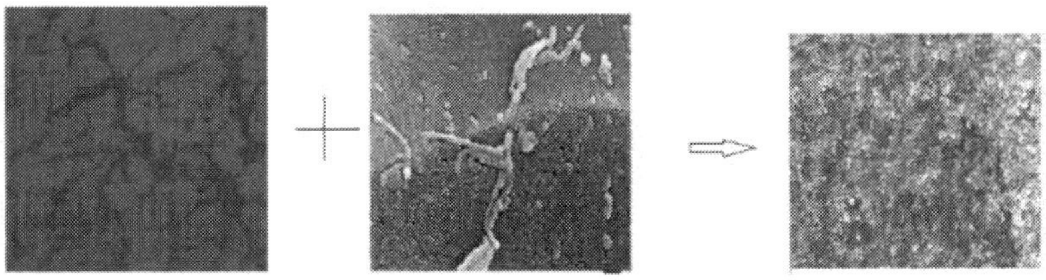

A-356 | Coated Silicon carbide Particle | Coated Silicon Carbide Particle mixed with A 356

Figure 4. Mixing of aluminium, coated silicon carbide particle and the resultant product of stir casting

3.3 Modelling and Measurement of Burr Height

Burr height is observed and studied. Data analysis and interpretation using suitable modelling and optimization methodologies are made using Math modelling.

Table 2. Output response - Burr height obtained in the experiment with three input parameters designed based on l_{27} array

Speed(N)-rpm	Feed(f)-mm/rev	Diameter(d)-mm	Burr height-mm
600	25	4	0.004
600	25	8	0.005
600	25	12	0.008
600	50	4	0.004
600	50	8	0.004
600	50	12	0.006
600	75	4	0.009
600	75	8	0.031
600	75	12	0.159
1260	25	4	0.006
1260	25	8	0.08
1260	25	12	0.013
1260	50	4	0.004
1260	50	8	0.005
1260	50	12	0.012
1260	75	4	0.009
1260	75	8	0.005
1260	75	12	0.012
1860	25	4	0.003
1860	25	8	0.002
1860	25	12	0.009
1860	50	4	0.002
1860	50	8	0.003
1860	50	12	0.005
1860	75	4	0.003
1860	75	8	0.005

1860	75	12	0.009

Table.2 listed the output response burr height and they are measured using Profile projector in North Laboratory, Chennai.The second order polynomial response surface model for burr height is given below.

BURR HEIGHT =-0.010905+8.02344e005 * speed-1.99650e-003* feed rate+2.74018e-003 * drill diameter-9.74152e-007 * speed * feed rate-4.72998e-006* speed * drill diameter+1.18333e-004 * feed rate * drill diameter-4.20875e-009* speed^2+2.50667e005 * feed rate^2-1.04167e-005* drill diameter^2 (1).

3.4 Optimized Parameter Using Math Modelling

In this paper math modelling is used to determine the optimized values of process parameters. At this input conditions the reduced value of burr height is obtained. Tables 3 and 4 describes the goals and limits used for optimization and table displayed the best goal solution for optimization. The figures 5 and 6 describes the interaction of input responses plot and the main effects for burr height

Table 3. Goals Set and Boundary used for Optimization

Parameter and Response	Goal	Lower limit	Upper limit	Lower Weight	Upper Weight	Importance
Spindle Speed	Is in Range	600	1860	1	1	3
Feed rate	Is in Range	25	75	1	1	3
Drill diameter	Is in Range	4	12	1	1	3
Burr Height	Minimize	0.002	0.159	1	1	3

Table 4. Best Goal Solution for Optimization

No	Speed	Feed	Diameter	Burr	Desirability
1	1854.07	28.30	11.92	0.00164	1.000
2	813.35	33.20	4.01	0.00077	1.000
3	1442.68	64.02	4.01	0.00046	1.000

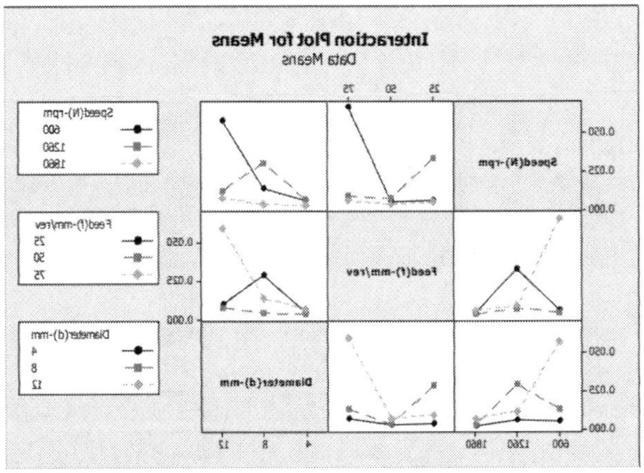

Figure 5. Interaction Plot for Means

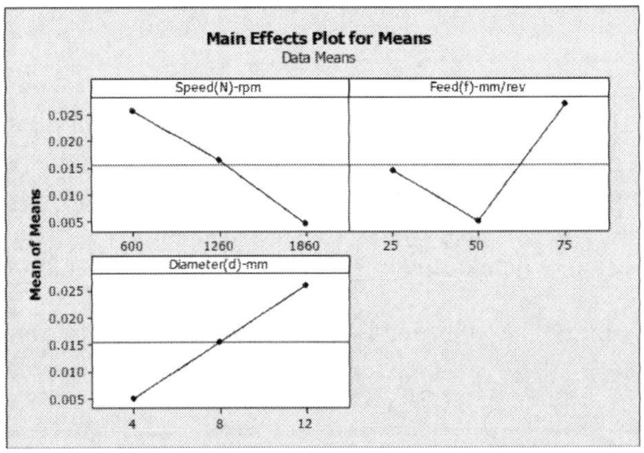

Figure 6. Main effect plot for Means

Table 5. Response table for means (Burr height)

Level	Speed(N)-rpm	Feed(f)-mm/rev	Diameter(D)-mm
1	0.025556	0.014444	0.004889
2	0.016222	0.005000	0.015556
3	0.004556	0.026889	0.025889
Delta	0.021000	0.021889	0.021000
Rank	3	1	2

Response table for means can be obtained using Minitab software and it describes that burr height is influenced mostly by feed rate followed by diameter and it is shown in table 5 The response table and the main effects graph describes that the output response burr height is greatly influenced by feed followed by drill diameter.

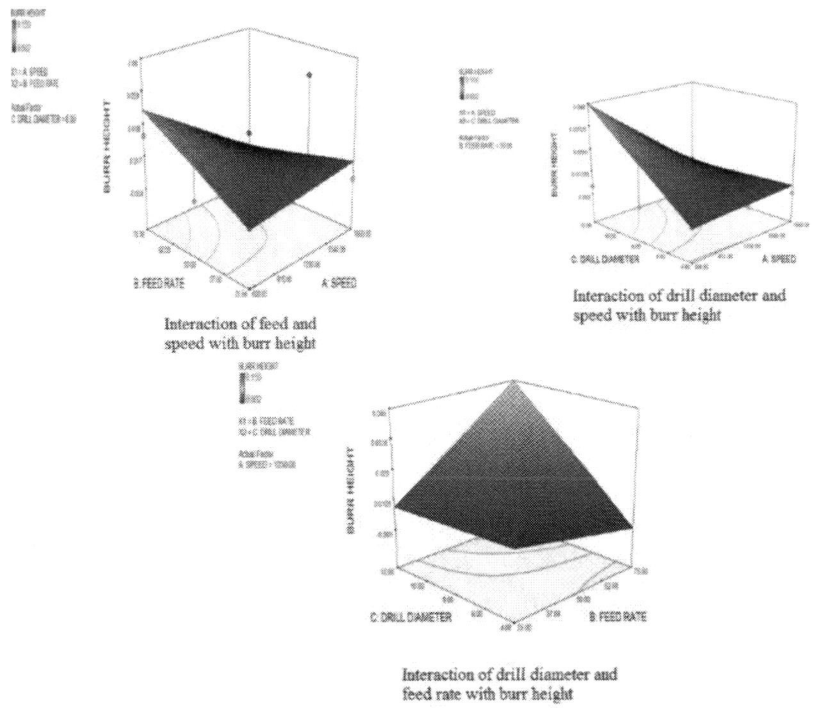

Figure 7. Interaction of input parameters with the Burr height

The figure 7 describes the variation of output, burr height with the interaction of input parameters. There is a rise of burr height with the increase of feed rate and drill diameter when compared to the spindle speed.

4. Conclusion

- Due to sonication process the clustering of nano particle is avoided
- Multi wall carbon nano tubes are evenly spread over the surface of ceramics

- Due to semi solid stir casting process there is a good interfacial bonding between the reinforcement and base alloy.

The output response burr height is obtained using Profile projector and analysed and this burr heightis highly affected by feed rate and drill diameter. As the feed increases the burr height increased and it decreased as the spindle speed increased.

5. References

1. T.Rajmohan, K.Palanikumar, Madhavan and G.Harish, "Optimizing the Machining Parameters for Minimum Burr Height in Drilling of Hybrid Composites", *Procedia Engineering*, (2012), Vol-38, pp-56-65.
2. Domingo, R.; García, M.; Sánchez, A.; Gómez, R. A sustainable evaluation of drilling parameters for PEEK-GF30. *Materials*. 2013, 6, 5907-5922. doi: 10.3390/ma6125907.
3. Jayaraman,P.; and Mahesh Kumar,L. Multi-response Optimization of machining parameters of turningAA6063T6 Aluminium alloy using Grey Relational Analysing (GRA) Taghuchi Method. *Procedia Engineering*.2014, 97,197-204.
4. Manjunatha,L.H and P.Dinesh, Fabrication and Properties of dispersed carbon nanotube–Al6061 composites, *International Journal of Innovative Research in Science, Engineering and Technology* Vol. 2, 2013, pp 500-507.
5. A. Saravanakumar and P. Sasikumar "Assessment of factors influencing burr height on the machining of particle reinforced hybrid composites"*Journal of Material Environmental Science*,(2015) Vol 6, Issue 5 pp 1638-1645.
6. Eder Silva Costa, MarcioBacci da Silva and Alisson Rocha Machado, "Burr Produced on the Drilling Process as a Function of Tool Wear and Lubricant-Coolant Conditions", *Journal of the Brazillian Society of Mechanical Sciences & Engineering*, 2009, Vol. XXXI, pp 57-63.
7. Shanti Parkash, MukeshVerma and Sarabjot Singh, "Modeling and optimization of burr height in drilling of Al-Fly ash composite using Taguchi method", *International Journal of Engineering, Research and Applications (IJERA)*, Vol. 2, Issue 1, 2012, pp. 383-390.
8. Ravinder Kumar Chauhan, "Experimental Investigation and Optimization of Burr Height in Drilling Stainless Steel AISI 202 Using Taguchi Method", *SSRG International Journal of Mechanical Engineering (SSRG-IJME) – EFES*, 2015,ISSN: 2348 – 8360, pp74-80.
9. NibirSaha, DebajitTalukdar, Kamal Upadhaya, ChayadBarbhuiya, and ManjuriHazarika, "An Experimental Study on Burr Formation in Turning, *International Journal of Innovative Research in Science, Engineering and Technology*,2016Vol. 5, Issue 3, 3215-3222.
10. G.Vijaya Kumar and P.Venkataramaiah, "Minimization of Burr Height and Tool Wear in Drilling, *Proceedings of the World Congress on Engineering,London*, U.K. (2013) Vol III, 2013, ISBN: 978-988-19252-9-9.

IOP Conf. Series: Materials Science and Engineering **197** (2017) 012071 doi:10.1088/1757-899X/197/1/012071

BUILDING BRAND POWER

Lakshmi S[1] and Dr.Muthumani S[2]

[1]Research Scholar, Sathyabama University, Chennai, Tamil Nadu, India

[2]Research Guide, Sathyabama University, Chennai, Tamil Nadu, India

Email: messagetolakshmi@gmail.com

E.mail:dr.s.muthumani@gmail.com

ABSTRACT

Brand power is established through brand awareness. It's all about making consumers familiar about their products and services. Marketing strategies should make the customers extend the positive approach towards brand and continue through repeated purchases. There is a triple perspective approach to investigate the brand awareness in this research. The brand awareness and brand equity are studied and the relationship between those are analyzed. This also drills down about the brand performance and knowledge with awareness which tries to find out the brands value and utility among the public. Continuous improvement on package design, quality and buying experience will lead to customer loyalty and preference. Branding should happen though creative ads, eye catchers and special campaigns. Brand awareness is the extent to which consumers are familiar with their product or services. Power of a brand is resides in the minds of the customers. To build a strong brand, it is one of the great challenge for the marketers to ensure that customers have the right experiences with products and services and various marketing programs. So that tenderness, beliefs, perspective, perception and so on linked to the brand. If we are presenting the brand with no enthusiasm or spunk, people are going to forget about our brand. Even though that may seem harsh, it's the naked truth in today's marketing world. Brand must reach out to the community by special events, creating campaigns to keep the brand relevant also offer customer a unique experience. Here we study about the brand consciousness and to identify the cohesion between brand awareness with knowledge and performance and also to assess the effect of brand awareness on consumer purchase. In this study we necessary statistical tools like chi-square test ad t- test has been used to analyse the collected data. It is highly recommend to increase brand awareness, the marketers are constantly required to build brand awareness both economically and efficiently in the minds of customers at a competitive environment. Generating brand power begins with building healthy brands. So that consumers are able to identify a brand through brand recognition or recall performance. This article contains the following sub headings.

1. Introduction 2. Objectives 3. Research questions 4. Research methodology 5. Data Analysis 6. Conclusion

Key words: awareness, recall, repeated purchase, attention, brand knowledge.

1. INTRODUCTION

In today's fast paced modern era, one cannot be lethargic and cannot take rest on their old laurels. Everyone's in this world they feel that they are short on time. Customers takeout little bit of time from their so called busy schedule to explore better quality products and services at a competitive price rather much cheaper price. Hence, now a days companies are facing stiff challenge in sustaining their business edge by promoting their brand tirelessly through various means at a highly competitive environment. Companies should device a clear strategy to integrate their response towards marketing the brand. Ultimately the sales numbers will be the yardstick for tangible and sustainable business environment. In today's world wide buyers' market, competition is becoming stiffer and stiffer. It becomes highly difficult for products and services to differentiate themselves from other customers'. Awareness of their buying power puts everyone on their toe to take care of them. Clash in the market takes place not between the companies but between the brands. Developing a strong awareness about the brand is a major responsibility of the firm. The brand only tie a customer to the firm by rational and emotional hooks, so it is a loyalty generator.

Brand responsiveness is how our products stand out from other products, how the appearance creates consumer awareness, it ramps up our attention, it leads to repeated purchase due to free gifts and sales offer and it paves way for major decision.

Brand accomplishes various functions like Top-of-mind awareness, recall of past experiences, quality sign and identity and it pulls the people towards brand loyalty. Xiaoling guo, Andy wei Hao, Xia oyan shang, (2011). Henry Yu Xie, David J.Boggs, (2006) stated to build a framework for the development of branding stratergy for international markets. Johan Anselmsson, Nikias vestman Bondesson, Ulf Johansson, (2014) mentioned in his study customers pay a premium price on the basis of brand image, quality, social image and uniqueness and awareness. Firm should be aware of what drives price premium in addition to the dimensions of quality and loyalty. Scott M.davis, (2000) mentioned brand occupies a space in the minds of the consumer's.

OBJECTIVES OF THE STUDY:

1. To study the brand consciousness
2. To identify the cohesion between brand awareness with brand knowledge and performance.
3. To assess the effect of brand awareness on consumer purchase.

STATEMENT OF THE PROBLEM:

The study has been undertaken to examine some issues pertaining to create brand awareness among people in Chennai city. The study gains relevance and importance in view of the following. The firm supplies consumers around the world with healthy products designed to promote health, vitality, beauty and overall good living. The concern focus on the continuous development and introduces of new products and programs to ensure the growth of their sales and profits while exceeding consumer expectations. Build brand awareness is equal importance to sustain business edge.

RESEARCH QUESTIONS

1. How it ramps up our attention?

2. How it induced us for repeated purchase?

3. Constant advertisement and association coaxes us infavour to the brand.

RESEARCH METHODOLOGY

Branding is all about creating a position in the market. The outcome depends on how value is being added almost at the same cost. It is a way to measure how well brands rank in the minds of consumers. In order to address the above questions an exploratory study was conducted. The idea was to probe and get deeper insight into brand equity of various products. A methodology has been framed in relation to the data collection, sampling, hypothesis and statistical tools. The objective of the study was to ascertain how to build brand power among the customers. For this best suited method was survey method. Hence the questionnaire method is chosen which is free from bias of the interviewer and large sample can be made use of and thus the result can be more dependable and reliable. For the purpose of the study, both the primary and secondary data were collected. Primary data is obtained through well framed questionnaire, which consist of optional type questions and sealing type questions. Scales are given in Likert' 5 point scale. Secondary data were collected from the published records, related books, journals, reports, broaches, magazines and websites. The study is focused on personal care products in Chennai city. Sample size consist of 320 customers. Responses from all class of people like teenage, housewives, private employees and from others. Necessary statistical tools like chi-square test and t-test has been used to analyses the data collected and the findings are as follows.

2. DATA ANALYSIS

2.1. Hypothesis:

The following hypothesis were formulated:

2.1. HO – Repeated purchase is independent to the perpetual advertisement.

TABLES

Chi-Square Tests

Table no.2.1.a

Constant advertisement * Awareness Cross tabulation

Count

		Awareness			Total
		High	moderate	low	
Constant advertisement	Often	132	67	27	226
	Rare	64	23	7	94
Total		196	90	34	320

IOP Conf. Series: Materials Science and Engineering **197** (2017) 012071 doi:10.1088/1757-899X/197/1/012071

Table no.2.1.b

Chi-Square Tests

	Value	Df	Asymp. Sig. (2-sided)
Pearson Chi-Square	2.913[a]	2	.233
Likelihood Ratio	3.001	2	.223
Linear-by-Linear Association	2.877	1	.090
N of Valid Cases	320		

a. 0 cells (.0%) have expected count less than 5. The minimum expected count is 9.99.

The Chi-square value is 2.913 for 2df and significance value is 0.233, (p value). Here the p value is less than 0.05, the difference between observed frequencies and expected frequencies is insignificant. Therefore the Alternate hypothesis (H1) is rejected.i.e. There is no significant relationship between perpetual advertisement and repeated purchase among the people. Advertisement is a way of communication for launch and strengthen the brand identity. This help pushes bond with customers and sales progress. Marketing keeps the customers informed about the choice and its worth. The advertisement should effectively improve the brand awareness and should have a impact on customers decision but on the other hand Perpetual advertisement may not be effective in case of luxury items and high end products Where customers buy based on their own experience and knowledge.**2.2.Paired sample t-test**

Ho: There is no significant relationship between market recognition and brand performance.

H1: There is a strong relationship between market recognition and brand performance.

Paired sample t-test

Table no.2.2.a

Paired Samples Statistics

		Mean	N	Std. Deviation	Std. Error Mean
Pair 1	Before market recognition (MS)	10.65	15	.814	.210
	After Market recognition (MS)	11.23	15	.942	.243

Refer Table.2.2.b

Paired Samples Test

	Paired Differences					t	df	Sig(2-tailed)
	Mean	Std. Deviation	Std.Error Mean	95% confidence interval of the difference				
				Lower	Upper			
Pair 1 Before- Market recognition-After- Market recognition	-.587	.391	.101	-.803	-.370	-5.815	14	.000

From the above table the mean market recognition level among the individual before the brand performance is 10.65 with a standard deviation of 0.81(market share) and 11.23 with a standard deviation of 0.94 (market share) after the brand performance. From the paired sample test, the t-value of -5.815 for 14 degrees of freedom(df) is highly significant as significant value for two tailed test is 0.000 (output 2). Therefore, we reject the null hypothesis. Consumer are ready to pay high price or best price once they recognize the performance of the product. Customers always want to pay as little as possible for true quality. Customers often eagerly pay more for a product even when they can get a functionally similar or even identical product elsewhere for less because easier to buy, it arrives more quickly. If we can indulge our customer's desires preferably than the competitors, they will usually pay a premium price.

CONCLUSION:

This findings state that how the necessity of brand building becomes more important and plays vital role in today's rapidly changing scenario. It is only way to sustain the entire business. Developing a strong brand is a major responsibility of the firm and it is more challenging. Brand is acting as a bridge between customers and firm. Market value of a business is determined by the type of brands one has, and to what extent the awareness is created among the people. Brand awareness is not only about how many people are aware of a brand, it also refers to the extent and ease with which consumers recollect and recognize the brand and can classify the products and services with which it is associated. From this result it is highly recommend to increase brand awareness, the marketers are constantly required to build brand awareness both economically and efficiently in the minds of customers at a competitive environment. Generating brand power begins with building healt brands. So that consumers are able to identify a brand through brand recognition or recall performance.

REFERENCE:

[1] Chris macrae (1996). Re-thinking brand management: the role of brand chartering. Marketing Intelligence and Planning 14(7), 46-55
Greyser, S.A.(2009). Corporate brand reputation and brand crisis management. Management decision, 14(7), 590-602
[2] brightblue.com.al/index.php/blog/article/consumers.
[3] https:://brandvision2009.wordpress.com
[4] https://en.m.wikiboks.org/wiki/marketing/introduction
[5] The effect of price bundling on consumer perception of value", journal of service marketing, vol.15, Issue: 4, pp.270-281.
[6] www.palgrave –journal.com/bm/journal/v16/n3/2550139a.htm.

IOP Conf. Series: Materials Science and Engineering **197** (2017) 012072 doi:10.1088/1757-899X/197/1/012072

Experimental Investigation on The Electromagnetic Clutch Water pump and Pneumatic Compressor for Improving the Efficiency of an Engine

Kumarasubramanian.R[1*], Goldwin Xavier[2], Mary Nishanthi.W[3], Rajasekar.R[4]
[1,4] Assistant Professor, Department of Automobile Engineering, Sathyabama university, Chennai.
[2]Assistant Professor, Department of Mechanical Engineering, Loyola Institute of Technology, Chennai.
[3]Assistant Professor, Department of Electrical & Electronics Engineering, St. Joseph's College of Engg., Chennai.
Email id: rkscad@gmail.com

Abstract - Considering the fuel crises today many work and research were conducted to reduce the fuel consumption of the internal combustion engine. The fuel consumption of an internal combustion engine can be relatively reduced by use of the electromagnetic clutch water pump and pneumatic compressor. Normally in an engine, the water pump is driven by the crankshaft, with an aid of belt, for the circulation of the water for the cooling process. The circulation of coolant is resisted by the thermostat valve, while the temperature inside the coolant jacket of the engine is below 375K the thermostat is closed only above 375K it tends to open. But water pump run continuously even when thermostat is closed. In pneumatic braking system, pneumatic or air compressor purpose is to compress the air and stored into the storage tank for the brake operation. When the air pressure of the storage tanks gets increases above its storage capacity pressure is regulated by governor, by passing them to atmosphere. Such unnecessary work of this water pump and air compressor can be minimized by use of the electromagnetic clutch water pump and air compressor. The European Driving Cycle is used to evaluate the performance of this water pump and air compressor when used in an engine. The result shows that the fuel economy of the engine while using electromagnetic water pump and pneumatic compressor were improved by 8.0% compared with conventional types which already exist. The application of these electromagnetic water pump and pneumatic compressor are expected to contribute for the improvement of engine performance because of their effect in reduction of the rate of fuel consumption.

Keywords - New European driving cycle test; Electromagnetic clutch; Water pump; Air compressor

1. Introduction

The role of the internal combustion engine is to provide tractive power to the driving wheels via drive train. The most commercial fuel used in internal combustion engine are petrol, diesel, CNG, LPG, biodiesel etc. Nowadays, alternate and hybrid technologies are used to reduce the emission from combustion of fossil fuel and also as an alternate source of energy for the internal combustion engine. Normally the combustion efficiency of an internal combustion engine are optimized, approximately 98% of the energy contained by combustion of the diesel fuel and 95-98% of the energy from the gasoline. However it may be, but only less than 40% of energy is transfer to drive the wheels and the fraction of energy are been used by ancillaries such as water pump, alternator, and air compressor [4]. The inability to convert all the chemical energy into brake power energy is termed as gross indicated thermal efficiency. From the definition of the engine fuel conversion efficiency is expressed as

$$\text{Fuel conversion efficiency} = \text{Combustion efficiency} \times \text{Gross indicated thermal efficiency}$$

For the improvement of the energy efficiency, electric and hybrid techniques, which can enhances the efficiency of the internal combustion engine these technology were been considered as valuable in recent era of development. The advancement in the science and technology, which were been contribute for the improvement in the efficiency in heavy-duty engines. High-efficiency technology like reduction in weight, air resistance, improving thermal efficiency and energy optimization were been used to reduce fuel consumption in the conventional vehicles. In most Commercial engines the coolant system consists of water jacket which is the outer cover of engine, thermostat and radiator. Generally water pump are operated in proportional to the engine speed to circulate the coolant throughout the water jacket of the engine block and radiator, in this configuration of direct connection of the engine crankshaft and the water pump results in excessive mass flow rate of coolant at cold start and part load condition the flow is regulated by thermostat. To optimize the coolant pump effect and to meet the efficiency of the coolant, alternative methods have been proposed by electrification of coolant components.

As per the effect, the electromagnetic water pump and intelligent thermostat have been used in vehicle cooling system. The mass flow rate of coolant in heavy duty diesel engine is relatively high at low engine speed [6] which can be optimized by this technique. In current study we have suggest a method for the efficiency improvement technique, which not only can be applied to the water pump but also to the air compressor of the pneumatic brake system. In pneumatic brake system air compressor process is to compression of the air for the brake operation is stored to the storage tank. The air pressure of the storage tank is regulated by the pressure relief valve. When the brake pedal is pressed, for the application of the braking process the pressurized air from the storage tank is flow through the hose to the brake cylinder which pushes out connecting rod of the brake cylinder to actuate s-cam and pushes the brake shoe outward. The air compressor, purpose is to pressurize the air and stored in the storage tank when the pressure accumulation reaches beyond the capacity of the storage tank, pressure is regulated by pressure relief valve [2]. The electromagnetic clutch type air compressor is used for the energy optimization technique as an alternative of the conventional type.

From our study, we suggest a methodology which improves fuel economy of an internal combustion by using electromagnetic clutch type water pump and air compressor. The performance of these water pump and compressor result in decrease in fuel consumption and co_2 emission were demonstrated by the experimental mean results

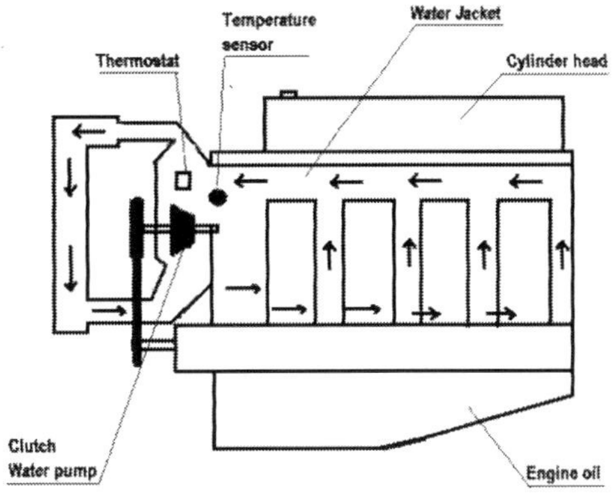

Figure 1. Layout of the coolant system with electromagnetic water pump

The Figure-1 gives the complete layout of the coolant system in an engine and with the electromagnetic water pump. The engine specification is given in the table 1.

IOP Conf. Series: Materials Science and Engineering **197** (2017) 012072 doi:10.1088/1757-899X/197/1/012072

Table 1. Specification of the Engine

Items	Value
Engine type	Water cooled type
Displacement	796 cc
Bore dia. X stroke	68.5 X 72.5 mm
length	37 KW @5000rpm
Max power	15 Nm @2500 rpm
Max torque	Petrol
Fuel used	Solex carburetor
Fuel system	

From the layout it is clearly illustrated the process of the coolant from the cylinder jacket to the radiator, the red arrows indicates high temperature coolant and blue arrows indicates the low temperature coolant. Figure-2 indicates the complete layout of pneumatic braking system, it consist of an air dryer, storage tanks, dual operated brake valve and brake cylinder. The air dryer purpose is to remove the moisture and water content in the compressed air and storage tanks is to store the pressurized into it. The dual valve used for control the air flow from storage tanks to the brake cylinder. The brake cylinder is to transfer kinetic energy of the pressurized air into mechanical energy by piston, connecting rod which connect the s-cam or wedge type for expanding the brake shoe inside the wheel drum.

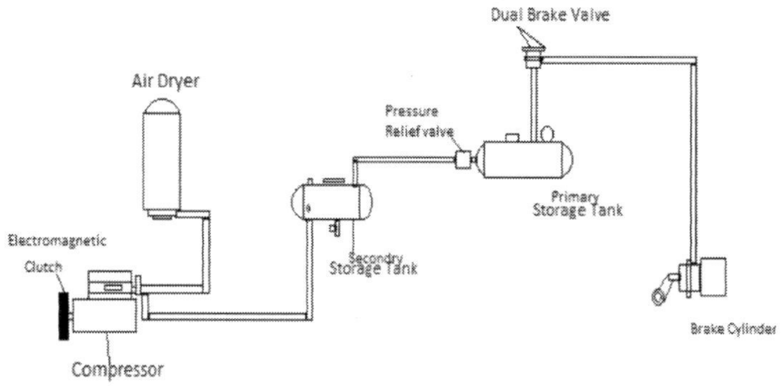

Figure 2. Layout of pneumatic braking system with electromagnetic clutch compressor

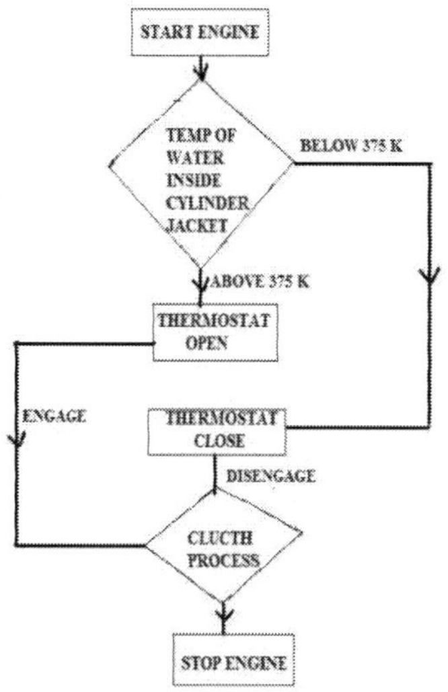

Figure 3. Flow Chart of Water Pump

IOP Conf. Series: Materials Science and Engineering **197** (2017) 012072 doi:10.1088/1757-899X/197/1/012072

The belt connects the pulley of the water pump as well as the air compressor to the crankshaft for transfer of the rotation motion for the necessary operational process. However, with the electromagnetic clutch the rotation speed is controlled by disengage and engage of the clutch and to transfer the crankshaft rotational motion to the water pump and the air compressor. The consumption of the electromagnetic clutch is 20 W under all operating condition. And the temperature sensor is installed near the cylinder water jacket for monitoring the temperature output of coolant. For evaluation of fuel consumption we use the New European Driving Cycle test.

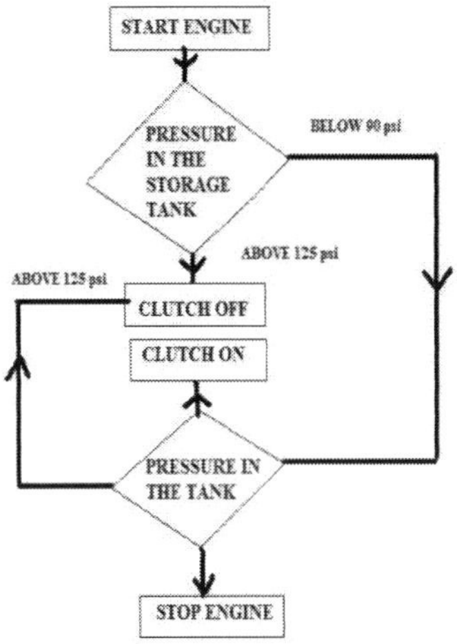

Figure 4. Flow chart of Air Compressor

Figure-5 shows about the schematic layout of the NEDC test process. In NEDC test [1], it consists of constant volume sample, dynamometer, etc. The constant volume purpose is to collect the exhaust for the measuring the emission parameters by gas analyzer method. The figure-3 shows the electromagnetic clutch type compressor. The figure-4 shows the model of the 'experimental setup of the engine with the electromagnetic clutch type water pump and Air compressor and the storage tank with relief valve.

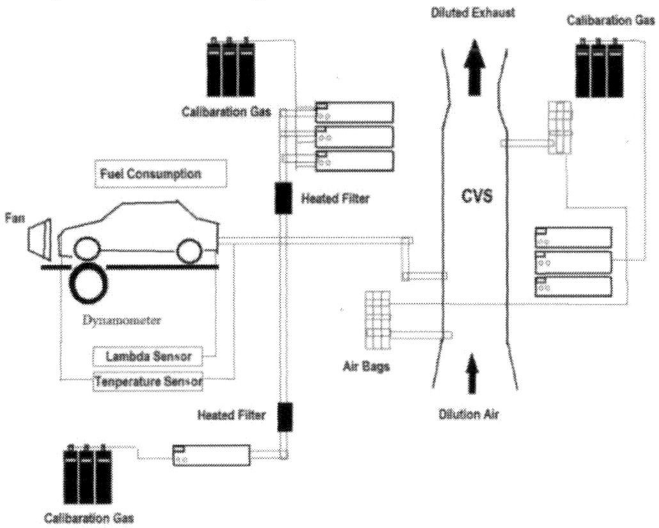

Figure 5. Schematic layout of testing process

IOP Conf. Series: Materials Science and Engineering **197** (2017) 012072 doi:10.1088/1757-899X/197/1/012072

2. Results And Discussions

The New European Driving Cycle Test (NEDC)

The NEDC is driving cycle designed to access the fuel consumption of the engine. The driving cycle used for emission type - approval of all light-duty vehicle models and Euro 3 in Europe. The emission limits (expressed as mass of pollutant emitted per kilometer driven) refer to the emissions over NEDC. From the NEDC evaluation test we obtained the results for fuel consumption and emission levels. We had conducted experiment on medium sized petrol engine to compare the performance of the clutch type water pump and pneumatic compressor with conventional types, the experimental setup consist of dynamometer for performance testing and co_2 emission parameter in the exhaust is determined from constant sampling process and gas analyzer test in NEDC. Figure-shows about the difference between conventional and electromagnetic water pump at cold start phase, the reduction in fuel consumption is due to the optimization of the water pump, the unnecessary work is reduced at this phase. From the figure-5 it is clearly illustrated that reduction in co_2 and fuel consumption level when compared with conventional type. The level of co_2 and fuel consumption has been decreased because at cruising phase the application of the brake is very less due that air pressure is maintained constant inside storage tank.

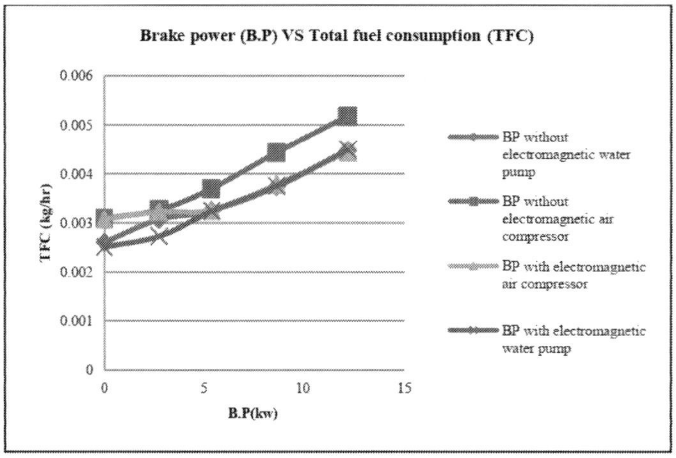

Graph 1. BP vs TFC

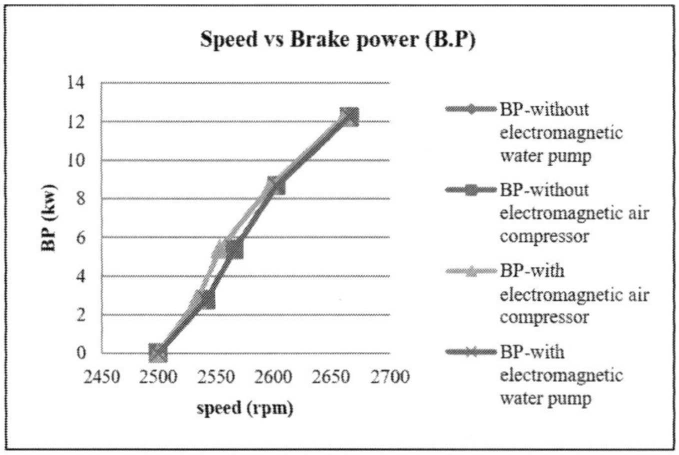

Graph 2. BP vs Speed

IOP Conf. Series: Materials Science and Engineering **197** (2017) 012072 doi:10.1088/1757-899X/197/1/012072

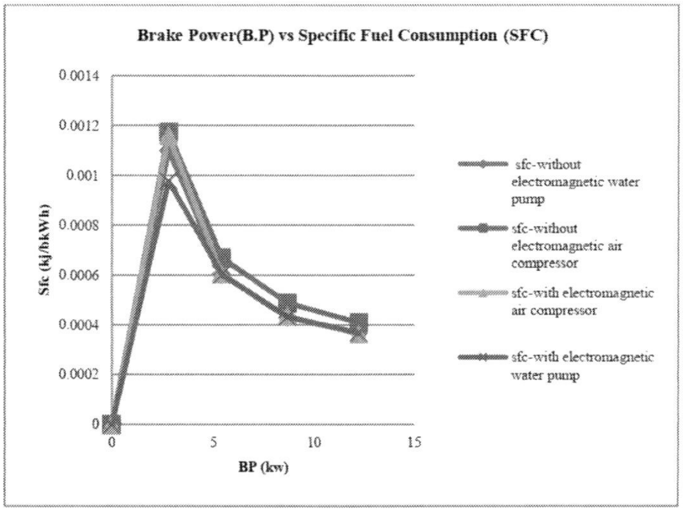

Graph – 3 BP vs SFC

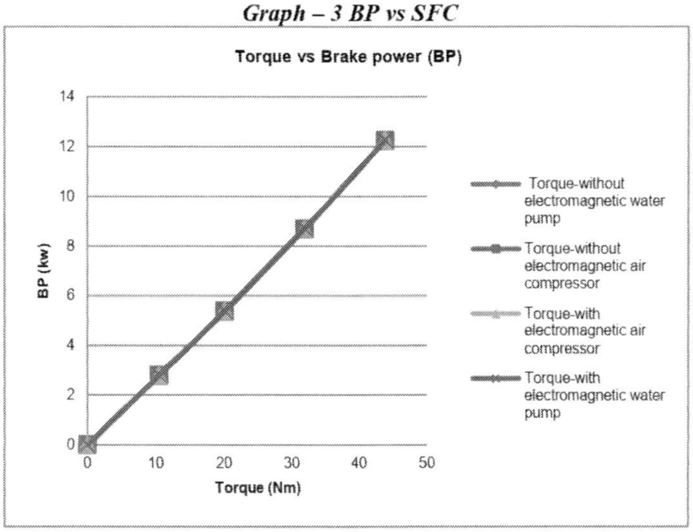

Graph – 4 Torque vs BP

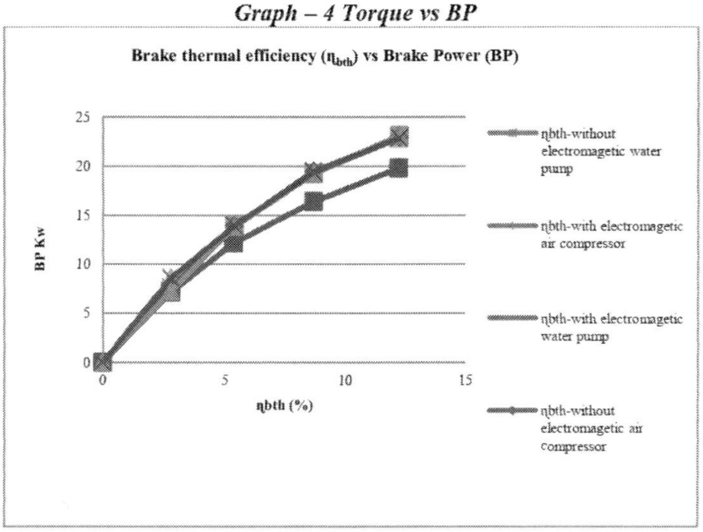

Graph – 5 Brake thermal efficiency (η_{bth}) vs Brake Power (BP)

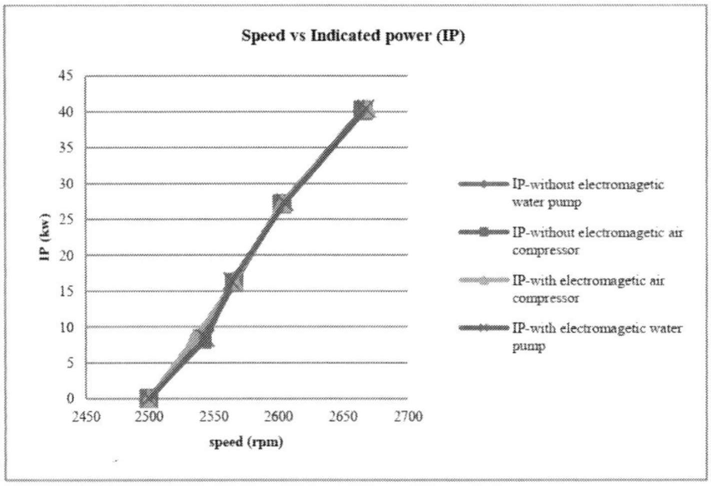

Graph – 6 Speed vs Indicated power (IP)

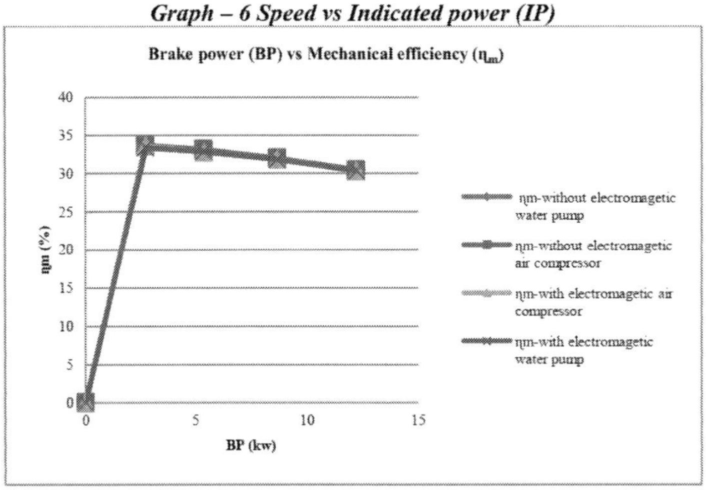

Graph – 7 Brake power (BP) vs Mechanical efficiency (η_m)

3. Conclusion

The paper study is to develop electromagnetic clutch water pump and air compressor to reduce the fuel consumption and CO_2 emission. The electromagnetic pulleys are installed and performance evaluation test were conducted using NEDC. From this test varies output results like fuel consumption, brake thermal efficiency and mechanical efficiency were obtained. New finding from the results are the decrease in the fuel consumption is upto 2.8% with electromagnetic water pump and upto 7.7% with electromagnetic air compressor when compared with the conventional type water pump and air compressor. By this method, the unnecessary work of the water pump and air compressor can be avoided from this we can expect considerable improvement in fuel consumption.

References

[1] Augusto F. Pacheco, Mario E.S. Martins New European Drive Cycle (NEDC) simulation of a passenger car with a HCCI engine: *Emissions and fuel consumption results*. Paper No.111 (2013) 733-739.

[2] Automotive and Off-Highway Air Brake Reservoir Performance and Identification Requirements, *Truck and Bus SAE*; Paper No.10 (2013)12.

[3] Christos Dardiotis , Georgios Fontaras, Alessandro Marotta, Giorgio Martini, Urbano Manfredi, Emissions of modern light duty ethanol flex-fuel vehicles over different operating and environmental conditions, *Fuel 140 (2015)* 531–540

[4] Heywood J. Internal combustion engine fundamentals. McGraw-Hill; 1988. ISBN 0-07-1004998.

[5] Krafft R, Faller W, Wolf A. Electromagnetic water pump clutch: working principle, design strategies and applications for heavy-duty vehicles. SAE;Paper No. 2007-01-4260.

[6]. R. Suarez-Bertoa , A.A. Zardini, C. Astorga ; Ammonia exhaust emissions from spark ignition vehicles over the New European Driving Cycle, Paper No.97 (2014) 043-053.
[7] Torregrosa AJ, Broatch A , Olmeda P, Romero C. Assessment of the influence of different cooling system configurations on engine warm-up, emissions and fuel consumption. Automotive Technology; Paper No. 9(2008) 447-58.
[8] Use of electromagnetic clutch water pumps in vehicle engine cooling systems to reduce fuel consumption 2013. Paper No. (2013)-04-073.
[9] Vicente Bermudez , José Manuel Luján, Santiago Ruiz, Daniel Campos, New European Driving Cycle assessment by means of particle size distributions in a light-duty diesel engine fuelled with different fuel formulations. Paper No.15 (2014) 649-659.

IOP Conf. Series: Materials Science and Engineering **197** (2017) 012073 doi:10.1088/1757-899X/197/1/012073

Experimental Investigation of Thermal Performance in a Vehicle Cabin Test Setup With Pcm in the Roof

M.Purusothaman[1,*], Saichand kota[2], Sam Cornilius.C[3], Siva.R[4]

[1,4]Asst Professor, [2,3]Student, School of Mechanical Engineering,
Sathyabama University, Chennai-600119, Tamilnadu, India.

Email.Id: purusothmani@gmail.Com, Contact No:+91 9042220783.
[*]Corresponding author

Abstract. Heat flow from the roof with radiation through glass windows obviously high level that contributes to the total heat gained of a vehicle cabin. The cabin temperature of closed stationary vehicles in direct sunlight can quickly rise to a very level that may damage property and harm children or pets left in the vehicle. The problem that is faced by many car users today is very hot interior after certain minutes or hours of parking in open or un-shaded parking area. The heat accumulated inside the vehicle with undesired temperature rise would cause the parts of the car's interior to degrade. Even the passengers are affected with the thermal condition inside the vehicle itself. The passenger has to wait for a certain time before getting into the car to cool down the interior condition either by lowering down the window or switching on the air conditioner at high speed that really affect the fuel consumption. A new roofing structure to improve its total thermal resistance is developed. Its uses phase change material properties to trap the heat from solar radiation and then release it back to the outer atmosphere by external convection when the vehicle is in use or during the nocturnal cycle. Phase change material, which has become an attractive means to store. Thermal energy, which has a wide range of applications, has been used. Phase change material has a high heat of fusion which is able to store and release large amount of energy. This PCM has been insulated in the roof of the vehicle to arrest the heat entering into the vehicle cabin. Experimental and numerical analyses have been conducted to compare the thermal performance of the new roofing structure and the normal roofing. By this experiment, the cooling process of the cabin could be much lower. The experimental investigation revealed that, on a hot day, the interior temperature of the vehicles cabin was approximately 22°Ce higher than the ambient temperature. The results show that the new roofing structure could effectively reduce the inlet of heat from the roof into the cabin. As a result, the interior temperature of the cabin could be much lower.

1. Introduction

When a vehicle is parked in an un shaded area the heat present inside the vehicle cabin increases to an average of 80°C. The increase in the internal temperature of the cabin can be so high that it may damage the interior parts or the children or pets left in the cabin. The vehicles roof is like a sheet metal roof which receives a incident solar radiation up to 1000 W/m k. in clear sky conditions, from the incident radiation 20 to 95% may be absorbed as heat. As a fact the interior temperature will be almost 11° C higher than the outside temperature when parked even for ten minutes. The hot interior is the main problem that is faced by many people after certain time of parking in an unshaded area. The part inside the interior of the car such as the dashboards and many plastic parts may degrade because of the excessive temperature of the car interior. The user has to wait for some time so as to let the interior temperature of car to cool down by either rolling down the window or by switching on the a/c. The increase in the temperature of the vehicle may make the passengers and the users very uncomfortable. Moreover the vehicle cabin may face aging problem and may damage the goods and materials kept in the vehicle. According to the finding in USA, every year has recorded many children deaths due to heatstroke due to being left inside the vehicles. In years 1998-2002, the number of children deaths recorded was 29 persons per year. In 2003, this number has increases to 42 and 35 persons in 2004. annually hundreds of children experience many heat illness for being left inside the vehicles [9].The problem arises when the vehicle is parked in a direct heat and sun the temperature rise in parked vehicle is mainly due to the solar radiation and green house effect the solar radiation enters the car through the glasses and is partially trapped within the car the measured temperature inside the car in a summer days could reach up to 120°C where the outside temperature ranges from 35°C to 45°C. The amount of irradiative heating from 12 am to 3pm is measured it could be seen that the irradiation was 1106W/m^2 approximately at 1pm. so to reduce this condition of thermal heat inside a vehicle we have decided to stop the heat entering into the cabin by arresting the heat at the roof itself. For that using a PCM has been used which has the ability to absorb the heat and store it within the material itself. Phase change materials can change phase from solid to liquid and vice versa. Phase change material has the ability to absorb heat and convert into liquid. And when it releases heat it converts back into solid state.

A new design is developed from the normal roof structure of the vehicle available in modern markets. A phase change material is inserted between the insulator layers and the upper metal sheet layers it could be found that the phase change properties of PCM couldn't only be utilized to store thermal energy as in heat storage application but also to improve thermal insulation effect of the PCM combined roofing structure. The vehicle roofing material is standard roof and the real roof of Tata sierra. The insulated roof contains the PCM in it. The special ability of a PCM arrests the heat at the roof itself without allowing the external heat from entering into the cabin this layer is very much useful in decreasing the temperature in vehicle cabin when it is parked.

The special ability of the phase change material is that is absorbs the heat radiation and stores within itself. The PCM changes its phase from solid to liquid as it stores the heat radiation within itself. Its a substance with high heat of fusion which melts and solidifies at certain temperature and it is capable of storing and releasing large amount of heat energy. The PCM is initially at solid state. As the environmental temperature increases, it absorbs energy in the form of heat. When the ambient temperature reaches the melting point, the PCM absorbs large energy of heat at an almost constant temperature. This continues until all the materials is then converted into liquid phase. In this way, heat is stored in a PCM and the temperature is maintained in a level that is optimum. When the temperature around the material gets low, it gets solidified as it releases the heat back into the atmosphere.

The aim is to design and fabricate the roof of a four wheeler which has the ability to resist the heat from flowing into the vehicle cabin so as to maintain a moderate temperature in a vehicle cabin when the vehicle is parked at an un shaded area. The main scope of the design is to make the cabin temperature much more comfortable to the passengers as they get into the parked vehicle. The installation of this roof design will reduce the interior temperature of the vehicle cabin. Due to this roofing structure deterioration of plastic parts can be decreased. The driver doesn't need to keep the windows of the car lower so as to ventilate the interior hot air. When the children and pets are left in the cabin of the car they would not be affected by the increasing temperature of the car cabin.

2. Materials and methods

IOP Conf. Series: Materials Science and Engineering **197** (2017) 012073 doi:10.1088/1757-899X/197/1/012073

2.1 materials used

- PCM -- 1-dodecanol
- Roof -- Roof of TATA sierra
- Side glass -- Window glass of Maruti Omni van
- Temperature Sensor -- DIGITEMP 9100
- Base -- Wood

2.2 PCM 1-dodecanol

1-Dodecanol is a phase change material manufactured by PCM product having its phase change temperature at 24^{0}c to 27^{0}c. The product has density of 831 kg/m³. The maximum operating temperature is 259^{0}c.It is a tasteless, colourless solid with a floral smell. It is soluble in ethanol and diethyl ether. The properties of the selected are shown in the table 1.

Table 1. Properties of pcm- 1 dodecanol

S. No	CONTENTS	PROPERTIES
1	Temperature	24°C to 27°C
2	Density	831 kg/m³
3	Max Temperature	259^{0}C
4	Taste	Tasteless
5	Colour	White crystals
6	Odour	Floral smell

The costs of different PCMs were compared to check cost efficient material. The cost comparison of those PCMs is shown in table 2.

Table 2. Cost of different PCM

S.No.	PCM MATERIAL	COST/KG
1.	HEPTADECANE	Rs. 75,480/-
2.	1-DODECANOL	Rs 900/-
3.	METHYL PALMITATE	Rs 18,460/-
4.	METHYL STEARATE	Rs 3,59,800/-

2.3 Roof

The roof is a basic structure made up of steel. Here the roof is made up of roof of Tata sierra. The roof is cut into dimension of 400×400 mm and painted in black colour as the heat absorbed by black is more. There are two roofs made for the experiment one of the roof is insulated with the PCM and the other roof is kept as it is. The amount of the heat entered the cabin is less in the insulated cabin and the when the roof without the insulated PCM is kept the amount of heat entered is more and thus the cabin's temperature increases.

2.3.1 Roof without PCM

The roof without PCM is made up of steel of the car TATA SIERRA. It doesn't have any insulated PCM so that the heat entering the vehicle cabin is more and the vehicle cabin's temperature increases rapidly. The plain roof which is made is shown in Figure 1.

Figure 1. Roof without PCM

2.3.2 Roof with insulated PCM

The roof with PCM is also made up of the same material that is TATA SIERRA steel roof. The roof contains the insulated material. The material used is a PCM which has the special ability to arrest the heat within itself. The process thus lessens the amount of heat entering into the vehicle cabin. As the heat entering the cabin decreases, the temperature inside cabin is low when compared to the roof without the insulation. The below shown figure.2 shows the constructed and the insulated roof it contains the PCM between the two layers of the roof metal and insulation.

Figure 2. Roof with PCM

3. Fabrication of experimental Setup

IOP Conf. Series: Materials Science and Engineering **197** (2017) 012073 doi:10.1088/1757-899X/197/1/012073

The test setup is constructed with the above mentioned materials. The laminated glasses were joined together with help of liquid sealant such that the heat from the cabin doesn't escape from the setup. The roofs were cut from a car roof and another layer of sheet metal is cut and placed between the roof and the insulation to place the PCM in position. Another roof of the same material was made without the insulating PCM so as the difference can be seen in temperature. The construction was completed as the designed setup as shown in figure 3.

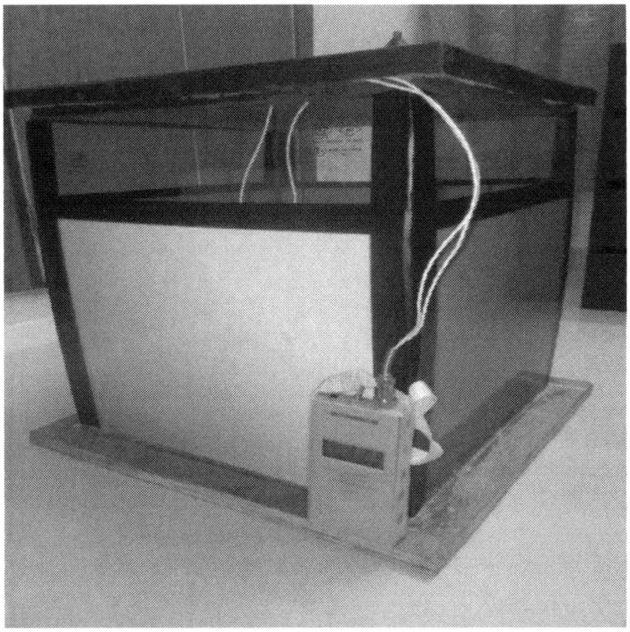

Figure 3. Fabricated setup of pcm

The vehicle cabin is represented by a test setup made up of exactly the same materials present in a vehicle. It consists of the laminated glass, steel roof and the base wood. The glasses are stuck together in the shape of a cube by using the liqiud seal which has the property to stick glass together. The glass is stuck and kept to dry for a day or two.the glass is again stuck to a base wood to give it a base stand. The steel roof is cut from the top of a car-TATA SIERRA. The top of the car is made up of hard steel. The insulation material used is a PCM the PCM is a phase change material which has the ability to absorb heat and convert is phase from its solid state to liquid state. And when the material releases heat it converts its phase back into solid state. The PCM is kept at a bottom steel plate made of the same material as the roof. The two different plates are joined together the design of the basic setup was created using solid works software and the basic structure was designed before it was made. After the design was finalised the fabrication procedure was started.

4. Results and discussion

4.1. Temperature variation with and without PCM

The table 3 shows the temperature of PCM in an insulated cabin and non-insulated cabin measured during four different days. The temperature was measured at peak hours (around 2 pm) of each day. The amount of heat present inside the cabin setup is almost the same ranging from 49°C to 55°C in a non-insulated roof.

Table 3. Temperature at different dates

TEST DATE	TEMP WITHOUT PCM	TEMP WITH PCM
12.03.2016	55°C	41°C
14.03.2016	53°C	42.3°C
17.03.2016	55.8°C	42°C
18.03.2016	49°C	39°C

When the insulated roof was installed in the setup the range of temperature of the cabin was noted much low than the previous experiment with the non-insulated roof, and the results were much better as it could decrease the temperature for at least 10°C. The temperature sensors were kept at two different locations one at the bottom corner and one at the top corner. The temperature range of the complete cabin setup can be measured with the help of this arrangement. The temperature at different location is noted.

Table 4. Temperature at different times

	TEMPERATURE (°C)			
	ROOF WITHOUT PCM		ROOF WITH PCM	
	Probe	Probe	Probe 1	Probe
11 AM	36	35.8	33.5	33.2
12 AM	43	44.2	34.8	35.3
1 PM	45.1	45.6	36.3	36.2
2 PM	47.6	49	35.9	36.8
3 PM	48	47.3	34.4	35
4 PM	44	45	31	32.4
5 PM	37.3	39.8	31.8	31

The table 4 shows the different types reading of the vehicle cabin at two particular days and of two different roof types. The readings were taken for a complete day with the non-insulated roof, the amount of heat and temperature present inside the cabin was noted and recorded. It was noted that the temperature is less at the early hours of the day and it gradually increases as the time increases. The cabin temperature is very high at peak hours and it gradually decreases as the external temperature decreases. And the next day the readings were noted for the cabin temperature with the insulated roof. The temperature was noted at two different locations of the vehicle cabin. The temperature sensors were kept at the top corner of one end and the bottom corner of the other end so as to find the complete temperature range of the vehicle cabin test setup. It was noted that the temperature variations was a gradual increase and decrease, but the temperature inside the cabin is much low than the earlier. The efficiency of the roof was seen at the peak hours the PCM could arrest most of the heat at the peak time. This shows the ability of the PCM chosen to arrest more heat at most of the time without allowing the heat from entering into the vehicle cabin.

4.2 Temperature variation with time

The graphs of the results obtained by the experiment are represented as shown in Figure 4 and Figure 5. The below graph shows the different variations of the temperature present inside the car cabin with and without the PCM roof. The variations were noted for a complete day. The variations show a gradual increase and a gradual decrease of the temperature of the vehicle cabin. The temperature was noted the maximum at the peak hours. The same was noted for the cabin with the PCM roof, the temperature was maximum at the peak hours. But, the PCM insulated roof had a difference of around 13°C between the two different cabin setup at the hottest hours of the day.

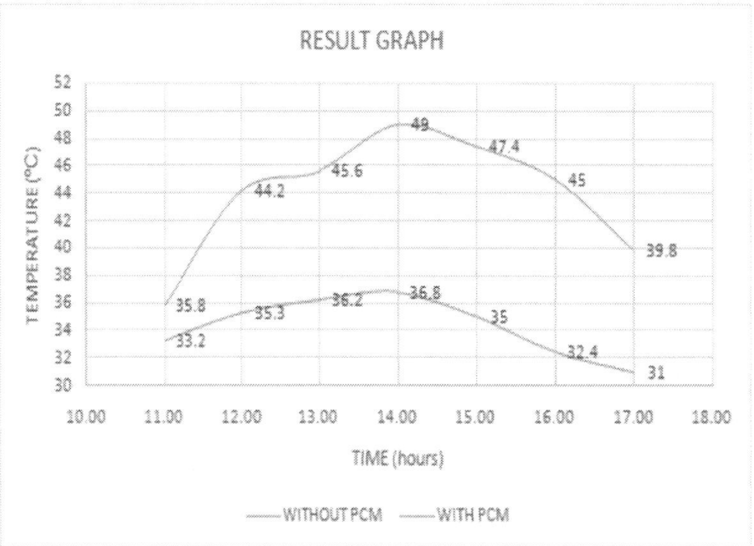

Figure 4. Temperature variation with time

The other graph shows the maximum temperature of the vehicle cabin, the temperature was measured at 1-2pm of each day so as to find out the maximum temperature range of the cabin test setup ranging at the peak temperatures. The temperature of the cabins were noted much high with the normal roofing structure than the temperature of the cabin with the insulated roof structure the insulated roof had arrested most of the heat from entering into the cabin.

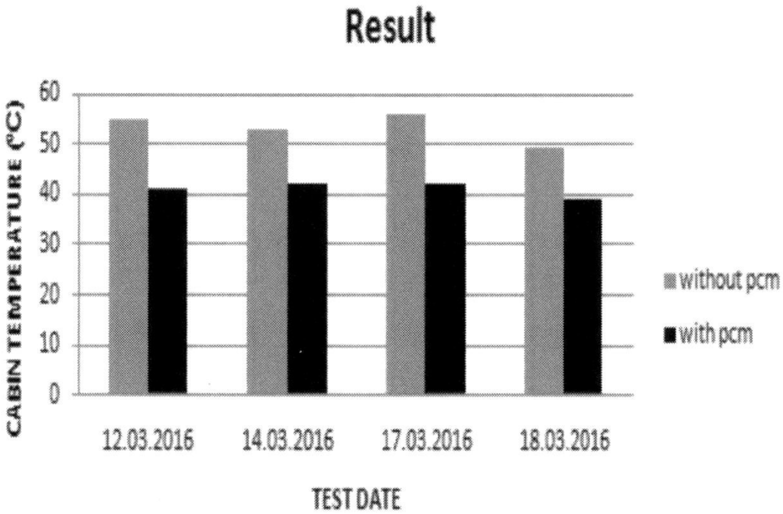

Figure 5. Temperature variation for different dates at peak hours

5. Conclusion

This project shows a new design of vehicle roof structure, which has been done by adding a layer of PCM in the roof of the vehicle. The result shows a decent decrease in the temperature of the vehicle cabin when a PCM insulated roof is attached to the vehicle roof. The experiment has full filled the purpose of decreasing in the amount of temperature of the vehicle cabin when a vehicle is parked in a sunny place or an un-shaded area.

The results obtained are favourable to the experiment. When the temperature measured of both the non-insulated and the insulated cabin, the insulated cabin had very less internal temperature than that of the non-insulated cabin. The results were calculated for three types of readings, one reading was taken at peak temperature around 12:30 pm to 2:00 pm of four different days the, another reading was taken for whole day from 10 am to 5 pm with the non-insulated roof, other reading was taken the next day with insulated roof for the same time period and the readings were noted. Experimental results have shown that the new design has a better thermal performance than the normal roof structure of the available vehicles. The new design may help to reduce up to about 30% of the energy amount required for cooling down the heat entering into the cabin from the roof. The saving rate could be even much higher in case of natural wind and movement of the car in use. Thus, the PCM insulated roof test cabin had the lowest cabin temperature when compared with the regular cabin, the PCM had solved the purpose of arresting the external heat from entering into the vehicle adding to the increase of the cabin temperature and if all the vehicles are insulated with the PCM the cabin's temperature would be very less and it would decrease deterioration of the vehicle plastic parts. This is a promising design which could be massively produced due to its simple structure and reasonable price.

References

[1]. Akanksha Mishra, Atul Sharma, Shukla.A "latent heat storage through phase change materials"
[2]. Abhat.A, "Low temperature latent heat thermal energy storage, Heat Storage materials", solar energy, 1983; 30, pp. 313-32.
[3] Atul Sharma, Tyagi.V.V, Chen.C.R And Buddhi.D, "Review on Thermal energy storage with phase change materials and applications, Renewable and Sustainable Energy Reviews", Vol.13, pp.318–345, 2009.

[4]. Aumboock, Braxmeier.S, Beck.A, Hellmann, Junghanss.T,Pluschke.G, Weinlaeder.H, Phase change material for Thermo Therapy of Buruli ulcer: modeling as an aid to implementation, Journal of Medical Engineering& Technology, Vol.33, No.7, pp.559–566, 2009.
[5]. Dincer.I, Rosen.M.A."Thermal energy storage, systems and Applications", John Wiley and sons, chickester (England) 2002.
[6]. Eunsoo Choi, Harold C. Lorsch, Young I. Cho, Forced convection heat transfer with phase change material slurries: Turbulent Flow in a circular tube, vol.37, No.2, pp 207-215, 1994
[7]. Fok S C, Tan F L, Cooling of helmet with phase change material, Applied Thermal Engineering, Vol.26, pp.2067–2072, 2006.
[8]. Husasain H.Al-Kayein, Firdaus Bin.M, Sidik Anr.Yuganthira.M, Munusammy A.L. "Study on the thermal accumulation and distribution inside a parked car".
[9]. Hsiao Y-F, Leong J.C, Tseng And C-Y, Tsai B-D, cabin heat removal From an electric car, Vol. 4, Nov 5-9, 2010
[10]. JAMEKHORSHID.A,SADRAMELI.M,"Application of phase change material (PCMs) in maintaining comfort Temperature inside an automobile, vol.6, No.1, 2012-01-25.
[11]. Jalal Marhoon Al Dhali,Sudhir Chitrapady Vishweshwara, Study of excessive cabin temperatures of the car parked in Oman and its Mitigation, International journal of multi-disciplinary sciences and Engineering, vol. 4, no. 9, October 2013.
[12]. Lane G.A, "Solar heat storage: Latent heat Material", Background and Scientific principles ,vol I, CRC press, Florida 1983.
[13]. Lane G.A, "Solar heat storage: Latent heat Material", Technology ,vol II, CRS Press, Boca Raton, FL, 1986.
[14]. CHANG Ren Chen, Huann Ming Chou, Nguyen Vu Lan, the use of PCM for ameliorating thermal performance of vehicle cabin.
[15]. Ravikumar.M, Dr.Srinivasan, Pss, phase change material as a Thermal energy storage material for cooling of building, Journal of Theoretical and Applied Information Technology 2005 - 2008.
[16]. Soubhi Ali Hassanein, Waleed A.Abd-Fadeel temperature variations in a parked car exposed to direct sun during hot and Dry climates.Vol.3,issue 1, pp.75-80, March 2013.

IOP Conf. Series: Materials Science and Engineering **197** (2017) 012074 doi:10.1088/1757-899X/197/1/012074

CFD AND THERMO MECHANICAL ANALYSIS ON EFFECT OF CURVED Vs STEP SURFACE IN IC ENGINE CYLINDER HEAD

S. BALAJI
N. GANESH
A. KUMARASAMY
SCIENTISTS, ENGINE DIVISION, CVRDE, DRDO, CHENNAI.
E-mail: BALAJI.S@CVRDE.DRDO.IN

Abstract

Current research in IC engines mainly focus on various methods to achieve higher efficiency and high specific power. As a single design parameter, combustion chamber peak firing pressure has increased more than before. Apart from the structural aspects of withstanding these loads, designer faces challenges of resolving thermal aspects of cylinder head. Methods to enhance the heat transfer without compromising load withstanding capability are being constantly explored. Conventional cylinder heads have got flat inner surface. In this paper we have suggested a modification in inner surface to enhance the heat transfer capability. To increase the heat transfer rate, inner flame deck surface is configured as a curved and stepped surface instead of flat. We have reported the effectiveness of extend of curvature in the inner flame deck surface in a different technical paper. Here, we are making a direct comparison between stepped and curved surface only. From this analysis it has been observed that curved surface reduces the flame deck temperature considerably without compromising the structural strength factors compared to step and flat surface.

I. Introduction

In present scenario, competitive automotive segment is in demand for higher power output and the demand is yet growing stronger. From the future engine trends [2], it is observed that specific power of the engine is likely to grow from 50 kW/l to 65 kW/l for CI engines and corresponding operating pressure and temperature also increases from 160 bar - 200 bar and temperature from 260°C to 280°C at the bridge between valve ports. In IC engines, cylinder head encounters end to end effect of combustion cycle i.e. from intake of combustion gas to exhaust of combustion products. It comprises of passage for intake, exhaust, coolant flow, valve and fuel injection arrangements. Since it acts as the load path, cylinder heads experience high operating temperature. Pulsating high peak pressure and temperature loads demand a rugged heavy cylinder head, whereas high specific power (kW/kg) needs a light weight design. This conflicting technical requirement stresses the designer with limited options in terms of material, space, coolant flow design. Cast iron in form of lamellar or Compacted Graphite Iron (CGI) are the most commonly used cylinder head materials. As material property is a function of body temperature. Hence, controlling the body temperature ensures longer service life of the component. The possibility of enhancing the heat transfer rate over the flame deck is studied by varying the thickness and shape of the inner flame deck surface.

The operational temperature range of cast iron cylinder heads could be specified up to approximately 420°C , 450°C, and 480°C for gray cast iron, compacted graphite iron, and spheroidal graphite iron. When operated at higher temperatures crack is likely to occur at nose bridge and ultimately lead to fatigue damage. Thus, effective coolant flow to maintain low body temperature, methods of transferring heat from combustion chamber reducing the thermal loads are explored by engine designers.

This paper reports the effect of introducing a curved and stepped inner surface to enhance heat transfer rate of the cylinder head and thereby reducing the body temperature. Thus, FEM is considered as starting step in development of new product and experimenting with new modifications. A hypothetical high performance engine cylinder head with step and curved surface is designed to study the effect of inner profile on heat reduction and thus, the best suited configuration is reported.

Figure 1: *GEOMETRY AND FE MODEL OF THE CYLINDER HEAD*

Table 1: *material property*

Parameters	Coolant	Grey Cast Iron
Density, kg/m^3	998.2	7200
Specific Heat in kJ/kg-K	4.18	0.515
Thermal Conductivity, W/m-K	0.6	52
Viscosity, kg/m-sec	0.001	–

II. Analysis Approach

I. Analytical model

Basic 3D model is created in CREO 2.0 and is shown in fig 1. Tetrahedral elements were used to mesh the model and it took 1950663 nodes and 10154632 elements to arrive at a considerable mesh quality. The material properties considered for analysis is given in table 1. CFD analysis is carried out as per the standard analysis method using ANSYS FLUENT. Coolant flow in the coolant passage was assumed to be 3D flow, steady state, incompressible turbulent flow, and the viscosity in the near wall region was also taken into account. CFD analysis was carried out using coolant at a constant inlet temperature of 120°C. Body Temperature is obtained as output from CFD. For ThermoMechanical analysis, load due to the combustion gas peak pressure is applied on the flame deck surface considering the body temperature into account.

II. BOUNDARY CONDITIONS

Energy and turbulence model was considered for CFD analysis. Cast Iron is considered as Solid body and Water as the coolant domain. 7 m/s water inlet velocity and 393°K is taken as input parameter for water. At outlet 1bar abs. pressure is taken as boundary condition. Simplex algorithm was used as solution method and first order up wind scheme was used. Boundary surfaces are shown in fig 2.

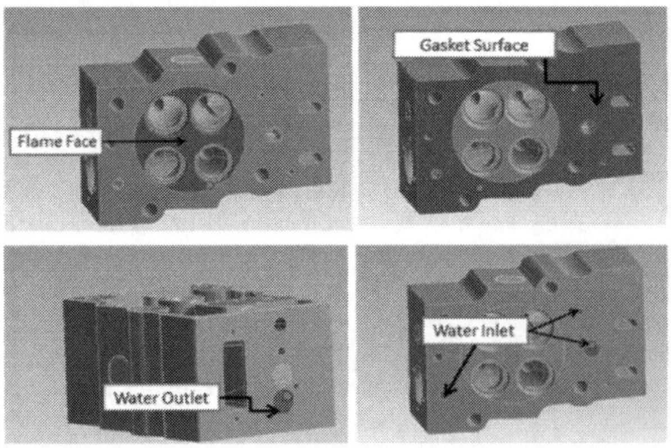

Figure 2: *BOUNDARY SURFACES*

Table 2: *HEAT TRANSFER COEFFICIENTS*

PART(COMBUSTION GAS CONTACT)	TEMPERATURE OF COMBUSTION PRODUCT(K)	HEAT TRANSFER COEFFICIENT(W/m^{2}K)
FLAME FACE	1262	858
EXHAUST PORT	1024	530
INTAKE PORT	379	230
OIL SURFACE	425	200

III. Heat Transfer Coefficient

Heat transfer coefficient is not a material property and depends on a number of factors such as the material conductivity, velocity of the fluid, temperature of the fluid, turbulence, surface finish etc. These complications have made it difficult to find the value of heat transfer coefficient (HTC) by experimental methods. Woschni empirical equation is used for calculation of HTC. Input values for this equation come from basic system specification. Value of the heat transfer coefficient can be found from Nusselt number - Reynolds number type correlation. Woschni model is preferred over other model and the assumptions made are given below:
1. Universally applicable equation for the instantaneous heat transfer coefficient in the internal combustion engine.
2. Calculates the amount of heat transferred to and from the charge.
3. It is the most commonly used heat transfer model and applied to all cylinder elements.

Assumptions:
1. On all surfaces of the cylinder heat transfer coefficient and velocity of charge is uniform.
2. This model accounts the increase in gas velocity in the cylinder during combustion.
3. Woschni's proposal that the average gas velocity should be proportional to the mean piston speed.
4. Air is the working medium.
5. Heat transfer by conduction through the walls is one-dimensional

Thermo dynamic simulation model is also considered for the HTC calculation and the values obtained are given in Table 2.

These heat transfer coefficient are validated using mathematical calculation based upon Woschni equation [2].The equation is given as below:

$$h = 0.820 * D^{-0.2} * P^{0.8} * W^{0.8} * T^{-0.53} \quad (1)$$

$$where, \quad (2)$$

$$h = Heat\ Transfer\ Coefficient \quad (3)$$

$$D = Bore\ dia\ in\ m$$

$$L = Stroke\ in\ m$$

$$N = Speed\ in\ rpm$$

$$P = Combustion\ Pressure\ in\ MPa$$

$$P_0 = Motor\ Pressure$$

$$W = Characteristic\ Speed\ in\ mps$$

IOP Conf. Series: Materials Science and Engineering **197** (2017) 012074 doi:10.1088/1757-899X/197/1/012074

$$W = C_1 * Sp + C_2 * V_s * \frac{T_1 * (P - P_0)}{(P_1 * V_1)} \quad (4)$$

$$\begin{aligned}
&where, && (5)\\
C1 &= 6.18; C2 = 0;\\
C1 &= 2.28; C2 = 0;\\
C1 &= 2.28; C2 = 0.00324 * Po;\\
Sp &= Mean\ piston\ Speed\ [m/s];\\
Tg &= Average\ gas\ temperature\ [K]\\
Pt &= Pressure\ at\ start\ of\ combustion\ in\ MPa\\
Vt &= Volume\ at\ start\ of\ combustion\ in\ m^3\\
Ti &= Temperature\ at\ start\ of\ combustion\ in\ m^3\\
Vs &= Swept\ volume\ of\ cylinder\ in\ m^3
\end{aligned}$$

The heat transfer coefficient on the gas side, h Vs θ for the engine can be represented by the Fig 3. Average of the heat transfer coefficient calculated is found matching with values obtained through thermodynamics simulation. Average heat transfer coefficient considered on the gas side is 800.87 W/m^{2}K.

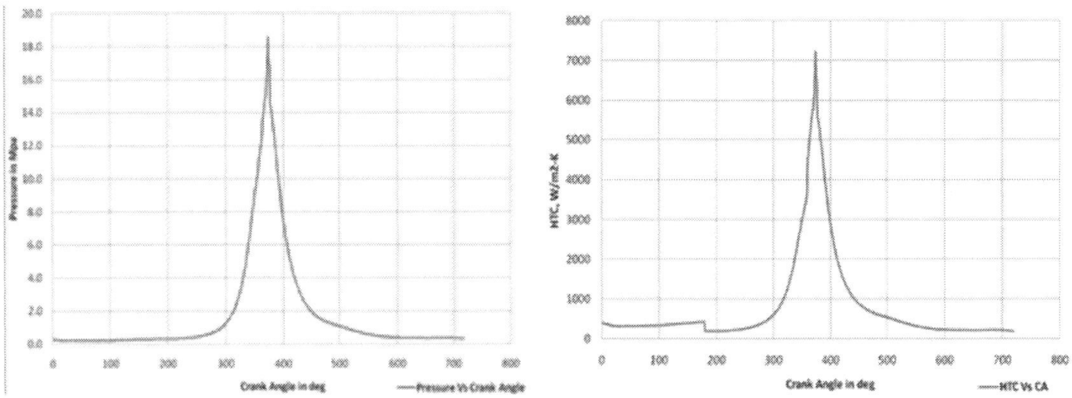

Figure 3

IV. Steady State Thermal Analysis

In IC engines heat from the combustion fluctuates periodically.This high rpm makes the temperature fluctuations to penetrate about a millimeter. Depth of heat penetration is governed by the Thermal diffusivity α. Considering rated speed engine as 2600 rpm calculation for depth of heat penetration is shown in fig 4.

$$\begin{aligned}
Engine\ rated\ speed &= 2600\ rpm\\
corresponding\ time\ scale &= 0.023\ sec\\
Thermal\ diffusivity\ of\ cast\ iron\ \alpha &= 1e^{-5} m^2/sec\\
Penetration\ Depth\ X &= \sqrt{\alpha * t} \quad (6)\\
X &= 4.87e^{-4} m
\end{aligned}$$

Since the depth of heat penetration is small it is assumed that the temperature profile does not change with reference to time. Hence steady state thermal analysis is carried out. The heat distribution is shown Table 3. Internal heat distribution is shown for each case.

IOP Conf. Series: Materials Science and Engineering **197** (2017) 012074 doi:10.1088/1757-899X/197/1/012074

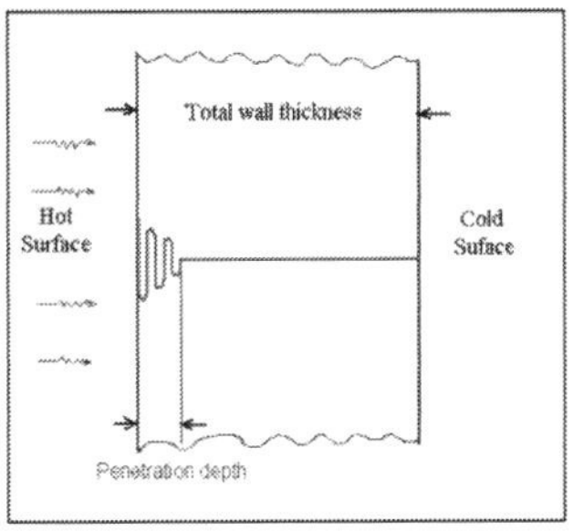

Figure 4

III. Grid Independent Test

One of the essential aspects of CFD simulation study is to check the grid sensitivity. The effect of grid size on the results has been studied for case-2 to prove the grid independent results. Number of grid has been reduced from 2751925 to 305030 as shown in Fig 5. Velocity and Temperature were compared for both fine and coarse mesh. It can be inferred from the figures 6 & 7 that there is no significant change in the values of velocity, temperature, which shows that the results are not sensitive to the grid.

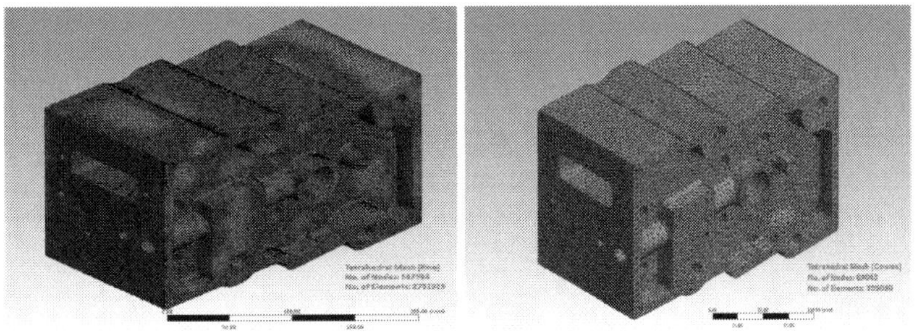

Figure 5: *COARSE AND FINE GRID*

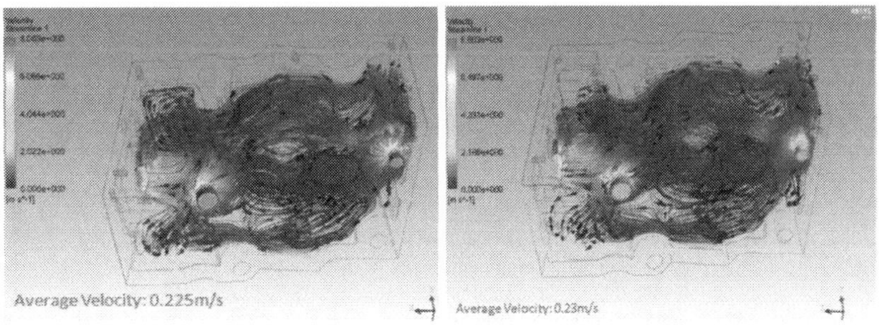

Figure 6: *VELOCITY PROFILE COMPARISION FOR FINE AND COARSE GRID*

IV. Results and Observation

CFD and thermo mechanical analysis was carried out for three different configurations. Coolant inlet velocity is taken as 7m/sec. Boundary conditions are maintained same for all the three cases. From the structural analysis it is found to withstand the peak firing pressure. Summary of CFD results for each case; velocity profile and temperature an stress distribution is given in Table 3. Configuration 3 performs better compared to other two cases in terms of low body temperature and stress distribution.

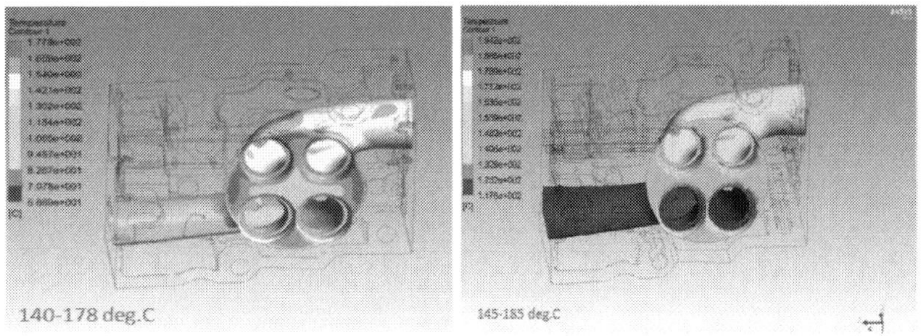

Figure 7: *TEMPERATURE PROFILE COMPARISION FOR FINE AND COARSE GRID*

References

[1] Prof.V. Ganesan, -INTERNAL COMBUSTION ENGINE

[2] Amit Paratwar, -SURFACE TEMPERATURE PREDICTION AND THERMAL ANALYSIS OF CYLINDER HEAD OF DIESEL ENGINE. IJERA, Vol 3. Aug 2013,pp. 892-902.

[3] Stefan Trampert, -THERMO MECHANICAL FATIGUE LIFE PREDICTION OF CYLINDER HEADS IN COMBUSTION ENGINES.INTEL., JOURNAL OF GAS TURBINE AND POWER, Vol 130. JAN 2008,012806-1 TO 10.

[4] Dr.-Ing Franz J Massen, -LIGHT WEIGHT DIESEL COMPONENTS

[5] Metals Databook , Alok Nayar .

Acknowledgement

Authors thank Dr. P. Sivakumar, Distinguished Scientist, Director, CVRDE for the support and encouragement in publishing this paper.

IOP Conf. Series: Materials Science and Engineering **197** (2017) 012074 doi:10.1088/1757-899X/197/1/012074

CONFIGURATION	VELOCITY PROFILE [m/s]	TEMPERATURE [°C]	STRESS[MPa]

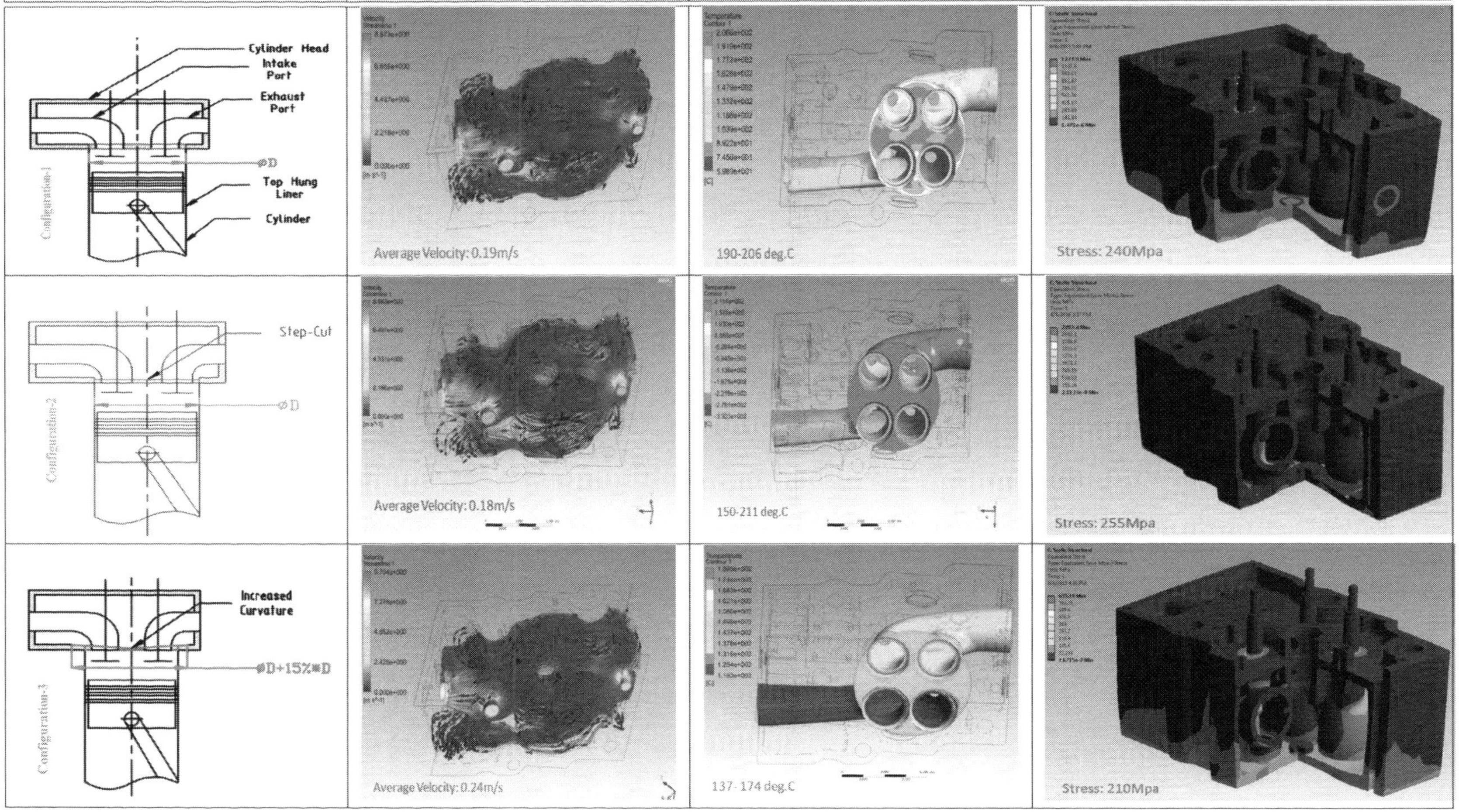

Table 3: *SUMMARY OF CFD RESULTS FOR 7 m/s AVERAGE INLET VELOCITY*

IOP Conf. Series: Materials Science and Engineering **197** (2017) 012075 doi:10.1088/1757-899X/197/1/012075

Analysis of Effect of Inlet Swirl In Four Stroke Single Cylinder Diesel Engine With Different Inlet Valve Geometries Using CFD

R.GOBINATH[1], G.MATHISELVAN[2], R.KUMARASUBRAMANIAN[3]

[1,2,3] Assistant Professors, Department of Automobile Engineering, Sathyabama University.
Email id:[1]gobinath32@ymail.com,[2]mathiselvan@yahoo.com,[3]rkscad@gmail.com

Abstract. Flow patterns are essential to ensure that the engine can produce high performance with the presence of swirl and tumble effect inside the engine cylinder. This paper provides the simulation of air is simulated in the software to predict the flow pattern. The flow pattern is simulated by using the steady state pressure based solver. The domain used for the simulations predicated on the particular engine parameters. Mistreatment the CFD problem solver ANSYS FLUENT, the CFD simulation is earned for four totally different geometries of the valve. The geometries consist of Horizontal, Vertical, curve and arc springs. In this simulation, only the intake strokes are simulated. From this results show that the velocity of the air flow is high during the sweeps the intake stroke takes place. This situation is produced more swirls and tumble effect during the compression, hence enhancing the combustion rate in a whole region of the clearance volume of the engine cylinder. This will initiate to the production of tumble and swirl in the engine cylinder.

1. Introduction:

Due to heterogeneous combustion the thermal efficiency of diesel Engine is less using a large excess air. Therefore to increase the thermal efficiency of C.I engine mixing of fuel and air is important factor. To improve the mixing of air and fuel, the air swirl is created inside the combustion chamber. If the swirl is weak the products of combustion don't seem to be hyped up quick from the surface of the burning drop plate. This will further suffocate burning of droplet. Hence to optimize swirl, inlet valve with different shapes on its periphery can be made first and simulated through computational fluid dynamic. Results are compared with the help of velocity and pressure.

Here five varieties of inlet valves are used,

1. Conventional inlet valve
2. Valve with horizontal blades
3. Valve with vertical blades
4. Valve with curve springs
5. Valve with arc springs

2. Valve Designs:

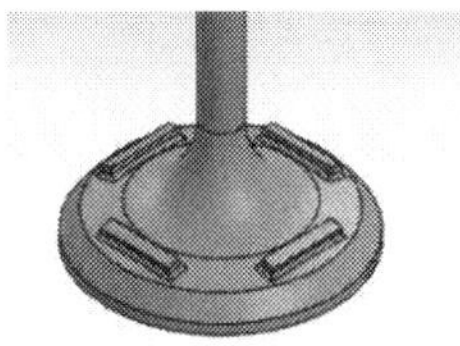

Figure 1. Valve with horizontal blades

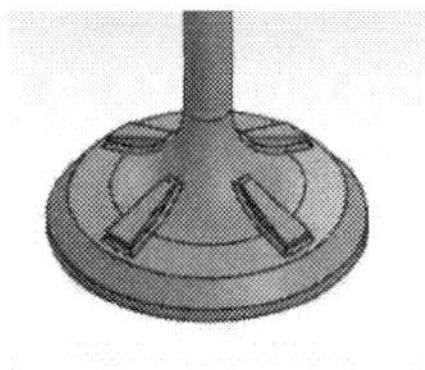

Figure 2. Valve with vertical blades

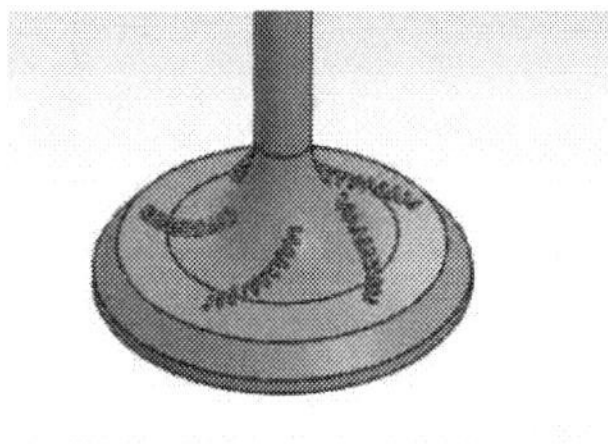

Figure 3. Valve with curve springs

IOP Conf. Series: Materials Science and Engineering **197** (2017) 012075 doi:10.1088/1757-899X/197/1/012075

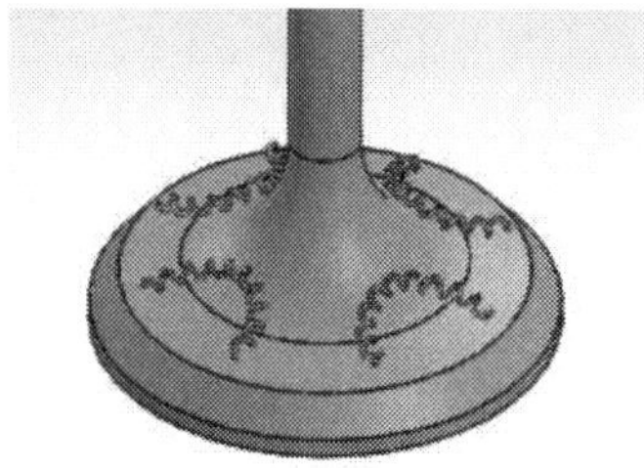

Figure 4. Valve with arc springs

3. Analysis of Normal Piston Ring:

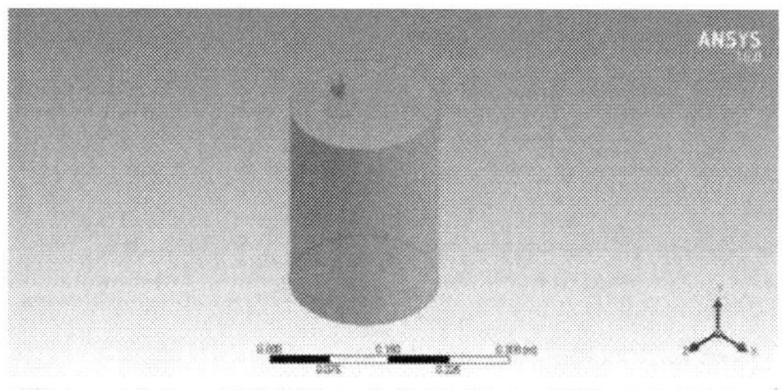

Figure 5. Created component is imported and edited for flow analysis in ansys workbench 2016

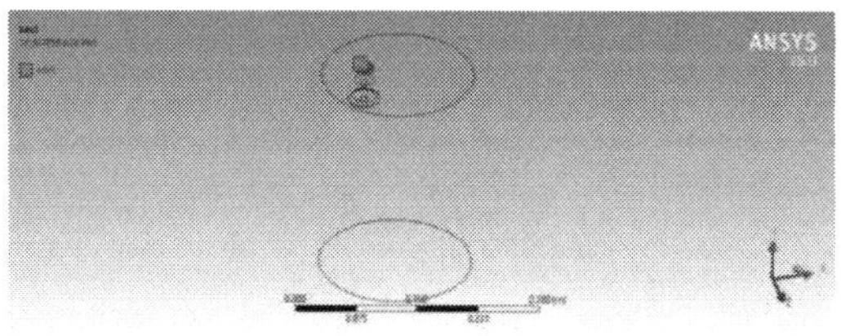

Figure 6. Assigning inlet conditions for flow to enter

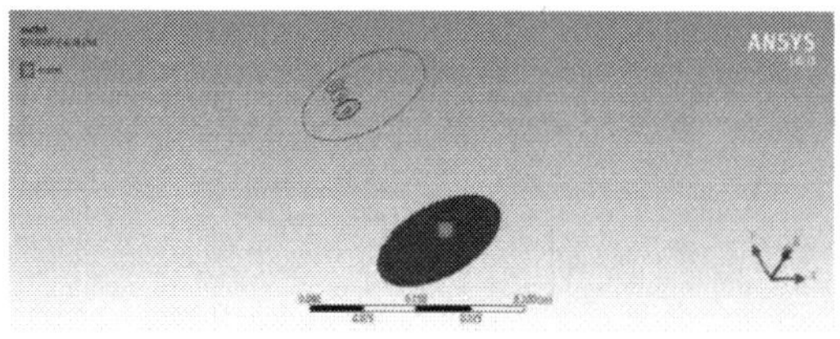

Figure 7. Assigning outlet condition for exit of the flow

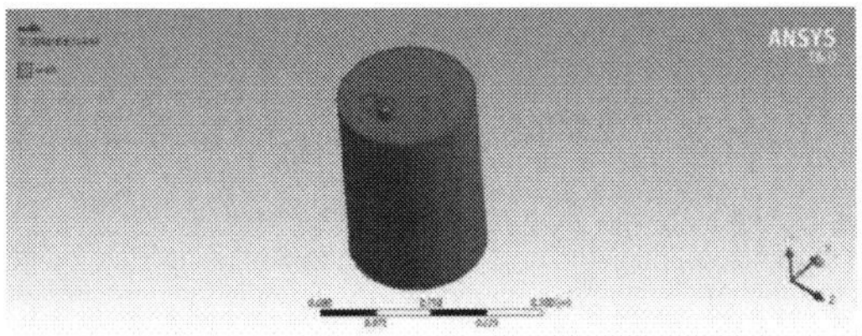

Figure 8. Assigning walls for controlling flow region

4. Velocity and Pressure Analysis:

In physics, mass–energy equivalence may be an idea developed by Einstein that explains the link between mass and energy. It states each mass has an energy equivalent and vice versa—expressed mistreatment the formula

$$E = mc^2$$

Where E is that the energy of a physical system, m is that the mass of the system, and c is that the speed of sunshine during a vacuum (about 3×10^8 m/s). In words, energy equals mass increased by the speed of sunshine square. as a result of the speed of sunshine may be a terribly sizable amount in everyday units, the formula implies that any touch of matter contains a awfully great deal of energy. a number of this energy could also be discharged as heat and lightweight by chemical or nuclear transformations. This additionally serves to convert units of mass to units of energy, irrespective of what system of activity units used.

Fluid medium used: Cynamidyl-radical

Boundary conditions:

Inlet

Type: velocity inlet (5m/s)

Outlet

Type: pressure outlet

No of iterations used: 10

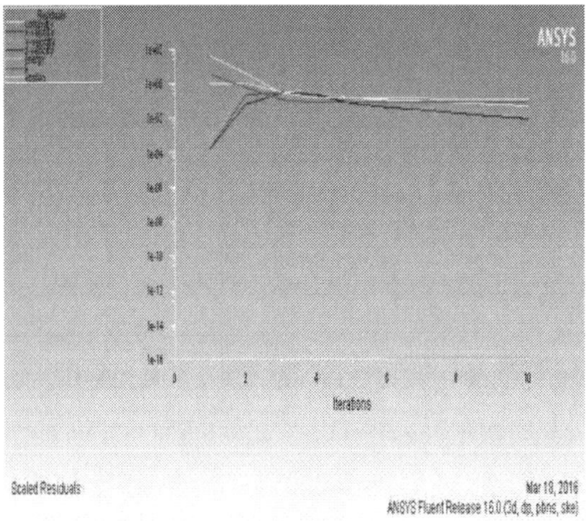

Figure 9. Graphical data for the solved results

5. Velocity flow over a normal valve design:

Figure 10. Velocity flow over a normal valve design

Figure 11. Pressure distributions due to the flow of fluids

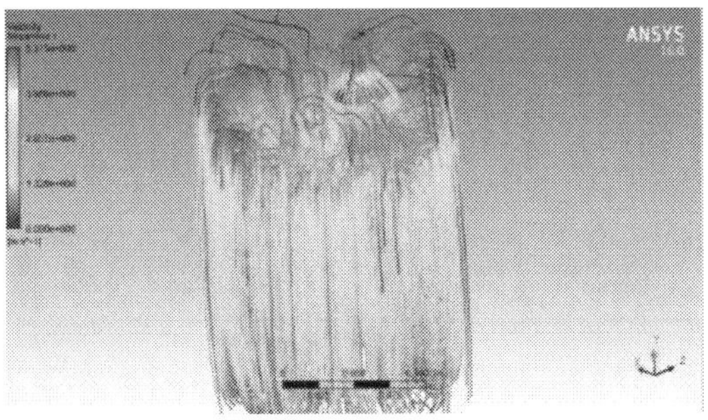

Figure 12. Stream line flow over a normal valve design

6. Analysed Results of Various Geometries of Engine Valve:

6.1 Results for engine valve with horizontal blade structures:

The first design which has modified with the horizontal blade structure where else this design helps to know and improve the results and take further steps in the valve modifications.

Table 1. Velocity and pressure for engine valve with horizontal blades

CONTENT	MINIMUM	MAXIMUM
Velocity (m/s)	1.286	5.143

Pressure (pa)	-5.392	8.199

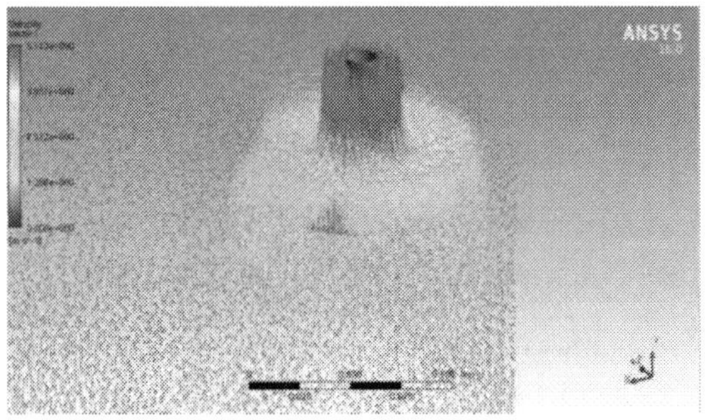

Figure 13. Velocity flow over a valve with horizontal blades

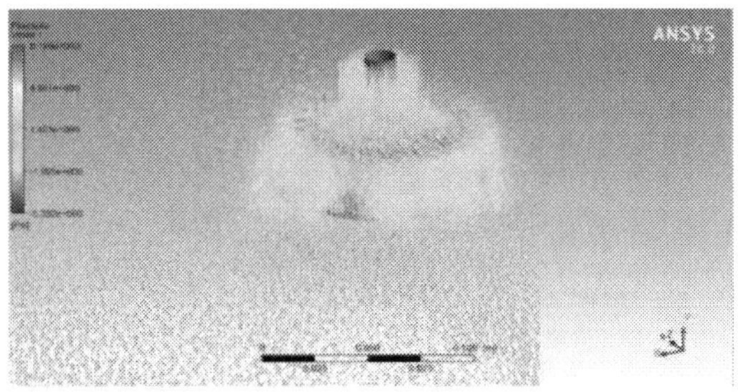

Figure 14. Pressure distributions due to the flow of fluids

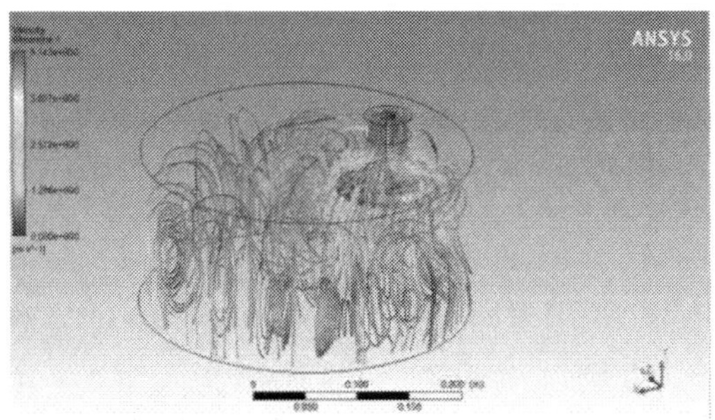

Figure 15. Stream line flow over a Horizontal valve blade design

6.2 Results for engine valve with vertical blade structure:

The second model of valve which as vertical blades on the surface of the valve this model does not supports much better increase of velocity according to design which is done but a 0.621m/s of velocity is

IOP Conf. Series: Materials Science and Engineering **197** (2017) 012075 doi:10.1088/1757-899X/197/1/012075

increased where the inlet air velocity is 5m/s.

Table 2. Velocity and pressure for engine valve with vertical blades

CONTENT	MINIMUM	MAXIMUM
Velocity (m/s)	1.405	5.621
Pressure (pa)	-1.106	$8.562e^{1}$

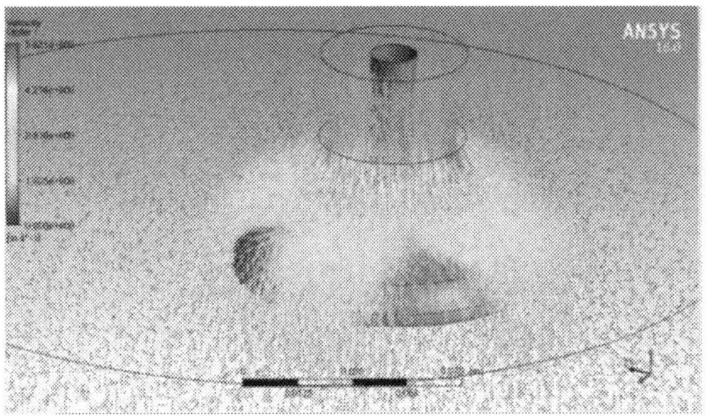

Figure 16. Velocity flow over a valve with vertical blades

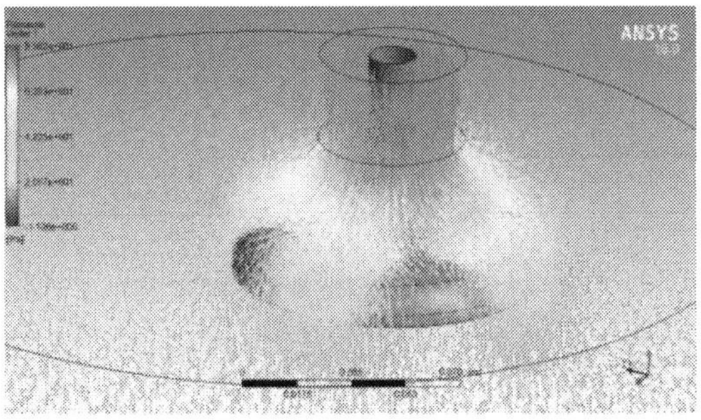

Figure 17. Pressure distributions due to the flow of fluids

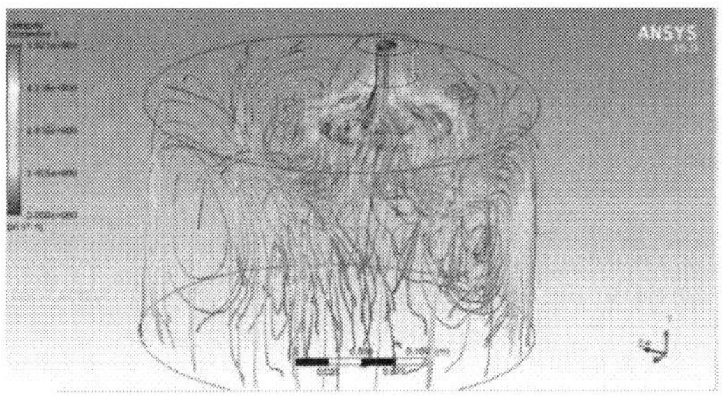

Figure 18. Stream line flow over a vertical valve design

6.3 Results for engine valve with curve springs:

The third result with the model curved springs which a springs fixed in a slight curved structure at the top of the valve where else a another approaches of the swirl motion in the valve where comparing to result of previous model of valve it is slightly about 0.373m/s this not much but a good improved for applied inlet velocity of 5m/s.

Table 3. Velocity and pressure for engine valve with curve springs

CONTENT	MINIMUM	MAXIMUM
Velocity (m/s)	1.499	5.994
Pressure (pa)	$-5.785e^2$	$9.309e^2$

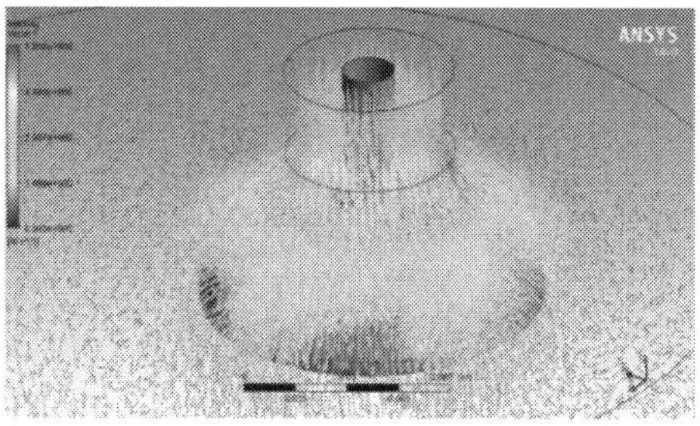

Figure 19 Velocity flow over a valve with curve springs

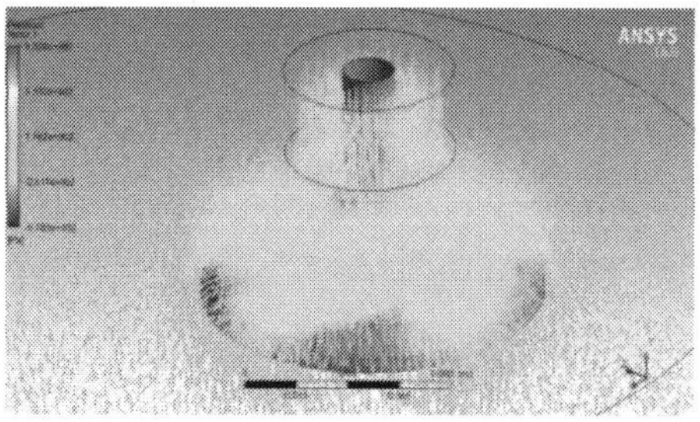

Figure 20. Pressure distributions due to the flow of fluids

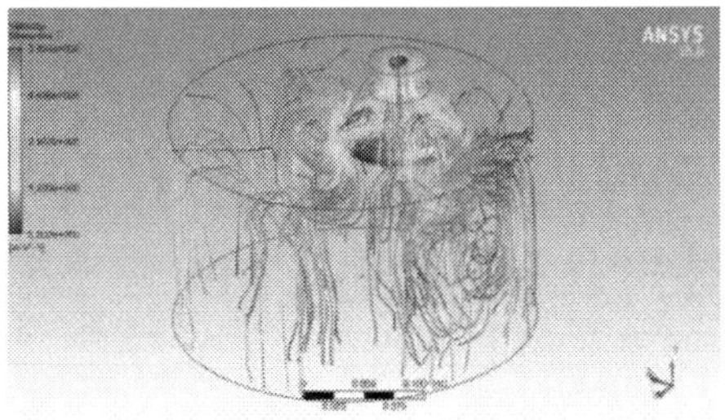

Figure 21. Stream line flow over a curve spring valve design

6.4 Results for engine valve with arc springs:

The result with the curved springs it the fourth model of the approached model it a fine springs arranged with a arc shape comparing to the all result the fourth model with arc springs gives good result where the velocity at the inlet in 5m/s and it results and increased the velocity to 1.212 where the good velocity changes happens to be good swirl motion in the chamber.

Table 4. Velocity and pressure for engine valve with arc springs

CONTENT	MINIMUM	MAXIMUM
Velocity (m/s)	1.553	6.212
Pressure (pa)	$-3.365e^2$	$1.462e^2$

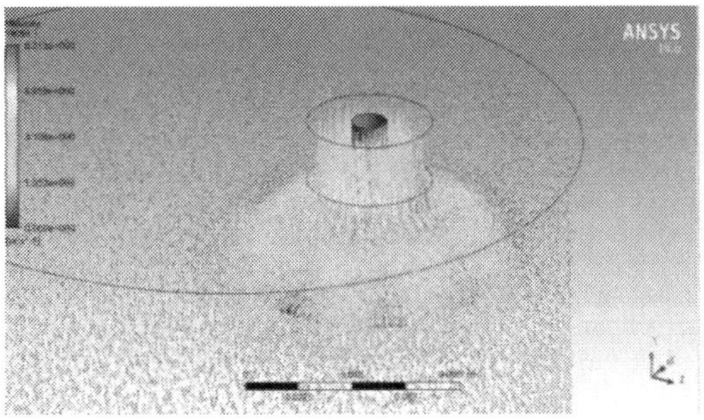

Figure 22. Velocity flow over a valve with arc springs

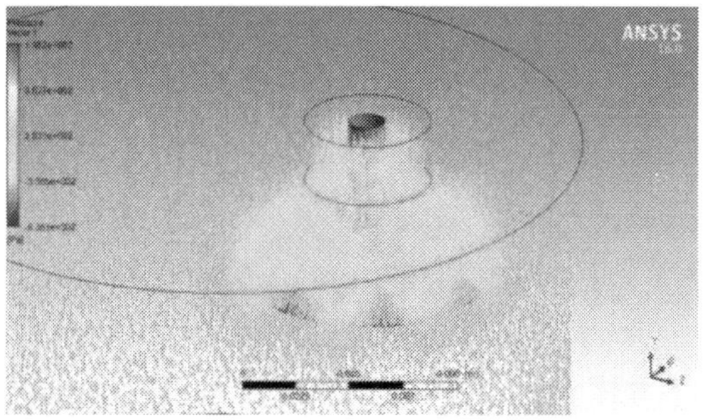

Figure 23. Pressure distributions due to the flow of fluids

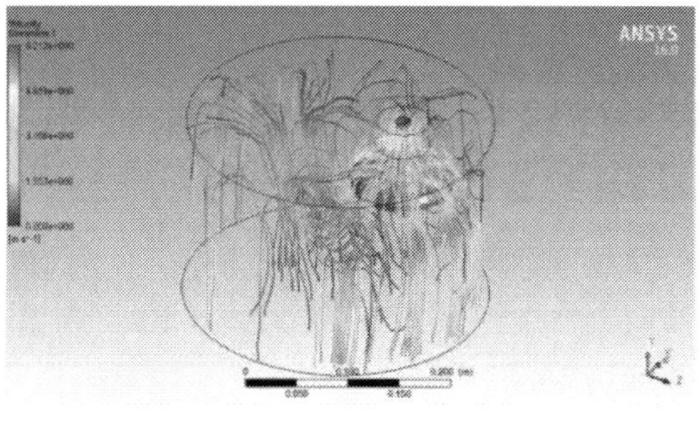

Figure 24. Stream line flow over a arc springs valve design

7.Conclusion:

Design optimisation of the intake/exhaust valves and piston bowl is crucial to understand the on top of mentioned needs. The utilization of CFD analysis in conjunction with optimisation tools will facilitate shorten the look optimisation cycle time. If we have a tendency to approach in experimental testing, it's terribly pricey similarly as time intense. The CFD analysis tool permits insight into the minute flow details that otherwise don't seem to be capture mistreatment flow bench tests. during this project work, the inner flow characteristic within the combustion chamber is investigated computationally for the various geometries of valve. The governing equations for unsteady, three-dimensional, compressible, flow ar resolved with the 2 equation RNG k-ε model to think about the quality of the pure mathematics and fluid motion. The general flow field within the combustion chamber and varied quantities, comparable to pressure, rate distribution were examined for all geometries of valves. While examination all the results of

IOP Conf. Series: Materials Science and Engineering **197** (2017) 012075 doi:10.1088/1757-899X/197/1/012075

various geometries of engine valve it's found that distribution of engine valve with arc spring attachments provides most velocity that tends to activates the most swirl motion that the correct quality of charge is discharged at correct content.

References:

[1] Abdullah .N.R, Shahruddin. N.S, Mamat .R, Mohd .A, Mamat .I and Zulkifli .A 2013. Effects of Air Intake Pressure on the Engine Performance, Fuel Economy and Exhaust Emissions of a Small Gasoline Engine. Journal of Mechanical Engineering and Science. Vol. 6, pp. 949-9

[2] Alfaiz .M, Musthafah .M.T, Rosli .A.B, Ali .M.S and Muhaimin .A. 2013. New Design of a Low Cost Small Engine Dynamometer for Engine Testing. In: 3rd International Conference and Exhibition on Sustainable Energy and Advanced Material. Trans Tech Publ. Melaka. pp. 642-647.

[3] Bari .S and Saad .I. 2013. CFD modelling of the effect of guide vane swirl and tumble device to generate better in-cylinder air flow in a CI engine fuelled by biodiesel. Computers & Fluids. Vol. 84, pp. 262-269.

[4] Hiregoudar Yerrennagoudaru .Dr, Shiva Prasad desai 2014 . Inlet Air Swirl On Four Stroke Single Cylinder Diesel Engine Performance. International Journal of Recent Development in Engineering and Technology Volume 2, Issue 6, ISSN 2347-6435.

[5] Huang .R, Yang .H and Yeh .C.N. 2008. In-cylinder flows of a motored four-stroke engine with flat-crown and slightly concave-crown pistons. Experimental Thermal and Fluid Science. Vol. 32, pp. 1156-1167.

[6] Fontanesi .S, Cicalese .G and Severi .E. 2013. Analysis of Turbulence Model Effect on the Characterization of the In-Cylinder Flow Field in a HSDI Diesel Engine. SAE Technical Paper.

[7] Kumar .V, Sivagaminathan .N, Gopalakrishnan .N, Morton .S, Marine .M and Radavich .P. 2000. Air flow and charge motion study of engine intake port. In: Proc. 37th National and 4th International Conference on Fluid Mechanics and Fluid Power. Number FMFP10-CR-10. pp. 26-30.

[8] Pandey .K and Roy .B. 2012. CFD Analysis of Intake Valve for Port Petrol Injection SI Engine. Global Journal of Researches in Engineering. Vol. 12.

[9] Punekar .H and Das .S. 2013. Numerical Simulation of Subcooled Nucleate Boiling in Cooling Jacket

of IC Engine. SAE Technical Paper.

[10] Ramasamy .D, Aik Soon .K, Walker-Gitano Briggs .H and Zainal .Z.A. 2013. Variation of Airflow Pattern through Dissimilar Valve Lift in a Spark Ignition Engine. Journal of the Chinese Institute of Engineers. Vol. 36, pp. 1083-1096.

[11] Ramasamy .D, Zainal .Z, Bakar .R and Kadirgama .K. 2014. Mass fraction burn comparison of compressed natural gas and gasoline.In: Applied Mechanics and Materials, Trans Tech Publ. pp. 442-446.

IOP Conf. Series: Materials Science and Engineering **197** (2017) 012076 doi:10.1088/1757-899X/197/1/012076

Aerodynamic Analysis Over Double Wedge Airfoil

U S Prasad[1], Ajay V S[2], Rajat R H[3], Samanyu S[4]

[1]Sr. Assistant Professor, Institute of Aeronautical Engineering, Hyderabad, India.
[2 3 4] UG scholar of Aeronautical Engineering, MIT, Manipal, Manipal University, India.

Email: shivaprasad047@gmail.com

Abstract. Aeronautical studies are being focused more towards supersonic flights and methods to attain a better and safer flight with highest possible performance. Aerodynamic analysis is part of the whole procedure, which includes focusing on airfoil shapes which will permit sustained flight of aircraft at these speeds. Airfoil shapes differ based on the applications, hence the airfoil shapes considered for supersonic speeds are different from the ones considered for Subsonic. The present work is based on the effects of change in physical parameter for the Double wedge airfoil. Mach number range taken is for transonic and supersonic. Physical parameters considered for the Double wedge case with wedge angle (ranging from 5 degree to 15 degree.

Available Computational tools are utilized for analysis. Double wedge airfoil is analysed at different Angles of attack (AOA) based on the wedge angle. Analysis is carried out using fluent at standard conditions with specific heat ratio taken as 1.4. Manual calculations for oblique shock properties are calculated with the help of Microsoft excel. MATLAB is used to form a code for obtaining shock angle with Mach number and wedge angle at the given parameters. Results obtained from manual calculations and fluent analysis are cross checked.

1. Introduction

Double wedge airfoils (DWA) are very rarely used in physical applications because of its complex geometrical parameter variations. DWA have different types of shocks at higher speeds which make studies interesting to analyse the flow parameters along the body. Supersonic flow over different contours is a leading study in today's aeronautical field. Since any disturbance (either physical or shock related) takes up great deal of importance in supersonic flow, it becomes a necessity to analyse flow over different contours to get higher and efficient performance. Detailed analysis of change in flow properties over airfoils is done with the aid of CAD Modelling and Fluent for analysis.

Supersonic characteristics for a Double wedge airfoil are associated with shocks and expansion waves. This results in increase in Drag. Manual and Fluent analyses over Double wedge airfoil was done to minimize this drag. Oblique shock and expansion waves for Double wedge airfoil were manually calculated for different wedge angles and the same was carried using ANSYS Fluent software. This was done at 0, 5, 10, 15degree angle of attack (AOA) for a 5deg half wedge angle. Also another model with 10 degree half wedge angle was analysed at 0 degree AOA.

Using coding software MAT lab shock angles for corresponding Mach number and Wedge angle is found. These angles are used for further calculation of flow properties after oblique shock and expansion wave are found. Modelling of Double wedge airfoil was carried out using CAD Software. Analysis was carried out in Fluent Software, Standard Flow conditions was used with change in Mach number and wedge AOA. Pressure coefficient, drag Coefficient and other important plots are considered for the study. Conclusions based on these plots were made.

Oblique shock and expansion waves theory was studied [1] and the implementation of shock and expansion waves over Double wedge airfoil was carried further. Manual calculations for oblique shock properties and expansion wave are formulated in Excel. Lift and drag are analyzed numerically by solving Navier stokes equations for variable density flow field on double wedged aerofoil by varying the angle of attack and thickness to chord ratio [2].

2. Modeling of double wedged airfoils

Modeling is carried out using CAD modeling software. Model varies in physical dimensions, the chosen value for half wedge angle is 5 degree and 10 degree at 0°, 5° , 10° and 15° AOA. Model Specifications: Domain Size 22 x 10 m (For 5° half angle model) Faces: 4, Boundary Conditions: Wedge surface = Wall, Walls of Domain: Pressure far field.

Second model specifications: Domain Size 22 x 10 m. (For 10o half angle model) Boundary Conditions: Wedge surface = Wall, Walls of Domain: Pressure far field Meshed model files was imported to FLUENT. Two equation model used for turbulence are K-epsilon model amongst viscous models [4]. Epsilon model is recommended for flows with separation, recirculation and whose scales are smaller than cell size. Two extra transport equations have been used for simulating the turbulent properties of the flow which allow two equation model to account for history effects like convection and diffusion of turbulent energy.

Figure 1a. 5° half angled aero foil **Figure 1b.** 10° half angled aero foil

3. Results and analysis

TheMeshed Model is imported in fluent software for analysis. Analysis is carried out for Mach number range of 1.2-1.8 at 0° and 5° AOA. Solver used is Density based with K-epsilon viscous model. Operating conditions are set to 0. Iterations are carried out until convergence of Standard Equations. Changes in Mach number was made and plots were analyzed. Changes in AOA and half wedge angle was made.

Further changes in the Angle of attack were made in fluent to 5, 10, 15 degree angle of attack and similar analysis was carried out at different Mach number. It was observed that as half wedge angle equals the AOA, No shock was visible at the leading edge. This mainly due to the fact that the upper first half of the plane becomes parallel to free stream direction. Hence no flow tends to turn into itself.

AOA was further increased to 10 degree and 15 degree. When the AOA is increased to value higher than half wedge angle, An Expansion wave is seen at top leading edge as show in Figure 7. For the same Mach number but different AOA it was observed that the expansion wave on top and the shock wave at lower surface got stronger as shown in Figure 8 and Figure 9. This was concluded based on velocity contours and oblique shock relations.

3.1 Important plots with respect to change in angle of attack

Properties Such as dynamic pressure, Mach number, Temperature and turbulence over the surface are compared for study. C_l, C_d Lift and Drag Vs Mach number was tabulated manually and plotted using post processing software.

3.2 Change in wedge angle

CAD Model made in modeling software to 10° half wedge angle (deflection angle) was read in fluent. Analysis at 0° AOA was carried out with Mach number ranging from 1.2 to 3.5. Contours and plots were studied.

Figure 2. Velocity Contour on 5° half wedge angle, 0° AOA at Mach 1.2

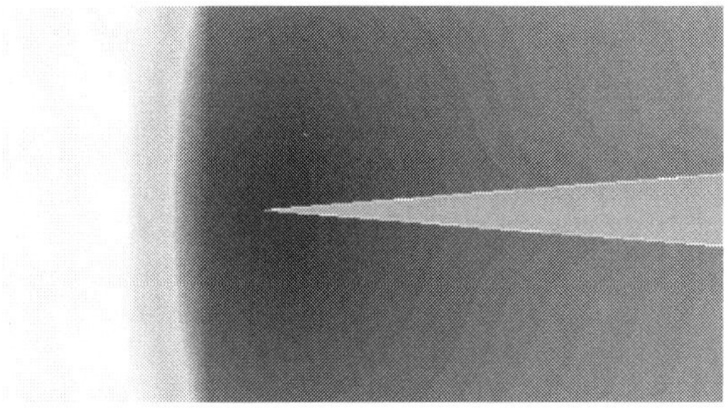

Figure 3. Detached shock wave (magnified image) on 5° half wedge angle

Figure 4. Mach 1.6velocity contour over 5° half wedge angle

Figure 5. Mach 1.6 attached Shock (magnified image)

Figure 6a. Velocity contours at 0° AOA respectively with 5° half wedge angle at Mach 2 freestream velocity.

Figure 6b. Velocity contours at 5° AOA respectively with 5° half wedge angle at Mach 2 freestream velocity.

Figure 7. Expansion wave observed at top Leading edge when the AOA exceeds the half wedge angle.

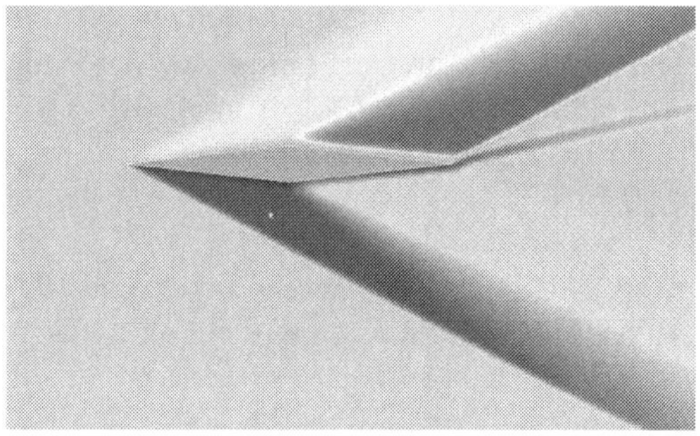

Figure 8.Velocity contours at Mach 3, AOA at 10° Angle of attack

Figure 9.Velocity contours at Mach 3, AOA at 15°.

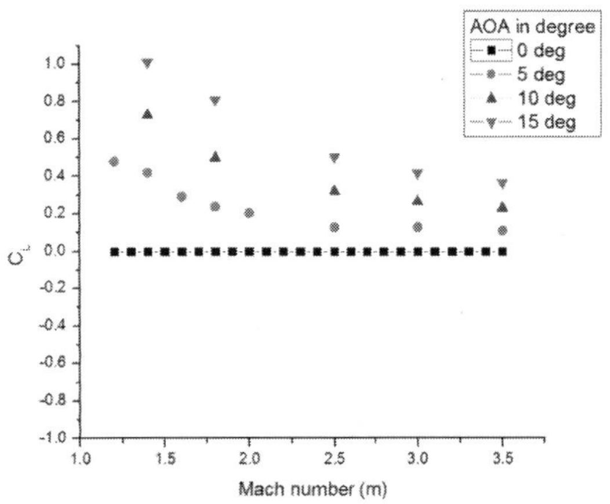

Figure 10. Coefficient of lift at different AOA with respect to Mach number.

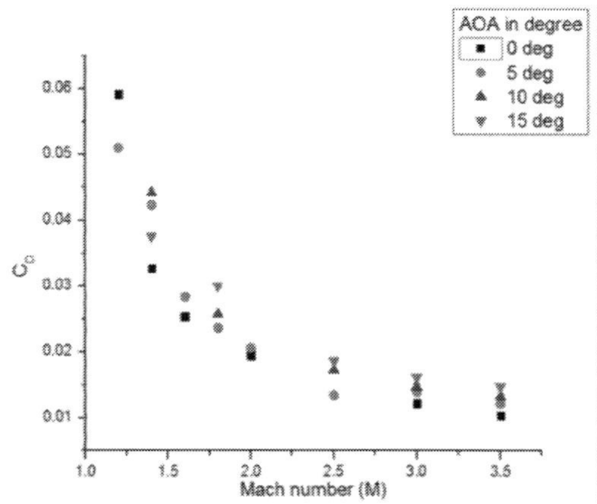

Figure 11. Coefficient of drag at different AOA with respect to Mach number.

Figure 12a. Velocity contours at Mach 3.5 with 5° Half wedge angle

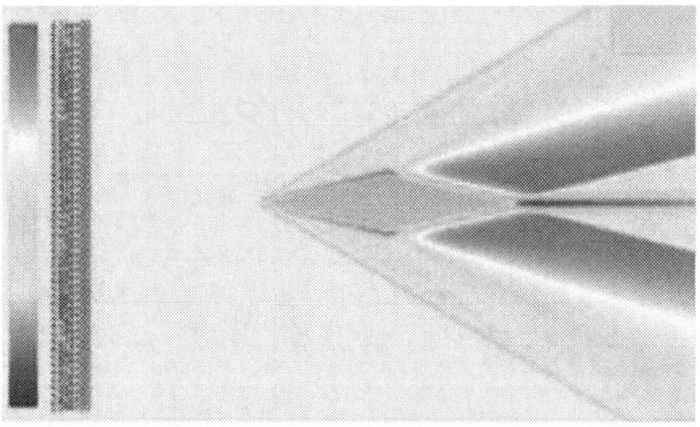

Figure 12b. Velocity contours at Mach 3.5 with 10° Half wedge angle

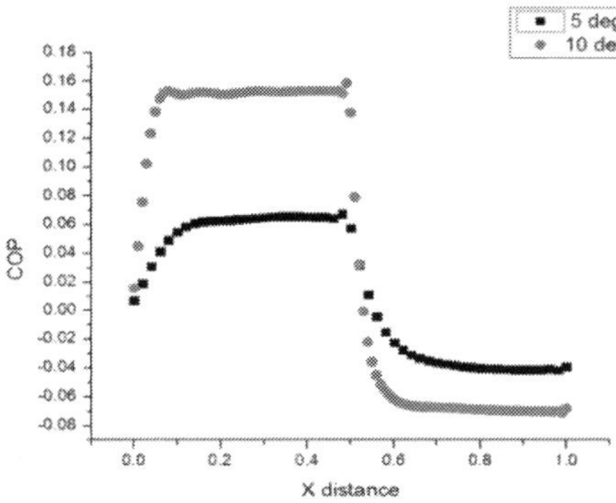

Figure 13. Coefficient of pressure Vs length(m) of aerofoil

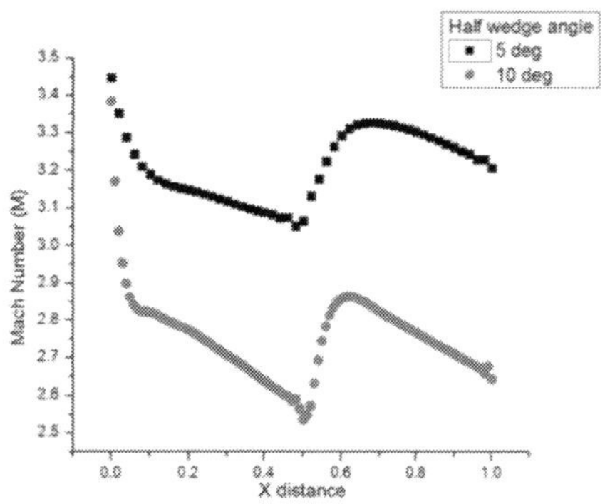

Figure 14. Mach number over surface Vs Length of the airfoil (m).

4. Conclusions

The use of sections to divide the text of the paper is optional and left as a decision for the author. Where the author wishes to divide the paper into sections the formatting shown in table 2 should be used.

4.1 *Double Wedge Airfoil*

4.1.1 *At 0 degree AOA* Manual calculations agrees well with fluent results. The de-attached shock for shock angle above 90° is clearly seen in fluent. Pressure Coefficient is seen to decrease with increase in Mach number after shock. Peak Mach number and turbulent kinetic energy above surface is seen to increase with inlet Mach number. Total temperature gradually decreases along wedge. It is higher at higher inlet Mach.

4.1.2 *At different AOA* No shock observed when half wedge angle equals AOA. An Expansion wave is seen at top leading edge at AOA>half wedge angle. The strength of Expansion wave and Shock wave at top and bottom leading edge increases with AOA provided AOA exceeds the half wedge angle.

4.2 *For AOA half wedge angle*

As the AOA increases, Patterns of plots remain same but the Pressure coefficient at the top surface increases and at bottom surface decreases. The same pattern is observed in case of Mach number, Dynamic pressure and turbulent kinetic energy plots.

C_l and C_d (Coefficient of Lift and Drag) seen to decrease with increase in M and decrease in AOA. Lift and Drag increase with increase in M and AOA.

IOP Conf. Series: Materials Science and Engineering **197** (2017) 012076 doi:10.1088/1757-899X/197/1/012076

Table 1: All calculations are tabulated in an Excel sheet as shown below and the Oblique shock pattern study is carried out.

γ	Θ	β(Radians)	M_1	P_1	M_{n1}	M_{n2}	M_2	T_1	P_2	ρ_1	ρ_2	T_2	P_{01}
1.4	0.0873	1.578	1.2	1	1.199969	0.84219	0.844899	288	1.513246	1.225	1.64341	324.857	2.42497
1.4	0.174	1.587	1.2	1	1.199842	0.842269	0.852865	288	1.512892	1.225	1.64314	324.834	2.42497
1.4	0.261	1.587	1.2	1	1.199842	0.842269	0.868151	288	1.512892	1.225	1.64314	324.834	2.42497
1.4	0.348	1.587	1.2	1	1.199842	0.842269	0.890857	288	1.512892	1.225	1.64314	324.834	2.42497
1.4	0.087	0.92	1.4	1	1.113842	0.90117	1.217875	288	1.280752	1.225	1.46119	309.235	3.18227
1.4	0.174	1.583	1.4	1	1.399896	0.739753	0.749543	288	2.119659	1.225	2.06961	361.332	3.18227
1.4	0.261	1.587	1.4	1	1.399816	0.739787	0.76252	288	2.1194	1.225	2.06944	361.318	3.18227
1.4	0.348	1.587	1.4	1	1.399816	0.739787	0.782463	288	2.1194	1.225	2.06944	361.318	3.18227
1.4	0.087	0.767	1.6	1	1.110365	0.9038	1.437357	288	1.27173	1.225	1.45388	308.6	4.25041
1.4	0.174	0.889	1.6	1	1.242307	0.816984	1.246129	288	1.633881	1.225	1.73359	332.508	4.25041
1.4	0.261	1.576	1.6	1	1.599978	0.668444	0.690925	288	2.819919	1.225	2.48884	399.731	4.25041
1.4	0.348	1.587	1.6	1	1.59979	0.668501	0.707065	288	2.819216	1.225	2.48846	399.693	4.25041
1.4	0.087	0.67	1.8	1	1.117775	0.89822	1.631553	288	1.290991	1.225	1.46945	309.953	5.7458
1.4	0.174	0.768	1.8	1	1.250456	0.81238	1.451508	288	1.657582	1.225	1.75097	333.983	5.7458
1.4	0.261	0.889	1.8	1	1.397596	0.740724	1.260748	288	2.112152	1.225	2.06472	360.905	5.7458
1.4	0.348	1.582	1.8	1	1.799887	0.616527	0.653226	288	3.612859	1.225	2.88983	441.07	5.7458
1.4	0.087	0.598	2	1	1.125981	0.89215	1.824252	288	1.312473	1.225	1.48673	311.449	7.82445
1.4	0.174	0.685	2	1	1.265346	1.644338	1.644338	288	1.701284	1.225	1.78275	336.679	7.82445
1.4	0.261	0.79	2	1	1.420707	1.448766	1.448766	288	2.188142	1.225	2.11377	365.213	7.82445
1.4	0.348	0.929	2	1	1.602043	1.216746	1.216746	288	2.827634	1.225	2.49309	400.141	7.82445

4.3 Change in Wedge angle

For same Mach number Increase in half wedge angle increases the strength of shock and expansion waves at 0o AOA. C_p increases in the first half region and decreases in second half region of wedge with increase in wedge angle. Pattern remains similar with respect to Dynamic pressure plots. Mach plots shifts down with increase in wedge angle representing a decrease in value.

Turbulent kinetic energy increases with increase in wedge angle.

Acknowledgments

Authors would like to express sincere gratitude to all those who have directly or indirectly contributed to this research work. Authors would like to thank Prof. Dr. Satish Shenoy B. Head of the Department Aeronautical & Automobile Engineering, Research Director (Technical), and MIT-Manipal University, Manipal, for the computational support given.

References

[1]. John D Anderson Jr, 2010 Fundamentals of Aerodynamics, 5 Edition.

[2]. Ramesh Kolluru and Vijay Gopal 2012 Proc. Excerpt Comsol p 1.

[3]. Pritam Kumar Patel and M Premchand CD adapco, *CFD basics*, American Agency, Training Document, P 38.

[4]. Colleen D. Scott-Pomerantz 2004 The k-epsilon model in the theory of turbulence. Tech. Rep. Univ. of Pittsburgh. P 10.

IOP Conf. Series: Materials Science and Engineering **197** (2017) 012077 doi:10.1088/1757-899X/197/1/012077

Aerodynamics and flow characterisation of multistage rockets

Srinivas G[1], M V S Prakash*[2]

[1]Assistant Professor, Dept of Aero & Auto Engineering, Manipal Institute of Technology, Manipal University, Manipal, Udupi, Karnataka, India-576104.
*[2]UG student, Dept of Mechatronics Engineering, Manipal Instituteof Technology, Manipal University, Manipal, Udupi, Karnataka, India-576104.

Abstract. The main objective of this paper is to conduct a systematic flow analysis on single, double and multistage rockets using ANSYS software. Today non-air breathing propulsion is increasing dramatically for the enhancement of space exploration. The rocket propulsion is playing vital role in carrying the payload to the destination. Day to day rocket aerodynamic performance and flow characterization analysis has becoming challenging task to the researchers. Taking this task as motivation a systematic literature is conducted to achieve better aerodynamic and flow characterization on various rocket models. The analyses on rocket models are very little especially in numerical side and experimental area. Each rocket stage analysis conducted for different Mach numbers and having different flow varying angle of attacks for finding the critical efficiency performance parameters like pressure, density and velocity. After successful completion of the analysis the research reveals that flow around the rocket body for Mach number 4 and 5 best suitable for designed payload. Another major objective of this paper is to bring best aerodynamics flow characterizations in both aero and mechanical features. This paper also brings feature prospectus of rocket stage technology in the field of aerodynamic design.

1. Introduction

The design of rocket model shape is very difficult job to the aerodynamic researcher because of its complexity in calculation of structural mass, inert mass and payload mass. However the designer also need to concentrate on the efficiency of the rocket in terms of thrust produced per unit time. Rocket model can make a simple way like cardboard tube completely bursting with black powder but to create an efficient and correct rocket implicates more number of complicated issues [1]. Some of the complicated difficulties like maintaining the desired combustion heat, maintaining the appropriate pressure of the fuel supplying into the combustion chamber, controlling the thrust levels producing from the rocket engine and most importantly the controlling the direction of motion. The rocket direction of motion again directly proportional to the aerodynamic shape of the rocket. This paper more concentrate on the desired shape to get better aerodynamic results in both single stage and double stage rocket model.

2. Literature survey on multi stage rockets

2.1 Forces on a rocket body

Aerodynamic study of the rocket body is the part of ballistics filed, whereas the spacecraft are studied in subdomains of the astrophysics. According to the rocket theory four forces lift, drag, gravity and thrust commonly act on the flying rocket [7]. Figure 1 shows four forces acting on flying rocket model. Any rocket travelling in the air must have thin in size and tall for providing high ballistic coefficient value as it makes very minimum drag losses. When high adequate speed in the exact direction and altitude are reached a stable escape velocity or orbit is obtained by addition centrifugal pseudo force and inertia can be significant due to path of the rocket around the centre of a celestial body [2-3].

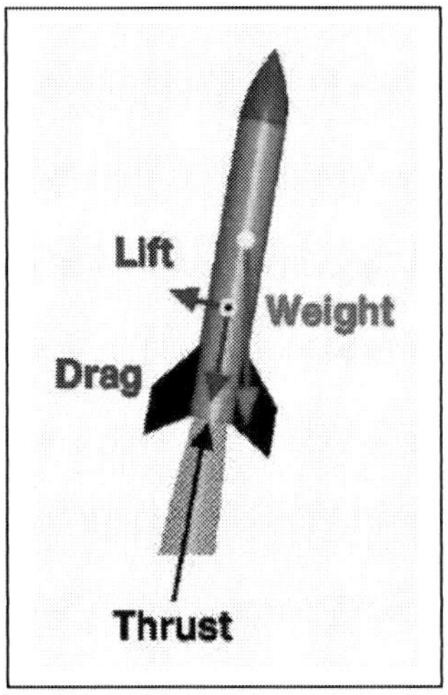

Figure 1. Forces acting on a flying rocket body [Reference: wikipedia.org]

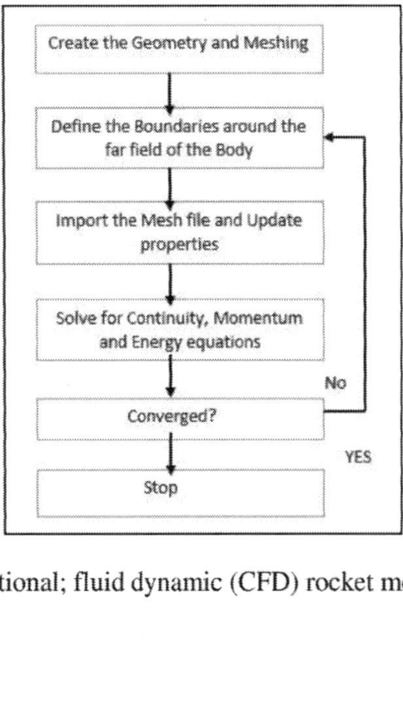

Figure 2. Computational; fluid dynamic (CFD) rocket model solving flow chart

Rocket body when its flying at higher altitudes obviously it follows parabolic trajectory termed a gravity turn and this trajectory is frequently used at least during the initial part of a launch [4]. Rocket body can thus maintain even zero or low angle of attack which minimizes the stress due to the transverse forces on the launch vehicles, permitting a very frailer and later lighter, launch vehicle. Opposite to the direction of rocket body motion is drag force. This force decreases by increasing the acceleration of the rocket body and structural loads. Drag equation used to calculate the deceleration forces for fast moving rockets. The drag equation is showing in equation (1) where the term A is the reference area of the rocket body, v is the flow velocity, ρ is the mass density and F_D is the drag force.

$$F_D = \frac{1}{2}\rho U^2 C_D A \tag{1}$$

The ambient atmospheric pressure almost always different than exit pressure of the rocket jet. After rocket engine ignited on the ground the atmospheric pressure is maximum than pressure in the rocket jet. As the rocket body moves to higher altitude the jet pressure remains constant while the atmospheric pressure decreases, the rocket exit jet becomes slowly start growing, very concisely becoming idyllically expanded although on its way to underexpansion. As the rocket body continues to increase the altitude, as the rocket is migrant into a vacuum, the rocket exit jet diameter increase to very high proportions. That is the reason as the rocket gains altitude rocket jet diameter increases. The jet diameter is again directly proportional to the area covered by the body miscellaneous component like nozzle, fins, engine location etc. This paper also discuss the entry and exit portion of the rocket body aerodynamic behaviour.

2.2 Rocket body staging

Rocket total mass includes structural, propellant, payload and also miscellaneous mass. Carrying this complete mass by the rocket to the higher altitude and reaching to the destination is very complicated to task to the researcher. There it helpful the concept called 'staging'. The rocket leaves the excess weight like empty tanks, related engines during launch. Comparing with the speed of light, the extreme speeds that can be achieved with the staging is ideally limited only. To achieve better incremental velocity of the staging carried goes down geometrically with each additional stage required. This paper also discuss about the aerodynamic behavior of such additional stage to get incremental velocity using ANSYS software.

2.3 Safety and reliability of rocket

The safety and reliability of the rocket that depends on the design and materials used. Rocket reliability as per all physical components is based on the quality of engineering proportions in design and construction also. Due to enormous chemical energy release from exit of the rocket propellant consequences of coincidences can be austere. Generally most of the space missions have some issues either with the geometry or due to improper assembly of the sub domains of the rocker or propellant used in the rocket motor [9-10]. Therefore the body aerodynamic behavior also plays a vital role in the field of rocket science, in this paper focus is given to aerodynamic behavior for better safety and reliability of the rocket [5].

3. Role of computational fluid dynamics on rocket body

Experimental simulations on rocket models using the wind tunnel are more expensive than solving numerically using CFD. Experimental work involves higher cost for set up, complexity is more in construction and also difficult to set all the boundary conditions. Time for the analysis also more, need to undergo several prototypes to predict the accurate results through experimental work. Nowadays to overcome all these problems the numerical simulations are very much handy to researcher [6, 7].

To solve the aerodynamic behavior of the rocket model the well-known computational fluid dynamics (CFD) tool was used to predict the flow behavior properties. In this work computers are used to perform the aerodynamic as mentioned in the above theory calculations required to simulate the interaction of the air with rocket surface defined by boundary conditions. With the speed computational facility good results can be achieved. In this paper research yields CFD tool that improves the speed and accuracy of complications like turbulent or transonic flows. The basic theory and basic of all Turbulent flow are the Navier stokes equation see in equation(2) where the term 'w' indicates specific thermodynamic work, 'v' kinetic viscosity, 'u' velocity profile which helpful in many single phase fluid flows[8].

$$\frac{\partial u}{\partial t} + (u.\nabla)u - v\nabla^2 u = -\nabla w + g \tag{2}$$

4. Methodology

In all the CFD methods same basic procedures are followed. During the pre-processing geometry defined based on the problem statement. After successful completion of geometry the body volume is divided into number of discrete elements or cells or mesh. In this paper for the better flow solutions structured grid was created. Consequently the physical modelling mentioned in the software for a given rocket model like enthalpy, species conservations and equation motion. The most important step in the CFD methodology is giving suitable boundary conditions for the rocket single stage and double stage involves specifying the fluid properties and behaviour at the boundaries of the problem statement. Once the setting of boundary conditions are done then simulations are started through solving the equations iteratively till solution get convergence under steady state or transient conditions. Lastly, the post processor is used for visualizing of the iterated solution and analysis of the rocket staging.

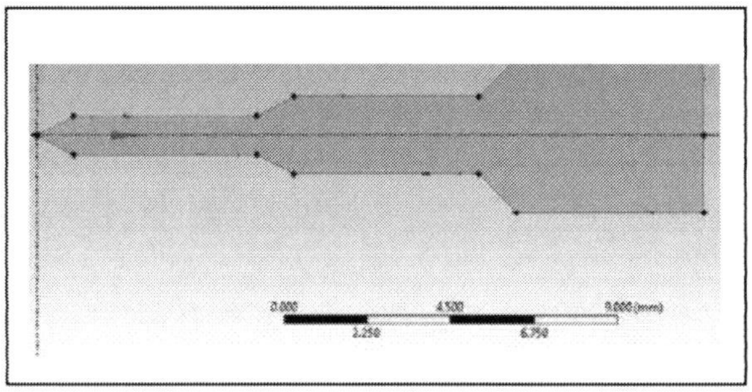

Figure 3: ANSYS workbench three stage rocket model.

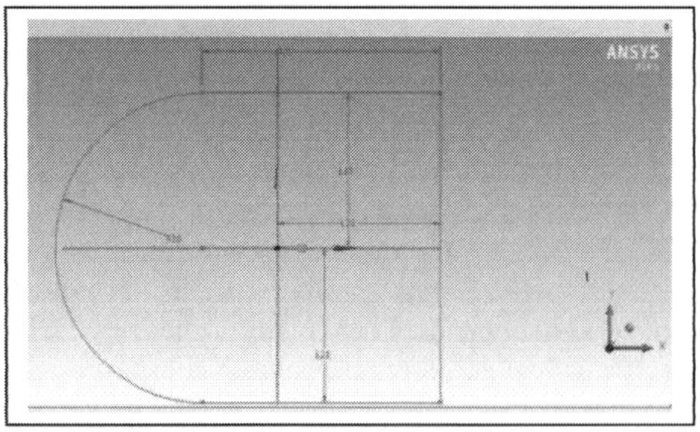

Figure 4 : Representation of farfield around the rocket model in ANSYS workbench.

Through the public literature survey the standardized basic rocket model was selected. After completing the successful convergence on the profile the optimized rocket model was created. Figure 3 shows three stage rocket model. The same geometry was imported into the ANSYS workbench where the geometry was scale down accordingly. As like experimental test section set up, the similar geometry called 'farfield' created around the rocket model see in the figure 4. The structured grid was created around the rocket model 71 million cells for east stage see in figure 5. Near the body surface the density of the mesh elements are more to predict the accurate results like shock wave propagation and aerodynamic properties. After setting appropriate elements attributes, mesh controls selection the required mesh created. The CFD methodology broad perspective flow chart shown in figure 2.

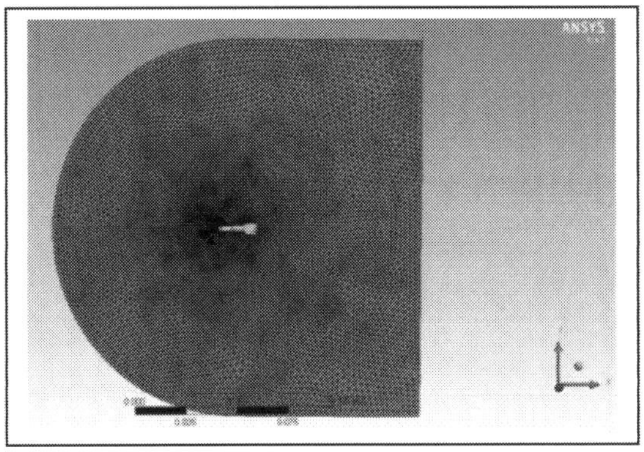

Figure 5: Mesh generation on rocket model in ANSYS.

The most important dimension in the numerical simulation is setting the suitable boundary conditions to single stage, double stage and three stage rocket. The flow simulations was run for different Mach numbers like M=3, 4 and 5. Till the fuel consumed by the rocket each stage separation time and also the corresponding altitude also calculated. From the calculated altitude, the density and mass flow rate also given to the farfield using atmospheric data. By the time all three stages separated from payload full propellant mass consumed for additional incremental velocity to enter into orbital velocity. For all three different stages boundary conditions remains same only the Mach number entry velocity and exit pressure only changes every time. All these calculations theoretically very strong since the experimental data over multistage rocket very inevitable.
Comparison also made in this analysis especially at the nose tip notable changes also observed, very maximum at the inclined profile and that too end corner of the rocket model also. The assumption in this analysis is flow pattern similar for all models shapes as the length of the each stage consider to be same.

5. Results and Discussions

The simulations were performed for single, double and multistage from mach speed 4, 5 and 6. All cases shown from figure 5-32. However the numerical results were presented only mach speed 3 double stage rocket model met the design criteria because the boundary conditions are well matched with theoretical calculations see in figure 3-8. Comparatively the rocket body from double stage to multi stage mass and propellant varied independently by ± 10% for a nominal value resulting in a combination of 16 cases for each parameter. Apart from these results it may note that the nominal mach speed value 3 nose tip shockwaves are capable of handling losses due to the aerodynamic flow interactions to keep the body accurate incremental velocity, density, temperature, pressure within the design data base limit shown from figure 9-21.

The small values of rocket path angle at three different stage separation are 42 deg, 52 deg and 65 deg and altitude $78x10^3$ ft , $82x10^3$ ft and $86x10^3$ ft respectively. The values were varied ±4000 ft and ±13 deg respectively. It is noted that these number differences can be controlled by the normal mach speed 3 and 4 satisfactorily and maintaining the rocket body incremental velocity, density, pressure and temperature with the design data base shown from figure 22-33.

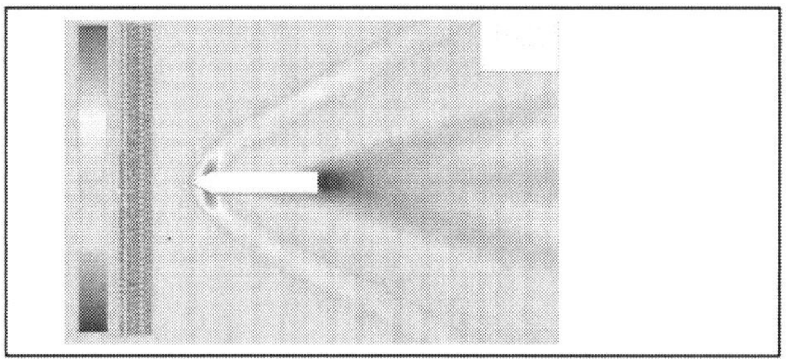

Figure 6: Single stage rocket density contour at mach speed 3

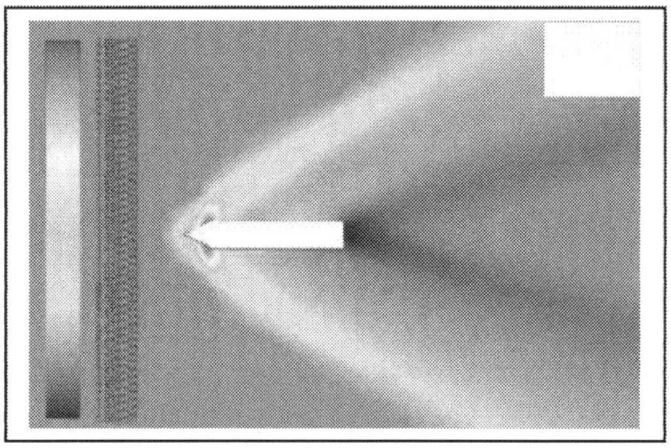

Figure 7: Single stage rocket static pressure at mach speed 3

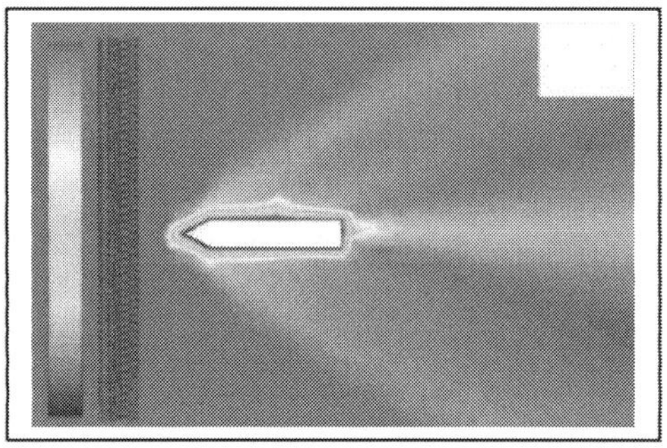

Figure 8: Single stage rocket velocity contour at mach speed 3

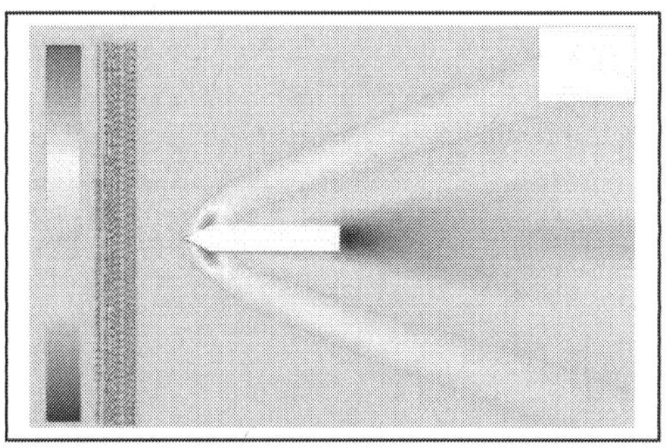

Figure 9: Single stage rocket density contour at mach speed 4

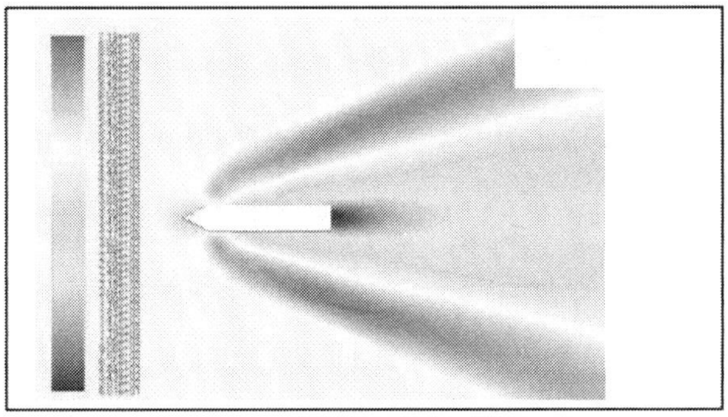

Figure 10: Single stage rocket static pressure at mach speed 4

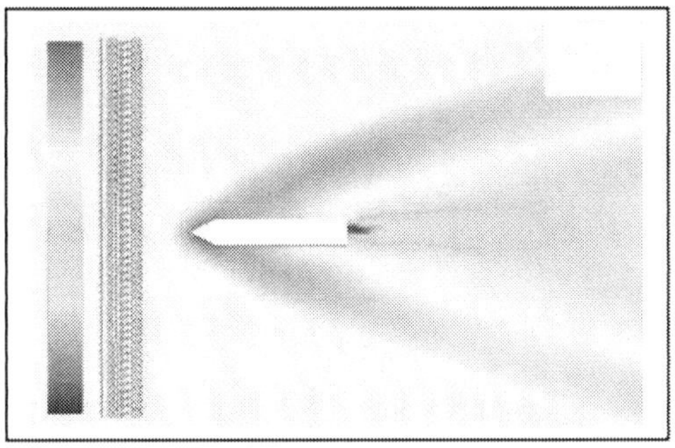

Figure 11: Single stage rocket temperature contour at mach speed 4

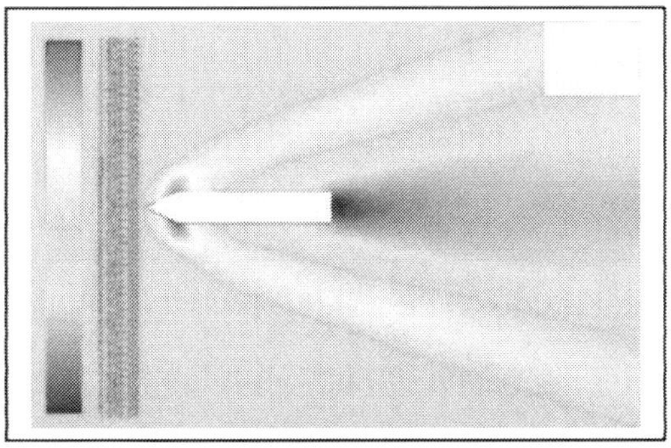

Figure 12: Single stage rocket density contour at mach speed 5

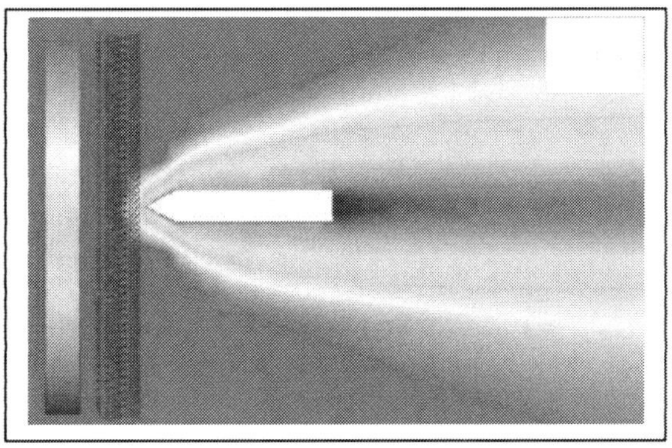

Figure 13: Single stage rocket pressure contour at mach speed 5

IOP Conf. Series: Materials Science and Engineering **197** (2017) 012077 doi:10.1088/1757-899X/197/1/012077

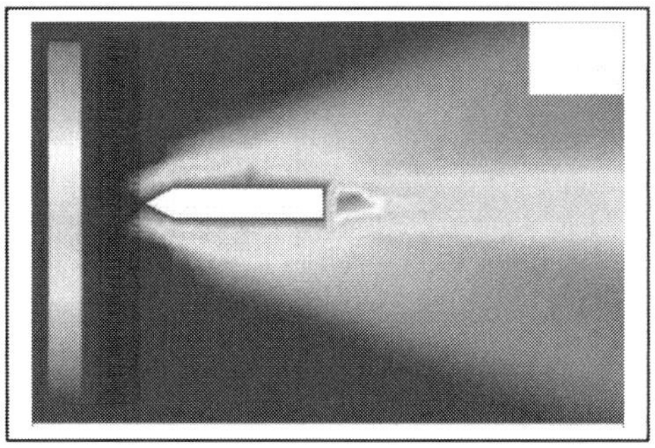

Figure 14: Single stage rocket static temperature contour at mach speed 5

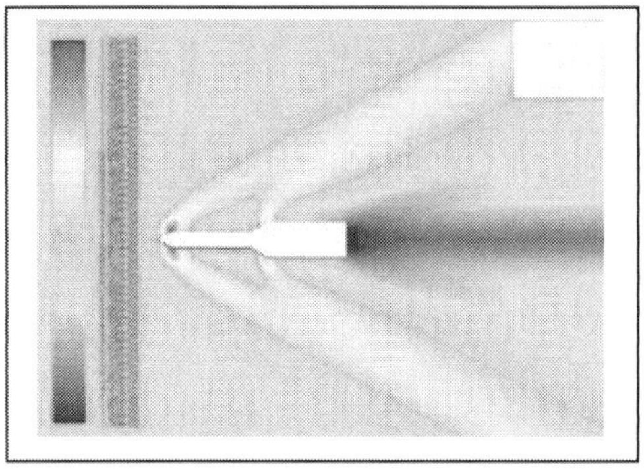

Figure 15: Double stage rocket density contour at mach speed 4

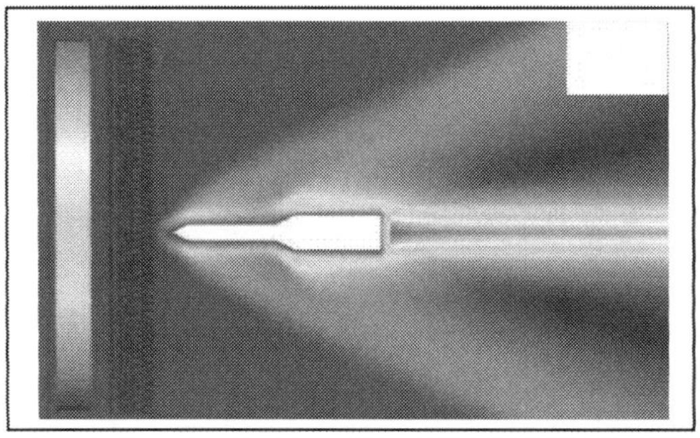

Figure 16: Double stage rocket density contour at mach speed 4

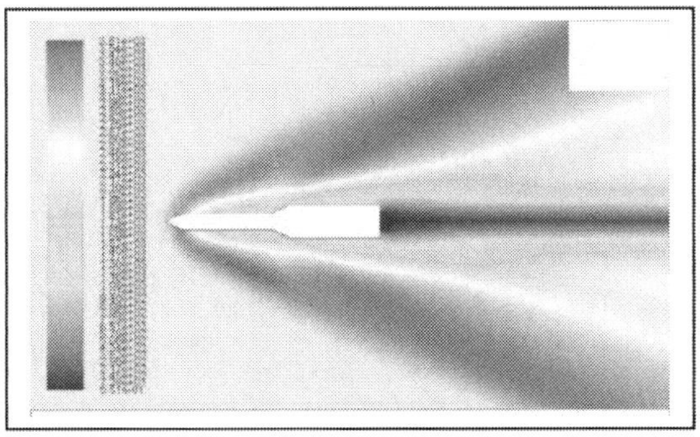

Figure 17: Double stage rocket pressure contour at mach speed 4

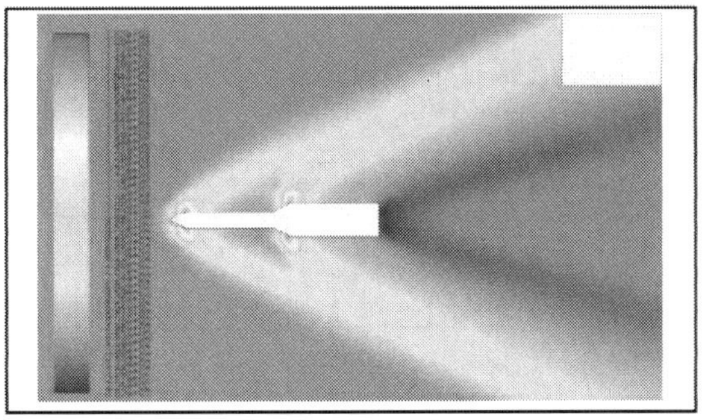

Figure 18: Double stage rocket static pressure contour at mach speed 4

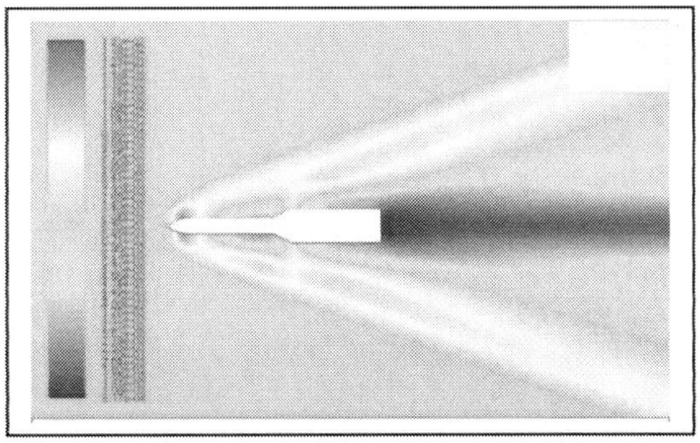

Figure 19: Double stage rocket static pressure contour at mach speed 4

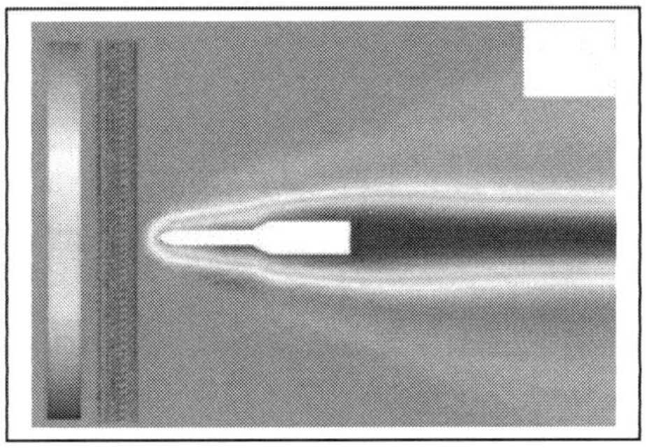

Figure 20: Double stage rocket total pressure contour at mach speed 4

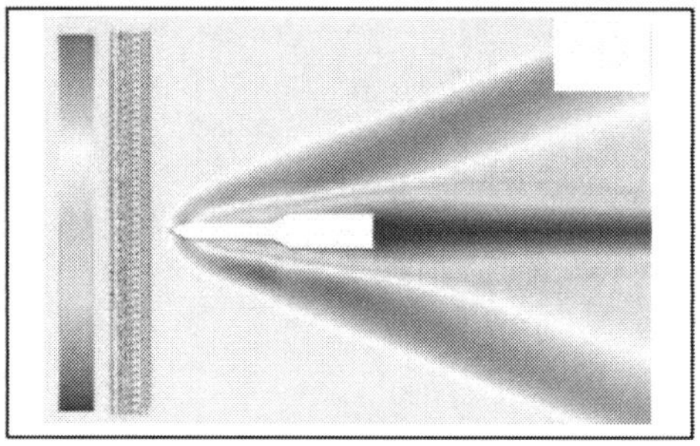

Figure 21: Double stage rocket pressure contour at mach speed 4

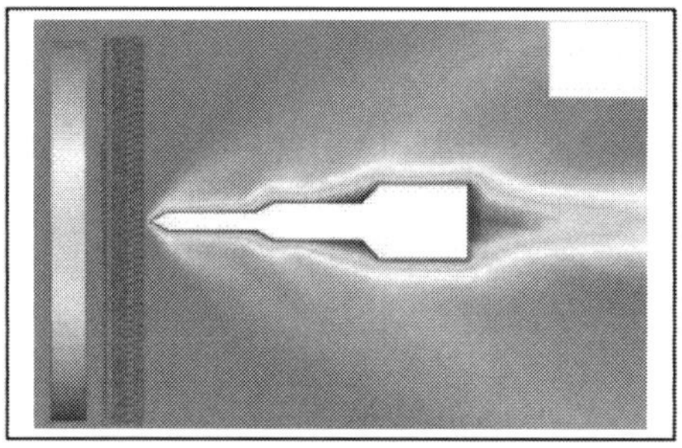

Figure 22: Three stage rocket velocity contour at mach speed 3

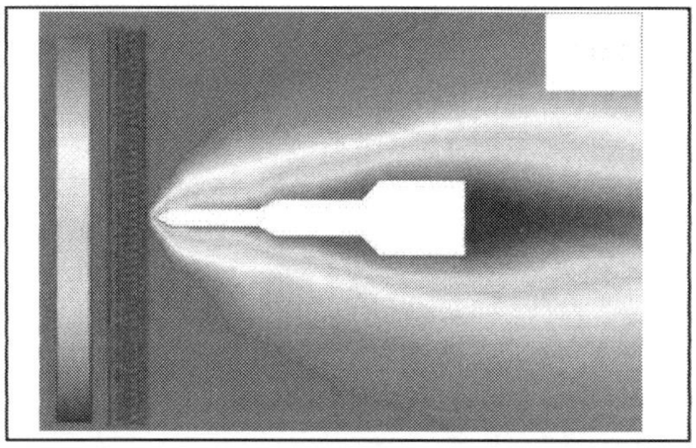

Figure 23: Three stage rocket velocity contour at mach speed 3

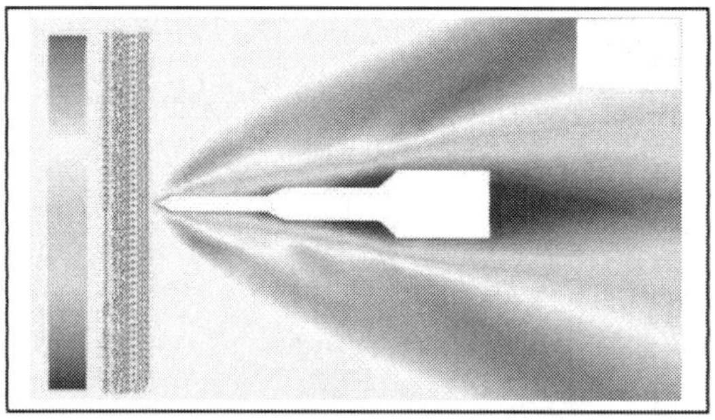

Figure 24: Three stage rocket dynamic pressure contour at mach speed 3

IOP Conf. Series: Materials Science and Engineering **197** (2017) 012077 doi:10.1088/1757-899X/197/1/012077

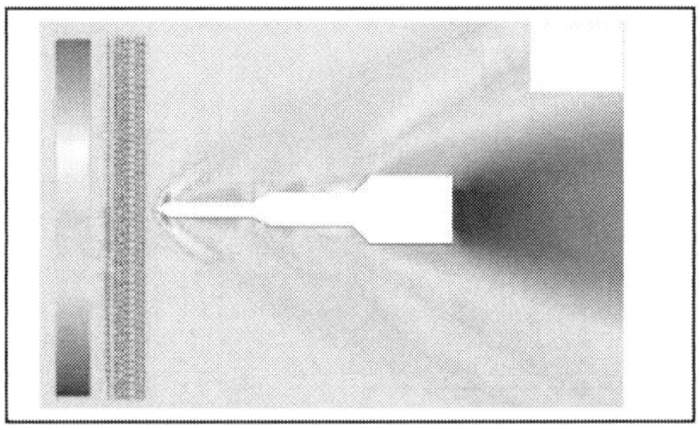

Figure 25: Three stage rocket density contour at mach speed 3

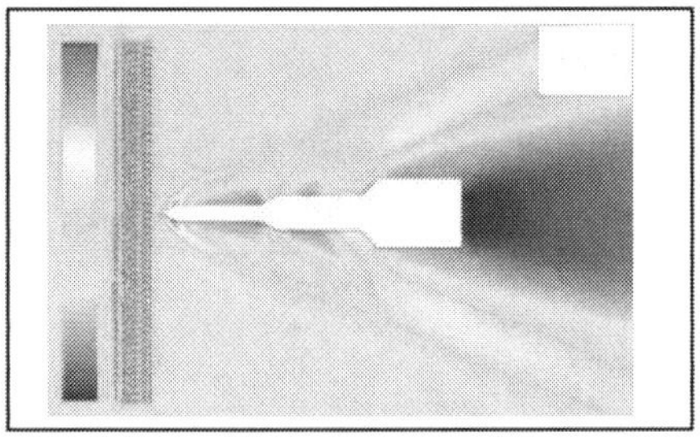

Figure 26: Three stage rocket density contour at mach speed 4

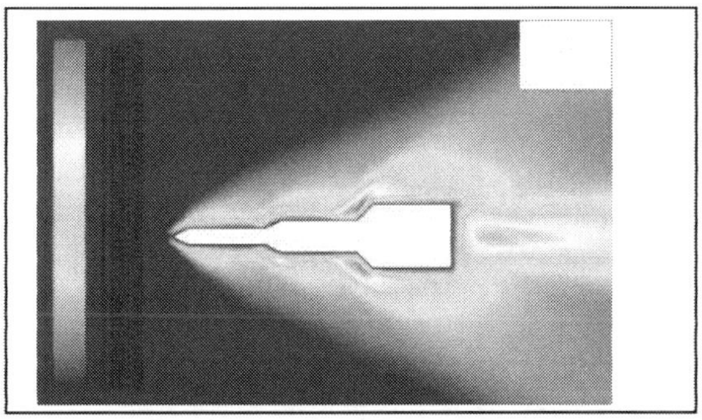

Figure 27: Three stage rocket static temperature contour at mach speed 4

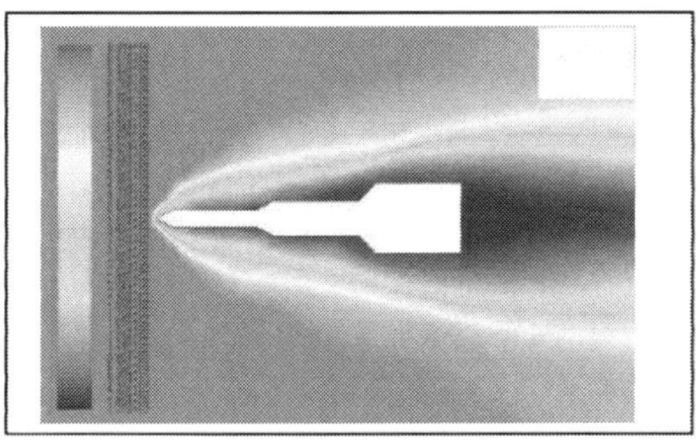

Figure 28: Three stage rocket total pressure contour at mach speed 4

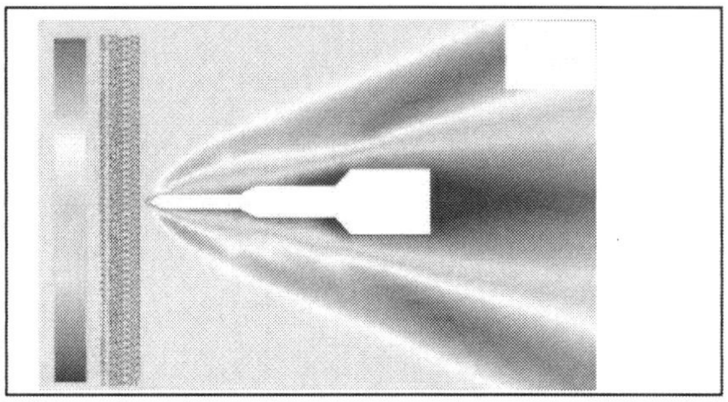

Figure 29: Three stage rocket dynamic pressure contour at mach speed 4

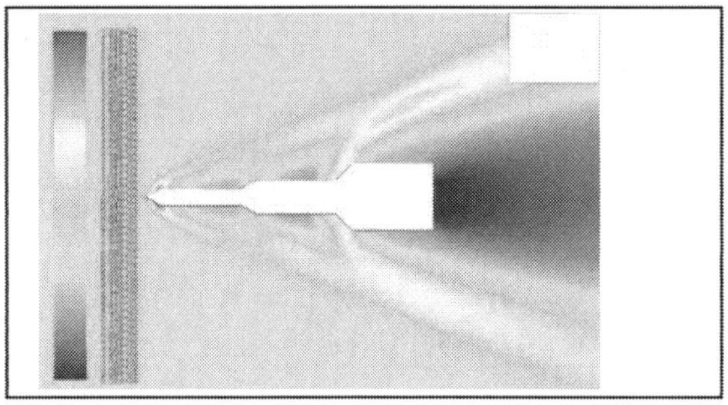

Figure 30: Three stage rocket density contour at mach speed 5

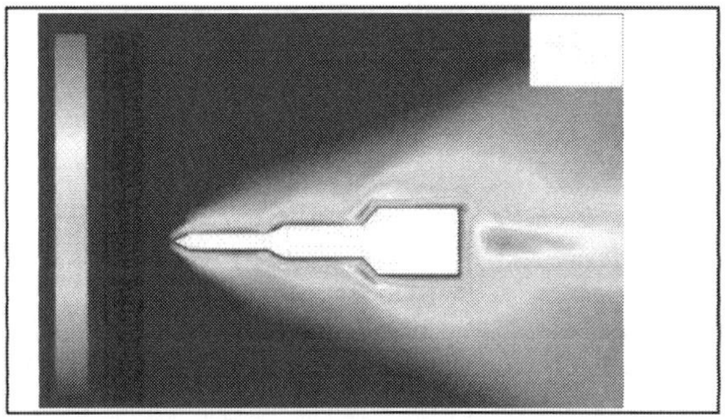

Figure 31: Three stage rocket static pressure contour at mach speed 5

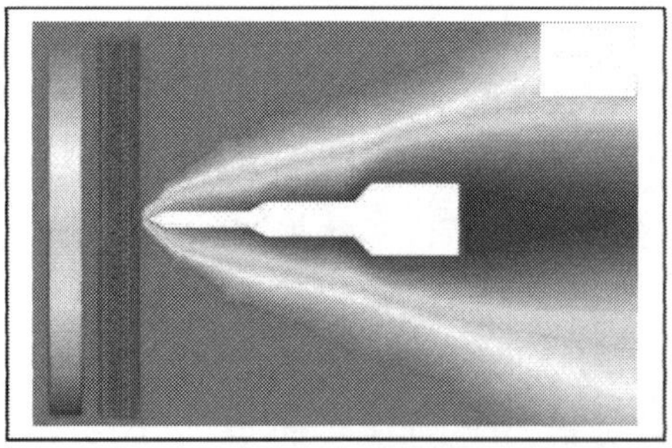

Figure 32: Three stage rocket total pressure contour at mach speed 5

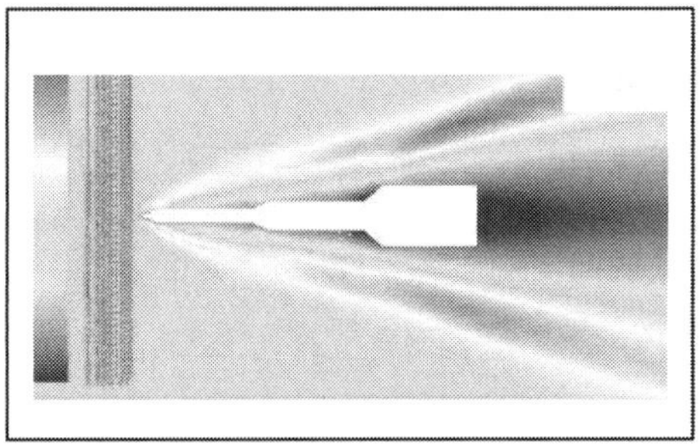

Figure 33: Three stage rocket dynamic pressure contour at mach speed 5

6. Conclusions

The analysis and flow simulation over the multistage, double stage and single stage rocket body vehicle concepts, single stage at mach speed 4 and double stage at mach speed 6 were performed outstandingly to the real-time criteria application. A passive agreement between the single stage and multistage rocket body mach speed 4 and 6 staging does not result in reasonable stage separation because the vehicle propellant mass and relative aerodynamic losses went very much outside the design data limits. However these combinations well suitable for mach speed 3 due to lower shock propagation strength at the entry of the nose tip. Further mach speed 6 combination with three rocket configurations well feasible at aerodynamic behavior to have better aeromechanical features. In all the 16 cases it is also investigated the best flow properties like pressure, temperature, density and velocity at entry and exit face of the rocket body.

References

[1] Sivaraj G., K. M. Parammasivam, M. Gokulraj, 2013 Computational Analysis of Flow Behaviour over the Multistage Launch Vehicle with Strapons, *Journal of Mechanical Engineering and Technology*. Volume **1**, Issue 1, pp. 54-65.

[2] N. Sreedhar, G. Ravindra Reddy 2013, Flow simulation over a Multi-Stage Launch Vehicle with Strapons using CFD techniques *International Journal of Computer Trends and Technology*, volume **4** Issue7, pp 2246-2249.

[3] 2015, Computational Fluid Dynamics (Cfd) In Launch Vehicle Applications Practice No. Pd-Ap-1311, http://llis.nasa.gov/lesson/758.

[4] Soon-Heum Koand Chongam Kim 1997 Separation Motion of Strap-On Boosters with BaseFlow and Turbulence Effects *Journal of Guidance control and dynamics*, Volume **20**, No.4.

[5] S.K.Chakrabartty,K. Dhanalakshmi an J. S. Mathur Computation of flow past aerospace vehicles http://www.iisc.ernet.in/currsci/nov25/articles18.html

[6] Robert E. Bartels Pawel Chwalowski, Steven J. Massey, Jennifer Heeg, Carol D. Wieseman 2012 Computational Aeroelastic Analysis of the Ares Launch Vehicle During Ascent American Institute of Aeronautics and Astronautics, PP 1-20.

[7] Arash Naghib-Lahouti, Amir Nejat, Taravat Khadivi. 2012 Parametric Analysis of Aerodynamic Characteristics of Launch Vehicles with Strap-On Boosters ICAS 2002 CONGRESS, pp 264.1-264.12.

IOP Conf. Series: Materials Science and Engineering **197** (2017) 012077 doi:10.1088/1757-899X/197/1/012077

[8] Soon-Heum Ko, Chongam Kim, Kyu Hong Kim 2006 Separation Analysis of Strap-Ons in the Multi- Stage Launch Vehicle Using the Grid Computing Technique 44th AIAA Aerospace Science Meeting and Exhibit, Reno, Novada.

[9] R. Venkatraman, S. Sankaran, G. Krishnaiah, P. Malleswara Rao, P. K. Vivekanand and T. Sundararajan 2015, Understanding Ignition over Pressure During Start Up of Typical Scaled Down Solid Rocket Motors, *Indian Journal of Science and Technology,* Vol **8(5)**, 481–488.

[10] Valeriy 2016 Developing a Test Stand to Verify Accuracy of Conducted Research at the Early Design Stages of Active Descent Systems of Spent Launcher Stages" Valeriy Ivanovich Trushlyakov and Kseniya Alekseevna Rozhaeva, Indian Journal of Science and Technology, Vol **9(36)**, September 2016. DOI: 10.17485/ijst/2016/v9i36/102037.

IOP Conf. Series: Materials Science and Engineering **197** (2017) 012078 doi:10.1088/1757-899X/197/1/012078

Recent developments of axial flow compressors under transonic flow conditions

Srinivas G[1], Raghunandana K*[2] and Satish Shenoy B[3]

[1]Assistant Professor, Dept of Aero & Auto Engineering, Manipal Institute of Technology, Manipal University, Manipal, Udupi, Karnataka, India-576104.

[2]Profess or, Dept of Mechatronics Engineering, Manipal Institute of Technology, Manipal University, Manipal, Udupi, Karnataka, India-576104.

[3]Professor, Dept of Aero & Auto Engineering, Manipal Institute of Technology, Manipal University, Manipal, Udupi, Karnataka, India-576104.

Abstract. The objective of this paper is to give a holistic view of the most advanced technology and procedures that are practiced in the field of turbomachinery design. Compressor flow solver is the turbulence model used in the CFD to solve viscous problems. The popular techniques like Jameson's rotated difference scheme was used to solve potential flow equation in transonic condition for two dimensional aero foils and later three dimensional wings. The gradient base method is also a popular method especially for compressor blade shape optimization. Various other types of optimization techniques available are Evolutionary algorithms (EAs) and Response surface methodology (RSM). It is observed that in order to improve compressor flow solver and to get agreeable results careful attention need to be paid towards viscous relations, grid resolution, turbulent modeling and artificial viscosity, in CFD. The advanced techniques like Jameson's rotated difference had most substantial impact on wing design and aero foil. For compressor blade shape optimization, Evolutionary algorithm is quite simple than gradient based technique because it can solve the parameters simultaneously by searching from multiple points in the given design space. Response surface methodology (RSM) is a method basically used to design empirical models of the response that were observed and to study systematically the experimental data. This methodology analyses the correct relationship between expected responses (output) and design variables (input). RSM solves the function systematically in a series of mathematical and statistical processes. For turbomachinery blade optimization recently RSM has been implemented successfully. The well-designed high performance axial flow compressors finds its application in any air-breathing jet engines.

1. Introduction

Axial flow compressors are mostly used in air breathing propulsion especially in aerodynamic applications. The design of axial flow compressor is not an easy task to the researcher because of its complexity in profile and its flow characteristics. Compressor size can vary from few meters to tens of meters in diameter, based on their applications. Till today recently the design of compressor two-dimensional compressor under real viscous conditions or three-dimensional meridional flow or inviscid three-dimensional flow analysis was done. With the use of modern sophisticated computational techniques, it is possible to solve three-dimensional viscous flow analysis and also potential to forecast better results of three dimensional multistage compressors. Therefore, it is required to retool the design procedures to have efficient results and physical dependability of these modern techniques.

In [1-2] discussed that the basic operations of an axial flow compressor were known and shown in the year 1853 to the FAS (French Academia des Sciences). Since then the development of axial flow compressor has been studied comprehensively and also slowly staging of compressor progressed meaningfully. To achieve high performance of the compressor the aerodynamics of each blade and stage(s) need to be understood fully. In this paper[1] main focus is given to understanding the aerodynamic flow characteristics and recent design techniques followed to improve the performance of the multistage compressor by referring to various research articles.

2. Understanding the Axial Flow Compressor Operation

The compressor is the first component in the core jet engine among various parts shown in the figure 1. The main purpose of the compressor is to increase the pressure ratio per unit stage working fluid using rotating shaft work. The axial flow compressor consists two set of blades one rotor and stator. The combination of rotor and stator together is called stage [3]. The working fluid which is entering from the earth atmosphere region accelerated through the rotor blade where the kinetic energy increases, then high kinetic energy of the fluid is further decelerated through the stator blades to form the high pressure. Stator blade is important to have proper pressure flow, by changing the energy associated with the swirl into pressure.

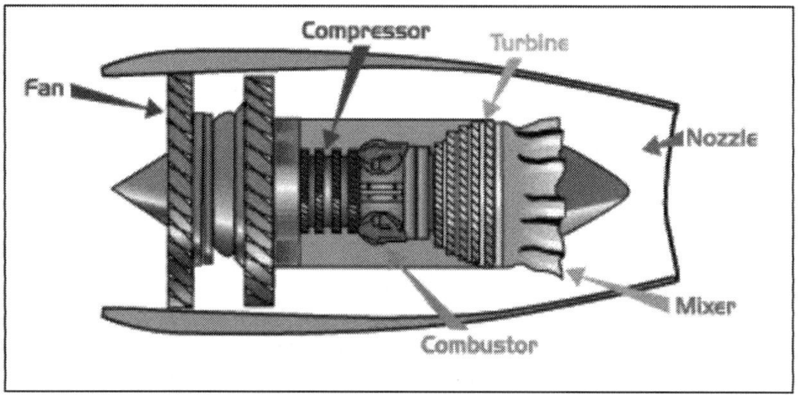

Figure 1: Various components of a commonly used turbo-engine [63].

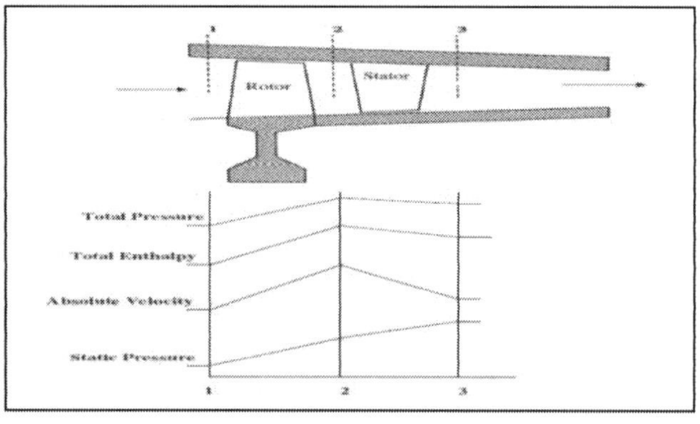

Figure 2: Various components of a commonly used turbo-engine [63]

Figure 2 represents a single stage axial flow compressor having a set of one rotor blade and one stator blade. This also represents the direction of working fluid which passes from left to right and for the rotor blades direction of rotation from bottom to top. Generally, every aircraft engine axial compressor has inlet guide vanes. These vanes will ensure that working fluid flow streamlines should be straight to the rotor blades to achieve high acceleration of mass flow rate in the blade rows when high accelerated working fluid leaves from the rotor blade exit, the flow is decelerated in the stator blade row and further enters into the next stage of the axial compressor.

Depending on the application of the aircraft engine specifications the number of stages in a compression system varies. In any axial flow compressor increasing the number of stages increases the weight, cost of the system, decreases the overall efficiency but increases the total pressure ratio for the system. In the working axial compressor, the fluid enters into gradually smaller volumes resulting in an increase in stagnation enthalpy of air and an increase in the stagnation pressure.

3. Axial flow compressor design

The compressor undergoes wide variety of operating conditions during the flight. During the take-off operation of the aircraft jet engine it requires more pressure ratio through the compressor to generate enough thrust. To control the thrust the compressor operates at different mass flow rates to vary the compressor and turbine speeds [4]. Here the staging plays a vital role in maintaining the desired efficiency. Through literature review it is evident that early designs of compressor at transonic and supersonic conditions were failure[4]. In [5] discussed more about subsonic compressor designs. It is found that low reliability and poor efficiency of the compressor blade steered to bad designs. In the early designs the researchers believed that poor efficiency was because of shock patterns alone, but later after successful failure of designs it was renowned that losses traits to flow blockages those are because of shocks. Since then drastic improvement have been made in compressor blading profile design, hub to tip design and also tip to hub design. In [6] discussed briefly about the early designs of compressor blade.

Through the existing designs it is evident that by increasing the number of stages in the compressor results in higher overall pressure rise, but at the same time the overall length of the compression system and weight also increases. Comparing with last stage the first stage of the compressor volume is larger, therefore the typical modern compressors are wide at the inlet. They generally have a conical shape and have somewhere amid 9 to 15 stages. Figure 3 illustrates the straightforward compressor design difficulty. The main components that need to be designed are the geometry of the blade rows and geometry of the blade end-wall contour. To accomplish design study on a mean stream surface (shown in figure 3) in an axisymmetric three dimensional multi stage compressor designs approach generally used. To achieve both the end wall geometry and blade geometry the mean stream surface was used as a baseline. This type of methods and findings may be found in reference [5].

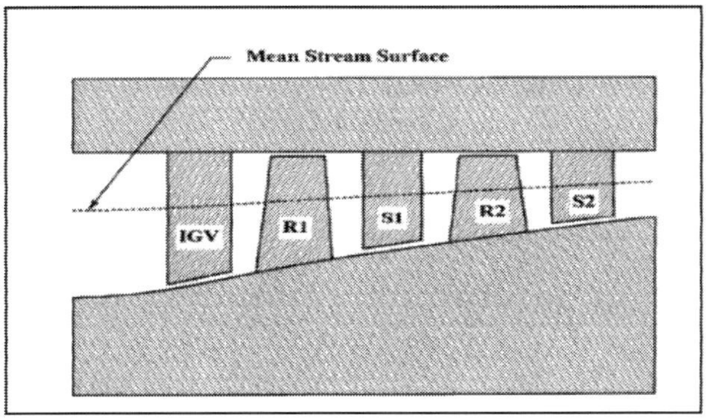

Figure 3: Compressor geometry design with endwalls and blade rows [1].

Considering from an aerodynamic perception, a more accurate three-dimensional design of compressor blade is very much essential as this safeguards the maximum possible blade loading. Structural and aeroacoustics responses are closely connected to aerodynamic design of a compressor blade. Through the literature it is evident that, the compressor blade load increases eventually structural deformation on the blade also increases ultimately leading to structural failure. An ineffective design of blades can naturally lead to greater than before acoustic response of the blade, from shock formations and rotor-stator interactions. Finally, it clears that the design of compressor system is a multidimensional problem.

4. CFD studies on axial flow compressor performance

Compressor flow solver is the turbulence model used in the CFD to solve viscous problems. In order to improve compressor flow solver and to get agreeable results careful attention need to be paid viscous relations, grid resolution, turbulent modeling and artificial viscosity. Figure 4 shows the harmonic balance simulation of complete axial flow compressor. Initially turbomachinery compressor performance calculations using CFD were frequently done in two dimensions. Such two-dimensional cascade analysis for both turbines and compressors were calculated by many authors [7-10]. By nature, the flow in an axial flow compressor is unsteady and vertical [11-14] and therefore huge turbulence and perfect modeling of the viscous effects is imperative.

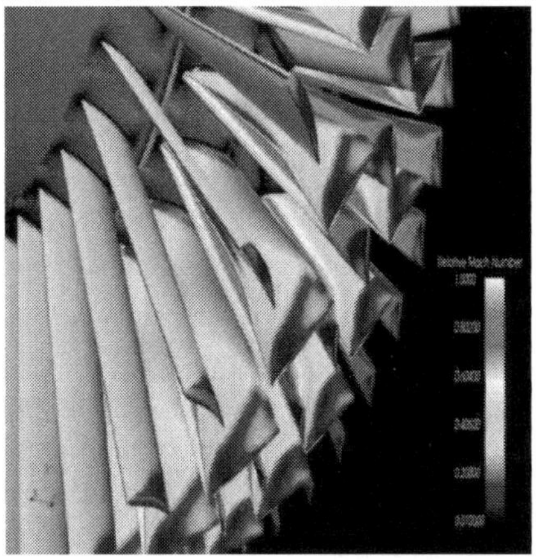

Figure 4: Harmonic balance simulation of multistage axial flow compressor relative mach number [64].

Ample of work done in progressing phenomenological and sophisticated simulation techniques to understand in addition simulate the stream flow behavior in compressor over the past two decades. The area of computational fluid dynamics during this time has undergone significant changes, giving engineers and designers having useful competence to model and study the intrinsically three-dimensional flow in axial flow compressor. Most recently having increased computational power and developments in visualization tools and post processing three dimensional simulations of axial flow compressor rotor machines are feasible. Also researchers now have access to a plethora of data for justification of compressor codes with the accessibility of dependable experimental test data by [15]. Since then the analysis of single and multistage axial flow compressor blade passage flow has been developed using a number of three dimensional CFD codes [16-23]. Computational codes for turbomachinery multistage three dimensional compressor configuration have been developed by different authors [24-26]. Very recently nitially developed compressible unsteady flow solver for an

IOP Conf. Series: Materials Science and Engineering **197** (2017) 012078 doi:10.1088/1757-899X/197/1/012078

isolated three-dimensional blade row, which was altered effectively to add in multistage competencies [27-28].

5. Background of axial flow compressor design methods

The axial flow compressor blade design is broadly classified into two approaches, direct and inverse methods. In the direct design method, the edge geometry is analyzed directly by experiments and / or by a computational analysis. In this method either the existing coordinates (x, y, z) of the edge are changed or performance constraints that are directly affect the blade profile are changed. The profile of the blade is consequently analyzed and the effect of performance variables on its overall efficiency is evaluated.

Introducing in the year 1970s the poplar techniques like Jameson's rotated difference scheme [29-30] and used method were possible to solve potential flow equation in transonic condition for two dimensional aerofoils and later three dimensional wings. These advance techniques had most substantial impact on wing design and aerofoil. The same techniques also had significant impact on axial flow compressor design as the new schemes may possibly be coupled with early established potential flow solvers.

In papers [31] and [32] investigated transonic direct method into inverse design methods. CAS22 a programme developed by Dulikravich which has capable of shock free aerofoil cascades designing and valid to transonic shock free redesign of existing cascade two dimensional aerofoil and also to aerodynamic analysis. Meauze was developed some of other applications of inverse design approach, thereafter Dulikravich passing few years implemented same technique for three dimensional blades.

In the year 1990's drastically improvements in computational fluid dynamics, the stage three-dimensional design and the flow analysis of compressor geometry configuration has been made possible. Significant work has been done on a four stage compressor through direct design analysis and showed that reiterating stage phenomenon where the blockage raises across the flow [33]. Complete three-dimensional inverse design approach for compressor blades in transonic flow was developed in 1993 [34]. After passing few years Dang et al redesigned NASA 67 configuration using three-dimensional inverse design method. In another investigation done on multi objective optimisation of compressor blades to find best blade profile with respect to working range and loss by [35].

In the recent years' researchers introducing the concept of 'sweep lean'. Where the term sweep describes about the moving blade in axial flow direction, whereas the lean explains the moving blade in circumferential direction. This concept is applying by many researchers [36-38] have applied the sweep and / or lean idea to multistage and single stage compressor configuration and found positive results.

When the design constraints are selected in an ad-hoc manner direct design method can sometimes be trial and error. Provided it has own benefits compared with the inverse design method, which normally required more number of inputs out of which few were always unknown (like pressure distribution over 3D blade) so as to produce the preferred flow features. Hence, the inverse direct design method always advisable and usually preferred for finding the unknown performance parameters to acquire detailed knowledge of the flow.

In the recent years' researchers have described the importance of compressor blade aerofoil curvature scattering on its performance and showed the potential methods that can be used to design extremely differentiable blade planes [39-41]. A set of Chained G Bezier polynomials capable of generating compressor aerofoils developed by corral [42] and pastor [48].

6. Aircraft system level engine design overview

This study describes about an axial flow compressor blade design analysis component level. However, the design studies on turbomachinery configuration can also be done at an overall engine system level, where the focus is not on individual component[42]. At the Aerospace System Design Laboratory (ASDL) at Georgia tech has been extensively performed such system level design studies. In paper [43] proposed a probabilistic design methodology for commercial aircraft engine cycle selection. In paper [44] investigated an assessment of lost thrust method for analysis of gas turbine engine thermodynamic performance. In paper [45] also done another study on probabilistic methods to systematically make the decisions under uncertainty and rationally to solve the preliminary design calculations [62]. Another probabilistic sensitivity in the gas turbine engine system level design can be found in references [46-48].

7. Axial flow compressor optimization techniques overview

For a variety of transonic flow design problems numerical optimization techniques have been used successfully. The design of compressor blade under aerodynamic perspective itself is a challenging task to a researcher. Initially, efficiency of the transonic compressor blade is most sensitive to its shape and hence the shape of the blade must be parameterized with required number of performance parameters are to be optimized. Provided in addition aerodynamic optimization design objective function problem is repeatedly multidimensional and non-linear reason is flow field directed by a system of non-linear partial differential equations. The optimization also has constraints such as an acoustic characteristic, operating mass flow range by varying the rotor speed and others make the aerodynamics blade profile optimization an effective problem. Many more optimization techniques have been reported with degree of success such as artificial intelligence methods (neural network and evolutionary algorithms EA's), orthogonal array method, response surface method (RSM), gradient based method.

The gradient base method is most popular method especially for compressor blade shape optimization. This gradient based algorithm in which the optimum is investigated by estimating the local gradient information. Constrained minimization (CONMIN) uses gradient algorithm and has been efficiently used for wing aerofoil design by vanderplaats [48]. Gradient based method design widely used in subsonic and supersonic wing design [49-50]. Later further studied [51-52] detailed manner and implemented same gradient based method on centrifugal compressor and design of compressor aerofoil.

Another type of optimization technique is Evolutionary algorithms (EAs). Evolutionary algorithms are quite simple than gradient based technique because it can able to solve the parameters simultaneously search from multiple points in the given design space. Recently EAs have been successfully used for solving the turbomachinery blade optimization problems found in references [53-55].

Response surface methodology (RSM) is another type of optimization algorithm. This method basically to design empirical models of the response that were observed and to study systematically the experimental data. This methodology analyses the function denotes the correct relationship between expected responses (output) and design variables (input). RSM solve the function systematically in a series of mathematical and statistical processes. For turbomachinery blade optimization recently RSM has been implemented successfully. Stacking line of the axial flow compressor rotor blades optimization was successfully carried out by Ahn [56]. To perform the shape optimization in a single stage transonic axial flow compressor stator blade successfully used RSM by Jang in 2006 [57].

8. Literature of summary of recent developments on axial flow compressor under transonic flow conditions

There is very little literature available in the unsteady and instability area. Very little is known about its behavior and its possible configurations as function of field variables. Experimental and Computational Analysis conducted in this area are less and a lot more is to be done to conceptually understand the flow behavior and its implications on rotor. This regime itself is non-uniformity associated, and is different at different regions and time. The solution is basically guided by the type of non-uniformity. Much is required to be understood about the impacts of different possible discontinuities, on the flow structure through the rotor. Some region of cross section subjecting to discontinuities will have its implications on remaining blades flow passages and therefore full rotor analysis is very much inevitable. From the literature it is understood that there were no attempts made to generate these discontinuities at the entry face of the compressor rotor.

There are varieties of configurations on non-uniformities possible at the entry face to rotor, like the one associated with pressure field or velocity and temperature and so on. Point to note here is they are not independent but are coupled effects. One discontinuity leads to the deviation of others. Though there is some theoretical data at hand very little is on the experimentation side. Since it is stall associated which is actually rotating by the rotor of the compressor in the aircraft engine. There were few literatures on Numerical analyses done towards understanding the effects of the tip sensitivity driven Distortion flow development in the rotor, by using some circumferential grooving and also there are few papers discussing about the stall inceptions and associated distortion.

Attempts have been made to find out the characteristics of distortion associated in the compressor rotor with rotating stall in multistage compressor but not exactly to understand the distorted flow field through compressor. It was all about tip vortex and separation implications[58] and it's as such flow disturbance effects were experimented[60]. Therefore, it is required that first to understand the flow structure under distorted entry flow field and its development through the rotor. In this process, simulations come very handy providing more comfort and relief, with ease of analyzing. Moreover, this area is very vast; there are lots of things to experiment and analysis, therefore in this research literature steady state simulation on axial flow fan, for which there is experimental work available have been carried out[59]. So to understand its effects turbulent kinetic energy, its development and growth is also reported[61] to some extent. There is still a lot more to understand with regard to vortex effects, unsteady disturbances and stall propagation, which guides passage flows. This served as the very Motivation to carryout distortion analysis of axial flow fan, and to find out the consistency of simulations and their ability to predict the unsteady flow field [58-59].

9. Conclusions

Review of literature revealed that implications of nonuniformities were studied to some extent, by specifying conventional one dimensional(axial), inaccurate profile at the entry, and no attempt was made towards possibility of velocity fields in the radial and circumferential directions, in distorted region at the entry face.

There are varieties of configurations on non-uniformities possible at the entry face to rotor, like the one associated with pressure field or velocity and temperature and so on. Point to note here is they are not independent but are coupled effects. One discontinuity leads to the deviation of others. Though there is some theoretical data at hand very little is on the experimentation side. Since it is stall associated which is actually rotating by the rotor of the compressor in the aircraft engine. There were few Numerical analyses done towards understanding the effects of the tip sensitivity driven distortion flow development in the rotor, by using some circumferential grooving and also there are a few papers discussing about the stall inceptions and associated distortion.

So to understand its effects turbulent kinetic energy, its development and growth is also reported to some extent. There is still need to understand with regard to vortex effects, unsteady disturbances and stall propagation, which guides passage flows. This served as the very Motivation to carryout unsteady flow analysis of axial flow fan, and to find out the consistency of simulations and their ability to predict the unsteady flow field.

Some region of cross section subjecting to discontinuities will have its implications on remaining blades flow passages and therefore full rotor analysis is very much inevitable. From the literature it is understood that they were no attempts made to generate these discontinuities at the entry face. This regime itself is non-uniformity associated, and is different at different regions and time. The solution is basically guided by the type of non-uniformity.

The computational work is to initiate with the intent to simulate and find out the effects and implications of the unsteady non uniform flow at the compressor inlet. Unsteady flow is not a fixed configuration problem. It is possible to understand it well upon investigation of flow through the rotor with different inlet conditions and varied problem specifications.

10. References

[1] H.K.Ozturk 2000 Stator well flows in axial compressors *International Journal of Heat and Fluid Flow* 21 PP 710-716.
[2] Gostelow, J. P. 1984 Cascade Aerodynamics, Pergamon Press, New York, NY.
[3] Mattingly, J. D., Heiser W. H., Daley, D. H., 1987 Aircraft Engine Design, AIAA, New York,
[4] Lakshminarayana, B. 1996 Fluid Dynamics and Heat Transfer of Turbomachinery, John Wiley and Sons. Inc,
[5] Hawthorne, W.R. 1964, Aerodynamics of Turbines and Compressors, Princeton University Press, Princeton, NJ.
[6] Aungier, R. H., 2003 Axial-Flow Compressors, ASME Press, New York,
[7] Weinberg, B. C., Yang, R.-J., McDonald, H., and Shamroth, S. J., J. 1986 Eng Calculations of Two- and Three-Dimensional Transonic Cascade Flow Fields Using the Navier- Stokes Equations,: *Gas Turbines and Power*, Vol. **108**, pp. 93-102.
[8] Davis, R. L., Hobbs, D. E., and Weingold, H. D., 1988 Prediction of Compressor Cascade Performance Using a Navier-Stokes Technique, *ASME*, International Gas Turbine Conference, Amsterdam, The Netherlands, Paper 88-GT- 96.
[9] Gundy-Burlet, K. L., Rai, M. M., and Dring, R. P., 1989 Two-Dimensional Computations of Multi-stage Compressor Flows Using a Zonal Approach *AIAA* Paper 89-2452, Monterey, Calif.
[10] Chevrin, P. A., and Vuillez, C. 1990 Viscous Flow Computations in Turbomachine Cascades *ASME* Paper 90-GT-76,
[11] Cambier, L., Escande, B., and Veuillot, J. P., 1986 Computation of Internal Flows at High Reynolds Number by Numerical Solution of the Navier-Stokes Equations *Rech. Aerosp.*, No.6, pp. 27-44.
[12] Boyle, R. J. 1990 Navier-Stokes Analysis of Turbine Blade Heat Transfer *ASME* Paper 90- T-42.
[13] Suder, K, L., and Celesttina, M. L., 1994 Experimental and Computational Investigation of the Tip Clearance Flow in a Transonic Axial Compressor *ASME* Paper 94-GT-365.
[14] Chima, R. V., 1996 Calculation of Tip Clearance Effects in a Transonic Compressor Rotor *NASA* TM 107216.
[15] AGARD, 1989 Test Cases for Computation of Internal Flows in Aero Engine Components *AGARD Propulsion and Energetics Panel*, Working Group 18, AGARD-AR-275.
[16] Srivastava, R., and Sankar, Lakshmi N. 1993 Efficient Hybrid Scheme for the Analysis of Counter-Rotating Propellers, *Journal of Propulsion and Power*, Vol. **9**, No. 3, pp. 382-388.
[17] Dawes, W. N. 1987 Numerical Study of the 3D Flowfield in a Transonic Compressor Rotor with a Modeling of the Tip Clearance Flow *AGARD Conference Proceedings*, No. 401, Neuilly Sur Seine.
[18] Adamczyk, J. J., Mulac, R. A., and Celestina, M. L., 1986 A Model for Closing the Inviscid Form of the Average Passage Equations, *ASME* Paper 86-GT-227.

[19] Hathaway, M. D., and Wood, J. R., 1997 Application of a Multi-Block CFD Code to Investigate the Impact of Geometry Modeling on Centrifugal Compressor Flow Field Predictions, *Transactions of the ASME*, Vol. **19**, pp. 820-830.
[20] Niazi, S., Stein, A., and Sankar, L. N. 1998 Development and Application of a CFD solver to the Simulation of Centrifugal Compressors, *AIAA* paper 98-0934.
[21]Chima. R. V., 1987, Explicit Multigrid Algorithm for Quasi-Three-Dimensional Viscous Flows in Turbomachinery, *AIAA Journal of Propulsion and Power*, Vol. **3**, No. 5, pp. 397-405.
[22] Chima, R. V., and Yokota, J. W., 1990 Numerical Analysis of Three-Dimensional Viscous Internal Flow, *AIAA Journal,* Vol. **28**, No. 5, pp. 798-806.
[23] Niazi, S., July 2000 Numerical Simulation of Rotating Stall and Surge Alleviation in Axial Compressors, Ph.D. Dissertation, Georgia Institute of Technology, Aerospace Engineering.
[24]Denton, J. D. 1990 The Calculations of Three-Dimensional Viscous Flow Through Multistage Turbomachines, *ASME Paper* 90-GT-19.
[25] Hall, E. J., 1997, Aerodynamic Modeling of Multistage Compressor Flow Fields - Part 1: Analysis of Rotor/Stator/Rotor Aerodynamic Interaction, *ASME* paper 97- GT- 344.
[26] Chima, R. V., 1998, Calculations of Multistage Turbomachinery Using Steady Characteristic Boundary Conditions, *AIAA* Paper 98-0968,. Also NASA TM-1998-206613.
[27] Niazi, S., Stein, A., and Sankar, L. N., 2000. Numerical Study of Surge and Stall Alleviation in a High-Speed Transonic Fan Rotor, *AIAA* Paper 2000-0225.
[28] Zaki, M., Iyengar, V., Sankar, L. N., 2006.Assessment of Rotor-Stator Interface Boundary Condition Techniques for Modeling Axial Flow Turbines, *AIAA* Paper 2006-4619.
[29] Jameson, A., 1973, Numerical Calculations of the Three-Dimensional Flow over a Yawed Wing, *Proceedings of the AIAA Computational Fluid Dynamics Conference*, Palm Springs, pp. 18-26.
[30] Murman, E. M., Cole, J. D., 1971. Calculation of Plane Steady Transonic Flows, *AIAA journal* PP 114-21.
[31] Dulikravich D. S. 1980 CAS2D – FORTRAN Program for Nonrotating Blade-to-Blade, Steady, Potential Transonic Cascade Flows," *NASA* TN 1705.
[32] Meauze, G., 1982 An Inverse Time Marching Method for the Definition of Cascade Geometry, *Transaction of the ASME: J. of Engineering for Power,* Vol. **104**, pp650- 656.
[33] Bolger, J. J., and Horlock, J. H., 1995 Predictions of the Flow in Repeating Stages of Axial Compressors Using Navier-Stokes Solvers, *ASME* Paper 95-GT-199.
[34] Dang, T. Q., 1993 A Fully Three-Dimensional Inverse Method for Turbomachinery Blading in Transonic Flows, *ASME Journal of Turbomachinery*, vol. **115**, no. 2, pp. 354-361.
[35] Keshin, A., Dutta, A. K., and Bestle, D., *2006* Modern Compressor Aerodynamic Blading Process Using Multi-Objective Optimization, *Proceedings of GT2006, ASME Turbo Expo 2006*, GT2006-90206.
[36] Gallimore, S. J., Bolger, J. J., Cumpsty, N. A., Taylor, M. J., Wright, P. I., Place, J. M. M., 2002 The Use of Sweep and Dihedral in Multistage Axial Flow Compressor Blading—Part I: University Research and Methods Development" ASME Journal of Turbomachinery, Vol. **124**, Issue 4.
[37] Benini, E., and Biollo, R., 2006 On the Aerodynamics of Swept and Leaned Transonic Compressor Rotors, Proceedings of GT2006, *ASME* Turbo Expo 2006, GT2006- 90547.
[38] Jang, C. M., Samad, A., and Kim, K. Y., 2006 Optimal Design of Swept, Leaned and Skewed Blades in a Transonic Axial Compressor, *Proceedings of GT2006*, *ASME* Turbo Expo 2006, GT2006-90384.
[39] Korakianitis, T., 1989 Design of Airfoils and Cascades of Airfoils, *AIAA Journal*, Vol. **27**, No. 4, pp. 455-461.
[40] Korakianitis, T., 1993 Prescribed-Curvature-Distribution Airfoils for the Preliminary Geometric Design of Axial Turbomachinery Cascades, *Journal of Turbomachinery*, Vol. **115,** pp. 325-333.
[41] Corral, R., and Pastor, G., 1999 A Parametric Design Tool for Cascades of Airfoils, International Gas Turbine and Aeroengine Congress and Exhibition, *American Society of Mechanical Engineers*, Paper 99-GT-73, Indianapolis, Indiana.
[42] Corral, R., and Pastor, G., 2004 Parametric Design of Turbomachinery Airfoils Using Highly Differentiable Splines, *Journal of Propulsion and Power* Vol. **20**, No. 2.
[43] Marvis, D. N., Mascotai, N. I., Roth, B., 1998 A Probabilistic Design Methodology for Commercial Aircraft Engine Cycle Selection, *AIAA*-98-5510, 3rd World Aviation Congress and Exposition, Anaheim, CA.
[44] Roth, B., De Luis, J., 2005 Lost Thrust Methodology for Gas Turbine Engine Performance Analysis, GT2005-68200, *Proceedings of ASME Turbo Expo 2005*, Reno, NV, USA.
[45] Marvis, D. N., Roth, B., and Mascotai, N. I., 1999 A Method for Probabilistic Sensitivity Analysis of Commercial Aircraft Engines, *ISABE*-99-BR, 14th ISABE, Florence, Italy.

[46] DeLaurentis, D., Mavris, D. N., and Schrage D. P., 1996 System Synthesis in Preliminary Aircraft Design Using Statistical Methods, ICAS-96 20th *International Council of the Aeronautical Sciences (ICAS) Congress*, Sorrento, Italy, September 8-13.

[47] Roth, B., Mavris, D. N., 2001 A Work Availability Perspective of Turbofan Engine Performance, *AIAA*-2001-0391, 39th AIAA, Aerospace Sciences Meeting and Exhibit, Reno, NV.

[48] Vanderplaats, G. N., 1973 CONMIN A FORTRAN program for constrained function minimization User's manual, NASA -TM-X-62282.

[49] Hicks, R. M., and Henne, P. A., 1978 Wing Design by Numerical Optimization *Journal of Aircraft*, Vol. **15**, pp. 407-412.

[50] Reuther, J. J., and Jameson, A., 1995 Supersonic Wing and Wing-body Shape Optimization Using an Adjoint Formulation, Technical Report, The Forum on CFD for Design and Optimization, (IMECE95), San Francisco, California, November.

[51] Lee, Y.-T., Luo, L., and Bein, T. W. 2001 Direct Method for Optimization of a Centrifugal Compressor Vaneless Diffuser, *Journal of Turbomachinery*, Vol. **123**, pp. 73-80.

[52] Koller, U., Monig, R., Kusters, B., and Schreiber, H.-A., 2000 Development of Advanced Compressor Airfoils for Heavy-Duty Gas Turbines- Part I: Design and Optimization, *Journal of Turbomachinery*, Vol. **122**, pp. 397-405.

[53] Yi, W., Huang, H., and Han, W. *2006* Design Optimization of Transonic Compressor Rotor Using CFD and Genetic Algorithm, *Proceedings of GT2006, ASME Turbo Expo 2006*, GT2006-90155.

[54] Oyama, A., Liuo, M., and Obayashi, S. 2004Axial-Flow Blade Optimization: Evolutionary Algorithms/Three-Dimensional Navier-Stokes Solver, Journal of Propulsion and Power, vol. **20**, No.4.

[55] Uelschen, M. and Lawerenz, M., 2001 Design of Axial Compressor Airfoils with Artificial Neural Networks and Genetic Algorithms, *AIAA* Paper 2000-2546.

[56] Ahn, C. S., and Kim, K. Y., Aerodynamic Design Optimization of a Compressor Rotor with Navier-Stokes Analysis, *Part A- Journal of Power and Energy*, Proceeding of The Institution of Mechanical Engineers, **217**, No. 2, pp. 179-184.

[57] Jang, C. M., and Kim, K. Y., Optimization of a Stator Blade using Response Surface Method in a Single-Stage Transonic Axial Compressor, *Part A- Journal of Power and Energy 2005,* Proceeding of The Institution of Mechanical Engineers, vol. **219**, no 8, pp. 595-603.

[58] Lingen Chen et al 2005 Optimum design of a subsonic axial-flow compressor stage, *Journal of Applied Energy*, 187-195.

[59] Akin Keskin et al 2006 Application of multi-objective optimization to axial compressor preliminary design *Journal of Applied Energy*, pg 581-589.

[60] CHEN Fu et al 2006 Experimental Study of Stator Clocking Effects in an Axial Compressor, *Chinese Journal of Aeronautics*, Vol.**19**.

[61] Babak Aryana 2008 Preliminary study to raise axial flow compressors up to the context-awareness standards *Journal of Aerospace Science and Technology*, Vol.**12** pg.515 -523.

[62] Rashidzadeh-Hiva et al 2016Industrial Twin-Shaft Gas Turbine Thermodynamic Modelling for Power Generation Application at Design Point and Off-Design Condition, *Indian Journal of Science and Technology*, Vol **9(3)**, pp 1-8.

[63] Turbo engine article. 2016. http://inventors.about.com/library/inventors/blhowajetengineparts.html. Date accessed: 2016 3rd april.

[64]Simulation on a multistage axial compressor article.2016 http://mdx2.plm.automation.siemens.com/cfdImage/simulation-multistage-axial-compressor#. Date accessed: 2016 03rd April.

IOP Conf. Series: Materials Science and Engineering **197** (2017) 012079 doi:10.1088/1757-899X/197/1/012079

Fault diagnosis of helical gearbox using acoustic signal and wavelets

Pranesh SK[1], Siju Abraham[2], V. Sugumaran[3], M.Amarnath[4]

[1]School of Electronics Engineering VIT University Chennai Campus Chennai,Indiaskpran@gmail.com

[2,3]School of Mechanical and building Sciences VIT UniversityChennai Campus Chennai, India v_sugu@yahoo.com

[4]Assistant Professor, Dept. of Mechanical Engineering IITDM, Jabalpur amaranth.cmy@gmail.com

Abstract: The efficient transmission of power in machines is needed and gears are an appropriate choice. Faults in gears result in loss of energy and money. The monitoring and fault diagnosis are done by analysis of the acoustic and vibrational signals which are generally considered to be unwanted by products. This study proposes the usage of machine learning algorithm for condition monitoring of a helical gearbox by using the sound signals produced by the gearbox. Artificial faults were created and subsequently signals were captured by a microphone. An extensive study using different wavelet transformations for feature extraction from the acoustic signals was done, followed by waveletselection and feature selection using J48 decision tree and feature classification was performed using K star algorithm. Classification accuracy of 100% was obtained in the study

Keywords: K star algorithm, Daubechies wavelet, Acoustic signals, Helical gearbox, J48.

1. Introduction:

Machines are the source of production of all the vital components that is required. Gears are responsible for the supply and transmission of the power for these machines making them a very essential part.Gears have cut teeth or cogs which mesh with each other to transmit torque. The gear tooth takes all the load showing its significance. Each and every tooth of the gear contributes for the transmission of the load. It is essential to ensure the proper maintenance of the gear tooth through real time condition monitoring system as the failure of even a single tooth can result in tremendous damage which can result in decreasing the efficiency for transmission or disruption. Gear failures can be caused by several factors such as incorrect design, installation, acid corrosion, poor lubrication etc. The effects of a faulty gearbox can be catastrophic as it can damage the machinery since the broken tooth may disrupt the rotatory motion of the gear. It is therefore essential to reduce the risk of gear failures. The situation can be monitored by examining the sound or vibration signals coming out of the box as the gears are enclosed inside the gearbox. The aim of this study is to reduce the risk of gear failures by real time monitoring of acoustic signals.

Acoustic signals were chosen for the experiment as the accelerometers are more expensive than microphones which would be used for recording the acoustic signals. Artificial fault conditions in gears were created for obtaining acoustic signals of different fault conditions. This was done so that the detection of faults could be done at early stages to avoid mishaps and thus resulting in better efficiency and transmission.

The processes involved were feature extraction, wavelet selection, feature selection and feature classification. Statistical features [1], histogram features [2] or wavelet features are the usually used features. FFT [5] (Fast Fourier Transform)was a very useful tool as the computation time for the Fourier Transform was reduced significantly. It was not suitable for the study as it was inconvenient for non-stationary signals. Though STFT[6] [18] (Short Time Fourier Transform) proved to be better than FFT for non-stationary signals, wavelet transformwere preferred as it provided useful information for fault identification and also provides multiresolution as performed by M. Lang et al . Discrete wavelet transform [3] [4] was preferred over continuous wavelet transform as the processing of data took place faster in discrete wavelet transform despite the provision of accurate data. With further analysis Daubechies 5 was found to produce the best feature set which was used for selection and classification.

The most popular techniques for feature selection are artificial networks [7] [8], fuzzy [9], decision tree [10] [11] [19], principal component analysis and genetic algorithm. J48 decision tree [10] [11] [19] was used for wavelet selection and feature selection as it is very compact, simple and a fast computation speed.

Commonly used classifiers for feature selection are Support vector machine (SVM)[12] [20], Proximal support vector machine (PSVM), artificial neural network (ANN) [13], Naïve Bayes (NB)and Bayes Net, Fuzzy, Decision tree (DT)[14] [15] etc. The SVM based models suffers from higher training time and computational complexity when the number of patterns increases. Despite good results provided by artificial neural network classifier, the training of artificial neural network classifier is complex and time consuming. For a real time system, the classifier used should consume least time with high classification accuracy. J48 decision tree was thus used.

Artificial faults were created in the gears by chipping off 20%, 40%, 60% and80% of the gear tooth and also by removing one entire gear tooth (100%) and another half of the next gear tooth (150%) all at full load capacity of the machine. Using J48 decision tree a set of wavelet features were extracted using 'db5' wavelet feature selection and classification. Simulations were performed and results were obtained.

2. Experimental studies

The experimental set up which is shown in Fig. 1 consists of a 5 HP two stage helical gearbox [17] [21] which is driven by 5.5 HP 3 phase induction motor at 1200 rpm. The DC generator is driven by the mechanical output from the gearbox. A load needs to be provided to the DC motor for the functioning. A resistor bank is connected to the DC motor such that the power can be dissipated in it. Resistor bank is preferred over dynamometers as additional torsional vibrations may occur due to torque fluctuations resulting in formations of inaccurate acoustic signals. Additional vibrations to the test rig are thus eliminated in the generator and resistor bank arrangement. The gearbox, generator and motor are mounted on stiff I-beams which are anchored to a big concrete block. To measure the vertical vibration signals which are generated on the bearing housing of the 16 teeth pinion, B&K 4332 accelerometer is used. Meshing gear frequencies are calculated as 320 Hz and its multiples. Data sets were collected when the helical gear was working at normal conditions and also when artificial faults were created by chipping off 20%, 40%, 60%, 80%, 100% and also by removing one entire gear tooth and another half of the next gear tooth (150%). The signals were curtailed to 3 kHz when a low pass filter was used and was sampled at 8kHz. B&K TYPE 2626 charge amplifier was used to condition the microphone output.

A pinion is connected to a DC motor to generate 2kW power, which dissipates power in the resistor bank. Subsequently the actual load on the gearbox is only 2.6 HP which is 52% of its rated 5 HP. Utilization of the load varies from 50% to 100% in the industrial environment. Tyre couplings are installed between the electrical machines and gear box so that the backslash can be restricted to the gears. The experimental setup is shown in Fig. 1

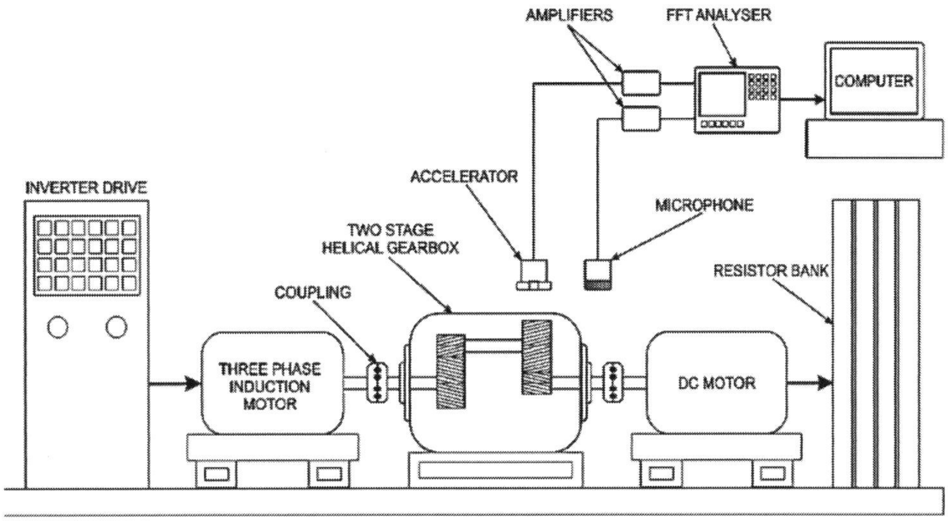

Figure:1 Experimental Setup

Table 1 Specifications of helical gearbox.

	First stage	Second stage
Number of teeth	43/13	73/16
Pitch circle diameter (mm)	198/65	202/48
Pressure angle	20	20
Helix angle	20	15
Modules	4.5/5	2.75/3
Speed of shafts	80 rpm (input)	1200 rpm (output)
Mesh frequency	59 Hz	320 Hz
Step-up ratio	15	
Power transmitted	5 HP	

3. Feature extraction

The acoustic signal recorded from the gearbox by the microphone was used to perform the fault diagnosis. The conversion of acoustic time-domain signal into time-frequency-domain data is done using Discrete Wavelet Transform (DWT) through wavelet decomposition. The wavelet decomposition forms a trend. The obtained trend is again decomposed into next level trend and details. The same process is repeated for several levels of trends to give multiple levels of details. For this study, a signal length of 2048 (2^{11}) is chosen and subsequently the signals can be decomposed into 11 levels.

At each level, the detail co-efficients were used to compute energy content using the following formula.

$$V_i = \sum_{i=1}^{n} X_i^2$$

Where x_i = details coefficients

n=number of details coefficients

Then, the features were defined as the energy content at each level. The feature vector is defined as

$V= (V_1, V_2, V_3, \dots\dots\dots, Vm)$

When m – (number such that length of signal) = 2^m

$V_1, V_2, V_3\dots$ are energy content at given level

The following discrete wavelet transformations were used in this study.

(1) Coiflet : coif – 1, 2, 3, 4, 5

(2) Reverse biorthogonal: rbio – 1.1, 1.3, 1.5, 2.2, 2.4, 2.6, 2.8, 3.1, 3.3, 3.5, 3.7, 3.9, 4.4, 5.5, 6.8

(3) Daubechies wavelet : db – 1, 2, 3, 4, 5, 6, 7, 8, 9, 10

(4) Discrete Meyer : dmey

(5) Sym wavelet : sym – 2, 3, 4, 5, 6, 7, 8

(6) Biorthogonal wavelet: bior – 1.1, 1.3, 1.5, 2.2, 2.4, 2.6, 2.8, 3.1, 3.3, 3.5, 3.7, 3.9, 4.4, 5.5, 6.8

(7) Haar

4. Wavelet selection

Processing of the time domain signals was done using 54 different discrete wavelet transforms from the seven wavelet families. The features which were extracted from the wavelet transform are passed to the J48 algorithm to find the maximum classification accuracy. The features extracted using Daubechies 5 gave the maximum classification accuracy among all the DWTs mentioned above. Daubechies 5 was thus selected for subsequent operations. Daubechies wavelet which is represented as 'db n' is a family of orthogonal wavelets which are characterised by highest number of vanishing points(n) for a given support width of 2^{n-1} . Of all these possible solutions for the point and orthogonality conditions, the solution whose scaling filter produces the maximum phase is selected.

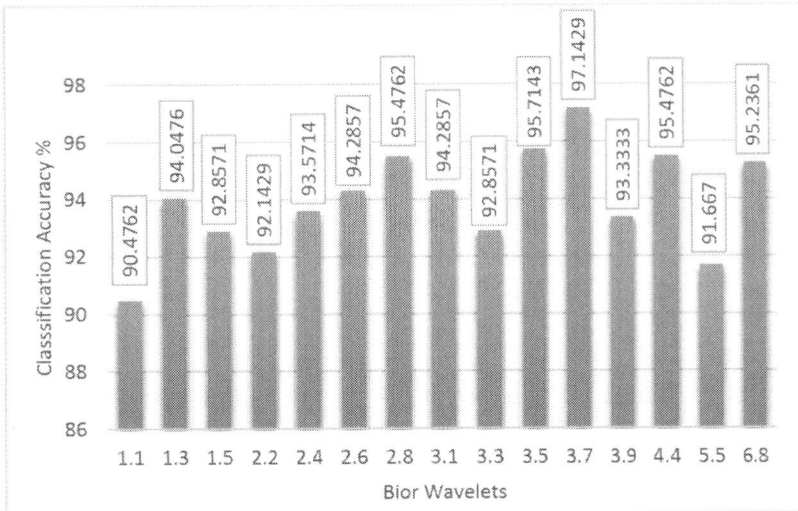

Figure: 2 Classification accuracy of bior wavelets

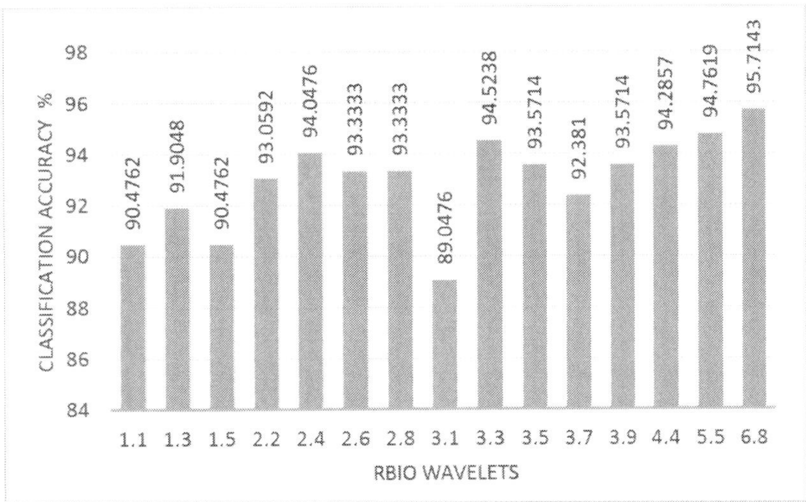

Figure:3 Classification accuracy of rbio wavelets

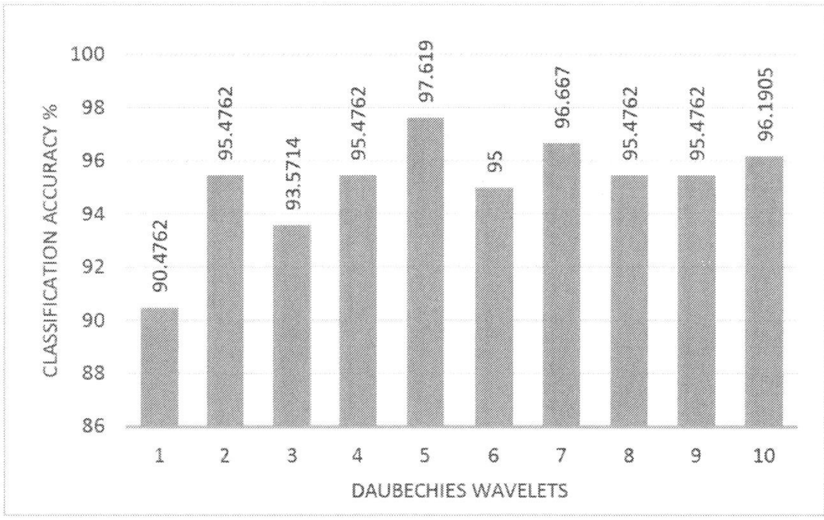

Figure:4 Classification accuracy of Daubechies wavelets

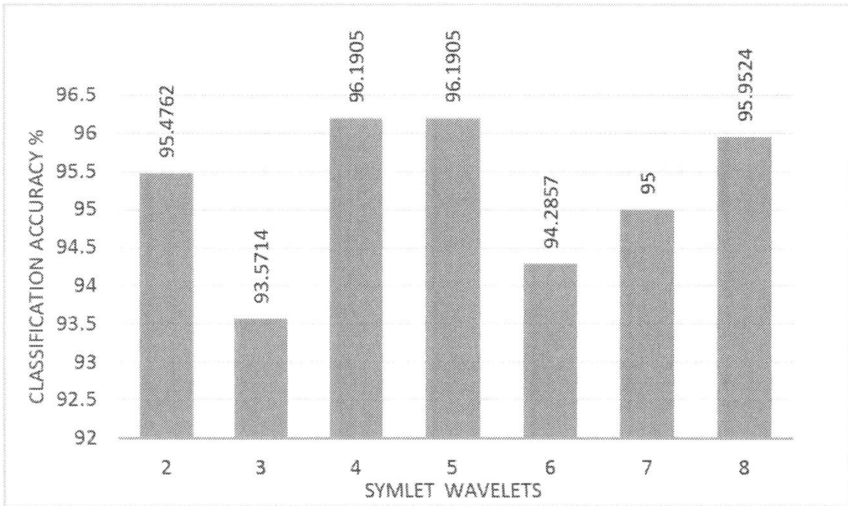

Figure:5 Classification accuracy of Symlet wavelets

IOP Conf. Series: Materials Science and Engineering **197** (2017) 012079 doi:10.1088/1757-899X/197/1/012079

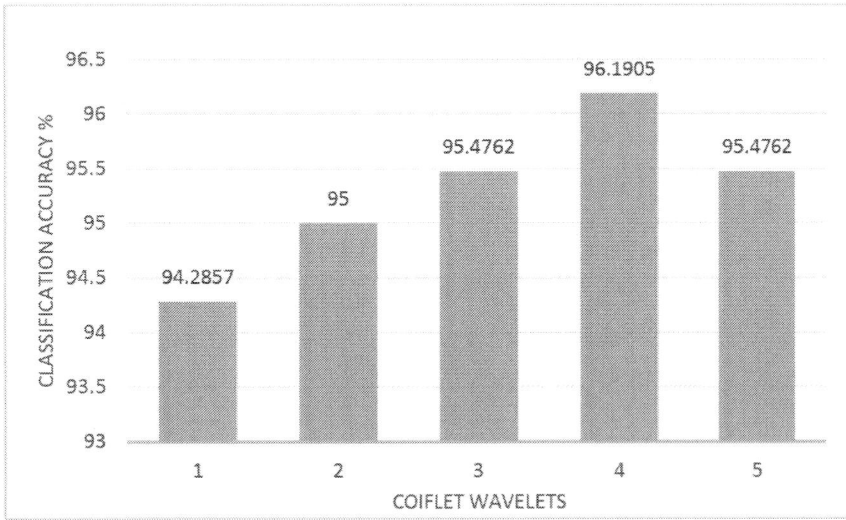

Figure:6 Classification accuracy of coiflet wavelets

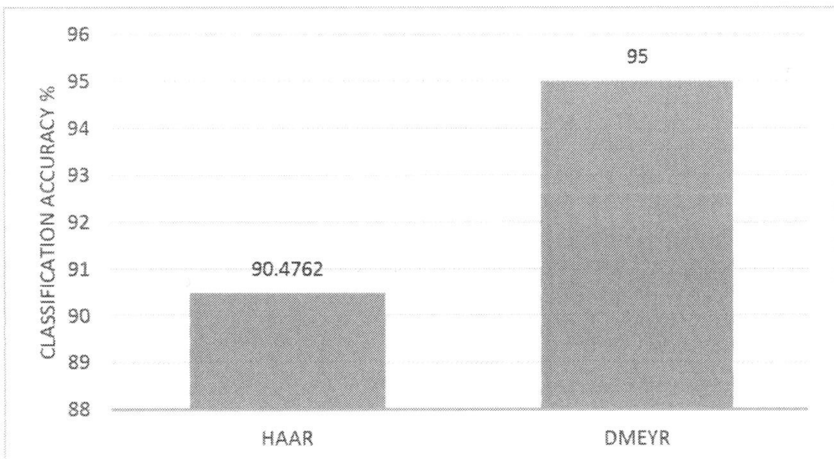

Figure:7 Classification accuracy of Haar and dmeyr wavelets

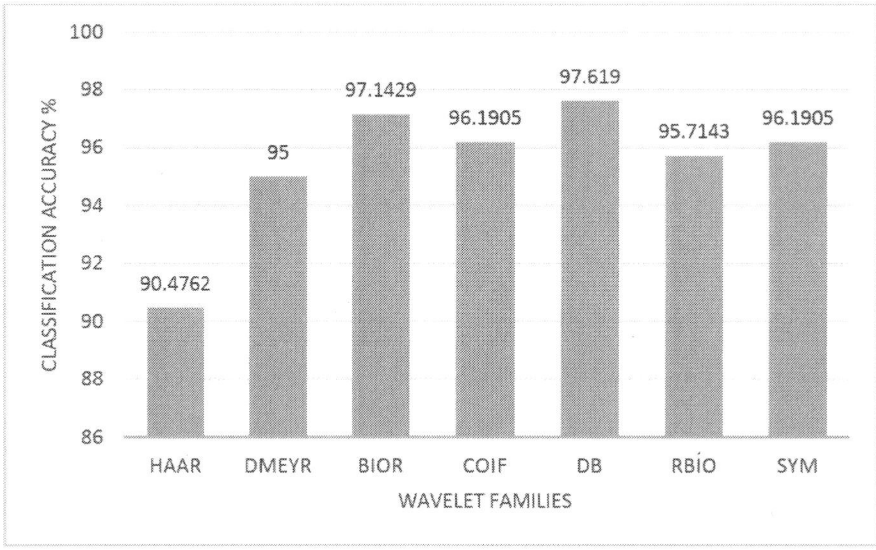

Figure:8 Classification accuracy of various wavelet families

5. Feature selection

J48 classifier is a simple decision tree used for classification. It creates a binary tree. The decision tree approach is most useful in classification problem. With this technique, a tree is constructed to model the classification process. Once the tree is built, it is applied to each tuple in the database and results in classification for that tuple.

While building a tree, J48 ignores the missing values i.e., the value for that item can be predicted based on what is known about the attribute values for the other records. The basic idea is to divide the data into range based on the attribute values for that item that are found in the training sample. J48 allows classification via either decision trees or rules generated from them.

J48 uses entropy based information gain as the selection criteria. As per information theory, entropy is a measure of the uncertainty in a random variable. The expected reduction in entropy due to the partitioning of the examples according to the given feature gives the information gain. It is a measure of the capability of a given attribute to separate its training examples according to the target function.

Information gain (S, A) of a feature A relative to a collection of examples S, is defined as $Gain(S, A) = Entropy(S) \ - \sum_{v \in Values(A)} \frac{|S_v|}{|S|} Entropy(S_v)$ (1)

Where, $S_v = (\{s \in S | A(s) = m\})$

Entropy is a measure of homogeneity of the set of examples and it is given by

$$Entropy(S) = \sum_{i=1}^{c} - P_i log_2 P_i \qquad (2)$$

where 'Pi' is the proportion of 'S' belonging to the class 'i' and 'c' is the number of classes.

IOP Conf. Series: Materials Science and Engineering **197** (2017) 012079 doi:10.1088/1757-899X/197/1/012079

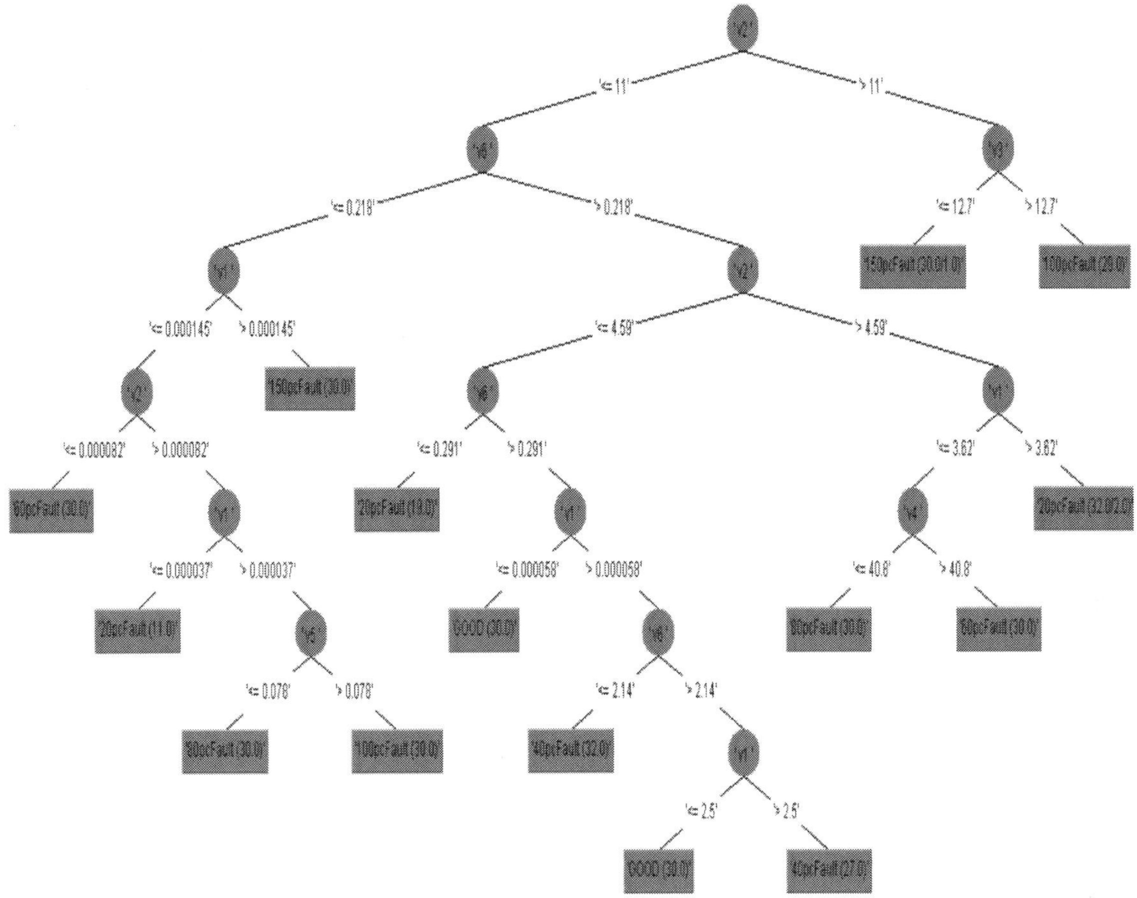

Figure:9 Feature selection

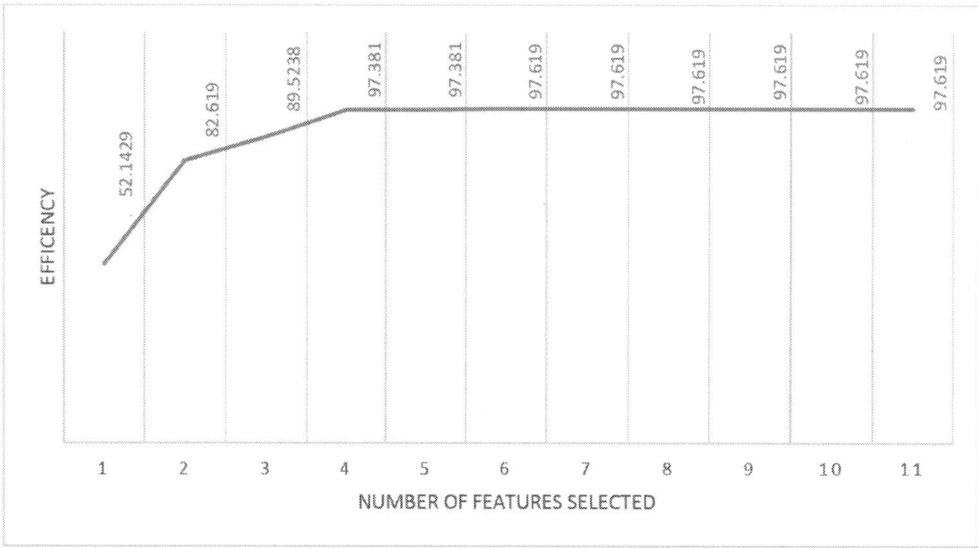

Figure: 10 Effect of features on classification accuracy

The features selected are *V1, V2, V3, V4, V5 and V6*as this combination gave the maximum classification accuracy.

6. Feature classification

The K* algorithm [16] is a method of cluster analysis which mainly aims at the partition of "*n*" observation into "*k*" clusters in which each observation belongs to the cluster with the nearest mean. K* algorithm can be described as an instance based learner which uses entropy as a distance measure. The benefits are that it provides a consistent approach to handling of real valued attributes, symbolic attributes and missing values. K* is a simple, instance based classifier, similar to KNearest Neighbour (*K-NN*). New data instances, *x*, are assigned to the class that occurs most frequently amongst the k-nearest data points $y(j)$, where $j = 1, 2 \ldots k$. Entropic distance is then used to retrieve the most similar instances from the data set. By means of entropic distance as a metric has a number of benefits including handling of real valued attributes and missing values

$$K^*(y_i,x) = -\ln P^*(y_i,x)$$

Where P^* is the probability of all transformational paths from instance x to y. It can be useful to understand this as the probability that x will arrive at y via a random walk in IC feature space. It will performed optimization over the percent blending ratio parameter which is analogous to KNN "sphere of influence", prior to assessment with other Machine Learning methods.

7. Results and discussions

Analysis of the acoustic signals from the gearbox under good and faulty conditions (20%, 40%, 60%, 80%, 100% and 150%) was done followed by the feature extraction, selection and classification using discrete wavelet transform and K star algorithm. Using Daubechies 5 wavelet transform ($v1 - v11$) 11 features were extracted from the acoustic signals. The features which contribute for the feature classification were selected using J48 decision tree. The maximum classification accuracy was obtained when the features *V1, V2, V3, V4, V5 and V6* were selected for training and testing. The remaining features were ignored as they reduced the classifier's accuracy.

7.1 Effect of number of features

(i) The decision tree gave a preview of the relative importance of the features extracted using 'db5' wavelets. The feature at the top gives the maximum contribution, *i.e.V2*, the contribution reduces down the tree.

(ii) The information gained as the entropy reduced gave the measure of the discriminating capability of the feature in a given data set and thus useful features were selected.

(iii) The classification accuracy with J48 algorithm is lower with less features and increases as the number of features increases, reaching a maximum value at 6 features and thereafter it remains constant at 97.619%.

7.2 Feature classification using K star algorithm

(i) Features *V1, V2, V3, V4, V5 and V6*which contributed for the classification were selected for training and testing from the J48 decision tree

(ii)When the number of instances from the data set of 60 increased, the classification accuracy was 100% initially and was constant for some range of values of the global blend and further diminished as the value of the global blend decreased. The reduction of accuracy may be due to increase in complexity of the problem or unnecessary confusion when the number of features increases.

(iii)The summary of stratified cross validation is illustrated in Table 2. The confusion matrix gives an insight into the details of misclassification. In confusion matrix, the first row corresponds to the total number of data points for the ''GOOD'' condition of the gearbox and the first column in first row corresponds to the correct classification as ''GOOD''.

(iv)The second row in the confusion matrix represents the total number of 60 data points for 20% fault condition and the first column represents misclassification of those data points as ''GOOD''. The cell corresponding to the second row and second column in the confusion matrix represents the number of data points in 20% fault condition that have been correctly classified as 20% fault condition.

(v) As shown in the confusion matrix, it is observed that all the data points have been classified correctly. Thus giving 100% accuracy.

	Good	20%fault	40%fault	60%fault	80%fault	100%fault	150%fault
Good	**60**	**0**	**0**	**0**	**0**	**0**	**0**
20%fault	**0**	**60**	**0**	**0**	**0**	**0**	**0**
40%fault	**0**	**0**	**60**	**0**	**0**	**0**	**0**
60%fault	**0**	**0**	**0**	**60**	**0**	**0**	**0**
80%fault	**0**	**0**	**0**	**0**	**60**	**0**	**0**
100%fault	**0**	**0**	**0**	**0**	**0**	**60**	**0**
150%fault	**0**	**0**	**0**	**0**	**0**	**0**	**60**

Table 2 Confusion matrix using K star algorithm

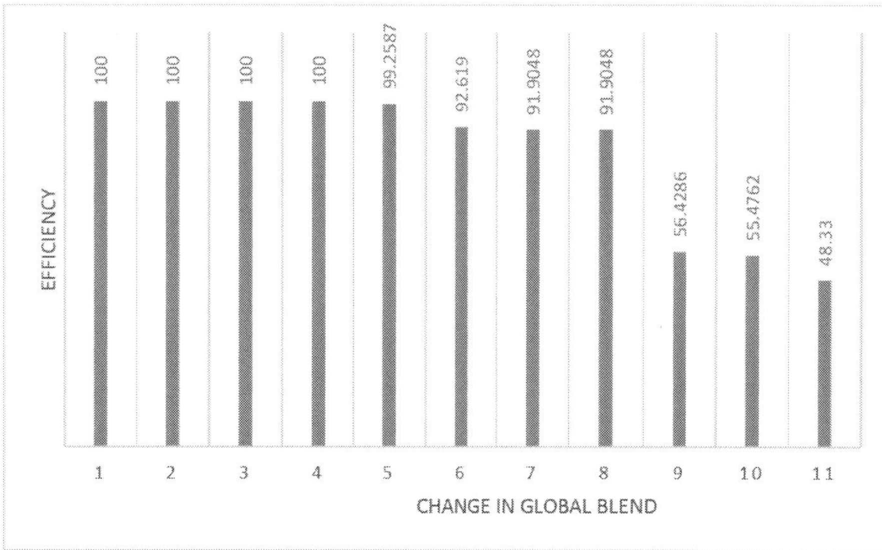

Figure: 11 Efficiency with variation of global blend

Total Number of Instances 420

Correctly Classified Instances 420 100%

Incorrectly Classified Instances 0 0 %

Kappa statistics = 1

Mean Absolute error= 0.0017

Root mean squared error= 0.0077

Relative absolute error= 0.6744%

Root relative square error= 2.2044%

The True Positive (TP) rate is a measure of instances which are considered as true instances in the same class and should be 1 in ideal cases. If a class is declared 'GOOD' all the instances in the same class are classified correctly and the TP rate is 1. Classes with artificial faults of 20%, 40%, 100% and 150% show TP rate as 1 implies that all the instances were classified correctly. The False Positive (FP) rate is a measure of non-satisfying conditions which were interpreted as satisfying conditions for perfect classification. From Table 2, it can be implied that FP is zero. Precision is the fraction of the retrieved instances that are relevant while recall is the fraction of relevant instances that are retrieved and both should be ideally one. In the present study they both confirm to an ideal condition.

8. Conclusion

The classification of features of the faults of the gearbox was done by extracting the discrete wavelet features from the acoustic signals. K star algorithm which was used for classification gave 100% accuracy. The proposed system of using J48 decision tree for selection along with the classification using K star algorithm is extremely effective and can be used for the real time condition monitoring of gearbox with minimal expenditure on the microphone and various components. This system can be implemented properly as it is cost effective and also reduces the risk of failure of the gearbox. This also enables prediction of the faults in the gearbox. Further developments can be made to increase the speed of the process of computation as K star algorithm is a time consuming process.

9. References

[1]Lambrou T, Kudumakis P, Speller R, Sandler M and Linney A1998Classification of Audio Signals Using Statistical Features on Time and Wavelet Transform Domains*Proceedings of the IEEE International Conference on Acoustics, Speech, and Signal Processing (ICASSP 1998)***Vol. 6**pp 3621-24

[2]Sakthivel N, Indira V, Nair B andSugumaran V 2011 Use of histogram features for decision tree based fault diagnosis of monoblock centrifugal pump *International Journal of Granular Computing, Rough Sets and Intelligent Systems (IJGCRSIS)* **Vol 2** pp 23–36.

[3]Lang M,Guo H, Odegard J, Burrus C and Well R 1996 Noise reduction using an undecimated discrete wavelet transform *IEEE Signal Processing Letters***Vol. 3** Issue 1 pp 10-12

[4]Shensa M 1992 The discrete wavelet transform: wedding the a trous and Mallat algorithms*IEEE Transactions on Signal Processing***Vol. 40** Issue 10 pp 2464-82

[5]Bergland G 1969 A guided tour of The Fast Fourier Transform *IEEE Spectrum***Vol. 6** Issue 7 pp 41-52

[6]Kwok H and Jones D 2000 Improved instantaneous frequency estimation using an adaptive short-time Fourier transform*IEEE Transactions on Signal Processing***Vol. 48** Issue 10 pp 2964-72

[7] Lipo W and Daniel L 1993 Artificial Neural Networks: Oscillations, Chaos and Sequence Processing *IEEE Computer Society Press*

[8] Kazuo T and Michio S 1992 Stability analysis and design of fuzzy control systems *Fuzzy Sets and Systems Elsevier***Vol. 45** Issue 2 pp 135-56

[9]Saravanan N andRamachandran K 2009 Fault diagnosis of spur bevel gear box using discrete wavelet features and Decision Tree classification*Expert Systems with Applications: An International Journal* **Vol. 36** Issue 5pp9564-73

[10]Sugumaran V, Ramachandran K andMuralidharan V 2007 Feature selection using Decision Tree and classification through Proximal Support Vector Machine for fault diagnostics of roller bearing *Mechanical Systems and Signal Processing Elsevier* **Vol. 21** Issue 2 pp 597-1196

[11]Glenn M andMangasarian O 2005 Multicategory Proximal Support Vector Machine Classifiers *Machine Learning Springer***Vol. 59** Issue 1pp 77-97

[12]Park D,El-Sharkawi M, Marks R andAltas L 1991 Electric load forecasting using an artificial neural network*IEEE Transactions on Power Systems***Vol.6** Issue 2 pp 442-49

[13] Cristina O, Louis W 2003 A complete fuzzy decision tree technique*Fuzzy Sets and Systems Elsevier***Vol. 138** Issue 2 pp 221-54

[14] Friedl M andBrodley C 1997 Decision tree classification of land cover from remotely sensed data*Remote Sensing of Environment Elsevier***Vol. 61** Issue 3 pp 399-409

[15] Deeman Y and Mohammed A2013 Intrusion Detection System Based on K-Star Classifier and Feature Set Reduction *IOSR Journal of Computer Engineering***Vol. 15** Issue 5pp 107-12

[16]Lin J andZuo M2003 Gearbox Fault Diagnosis using Adaptive Wavelet Filter *Mechanical Systems and Signal Processing Elsevier***Vol. 17** Issue 6 pp 1219-35

[17] Griffin D and Jae L 1984Signal modification from Short Time Fourier Transform*IEEE Transactions on Acoustics, Speech and Signal Processing* IEEE**Vol. 32** Issue 2 pp 236-43

[18]Tina R andSherekar S 2013 Performance Analysis on Naïve Bayes and J48 Classification algorithm for data classification*International Journal of Computer Science and Applications***Vol. 6** Issue 2

[19] Margaret H and Sridhar S 2006 Data mining, Introductory and Advanced Topics *Pearson education* 1st Edition

[20] Amarnath M and Praveen K 2014 Local fault detection in helical gears via vibration and acoustic signals using EMD based statistical parameter analysis*Measurement***Vol.58**pp 154–64

Numerical simulation of dimples in airfoil using MATLAB

P. Booma Devi[1] and Dilip A Shah[2]
[1, 2] Department of Aerospace Engineering,
Hindustan University, Chennai, India.

E-mail: [1]boomajeeva@gmail.com, [2]aero@hindustanuniv.ac.in

ABSTRACT: The Aircraft wing is a point of important research which poses greater challenge in terms of aerodynamic efficiency. The flow separation control method is addressed in classical aerodynamics methods. This study focuses on influence of dimples on controlling the flow and also increasing the aerodynamic efficiency. The periodic process of placing the cavities on the wing starting from root to tip controls the flow separation. The linear variation of characteristic curve provides the information about the flow separation and control of flow on upper surface of the airfoil.These different shapes are utilized viz., Square, Rectangle and Triangle. The numerical simulation is carried out in using MATLAB package. Preliminary analysis on the flow separation is carried out focuses on laminar flow separation, which has the influence on the overall lift generation and drag generation.

Keywords: Flow separation, dimples, laminar flow, lift and drag.

1. INTRODUCTION:

The laminar flow under low Reynolds number aerodynamics is of greater significance for civil aircraft. Typical performances of these airfoils are characteristics by the aerodynamic efficiency. It is a well-known fact that being aerodynamic problem have greater factor for low Reynolds number. In general, the range of the aircraft angle of attack to be aerodynamic efficiency and also the overall weight characteristics of the aircraft. However for low Reynolds number less than 5×10^5. The non-linear features are responsible for the flow separation. Hence at lower speed, the greater increase in angle of attack is mandatory to produce the constant lift. The phenomenon of laminar flow separation complies with the loop found in the lift curve which further explains the reasons contributing towards flow separation. The application of various models creates region of vortices by reducing the pressure drag of the cavities.

In general, the interaction between the forces on the motion of aircraftin air is noted as aerodynamics. Lift to drag ratio is one of the important aerodynamic parameter which determines the total weight and the cost of the aircraft.[1]

In general, the range of aircraft is directly proportional to its aerodynamic efficiency at a constant fuel consumption. Increase in the efficiency of a wing for commercial usage results in reduction of the operational cost. Stalling is a phenomenon associated with flow seperation on the upper wing surface. It is know that C_L is directly proportional to Angle of Attack. Hence at lower speed,high Angle of Attack is required to have a constant lift.[2]

The concept of introducing cavities in the wing creates turbulence which in turn delays the flow seperation(boundary layer). This reduces the region of vortices, there by reducing the pressure drag. Cavities are also called dimples, which are effective with different Angle of attack and also effectively increases the stalling angle. This can be verified by introducing different geometrical cavities over the NACA 0012 airfoil, at the flow of seperation point.[3]

Accelerated flow on the dimple surface converts the laminar flow into turbulent flow. This continuous flow translation results in the reduced dreg and also increases stalling angle. The flow separation over NACA0012 airfoil is observed and results indicate enhanced lift coefficient.[4]

To reduce the flow separation, boundary layer is energized using either leading edge slot or dimple effect. This study aims to validate, document and support the incrasein lift to drag ratio, reduction in the pressure drag and increase in stalling angle. Care is taken to ensure proper pressure coefficient distribution over the airfoil surface.[5]

The introduced cavities converge the laminar to turbulent which as the result of flow translation. The symmetrical of NACA0012 airfoil helps this transition to reduce the drag curve increase the stalling angle. This paper conducts numerical simulation on this airfoil, which incorporates dimples, using the MATLAB package. The subsequent results are then plotted for various aerodynamic parameters.

2. RESEARCH METHODOLOGY - COMPUTATION SCHEME

The unsteady and incompressible flow which is in 2D can be expressed in Cartesian form without taking in account the gravity and body force.

$$\frac{\partial u_i}{\partial x_i} = 0 \tag{1}$$

$$\frac{\partial u_i}{\partial x} + u_j \frac{\partial u_i}{\partial x_j} = - \frac{\partial \rho}{\partial x_i} + v \frac{\partial u_i}{\partial x_i \partial x_j} - \frac{\overline{\partial u_i v_j}}{\partial x_i} \tag{2}$$

Where

U_i ----- the mean velocity
γ ---- Kinematic viscosity
ρ ---- Density
P ---- Pressure
$U_i^{'}U_j^{'}$ -----Reynolds stress

2.1 TURBULENCE MODEL:

The SST k-ω turbulence model is a two- equation turbulence model which can predict precisely the adverse pressure gradient flows and airfoil flows. This effectively combines the accurate combination of turbulence.

$$\frac{\partial}{\partial t}(\rho k) + \frac{\partial}{\partial x_i}(\rho k u_i) = \frac{\partial}{\partial x_j}\left(\gamma_k \frac{\partial k}{\partial x_j}\right) + \widetilde{H_k} - I_k \quad (3)$$

$$\frac{\partial}{\partial t}(\rho \omega) + \frac{\partial}{\partial x_i}(\rho \omega u_i) = \frac{\partial}{\partial x_j}\left(\gamma_\omega \frac{\partial \omega}{\partial x_j}\right) + \widetilde{H_\omega} - I_\omega + G_\omega \quad (4)$$

Where,

$\widetilde{H_k}$ ---- the generation of turbulent kinetic energy due to mean velocity gradients.
H_ω ----- the generation of ω, $\gamma_k$and γ_ω represent the effective diffusivity of k and ω.
I_k andI_ω ----- the dissipation of k and ω due to turbulence.
G_ω ----- the cross-diffusion term.

The simulation representing laminar flow correction can be represented in terms of Reynolds number as follows

$$\mu_t = \alpha^* \frac{\rho k}{\omega} \quad (5)$$

$$G_\omega = \alpha \frac{\omega}{k} . \tau_{ij} \frac{\partial u_j}{\partial x_j} \quad (6)$$

$$Y_k = \rho \beta^* k\omega \quad (7)$$

The above equations are converted into MATLAB codes. Dimples of volume, $6x10^{-4}m^3$is made along the entire wing span of 3m long for both the square and triangle dimples.

Figure 1: plain wing section of NACA-0012

In figure 1 shows the base or plain airfoil that is NACA 0012 which is imported model from CATIA V5R20.

IOP Conf. Series: Materials Science and Engineering **197** (2017) 012080 doi:10.1088/1757-899X/197/1/012080

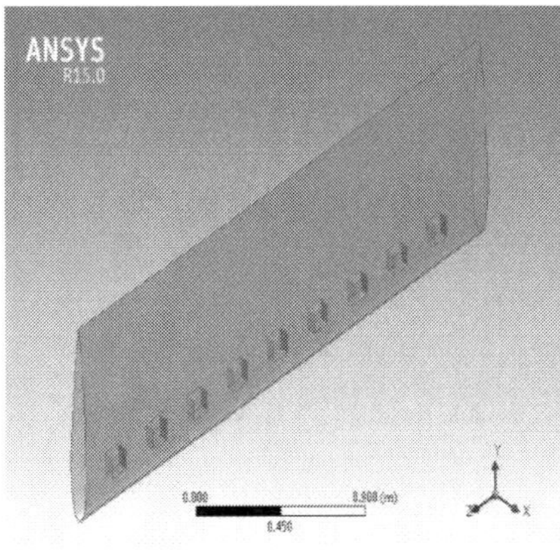

Fig 2: Square cavity pattern over the wing section

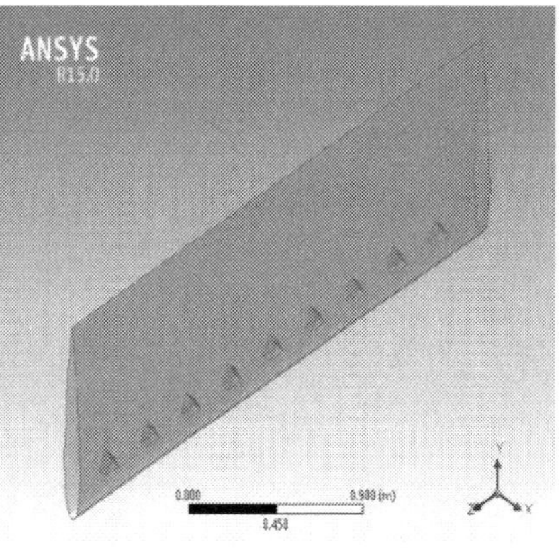

Fig 3: Triangle cavity pattern over the wing section

In fig2, the dimples are arranged in a pattern on the surface of aerofoil at a distance of 0.1m from leading edge. The domain length is considered as 20 times the chord length of the aerofoil. For a square dimple the area is $0.01m^2$. NACA-0012 airfoil with chord 1m and span 3m is considered for the analysis. In fig3, for the same location of a triangular dimple is considered at the same location as the square one. The base dimensions of the triangle dimples are 0.03m long and 0.03m height, with one of its sharp edges facing the trailing edge. The depth of the dimples is 0.06m.

3. PRESENTATION OF RESULTS:

The simulation results for NACA0012 show the time dependent lift curve of various angles of attack. The simulation result agrees with the previous numerical work. The angle of attack is varied from 0^0 to 20^0 and it observed that the slope of the curve is closure to 2π. The inward cavities which as placed are wing span reduce the drag and the cumulative average of drag reduction is considerably low. For overall lift co-efficient the triangular cavities found to be 1.45, which is closure to 1.47 found using previous numerical results. The both cases the stalling angle is increases to 15^0 and 18^0. This is established in correlation with low Reynolds number using standard SST model.The factors which affect the boundary layer are fluid viscosity at adverse pressure gradient. As the angle of attack increases the flow pattern remains until the stalling angle. Hence the overall characteristics agree with the simulated MATLAB analysis.

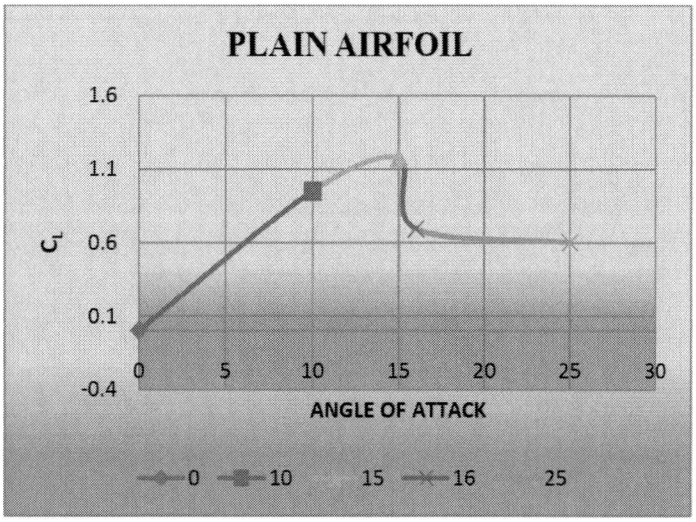

Figure 4: Co-efficient of lift Vs Angle of attack for plain NACA0012 airfoil

In fig 4, the above simulated results show positive increase the co-efficient of lift and the information recording the angle of attack which is at 15^0.

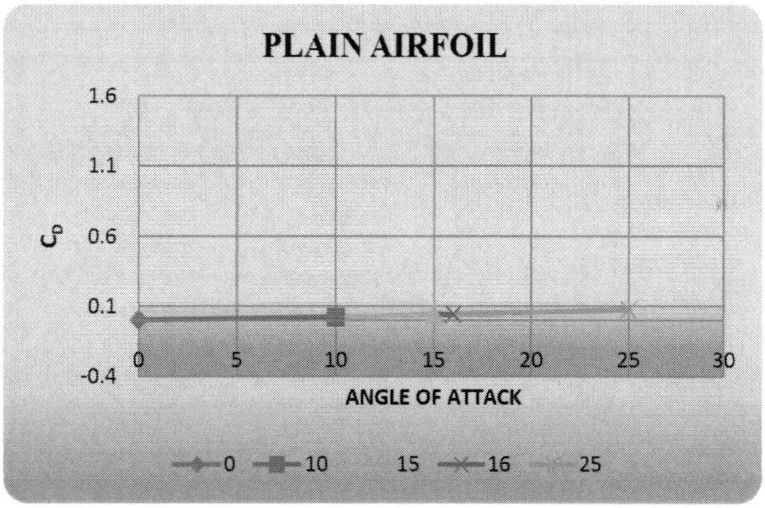

Figure 5:Co-efficient of Drag Vs angle of attack for plain NACA0012 airfoil

In fig 5, the subsequent drag profile for the plain airfoil indicates that for varying angle of attack the drag polar increases with increase in lift.

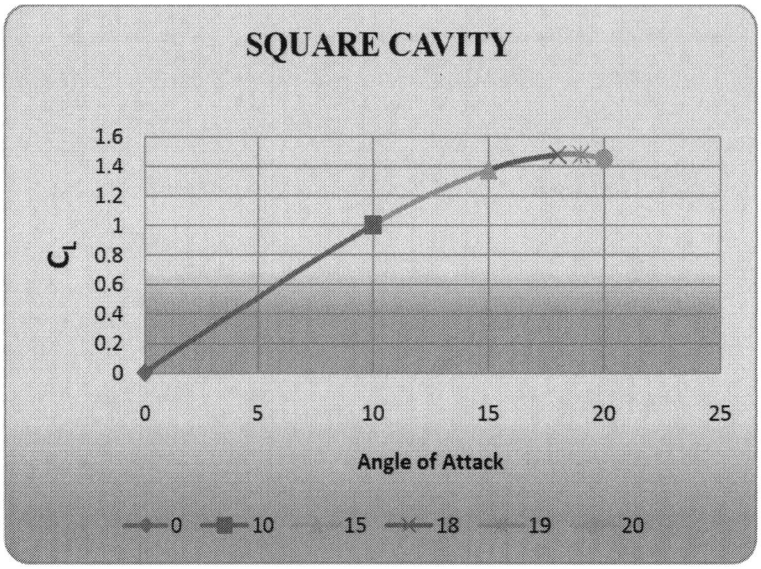

Figure 6:Co-efficient of Drag Vs angle of attack for square dimple

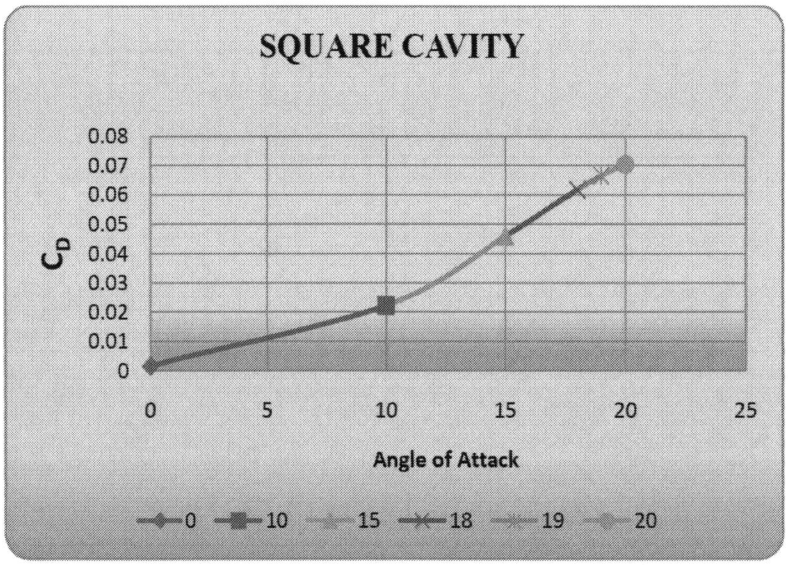

Figure 7: Co-efficient of lift Vs angle of attack for square dimple

In fig 6 and fig 7, the above graph for square dimple represents increase in stalling angle from 15^0 to 18^0. The drag polar explains the drag profile which is correspondingly less comparative plane airfoil.the drag polar for such a cavitiy is parbolicaly less than the angle of attack profile . this shows that the total drag acting on the airofoil is maintained at optimum level in order to provide the required aerodynamic characteristics. the angle of attack profile also increases from 15 degree to 18 degrees.

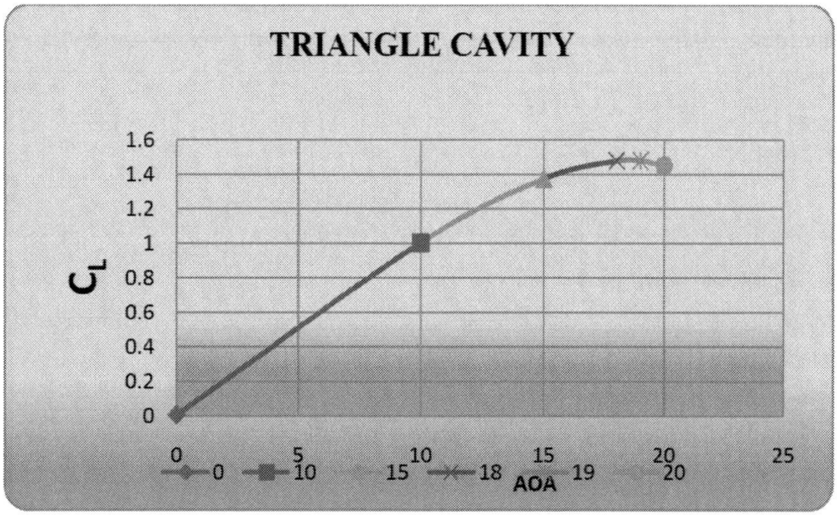

Figure 8:Co-efficient of lift Vs angle of attack for triangular dimple

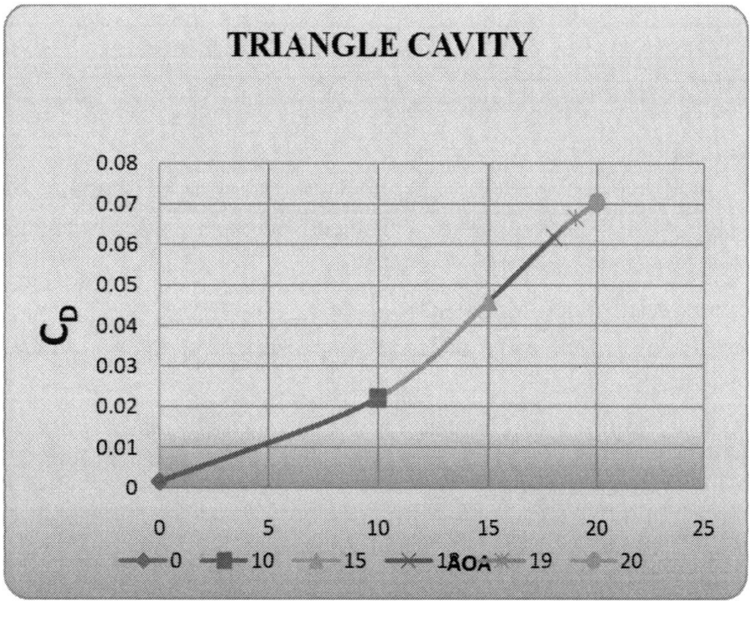

Figure 9:Co-efficient of Drag Vs angle of attackfor triangular dimple

In fig 8 and fig 9, the above graph for triangle dimple represents increase in stalling angle from 15^0 to 18^0. The drag polar explains the drag profile which is correspondingly less comparative plane airfoil.the drag polar for such a cavitiy is parbolicaly less than the angle of attack profile . this shows that the total drag acting on the airofoil is maintained at optimum level in order to provide the required aerodynamic characteristics. the angle of attack profile also increases from 15 degree to 18 degrees.

IOP Conf. Series: Materials Science and Engineering **197** (2017) 012080 doi:10.1088/1757-899X/197/1/012080

The above set of simulation using MATLAB agrees with the previous work and hence it is very clear that the equation taken for evaluating the effect of dimples is inperfect match with the analytical results.

4. CONCLUSION:

1. The non-linear charactersied of symmetrical airfoil NACA 0012 are characterised by the inward cavities which placed root to tip.
2. At very low reynolds number the laminar flow separation occurred as the angle of attack increased periodically.
3. The overall stalling angle exhibits and increase of 18^0 from 15^0.
4. The numerical simulation also indicates the dimple airfoil configuration gives better aerodynamic efficiency compare to plain aerofoil.
5. Furthur more the periodical flow separation on the upper surface of airfoil reduces the stalling effect on the dimple case.
6. The intensity of the cavity increases the lift co-efficient and decreases the drag c0-efficient.

REFERENCES

1. Srivastave.D, "Flow controls over airfoils using different shaped dimples" BITS Bilani, k.k Birla Goa, India-403726, 2012 pp 92-97.
2. Livya.E, Anitha.G and Valli.P " Aerodynamic analysis of dimple effect over aircraft wing" international journel of mechanical aerospace industrial and mechatronics engineering, Vol-9. No 2, 2015 pp 350-353.
3. Md.AmzadHossain and Md.Nizzam Uddin ,"Experimental study of aerodynamic characteristics of airfoils using different shaped dimples" ISSN (e):2319 The IJCE Vol-4.No1,2015 pp13-17.
4. "Aerodynamic effects of dimples on aircraft wing" Intl. Conf. On Advances in Mechanical, Aeronautical and Production Techniques - MAPT 2015, ISBN: 978-1-63248-072-9.
5. Makwana.P.B,Makadiya.J.J, "Numerical simulation of flow over airfoil and different techaniquies to reduce flow seperation alongwith basic CFD model: A review study" ISSN: 2278-0181, 2014 PP 399-404.
6. Gad-el-Hak, Flow Control Passive: Active and Reactive Flow Management, First ed., Cambridge University Press, 2000.
7. D.J. Butter, "Recent Progress on Development and Understanding of High Lift Systems," in Improvement of Aerodynamic Performance Through Boundary Layer Control and High Lift Systems, Brussels, Belgium, pp. 1.1-1.26,1984.
8. J.D. Anderson, Fundamentals of Aerodynamics, McGraw Hill Series, Third Edition, pp. 41-47, 2001.
9. A.M.O. Smith, "High-Lift Aerodynamics,"Journal of Aircraft vol. 12, pp. 501-530, 1975.

10. R.H. Liebeck, Smyth, D.N., "Study of slat-airfoil combinations using computer graphics," Journal of Aircraft,vol.10, April, pp. 254-256, 1973.
11. M. Sajben, et al., "A New, Passive Boundary Layer Control Device," AIAA-1976-700, presented at the American Institute of Aeronautics and Astronautics and Society of Automotive Engineers, Propulsion Conference, 12th, Palo Alto, Calif, 26-29 July, 1976.
12. F.G. Howard and W.L. Goodman, "Drag Reduction on a Bluff Body at Yaw Angles to 30Degrees," Journal of Spacecraft and Rockets, vol. 24, pp. 179-181, 1987.

IOP Conf. Series: Materials Science and Engineering **197** (2017) 012081 doi:10.1088/1757-899X/197/1/012081

Experimental and numerical analysis of convergent nozzle

Srinivas G[1], Bhupal Rakham*[2]

[1]Assistant Professor, Dept of Aero & Auto Engineering, Manipal Institute of Technology, Manipal University, Manipal, Udupi, Karnataka, India-576104.
[*2]Assistant Professor, Dept of Aeronautical Engineering, MLR Institute of Technology-JNTU, Dundigal, Rangareddy District, Telangana State, India-500043.

Abstract. In this paper the main focus was given to convergent nozzle where both the experimental and numerical calculations were carried out with the support of standardized literature. In the recent years the field of air breathing and non-air breathing engine developments significantly increase its performance. To enhance the performance of both the type of engines the nozzle is the one of the component which will play a vital role, especially selecting the type of nozzle depends upon the vehicle speed requirement and aerodynamic behavior at most important in the field of propulsion. The convergent nozzle flow experimental analysis done using scaled apparatus and the similar setup was arranged artificially in the ANSYS software for doing the flow analysis across the convergent nozzle. The consistent calculation analysis are done based on the public literature survey to validate the experimental and numerical simulation results of convergent nozzle. Using these two experimental and numerical simulation approaches the best fit results will bring up to meet the design requirements. However the comparison also made to meet the reliability of the work on design criteria of convergent nozzle which can entrench in the field of propulsion applications.

1. Introduction

Today's all the modern aircraft engines are using highly sophisticated components to achieve better fuel efficiency and thrust. These components are highly reliable and giving better service to the aerospace industry. However there are some technical problems which even today facing by all air breathing engines in each and every component like engine intakes, compressor, combustion chamber, turbine and most importantly the nozzle. Nozzle has a vital role in every air breathing and non-air breathing engines, it makes the complete expanded hot gas coming from the turbine to expel with the high acceleration [1]. The more massive expulsion of hot gas gives more forward motion to the aircraft or space craft as per the well-known theory Newton's third law of motion [3].

In general discussion the word propulsion meaning gives the net force that results from imbalanced pressure. To understand more clearly can take the sealed container, gas or air under pressure in a sealed container applies equal forces or pressure on all sides of the container [2]. One more observation that all forces are equally balanced and there is no possibility of container to move.

The purpose of this paper is to give straightforward information about the convergent nozzle basic engine operational theory and discussion on experimental results. Apart from this information this paper also describes pertinent information about problems that may phase during the experiment on

IOP Conf. Series: Materials Science and Engineering **197** (2017) 012081 doi:10.1088/1757-899X/197/1/012081

nozzle especially those cannot be solved through the simulations well and therefore cause confusion. However the consistent numerical simulation also done for better reliability of convergent nozzle [4]. In common the purpose of the nozzle in terms of the aeronautical applications it converts the chemical thermal energy formed in the combustion chamber into the required kinetic energy [6]. The nozzle makes the high velocity gas of lower temperature and pressure from high pressure, high temperature and low velocity of hot gas. According to the literature survey it is found that the nozzle generally gives exhaust velocity range from 2-4.5 km/sec [5]. If we closely observe the properties of the real fluid in the nozzle the inlet Mach number is less than one and near the throat section the flow is sonic. If we further attach the divergent section we may found that supersonic velocities depends upon the boundary conditions set in the simulation or experiment.

2. Literature on theoretical analysis on convergent nozzle

The convergent-divergent nozzle, the speed of the gas can increase from end to end, the mean speed of the molecules in the molecular structure of the gas will fall from end to end and at some point the speed of flow will equal the mean speed of the molecules [9]. We have seen that this is a special point where the duct must change from convergence to divergence. Then the flow in the convergence depends on expansion in the direction of flow and in the divergence on expansion across the flow.

Thespeed at the junction, that is, in the throat, is $\sqrt{\gamma RT}$ where $T$the absolute temperature at the throat is of course. This immediately raises the issue of the use of the Mach number in connection with nozzles [10]. In the early part of the 20th century there was great optimism that the use of non-dimensional groups of physical properties would lead to significant improvements in the ways in which we store experimentally gathered data. One of the important groups is the Mach number denoted M and named in honor of Ernst Mach [7, 8]. It should be noted that in order to draw graph shown below to start with values for the stagnation conditions. At one time this would have meant that the calculations for drawing the graphs would have been very time consuming. That problem has disappeared with the emergence of mathematics packages on computers. We can draw as many graphs as we like almost instantly like figure 1.

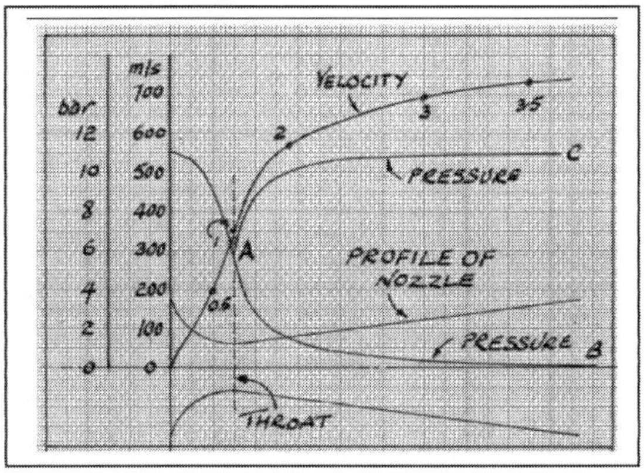

Figure 1. Representation of stagnation values of convergent nozzle.

3. Experimental study on convergent nozzle

For a convergent nozzle where the area of cross-section decreases along the length of the nozzle, velocity increases according to the continuity equation as well as the Bernoulli's equation. It is essential to demonstrate the working of the nozzle and study the velocity and pressure variations as the flow passes through the nozzle.

A convergent nozzle with surface/static pressure taps along the length of the nozzle is provided to measure the static pressure variation as the flow passes through the nozzle shown in figure 2-3. An orifice is provided in the upstream direction to measure the volume flow rate.

Using the flow rate measurement and the geometry of the nozzle at various locations, it is observed that average velocity is varied along the nozzle. Comparisons can be made theoretically by obtaining the static variation using Bernoulli's and continuity equations and practically using the measurements. Hence the working of the convergent nozzle can be demonstrated.

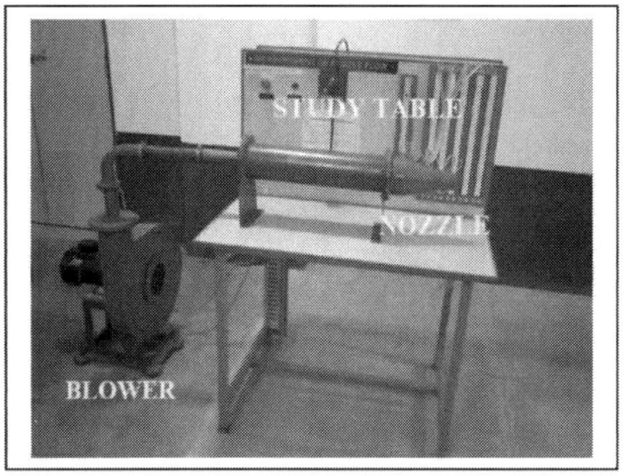

Figure 2. Experimental setup for convergent nozzle flow analysis.

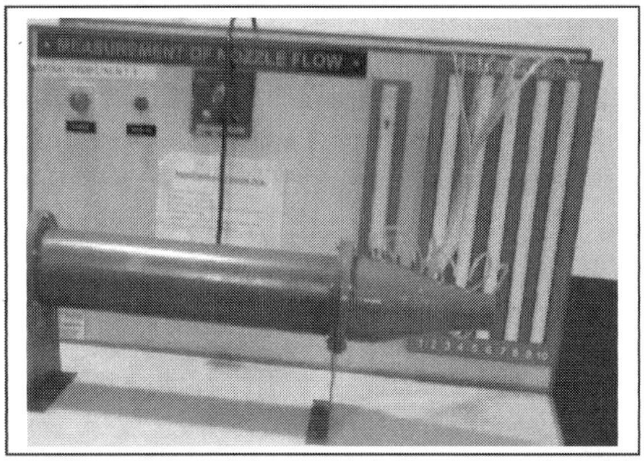

Figure 3. Representation of convergent nozzle experimental setup manometer arrangement.

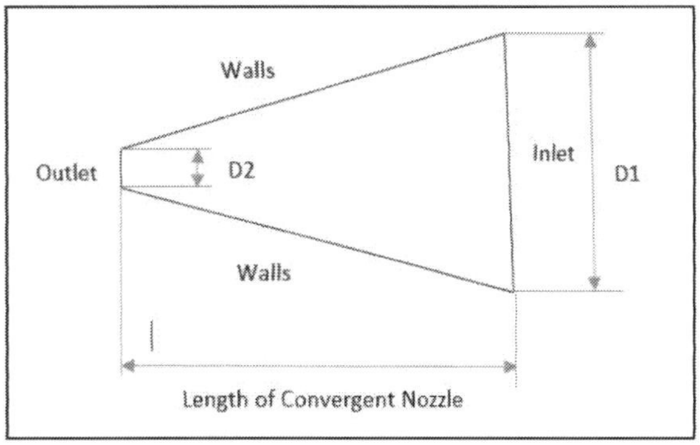

Figure 4. Geometrical details of convergent nozzle inlet and outlet diameters.

The set up consists of a blower unit coupled to D.C. motor and is connected to a convergent nozzle through a settling chamber with a hose. The discharge can be controlled by controlling the speed of the D.C. motor. The pressure tap rings (9 nos) are made in the nozzle surface and are connected to the multi-bank manometer. The orifice plate is fitted in the pipeline of the blower outlet, to measure the discharge of flow and is connected to differential manometer. The control panel consists of the mains-on indicator, control switch, D.C. motor speed controller, differential manometer and multi-bank manometer. The whole instrument is mounted on a self-contained sturdy table and is isolated from the blower unit so that vibration should not transfer to the study table.
Formulas used:
Q = Discharge = $C_d A\sqrt{2gh}$
g : Acceleration due to gravity = 9.8m/s^2
ρ_w: Density of water = 1000kg/m^3
ρ_a: Density of air = 1.22 kg/m^3
h : Differential manometer reading in meters of H_2O column
d : Diameter of orifice = 25mm
Where C_d – Co-efficient of orifice = 0.64

$$Area\ of\ orifice = \frac{\pi d^2}{4} = \frac{\pi(0.025)^2}{4}\ m^2$$

V - Velocity of nozzle at particular location = $\frac{Q}{A_n}$
Where n – the nozzle number
A_n – Cross technical area at point n of the nozzle.

After successful iteration of the nozzle requirement in design point of view, numerically and experimentally the appropriate tabulation was formed. After conducting the experiment the observations has tabulated as shown in the table 1 and 2. From the figure 5 it is clear that the port number between 7 and 8 all the pressure lines meeting at one point because the area contacting in the nozzle has more influence. According to the theoretical understanding the exit velocity flow speed and discharge flow speed vary proportionally, same was observed in the experimental graph showed in figure 6 and figure 7. Similar performance also given from the figure 8 and figure 9 that velocity is directly proportional to the pressure head also discharge is proportional to the pressure head because of the area contraction at desired design point is quite good enough to achieve better velocity.

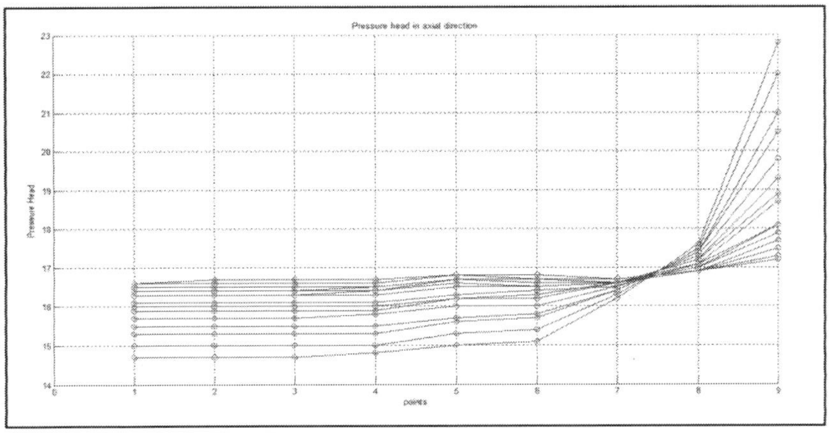

Figure 5. Representation of pressure head in axial direction.

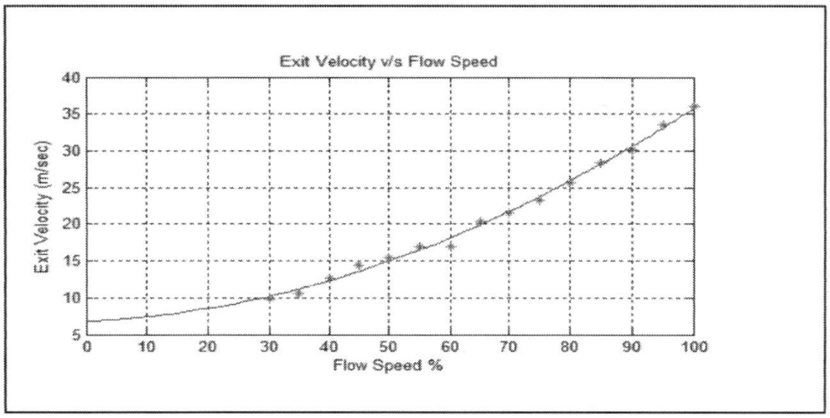

Figure 6. Representation of exit velocity vs flow speed

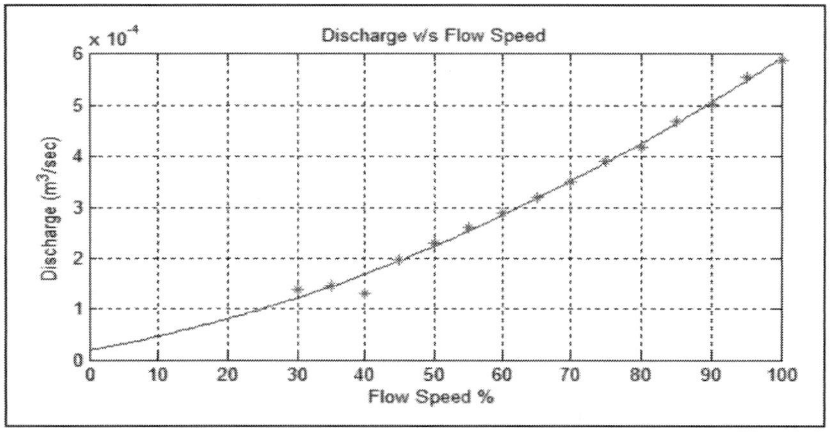

Figure 7. Representation Discharge vs flow speed

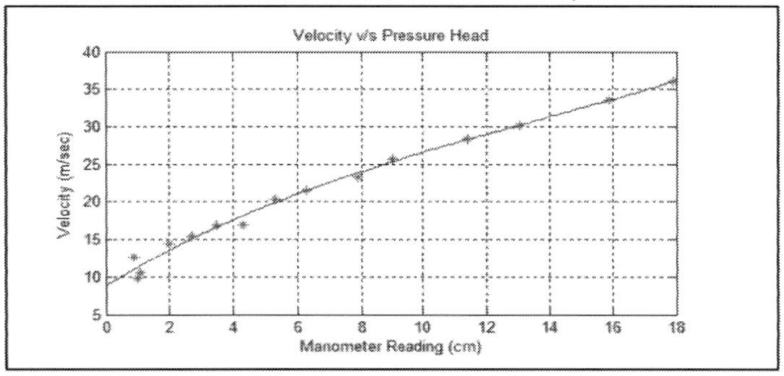

Figure 8. Representation of velocity vs pressure head

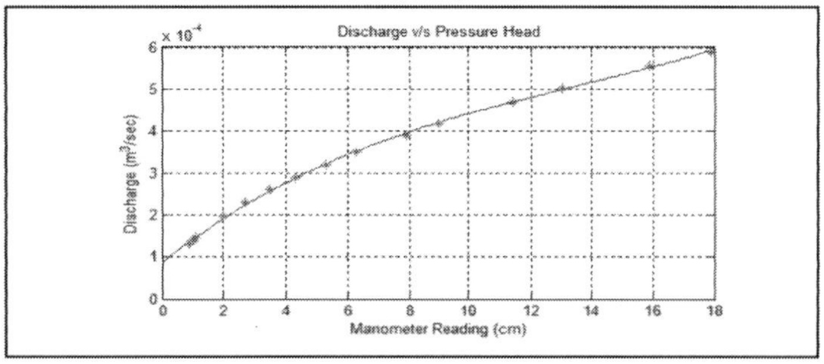

Figure 9. Representation of discharge vs pressure head.

Table 1. Experimental tabulation static pressure values across the convergent nozzle.

Speed	Static pressure head									
%	1	2	3	4	5	6	7	8	9	H_ref
30	16.6	16.6	16.6	16.6	16.8	16.7	16.6	16.9	17.2	17.2
35	16.6	16.7	16.7	16.7	16.8	16.8	16.7	16.9	17.3	17.3
40	16.5	16.5	16.5	16.5	16.7	16.7	16.7	16.9	17.5	17.5
45	16.4	16.4	16.4	16.5	16.7	16.7	16.6	16.9	17.7	17.7
50	16.4	16.4	16.4	16.4	16.7	16.6	16.6	17	17.9	17.9
55	16.3	16.3	16.3	16.4	16.6	16.5	16.6	17	18.1	18.1
60	16.3	16.3	16.3	16.3	16.5	16.5	16.6	17.1	18.1	18.1
65	16.1	16.1	16.1	16.1	16.3	16.4	16.5	17.1	18.7	18.7
70	16	16	16	16	16.2	16.3	16.6	17.2	18.9	18.9
75	15.9	15.9	15.9	15.9	16.2	16.2	16.6	17.2	19.3	19.3
80	15.7	15.7	15.7	15.8	16	16	16.5	17.3	19.8	19.8
85	15.5	15.5	15.5	15.5	15.7	15.8	16.4	17.4	20.5	20.5
90	15.3	15.3	15.3	15.3	15.6	15.7	16.4	17.4	21	21
95	15	15	15	15	15.3	15.4	16.3	17.5	22	22
100	15.1	14.7	14.7	14.8	15	15.1	16.2	17.6	22.8	22.8

Table 2. Experimental convergent nozzle tabulation values of manometer readings.

Speed	Manometer head			Velocity	Discharge
%	L	R	H (cm)	m/sec	m^3/sec
30	-0.6	0.4	1	9.80	1.39E-04
35	-0.6	0.5	1.1	10.59	1.46E-04
40	-0.8	0.1	0.9	12.66	1.32E-04
45	-1	1	2	14.43	1.97E-04
50	-1.4	1.3	2.7	15.50	2.29E-04
55	-1.8	1.7	3.5	16.98	2.60E-04
60	-2.2	2.1	4.3	16.98	2.89E-04
65	-2.7	2.6	5.3	20.41	3.20E-04
70	-3.2	3.1	6.3	21.55	3.49E-04
75	-4	3.9	7.9	23.34	3.91E-04

80	-4.6	4.4	9	25.63	4.18E-04
85	-5.8	5.6	11.4	28.30	4.70E-04
90	-6.5	6.5	13	30.21	5.02E-04
95	-8	7.9	15.9	33.48	5.55E-04
100	-8.9	9	17.9	36.02	5.89E-04

4. Numerical simulation of convergent nozzle

To solve the numerical simulation on nozzle suitable turbulent model needed. The final solution of final step in the navier stokes equation is flow velocity. Equation (1) represents the navier stokes equation where the term 'w' indicates specific thermodynamic work, 'v' kinetic viscosity, 'u' velocity profile which helpful in many single phase fluid flows. Once the flow velocity is known other quantities such as temperature or pressure found easily. The Cauchy momentum equation derived originally from navier stoke momentum equation. Equation (2) shows conservative form of cauchy momentum equation, the left side of the equation designates acceleration and right side of the equation combination of body forces.

$$\frac{\partial u}{\partial t}+(u.\nabla)-v\nabla^{2}u=-\nabla w+g. \tag{1}$$

$$\frac{\partial}{\partial t}(\rho u)+\nabla.(\rho u\Theta u+pI)=\nabla.\tau+\rho g \tag{2}$$

The same convergent nozzle used for the experimental work, the numerical simulation has been conducted by creating the similar artificial environment in the ANSYS software. The basic geometric details are shown in figure 4. Once the geometry is created by using proportionate nozzle profile the boundary conditions are given before the meshing. The boundary conditions for the convergent nozzle is shown in table 3. The exit diameter of the nozzle obviously smaller comparatively the inlet diameter, therefore the velocity inlet was given to entry of the nozzle and pressure outlet was given to exit of the nozzle to find the accurate convergence solution.

Table 3. Representation of convergent nozzle analysis procedure using ANSYS.

Description	Setup Details
General values	Fluid type: Density type
Fluid Description	Viscous: Inviscid, Energy: on
Materials	Fluid: Air Density: Real gas
Start-up cell zone condition	Operating condition : 40 pa (Equivalent to experimental)
Boundary conditions	Inlet: velocity Outlet: pressure Temperature: 240 K
Solution control	Courant number: 6

IOP Conf. Series: Materials Science and Engineering **197** (2017) 012081 doi:10.1088/1757-899X/197/1/012081

	Solution initialisation: Standard compute inlet Run calculation: Enter the number of iteration (50,000), select calculation.

The surface of the nozzle taken as walls so the interference results predicted easily. The meshing plays a vital role in the simulation where the accuracy of the flow parameters like static pressure can found at each location. The mesh elements maintained almost very high density to have better flow characteristics across the nozzle. The mesh density representation over the walls shown in figure 10-11.

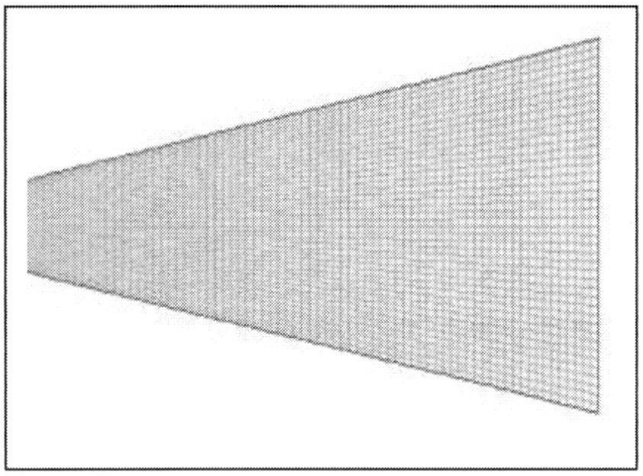

Figure 10. Representation of convergent nozzle grid along XZ full symmetry in ANSYS.

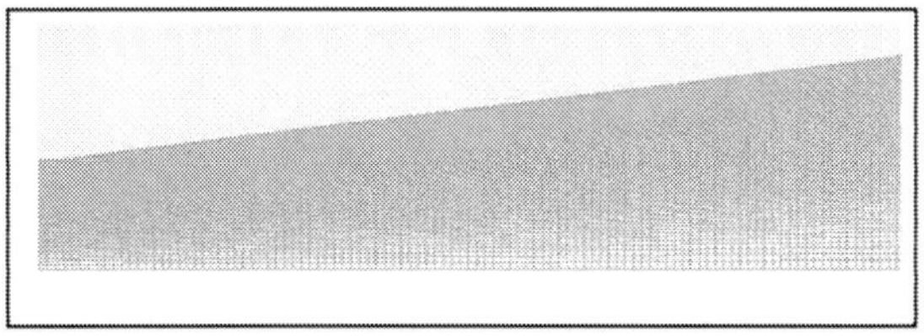

Figure 11. Representation of nozzle wall mesh density high level contrast.

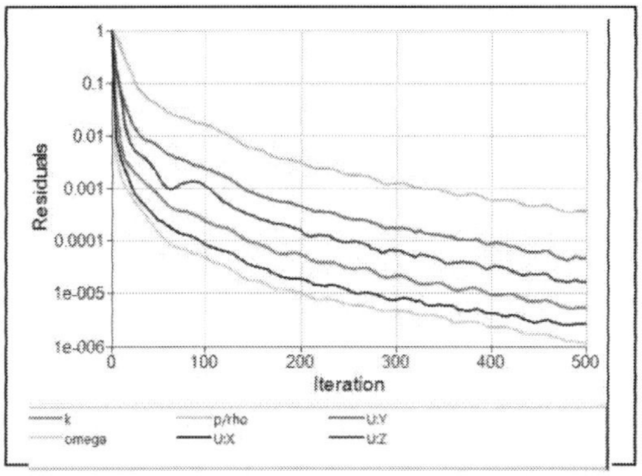

Figure 12. Representation of ANSYS fluent solution convergence plot at convergence criteria 10^{-5}.

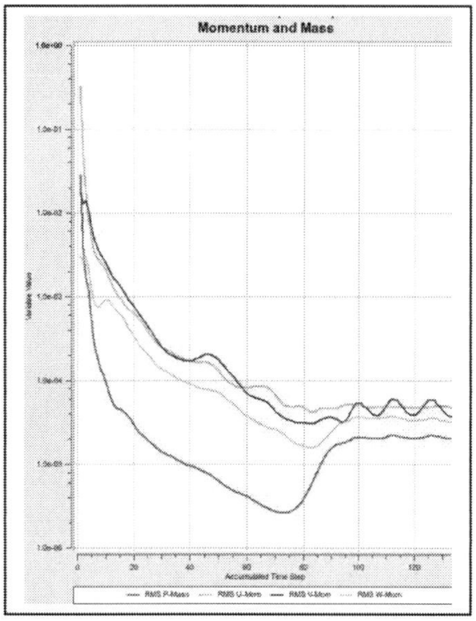

Figure 13. Representation of ANSYS solution momentum and mass along the axis of convergent nozzle.

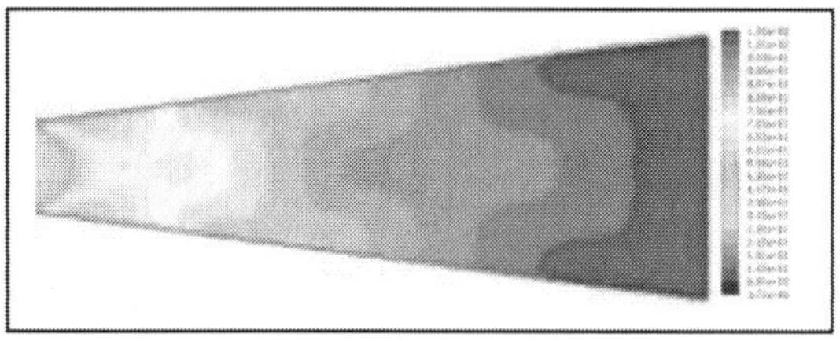

Figure 14. Representation of convergent nozzle mach contour.

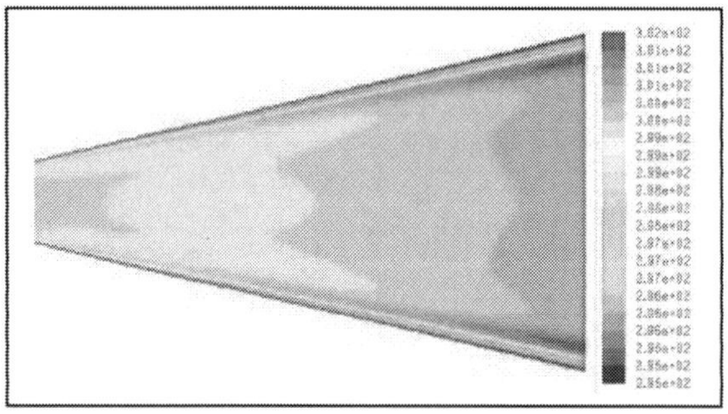

Figure 15. Representation of convergent nozzle Total temperature.

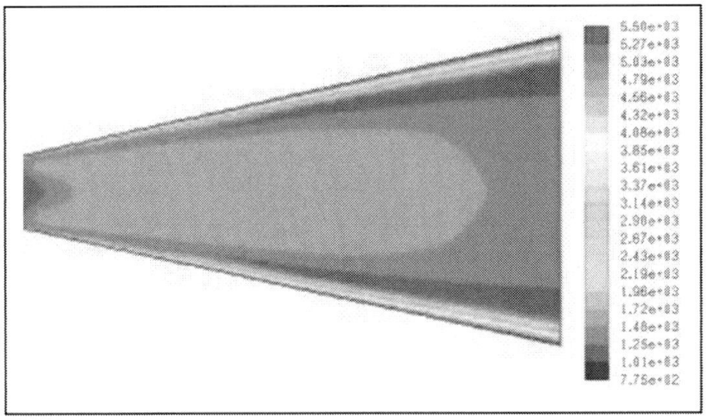

Figure 16. Representation of convergent nozzle turbulent intensity.

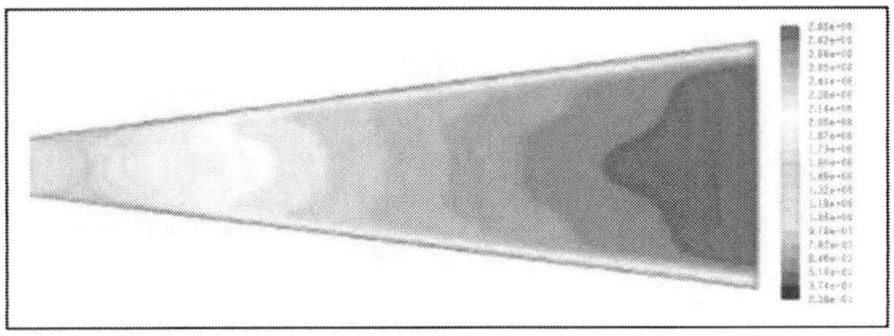

Figure 17. Representation of convergent nozzle velocity contour.

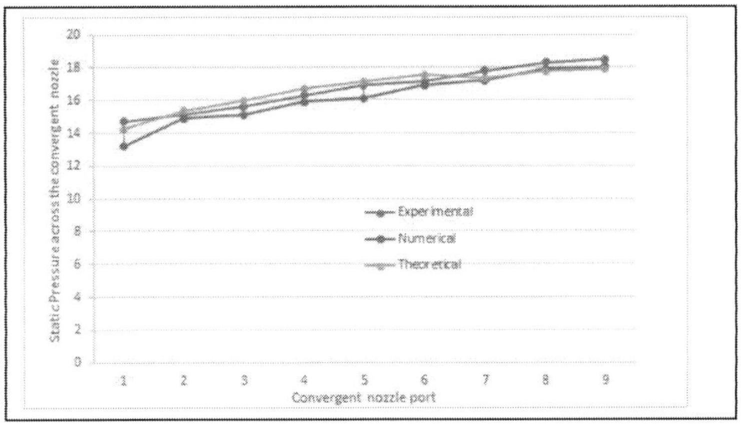

Figure 18. Representation of convergent nozzle static pressure vs nozzle port.

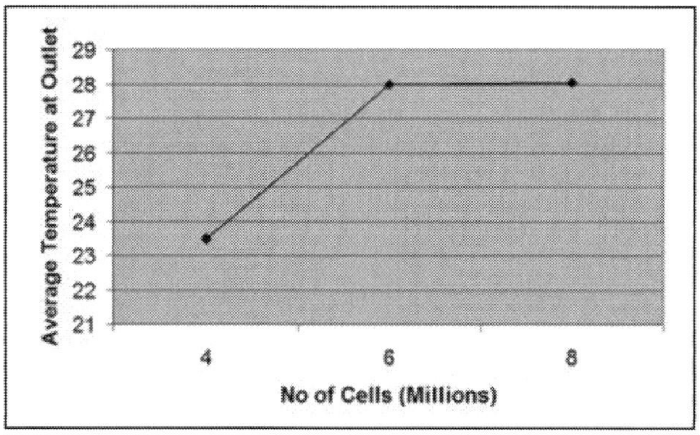

Figure 19. Representation of mesh independence check of convergent nozzle average temperature at exit.

The final most important step in the simulation giving boundary conditions. Based on subsonic standardized theory the flow boundary conditions given to convergent nozzle solver. The convergence criteria has been maintained 10^{-5} value so that accuracy can be improved. The convergent nozzle for the given set of mesh 2 million cells convergence solution obtained within the 500 iterations. The solution convergence plot as shown in figure 12. For the same boundary conditions the mass and momentum curves are further captured through ANSYS shown in figure 13. The flow aerodynamic behavior in convergent nozzle further explained through the Mach number, total temperature, turbulence model and velocity contour are shown from figure 14-17. The maximum total temperature found to be $3.02x10^2$ K, the velocity at the exit of nozzle found to be $2.85x10^3$ m/sec are optimum values comparatively with the experimental and theoretical results. From experimental, theoretical and numerical approached it is found that the port six values are very much optimum to the design point of view. The same representation of three approaches the desired values of convergent nozzle static pressure vs port number shown in figure 18. The mesh independence check also performed for optimum temperature values that has been shown in figure 19. The average temperature at outlet maintained to be 28 K which is best suitable for subsonic flow conditions in air breathing propulsion.

5. Conclusions

According to the theory studied on the nozzle and experiment conducted to the convergent nozzle it is understood that the area rule concept perfectly agreed with the numerical simulation results. After successful completion of numerical simulation, convergent nozzle experiment and theoretical calculation the desired mach value achieved at the exit of the convergent nozzle section found to be $1.8x10^2$ and also the total temperature also found to be $3.02x10^2$ K . Similarly the turbulent intensity was found that $5.50x10^3$ which is optimum velocity value for all subsonic air breathing propulsion. Because as the area decreases it increases the velocity especially at throat the flow is sonic and the flow further accelerates to supersonic if we attach additional divergent section. However the given boundary conditions are meeting by conducting the experiment on the convergent nozzle. It is also further understood that there is a scope to improve the same experimental setup to understand the convergent and divergent section nozzle three dimensional flow analysis for better aero mechanical features.

6. References

[1]. Pandey,K.M.; Singh, A.P. 2010 CFD Analysis of Conical Nozzle for Mach 3 at Various Angles of Divergence with Fluent Software, *International Journal of Chemical Engineering and Applications*, Vol. **1**, No. 2, ISSN: 2010-0221.

[2]. Natta, Pardhasaradhi.; Kumar, V.Ranjith.; Rao, Dr. Y.V. Hanumantha; 2012 Flow Analysis of Rocket Nozzle Using Computational Fluid Dynamics (Cfd), *International Journal of Engineering Research and Applications (IJERA)*, ISSN: 2248-9622,Vol. **2**, Issue 5, pp.1226-1235.

[3]. P. Padmanathan, Dr. S. Vaidyanathan, 2012 Computational Analysis of Shockwave in Convergent Divergent Nozzle, *International Journal of Engineering Research and Applications (IJERA)*, ISSN: 2248-9622, Vol. **2,** Issue 2, , pp.1597-1605.

[4]. Biju kuttan. P, M. Sajesh., 2013, Optimization of Divergent angle of a rocket engine nozzle using Computational Fluid Dynamics The *IJES*, vol.**2**, Issue2, Pages 196-207.

[5]. C.Satheesh, A. Arulmurugu., 2015 Design and analysis of C-D Nozzle increase the efficiency using CFD IJMTER –PP 490 - 494.

[6]. Mayur Chakravarti 2011 Design analysis of flow in convergent divergent rocket nozzle – University of Petroleum and energy studies, Department of Aerospace Engineering, Dehradun, India,.

[7]. Vipul Sharma, Ravi Shankar, Gaurav Sharma, 2011, CFD Analysis of Rocket Nozzle – University of Petroleum and energy studies, Department of Aerospace Engineering, Dehradun, India,.

[8]. Yasuhiro Tani, 2005, Fundamental study of supersonic combustion in pure air flow with use of shock tunnel‖ Department of Aeronautics and stronautics, Kyushu University, Japan, Acta Astronautica 57, PP 384 – 389.

[9]. Patankar, S.V. and spalding. D.B. 1974, A calculation for the transient and steady state behaviour of shell- and- tube Heat Exchanger‖. Heat transfer design and theory sourcebook. Afghan A.A. and Schluner E.U. Washington. PP. 155-176.

[10] M. Sundararaj and S. Elangovan 2013, Computational Analysis of Mixing Characteristics of Jets from Rectangular Nozzle with Internal Grooves" Supplementary Article, *Indian Journal of Science and Technology*, Print ISSN: 0974-6846, Online ISSN: 0974-5645. Vol **6 (5S).**

IOP Conf. Series: Materials Science and Engineering **197** (2017) 012082 doi:10.1088/1757-899X/197/1/012082

Numerical investigation of mixing characterstics of chevron nozzle by passive controls method

J.Devipriya[1], Dr.Kanimozhi[2]

[1, 2,3] Department of Mechanical Engineering, Sathyabama University, India

Abstract.

This paper deals with the Reduction of noise in the aircraft exhaust is done by installing Chevrons with particular parameters in the Nozzle section. Numerical investigations have been carried out on chevron Nozzles to evaluate the importance of Chevron parameters by adding number of Chevrons and the mixing characteristics of jet. After assessing the Chevron parameters we vary the Chevron shapes at the exit by installing the triangular wedge in order to regulate maximum noise reduction along with a negligible thrust loss. Finally the results is compared with free jet Nozzle with Chevron and Chevron with wedge has been analysed using CFX CFD software and the results of potential core decay of these Nozzles has been measured from the analysis.

1. Introduction

A Nozzle is designed to control the direction of a fluid flow as it enters and exit an enclosed chamber or pipe via orifice. A nozzle is a tube with varying C-S area and it can be used vary the flow of a fluid. Nozzles are used to control the rate of flow, speed, direction, mass, shape, and the pressure of the stream that comes from mid engine. The goal is to increase the kinetic energy of the flow medium with respect to energy. Nozzles can be classified as convergent or divergent. A de Laval nozzle has a convergent section followed by a divergent section and is often called a nozzle[2]. Convergent nozzles have ability to accelerate subsonic fluids. If the nozzle pressure ratio is high enough the flow will reach sonic velocity at the nozzle throat, that situation the nozzle is said to be choked.
Jet is a free shear flow driven by momentum introduced at the nozzle exit of, usually, a nozzle. As the jet fluid travels further away from its origin, it slows down due to mixing with slower speed ambient fluid. This is due to boundary layer at the nozzle exit which develops roll up structure, or ring vortices[1]. Thus, mass flow at any cross-section of the jet progressively increases along the downstream direction. Hence, to converse momentum the centerline velocity decreases with downstream distance. The resulting centerline velocity decay as proportional to gradient across the shear layer and is a strong function of distance downstream.

IOP Conf. Series: Materials Science and Engineering **197** (2017) 012082 doi:10.1088/1757-899X/197/1/012082

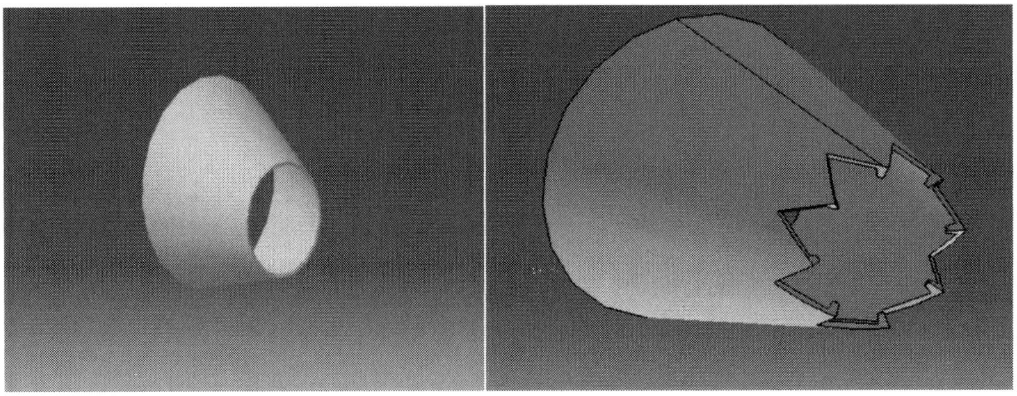

Figure1. Three Dimensional Freejet Nozzle

2. Computational work

2.1 Three Dimensional modelling

Designing three dimensional nozzles with different configurations at exit we used "CATIA" modelling software which is flexible one and user friendly. Initially the sketch has been created with the given dimension, and at the exit the domain for the visualization of the potential core region has been designed with the respective dimensions. The computational approach is applied with similar basic boundary conditions[6][3].Surfaces on aircraft are embedded with millions of rivets. The model considered, Cessna 152 consists, in total of 2 million rivets. The horizontal stabilizer alone consists of about 200 rivets. The common types of rivets used are brazen head rivets and counter-sunk rivets. Some tests were conducted regarding the drag produced due to rivet heads way back in 1950's and with some positive results. Tests have been conducted in the NACA 8-foot high-speed wind tunnel to find the effect of rivet heads and spot welds on the wing drag. The tests were made of 5-foot chord airfoil. The velocity of the air was varied from 80 to 500 miles per hour and the lift coefficient from 0 to 0.30. The increment of the drag force on the 5-foot airfoil varied from 6%, due to countersunk rivets, to 27%, due to the 3/32-inch brazier rivet head, with the rivets in a arrangement. The drag increases caused by protruding rivet heads is directly related to the height of the heads. With the front row of rivets well forward, changes in span-wise pitch having insignificant effects on drag unless the pitch was more than 2.5% of chord. Data are presented for evaluating the reduction in drag attained by eliminating rivets from the forward part of the wing surface body; for example, it is shown that over 60 to 75% of the rivet drag is caused by the rivets on the forward 30 to 35% of the airfoil in a typical case.

Table 1. Nozzle specifications

Specifications	Dimensions
Length	30mm
Inlet diameter	30mm
Exit diameter	16mm
Chevron length	5.44mm
Wedge	3.08mm

IOP Conf. Series: Materials Science and Engineering **197** (2017) 012082 doi:10.1088/1757-899X/197/1/012082

2.2 Boundary Conditions

The below diagram illustrates the boundary conditions of 3dimensional models. In ICEM CFX select the pre-processor modules and feed the input for the respective Mach number. It contains elements of 556208.

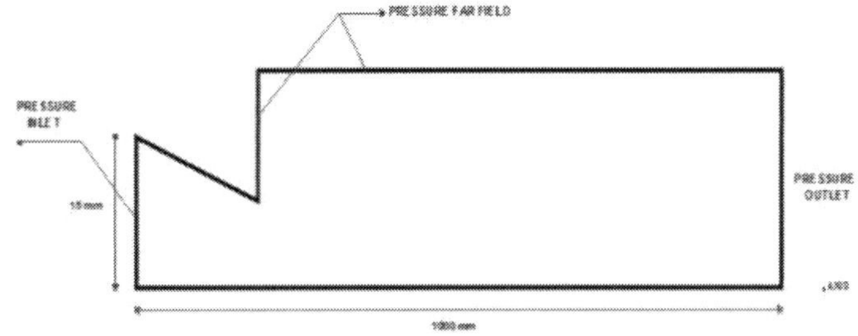

Figure 2. Boundary conditions

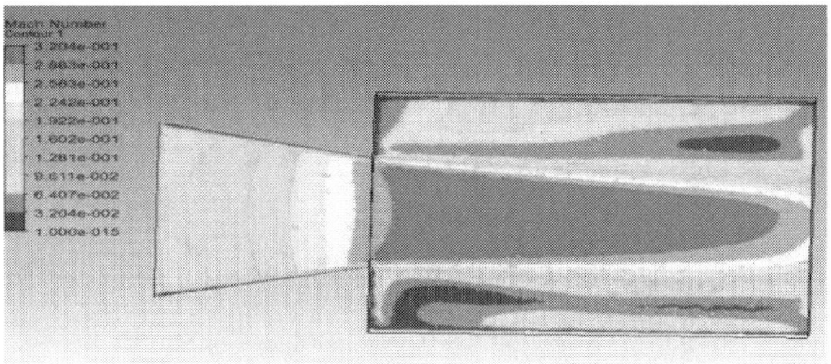

Figure 3. Mach number plot for free jet M= 0.4

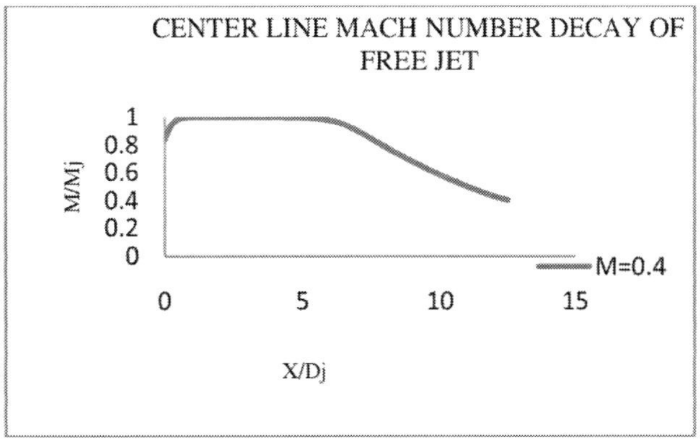

Figure 4. Centerline Mach number decay of free jet

The above graph shows the centerline Mach number decay for the Mach number 0.4 and is plotted between X/Dj and M/Mj. Here the potential core decays at 5.5D.

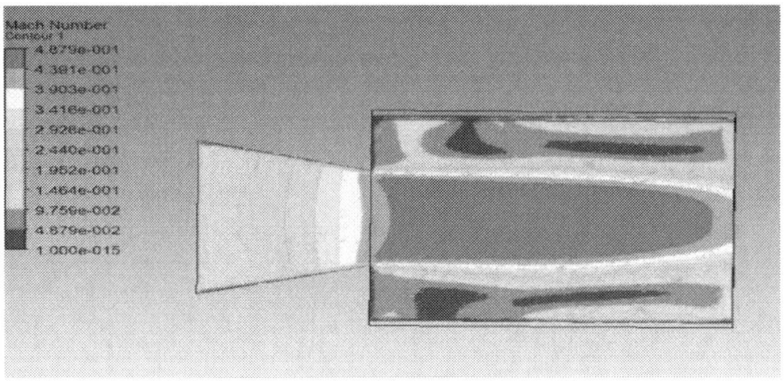

Figure 5. Mach number plot for free jet M= 0.6

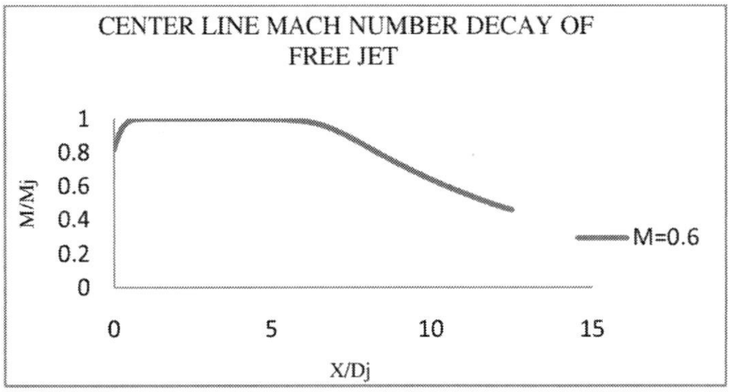

Figure 6. Centerline Mach number decay of free jet

The above graph shows the centerline Mach number decay for the Mach number 0.6 and is plotted between X/Dj and M/Mj. Here the potential core decays at 5.81D. When comparing with 0.4 Mach number of free jet the potential core length is high.

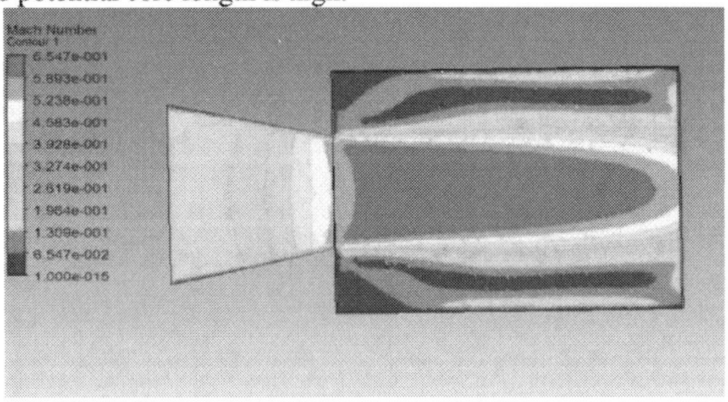

Figure 7. Mach number plot for free jet M= 0.8

The above graph shows the center line Mach number decay for the Mach number 0.8 and is plotted between X/Dj and M/Mj. Here the potential core decays at 6.12D. When comparing with 0.4 and 0.6 Mach number of free jet the potential core length is high.

4. Center line mach number decay of chevron

The above centreline Mach number decay graph is plotted between M/Mj and X/Dj for the chevron nozzle for the Mach number 0.4. The potential core decay starts from 3.88D. When comparing with 0.4 Mach number of free jet the potential core length is 30 % reduced.

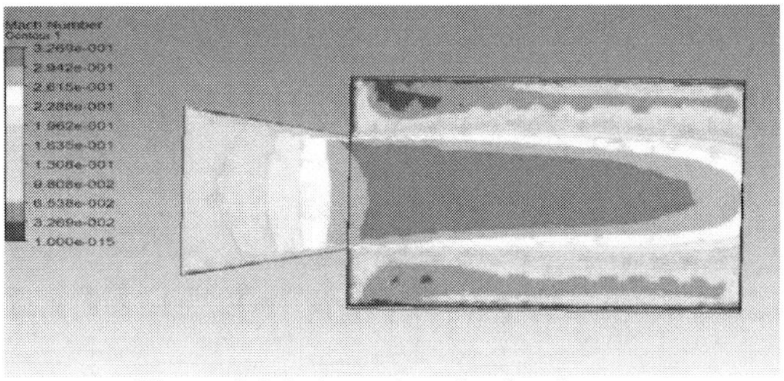

Figure 8. Mach number plot for chevron

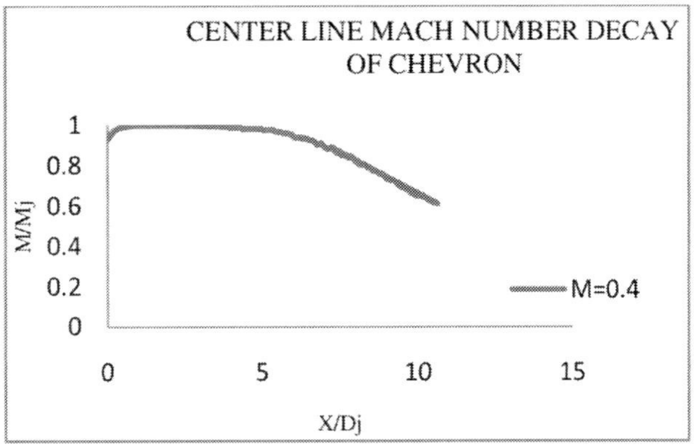

Figure 9. Centerline Mach number decay of chevron

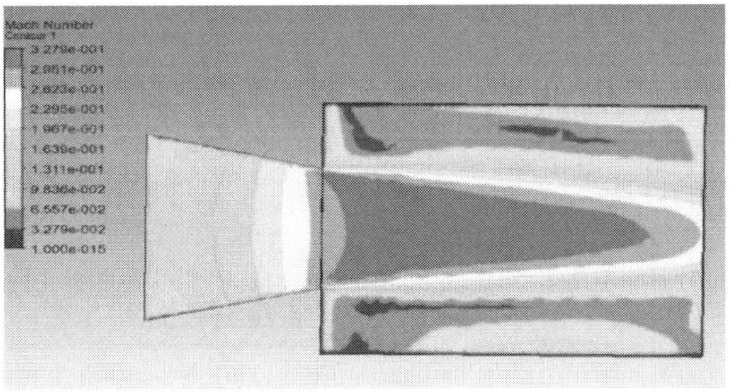

Figure 10. Mach number plot for chevron with wedge thickness

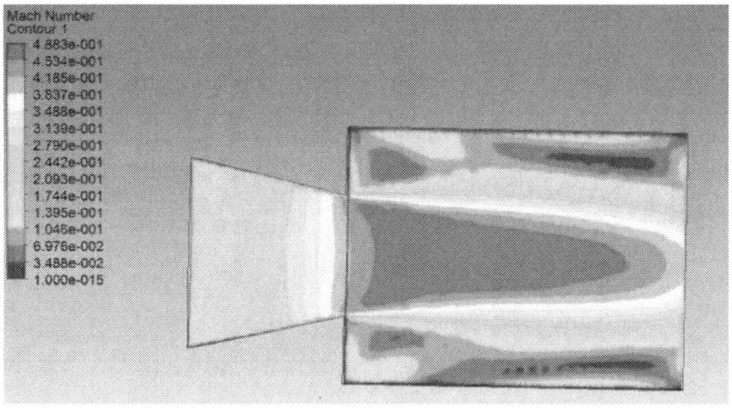

Figure 11. Mach number plot for chevron with wedge thickness

3. Conclusion

The Wedges and chevrons were used in the presents study, namely chevron, chevron with 1mm wedge thickness, chevron with 2mm Wedge thickness were found to be effective in distorting the jet structure. From the figure 6.1, fig 6.2, fig 6.3 among the three configurations, chevron with 2mm Wedge was found to be more efficient in reducing the potential core length and distorting the jet structure compared to the other two configurations in all subsonic mach numbers. From the above figures, Nozzle with the chevron with 2mm Wedge , the maximum core length reduction achieved was 67 % of uncontrolled jet , followed by rapid decay of jet center line mach number, where the chevron with 1 mm Wedge reduce the core length to 50 % and chevron reduce the core length to 40 % at all subsonic mach numbers. Chevron with 2 mm Wedge is showing consistently better performance for all mach numbers compare to other two configurations. It may due to that fact the chevron with 2mm Wedge creates stronger pressure gradient shedding stronger streamwise vortices causing reduction in potential core and enhanced mixing. Mach number profiles at different X/D locations shows the chevron with 2mm Wedge is more effective spreading of the jet normal to the wedge compare to other two configurations.

Chevron technology has given a modest relief. Unfortunately, total understanding of jet noise mechanisms is still not in our grasp. The insight of fundamental experiment and application of CFD allowed the development of the subject technology to the next level of analysis.Hope for further control and reduction of jet noise hinges on advancement of our understanding of the relevant mechanisms.

3.1. Comparisons Of Centerline Mach Number

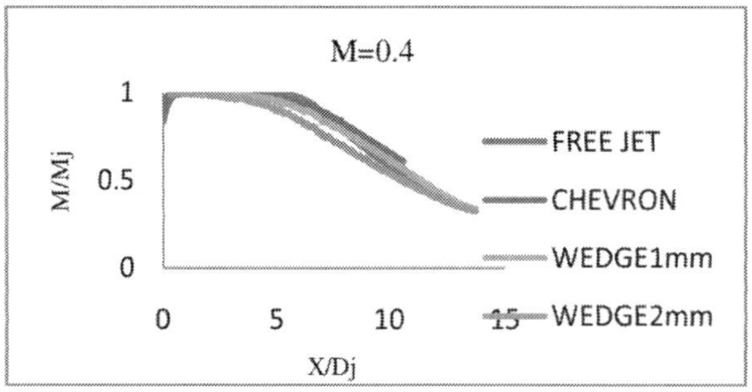

Figure 12. Comparison of Centerline Mach number decay

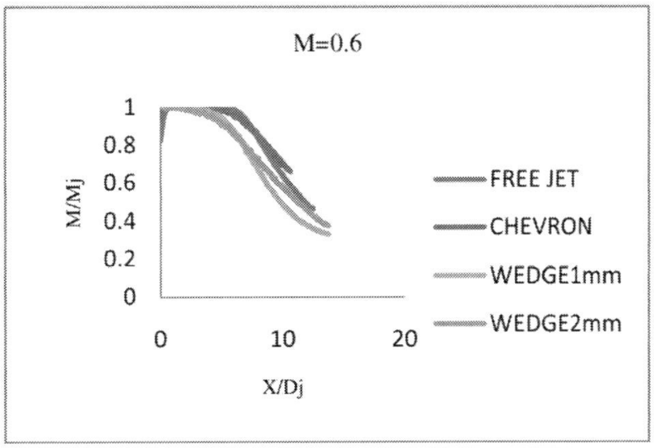

Figure 13. Comparison of Centerline Mach number decay

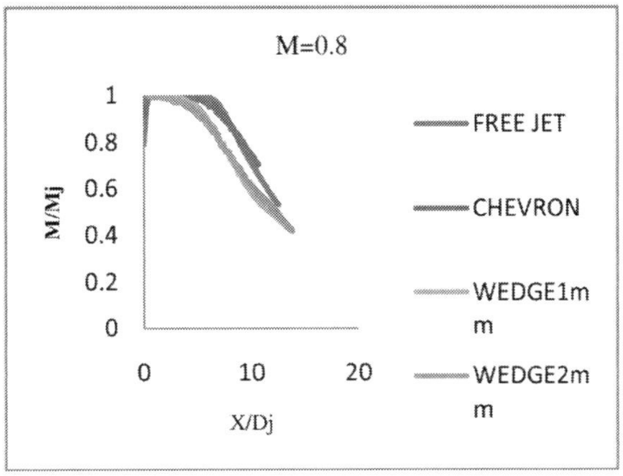

Figure 14. Comparison of Centerline Mach number decay

References

[1] Hao Xia, Paul G. Tucker, Simon Eastwood, "Large-eddy simulations of chevron jet flows with noise predictions" International Journal of Heat and Fluid Flow. International Journal of Heat and Fluid Flow, 2009.

[2] M.L. Shur, P.R. Spalart, M.Kh. Strelets, A.K. Travin, "Towards the prediction of noise from jet engines: International Journal of Heat and Fluid Flow", Volume 24, Number 4, August 2003.

[3] P.S.Tide, K.Srinivasan, "Effect of chevron count and penetration on the acoustic characteristics of chevron nozzles", 2009.

[4] Manigandan, S., P. Gunasekar, J. Devipriya, A.Anderson and S. Nithya. " Energy saving potential by changing window position and its size in isolated building." International Journal of Ambient Energy, 2017.

IOP Conf. Series: Materials Science and Engineering **197** (2017) 012082 doi:10.1088/1757-899X/197/1/012082

[5] P.S. Tide, V. Babu, "Numerical predictions of noise due to subsonic jets from nozzles with and without chevrons" Applied Acoustics, 2009.
[6] Manigandan, S., P. Gunasekar, J. Devipriya, and W. S. R. Saravanan. "Reduction of greenhouse gases by the effect of window position and its size in isolated building."Journal of chemical and pharmaceuticals sciences, 2016.
[7] Manigandan, S., P. Gunasekar, J. Devipriya, and S. Nithya. "Determination of heat flux on dual bell nozzle by Monte carlo method." Journal of chemical and pharmaceuticals sciences, 2016.

IOP Conf. Series: Materials Science and Engineering **197** (2017) 012083 doi:10.1088/1757-899X/197/1/012083

Aerodynamic analysis of seamless horizontal stabilizer

S.Nithya[1] S.Kanimozhi[2]

[1, 2]Department of Mechanical Engineering, Sathyabama University, India

Abstract. This project presents an investigative view into the concept of seamless aeroelastic wing and hingeless flexible trailing edge. Wings are designed to provide maximum lift and minimal drag and weight. But with conventional wings where rivets are used and the control surfaces are separately hinged, parasite drag comes into play. This project is about analysing a smooth seamless wing with hinge-less flexible trailing edge. This type of wing reduces the drag considerably and the hinge-less trailing edge leads to a minimal control demand and reduces the noise produced when the aircraft comes for landing. Seamless aeroelastic wing will function as an integrated one piece lifting and control surface. It has been designed to enhance a desirable wing camber for control by deflecting a hinge-less flexible trailing edge part instead of a traditional hinged control surface. This kind of flexible wing can be achieved either by a curved beam and disc actuation mechanism or by piezo-electric materials, whose shape change can be achieved by electricity. The intent of this project is to analyze the effects of introducing the concept of Seamless Wing to the horizontal stabilizer. While the removal of rivets and serrations that hinge the elevators to the stabilizer reduces the overall drag by a reasonable value, the overall concept of a control surface-less stabilizer where the maneuvers are done by deflecting the trailing edge offers better maneuverability.

1. Introduction

Surfaces on aircraft are embedded with millions of rivets. The model considered, Cessna 152 consists, in total of 2 million rivets. The horizontal stabilizer alone consists of about 200 rivets. The common types of rivets used are brazen head rivets and counter-sunk rivets. Some tests were conducted regarding the drag produced due to rivet heads way back in 1950's and with some positive results. Tests have been conducted in the NACA 8-foot high-speed wind tunnel to find the effect of rivet heads and spot welds on the wing drag. The tests were made of 5-foot chord airfoil. The velocity of the air was varied from 80 to 500 miles per hour and the lift coefficient from 0 to 0.30. The increment of the drag force on the 5-foot airfoil varied from 6%, due to countersunk rivets, to 27%, due to the 3/32-inch brazier rivet head, with the rivets in a arrangement. The drag increases caused by protruding rivet heads is directly related to the height of the heads. With the front row of rivets well forward, changes in span-wise pitch having insignificant effects on drag unless the pitch was more than 2.5% of chord. Data are presented for evaluating the reduction in drag attained by eliminating rivets from the forward part of the wing surface body; for example, it is shown that over 60 to 75% of the rivet drag is caused by the rivets on the forward 30 to 35% of the airfoil in a typical case.

2. Modeling.

Modeling is done using CATIA V5R20. A basic model of riveted horizontal stabilizer is made. The dimensions are taken from Cessna 152 aircraft. The dimensions of the rivets are as specified below.
The dimensions of the rivets are:
Diameter:5.486mm
Height:2.134mm
The distance between the rivets are calculated accordingly Edge distance= 2 X Rivet diameter
Rivet pitch (Spacing) = rivet diameter X 4

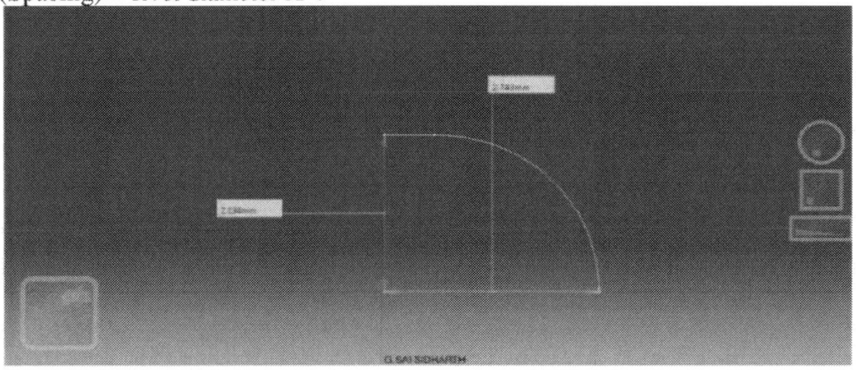

Figure 1. Rivet Dimensions.

2.1 Modelling of the Horizontal stabilizer

Horizontal stabilizer of cessna 152 is used for modelling, as it was available at the Sathyabama University Aero-Hangar and was available to be measured. A total of 24 models are meshed and analyzed to establish a set of results.

Root chord : 4feet
Tip chord : 2feet 7inches
Span length : 4feet 7inches
Elevator at root : 1foot 8inches
Elevator at tip :1foot 2inches

Figure 2. Regular surfaces with rivets.

Figure 3. Devoid of rivets and serrations.

Figure 4. Meshed model of Regular surface.

Figure 5. Seamless surface.

3. Analysis and Results

3.1 Method Of Analysis

3.1.1 Effect of span-wise location of rivets

To find the effect of span-wise location of rivets, four rows of rivets have been considered, one each at 0.275 feet, 1.2 feet, 2.4 feet and 3.9 feet from the root chord leading edge.

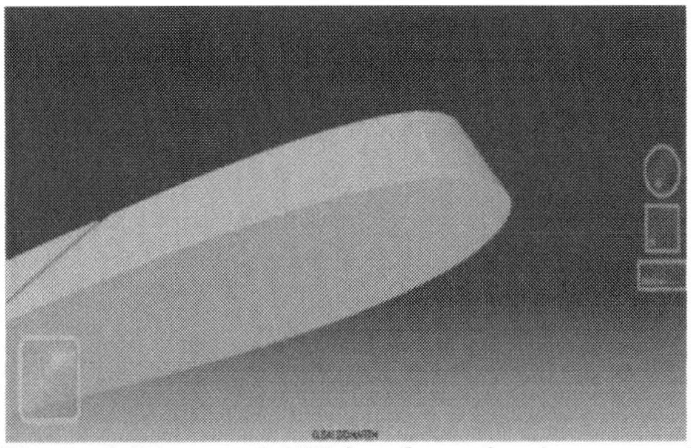

Figure 6. Rivets at 0.275 feet from chord.

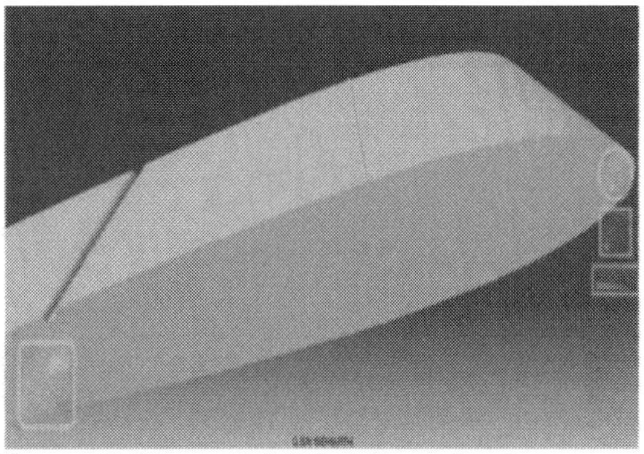

Figure 7. Rivets at 1.2 feet from chord.

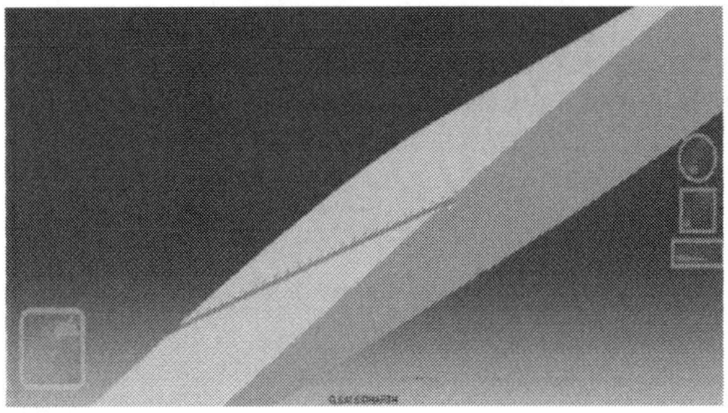

Figure 8. Rivets at 2.4 feet from chord.

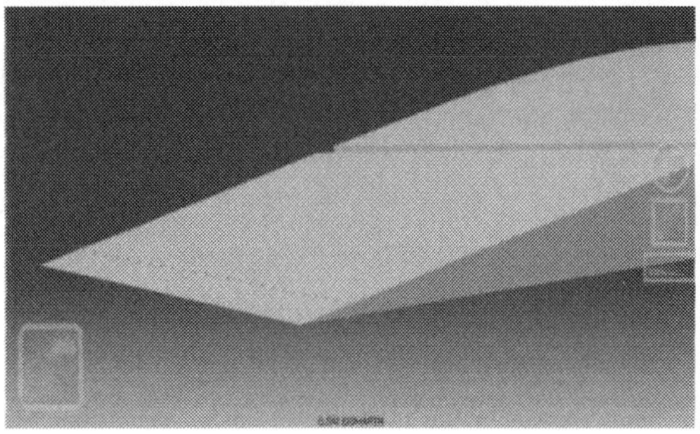

Figure 9. Rivets at 3.9 feet from chord.

The above three models had rivets at 0.275 feet, 1.2 feet and 2.4 feet respectively, the former two setup is located within 30% of the leading edge of the 4 foot chord airfoil, where the flow separation initially takes place. The fourth model is setup to have rivets at its trailing edge.

IOP Conf. Series: Materials Science and Engineering **197** (2017) 012083 doi:10.1088/1757-899X/197/1/012083

3.1.2 C_L vs α

To find the overall increase of Lift coefficient due to the use of seamless aerofoil over regular airfoil, 12 models are considered. -3', 3', 6', 9', 12', 15' for each, regular and seamless surface.A graph depicting the angles of attack and C_l values of both the types are plotted.

3.1.3 δ_e (angle of deflection of elevator) vs C_L

CATIA models of both, regular surface and seamless surface with elevator deflection depicted at angles of 0' , 5' , 10' , 15' are made, meshed and analyzed. The resulting graph is used to arrive at the elevator efficiency, which is the parameter used to differentiate the stabilizers considered.

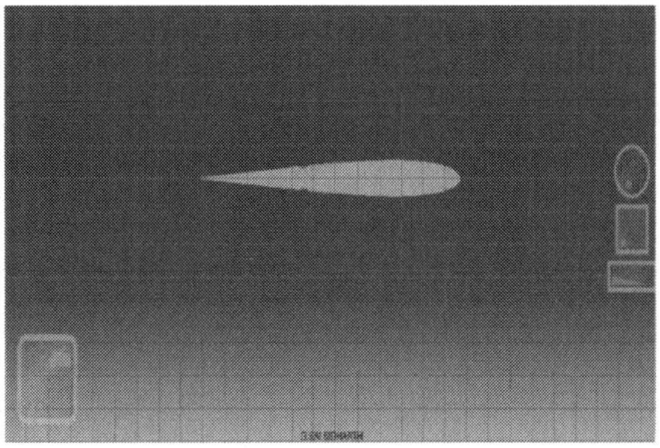

Figure 10. Zero degree deflection.

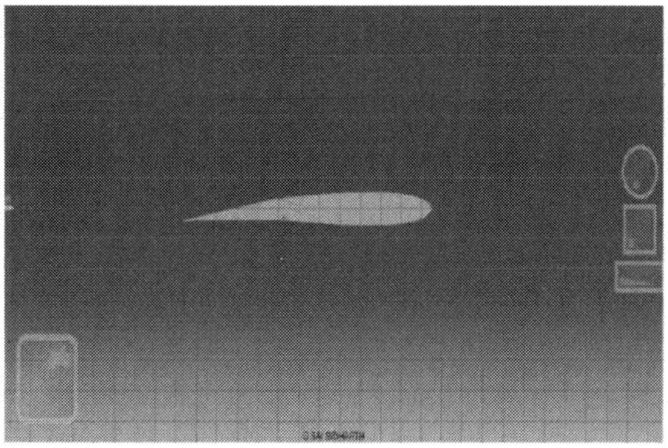

Figure 11. Five degree deflection.

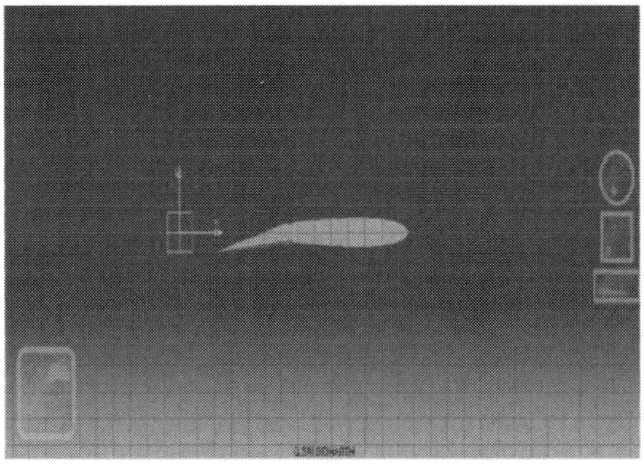

Figure 12. Ten degree deflection.

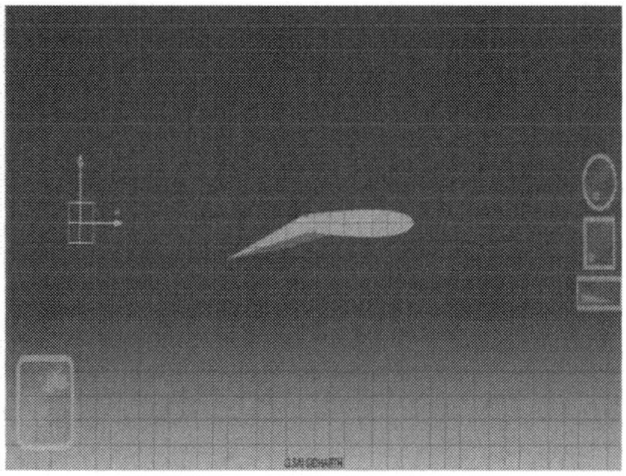

Figure 13. Fifteen degree deflection.

3.2 Analysis

Meshing is done in ICEM-CFD and the analysis is done in CFX, both are a part of the ANSYS workbench. The analyses are all carried out in the same boundary conditions.

The inlet and outlet of the domain is maintained at subsonic as the aircraft considered is a subsonic aircraft and supersonic analysis does not offer any further solutions to the horizontal stabilizer problem at hand.

The density of the atmosphere considered is 1.27 kg/m^3 , which indicates the altitude to be sea level and the temperature is at 25'C.

- Subsonic inlet and outlet

- Density: 1.27 kg/m^3
- Relative Velocity: 200m/s
- Wall: No slip wall
- Air at 25'C
- Altitude: Sea level

This test was generic in nature and carried out to check the drag caused due to the span-wise location of rivets on a lifting surface.

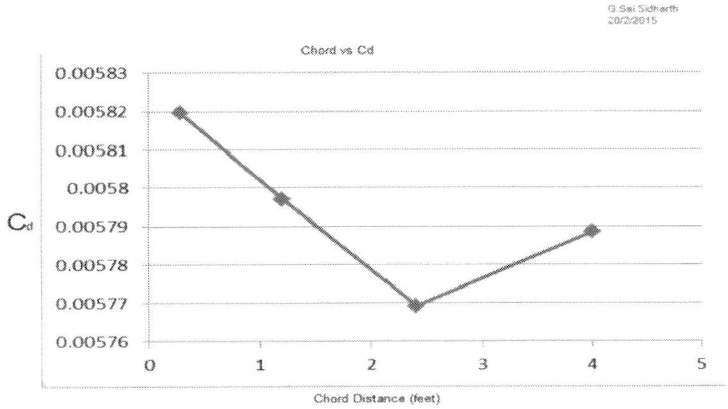

Figure 14. Chord vs C_d.

There is a high drag coefficient at the leading edge, which seems to reduce at 1.2 feet and 2.4 feet of chord and then there is a surge in the drag at the trailing edge.

Elevator Effectiveness

Elevator effectiveness is a parameter largely used to design horizontal stabilizers to suit an aircraft design.

The primary design of elevator in an aircraft is for the pitch rate. The non-dimensional derivative which represents the longitudinal control power derivative is the rate of change of pitching moment coefficient of aircraft with respect to elevator deflection (Cmδe).

3.3 Contribution of elevator to tail lift

"dCl/dδe" } Contribution of elevator to tail lift.

It is determined from the slopes of the graph Cl vs δeof both, regular surface and seamless surfaces.

Slopes of CL vs δe = "dCl/dδe"

The slopes from both the graphs are taken separately to figure out the contribution of elevator to tail lift.

"dCl/dδe" (For regular surface)=0.01351

"dCl/dδe" (For seamless surface)=0.01395

3.4Longitudinal Control Power

Longitudinal control power is defined as the coefficient of pitching moment produced by the horizontal stabilizer for a given deflection of the elevator.

Cmδe= "(-VH η)dCl/dδe"

Where VH is defined as the horizontal stabilizer volume ratio. It changes with a change in pitching moment. η is the pressure ratio and the value is a constant for given aircrafts, for Cessna 152, the value is 0.85.
VH= 0.50
Cmδe = -0.00574175

3.5 Seamless Surface

Cmδe="(-VH η)dCl/dδe"
VH= 0.50
Cmδe = -0.00592875

Table 1. C_L vs δ_e for seamless surface.

Angle	Stress N/m^2
0'	0.00029557
5'	0.0667603
10'	0.136463
15'	0.209329

4. Conclusion

Seamless surface offers 3.25% more efficiency in control power and elevator tail lift. This can be observed due to the entire horizontal stabilizer acting as a one piece lifting surface. There is an increase in the span area of the stabilizer due to the absence of the gap where the elevator is attached to the horizontal stabilizer.The reduction in drag varies from 3% to 8% depending upon the angle of attack of the horizontal stabilizer and the location of rivets. The rivets placed at the leading and trailing edges cause the most of the drag whereas the rivets positioned at the center of the airfoil produce minimal or even negligible drag.Introduction of seamless surface to elevators may initially prove to be a bit costly on the economic aspects. But with usage, the economic returns start turning profitable to the airlines. A better control power, longitudinally can lead to shorter runways with advanced research. Due to the removal of the serration, there is no flow separation and no build up in pressure at the gap. This leads to a smoother flow over the surface.

References

[1] Abbott & Ira H 1932 The Drag of Two Streamlined Bodies as Affected by Protuberances and Appendages T.R.No.451, N.A.C.A.
[2] Allegri.G, Guo.S&Trappani.M, 2007 Design of Aeroelastive Fin for a Flight Demonstration.
[3] Bergmann, A. and Huebner, A., Loeser,T 2008 Experimental and Numerical Research on The Aerodynamics of Unsteady Moving Aircraft Progress in Aerospace Sciences.
[4] Dang, T.Q., and Bushnell, P.R 2009 Aerodynamics of Cross-Flow Fans and Their Application to Aircraft Propulsion and Flow Control Progress in Aerospace Sciences.
[5] Eastman N.Jacobs& Albert Sherman 1932 Wing Characteristics as Affected by Protuberances of Short Span.
[6] Filippone,A. 2008 Comprehensive Analysis of Transport Aircraft Flight Performance, Progress in Aerospace Sciences.
[7] G.Suttera, G.Lista 2013 Very high speed cutting of Ti–6Al–4V titanium alloy – change in morphology and mechanism of chip formation, International Journal of Machine Tools and Manufacture.

Analysis and Optimization of Central Processing Unit Process Parameters

Kaja Bantha Navas R [1] Budi Venkata Chaitana Vignan[2] Margani Durganadh [3], Chunduri Rama Krishna [4]
School of Mechanical Engineering, Sathyabama University, Chennai, India

E-mail: [1]sumai244@gmail.com, [2] vignanchaitanya@gmail.com, [3]nadhu511@gmail.com

Abstract. The rapid growth of computer has made processing more data capable , which increase the heat dissipation. Hence the system unit CPU must be cooled against operating temperature. This paper presents a novel approach for the optimization of operating parameters on Central Processing Unit with single response based on response graph method. These methods have a series of steps from of proposed approach which are capable of decreasing uncertainty caused by engineering judgment in the Taguchi method. Orthogonal Array value was taken from ANSYS report. The method shows a good convergence with the experimental and the optimum process parameters.

1. Introduction

Computer cooling is the process of removing hot air from the system unit components. In earlier days, it was possible to cool most of the components of personal computer using natural convection, but later modern components require more effective active cooling. To cool these component fans are used to move hot air away from the components and draw cool air over them. In addition to it, the normal operation of system unit can be disturbed by other causes such as dust, poor air flow and poor heat transfer. Dust acts as a thermal insulator and impeding air flow there by reducing heat sink and fan performance. The poor air flow is due to friction against system unit components and improper orientation of fan can reduce the amount of air flowing through a case. The air is the most commonly used cooling medium. The cooling of system unit can be improved by supplying air directly to the components thereby improving air exchange in the system unit. The forced convection cooling is commonly used for system unit. System unit is also known as the computer data or central processing unit. The system unit is the most important part of a desk top personal computer. It may be known as the brain behind every action of a PC since in executes all the function performed by the personal computer. The system unit consists of several devices. The following components are listed as CPU/Heat Sink, Accelerated Graphical Port, SMPS, Hard Disk Drive, RAM,CD/ DVD, Mother Board and Processor.

The rapid development in desktop personal computer is has made processing more data possible which increase loads , which in turn increases the heat dissipation and temperature. Unless the heat is removed the temperature of the device will keep on increasing which leads to decrease in efficiency and caring house to the computers of system unit. The system unit must be cooled to satisfy the operating temperature unit hence keeping the temperature of heat source. The heat transfer process with heat generating components such as CPU - Heat Sink, Hard Disk Drive, RAM, AGP, CD/DVD and adaptive cards generates more heat. In computer the processor, AGP, RAM are increasing the speed. Also these components need to be kept within a specified temperature range to prevent overheating and instability. Many components are required to be kept within a specified temperature range to prevent overheating and instability. The major heat generating components in CPU and its capacity (in Watts) are CPU-70, AGP-25, HDD-20, CHIPSET-10, SMPS-75 and FDD-10.

2. Literature Survey

Several researchers have worked on heat transfer in system via CFD among other , Yu and Webb (2001) simulated a complete pc which used an 80 W CPU with the addition of other components (AGP, PCI) a total of 313 W of heat was dissipated in to the system. This domain was solved with software ANSYS icepack. To decrease complexity, CPU heat sink was modelled as a volume resistance having the same time impedance as the detailed geometry. The cooling of PCI cards and baffle were improved by the analysis. Argento, Joshi and Oysterman (2001) has studied and verified system level electronic packaging in thermal design. In a complex computer CFD is a good approach to explore various design alternatives quickly with accuracy. There has been much success in the thermal design of complex electronics system using CFD Lech et al used. Flotherm code to simulate detailed flow and temperature fields within a computer chassis with two fans. Emrezturk and iiker Tari (2007) had drawn a CFD road map for forced cooling heat transfer analysis in a computer chassis and the main sources of error in CFD analysis arise in numerical models including turbulence model. Icepack is used for pre processing and fluent is used for simulation and post processing. Biswas et.al(2006) studied the air flow in a compact electronic enclose and investigated the pressure loss due to the inlet and outlet and use of fan curves obtained from the manufacturing was considered and the fan curve may need to be modified if the fan is not closely ducted. Eveloy vet.al (2003) used Flotherm software has a perspective on the current CFD as a design tool to predict component temperature on PCB. Their computation predicted on the component operating temperature in an accuracy range of 30c to 220c with up to 35% error. They suggested that components junction temperature would need to be experimentally measured.

3. Parametric Study for System Units

The main objective of the parametric study is to investigate the effect of various parameters in the CPU case. There are power dissipation of the processor, fan speed, Ambient air temperature and air intake area.The range of parameter which the Anandakrishnan .M and Balaji.C, (2009) considered is given in table 1. The power dissipation in processor is between 60 w to 130 w and the ambient temperature of a system unit in desktop personal computer is between 240 ^{0} c and 390 ^{0} c .

3.1 Power Dissipation of CPU

The processor may not dissipate its maximum amount of heat all the time. It will vary depending upon the computational work it performs. The profile is obtained by varying the power dissipation value of the processor over the range of 60 to 130 W. while other parameters are kept constant. It can be seen that maximum CPU case temperature has a linear dependence on the power dissipation rate and the order temperature are less sensitive to the power dissipation of the processor. The thermal design power and maximum case temperature of a processor are the maximum values of the thermal profile. By design the thermal solution must meet the thermal profile for all system operating conditions and processor power levels.

3.2 Fan Speed

Since the processor is cooled by forced convection. The fan speed will exert a significant influence on the heat transfer characteristics. An increase in the fan speed will increase the velocity of the air flow over the heat sink and hence the heat transfer coefficient. This will improve the heat transfer from the processor and reduce its temperature. Simulation was done for different fan speed that was varied over the range of 2850 rpm to 5000 rpm. The corresponding flow velocities were 2.33 m/s and 4.29 m/s respectively. For each fan speed the corresponding fan performance curve was given as the heat sink fan boundary conditions. The nature of the plot shows that maximum CPU case temperature has a non linear dependence on fan speed. The temperature of other components also shows a non linear variation with fan speed.

Table 1. Power dissipation in Processor

Parameters	Range
Power Dissipation of the processor (watts)	$60 \leq Q \leq 130$
Fan speed (rpm)	$2850 \leq N \leq 5000$
Ambient temperature (^{0} c)	$24 \leq T \leq 39$
In take area (m^2)	$0.0036 \leq A \leq 0.0084$

3.3 Ambient Air Temperature

The ambient air temperature at which the system has to work may vary depending upon the weather conditions or geographical variations. The effect of ambient temperature on CPU case temperature was studied by varying the air inlet temperature from 2400^0 c to 3900^0 c. The air properties were changed accordingly. Maximum CPU case temperature and the temperature of the other components with the intake air temperature .It can be seen that the variation is more or less linear. Usually air circulates through the system cabin by induced draught produced by an exhaust fan. Air vents on cabin walls are provided at appropriate locations to make air in to the cabin cool, proper vending is a key element in any good thermal design providing an insufficient amount of venting does not allow enough air into the system for adequate cooling. Providing too much venting can decrease the air velocity across the system components resulting in less heat transfer through forced convection. In order to study the effect of air intake area on the CPU case temperature, the former was varied over the range 0.0036 m^2 to 0.0084 m^2 the maximum temperature decreased linearly from 77 ^{0} c to 70.50 ^{0} c .

4. Simulation of System Units

By experimental method it is very difficult to predict the performance of air velocity inside the system unit. So by the Simulation it is easier to carry out the air velocity and analysis of heat transfer in system unit. The boundary conditions is used here are Inlet velocity - 1 to 6 m/s, Pressure-1atm, Temperature -300 K to 312 K. The Analysis is made for various parameters. Then the calculation activities are started for the given number of iterations. The load consideration taken is 70W/2.8 GHz processor load.

The desktops components standard sizes are selected as follows in meters. CD/DVD-0.051×0.175× 0.17, FD- 0.05 × 0.14 × 0.010, HDD- 0.024 × 0.155 × 0.098, CPU/HS-0.048 × 0.048 × 0.089, AGP- 0.082 × 0.145 × 0.005, RAM0.03 × 0.16 × 0.005, SMPS-0.085 × 0.140 × 0.150, CHASSIS-0.45 × 0.47× 0.21, MOTHER BOARD-0.26 × 0.23 × 0.005.Before starting simulation process we must provide an initial solution to be fixed for attaining the final solution. By selecting the entire flow field and the solution is computed from all zones. The meshed modelled is imported to ANSYS FLUENT. In model, the energy equation is enabled. The standard K- epsilon model is used in this analysis. The material used for this analysis is fluid- air.

The boundary conditions are
Inlet velocity – 1 to 6 m/s
Pressure – 1 atm
Temperature – 300 K To 312 K

The meshing of the Desktop computer components such as CD/DVD, HDD, AGP, CPU-Heat Sink & SMPS are done. The meshing of CD/DVD is shown in figure 1.

IOP Conf. Series: Materials Science and Engineering **197** (2017) 012084 doi:10.1088/1757-899X/197/1/012084

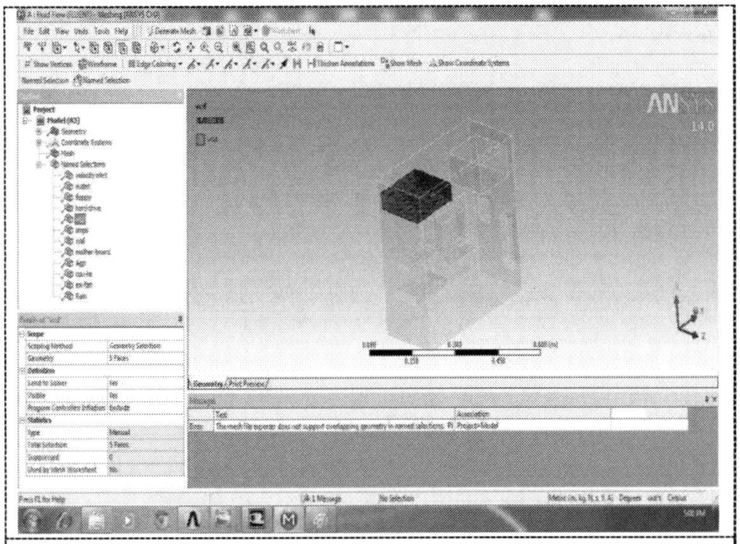

Figure 1. Meshing of CD/DVD

The Analysis is made for various parameters by varying the ambient temperature & air velocity at constant atmospheric pressure and heat dissipation is computed from all zones. Then the calculation activities and the calculation are started for the given number of iterations. The load consideration taken is 70W/2.8 GHz processor load. Before starting simulation process we must provide an initial solution to be fixed for attaining the final solution. There are two types of initialization, they are as follows: Initialize the entire flow field selected and the solution is computed from all zones. Then the calculation activities and the calculation are started for the given number of iterations. The various heat dissipation of components are studied and heat dissipation taken from earlier studies by Emre ozturk et.al. (2007) are given in the table 1 and 2.The system parameters path diagram is shown in figure 2.

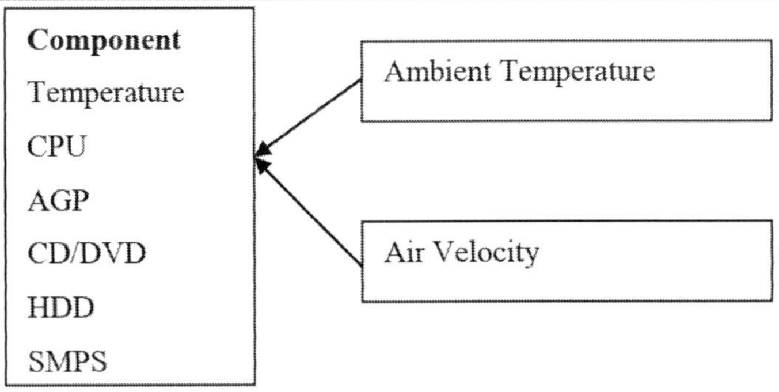

Figure 2. Path Diagram for System Parameter

Table 2. Desktop Component's Heat dissipation and Temperature

Components	Heat Dissipation (W)	Temperature (K)
CPU	70	406
AGP	25	320
HDD	20	311
SMPS	75	415
CD/DVD	20	311

5. Temperature Distribution in Various Components using ANSYS

The Temperature of various components plays a vital role in the performance of the system unit. The temperature variation is due to the ambient temperature, Heat dissipation of the components, velocity of air at inlet, etc. The influence of all these factors on temperature was studied. The 3D modelling and analysis of system unit has been carried out with the help of solid works and ANSYSIS. The analysis of operating conditions such as varying ambient temperature, air inlet velocity and convective heat transfer coefficient with the post processing facility in the software, the contour plots for temperature and vector plots for velocity has been generated for various operating conditions and the significance of various parameters considered in this study. In order to predict the flow of air inside the system unit, the velocity vector was analyzed. The air enters from the inlet and the flow is shifted downwards near the processor and then cooled by the convective walls and finally exists through the outlet. The heat transfer process with heat generating components such as CPU, Hard Disk Drive, AGP, HDD and Adaptive Cards have been generating more and more which impact on Performance of the system unit. The temperature variation is due to the ambient temperature, Heat dissipation of the components and velocity of air at inlet. To keep the these components CPU, Hard Disk Drive, AGP, HDD need to be within a specified temperature range, the study made with various air velocity, Ambient temperature & constant atmospheric pressure to prevent overheating and instability. The temperature distribution of AGP, CD/DVD, CPU-Heat Sink, Hard Disc Drive, SMPS are from ANSYS Result for SMPS and ACP is shown in figure 3a and 3b.

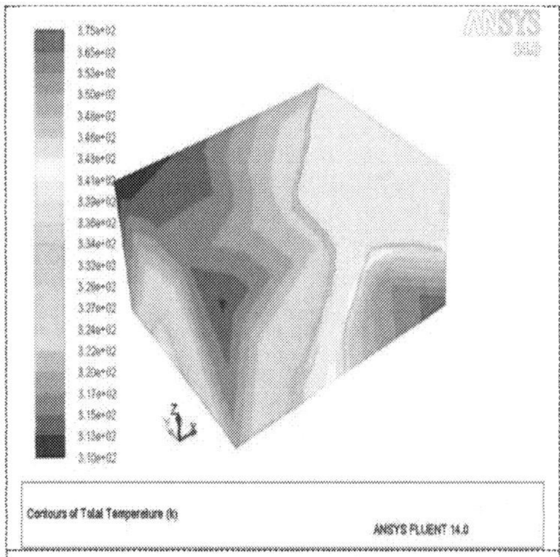

Figure 3a. ANSYS Result for SMPS

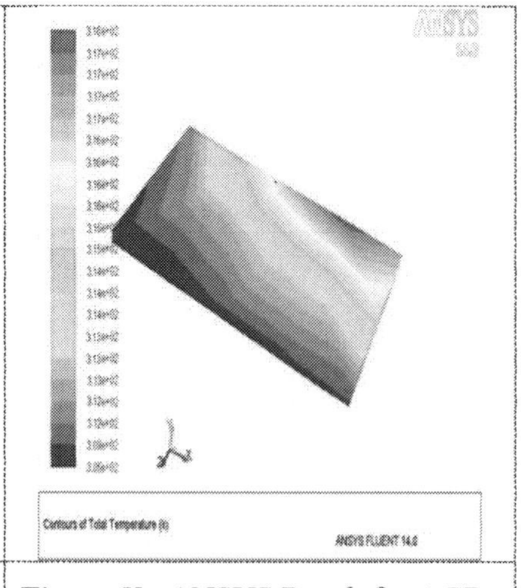

Figure 3b. ANSYS Result for AGP

IOP Conf. Series: Materials Science and Engineering **197** (2017) 012084 doi:10.1088/1757-899X/197/1/012084

6. Impact of Air Velocity in System Unit

The Variation in air inlet velocity influences the temperature distribution of system so temperature and atmospheric transfer are kept constant and the variation of component's temperature with respect to the inlet velocity has been analysed. It is observed that the temperature of various components decrease significantly. When the velocity is increased from 1 m/s to 6 m/s. The temperature range for various components such as CPU, AGP, CD/DVD, SMPS and HDD are 300 K, 306 K and 312 K.

7. Process Parameters Optimization

An experiment was conducted using array concept and data were collected from ANSYS report. The objective is to minimize the response (Components temperature), we have taken smaller –the better type of quality to compute S/N ratios. Here factors considered are ambient temperature & Air velocity; the response is component's (AGP, CD/DVD, CPU, HDD and SMPS) temperature. The components temperature achieved for the ambient temperature of 300 K, 306 K and 312K as well as the velocity of 1 m/sec, 3 m/sec and 6 m/sec with respect to components as shown in table 3. The table 4 represents operating parameters (Factors) and its level.

Table 3. Experimental Data From ANSYS Report

AGP (Y_1)	Velocity (m/sec)		
Temperature (K)	1	3	6
300	315.9	314.5	314
306	316	315	314
312	316.5	315.5	314.5

CD/DVD (Y_2)	Velocity (m/sec)		
Temperature (K)	1	3	6
300	308.5	307.5	307
306	307.5	306.5	306
312	308.9	308	307

CPU/Heat Sink (Y3)	Velocity (m/sec)		
Temperature (K)	1	3	6
300	362	358	350
306	362	358	350
312	370	365	358

IOP Conf. Series: Materials Science and Engineering **197** (2017) 012084 doi:10.1088/1757-899X/197/1/012084

HDD (Y_4)	Velocity (m/sec)		
Temperature (K)	1	3	6
300	310	309.5	309.3
306	310.2	309.7	309.3
312	310	310.3	309.7

SMPS (Y_5)	Velocity (m/sec)		
Temperature (K)	1	3	6
300	379	372	368
306	380	382	375
312	380	376	372

Table 4. Factors and Levels for Operating Parameters

Operating Parameters	Symbol	Unit	Low	Medium	High
Temperature	A	K	300	306	312
Velocity	B	m/sec	1	3	6

Computed S/N ratio with respect to listed operating parameters using the formula S/N ratio = -10 log {1/n ∑ Yi2 . Table 5 represents Signal to Noise Ratio for operating parameters. The total response values are tabulated in table 6. Average effects and Rank for operating parameters are listed in table 7

Table 5. S/N ratio for Operating Parameters

Trial	A	B	Y_1	Y_2	Y_3	Y_4	Y_5	S/N ratio
1	1	1	315.9	308.5	362	310	379	-50.53
2	1	2	314.5	307.5	358	309.5	392	-50.57
3	1	3	314	307	350	309.3	368	-50.37
4	2	1	316	307.5	362	310.2	388	-50.57
5	2	2	315	306.5	358	309.7	382	-50.52
6	2	3	314	306	350	309.3	375	-50.42
7	3	1	316.5	308.9	370	310	380	-50.57
8	3	2	315.5	308	365	310.3	376	-50.56
9	3	3	314.5	307	358	309.7	372	-50.59

Table 6. Total Response for Operating Parameters

Factor	Level 1	Level 2	Level 3
A	-151.47	-151.53	-151.58
B	-151.71	-151.62	-151.25

Table 7. Average Effect and Rank for Operating Parameters

Factor	A	B
Level 1	-50.49	-50.57
Level 2	-50.51	-50.54
Level 3	-50.53	-50.42
Difference	0.04	0.15
Rank	2	1

From the figure 4, Factor B has given main impact to reduction in component temperature.

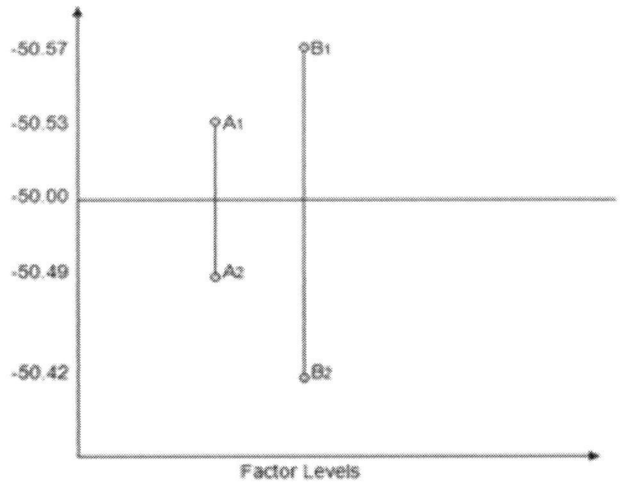

Figure 4. Factor Ranking with Response Graph

8. Predicting Optimum Condition

It can be seen that factor B is significant since it is Rank 1. Hence the predicted optimum response in terms of S/N ratio is given by

$\mu)_{optimum}$ = $\mu)_{mean}$ + { Average Effect Value for Factor$)_B$ - $\mu)_{mean}$ } + { Average Effect Value for Factor$)_A$ - $\mu)_{mean}$ }.............................. (1)

= - 50.51 + {-50.57 – (-50.51)} + {-50.53 (-50.51)}

= - 50.59

It is observed that from the table 5, that the S/N ratio is - 50.59 and it is correspond to experimental trail no. 9 which predicted the yield of - 50.59. From this we can 50.59 and it is correspond to experimental trail no. 9 which concludes the optimum condition obtained is at 6 m/sec.

9. Conclusion

Based on the analysis conducted on the system unit of desktop personal computer for various operating and boundary conditions,a good flow is developed in CFD software based on the detailed study of flow models, heat transfer models and solution controls. The CPU's component temperature is observed with respect to the change in operating parameters. In this paper, single response problem was taken with two independent variables such as ambient temperature and air flow and optimum component temperature values were determined by response graph method

10. References

[1] Anandakrishnan .M and Balaji.C, "CFD Simulations of thermal and Flow Flieldsinsidea Desktop Personal Computer Cabin with Multi core Processors" *Engineering Application of Computational Fluid Mechanics* Vol.3, No.2, pp.277-288, 2009.

[2] Biswas Rebecca, Raghu .B. Agarwal, Avijit Goswami, Vivek mansingh, "Evacluation of Air flow prediction methods in Compact Electronic Enclosures" ,*Fifteenth IEEE semi Thermo Symposium*, pp. 48-53,1999.

[3] Argento, Joshi and Oysterman, "Forced Convection air cooling of a commercial chassis; an experimental and computational case study", *IEEE Transaction on Components Packaging and Manufacturing Technology*, 19,2 248-257.

[4] ChauCFD K.W. "Simulations of Thermal and Flow Fields inside a Desktop Personal Computer Cabin with Multi-Core Processors more", *Engineering Applications of Computational Fluid Mechanics*, Vol.3, No.2, pp.277-288, 2007.

[5] Chiang J.S " Numerical simulation of Heat Transfer in a Desktop Computer with Heat Generating Components" *International Communication in Heat and Mass Transfer*, Vol. 32, pp.184-194, 2005.

[6] Emre Ozturk and iiker Tari, " CFD Modeling of Forced Cooling of Computer Chassis, *Engineering Applications of Computational Fluid Mechanics,* Vol. 1, No. 4, pp. 304-313, 2007.

[7] Emre Ozturk and iiker Tari, " Forced Air Cooling of CPU with Heat Sinks- A numerical Study"*IEEE Transactions on components and packaging technologies*, Vol.31, No.3, pp. 650-660, 2008.

[8] Yuc.W ,Webb R. "Thermal Design of a Desktop Computer system using CFD Analysis" *Seventeenth IEEE semi thermal symposium*, pp. 18-28, 2001.

[9] Eveloy Valerie, Peter Rodgers, and Hashmi M.S. "Numerical Prediction of Electronic Components Heat" *Anna industry perspective proceeding of 19th SEMI THERM,* sanjose USA, pp.14-26, 2003

Experimental and Transient Thermal Analysis of Heat Sink Fin for CPU processor for better performance

S.Ravikumar[1,*], Parisaboina Subash Chandra[2], Remella Harish[2], Tallapaneni Sivaji[2]

[1,*] Assistant Professor, [2] UG students of Department of Mechanical Engineering, Sathyabama University, Chennai, Tamilnadu, India. Pin: 600 119.
[1,*]mahailakumar@gmail.com, [2]subash.chandra000@gmail.com, [2]hremella5@gmail.com
[2]tallapaneni.sivaji@gmail.com

Abstract: The advancement of the digital computer and its utilization day by day is rapidly increasing. But the reliability of electronic components is critically affected by the temperature at which the junction operates. The designers are forced to shorten the overall system dimensions, in extracting the heat and controlling the temperature which focus the studies of electronic cooling. In this project Thermal analysis is carried out with a commercial package provided by ANSYS. The geometric variables and design of heat sink for improving the thermal performance is experimented. This project utilizes thermal analysis to identify a cooling solution for a desktop computer, which uses a 5 W CPU. The design is able to cool the chassis with heat sink joined to the CPU is adequate to cool the whole system. This work considers the circular cylindrical pin fins and rectangular plate heat sink fins design with aluminium base plate and the control of CPU heat sink processes.

Key words: Heat sink fin, CPU Processor, ANSYS, Transient Thermal Analysis

1. Introduction

In the present era, the trend to design electronic products becomes thinner, lighter, shorter, & smaller. Due to the actuality that shrinking in the dimension of these electronic components will consequence in a drastic increase in the heat generation rate when evaluating with previous products. For this reason, an efficient cooling system to remove the high heat generation and consequently maintains the reliability and stability of the products, have gained much attention. The heat sink component is the most common heat exchanger for CPUs and has been extensively exercised in order to provide cooling utility for electronic components. The conventional heat sink module utilized the natural as well as forced convection cooling technique; dissipate heat from CPUs to the ambient air. The combination of the heat sink and fan design usually involved in this forced convection cooling technique. Cheng-Hung et al [1] developed a three-dimensional heat sink design to estimate the optimum design variables. Temperature distributions are dignified by using thermal camera for the optimal heat sink modules and results are compared with the numerical solutions to validate the design. C.J. Kobus et al [2] investigate the effect of thermal radiation on the thermal performance of heat sink having pin fin array by theoretical and experimental approach. Dong-Kwon Kim et al [3] compared the thermal performances of two types of heat sinks i.e. plate-fin and pin-fin and used a volume averaging approach based model for envisaging the pressure drop and the thermal resistance. Emrana et al [4] performed a three-dimensional numerical simulation is in order to investigate the flow dynamics and heat transfer characteristics in a micro channel heat sink. Goshayeshi et al [5] conduct numerical studies on vertical fins, attached with the surface. Natural convective heat transfer find out from heated plane, which is kept into air with horizontal and vertical surface. Li et al [6] investigated the performance of plate fin heat sinks with cross flow. The effect of different parameters like the fin width, fin height, Re. number of cooling air on the thermal resistance and the pressure drop of heat sinks were studied. Mehran et al [7] examined numerically and experimentally, Steady-state external natural convection heat transfer from vertically-mounted rectangular interrupted fins. FLUENT software is employed to develop a 2-D numerical model of fin interruption effects. An experimental Numerical parametric study was performed to investigate the effects of fin spacing, and fin disruption.

IOP Conf. Series: Materials Science and Engineering **197** (2017) 012085 doi:10.1088/1757-899X/197/1/012085

Mahmoud et al [8] examined numerically and experimentally, Steady-state external natural convection heat transfer from vertically-mounted rectangular interrupted fins, fin height ranging from 0.25-1.0 mm and fin spacing from 0.5 to 1.0 mm was taken. Qarnia et al [9] using numerical approach investigates the heat transfer by natural convection during the melting of a phase change material. Sable et al [10] investigated the natural convection of a vertical heated plate with a multiple v- type fins having ambient air surrounding. In this paper, the CPU cooling performances of a computer chassis with rectangular and circular pin fin heat sinks were investigated using transient thermal analysis and the results were compared. An alternative model of heat fins has been designed to increase heat dissipation. These fins utilize thermal analysis to identify a cooling solution for a desktop computer, which uses in a 5W CPU. In ANSYS both the materials existing and proposed materials is analysed and the results of steady state and transient thermal analysis are taken for comparison.

2. Experimental set up

Experiments were conducted to investigate forced convective cooling performance of an air cooled parallel plate fin heat sink circular ,with coated and without coated rectangular pin fins between the plate fins. The original parallel plate heat sink was fabricated consist parallel plates of length 53 mm with cross-sectional area of 1.4 mm in width by 20 mm height for each plate. The second heat sink has the same geometry of original one but with some circular pins between the plate fins. Thermal and hydrodynamics performances of the heat sinks have been assessed from the results obtained for the pressure drop, thermal resistance and overall performance with the free stream air velocity ranging from 4.7 to 12.5 m/s. Fig 1(a),(b) and (c) represents circular fin , rectangular fin with coated and without coated.

(a) (b) (c)

Fig 1 Photo image of (a) Circular fin (b) Rectangular fin with coated
(c) Rectangular fin without coated

2.1 Modeling and Analysing of heat sink fins

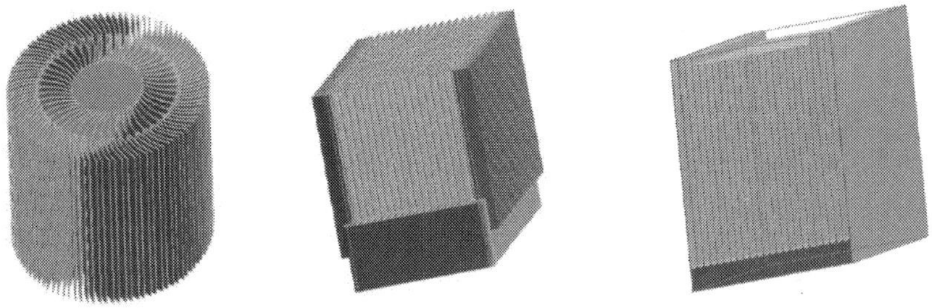

(a) (b) (c)

Fig 2 Model image of (a) Circular fin (b) Rectangular fin with coated
(c) Rectangular fin without coated

The flow of work used in each simulation. The steps are the following.

IOP Conf. Series: Materials Science and Engineering **197** (2017) 012085 doi:10.1088/1757-899X/197/1/012085

- Creation (reception) of 3D model
- Preparation of the model before taking it into FEM software such as elimination of bad geometry, simplification of unnecessary parts and improvement of contact regions
- Basic inputs in ANSYS workbench15.0. This information will allow the model to work properly during simulation and will define the behavior of the parts during the analysis
- FEM inputs. These parameters will control the computational time and the accuracy of results in the simulation. A special analysis of factors is performed in order to define them.
- Simulation is carried out by the software
- Results obtained need to be evaluated and judged as good or bad before presenting them in the report.

Fig 2 (a),(b) and (c) represents the model of circular fin , rectangular fin with coated and without coated.

2.2 Meshing of heat sink fins

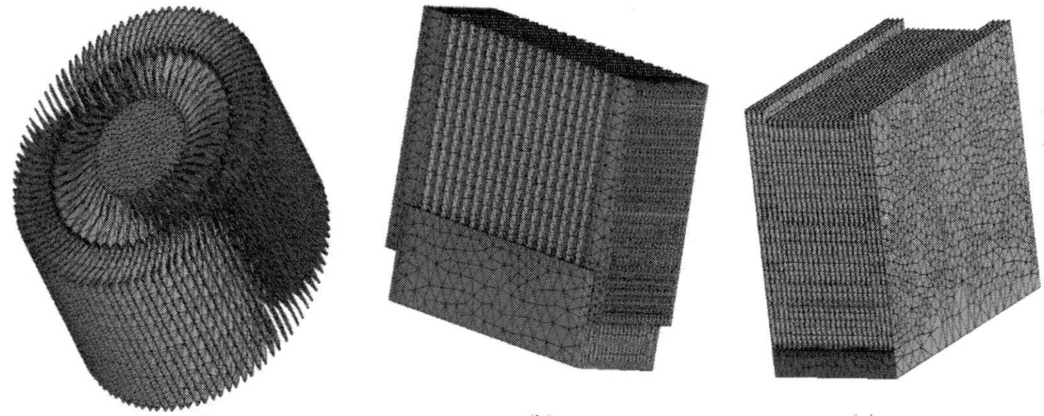

(a) (b) (c)

Fig 3 Meshed image of (a) Circular fin (b) Rectangular fin with coated (c) Rectangular fin without coated

Table 1 Mesh details for heat sink fins

S.No	Heat sink fin	Mesh size	Element type	No of nodes	No of elements
1	Circular fin	5mm	tetrahedron	73712	233998
2	Rectangular fin with coated	2mm	tetrahedron	41730	92289
3	Rectangular fin without coated	3mm	tetrahedron	53904	163653

Fig 3 (a), (b) and (c) represents the meshed image of circular fin, rectangular fin with coated and without coated.

2.3 Boundary conditions

Selection of the proper boundary conditions is important in the analysis of structures. For a static analysis, it is common to use a simpler assumption of supports without considering the soil system stiffness. However for dynamic analysis, representing the soil stiffness is most important. The boundary conditions used are corresponding to a double-pinned axle and the model is prevented from translation and rotation in all directions.

2.4 Transient Thermal Analysis

Transient thermal analysis shows temperature and other thermal parameters that vary over time; which are input in evaluating the structural analyses which is a steady-state thermal analysis. The induced loads in a transient analysis are functions of time that can separate the load versus time curve into load steps.

3. Results and Discussions

3.1 Analysis of Existing heat sink fins

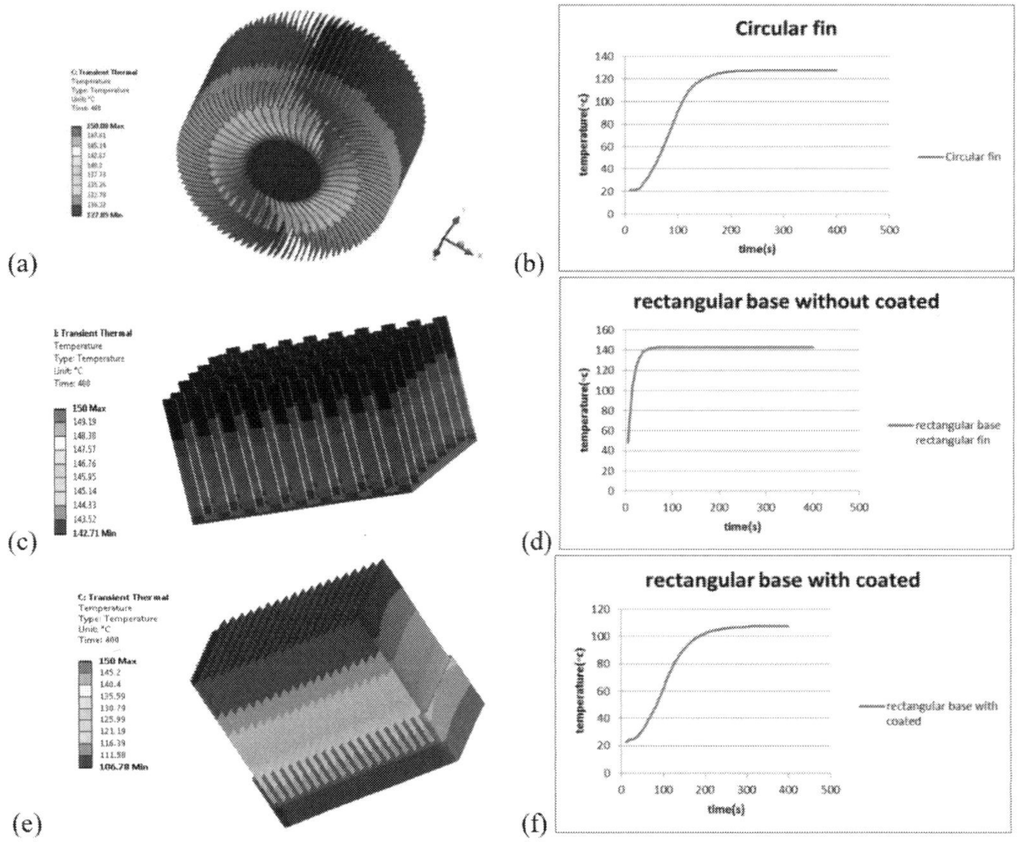

Fig 4 Meshed image and Transient thermal analysis of (a) Circular fin (b) Rectangular fin with coated (c) Rectangular fin without coated

3.2 Proposed models and its analysis

IOP Conf. Series: Materials Science and Engineering **197** (2017) 012085 doi:10.1088/1757-899X/197/1/012085

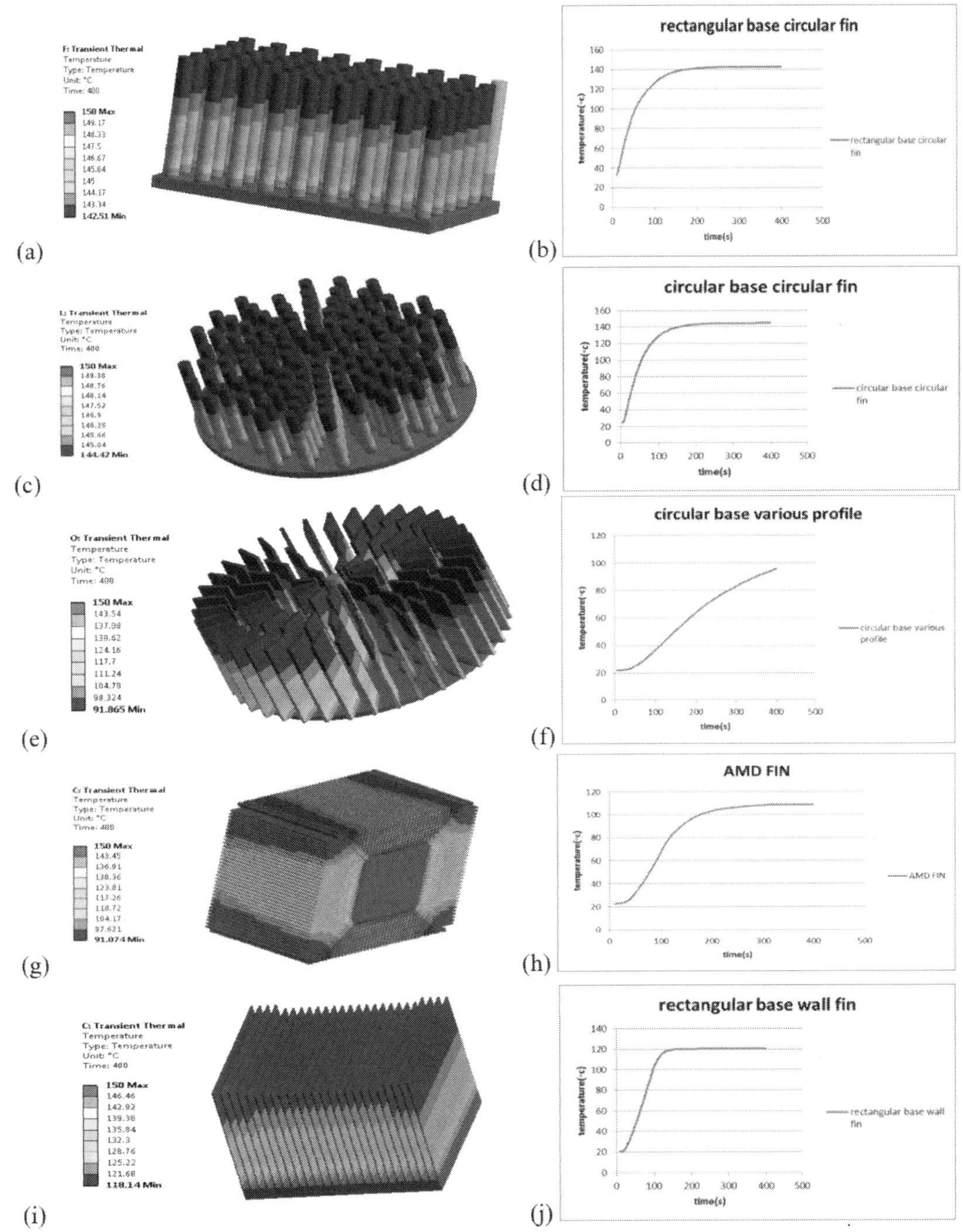

Fig 5 Meshed image of (a) Rectangular base circular fin (b) Circular base circular fin (c) Circular base various profile fin (d) AMD fin (e) Rectangular base wall fin

IOP Conf. Series: Materials Science and Engineering **197** (2017) 012085 doi:10.1088/1757-899X/197/1/012085

3.3 Proposed models comparison of analysis

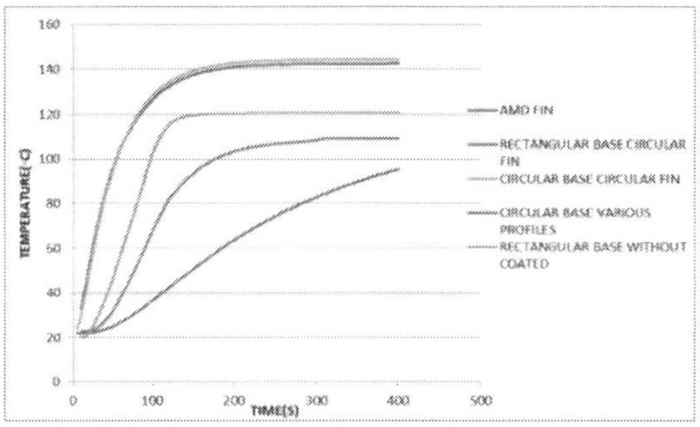

Fig 6 Comparison of analysis of proposed models

4. Conclusions

An alternative model of heat fins has been designed to increase heat dissipation. In ANSYS both the materials existing and proposed materials is analysed and the results of steady state and transient thermal analysis are taken for comparison. From the observations the following conclusions are made:

- CPU cooling performances of a computer chassis with rectangular and circular pin fin heat sinks were investigated using transient thermal analysis and the results were compared. The heat sink temperature difference results have been compared with an experimental result to find out best heat sink designs, and it shows the good correlation.
- Model 3 is proposed due to maximum heat dissipated, compare to other proposed models.

References

[1] Cheng-Hung Huang , Jon-Jer Lu , Herchang Ay 2011 A three dimensional heat sink module design problem with experimental verification, International Journal of Heat and Mass Transfer **54** 1482–1492.

[2] Kobus CJ, Oshio, T 2005 Predicting the thermal performance characteristics of staggered vertical pin fin array heat sinks under combined mode radiation and mixed convection with impinging flow, International Journal of Heat and Mass Transfer **48** 2684–2696.

[3] Dong-Kwon Kim, Sung Jin Kim, Jin-Kwon Bae 2009, Comparison of thermal performances of plate-fin and pin-fin heat sinks subject to an impinging flow, International Journal of Heat and Mass Transfer **52** 3510–3517.

[4] Emrana, Mohammad Ariful Islama 2014 Numerical investigation of flow dynamics and heat transfer characteristics in a micro channel heat sink Procedia Engineering **90** 563-568.

[5] Goshayeshi Ampofo 2009, Heat Transfer by Natural Convection from a Vertical and Horizontal Surfaces Using Vertical Fins Energy and Power Engineering, 85-89.

[6] Li and Chao 2009, Measurement of performance of plate-fin heat sinks with cross flow cooling International Journal of Heat and Mass Transfer **52** 2949–2955.

[7] Mehran Ahmadi, Golnoosh Mostafavi, Majid Bahrami 2014, Natural convection from rectangular interrupted fins International Journal of Thermal Sciences **82** 62-71.

[8] Mahmoud R, Al-Dadah DK, Aspinwall SL, Soo H Hemida 2011, Effect of micro fin geometry on natural convection heat transfer of horizontal microstructures Applied Thermal Engineering **31** 627-633.

[9] Qarnia and Lakhal 2013, Computation of melting with natural convection inside a rectangular enclosure heated by discrete protruding heat sources Applied Mathematical Modelling **37** 3968-3981.

[10] Sable SJ, Jagtap PS, Patil PR, Baviskar SB 2010 Barve Enhancement Of Natural Convection Heat Transfer On Vertical Heated Plate By Multiple V-FIN Array IJRRAS **5** 2.

IOP Conf. Series: Materials Science and Engineering **197** (2017) 012086 doi:10.1088/1757-899X/197/1/012086

Design and analysis of Air flow duct for improving the thermal performance of disc brake rotor

Raja.T[1] Mathiselvan.G[2], M Sreenivasulureddy[3], X.Goldwin Xavier[4]

[1,2]Department of Automobile Engineering, Sathyabama University, Chennai
[3] Department of Mechanical Engineering, Malla Reddy Engineering College, Hydrabad
[4]Department of Mechanical Engineering, Loyola Institute of Technology, Chennai

E-mail: raja.thiya@gmail.com

Abstract. safety in automotive engineering has been considered as a number one priority in development of new vehicle. A brake system is one of the most critical systems in the vehicle, without which the vehicle will put a passenger in an unsafe position. Temperature distribution on disc rotor brake and the performance brake of disc rotor is influenced by the air flow around the disc rotor. In this paper, the effect of air flow over the disc rotor is analyzed using the CFD software. The air flow over the disc rotor is increased by using a duct to supply more air flow over the disc rotor. The duct is designed to supply more air to the rotor surface and it can be placed in front of the vehicle for better performance. Increasing the air flow around the rotor will maximize the heat convection from the rotor surface. The rotor life and the performance can be improved.

1. Introduction

A braking system is one of the most important safety components of an automobile. It is mainly used to decelerate vehicles from an initial speed to a given speed. The friction based braking systems are the common device to convert kinetic energy into thermal energy, through friction between the brake pads and the rotor faces. The vehicle brakes are grouped into disc and the drum brakes. In the drum brakes, brake shoes are pushed in a radial direction against a brake drum whereas in the disc brakes the pads are pressed axially against a rotor or disc. The heat generated during braking needs to be dissipated effectively to attain satisfactory performance of brake as too high heat causes excessive wear of the rotor.

2. Computational Methodology

CFD is a commercially available code which is capable of giving accurate results for the flow field and cooling rate of brake discs. The CFD programs employed in this analysis are the *Fluent.* Fluent uses the finite volume method to solve the fluid flow transport equations in integral form and is known for its robustness in simulating many fluid dynamic phenomena.

3. Design of Air Duct

IOP Conf. Series: Materials Science and Engineering **197** (2017) 012086 doi:10.1088/1757-899X/197/1/012086

The duct is designed to supply more air to disc rotor, so that the air flow around the disc is maximized. Increasing of air flow around the rotor will maximize the heat convection and improve the performance of rotor. The inlet of the duct is wider than the outlet in order to supply more air. The duct is converging at the outlet, this leads to increase in the speed of the air. Solid Works is used to model the design a duct.

Figure 1. Design of Disc Rotor

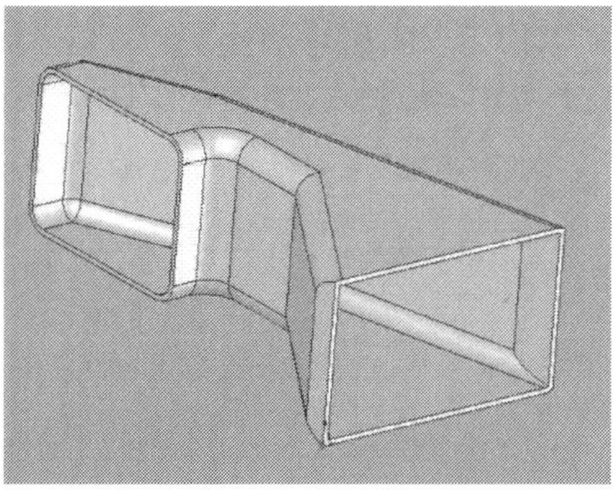

Figure 2. Design of Duct-Left Side View

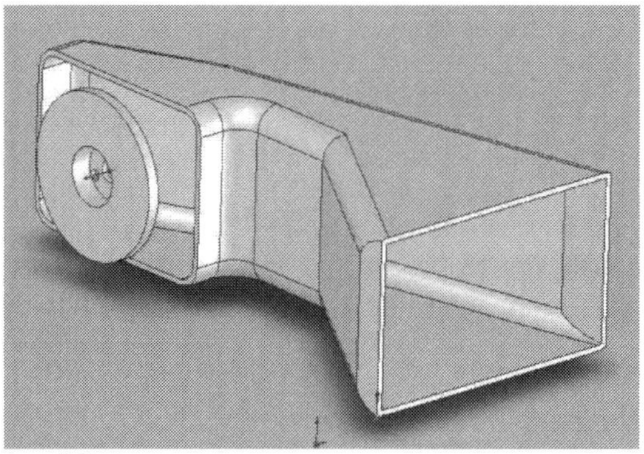

Figure 3. Assemble of Duct and Rotor

The assembled view shows the arrangement of the rotor and the duct. The disc rotor is placed in the outlet of the duct. The air from the inlet of the duct flows to the outlet and over the disc rotor.

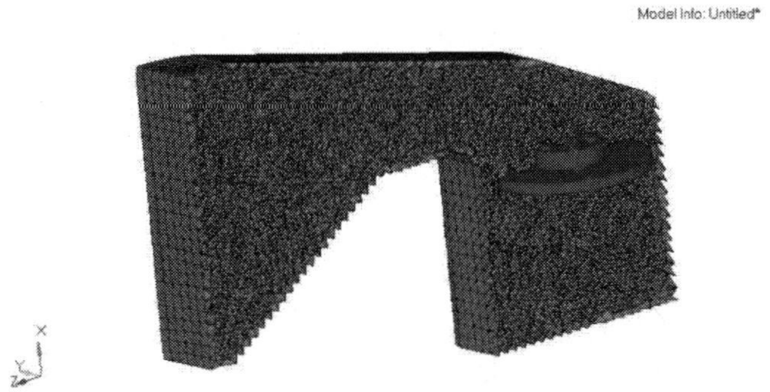

Figure 4. Hypermesh-Meshed Model

3.1 Boundary Condition for The Air Flow Over a Disc

The inlet velocity of the air flow over a disc is 27.7 m/s. and 13.85 m/s. Disc wall heat flux is 120,084 W/m2. The model of the duct is realizable model, Energy, K-epsilon

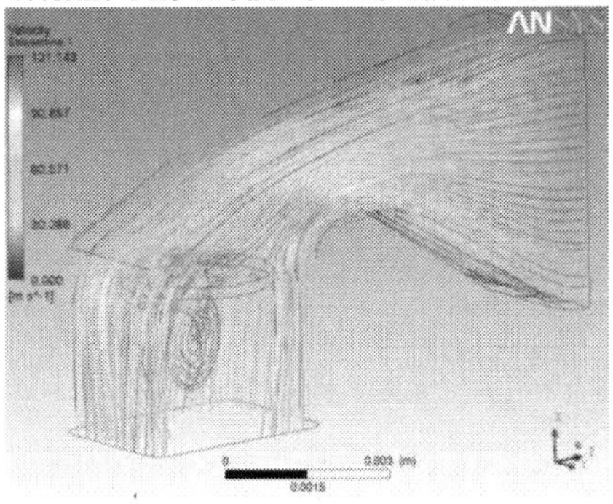

Figure 5a. Velocity streamline (100m/s)

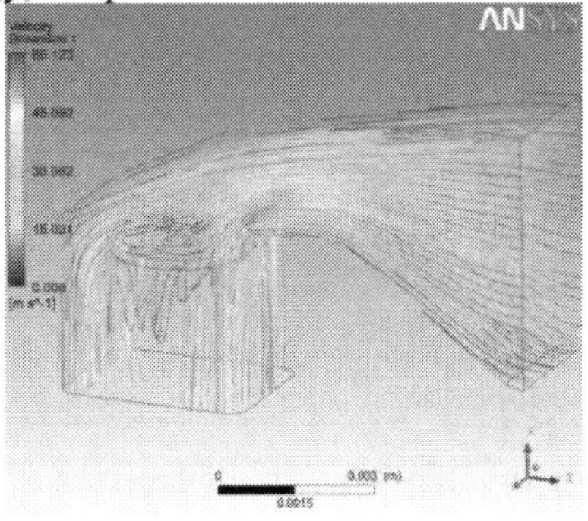

Figure 5b Velocity streamline (50m/s)

3.2 Velocity streamline:

Velocity distribution for the given condition is given in the Fig.5a &b. When the input velocity is 27.7m/s, the maximum velocity of the fluid is 121.143m/s which is produced in the region of the converging part of the duct. When the input velocity is 13.85m/s, the maximum velocity of the fluid is 60.123m/s Thus it is evident the average air flow rate over the rotor is increased.

Rotor Temperature:

The rotor temperature produced in the rotor surface is shown in Fig.6a & b. The maximum temperature produced on the rotor surface for the input velocity of 27.7m/s is 639.5K. The maximum temperature produced on the rotor surface for the input velocity of 13.85m/s is 642.K. The inlet air temperature is 300K

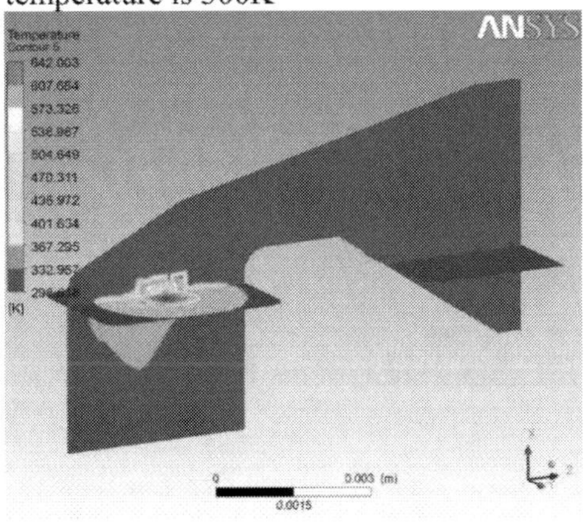

Figure 6a. Temperature on the Rotor (100m/s)

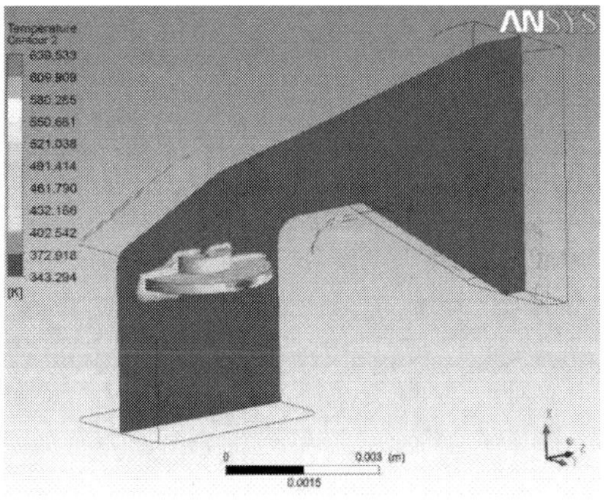

Figure 6b. Temperature on the Rotor urface(50m/s)

3.3 Convective Heat transfer:

The heat carried by the fluid is shown in the Fig.8a and Fig.8b. The temperature of the fluid at the exit for the input velocity of 27.7m/s is 318.73K. The temperature of the fluid at the exit for the input velocity of 13.85m/s is 309.38K.

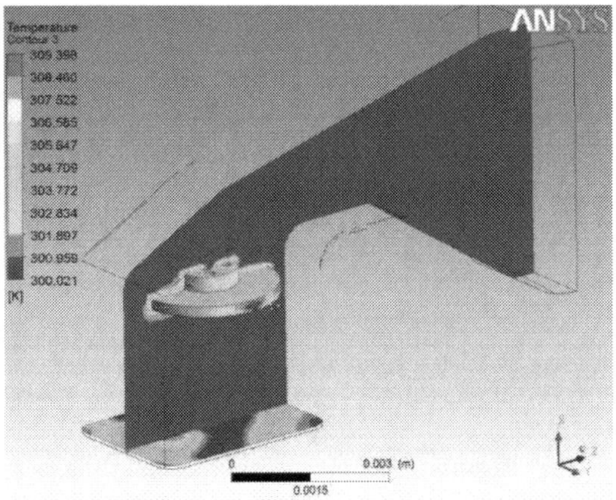

Figure 8a. Temperature of the fluid at outlet

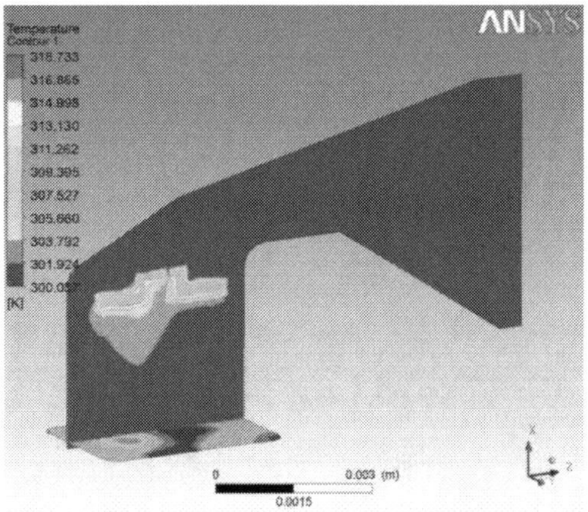

Figure 8b. Temperature of the fluid at exit

The cooling effect of the rotor is plotted in the Fig.8 and Fig.8b. The temperature of the fluid at the exit for the input velocity of 27.7m/s is 318.73K. The temperature of the fluid at the exit for the input velocity of 13.85m/s is 309.38K.

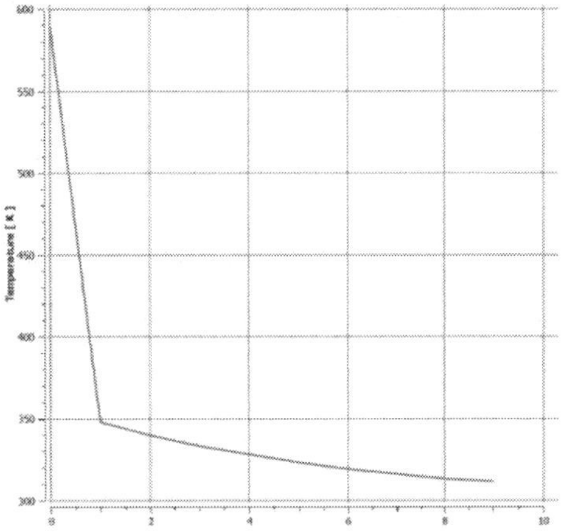

Figure 9a. Cooling effect of the Rotor (100m/s)

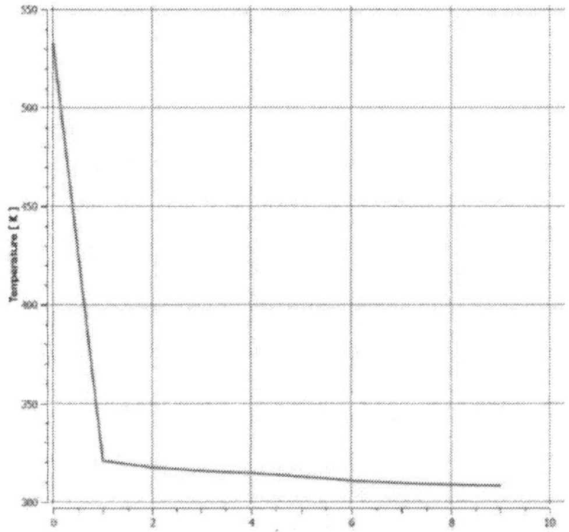

Figure 9b. Cooling effect of the Rotor (50m/s)

4 CONCLUSION

The present study can provide a useful design tool and improve the brake performance of disk brake system. Thus, the design of the duct is useful and it improves the heat convection of the disc rotor. The duct supply more air flow to the disc rotor and it reduce the heat of the disc rotor and it improves the life and the performance of the disc rotor.

5. REFERENCES

[1] Palmer, E., R. Mishra and J. Fieldhouse," 2007.*Comparison of air flow and heat dissipation from the front brakes fitted to a high performance GT car*" in 2007.

[2]. Wallis L, Leonardi E, Milton B, "*Air Flow and Heat Transfer in Ventilated Disc Brake Rotors with Diamond and Tear-Drop Pillars*" Numerical Heat Transfer, Part A; 41 pp643-655, 2001

[3]. Tirovic M, Galindo-Lopez C H, "*Analysis of Air Flow and Heat Dissipation from Wheel Mounted Railway Brake Discs*", IMechE, Brakes 2006 pp 125-134, 2006

[4]. Voller G P, Tirovic M, Gibbens P, "*Analysis of Automotive Disc Brake Cooling Characteristics*"

MRR and TWR evaluation on electrical discharge machining of Ti-6Al-4V using tungsten : copper composite electrode

J Prasanna [1], S Rajamanickam [2], O Amith Kumar [2], G Karthick Raj [2] and P V V Sathya Narayanan [2]

[1] Department of Mechanical Engineering, College of Engineering Guindy, Anna University, Chennai 25.

[2] Department of Mechanical Engineering, Vel Tech High Tech Dr. Rangarajan Dr. Sakunthala Engineering College, Avadi, Chennai-62

Prasanna@annauniv.edu , manic1327@gmail.com, amithombir01@gmail.com , karthikgowtham31@gmail.com, sathyanarayama4189@gmail.com

Abstract. In this paper Ti-6Al-4V used as workpiece material and it is keenly seen in variety of field including medical, chemical, marine, automotive, aerospace, aviation, electronic industries, nuclear reactor, consumer products etc., The conventional machining of Ti-6Al-4V is very difficult due to its distinctive properties. The Electrical Discharge Machining (EDM) is right choice of machining this material. The tungsten copper composite material is employed as tool material. The gap voltage, peak current, pulse on time and duty factor is considered as the machining parameter to analyze the machining characteristics Material Removal Rate (MRR) and Tool Wear Rate (TWR). The Taguchi method is provided to work for finding the significant parameter of EDM. It is found that for MRR significant parameters rated in the following order Gap Voltage, Pulse On-Time, Peak Current and Duty Factor. On the other hand for TWR significant parameters are listed in line of Gap Voltage, Duty Factor, Peak Current and Pulse On-Time.

Keywords: Ti-6Al-4V Electrical Discharge machining, TWR, MRR

1. Introduction

Electrical Discharge Machining (EDM) is on the pioneering machining process for machining hard to machine materials from lot of literature reviews. It comes under the category of non-contact machining process, in a dielectric medium workpiece and tools are impressed, by applying the electrical energy is to create a millions of isolated spark between the workpiece and tool for a micro seconds. This leads to the enormous heat development at the work zone in a dielectric fluid medium and resulted in material removal from both workpiece and tool. The detailed literature review is mentioned below:

The Electrical Discharge Machining finds important applications in drilling small holes in hard to machine materials [1]. J Marafona and C Wykes discovered the electric discharge machining of D2 tool steel using tungsten–copper electrode performed in two stages. The first stage was done using low current intensity and a long pulse duration to move carbon to the electrode thus preventing tool wear. The second stage was done using high current intensity and long pulse duration to increase the MRR with reduced TWR [2].

C H C Haron et al done MRR and TWR investigation on EDM of AISI 1045 steel using copper electrodes with diameter of 9.5 mm, 12 mm and 20 mm in two current setting of 3.5 A and 6 A. The MRR and TWR are depends on the diameter of tool and current. The (3.5 A) is suitable for small diameter workpiece. The (6 A) is suitable for large diameter workpiece [3]. The effect of TiC content varying from 5% to 40% in sintered copper-tungsten electrodes on their EDM performance during machining tool steel were studied. The 15% of TiC addition gives better surface finish, lower electrode wear and higher material removal rate has been recorded [4].

P M George et al employed Taguchi method for Electrical Discharge Machining of carbon-carbon composite under different machining conditions namely pulse current, gap voltage and pulse-on-time using copper electrode. Among these gap voltage is most significant and pulse current is ranked next to the gap voltage and pulse on time found to be insignificant on the performance characteristics such as MRR and TWR [5].

En-31 tool steel was electrical discharge machined with copper, copper-tungsten, aluminium and brass electrode materials, the results proved that the Copper tungsten provides better surface finish [6]. S Sivasankar and R Jeyapaul performed electric discharge machining of ZrB_2 using different tool materials namely Titanium, Graphite, Niobium, Tantalum, Tungsten, Stainless Steel, Molybdenum, Copper, Nickel and Silver by ANOVA, Grey and Regression analysis. The tool is having low melting produces irregular shape of the holes. From ANOVA the tool material is important next to that pulse on time. Grey analysis rated the EDM tool is the order of Titanium, Graphite, Niobium, Tantalum, Tungsten, Molybdenum, Nickel, Stainless Steel, Copper & Silver [7].

K M Kumar and P Hariharan did experimental investigation of austempered ductile iron EDM using copper electrode. They conclude that cross sectional area does not depend on the MRR and TWR. The MRR and TWR are increases with increasing the current and pulse on time [8]. T Muthuramalingam and B Mohan reviewed the electrical parameters of electrical discharge machining process and noted that, optimizing the electrical parameters will improves efficiency of electric discharge machining process [9].

With various literature reviews the tool material is one of noted parameter in EDM. Hence this work is intended to attain MRR and TWR Evaluation on Electrical Discharge Machining of Ti-6Al-4V Using Tungsten: Copper Composite Electrode.

2. Experimental procedure

The experiment was conducted using Electrical Discharge Machine, model Electrica Ratanparhki 550 x 350 Die Sink EDM as in figure 1 and figure 2 shows the enlarged view of work zone of Electrical Discharge Machine. Titanium Grade 5 (Ti–6Al–4V) alloy of 0.55 mm thickness was used as work material. The dielectric fluid used was EDM oil (specific gravity-0.763).

The work piece and tool material is captured in Figure 3 and 4 respectively. The composition of work material and tool is pointed in the Table 1and 2. The densities of the workpiece and tool material are tested as 4.415 g/cc and 15.2 g/cc. The hardness of the workpiece and tool material are experienced as 363 in HV10 kg and 235 in HV 10 kg.

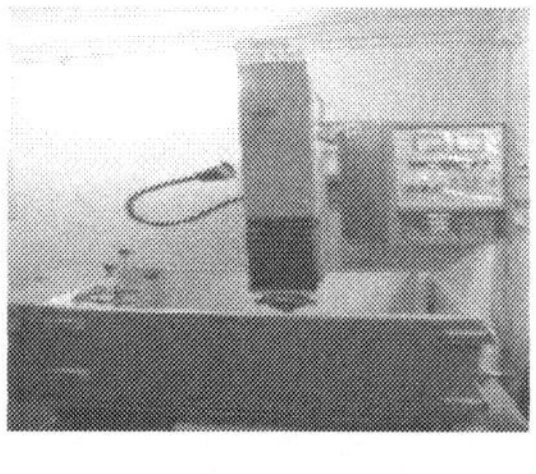

Figure 1 Electrica Ratanparhki 550 x 350 Die Sink EDM

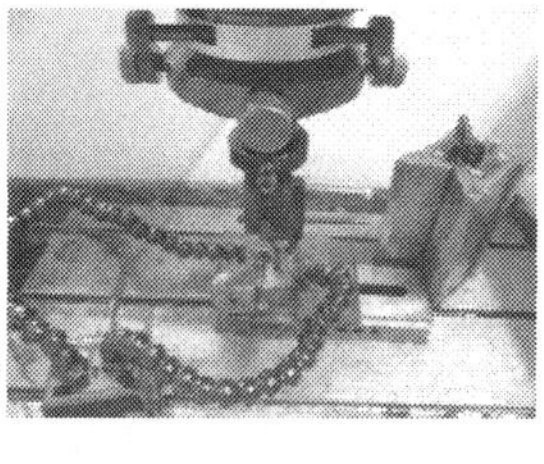

Figure 2 Work zone of EDM

IOP Conf. Series: Materials Science and Engineering **197** (2017) 012087 doi:10.1088/1757-899X/197/1/012087

Figure 3 Workpiece – Titanium (Ti-6Al-4V)

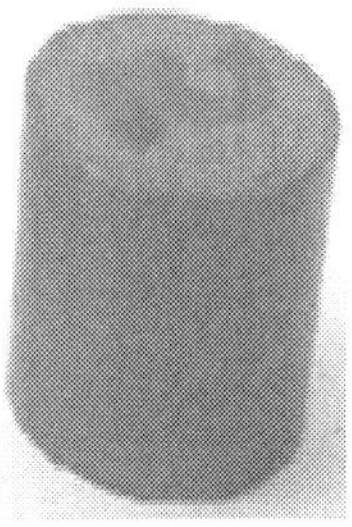

Figure 4 Tool- Tungsten:Copper (70:30)

Table 1 Composition of Work piece Material

Composition	Percentage (%)
Aluminium	5.710
Titanium	90.10
Vanadium	3.720
Others	Rest

Table 2 Composition of Tool Material

Composition	Percentage (%)
Tungsten	69.16
Copper	28.62
Nickel	0.67
Manganese	0.16
Vanadium	0.42
Zinc	2.04
Bismuth	0.02
Iron	0.49

The Ti-6Al-4V sheet of 50 x 50 mm of required no is cut by hand shearing machine and Tungsten:copper(70:30) tool is cut wire electrical discharge machine of 3 mm length are prepared. Using Electrical Discharge Machine experiment were conducted for 15 minutes of constant time by considering different levels of Gap Voltage, Peak Current, Pulse On-Time, Duty Factor to analyse the performance charecteristics such as MRR and TWR.

3. Taguchi optimization technique

Taguchi optimization technique offers organized, straightforward and clean methodology for the optimization of the parameters with only a minimum number experimental set that can operate consistently over a wide range of conditions. The L9 Taguchi orthogonal array is planned to carry out the experiment. The significant parameters were determined by using S/N graphs and Response table obtained from Minitab software trail version 15.

The aim of this research work was to determine the process parameters required to achieve higher material removal rate and lower tool wear rate. All machining process parameter was planned to have three levels which are presented in Table 3.

Table 3 Experimental Parameter and their Levels

Machining Process Parameter	Symbol	Levels		
		1	2	3
Voltage (V)	v	50	100	150
Peak Current (A)	I_p	5	10	15
Pulse on time (□s)	T_{on}	20	30	40
Duty Factors (%)	□	3	6	9

4. Results and discussion

The weight of workpiece and tool material is measured before and after machining. The constant time of 15 min set for each experiment. MRR and TWR for all the experiment were calculated using the equations 4.1.

$$Material\ removal\ rate = \frac{weight\ before\ machining - weight\ after\ machining}{machining\ time} \quad ....(4.1)$$

Results obtained were tabulated in table 4. The difference of MRR for various experiments is illustrated in figure 5 and 6. It is witness that the larger MRR of experiments are ranked in the following order 2, 3, 1, 6, 4, 9, 5, 8, 7. It is observed that the smaller TWR of experiments are graded in the following order 1, 7, 9, 8, 5, 2, 6, 4, 3.

Table 4 Results of the Experiment

V (v)	I_p (A)	T_{on} (s)	(%)	MRR (mm^3/min)	TWR (mm^3/min)
50	5	20	3	0.00933	0.00004
50	10	30	6	0.01227	0.00101
50	15	40	9	0.00951	0.00157
100	5	30	9	0.00319	0.00123
100	10	40	3	0.00290	0.00085
100	15	20	6	0.00322	0.00113
150	5	40	6	0.00069	0.00013
150	10	20	9	0.00145	0.00023
150	15	30	3	0.00311	0.00018

4.1 Analysis of Machining Performance on MRR

The main effects plot for SN ratios- MRR graphs, as shown in Figure 7, it can be seen that MRR decreases as the gap voltage decreases throughout the entire range. In case of pulse on time, the MRR increases in a similar fashion as the peak current increases. The MRR increases linearly along with the increase in Pulse On-Time up to second level then it starts decreases on the other side for duty factor decreases up to second level then it starts increases. The response table – MRR is created in Table 5 gives the most significant factor for MRR is Gap Voltage and the least significant for MRR is duty factor.

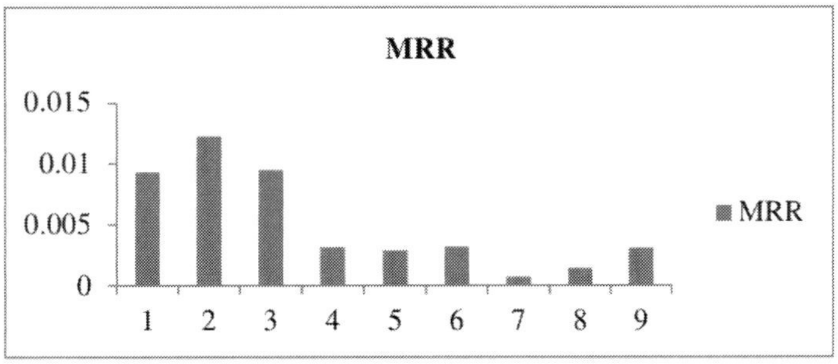

Figure 5 Difference of MRR for Various Experiments

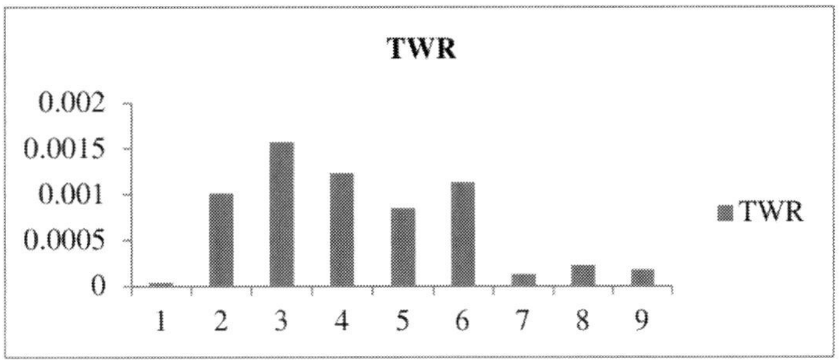

Figure 6 Difference of TWR for Various Experiments

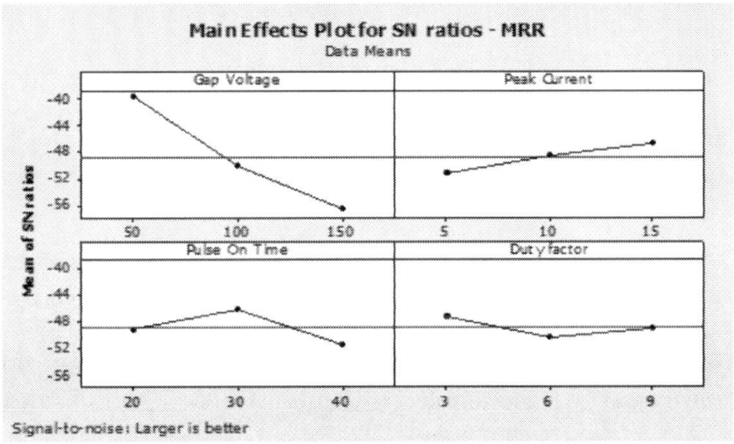

Figure 7 Main Effects Plots for SN ratios – MRR

Table 5 Response Table -MRR

Source	V	I_p	T_{on}	
1	-39.75	-51.25	-49.07	-47.17
2	-50.17	-48.58	-46.10	-50.43
3	-56.71	-46.81	-51.47	-49.04
Delta	16.96	04.44	05.37	03.26
Rank	1	3	2	4

4.2 Analysis of machining performance on TWR

The main effects plot for SN ratios- TWR graphs, as shown in Figure 8, it can be seen that TWR decreases as the gap voltage decreases TWR up to second level then it increases the TWR linearly. In case of pulse on time, the TWR decreases in a similar fashion as the peak current increases. The TWR decreases linearly along with the increase in Pulse On-Time and duty factor increases. The response table – TWR is formed in Table 6 furnish the most significant factor for TWR is Gap Voltage and the least significant for TWR is Pulse On-Time.

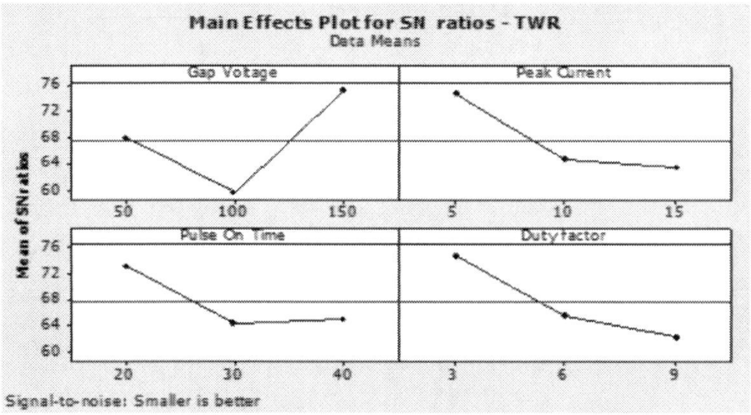

Figure 8 Main Effects Plots for SN ratios – TWR

Table 6 Response Table for TWR

Source	V	I_p	T_{on}	
1	67.98	74.63	73.22	74.75
2	59.52	64.70	64.34	65.52
3	75.13	63.30	65.07	62.35
Delta	15.61	11.32	08.88	12.41
Rank	1	3	4	2

5. Conclusion

The elaborate study on MRR and TWR Evaluation on Electrical Discharge Machining of Ti-6Al-4V Using Tungsten: Copper Composite Electrode has drawn the following conclusions.

The most advantageous combination of process parameters for obtaining maximum MRR are as detected at Gap Voltage 50 V, Peak Current 10 A, Pulse On Time 30 □Sec, Duty Factor 6 % on the other hand minimum TWR are founded at Gap Voltage 50 V, Peak Current 5 A, Pulse On Time 20 □Sec , Duty Factor 3, respectively.

References

[1] Jeswani M L,1979 *Int. J. Mach, Tool. Des.* **19.** (1979) pp. 165-169.

[2] Marafona J andWykes C 2000 *Int. J. Mach, Tool. Manu.***40** 153–164

[3] Haron C H C, Deros B M, Grinting A and Fauziah M 2001 *J. Mater. Process. Technol.* **116** 84-87.

[4] Li L, Wong Y S, Fuh J Y H and Lu L 2001 *J. Mater. Process. Technol.* **113** 563-567.

[5] George P M, Raghunath B K, Manocha L M and Warrier A M 2004 *J. Mater. Process. Technol.* **145** 66–71.

[6] Singh S, Maheshwari S, Pandey P C 2004 *J. Mater. Process. Technol.* **149** 272–277.

[7] Sivasankar S and Jeyapaul R 2012 *Procedia Eng.* **38** 3977 – 3991.

[8] Kumar K M and Hariharan P 2013 *Procedia Eng.* **64** 495 – 1504.

[9] Muthuramalingam T and Mohan B 2015 *J.acme.* **15** 87-94.